AF807101

SECOND EDITION

Monitoring and Control of Macrofouling Mollusks in Fresh Water Systems

SECOND EDITION

Monitoring and Control of Macrofouling Mollusks in Fresh Water Systems

GERALD L. MACKIE • RENATA CLAUDI

CRC Press
Taylor & Francis Group
Boca Raton London New York

CRC Press is an imprint of the
Taylor & Francis Group, an **informa** business

CRC Press
Taylor & Francis Group
6000 Broken Sound Parkway NW, Suite 300
Boca Raton, FL 33487-2742

© 2010 by Taylor and Francis Group, LLC
CRC Press is an imprint of Taylor & Francis Group, an Informa business

No claim to original U.S. Government works

International Standard Book Number: 978-1-4398-0050-8 (Hardback)

This book contains information obtained from authentic and highly regarded sources. Reasonable efforts have been made to publish reliable data and information, but the author and publisher cannot assume responsibility for the validity of all materials or the consequences of their use. The authors and publishers have attempted to trace the copyright holders of all material reproduced in this publication and apologize to copyright holders if permission to publish in this form has not been obtained. If any copyright material has not been acknowledged please write and let us know so we may rectify in any future reprint.

Except as permitted under U.S. Copyright Law, no part of this book may be reprinted, reproduced, transmitted, or utilized in any form by any electronic, mechanical, or other means, now known or hereafter invented, including photocopying, microfilming, and recording, or in any information storage or retrieval system, without written permission from the publishers.

For permission to photocopy or use material electronically from this work, please access www.copyright.com (http://www.copyright.com/) or contact the Copyright Clearance Center, Inc. (CCC), 222 Rosewood Drive, Danvers, MA 01923, 978-750-8400. CCC is a not-for-profit organization that provides licenses and registration for a variety of users. For organizations that have been granted a photocopy license by the CCC, a separate system of payment has been arranged.

Trademark Notice: Product or corporate names may be trademarks or registered trademarks, and are used only for identification and explanation without intent to infringe.

Library of Congress Cataloging-in-Publication Data

Mackie, Gerry L.
Monitoring and control of macrofouling mollusks in fresh water systems / Gerald L. Mackie, Renata Claudi. -- 2nd ed.
p. cm.
Includes bibliographical references and index.
ISBN 978-1-4398-0050-8 (hardcover : alk. paper)
1. Gastropoda--Monitoring--North America. 2. Bivalves--Monitoring--North America. 3. Marine fouling organisms--Control--North America. 4. Freshwater ecology. I. Claudi, Renata. II. Title.

QL430.4.M3235 2010
628.9'64--dc22 2009042042

Visit the Taylor & Francis Web site at
http://www.taylorandfrancis.com

and the CRC Press Web site at
http://www.crcpress.com

Contents

Preface

This book is the first to compile information on the most invasive, nuisance species of bivalves and gastropods in North America. Not even in Eurasia, South America, Australia, New Zealand, or other continents is there a compilation of invasive features, their current and historical distributions, monitoring for their presence and abundance, their major impacts to industry and agriculture, and how to control numerous species of mollusks. Fifteen species, five bivalves and ten gastropods are discussed. Selecting the species, especially the gastropods, was a task in itself, and some may wonder why we selected the species we did. First, they all had to have invasive and nuisance qualities (see below). Second, they had to be introduced from another country or from another state.

The dreissenids, zebra, and quagga mussels are the most obvious choices. We include another dreissenid, *Mytilopsis leucophaeata*, or Conrad's false mussel, which is native to Florida but is now spreading to other states and has the same invasive, nuisance features as zebra and quagga mussels. The Asian clam, *Corbicula fluminea*, is another obvious choice and has demonstrated its nuisance features in North American waters for over 70 years now. The golden mussel, *Limnoperna fortunei*, has the same invasive, nuisance features as dreissenids and is also included here; while it is not yet in North America, its presence and rapid northward spread in South America portend its eventual invasion into North America.

The selection of gastropods for discussion was slightly more difficult. The apple snails of the family Ampullariidae are an obvious choice, but which ones? There is still much taxonomic confusion in the family, but the recent publication of two books, one by Barker (2002) and another by Joshi and Sabastian (2006), helped in reducing the section to six species: *Marisa cornuarietis*, *Pomacea canaliculata*, *Pomacea haustrum*, *Pomacea insularum*, *Pomacea diffusa*, and *Pomacea paludosa*. All but *P. paludosa* are introduced; *P. paludosa* is native to Florida but is spreading to other states. Two thiarids, *Tarebia granifera* and *Melanoides tuberculata*, are included because they are hosts to dangerous human parasites and have ecological impacts. The bithyniid *Bithynia tentaculata*, or faucet snail, is included because it is a nuisance in some municipal water supplies. The New Zealand mudsnail, *Potamopyrgus antipodarum*, is a recent invader, very prolific, and is rapidly spreading throughout North America impacting the ecology of several ecosystems. The physid *Physa acuta* is also spreading quickly and is a nuisance in many industrial intakes, but it is a native species and is not discussed here.

Complete descriptions of the biology of all invasive macrofouling mollusks would require a tome and is beyond the scope of this book. Brief descriptions of the external and internal structures are provided to introduce the reader to the most common terminology used in describing macrofouling bivalves and snails. The overviews examine only those biological features that are relevant to the monitoring and control of the invasive macrofouling mollusks. Descriptions of their shell and soft parts are essential because they help both in explaining why mollusks are pests and in distinguishing these mollusks from native species of mollusks. Some explanation of their habits and habitat is important for knowing why they can live and be so prolific in pipelines, on intake structures, in canal systems, in aqueducts, and in lakes and rivers. Knowledge of the mollusk's reproductive potential helps in explaining their biofouling nature and why they are so much more prolific than our native species that also live in freshwaters. Information on their life cycles and population dynamics is crucial for successful control of the species. The efficient dispersal mechanisms employed by the nuisance mollusks help in explaining why they have spread so rapidly to inland lakes and rivers across continents, and also how they can invade virtually any part of a facility or water body that is not protected. An understanding of their physiology can explain why some natural physical (e.g., heat) and chemical properties (e.g., salinity, pH, alkalinity, calcium levels, dissolved oxygen levels) and nutrient (e.g., total phosphorus and chlorophyll) levels can be effective control agents. Finally, knowledge of their potential impacts on ecosystems, industries, and utilities, as well as of the many human-made physical and chemical mitigants that are being used to control the mollusks, is crucial for preserving the health of our raw water supplies.

Gerry Mackie
Renata Claudi

Acknowledgments

It has been more than 20 years since the zebra mussel arrived in North America and 15 years since the first book by Claudi and Mackie (1994) on monitoring and control of the species was published. But much has been learned about the distribution, ecology, and physiology of not only the zebra mussel but other dreissenids such as the quagga mussel and Conrad's false mussel as well. Largely ignored in the meantime were invasive gastropods, such as the faucet snail and the New Zealand mudsnail, although several apple snails and thiarids were gaining infamy as ecological and agricultural pests. Fifteen invasive mollusk species are discussed in this book, and because one person cannot be an expert on all of them we must rely on contributions from several people. We are especially grateful to Amy Benson for providing distribution maps and ecological information for many species. We are grateful for the timely submissions by several authors who provided information and their experiences on invasive bivalves in other countries, such as European Russia and adjacent countries (Dr. Marina Orlova, Chapter 10, Section 10.1), Ireland (Dr. Frances Lucy, Chapter 10, Section 10.2), Argentina and neighboring countries (Dr. Gustavo Darrigran, Chapter 10, Section 10.3), Brazil (Maria Rolla and Hélen Mota, Chapter 10, Section 10.4), and South America in general (Dr. Márcia Oliveira, Dr. Stephen Hamilton, and Dr. Claudia Jacobi, Chapter 3, Section 3.3.6). We appreciate the inclusion of the newest technology on polymerase chain reaction (PCR) detection of dreissenid larvae by Dr. Paul Rochelle, Anne Johnson, and Ricardo De Leon (Chapter 4, Section 4.4) and the use of navigational buoys as monitoring platforms by Dr. Bruce Conn (Chapter 4, Section 4.5). Tom Prescott provided engineering support, and David Miller and Einar Mackie helped greatly with keeping us on schedule by organizing and formatting references and text and providing some graphics. I am thankful to Dr. Andre Martel, who provided many specimens, especially apple snails and thiarids, and also photographic equipment for photos for this book. We appreciate the permission of the United States Geological Survey (USGS) and Cassie Thibodeaux, USGS National Wetlands Research Center, for use of their photo of *Pomacea insularum*. Dr. Philip Darby kindly gave permission to use his drawing, with some modifications, of a funnel trap for collecting ampullariids. Finally, we are grateful to Nancy Binnie and Jonathan Moore for their last-minute contribution on damage caused by quagga mussels on shipwrecks in Lake Ontario (Chapter 6, Section 6.2.6).

Authors

Gerry Mackie is professor emeritus in the Department of Integrative Biology at the University of Guelph, Guelph, Ontario, Canada, and has had his own consulting business, Water Systems Analysts, since 1995. He has supervised 29 graduate students and has authored four books (*The Cottage Bible*, *Biology of Freshwater Corbiculid and Sphaeriid Clams of North America*, *Applied Aquatic Ecosystem Concepts*, and *Practical Manual for Zebra Mussel Monitoring and Control*), 12 chapters in books, and more than 150 peer-reviewed scientific articles. He was the first to identify and report the presence of the zebra mussel in Lake St. Clair (between Lake Huron and Lake Erie) in June of 1988. He was honored with a lifetime achievement award by the Freshwater Mollusk Conservation Society in April 2009 for his singular accomplishments and long-term contributions that have advanced the conservation and science of freshwater mollusks at the national and international levels.

Renata Claudi is an aquatic biologist with over 20 years of diverse business and technical experience. At this time, she is the chief scientist of RNT Consulting Inc. This environmental consulting firm focuses on the various aspects of alien species invasions, including their economic impact, risk of introductions, and the selection of appropriate control options and their installation.

Ms. Claudi received her BSc and MSc degrees in marine biology from McGill University, Montreal, Quebec, Canada. From 1989 to 2000 she worked for a major Canadian electrical utility. There she was responsible for the assessment of the impact of dreissenid mussels on operating power stations, the development of mitigation plans, interaction with other utilities and industries, and the organization of major scientific meetings. She was also one of the chief organizers of numerous international conferences on aquatic nuisance species.

In 2002, Ms. Claudi was the lead editor of a volume entitled *Alien Invaders in Canada's Waters, Wetlands and Forests*. This book was published by Natural Resources Canada and has 24 contributors.

In 1999, she coedited *Fresh Water Alien Organisms in North America*, published by Lewis Publishers. This volume has 43 contributors from across Canada, United States, and Mexico.

In 1993, she coauthored a book entitled *Practical Manual for Zebra Mussel Monitoring and Control*, published by Lewis Publishers. This book summarizes the state of the art of zebra mussel control as well as the experience gathered by utilities and other industries dealing with zebra mussels on the Great Lakes.

Contributors

Nancy Binnie
Department of Canadian Heritage
Canadian Conservation Institute
Ottawa, Ontario, Canada

David Bruce Conn
North American Scientific Enterprises (NASCENT)
Monteagle, Tennessee

Gustavo Darrigran
Grupo de Investigación en Moluscos Invasores/Plaga
División Zoología Invertebrados
La Plata, Argentina

Stephen K. Hamilton
Kellogg Biological Station and Department of Zoology
Michigan State University
Hickory Corners, Michigan

Claudia M. Jacobi
Instituto de Ciências Biológicas
Universidade Federal de Minas Gerais
Departamento de Biologia Geral
Belo Horizonte, Brazil

Anne M. Johnson
Metropolitan Water District of Southern California
Water Quality Section
La Verne, California

Darren Keyes
ASI Group
St. Catherines, Ontario, Canada

Ricardo De Leon
Metropolitan Water District of Southern California
Water Quality Section
La Verne, California

Frances E. Lucy
Department of Environmental Science
Institute of Technology
Sligo, Ireland

and

Environmental Services Ireland
Carrick on Shannon, Ireland

Jonathan Moore
Underwater Archaeology Service
Parks Canada Agency
Ottawa, Ontario, Canada

Hélen Regina Mota
Centrais Eletricas de Minas Gerais (CEMIG)
Belo Horizonte, Brazil

Márcia D. Oliveira
Empresa Brasileira de Pesquisa Agropecuária
Centro de Pesquisa Agropecuária do Pantanal,
Embrapa Pantanal, Corumba, Brazil

Marina I. Orlova
Zoological Institute of the Russian Academy of Sciences
St. Petersburg, Russia

Paul A. Rochelle
Water Quality Section
Metropolitan Water District of Southern California
La Verne, California

Maria Edith Rolla
Centrais Eletricas de Minas Gerais (CEMIG)
Belo Horizonte, Brazil

General Introduction

The zebra mussel, *Dreissena polymorpha* (Pallas), was introduced into Lake St. Clair, probably in 1985 (Hebert et al., 1989). The species is native to Europe and appears to have been transported to Lake Erie in the freshwater ballast of a trans-oceanic ship. Until recently, most ships have been discharging some of their ballast water into the Great Lakes, releasing of either veliger larvae or young adults into Lake Erie possible. Adult shells, about 2 cm long, were first found in August 1988 near Belle River in Lake St. Clair (Hebert et al., 1989) and were probably carried there on the hull of a fishing boat traveling from Lake Erie to Lake St. Clair via the Detroit River. Assuming it takes the species about 2 years to grow to 2 cm, it is most likely that the species was introduced in either the spring or summer of 1986 as young adults or in the summer or fall of 1985 as veliger larvae. Since adults are benthic and larvae are pelagic, and ships normally take on ballast water from the pelagic zone, the species was probably introduced as veliger larvae in Lake Erie in the summer or fall of 1985. The mussels could have been introduced as adults only if the ships were taking on ballast water in a harbor where the water was so heavily stirred or mixed (e.g., by ship propellers) that adults were washed from nearby structures into the ballast water supply. The species was not found in an extensive survey of bottom sediments in 1983 (Griffiths, 1987) or in an extensive native clam survey in 1986 (Tom Nalepa, U.S. Fish and Wildlife, personal communication). However, the numbers and sizes of adults of the pioneering population were probably so small that they could easily have been missed in these surveys.

Most introduced species are rather benign and have very little apparent effect on the ecosystem or on socio-economics. In fact, many are beneficial; for example, *Nasturtium officinale*, or water cress, is an introduced plant that lives in cold, clear streams and is favored by many for use in salads. However, many other introduced species have negative impacts and some can have devastating consequences. The zebra mussel is one of them. In the past 20 years, it has not only become the most serious biofouling pest of any exotic species ever introduced to North American surface waters, it has also completely altered the ecology of the Great Lakes. For more than 100 years Europeans have witnessed its ability to (1) clog intakes and distribution pipelines in industrial and domestic facilities, (2) increase the corrosion potential of steel and iron pipes, (3) taint and contaminate potable water supplies, and (4) encrust the hulls of boats and ships (Morton, 1969; Mackie et al., 1989). In only 3 years, utilities, industries, water treatment plants, and boaters on Lake Erie and Lake St. Clair had already seen these effects and, in addition, had documented (5) fouling fishing gear and navigational aids (e.g., buoys) and rendering them useless; (6) reduction, or possibly elimination, of the recreational and aesthetic value of beaches and shorelines; (7) potential displacement of native species of bivalves;

(8) alteration or even destruction of fish habitats; and (9) changes in the physical limnology (e.g., Secchi depth) of infested lakes (Lakes Erie and St. Clair).

This book follows up on the information provided by Claudi and Mackie in 1994 and provides the newest information available (up to April 2009) on biology, impact, and control of zebra and quagga mussels. While writing the updated version, we realized that there are not only other invasive species of bivalves that were also causing havoc to industries and utilities, but also two freshwater gastropods, faucet snails and New Zealand mudsnails. At the same time, there were apple snails and thiarids impacting agriculture and/or altering the ecology of lakes and streams. Hence, the objectives of this book have changed dramatically from those of Claudi and Mackie (1994). However, we do retain the objective of providing sufficient information, much of it new, for professional engineers and biologists, as well as for plant supervisors and their staff, including technicians, pumphouse operators, and other "hands-on" personnel, (1) to understand the new fouling species, their biology, and their potential for harm and (2) to detect, monitor, and control the biofouling species. It is still a "how to" book to guide plant personnel through the normal sequence of events in dealing with biofouling bivalves. We follow our general philosophy by first determining if a facility is in danger of mollusk infestations then, if so, how to monitor to detect the first appearance of larvae and/or adults. This book provides new information for assessing the potential for invasion by invasive mollusks. If a facility, or farm for that matter, is in a high-risk area, information is provided to distinguish the mollusk from native mollusks and from other pest species. Monitoring to detect the first appearance of young and adults, and then monitoring the growth and abundance helps in determining the method(s) of control and the numerous control options that are available. Minimizing the cost(s) of the option(s) selected is also described.

Bivalves and clams are treated first, and then the gastropods. For bivalves, the book is written for users of all types of raw water systems, with the aim of keeping their facilities operational and avoiding downtime due to macrofouling. The book will be most useful for raw water users facing zebra or quagga mussel, or Asian clam, problems for the first time. Conrad's false mussel, *Mytilopsis leucophaeata*, a close relative of zebra and quagga mussels, and the golden mussel, *Limnoperna fortunei*, have very similar impacts and are controlled in much the same way as zebra and quagga mussels. This book provides information on the most recent treatment technologies, reviews the different strategies for use of the technologies, and examines some of the successes and failures. The book will be useful for mitigating clams and mussels in plants that are already impacted by providing plant operators with a quick guide to techniques for eliminating mussels in their facility.

Most industries have some systems that are common to all. For example, most industries have fire protection systems and service water systems (SWSs), and the mitigation strategies for these will be the same. However, even within these systems differences exist. For example, in some industries water for the fire protection system is drawn from the service water supply, in others from the circulating water system (CWS), and in still others from a separate devoted system. The mitigation strategy for each type of water supply may differ. Moreover, many industries have unique systems that require special consideration. For example, seal water systems, demineralized water systems, and water clarifying systems are common but are not present in all industries. Even in utilities such as power generating plants and water treatment plants, all systems are not common or they are common but have different elements (e.g., alum flocculation versus direct filtration). Although a summary of treatment strategies and technologies is given for a variety of industries and utilities, detailed discussions for every type of industry are beyond the scope of this book. Instead, we have designed a hypothetical industry that has most systems and elements present (Figure I.1). This will help one to relate to structures in their facility. The major systems and structures that have been subjected to severe biofouling in industries on the Great Lakes are listed in Table I.1 and are included in the hypothetical system shown in Figure I.1.

Of the numerous questions that arise when faced with a potential mussel infestation, four arise almost every time, at least after you have convinced yourself that mussels will arrive: (1) Will mussels survive and grow to be a problem in my facility? (2) Should we be doing something now (i.e., be proactive) or should we wait until they arrive (i.e., be reactive)? (3) In either case, what structures and systems do we need to protect? (4) How do I protect them? Both proactive and reactive treatment strategies are described, when appropriate. Proactive strategies are those that will prevent mussels from settling and establishing themselves in or on a structure; usually the proactive strategies are aimed at larval stages. Reactive strategies are those that allow mussels to settle but prevent them from growing and establishing themselves; usually reactive strategies are aimed at juveniles and adults. Reactive strategies are the first choice when large infestations are discovered for the first time.

GETTING STARTED FOR BIOFOULING BIVALVES

Step 1: "Will mussels and clams infest our facility?" Chapter 1 gives a review of the important biological features of mussels and clams, but the "physiology" of each biofouling species will largely dictate the success of the bivalves in your water supply. In Chapter 3, Tables 3.7, 3.10, 3.12, 3.14, and 3.16 summarize the limiting environmental factors for zebra mussels, quagga mussels, Conrad's false mussels, golden mussels, and Asian clams, respectively, and allow you to quickly evaluate the infestation potential of your facility on the basis of your raw water source. If the units of your water chemistry data are not the same as in the tables (e.g., values are in μeq/L instead of mg/L), use the conversions given in the appropriate sections of Chapter 3. If you do not have data on calcium levels, follow the procedures for calculating the calcium levels from other characteristics, for example, alkalinity, total hardness, or conductivity, and determine whether your water source has the appropriate threshold levels for survival, growth, and reproduction. Examples are given for several scenarios: one for water in which the temperature is ideal for the survival, growth, and reproduction of zebra mussels but the calcium levels are not; another for which the calcium levels are high enough to support their survival but not sufficient to support infestation levels of growth and reproduction; and a third example for which both temperature and calcium are ideal but the temperatures may be high enough for a short period to either curtail growth and reproduction or even eliminate part of the population. Although there are exceptions, a good rule of thumb to follow is that if native clams thrive around your facility, biofouling mussels and clams will probably thrive there as well. Once you have convinced yourself that zebra mussels will be a problem, go to Step 2.

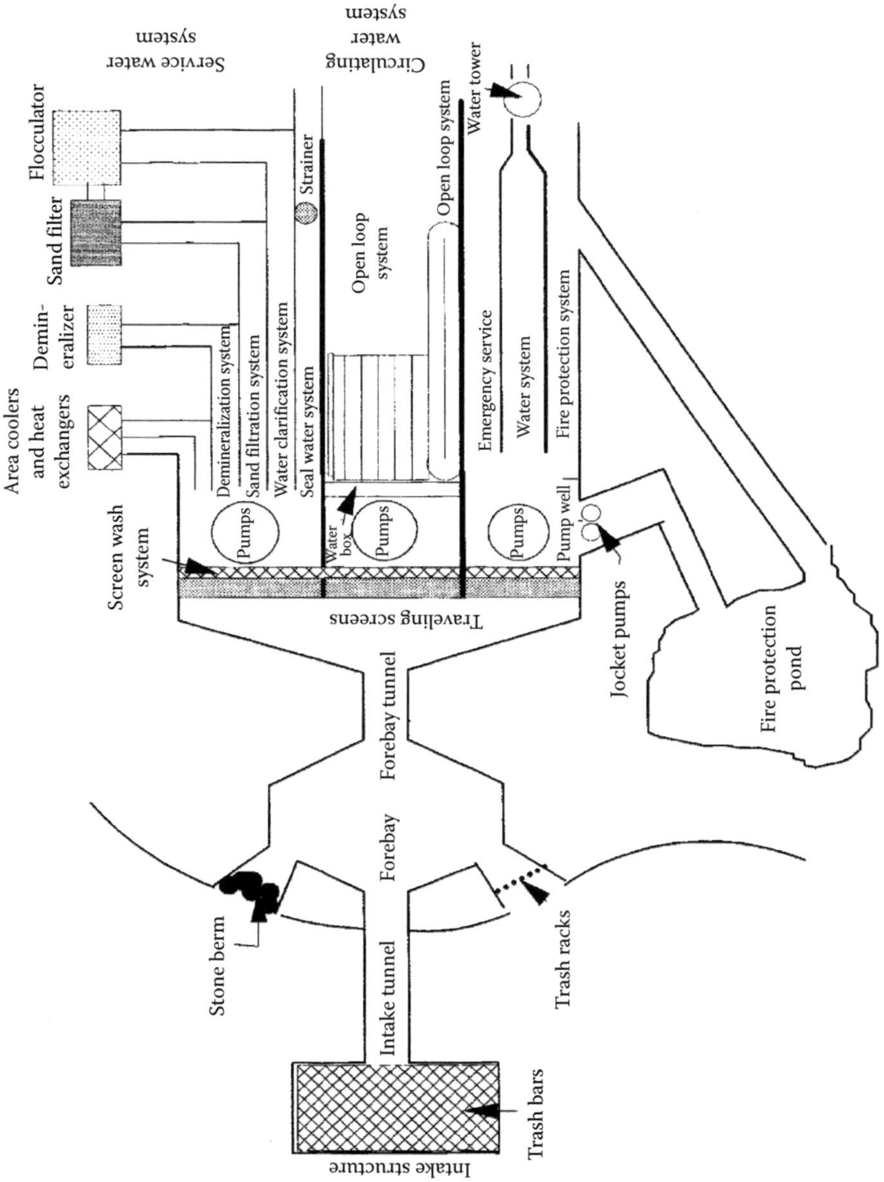

FIGURE I.1 A hypothetical facility containing systems and components of biofouling.

TABLE I.1
Typical Systems, Components, and Structures Subjected to Severe Biofouling by Zebra and Quagga Mussels in the Great Lakes

- Intake tunnel, trash racks, and small grates
- Forebays and holding ponds/storage tanks
- Traveling screens and support structures
- Pump suction chambers, pump wells, and wet wells
- Screen wash systems
- Low lift pumps and pump bell housings
- CWS
  - Condenser water boxes and condenser tubes
  - Instrumentation
    - Pressure taps, water sampling lines, and level and flow measuring equipment
- SWS
  - Motor air coolers and lube oil coolers
  - HVAC chillers and local air coolers
  - Air compressor intercoolers
  - Emergency or back-up heat exchangers
  - Seal water systems
  - Strainers and filters
  - Drain or return lines
  - Sand filtration system
  - Water clarification system
- Emergency SWS
- Fire protection system
  - Pumps, nozzles, and sprinklers
- Piping, vent lines, and air release valves
- Aqueducts

Step 2: "Should we be doing something now?" We advocate a proactive approach. Chapters 7 and 8 (for biofouling bivalves) and Chapter 9 (for biofouling gastropods) will provide you with information on solving problems that you are likely to experience if you allow biofoulers to colonize near your facility. If your budget allows, in the time remaining in your fiscal year, at the very least install a monitoring program. Perform a mainstream sampling program (Chapter 4) and find out if mussels are in your area or even in your facility. If you do not find mussels in your area, find out the closest recorded sighting. As a rule of thumb, in lakes mussels are known to disperse at a rate of at least 250 km/yr and in rivers at about 1000 km/yr in the downstream direction (see Chapter 2). Use these estimates as minimum dispersal rates. Once the mussels arrive at your facility, you can expect biofouling problems in eight months if you are in warm parts of the globe and in the second year if you are in a temperate region. The problems to expect are described in Chapters 5 and 6. However, the rate and magnitude of accumulation depend largely on the abundance of veligers (and adults) arriving at your facility and on the length of the breeding season. Remember, it is not necessarily the adults that produce the veligers in the immediate vicinity of your plant that will cause problems. It may be the populations upstream that produce veligers, which develop while drifting with currents for three to five weeks, that could be the primary source of mussels. The sidestream monitoring program will tell you the abundance and rate at which veligers enter and settle within your facility. A good detection program will cost between $2000 and $2500 for capital costs and probably between $5000 and $10,000 for operating costs (electricity for pumps, if necessary; labor for sampling and analyses; and expendables such as vials, jars, etc.). The more samples you take, the more costly the program, but the better informed you will be about the arrival, abundance, growth rate, and settling rate of the biofoulers. Now, you need to know the rest of Chapter 1, that is, what the adults and larvae look like (Section 1.2.1), the reproduction and the life cycle (Section 1.4.1), and population dynamics (Section 1.5).

Step 3: "What structures should I protect?" The most immediate concern is those structures that, if biofouled, will immediately restrict the flow of water. Fixed grates are of immediate concern, but various intake structures or openings, stone (or other materials) berms, and trash racks must also be monitored. These can become fouled sufficiently even in the first year of infestation, especially if the openings or spaces (e.g., between bars of the trash bar structures) are very narrow (e.g., <7.5 cm or 3 in.). Next, one should take care of the pipes and tunnels where the water velocity is chronically <1.5 m/s or 5 ft/s. Van Cott et al. (1993) found that zebra mussel settlement is proportional to flow rate (up to 5 ft/s) and that infestation along the length of the pipe is proportional to the duration of flow in the pipe. They also showed that hydraulic capacity (measured as the Hazen Williams coefficient of friction) can be decreased within four weeks after the settlement event begins. After about 2 years of heavy infestation, clumps (called druses) of mussels start breaking off and, through downstream transport, will plug small orifices, such as those of condenser tubes. Therefore, by the second year at the latest (i.e., after their first year of discovery), one should have operational a control program that will prevent the buildup of mussels in pump wells and/or suction chambers, pumps and pump bell housings, water boxes, structures on the submerged discharge side of traveling screens, and strainers. Traveling screens, if present, will remove any large clusters collected on their inflow side. Exposed parts of the traveling screen and associated structures are rarely, if ever, fouled, at least not enough to affect the efficiency of operation. Also by the second year, one should have in operation control programs that will prevent the biofouling of pipes, structures, and subsystems of the SWS; the CWS (if present); the fire protection system; and the emergency SWS (if present).

Step 4: "How will I protect my facility?" This is the most complex of the four steps. There is only one control option, the infiltration gallery that will prevent biofouling of all structures and systems within a facility. But it is expensive and requires a year or two to be installed and operational, *if* it can be utilized at your facility (see Chapter 7). For all other options, there are a minimum of two, but usually three, combinations of strategies required to completely protect your facility. One is invariably mechanical and requires man hours, such as periodically cleaning out strainers. The other (one or two) can be chemical or physical, depending on your circumstances. We advocate chlorination if you already have it on site. But there are some decisions to make before choosing chlorine as an option. Your regulatory agency (e.g., Department of Environmental Conservation, Ministry of Environment, Environmental Protection Agency, or similar authorities) has standards and guidelines that you will have to follow. You will also have to consider the volume of dilution water or some mechanism for dechlorination, the effect of chlorine on plant processes, increased corrosion rates, and so on. For details, see Chapter 8.

Step 5: "How do I know my various control programs are working?" You are probably using at least two options: controlling the mussels at the intake structure (e.g., coating the bar racks with antifoulant—see Chapter 7) and protecting the rest of the system (e.g., chlorination at the intake structure—see Chapter 8). You will learn that there are two basic kinds of control, continuous and periodic, with variations in each. But the procedures for verifying the effectiveness of your control program will be similar. The procedures are described in Chapter 4. They are the same procedures as for detecting the presence of mussels in your facility, except that emphasis is placed on calculating the percent mortality of mussels in a monitor receiving treatment and comparing it with the percent mortality of mussels in monitors not receiving any treatment. There are alternatives, and all of them are discussed.

Step 6: "How can I reduce the costs of control?" Once you have paid for the capital costs of installation, the only other way you can save money is to reduce the amount of treatment. This is accomplished with a "minimization control program." There are two ways of reducing the amount of treatment: one is to reduce the amount of chemical mitigant (or physical mitigant) at each application, and the other is to reduce the number of applications with the reduced amount of mitigants (see Chapters 7 through 9). Obviously, the latter can apply only to intermittent programs. Chapter 10 gives the control methods used in other countries.

INVASIVE GASTROPODS

Apple snails and thiarids have impacts on agriculture, ecology, and human health. The faucet snail and the New Zealand mudsnail have ecological impacts, mostly through competition with native species, but they also appear in intakes of industries and utilities and can be controlled using the same methods given above for biofouling bivalves. The impacts of invasive gastropods are not nearly as diverse as for bivalves, mainly because industries and utilities are more diverse systems than rice fields. Hence the monitoring methods described in Chapter 4 to detect the presence, and measure the abundance, of apple snails and thiarids are very different and resemble methods used for benthic studies. While the different kinds of impacts by invasive gastropods are not as diverse as for invasive bivalves, they can be just as much of a nuisance, and even more dangerous to human health, than biofouling bivalves, as described in Chapter 6. Most of the control options for nuisance gastropods are labor-intensive, involving handpicking for larger species such as the applesnail, or chemical control. Chemicals, especially molluscicides, are more commonly used to control apple snails than for biofouling bivalves. Because most of the snail impacts are in streams, lakes, or marshes, chlorination and other chemicals that impact other organisms are rarely used, unless the ecosystems are closed systems. Hence, Chapter 9 is devoted solely to the control methods used for invasive gastropods.

1 Biology

1.1 INTRODUCTION

Complete descriptions of the biology of all invasive macrofouling mollusks would require a tome and is beyond the scope of this book. Brief descriptions of the external and internal structures are provided to introduce the reader to the most common terminology used in describing the macrofouling bivalves and snails. The overviews examine only those biological features that are relevant to the monitoring and control of the invasive macrofouling mollusks. Description of their shell and soft parts is essential because they not only help to explain why the mollusks are pests but the descriptions are also important for distinguishing it from native species of mollusks. Some explanation of their habits and habitat is important for knowing why they can live and be so prolific in pipelines, on intake structures, in canal systems, in aqueducts, and in lakes and rivers. Knowledge of the mollusk's reproductive potential helps to explain their biofouling nature and why they are so much more prolific than our native species that also live in fresh waters. Information on their life cycles and population dynamics are crucial for successful control of the species. The efficient dispersal mechanisms employed by the nuisance mollusks help to explain not only why they have spread so rapidly to inland lakes and rivers across continents, but also how they can invade virtually any part of a facility or water body that are not protected. An understanding of their physiology helps to explain why some natural physical (e.g., heat) and chemical properties (e.g., salinity, pH, alkalinity, calcium levels, dissolved oxygen levels) and nutrient (e.g., total phosphorous and chlorophyll) levels can be effective control agents. Finally, knowledge of their potential impacts on ecosystems, industries, and utilities, as well as of the many human-made physical and chemical mitigants that are being used to control the mollusks, is crucial for preserving the health of our raw water supplies.

This chapter briefly examines those aspects of life and processes that contribute to the macrofouling nature of bivalves and snails. Bivalves belong to a class of mollusks (Bivalvia) with two calcareous valves and the snails (Gastropoda) that usually have a coiled calcareous shell. The two classes are discussed separately.

The freshwater macrofouling bivalves consist of two forms commonly referred to as mussels (e.g., zebra mussels, quagga mussels, golden mussel) and clams (e.g., Asian clams). The freshwater mussels are emulated by the marine blue mussel, *Mytilus edulis* (Figure 1.1a and b) and the freshwater clams by the Mapleleaf, *Quadrula quadrula* (Figure 1.1c and d). Typically, mussels are marine bivalves that have equal-size valves, unequal-size adductor muscles (i.e., are heteromyarian, the anterior adductor smaller than the posterior one), a byssal mass used to attach the mussel to solid surfaces (usually), and are epibenthic, or live on the surfaces of objects bathed by water. The surfaces can be objects on the bottom (e.g., rocks,

FIGURE 1.1 **(See color insert following page 286.)** (a) *Mytilus edulis*, outer view of left valve; (b) inner view of right valve. (c) *Quadrula quadrula*, outer view of left valve; and (d) inner view of right valve.

animals), objects projecting up from the bottom (e.g., plants, logs) of water bodies, floating objects (e.g., buoys, boat hulls), and walls of pipelines and intake structures. Clams have equal-size valves, equal-sized adductor muscles (i.e., are homomyarian), lack a byssus (in adult life), and are infaunal (spend most of their time buried in the sediments). Most of the native bivalves in North America are clams, especially the fingernail and pea clams of the family Sphaeriidae. Ambiguity in the use of the term "mussel" arises within the family Unionidae, commonly referred to as "freshwater pearly mussels" or "freshwater mussels" (or even unionids or naiads). Yet, these bivalves lack all the features that characterize marine mussels and have all the features characteristic of marine and freshwater clams. Since the term "mussel" is already pervasive in the freshwater molluskan literature, the usage of the term "mussel" in this book (see Glossary) is broad and includes the byssate bivalves as well as the large freshwater bivalves (i.e., unionids).

There are three families of macrofouling bivalves discussed here, the Dreissenidae, Mytilidae, and Corbiculidae. The Dreissenidae includes *Dreissena polymorpha* (zebra mussel), *Dreissena bugensis* (quagga mussel), and *Mytilopsis leucophaeata* (dark false mussel, Conrad's false mussel, platform mussel). *Limnoperna fortunei* (golden mussel) is mytilid and *Corbicula fluminea* (Asian clam) is a corbiculid.

The only other group of freshwater mollusks is the snails, which generally have a single coiled shell, although the limpets have a single uncoiled shell that looks like a single valve. All the macrofouling snails discussed below have a spirally coiled shell. The species selected as invasive macrofoulers include those that have spread quickly throughout North America in the last 10 years and have caused such major havoc that mitigants are required to control their population growth and spread. There are four gastropod families discussed, the Ampullariidae, which include six species, *Marisa cornuarietis* (giant ramshorn snail), *Pomacea canaliculata* (golden apple snail), *Pomacea haustrum* (Titan applesnail), *Pomacea insularum* (island applesnail), *Pomacea diffusa* (spiketopped applesnail), and *Pomacea paludosa* (Florida applesnail). Two species, *Tarebia granifera* (quilted melania) and *Melanoides tuberculata* (red-rimmed melania), from the family Thiaridae are also discussed because they are hosts to dangerous human parasites. Finally, the bithyniid *Bithynia tentaculata* (faucet snail) and the hydrobiid *Potamopyrgus antipodarum* (New Zealand mudsnail) are also included.

1.2 MORPHOLOGY AND ANATOMY

1.2.1 External Morphology

1.2.1.1 Invasive Bivalves

The three families of invasive bivalves (Dreissenidae, Mytilidae, and Corbiculidae) must first be distinguished from the two families of native bivalves, the Sphaeriidae and the Unionidae. The dreissenids, mytilids, and corbiculids differ from the native bivalves in (i) shell form, (ii) mode of life, (iii) reproductive potential, (iv) larval life cycle, (v) population dynamics, (vi) distribution, (vii) dispersal mechanisms, (viii) physiology, (ix) potential negative impact on the ecosystem, and (x) potential negative impact on utilities, industries, society, and the economy.

The shells of dreissenids and mytilids are distinctly different from corbiculids, sphaeriids, and unionids in that the umbone (beak) is lateral rather than dorsal as on *Corbicula* and all native clams. Mytilids differ from dreissenids in possessing a nacreous layer but lacking a myophore plate. Figure 1.2 is a flowchart that distinguishes the five families of freshwater bivalves on the basis of adult shell features alone. Only the five invasive species are identified in the chart.

1.2.1.1.1 Unionidae

Unionids are large, averaging about 10 cm in shell length (Figure 1.3f and g) and the shells of many are thick and robust and have heavy pseudocardinal teeth (Figure 1.1b and c). Others have rather thin shells and either vestigial pseudocardinal teeth (Figure 1.3g) or none. The shells of unionids are not only larger than the invasive bivalves, but also have distinctly different shapes and ornamentations on the shell (e.g., pustules/nodules, rays, heavy ridges, and sculpture on the umbones). Some are triangular in shape (e.g., the rocky mountain ridged mussel, *Gonidea angulata*, Figure 1.3f and g) like dreissenids but the umbone is dorsal on unionids. Unionids are dioecious and produce parasitic larvae called glochidia. Many glochidia are host specific requiring only one or two species of fish for development into juveniles. One species, the salamander mussel, *Simpsonaias ambigua*, requires the mudpuppy, *Necturus maculosus*, for development into juveniles. Most species have long life spans, usually greater than 10 years; some, like the Eastern pearlshell, *Margartifera margaritifera*, can live more than 100 years, making it the longest-living invertebrate on the planet (Nedeau et al., 2000). Details on form and function, ecology,

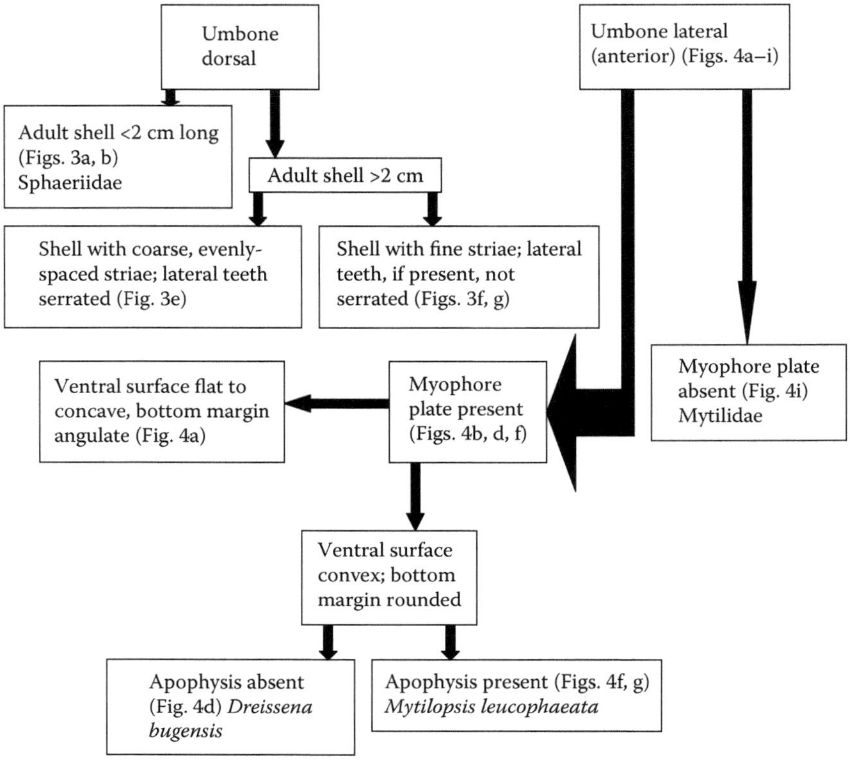

FIGURE 1.2 A flowchart for identifying the two native families (Sphaeriidae and Unionidae) and the three invasive families (Corbiculidae, Dreissenidae, and Mytilidae) and their species in freshwaters of North America, with illustrations in Figure 1.3.

physiology, and reproduction and development can be found in Purchon (1977), Wilbur and Yonge (1983a–g), and McMahon et al. (1993).

1.2.1.1.2 Sphaeriidae

The burrowing bivalves have central to sub-central umbones, the sphaeriids being the smallest in size, varying from 2 mm in shell length for some *Pisidium* species to 2 cm for some *Musculium* and *Sphaerium* species (Figure 1.3a and b); most adults average 5–8 mm in shell length. Sphaeriids have delicate cardinal and lateral teeth of diagnostic shapes (Figure 1.3b). Sphaeriids are hermaphrodites and brood their young in brood sacs on the inner gills of the adults (Mackie, 2007). Their life span typically varies from one to two years. For details on anatomy, life history, distribution, physiology, ecology, and taxonomy, see Mackie (2007).

1.2.1.1.3 Corbiculidae

1.2.1.1.3.1 Asian Clam, Corbicula fluminea *(Prashad, 1929)* The taxonomy within this genus is still unresolved (Kornuishin, 2006). There is no agreement

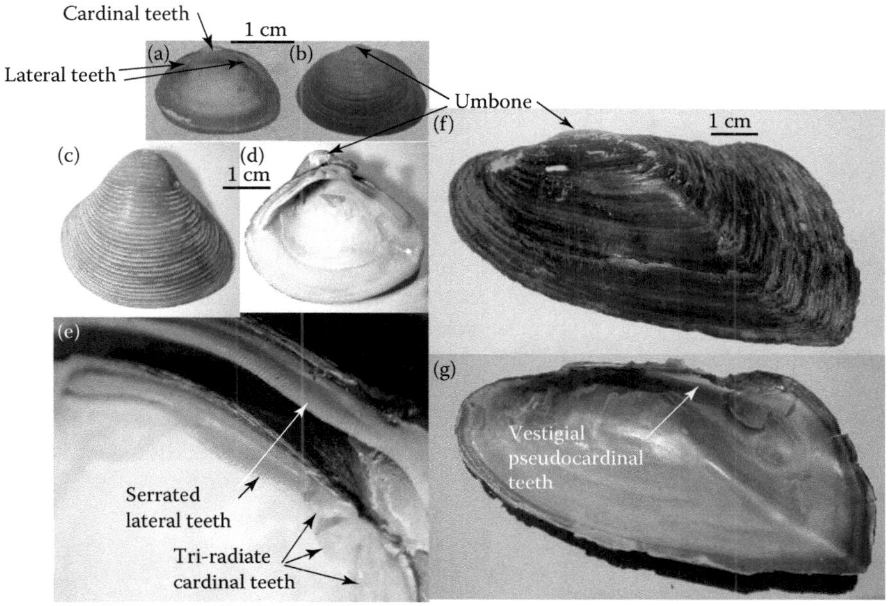

FIGURE 1.3 **(See color insert following page 286.)** (a–c) Infaunal (burrowing) bivalves. All upper images are outer views of left valves; all lower images are of inner views of valves. (a, b) Sphaeriidae; (c, d, e) Corbiculidae [(e) shows serrated lateral teeth]; and (f, g) Unionidae.

among investigators neither about the general number of valid species that vary from six in China (Prashad, 1929) to only two in the world (Morton, 1986), nor about the species identity of the two morphotypes in North America recorded one in South America (Ituarte, 1984, 1994; Morton, 1986; Kinzelbach, 1992; Araujo et al., 1993; Swinnen et al., 1998; Siripattrawan et al., 2000; Pfenniger et al., 2002).

Two modes of reproduction occur in Corbiculidae. Most Asian taxa, including *Corbicula japonicus*, are dioecious and produce free-swimming larvae. All North American species (morphs) are hermaphroditic and incubate their larvae. Morton (1986) assigned all nonincubating forms to *Corbicula fluminalis* in Asia and all incubating forms to *Corbicula fluminea*. However, three distinct shell morphotypes have been distinguished (Lee et al., 2005); two of them occur in North America (Hillis and Patton, 1982), a widespread "Form A" and a southwestern "Form B" (Britton and Morton, 1986). A third morphotype, "Form C," has been described in La Plata, Argentina (Ituarte, 1994). All three morphotypes are hermaphroditic (McMahon, 1991; Ituarte, 1994). Form A and Form B morphotypes represent two distinct allozymic (Hillis and Patton, 1982; McLeod, 1986) and mitochondrial (Siripattrawan et al., 2000) lineages. Burch et al. (1998) found that both morphs are triploid ($3n = 54$), lack meiotic metaphases, and produce unreduced sperm. Additional differences among morphs and dioecious and monoecious forms are provided later in Section "1.4.1.3.1 Reproduction" but the current accepted opinion is that there is only one species, *Corbicula fluminea*, currently living in North America.

Corbicula fluminea (Figure 1.3c–e) is of a medium size, averaging 3–4 cm in shell length. The shell is glossy and has heavy, evenly spaced ridges or striae. The cardinal teeth are tri-radiate with three heavy protuberances (Figure 1.3d and e). The lateral teeth are heavy and serrated (Figure 1.3e).

1.2.1.1.4 Dreissenidae

There are three species of dreissenids in North America. Two species belong to the genus *Dreissena* and one to *Mytilopsis*. The dreissenids can be distinguished easily from the mytilids by the presence of a myophore plate on the inner shell at the umbone (Figure 1.4b, d, and f) but lacking a nacre; mytilids lack a myophore plate (Figure 1.4i) but possess a nacre. Conrad's false mussel can be separated immediately from the dreissenids by the presence of an apophysis under the myophore plate of each valve (Figure 1.4f and g). The anterior adductor muscle attaches to the myophore plate and the apophysis of each valve. Kauffman (1969) suggests that *Mytilopsis* species are better suited for more turbulent conditions because the

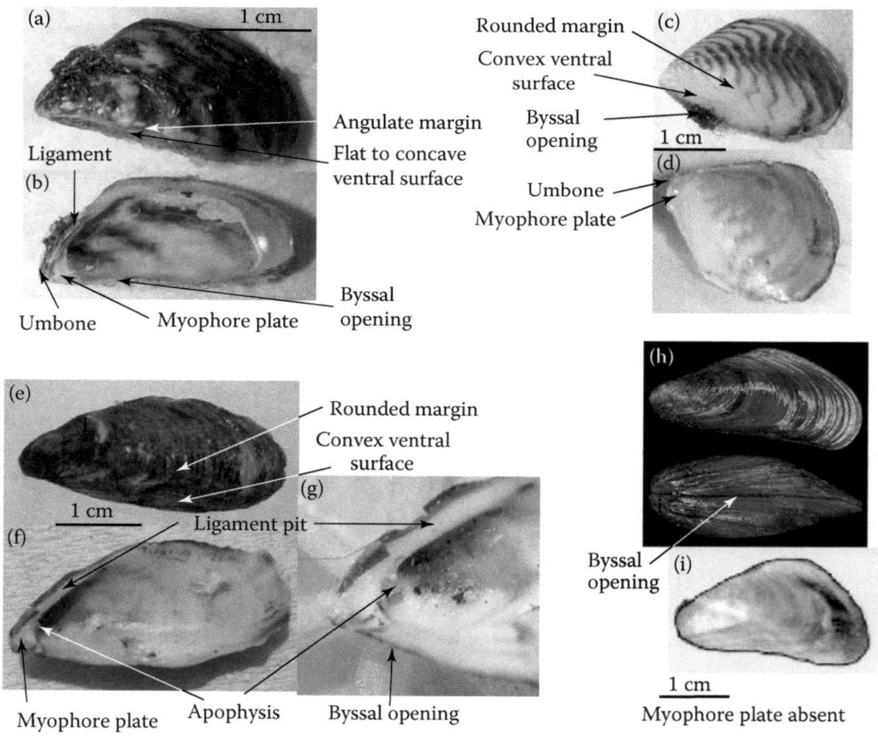

FIGURE 1.4 (See color insert following page 286.) (a–i) Epibenthic invasive bivalves. All upper images are outer views of left valves; all lower images are inner views of right valves. (a–g) Dreissenidae; (h, i) Mytilidae. (a, b) *Dreissena polymorpha*; (c, d) *Dreissena bugensis*; (e, f, g) *Mytilopsis leucophaeata* [(g) = 3× magnification of umbonal end]; and (h, i) *Limnoperna fortunei* in side view (upper) and ventral view to show byssal opening.

apophysis may result in a stronger byssal attachment to the substrate than in the zebra mussel (MacNeill, 1991c) and the quagga mussel.

The most common dreissenid is the zebra mussel, its common name being derived from the pattern of zebra stripes on the shell and its scientific name, *polymorpha*, from the many morphs or forms that occur in the shell's color pattern, including albino and solid black or brown. Most have alternating zigzag lines of brown and yellow (Figure 1.3g). There has been some confusion in distinguishing both *Mytilopsis leucophaeata* and the quagga mussel from *Dreissena polymorpha* in North America (O'Neill, 1990) because of the similar attachment mechanisms, shell shape, and coloring and color patterns. Conrad's false mussel is usually uniformly dark with no stripes (Figure 1.4e) but some specimens, especially juveniles, have distinct stripes or zigzag lines, much like *Dreissena polymorpha* (Figure 1.4a). In addition, not all zebra mussels are striped. Some are all black, brown, or white, occasionally making it difficult to distinguish *Mytilopsis leucophaeata* from *Dreissena polymorpha* (O'Neill, 1990). In practice, one could distinguish the zebra mussel from the quagga mussel by placing the shells with the ventral side down; the quagga mussel topples over, whereas the zebra mussel usually remains upright on its bottom surface. However, some zebra mussels are heavily arched (but still strongly angulate on the lateral–ventral margin) and topple over.

1.2.1.1.4.1 Zebra Mussel, Dreissena polymorpha *(Pallas, 1771)* Pathy and Mackie (1993) gave detailed descriptions of the three North American dreissenids. Only brief descriptions are provided here. The adult shells average 2.0–3.0 cm in shell length but may reach 4.0–4.5 cm. Externally, the umbone of the zebra mussel is acute and lies anterior in position (Figures 1.4a, b, and 1.5b). The ventral side of the shell of a typical zebra mussel specimen is flattened with an acute lateral–ventral angle (Figures 1.4a, b, and 1.5a). In cross-section the shell is tent shaped and the ventral surface may be flat to slightly concave (Figure 1.5a). In longitudinal section, the shell is elongate with a roundly angled ventral posterior margin (Figure 1.5b).

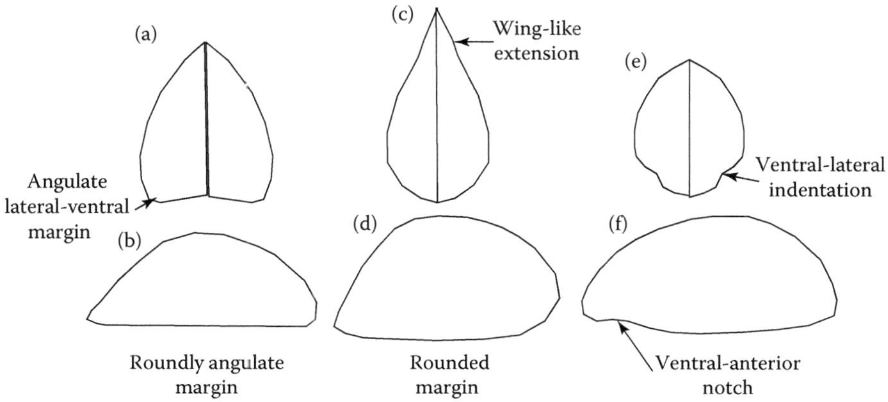

FIGURE 1.5 Cross (upper row, a, c, e) and longitudinal sectional views (b, d, f) of the shells of three dreissenids: (a, b) *Dreissena polymorpha*; (c, d) *Dreissena bugensis*; (e, f) *Mytilopsis leucophaeata.*

1.2.1.1.4.2 Quagga Mussel, Dreissena bugensis *(Andrusov, 1897)* There is considerable variation in shell morphology (e.g., shell length:shell height ratio, shell length:shell width ratio, shell height:shell width ratio) and shell patterns in the quagga mussel (Dermott and Munawar, 1993; Spidle, 1994; Claxton et al., 1998). The quagga mussels that occur in the Great Lakes are of two morphologically distinct forms, one that occurs in some deep sections of Lakes Erie and Ontario ("profunda") and one that is more common in shallow water ("epilimnetic") (Dermott and Munawar, 1993; Spidle, 1994; Claxton and Mackie, 1998). Genetic studies suggest that the two forms are phenotypic variants rather than genetically distinct taxa (Spidle, 1994; Spidle et al., 2005), but Claxton and Mackie (1998) recommend further analysis using faster evolving genes to show genetic differences.

Therriault et al. (2004) used DNA sequence analyses of two mitochondrial gene fragments to determine phylogenetic relationships among six dreissenids, including *Dreissena rostriformis*, *Dreissena bugensis*, *Dreissena polymorpha*, and *Mytilopsis leucophaeata*. "Sequence divergence between *Dreissena rostriformis* and *Dreissena bugensis* was relatively low (0.3–0.4%), suggesting that these two taxa constitute a single species." However, environmental differences suggest two races of *Dreissena rostriformis*, a brackish water race (*rostriformis*) and a freshwater race (*bugensis*) (Therriault et al., 2004). On this basis, the quagga mussel is herein referred to as *Dreissena bugensis* (Andrusov, 1897) and not a subspecies of *Dreissena rostriformis* (Desh, 1838).

Adult shells average 2.0–3.0 cm in shell length but may grow up to 3.5–4.0 cm. In the quagga mussel the ventral side is convex and the ventral–lateral margin is rounded (Figure 1.5c). Usually the upper portion of the posterior half of the shell is expanded, thin, and "winglike," giving the shell a fan shape in side view (Figure 1.5d).

1.2.1.1.4.3 Conrad's False Mussel, Mytilopsis leucophaeata *(Conrad, 1831)* The taxonomy of *Mytilopsis* is perplexing because of a wealth of synonyms and because the systematic relationships of species within *Mytilopsis* are vague (Marelli and Gray, 1985). Some authors (e.g., Morton, 1970) believe *Mytilopsis* is a junior synonym of the extinct genus *Congeria* (Partsch, 1835); others (e.g., Marelli and Gray, 1985) believe it to be a subgenus of *Congeria*. It is probable that both *Dreissena* and *Mytilopsis* evolved from the extinct genus *Congeria* (Morton, 1970; Mackie et al., 1989; Partsch, 1835). However, on the basis of shell morphologies (Marelli and Gray, 1985) and recent genetics (Therriault et al., 2004), it appears that *Mytilopsis* and *Congeria* are more closely related to each other than to *Dreissena*. Recent work treats *Mytilopsis* and *Congeria* as distinct genera. Marelli and Gray (1985) suggest that there are only five valid species of *Mytilopsis*, which in turn are separated into two distinct species groups based on conchological characteristics. Some authors refer to Conrad's false mussel as *Congeria leucophaeata* (Archambault-Guezou, 1982), while others refer to it as *Mytilopsis leucophaeata* (Morton, 1970; Mondadori, 1980; Britton and Morton, 1989; Therriault et al., 2004). For this book the mussel will be referred to as *Mytilopsis leucophaeata*.

Common names are also varied within the species. *Mytilopsis leucophaeata* is most commonly known as Conrad's false mussel but the dark falsemussel (also spelled as dark false mussel, false dark mussel) and platform mussel are also used (e.g., Morris, 1975; Mondadori, 1980; Howells et al., 1996). Conrad's false mussel is used here because (1) the species was first described by Conrad (1831); (2) spelling of dark false mussel is inconsistent; (3) platform refers to myophore plate (septum) that is present in other dreissenid species.

The adult shell averages 1.5–2.5 cm in shell length but may grow to 3.0–4.0 cm. The shell is elongate, somewhat rounded in cross-section (Figure 1.5e) and usually has a distinct ventral–lateral indentation (Figure 1.5f). The adult shell is brownish and lacks stripes, the striae are coarse and the periostracum is often loose or "flaky." The most diagnostic feature of Conrad's false mussel is the presence of an apophysis on the underside of the myophore plate (Figure 1.4g).

1.2.1.1.4.4 Golden Mussel, Limnoperna fortunei *(Dunker, 1857)* The mean shell length varies between 2.0 and 4.0 cm but may reach 6 cm in shell length under ideal conditions. Juveniles are considered to be between 2.0 and 8.0 mm in shell length. The shell periostracum is golden or yellowish, often glossy in young, healthy specimens. Like dreissenids, the inner shell lacks hinge teeth but unlike dreissenids they lack a myophore plate and they have a nacre that appears bluish or purplish in older specimens. The byssal opening is also more posterior in position than in dreissenids (Figure 1.4h).

1.2.1.2 Gastropoda

There are 15 families of freshwater gastropods in North America divided between two subclasses, Pulmonata and Prosobranchia. Pulmonates are represented by five families, Acroloxidae, Ancylidae, Lymnaeidae, Physidae, and Planorbidae, all of which can be distinguished from each other by features shown in Figure 1.6. Pulmonates have lungs and can utilize atmospheric oxygen, are hermaphroditic, and lack an operculum (a lid that closes the aperture) (Figure 1.7a–f). Prosobranchs have gills that derive oxygen dissolved in the water, are generally dioecious, and have an operculum that closes the aperture and protects the soft parts within the shell. The Prosobranchs are represented by 10 families, Ampullariidae, Bithyniidae, Hydrobiidae, Neritidae, Pleuroceridae, Pomatiopsidae, Thiaridae, Valvatidae, and Viviparidae, all of which can be identified by diagnostic features shown in Figure 1.7; Micromelaniidae is represented by a single species, *Antroselates spiralis*, a subterranean species that lives in caves and is not treated here.

The descriptions below are for species that are documented to be introduced and a nuisance and are extracted from the list in Keller (2006) and Keller et al. (2007). Of the four species of Ampullariidae that have been introduced (*Marisa cornuaurietis*, *Pomacea bridgesii*, *Pomacea canaliculata*, and *Pomacea haustrum*), two, the Ram's horn snail *Marisa cornuaurietis* and the golden applesnail *P. canaliculata*, are documented to produce economic and environmental impacts and are discussed below.

1.2.1.2.1 Invasive Gastropoda

There are species in five families of gastropods (Ampullariidae, Bithyniidae, Hydrobiidae, Physidae, and Thiaridae) that are purported to have been introduced and have been or are threatening to be an environmental and/or economic nuisance in North America. Within the Physidae, the European freshwater pulmonate, *Physa*

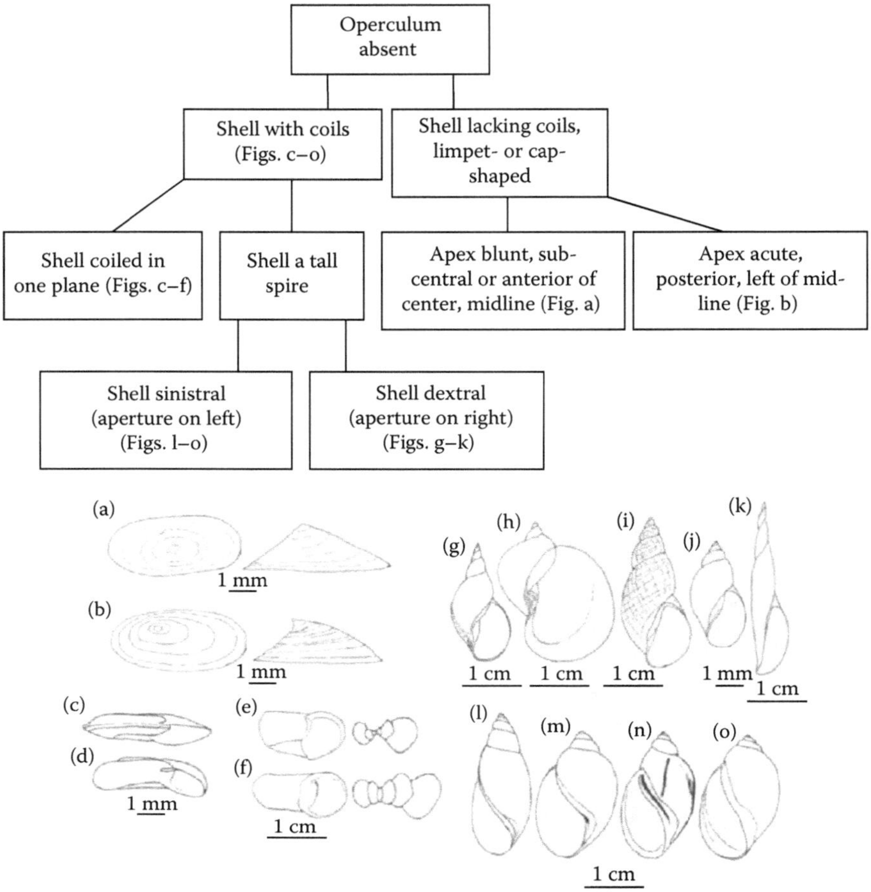

FIGURE 1.6 Flow diagram to identify five families of Pulmonata. Figure numbers correspond to those below the flowchart. (a) Ancylidae (left is dorsal view, right is lateral view); (b) Acroloxidae (left is dorsal view, right is lateral view); (c–f) Planorbidae [(c) *Promenetus*, (d) *Gyraulus*, (e) *Helisoma*, and (f) *Planorbella*]; (g–k) Lymnaeidae [(g) *Lymnaea*, (h) *Radix*, (i) *Stagnicola*, (j) *Fossaria*, and (k) *Acella*]; (l–o) Physidae [(l) *Aplexa*, (m–o) *Physa*].

acuta, is rapidly spreading throughout North America and is forming massive populations in some water treatment facilities. However, recent genetic evidence by Dillon et al. (2002) suggests that the European freshwater pulmonate is not reproductively isolated from American *Physa heterostropha* or *Physa integra* species and is probably native to North America. Therefore, the species is not considered further here. The ampullariids, the faucet snail, *Bithyinia tentaculata* of the family Bithyiniidae, the New Zealand mudsnail, *Potamopyrgus antipodarum* of the family Hyrobiidae, and the red-rimmed melania, *Melanoides tuberculata* of the family Thiaridae are also considered to have documented environmental or economic impacts in North America (Keller, 2006; Keller et al., 2007).

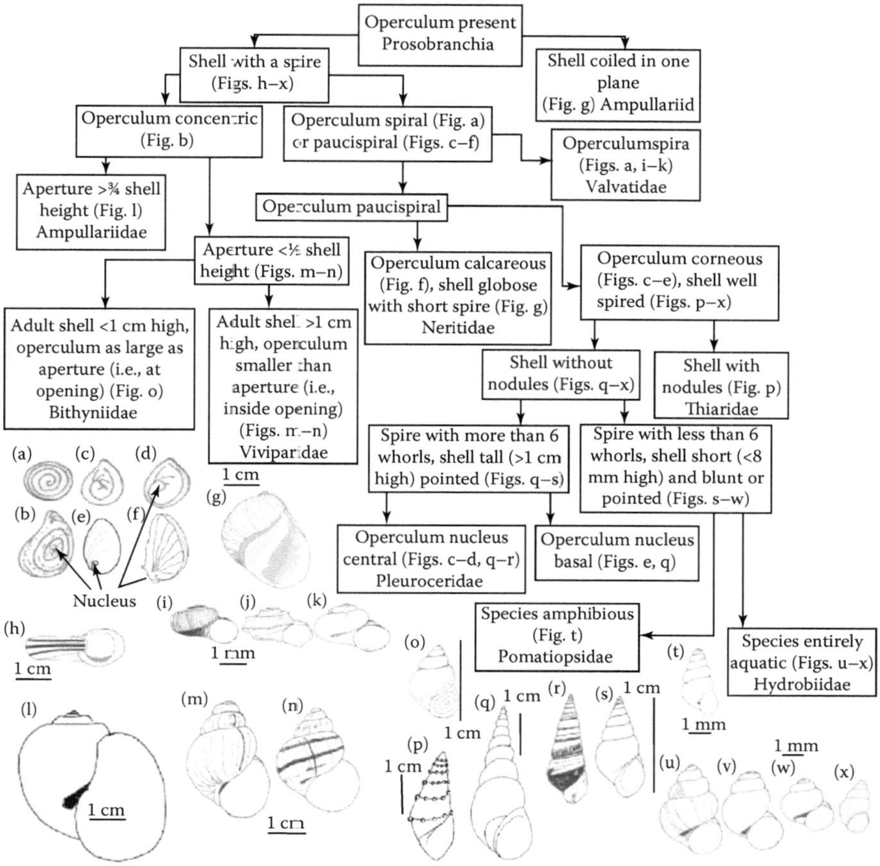

FIGURE 1.7 Flow diagram to identify nine families of Prosobranchia. (a–e) corneus opercula: (a) spiral; (b) concentric; (c, d) paucispiral typical of pleurocerids and hydrobiids; (e) paucispiral, of thiarids; (f) calcareous operculum of Neritidae. (g–x) aperture views of shells: *Neritina usnea*; (h) *Marisa cornuarietis*; (i–k) Valvatidae, (i) *Valvata sincera*, (j) *Valvata tricarinata*, (k) *Valvata piscinalis*; (l) *Pomacea*; (m–n) Viviparidae, (m) *Campeolma decisum*, (n) *Viviparus georgianus*; (o) *Bithyinia tentaculata*; (p) *Tarebia granifera*; (q) *Melanoides tuberculata*; (r–s) Pleuroceridae, (r) *Pleurocera acuta*, (s) *Elimia livescens*; (t) *Potamiopsis lapidaria*; (u–x) Hydrobiidae, (u) *Marstonia lacustris*, (v) *Probythinella lacustris*, (w) *Amnicola limosa*, (x) *Potamopyrgus antipodarum*.

1.2.1.2.1.1 Ampullariidae Rawlings et al. (2007) examined the phylogenetic relationships among 46 unique mitochondrial haplotypes from 95 individual apple snails for which complete sequences were generated; their analyses were based on combined *12S–16S* (~1900 base pairs) and *COI* (~658 base pairs) datasets. They concluded that five species of nonindigenous ampullariids have been introduced to North America. Based on comparison with native-range samples and type material, and knowledge of type localities, the three clades represented three applesnail species with channeled sutures (*Pomacea insularum*, *Pomacea canaliculata*, and *Pomacea haustrum*). "*Pomacea canaliculata* was strongly supported as sister to *Pomacea insularum*, with *Pomacea paludosa* robustly supported as sister to these two species. Discrete *Pomacea haustrum* and *Pomacea diffusa* clades formed a clade that was sister to the *Pomacea insularum*, *Pomacea canaliculata*, and *Pomacea paludosa* clade" (Rawlings et al., 2007). Of the three apple snail species with channeled sutures in North America, *Pomacea canaliculata* occurred only in California and Arizona, *Pomacea insularum* occurred in Texas, Georgia, and Florida, and *Pomacea haustrum* was restricted to Florida. *Pomacea diffusa* was formerly a subspecies in a *Pomacea bridgesii* complex but Rawlings et al. (2007) provided phylogenetic evidence that *diffusa* also deserves species status.

The web site, www.applesnail.net (accessed October 30, 2008) provides useful information on shell description, food, behavior, and distribution for each species. The web site, http://www.applesnail.net/content/species_genera.htm (accessed October 30, 2008), is useful for separating the three channeled snail species from *Pomacea diffusa* (formerly *Pomacea bridgesii*). Figure 1.8 is adapted from the apple snail web site but adds *Pomacea insularum* as a distinct species to the comparisons because of the mtDNA sequence data results of Cowie et al. (2006) and Rawlings et al. (2007).

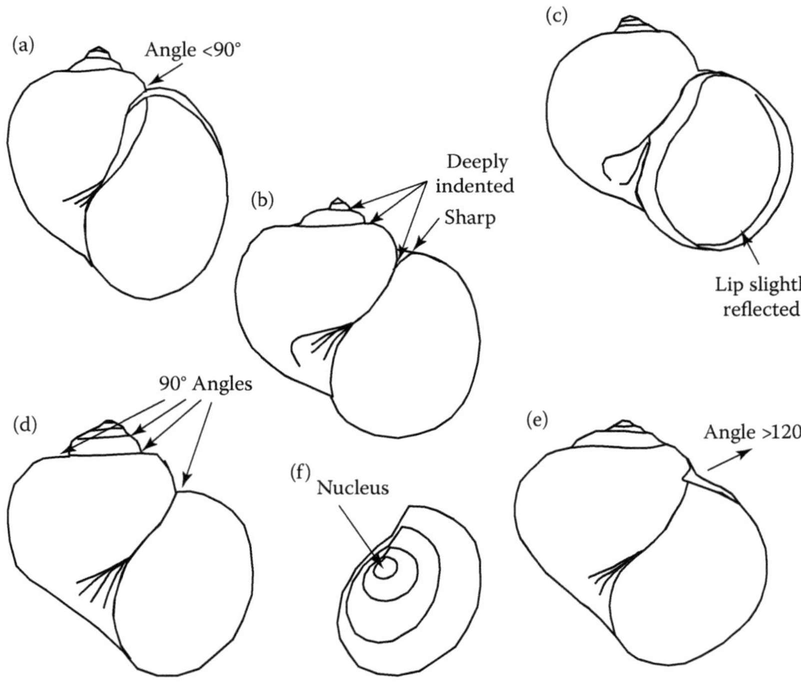

FIGURE 1.8 Diagnostic shell features of five species of *Pomacea*: (a–c) *Pomacea canaliculata* complex: (a) *Pomacea canaliculata*; (b) *Pomacea haustrum*; (c) *Pomacea insularum*. (d) *Pomacea diffusa*; (e) *Pomacea paludosa*. (f) Concentric operculum with offset nucleus.

Giant Ramshorn Snail, *Marisa cornuarietis* (Linnaeus, 1758) Is a planorbid-shaped shell (Figures 1.1c–f, 1.7h, and 1.9), the adults have 3.5–4 whorls with a slightly elevated spire, and the juveniles have fewer whorls and no spire. The adult shell is moderately large being 18–22 mm in diameter and 5–6 mm high. Transverse growth lines are pronounced on the adult shells near the aperture. Color and striae patterns are quite variable. The shell has three to six dark spiral bands that are yellow to dark brown or even black, of unequal width and are most pronounced near the umbilicus. Hence, the snails will often have a yellow and a black side. Some have a mutation in which the banding pattern is absent and the shell is completely yellow.

Golden Apple Snail, *Pomacea canaliculata* (Lamarck, 1822) The size and appearance of the shell of the golden apple snail, *Pomacea canaliculata*, is highly variable and Cazzaniga (2002) suggested that it should be more appropriately classified as *Pomacea canaliculata* complex. Within the complex he placed *Pomacea canaliculata*, *Pomacea haustrum*, and *Pomacea insularum*. Although the external differences in shape among the three species in the complex are subtle and the variation within the species complex is high, Rawlings et al. (2007) provided phylogenetic evidence that the three species with channeled sutures deserve species status.

FIGURE 1.9 (See color insert following page 286.) Invasive ampullariids of North America: (a) *Marisa cornuarietis*; (b) *Pomacea canaliculata*; (c) *Pomacea haustrum*; (d) *Pomacea insularum* (Modified from http://www.nwrc.usgs.gov/invasive_species/applesnail_images2.htm. With permission.); (e) *Pomacea paludosa*; and (f) *Pomacea diffusa*.

The golden apple snail is also spelled as the golden applesnail. The shell is globose, relatively heavy especially in older snails, has five to six whorls, a large oval to round aperture, and a large, deep umbilicus (Figures 1.8a and 1.9b). The species name, *canaliculata*, is derived from the deeply indented sutures between whorls. It can be distinguished from other *Pomacea* species by the >90° angle at the junction of the aperture whorl to the body whorl (Figure 1.8a). There is sexual dimorphism in that males tend to have a rounder aperture than females. The umbilicus is large and deep. The shell varies from 40 to 60 mm wide and 45 to 75 mm high, depending on the conditions. The periostracum color varies from completely yellow and green in cultivated forms to brown with or without dark spiral banding patterns around the whorls in wild forms. Shell growth occurs mainly in spring and summer; it aestivates in fall and winter.

Within the aperture is a moderately thick and corneous operculum, the growth lines being concentric about an offset nucleus (Figure 1.8f). The color varies from light yellow in young snails to dark brown in older ones. The operculum is slightly smaller than the aperture and can be retracted somewhat into the shell.

The color of the eggs can be useful in separating the species of Pomacea as well; the eggs of *Pomacea canaliculata* are bright orange (sometime bright reddish pink) and laid loosely in clutches of 200–600 eggs. The eggs vary in size from 2.20 to 3.5 mm (0.5–0.9 inch) diameter.

Titan Applesnail, *Pomacea haustrum* (Reeve, 1856) The shell is relatively large (9–12 cm in height), less round and higher than *Pomacea canaliculata*, and has heavy, dark-brown bands on a greenish to brownish color background. The sutures are deeply indented, forming greater than 90° angles at the junctions of the whorls (Figures 1.8b and 1.9c). The operculum is moderately thick with an offset nucleus (Figure 1.8f) and can be retracted within the shell. The eggs are greenish and deposited in clutches up to 230 (Cowie, 2002) above the water line on vegetation. The individual eggs are about 3–5 mm in size and compressed into polygonal shapes, giving the egg mass an irregular honeycombed appearance (Rawlings et al., 2007).

Island Applesnail, *Pomacea insularum* (D'Orbigny, 1839) This species has been confused with *Pomacea canaliculata* and *Pomacea haustrum* from Florida and Texas and as a consequence the original introduction of *Pomacea insularum* is unknown (Rawlings et al., 2007). Normal coloration typically includes bands of brown, black, and yellowish-tan, and color patterns are extremely variable. Albino and gold color variations exist (Benson, 2008c). The shell is globose with four to five whorls and can grow up to 15 cm in shell height, although 3–4 cm is more common. The species can be distinguished from *Pomacea canaliculata* by the presence of a strongly pleated columella (Figures 1.8c and 1.9d). The shell of this apple snail species is globose and relatively thick (especially in older snails) with a low spire. The umbilicus is large and deep. The operculum is moderately thick and has an offset nucleus (Figure 1.8f). The operculum can be retracted in the aperture (shell opening). The species lays 500–700 small pink eggs above the water surface.

Spiketop Applesnail, *Pomacea diffusa* (Blume, 1957) The *Pomacea bridgesii* complex consisted of two subspecies, *Pomacea bridgesii bridgesii* (Reeve, 1856), which is only known from the Rio Grande, Reyes River at Beni (bridges), and *Pomacea bridgesii diffusa* (Blume, 1957), which inhabits almost the entire Amazon river system (Pain, 1960; Cowie et al., 2006). It is the latter that inhabits several counties in Florida but as stated earlier each deserves full species status (Rawlings et al., 2007). The shell of *Pomacea diffusa* varies from 40 to 50 mm wide and 45 to 65 mm high, has a high, sharp spire hence the common name, spiketop applesnail (also called/spelled spike-topped, spiketopped, spike-top, golden mystery snail). The shell has about five to six whorls, the most diagnostic features being the square shoulders (flat at the top of the whorls) and almost 90° sutures (Figures 1.8e and 1.9e). The aperture is large and oval and the umbilicus is large and deep. The periostracum varies from yellow (often referred to as the golden mystery snail), green to brown, and usually has dark spiral bands. Cultivations have resulted in white and blue mutations, the white variation sometimes referred to as the ivory snail in the aquarium world. The operculum is moderately thick and corneous, it can be retracted in the aperture, and the nucleus is offset (Figure 1.8f). The color of the operculum varies from light to dark brown. The egg masses of *Pomacea diffusa* "have an irregular honeycombed appearance, like those of *Pomacea haustrum*, but are smaller and have a tan to salmon color" (Rawlings et al., 2007). The egg clutches range in size from 200 to 600 eggs, are pale pink to reddish, closely attached to each other, and deposited above the water surface. Their size varies from 2.20 to 3.5 mm (0.5–0.9 inch) in diameter.

Florida Applesnail, *Pomacea paludosa* (Say, 1829) The snail is native to Florida (and Cuba) but has recently spread to Hawaii as a few local populations (Rawlings et al., 2007). The shell is globose, 40–55 mm wide and 45–65 mm high, and has a yellow to green periostracum with reddish streaks and/or dark spiral bands. The shell aperture is large and oval with a relatively thick and corneous operculum that has an offset nucleus (Figures 1.8f and 1.9f). The shell has a large and deep umbilicus but the sutures are shallow, with greater than 120° angle (Figure 1.8e), which gives the snail a cone-shaped shell top (spire). The egg clutches are small (10 to 80 eggs) with whitish to slightly pinkish eggs that are loosely packed together and laid on emerging stems of vegetation. Although the egg clutches are small relative to other apple snails, the eggs of *Pomacea paludosa* are relatively large, about 3–6 mm in diameter.

1.2.1.2.1.2 Thiaridae

Quilted Melania, *Tarebia granifera* (Lamarck, 1822) The shell is tall (2–3 cm high), tapered, and sculptured with prominent nodules and overlapping sutures with crenulations (i.e., scalloped) (Figure 1.10). Chaniotis et al. (1980a) reported a maximum shell length of 35 mm for several thousand specimens measured in Puerto Rico. The base of the last whorl has prominent spiral ridges but lacking the nodules. The shell is always uniformly brown except on the top half of the spire where periostracum is grayish. It is approximately 0.40 times as wide as it is high (Thompson, 1984). There is some disagreement regarding whether or not the quilted

FIGURE 1.10 **(See color insert following page 286.)** *Tarebia granifera* showing aperture view.

melania established in the Gulf ecosystem compromises more than one species (Thompson, 1984) (http://nis.gsmfc.org/nis_factsheet.php?toc_id=163).

Red-Rimmed Melania, *Melanoides tuberculata* (Müller, 1774) Red-rimmed melania have an elongate conical shell with six to eight whorls that regularly decrease toward the nuclear whorl (Figure 1.11), which is often broken. The shell is light brown, usually mottled with rust colored streaks that may form a spiral band around the base of the body whorl. The spire is usually more than twice the length of the aperture and is sculptured with vertical, weakly curved ribs and much finer spiral striations (Morrison, 1954). Vertical ribs are prominent on the middle and upper whorls. The operculum is paucispiral, with the nucleus near the base (Figure 1.7e) (Morrison, 1954; Lee, 1973; Thompson, 1984). The shell at the aperture is thin and the umbilicus is lacking or vestigial.

1.2.1.2.1.3 Bithyniidae

Faucet Snail, *Bithynia tentaculata* (Linnaeus, 1758) The faucet snail is usually less than 10 mm high and 5–7 mm wide but may reach 15 mm in height under ideal conditions. The shell is oval in shape with five to six whorls and the periostracum is shiny pale brown. The whorls are separated by rather shallow sutures, the aperture is less than half the height of the shell, and the operculum is always located very close to the aperture of the shell (Figure 1.12). The umbilicus is vestigial or lacking.

FIGURE 1.11 **(See color insert following page 286.)** *Melanoides tuberculata*, aperture view.

FIGURE 1.12 **(See color insert following page 286.)** *Bithynia tentaculata.*

1.2.1.2.1.4 Hydrobiidae

New Zealand Mudsnail, *Potamopyrgus antipodarum* (Gray, 1843) *Potamopyrgus antipodarum* has an elongated shell with five or seven whorls separated by rather deep sutures (Figure 1.13). The average height of the shell in North America is about 5 mm (0.2 inch) but attains approximately 12 mm (1.5 inch) in its native range. The operculum is thin and corneus, paucispiral, and has an off-center nucleus (Figure 1.7a and b). The aperture is cval and its height is about 0.4 times the height of the spire. Some morphs, including many from the Great Lakes, have a keel in the middle of each whorl, the keel often being hirsute (with hairs). The shell periostracum can vary from gray and dark brown to light brown.

1.2.2 Internal Morphology

1.2.2.1 Bivalvia

1.2.2.1.1 Dreissenidae and Mytilidae

Internally, the shell of the three dreissenids and *Limnoperna fortunei* lack teeth known as cardinal (Figure 1.3a and d) and pseudocardinal teeth (Figure 1.3g) but

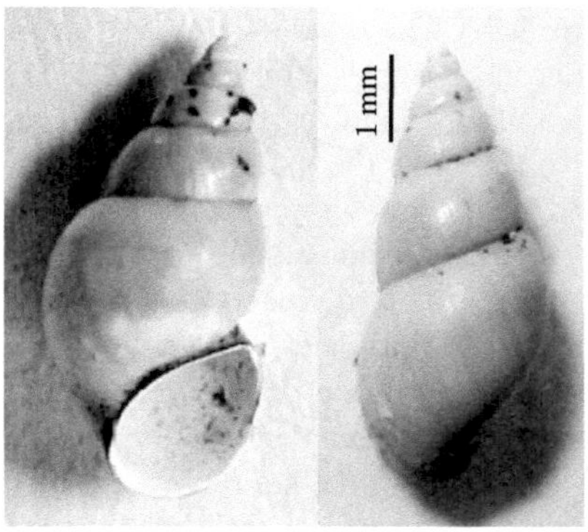

FIGURE 1.13 **(See color insert following page 286.)** *Potamopyrgus antipodarum* in aperture view (left) and back view.

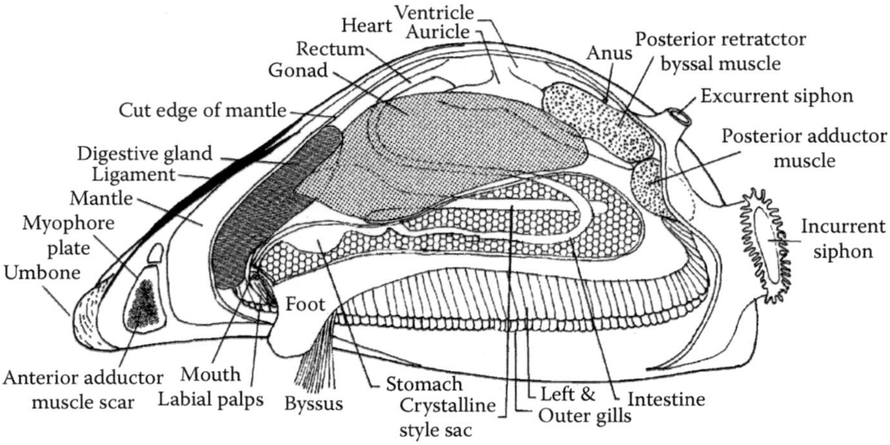

FIGURE 1.14 Major organ systems in dreissenids (*Dreissena polymorpha* shown).

a vestigial lateral tooth may be present. The ligament (or hinge) is internal, anterior (Figure 1.4b and f), and sunk in an elongated pit just posterior to the umbone (Figure 1.4b, f, and g). The umbonal end (pointed end) of the shell bears an apical septum or myophore plate (Figure 1.4b, d, f, and g) to which attaches the small anterior adductor muscle (Figure 1.14), one of the two muscles that help to close the two valves. The posterior end (broad, round end) of the shell bears a large posterior adductor muscle scar; both the posterior and anterior adductor muscles serve to close the valves. The ligament has an elastic hinge that serves to open the two valves when the two adductor muscles relax or when the animal dies, which explains why most dead shells are found gaping.

The major internal organs of dreissenids and mytilids are shown in Figure 1.14. The body is enveloped by a mantle (Figure 1.14), or a thin tissue that secretes the shell. The internal space enveloped by the mantle is known as the mantle cavity and is filled with water during life. The mantle forms two openings for the inhalant and exhalant siphons (Figure 1.14). The inhalant siphon is the larger opening of the two and is ringed with numerous tentacles that assist in selection of food particles. The exhalant siphon is conical, lacks tentacles, and is immediately dorsal to the inhalant siphon. The only other opening in the mantle is the pedal gape that allows for the extrusion of the foot and for the large byssus.

On each side of the body, between the mantle and the foot, are the gills (Figure 1.14). The gills (Figures 1.14 and 1.15) are divided into a series of water tubes by septa or filaments and the filaments make up sheets or lamellae. The lamellae have numerous small openings through which the lake water circulates. Cilia on the gills generate currents that pull water through the inhalant siphon, into the mantle cavity, and over the gills. Solid particles, including food, are removed by cilia from the water as they pass over the gills. Digestible food particles are directed by the cilia towards the mouth for ingestion. The unpalatable particles are wrapped in mucous secreted by cells in the gills and are then rejected as pseudofeces.

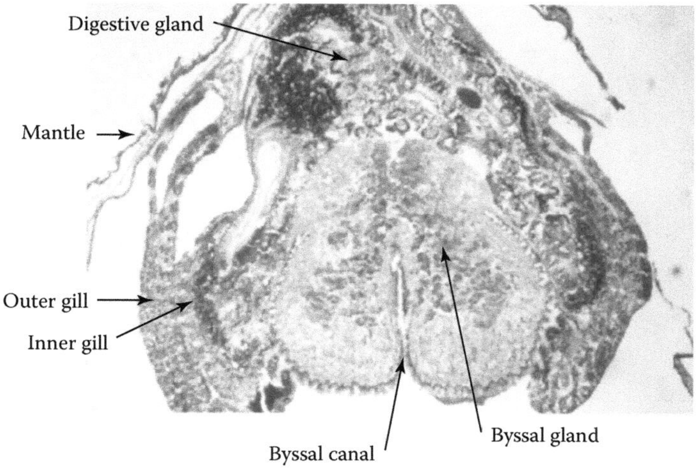

FIGURE 1.15 Cross section through *Dreissena polymorpha* showing foot with byssal gland and gills.

The mouth is an orifice at the anterior end of the body (Figure 1.14). On each side of the mouth is a pair of flaps called labial palps that assist in guiding and selecting food to be ingested. The mouth opens into a short esophagus, which leads into a large thin-walled stomach. From the stomach, food is passed to the digestive diverticula (liver). At the posterior base of the stomach lies the crystalline style sac that secretes a rod of gelatin called the crystalline style. The crystalline style rotates in the sac and by grinding against the walls releases digestive enzymes. Undigested food is passed by cilia from the stomach to the mid-gut and then proceeds to the hind gut, which eventually terminates at an anal papilla situated just within the exhalant siphon (Figure 1.14).

On the ventral side, the body forms a muscular foot that provides the means for crawling as well as housing the byssal gland that is responsible for the secretion of byssal threads. Secretions by the byssal gland are passed to a byssal canal (Figure 1.15), which directs the fluid to the base of the foot. Byssal threads are formed one at a time. The individual byssal threads branch from a central stem, each thread terminating in an adhesive pad or plaque (Figure 1.16). The thread itself has two parts. An interior cortex containing longitudinal fibers embedded in a matrix and an exterior sheath. In order to detach itself from a substrate, enzymes are secreted at the base of byssal mass and the entire mass of byssal threads is released. The mussel then secretes new threads. Mussels 2.5 cm in shell length may have up to 600 threads holding it in place (Figure 1.17).

1.2.2.1.1.1 Byssal Thread Production The biochemical composition of byssal threads is similar to that found in its marine relative, the blue mussel, *Mytilus edulis* (Rzepecki and Waite, 1993). Both have 3,4-dihydroxyphenylalanine (DOPA) as a primary component. However, the DOPA content is lower and more variable in zebra mussels than in blue mussels; also, cystine forms a major component of some

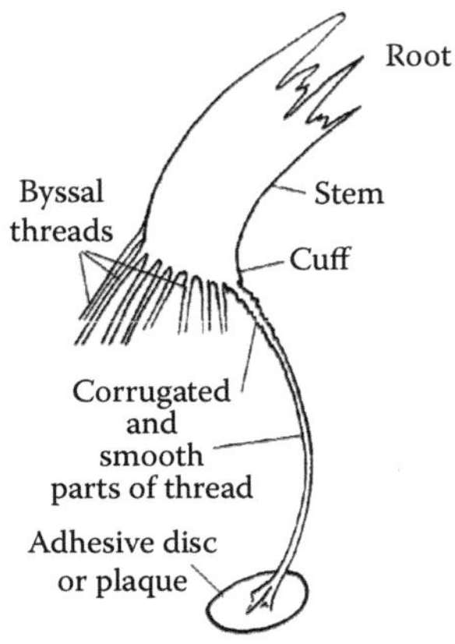

FIGURE 1.16 Major components of a byssal apparatus.

proteins and can therefore bind more readily to sulfides. The zebra mussel proteins serve analogously as glues, varnishes, and structural components of the byssus with some novel chemical properties, the intrinsic quinone chemistry of DOPA proteins conferring strength and durability on diverse composite materials (Rzepecki and Waite, 1993). Coyne et al. (1997) found that byssal threads contain collagenous proteins (preCol-P) and each thread is a "stiff tether at one end and a shock absorber with 160% extensibility at the end protruding from the mussel foot." The elastic extensible portion at the proximal end contains three major domain types, a central collagen domain, flanking elastic domains, and histidine-rich terminal domains. The byssal threads are five times tougher and 16 times more extensible than a human tendon (Coyne et al., 1997).

The rate of byssal secretion varies from about 1 to 9 per day (Ekroat et al., 1993), depending on factors such as age, temperature, and water quality. At 30°C, byssus

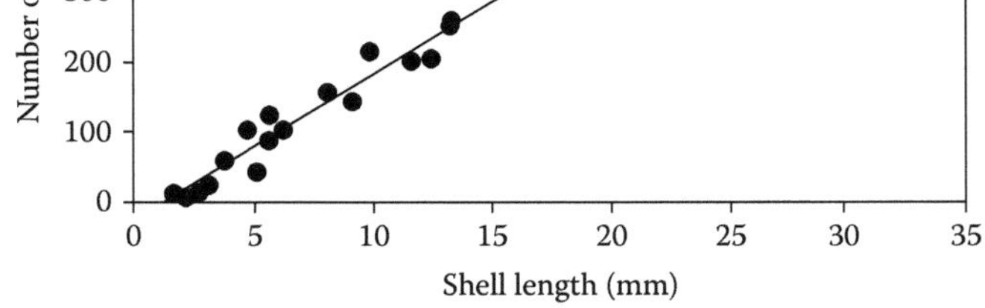

FIGURE 1.17 Relationship between size and number of byssal threads in *Dreissena polymorpha.*

reformation appeared to consist of two phases, an initial phase of rapid thread production (1–7 days), followed by a phase in which production rate slowed (8–21 days). At low temperatures, there was no evidence of a two-phase reformation of the byssal mass within 21 days, the rate of thread production being constant throughout (Clarke and McMahon, 1996a).

Clarke and McMahon (1996a) found that the rate of byssal thread production (byssogenesis) increases proportionally with increasing temperature, the slowest rate being at 5°C, the highest rate at 30°C, just 1°C below the reported upper lethal temperature for this species. The same authors reported that a significantly greater number of byssal threads were formed by mussels exposed to high test temperatures (15°C, 25°C, and 30°C) than those exposed to low temperatures (5°C and 10°C). In contrast, Rajagopal et al. (1996) found that byssogenesis was significantly reduced when temperature was increased beyond 20°C and decreased below 10°C. They also found that byssus production by mussels did not vary significantly at salinities up to 3 ppt but byssus production was reduced significantly above 3 ppt. Other factors affecting byssogenesis include agitation (increased byssus production with increasing frequency of agitation) and severed threads (mussels with severed threads produced a substantially increased number of threads and mussels with uncut byssus threads were comparatively more mobile) (Rajagopal et al., 1996); current velocity (increased current velocity significantly elevated rates of byssal thread production between 0.1 and 0.2 m/s and suppressed production at velocities of 0.27 m/s) (Clarke and McMahon, 1996b); oxygen tension (mussels exposed to a P_{O_2} of ≤15.4 Torr exhibited significant suppression of byssogenesis) (Clarke and McMahon, 1996c). While Rajagopal et al. (1996) examined frequency of agitation, Clarke and McMahon (1996c) examined rate of agitation and found that adult zebra mussels exhibited the highest rate of byssogenesis in still water [0 cycles per minute (CPM)]; the rate of byssogenesis progressively decreased as agitation rate increased up to 40 CPM.

1.2.2.1.1.2 Problems of Byssal Threads In any facility, large numbers of even very small mussels can leave enough byssal threads to disturb the laminar flow of water through pipelines. Both dreissenids and mytilids produce byssal threads that can affect the integrity of iron and steel pipelines through potentially increasing corrosion rates underneath the points of attachment. Bacteria invade the space between the plaque of the thread and the wall of the pipe and through anaerobic respiration produce acidic compounds, which may accelerate pitting and corrosion rates. The resulting increase in turbulence causes a decrease in head pressure. Veligers and juvenile mussels will settle in internal piping where the water flow is sufficiently slow to allow attachment (less than 1.5–2 m/s). Mussels colonizing such piping will result in a decreased supply of water to vital areas, obstruction of valves, and loss of heat exchange efficiency. There are enormous safety hazards if systems such as the fire protection system are invaded.

For an estimate of the numbers of threads probably attached to an intake structure (or other structure), see Figure 1.17. Do not let mussels grow too large or you will have enormous numbers of threads to deal with (e.g., approximately 500 threads for each mussel that is 25 mm in shell length).

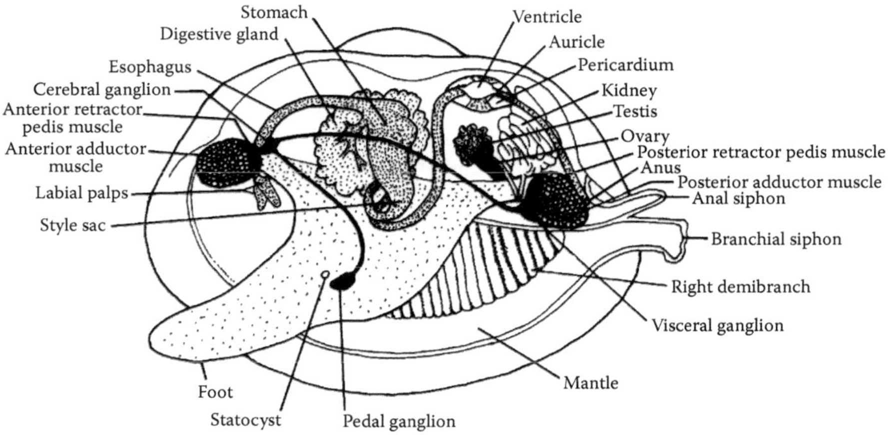

FIGURE 1.18 Internal anatomy of Sphaeriidae (*Musculium securis* shown). (Redrawn from Mackie, G.L., *Ohio Biological Survey, Bulletin New Series 15*, 2007.)

1.2.2.1.2 Corbiculidae

Burrowing bivalves have distinctive cardinal or pseudocardinal teeth and lateral teeth are present (Figure 1.1b). The teeth serve only to align the two valves during closure. Asian clams have heavy tri-radiate (three branches) cardinal teeth and serrated (saw-toothed) lateral teeth (Figure 1.3e–g), and most unionids have robust pseudocardinal teeth and large lamelliform lateral teeth (Figure 1.1b); some species, like those within the genus *Pyganodon*, are an exception and lack hinge teeth. All native bivalves and *Corbicula* have an external ligament and all lack a myophore plate.

The internal anatomy of Corbiculidae is similar to its relative, the Sphaeriidae (Figure 1.18). One major difference is in reproduction; sphaeriid larvae are brooded in brood sacs to juvenile stage but corbiculids release larvae as pediveligers (see Figure 1.22). Otherwise, both are homomyarian with similar-sized anterior and posterior adductor muscles; both have a complete digestive system with labial palps for food sorting, and a crystalline style containing enzymes for digesting food, like the dreissenids; both have a nervous system with pedal, cerebral, and visceral ganglia, a statocyst or balance organ for maintaining balance, and an osphradium or olfactory organ linked to the respiratory system for "testing" the water for silt; both have a closed circulatory system enclosed within a pericardium; both have an excretory system; and both have a respiratory system with two pairs of gills, each composed of inner and outer lamellae, the lamellae composed of filaments or water tubes, the filaments containing tiny pores or ostia enabling water to enter the tubes for exchange of respiratory gases. Details for the structure and functioning of each system are given by Mackie (2007).

1.2.2.2 Gastropoda

All invasive gastropods discussed here belong to the subclass Prosobranchia. During larval development the body undergoes torsion where the visceral mass, mantle, and shell rotate 180° with respect to the head and foot. This results in the mantle cavity

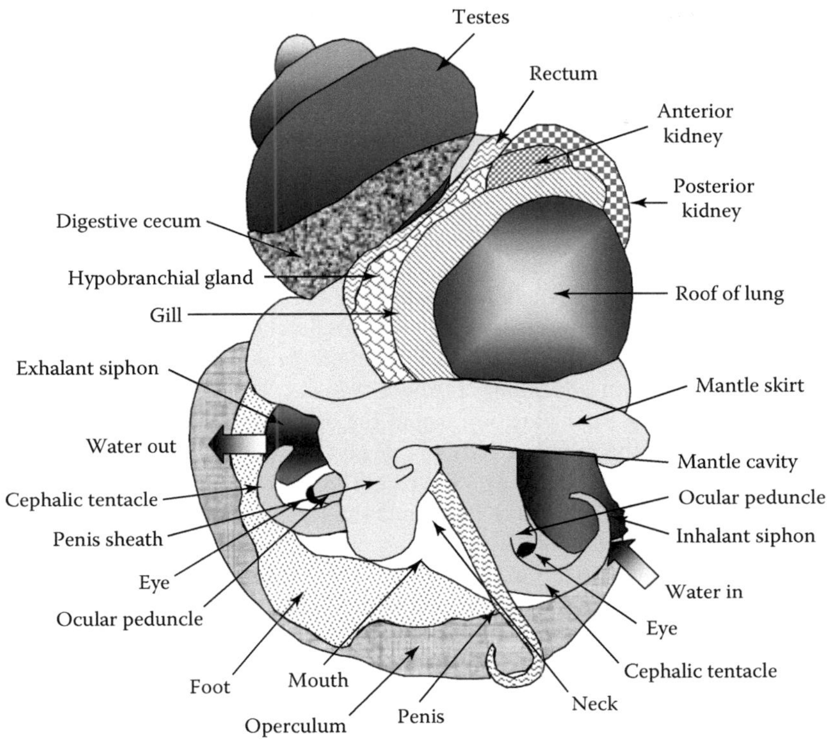

FIGURE 1.19 Soft anatomy of a snail (male *Pomacea* shown). [Redrawn from Richard Fox, web site, http://webs.lander.edu/rsfox/invertebrates/pomacea.html (accessed November 18, 2008).]

and anus moving to an anterior position above the head. Most are dioecious, some are monoecious (hermaphroditic), and still others are parthenogenetic. The reproductive strategy for each species is discussed under the life cycle section. All species discussed here have an operculum, only one gill in the mantle cavity and the right atrium is lost. Figure 1.19 shows the organs visible from above when the shell is removed.

The head, foot, and visceral mass are the main regions visible when the shell is removed (Figure 1.19). The foot is a large muscular mass that is ventral and posterior to the head. The visceral mass sits on the foot and is coiled and tapered to fit within the upper whorls of the shell. In order to fit within the aperture, it folds transversely on itself when retracted. A horny operculum is attached to the dorsal surface of the posterior half of the foot and is the last to enter the aperture to protect the soft parts behind it.

The head is dorsal and anterior to the foot and is connected to the visceral mass by a slightly narrower neck (Figure 1.19). The anterior end of the head is blunt and bears the mouth with a labial tentacle on either side of the mouth and a long cephalic tentacle on each side of the dorsal head (Figure 1.19). At the base of each cephalic tentacle is a short ocular peduncle that bears an eye at its distal end. On each side of the neck is a siphon; the left one is an inhalant siphon that draws in water or air (for amphibious species) or just water (for strictly aquatic species); the right siphon is the exhalant siphon through which water or air exits.

As in all mollusks, the mantle lies adjacent to the inner surface of the shell wall and secretes the shell. The free anterior edge of the mantle forms a collar, or mantle skirt, that completely encircles the neck (Figure 1.19) and secretes the growing edge of the shell. The mantle cavity opens above the head and is exposed to the water (or air). In ampullariids, the mantle cavity is divided into a lung on the left side and a larger branchial chamber containing the gill on the right (Figure 1.19). In other Prosobranchs, the mantle cavity is single and only a gill is present. An osphradium is attached to the roof of the mantle cavity on the left side near the base of the left siphon and monitors the chemistry and silt load of the inhalant water current. A muscular partition called the epitaenium can be raised from the floor to the roof of the mantle cavity to isolate the inhalant flow of air on the left side from the branchial chamber on the right. It channels air into the lung through a tiny pore, the pneumostome, and prevents the flow of water into the lung. When submerged, the epitaenium is lowered, water enters the inhalant siphon, passes over the osphradium, flows posteriorly in the left side of the mantle cavity, moves across the gill filaments to the right side of the mantle cavity, and then exits through the right siphon.

In male ampullariids, the right margin of the mantle skirt forms a penis sheath through which the penis can be extended during copulation (Figure 1.19). The penis can be extended through the lumen of the sheath to emerge at the tip when needed. In some individuals, the sheath may be folded inside the mantle, under the skirt, where it is not immediately apparent.

A hypobranchial gland is situated between the rectum and the gill and secretes mucus into the mantle cavity. The function of the gland is speculative but in the mucus of purple snails, Naegel and Alvares (2005) found substances with antibacterial properties which could serve as a defense against bacterial infections.

Most prosobranchs are dioecious, oviparous, and fertilization is internal. If amphibious, like ampullariids, females lay masses of large, yolky, shelled eggs above the waterline; if aquatic, the eggs are laid on rocks, plants, or anything solid as long as it remains submerged. The gonads are located in the upper whorls of the visceral mass. The male has a penis for transferring sperm directly into the female.

All prosobranchs have a kidney, or nephridium, but it is divided into anterior and posterior lobes in ampullariids (Figure 1.19) and undivided in most other prosobranchs. The nephridium filters impurities from the blood of the hemal (circulatory) system. The hemal system consists of a two-chambered heart situated within a pericardial cavity located on the left surface of the visceral mass posterior to the mantle cavity. The blood is circulated to all parts of the visceral mass through a series of sinuses, the largest of which is the cephalic hemocoel.

The digestive system consists of a mouth that opens into a muscular buccal mass, which contains a radula (series of teeth for scraping algae from surfaces), an esophagus, stomach (lacks a crystalline style), intestine, rectum, and anus. A large digestive cecum (Figure 1.19) secretes enzymes into the stomach to digest food particles.

The nervous system is well developed but asymmetrical as a result of torsion. The gastropod central nervous system consists of several pairs of ganglia (e.g., cerebral, buccal, pedal, pleural) joined by nerve cords called commissures (connect transverse pairs of ganglia) or connectives (connect pairs of ganglia longitudinally). Sense organs include labial and cephalic tentacles (both are mechanoreceptors, sensitive to tactile stimuli, and are probably also chemoreceptive), paired eyes (photoreceptors not capable of forming an image) each borne distally on an ocular peduncles (Figure 1.19), an osphradium, and a pair of statocysts in the foot.

1.3 MODE OF LIFE

1.3.1 Bivalvia

1.3.1.1 Dreissenidae

All three dreissenids and *Limnoperna fortunei* are byssate mussels and have a permanent opening through which the byssal apparatus extends. The openings are slit like and are very anterior in position in *Mytilopsis leucophaeata* (Figure 1.4g) and somewhat central in *Limnoperna fortunei* (Figure 1.4h). The byssal opening tends to be more anterior in position in quagga mussels than in zebra mussels (cf. Figure 1.4b and d). The shell shapes of the three dreissenids and *Limnoperna fortunei* are perfectly adapted to life on a variety of substrates: (1) the flat ventral surface (especially of *Dreissena polymorpha*, Figure 1.5a) allows the animal to be pulled tightly against the substrate by its byssal apparatus, making it difficult for predators to pry the shell from the substrate; (2) the umbone is lateral and adjacent to the substrate, giving the animal maximum upright stability at the surface of the substrate; (3) the shell is tapered dorsally (tent shaped in cross-section, Figure 1.5a, c, and e).

The differences in shell shape among the zebra mussel, quagga mussel, and Conrad's false mussel may also be important in determining their habitat preferences. Kauffman (1969) suggests that the more rounded shape in Conrad's false mussel than in the zebra mussel may be more efficient in directing water flow posteriorly over the ventral (incurrent) siphon and away from the excurrent (dorsal) siphon; the rounded shape would also help to reduce drag and uplift forces. The angular shape of the zebra mussel would also help to deflect water away from the incurrent siphons. The more flattened ventral surface of the zebra mussel would tend to increase drag and ventral lift (Kauffman, 1969). Combining the functions of shape and muscle attachment to both the myophore plate and the apophysis, one would think that Conrad's false mussel would inhabit more swiftly moving waters and powerful waves than zebra mussels. The less angular shape of quagga mussels than zebra mussels would suggest that quagga mussels would prefer even quieter waters than both zebra mussels and Conrad's false mussel. This might help to explain the quagga mussel's prevalence over zebra mussels in deeper waters (Claxton et al., 1998). Roe and MacIsaac (1997) found that quagga mussels occupied mostly muddy substrates in deep waters of Lake Erie, about ~70% at 37-m sites and between 63% and 90% at 55-m sites. The lack of a myophore plate and an apophysis in the golden mussel would suggest that it is even less adapted to fast waters than any of the dreissenids.

1.3.1.1.1 Zebra Mussel, Dreissena polymorpha

All dreissenid larvae generally need solid substrates (e.g., rocks, shelled animals, boat hulls, artificial structures, vegetation, woody debris, etc.) to settle on. Mellina and Rasmussen (1994) found a positive correlation between zebra mussel density and mean particle weight, explaining 38–91% of the variation in the Hudson and St. Lawrence rivers and Oneida Lake, and 75% of the variation in 72 other lake/sites reported in the literature, with the mussels being more abundant in coarser substrates. However, in lakes with meager hard substrates or large muddy/sandy areas between hard substrates, Mackie et al. (1989), Mackie (1990, 1991), Ramcharan et al. (1992), Claudi and Mackie (1994), Mellina and Rasmussen (1994), Nichols (1996), Berkman et al. (1998), Jones and Ricciardi (2005) found that groups (called druses) of mussels that slough off solid substrates can act as a substrate for settlement of larvae and form large aggregations. Hence, the correlation between mean particle weight and mussel density declines over time. For example, Jones and Ricciardi (2005) found that only 20% of the zebra mussel biomass was explained by mean particle weight and up to 900 g/m^2 of zebra mussel biomass was found on fine substrates.

Zebra mussels apparently preferentially settle on submersed aquatic plants, or on the underside of substrates (Ackerman et al., 1994), and later move to other substrates. Wainman et al. (1996) showed that a primary biofilm enhances recruitment, with up to 29,300 (±34,800) mussels/day on plates with biofilm compared to 4640 (±540) mussels/day on plates with biofilm removed. In addition, they found that dreissenid shell material (dead or alive) was a particularly attractive substrate for dreissenid recruitment in Lake Erie, which helps to explain why druses are effective substrates for settling larvae. Wainman et al. (1996) suggest that mussel recruitment can be reduced 10–20% by removal of the biofilm. Ackerman et al. (1994) found that translocation, or postsettlement relocation, occurs throughout the year, but peaks in the spring.

1.3.1.1.2 Quagga Mussel, Dreissena bugensis

Quagga mussels have displaced zebra mussels in much of Lake Erie and Lake Ontario and it may be due to the predominance of soft substrates in these lakes, especially deep portions where quagga mussels prevail. Mills et al. (1996) found that in these lakes, quagga mussels colonize sand and sandy silt between 10 and 30 m depth, and silty substrate below 40 m depth. As reported earlier (Section 1.2.1.1.4.2), Claxton and Mackie (1998) showed that the quagga mussels that occur in the Great Lakes are of two morphologically distinct forms, one that occurs in deep waters of Lakes Erie and Ontario ("profunda") and the other that is more common in shallow waters ("epilimnetic form") (Dermott and Munawar, 1993; Spidle, 1994). Genetic studies by Spidle (1994), Claxton and Mackie (1998), and Spidle et al. (2005) on these morphs suggest that the two forms are phenotypic variants rather than genetically distinct taxa.

Currently, quagga mussels have displaced zebra mussels even on hard substrates in Lake Erie (Ricciardi and Whoriskey, 2004). Jones and Ricciardi (2005) reported greater quagga mussel biomasses on larger substrate sizes (cobbles or boulders) than on smaller size classes (sand, silt, and mud), but substrate size explained only 11% of the variation in quagga mussel biomass. In European waters where quagga mussels have become more abundant than zebra mussels, zebra mussels remain the dominant species attached to aquatic vegetation (Karatayev et al., 2007b), but this has not been reported for Great Lakes waters yet.

Orlova et al. (2005) hypothesized that given sufficient time, quagga mussels will displace zebra mussels, in deeper, more oligotrophic habitats. They found four patterns in the times of establishment of *Dreissena polymorpha* and *Dreissena rostriformis bugensis* (here called *Dreissena bugensis*, see Section 1.2.1.1.4.2) in the Dnieper and Volga basins. First, the quagga mussel was not found in these systems prior to reservoir creation, while the zebra mussel was common to both systems prior to this. Second, quagga mussel populations established themselves 5–10 years following reservoir impoundment and several years after the extensive colonization by zebra mussels. Third, by the mid-1960s it was noted that in the older, deeper reservoirs along the Dnieper River quagga mussels largely replaced zebra mussels and became the dominant dreissenid species. In contrast, the younger, shallower reservoirs in this system contained extensive zebra mussel populations and limited quagga mussel populations. Fourth, in two, old deep reservoirs in the Dneiper basin, quagga mussels occupy deeper habitats than zebra mussels down to 28 m, where temperatures are 10°C lower than the epilimnion through the summer. The observed time needed for replacement of *Dreissena polymorpha* by *Dreissena bugensis* closely matches the 5–10 years interval required for stabilization of environmental conditions in newly created reservoirs along the Dnieper River. Similarly, the zebra mussel was the dominant species in the 1950s at most locations along the Volga River, but as of 2000 the quagga mussel dominates the river (Orlova et al., 2004).

1.3.1.1.3 Conrad's False Mussel, Mytilopsis leucophaeata

Like other dreissenids, *Mytilopsis leucophaeata* settles preferentially on natural or artificial solid substrates, with up to 100% coverage of stones and boulders in 1–3 m depths in Finnish waters, as well as a few on macroalgae like *Fucus vesiculosus* (Laine et al., 2006). Koch (1989) reported finding *Mytilopsis leucophaeata* attached to dams and locks as well as to a freshwater clam in the Mississippi River. The original description of the species was of a specimen attached to an oyster shell (Marelli and Gray, 1983).

Mytilopsis leucophaeata is a euryhaline species typically found in salinity concentrations ranging from freshwater (<0.5 ppt) to oligohaline water (brackish water between 0.5 and 5 ppt salinity) (Therriault et al., 2004). In Europe, *Mytilopsis leucophaeata* is found in mesohaline water (5–18 ppt) in estuaries with 12 ppt salinity (Verween et al., 2005). Siddall (1980) reports that the species is able to complete development in 32 ppt (euhaline).

Conrad's false mussel can also tolerate widely fluctuating salinities. Darr and Zettler (2000) reported fluctuations between 0.3 and 8 ppt for a population in the German Baltic Sea Coast, and Verween et al. (2005) reported salinity fluctuations between 0.8 and 10.3 ppt from beginning to end of the recruitment period in a harbor in Antwerp. The species has been reported to coexist with the zebra mussel in European oligohaine waters (Jenner and Janssen-Mommen, 1993; Van der Velde et al., 1998; Darr and Zettler, 2000; Rajagopal et al., 2005a,b).

1.3.1.2 Mytilidae

1.3.1.2.1 Golden Mussel, Limnoperna fortunei

Limnoperna fortunei can be found attached to solid substrates, including natural ones (e.g., shelled invertebrates such as gastropods, crayfish, crabs, and other golden mussels), under and on rocks, roots, driftwood, and on artificial substrates (e.g., concrete walls, docks, piers, etc.) (Darrigran and Pastorino, 1995; Darrigran and Ezcurra de Drago, 2000; Darrigran, 2002; Orensanz et al., 2002; Karatayev et al., 2007a). It will live on compacted bottoms of silt–sand or silt–clay, and occasionally found as druses on loose silt or mud substrates (Karatayev et al., 2007a). When juveniles or adults are disturbed or translocated, they prefer to resettle in crevices and on dark or shaded substrates. Morton (1975, 1977) found that *Limnoperna fortunei* preferred depths of 6–9 m (20–30 feet), settling mainly in fissures, on shaded horizontal surfaces if in shallow water, and on illuminated vertical surfaces if in deeper water in Plover Cover reservoir, Japan. Iwasaki and Uryu (1998) reported similar habitat preference, with *Limnoperna fortunei* being most abundant on the undersides of boulders in the Uji River in Japan. In contrast, Iwasaki (1997) found that juveniles in particular tend to move upward along vertical walls and resettle just below the air–water interface.

1.3.1.3 Corbiculidae

In burrowing freshwater bivalves, the umbone is rounded and dorsal in position (Figure 1.3) and adapted for life within sediments. The ventral margin is rounded and lacks a byssal opening because the adults do not have a functional byssal gland. Corbiculids and sphaeriids are related and both have a byssal thread during the larval stages but it is lost in the adults (Mackie, 2007).

1.3.1.3.1 Asian Clam, Corbicula fluminea

Corbicula is also a biofouler, but it does not have the ability to attach to structures. It will usually be found on the bottom of the structures, but beware juvenile forms of Asian clams (and even pearly mussels, and adults and juveniles of fingernail clams) can be found embedded in moss and mud on the walls of pump wells.

Due to differences in the life cycles (see below for details), only dreissenids, mytilids, and Asian clams are likely to be found in pipelines within the facility. Native bivalves present no danger to a facility and are often considered indicators of good water quality because they are quite sensitive to pollution.

1.3.2 Gastropoda

1.3.2.1 Ampullariidae

Pomacea inhabits rivers, springs, ponds, ditches, wetlands (mostly marshes, rarely swamps), and similar freshwater environments with low flow and large standing crops of aquatic macrophytes, especially rice and taro of which many species are voracious consumers.

1.3.2.1.1 Giant Ramshorn Snail, Marisa cornuarietis

Marisa cornuarietis lives in lakes, rivers, farm ponds, irrigation systems, and wetlands (Akerlund, 1973). They prefer shallow waters (<1 m) with a lot of vegetation and are usually seen close to the surface or on leafy margins in lakes and ponds

(Ferguson and Palmer, 1958; Akerlund, 1973). They can tolerate relatively high salt concentrations (30 ppt) and are occasionally found in light brackish waters (<8.5 ppt) but apparently they do not reproduce under highly brackish conditions (Hunt, 1961; Robin, 1971; Santos et al., 1987). The giant ramshorn snail can tolerate 12°C for short periods of time but will die under prolonged exposure to <12°C (Hunt, 1958). The ideal habitat for the giant ramshorn snail is one that has relatively stable water levels, a shoreline margined with vegetation and a fair percentage of surface area that is covered with leafy plants (Ferguson and Palmer, 1958). However, it has been found to thrive in both nonpolluted and heavily polluted waters in Puerto Rico and has been found in streams receiving effluent from hog abattoirs and outhouses (Ferguson and Palmer, 1958). Hunt (1958) has kept specimens alive for months in unaerated aquaria where the dissolved oxygen regularly dropped to under 0.5 ppm.

1.3.2.1.2 Golden Apple Snail, Pomacea canaliculata

In Japan *Pomacea canaliculata* is found in quiet waters with a high level of suspended matter and a high chemical oxygen demand. Near the northern boundary of its range in Japan, the snail can overwinter only in shallow canals with a high level of dissolved oxygen and a low pH (Ito, 2003). Oya et al. (1987) kept golden applesnails in storage incubators at 0°C, –3°C, and –6°C and observed total mortality within 25 days, three days, and one day, respectively. Young snails with a shell height of 2–3 cm seemed to have a greater tolerance to low temperature than mature ones with a size larger than 3 cm (Oya et al., 1987). These results suggest that the golden applesnail may limit their distribution to the lower portion of the warm temperate zone. However, in some parts of Japan and Korea, the golden applesnail burrows into the mud and closes its operculum to hibernate and provide some protection against freezing temperatures (Halwart, 1994a,b). It appears that the existence of different ecotypes may explain differences in overwintering abilities.

1.3.2.1.3 Titan Applesnail, Pomacea haustrum

This species was considered a synonym of *Pomacea canaliculata*, so much of the habitat requirements and tolerances are probably similar for the two species. The only established populations of *Pomacea haustrum* known in the United States are in Palm Beach County, Florida (Rawlings et al., 2007). The populations are located in canals and lakes of Loxahatchee National Wildlife Refuge and in some surrounding communities (Rawlings et al., 2007). *Pomacea haustrum* was collected in Florida as early as 1983 (Howells et al., 2006) but recorded as *Pomacea canaliculata*; however, Rawlings et al. (2007) believe that specimens collected in the 1970s and 1980s and attributed to *Pomacea canaliculata* are *Pomacea haustrum*. Therefore, *Pomacea haustrum* was probably introduced and successfully established by the late 1970s, but has failed to expand its range appreciably since then (Rawlings et al., 2007).

1.3.2.1.4 Island Applesnail, Pomacea insularum

Pomacea insularum is a tropical and subtropical snail that is native to slow-flowing rivers and canals in South America. It is also a voracious feeder of rice and taro. It was probably introduced to the United States in 1989, establishing populations in the rice belt region of southeastern Texas (Rawlings et al., 2007).

Ramakrishnan (2007) examined the resistance adaptations of *Pomacea insularum* to temperature, salinity, pH, emersion, and progressive hypoxia in order to predict its potential distribution in North America, and to evaluate its threat to rice and taro crops in the United States. He found that the main factor likely to limit its distribution was its lower lethal temperature of 15°C. This probably explains, at least in part, its current distribution being limited to Texas, Florida, and Georgia. Its incipient tolerated temperature range is 15.2–36.6°C (Ramakrishnan, 2007). Other factors that may play a role in habitat selection are its tolerance ranges of 0–6.8 ppt salinity (which should prevent it from invading estuarine habitats), 4.0–10.5 pH (which includes all freshwater bodies in North America), and 70 days at 30°C and >95% relative humidity (RH) to >308 days at 20–25°C and 75% to >95% RH incipient emersion tolerance; it was a moderate regulator of oxygen consumption when subjected to progressive hypoxia, maintaining a normal oxygen uptake rate down to a critical PO_2 of 80–120 Torr depending on temperature. These results indicated that within its tolerated temperature range, *Pomacea insularum* will be most successful in oxygenated, flowing water habitats (Ramakrishnan, 2007).

1.3.2.1.5 Spike-Topped Applesnail, Pomacea diffusa

Formerly known as *Pomacea bridgesii*, this apple snail's habitat preferences are much like those in the *Pomacea canaliculata* complex and also prefers lentic habitats (i.e., standing water) that may have alternating periods of drought and high rainfall. Like all pomaceans, the possession of both lungs and gills reflects an adaptation to low dissolved oxygen conditions that are often present in swamps and shallow water due to decay of organic materials like dead vegetation and high temperatures. When the dissolved oxygen levels of the water drops, the snails are forced to use aerial respiration and will float to remain at the surface using their lungs to adjust their buoyancy.

1.3.2.1.6 Florida Applesnail, Pomacea paludosa

Typical *Pomacea paludosa* habitat in Florida is wetlands with a variety of emergent and floating leaved macrophytes, including sawgrass (*Cladium jamaicense*), a dominant sedge in central and south Florida wetlands that often reaches 2–3 m in height (Steward and Ornes, 1975). Many wetlands are densely vegetated (Hann, 1999) and require specialized collecting equipment, such as funnel traps, to collect Florida applesnails (Darby et al., 2001). Juvenile and adult Florida applesnails can tolerate fairly high nitrate levels, at least up to 500 mg/L (Corrao et al., 2006). Given that snails exhibited little to no response to nitrate concentrations orders of magnitude greater than those found in Florida springs, Corrao et al. (2006) suggest that other factors, possibly including natural differences in habitat structure or changes in structure related to exotic plant invasions, may help explain the observed declines in apple snail abundance.

1.3.2.2 Thiaridae

1.3.2.2.1 Quilted Melania, Tarebia granifera

The quilted melania can be found in the shallow waters [30–120 cm (1–4 ft)] of almost any body of freshwater including rivers, streams, and lakes to drainage ditches, irrigation canals, cement ponds, and wetlands (Chaniotis et al., 1980a) but it seems to prefer riffles of shallow, fast-flowing freshwater streams (Lachner et al., 1970). It is capable of maintaining high abundances in most of these habitats (Prentice, 1983). The quilted melania cannot tolerate saltwater or temperatures less than 10°C (Abbot, 1952; Chaniotis et al., 1980c).

1.3.2.2.2 Red-Rimmed Melania, Melanoides tuberculata

The red-rimmed melania is typically found in shallow, slow-running water [60–120 cm (2–4 ft) in depth], on soft substrates of mud or sand but has been reported on rocks in relatively deep portions of freshwater pools (3–3.7 m deep, 3–12 ft) (Murray, 1975). Dundee and Paine (1977) reported high densities of up to 2700 m^{-2} in areas rich in detritus and silt behind overhanging stems and protruding roots of bank vegetation in Louisiana. It is very resistant to low oxygen levels (Neck, 1984) and high salinities (e.g., 30 ppt), reaching densities of up to 37,500 m^{-2} in areas of Florida dominated by red (*Rhizophora mangle*) and white (*Laguncularia racemosa*) mangroves (Russo, 1973; Roessler et al., 1978). The red-rimmed melania is a nocturnal species, active at night but hiding or burying themselves in mud beneath decaying plants and stones during the day (Livshits and Fishelson, 1983). As with most thiarids, red-rimmed melania are tropical snails believed to be restricted to waters between 18°C and 25°C (Murray, 1964, 1971, 1975), although Livshits and Fishelson (1983) reported the species hibernating in the mud in the winter in Israel.

1.3.2.3 Bithyniidae

1.3.2.3.1 Faucet Snail, Bithynia tentaculata

The faucet snail is commonly found naturally in freshwater rivers, streams, ponds, shallows (<5 m, 16 ft) of lakes, and canals and does extremely well in backyard fish ponds but does not do well in fast-flowing turbulent water (Harman, 1968). It seems to live equally well on gravel, sand, clay, mud or undersides of rocks, and on submersed aquatic macrophytes (Vincent et al., 1981; Jokinen, 1992; Mackie, 2000b; Smith, 2001). The faucet snail is quite tolerant of a variety of conditions and is often abundant in eutrophic systems. Jokinen (1992) and Harman (1974b) ascribe chemical requirement values for pH of 6.6–8.4, conductivity of 87–2320 μmhos/cm, Ca^{2+} of 5–89 mg/L, and Na^{+} of 4–291 mg/L, and dissolved oxygen of at least 4 mg/L. It is apparently tolerant of high levels of K^{+} and low concentrations of NO_3^- and inhabits intertidal zones in the Hudson River (Jokinen, 1992). It is one of the most abundant mollusks in the Great Lakes (Mackie, 1980, 2000a) and in the St. Lawrence River where it occurs mostly in relatively unpolluted (but organically enriched), nearshore areas (Vaillancourt and Laferriere, 1983) and among dreissenid mussel beds (Ricciardi et al., 1997), presumably feeding on particles in their pseudofeces. As for most gastropods, it is a grazer using its radula to selectively scrape algae off rocks (Brendelberger, 1997). But it can also filter feed using its gills to filter suspended particles such as algae from the water column (Jokinen, 1992). When filter feeding, particles are sucked in, packaged, and then passed out between the right tentacle and exhalant siphon in pellet-like bundles, which are then eaten (Jokinen, 1992). Jokinen believes that the species' ability to filter feed may play a role in allowing populations of the faucet snail to survive at high densities in relatively eutrophic, anthropogenically influenced water bodies (Jokinen, 1992). The faucet snail is reported to feed on black fly larvae in Eurasia (Pavlichenko, 1977).

1.3.2.4 Hydrobiidae

1.3.2.4.1 New Zealand Mudsnail, Potamopyrgus antipodarum

The New Zealand mudsnail lives at 4–45 m depths on silt and sand in Lake Ontario but in New Zealand it lives in lowland rivers, stony streams, creeks, ponds, lakes, springs, and estuaries where salinities can range up to 26 ppt (Zaranko et al., 1997; Mackie, 2000b; Levri et al., 2007). Numerous authors (Zaranko et al., 1997; Collier et al., 1998; Holomuzki and Biggs, 1999, 2000; Negovetic and Jokela, 2000; Richards et al., 2001; Weatherhead and James, 2001; Death et al., 2003; Schreiber et al., 2003; Suren, 2005) allude to its propensity for littoral habitats somewhat enriched with silt, organic matter, and nutrients. In Australian streams their success appears to be facilitated by human disturbance, as densities are positively correlated with human land use and flow disturbance (Schreiber et al., 2003). The species is euryhaline, the optimal salinity being near or below 5 ppt but capable of feeding, growing, and reproducing at salinities of 0–27 ppt and can tolerate 30–35 ppt for short periods of time (Jacobsen and Forbes, 1997; Zaranko et al., 1997; Leppakoski and Olenin, 2000; Costil et al., 2001; Gerard et al., 2003). It tolerates temperatures of 0–34°C (Zaranko et al., 1997; Cox and Rutherford, 2000). In the western United States, it reaches extremely high densities in geothermally influenced rivers (Kerans et al., 2005) and plays a large role in ecosystem processing of nitrogen and carbon (Hall et al., 2003). Although densities rarely exceeded 1000 snails/m^2 (Holomuzki and Biggs, 1999, 2000) but can reach 4000 snails/m^2 (Collier et al., 1998) in their native range, they usually numerically dominate the invertebrate assemblages in their preferred habitats (Duggan et al., 2002). Outside of their native range, secondary production of *Potamopyrgus* was 7 to 40 times higher than any other taxon in three rivers in the Greater Yellowstone area of Wyoming. In North America, densities of the New Zealand mudsnail are also larger than in their native range, from 5600 m^2 in the Great Lakes (Zaranko et al., 1997; Levri et al., 2007); 30,000 m^{-2} in Snake River Springs, Idaho (Richards et al., 2001); 300,000 m^{-2} in Madison River, Wyoming (Kerans et al., 2005), and 350,000–700,000 snails/m^2 in vegetated portions of Firehole Creek and in Polecat Creek in Wyoming (Hall et al., 2006). They also dominate native snail assemblages in Poland (Strzelec, 1999) where fecundities are positively correlated with water temperatures.

1.4 REPRODUCTION AND LIFE CYCLE

1.4.1 Bivalvia

1.4.1.1 Dreissenidae

1.4.1.1.1 Reproduction

Most dreissenids have separate sexes, usually with a sex ratio of about 1:1; one exception is the black-striped mussel, *Mytilopsis sallei*, for which a proportion of mussels in any population is present as hermaphrodites but during their lifespan about half the individuals change sex (Morton, 1989). Ovaries and testes ripen in the winter months. By spring, most mussels over 8 mm in shell length will have ripe gonads. For zebra mussels, eggs and sperm usually begin appearing when the surface water temperature reaches about 12°C (Neumann et al., 1993) but their numbers do not peak until water temperatures are near 15–17°C (Mackie, 1991). In Canada and northern states of the United States these surface water temperatures usually occur near the end of May to the first week of June. In southern states, especially south western states of the United States, these surface temperatures are reached earlier and last longer, for example, from March/April to November/December in Barttlett Lake, Saguaro Lake, Canyon Lake, and Lake Powell, Arizona (GLM, unpublished data). In contrast, quagga mussels begin reproduction at much lower temperatures. In the hypolimnion of Lake Erie at 4.8°C, 80% of female mussels contained mature oocytes (80%) or had spent gonads (20%) (Roe and MacIsaac, 1997).

The eggs are usually 40–50 μm in diameter. Recent studies in North America indicate that some zebra and quagga mussel females may produce nearly one million eggs per year [Keller et al. (2007) reported 960,000 eggs/female/year]. Neumann et al. (1993) found that fecundity in zebra musels was correlated with shell size, with 24-mm-long females producing 310,000–1,610,000 oocytes in the Upper and Lower Rhine River; they also found that the onset of spawning correlated with a threshold temperature of about 12°C. However, Garton and Haag (1993) found that zebra mussel spawning and veliger abundance were not correlated with temperature in Lake Erie; they reported two spawning events in 1989 and only one spawning event in 1990, with spawning occurring over a 5-week period (late July to late August).

The fecundity of *Mytilopsis leucophaeata* has not yet been reported but a closely related species, *Mytilopsis sallei*, produces "tens of thousands" of eggs per year (Morton, 1980). The sperm of *Dreissena polymorpha and Dreissena bugensis* are quite different. In the zebra mussel, the sperm is straight and short in appearance, averaging 4.0 ± 0.1 μm ($N = 5$) in length, and the acrosome is more bulbous and oval in shape than that of the quagga mussel; in the quagga mussel the sperm are curved and longer in appearance, averaging 4.7 ± 0.2 μm ($N = 7$) in length, and the acrosome is wide at the point of attachment to the rest of the sperm head, but tapers more sharply toward the tip (Denson and Wang, 1994).

For all three dreissenids, oogenesis (egg production) and spermiogenesis (sperm production) appear to occur at about the same time. Fertilization is external and occurs in the water (Franzen, 1983; Sprung, 1987). In Lake St. Clair, most zebra mussels are sexually mature by the time they reach 8–10 mm in shell length; a few mature at even 5.5 mm in shell length (Gillis, 1989). Peak reproduction normally occurs about the same time every year, as indicated by the large proportion of adults with partially spent gonads (Mackie, 1991), but the peak in reproduction appears to vary from one aquatic system to another. If larvae are born in the spring and water temperatures vary from 15°C to 20°C, they will settle about 28 days later (Neumann et al., 1993), or early to mid-summer; the juveniles may grow to reproductive size by the fall and contribute to a fall recruitment (production of new larvae). The rate of development increases with increasing temperatures. As a result, one can expect larvae born when water temperatures are 16–24°C to settle in 18 days (Neumann et al., 1993).

Apparently, most adults do not attain the reproductive state until the following spring. Only those that settle in June and July and grow quickly to reproductive size (8–10 mm) by August or September can reproduce in the year of birth, although the level of reproduction appears to be significantly lower than those that reproduce in the year following their birth. In northern latitudes, gametogenesis ceases in late September to mid-October, as indicated by the spent condition of gonads in all length classes. However, in southwestern states like Nevada, winter surface water temperatures are usually well above the reproductive threshold of 5°C (e.g., 4–16°C for Lake Mead and Lake Mohave) for quagga mussels so reproduction may occur all year, although growth and recruitment of quagga mussels apparently ceases (McMahon, 2007).

1.4.1.1.2 Life Cycle

The life cycle of dreissenids is shown in Figure 1.20. It is almost impossible to distinguish the appearance of each larval stage among the three species of dreissenids.

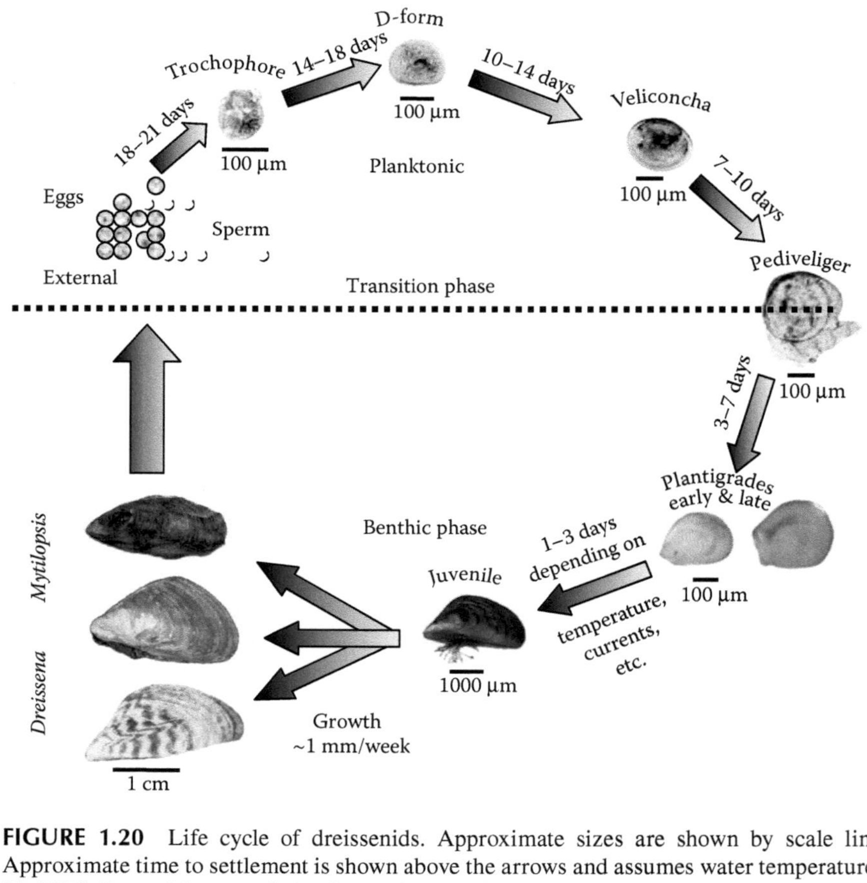

FIGURE 1.20 Life cycle of dreissenids. Approximate sizes are shown by scale lines. Approximate time to settlement is shown above the arrows and assumes water temperature is 18–20°C. Rate of descent of plantigrade larvae will depend on horizontal and vertical water currents, thermal profile, and other factors discussed in text.

The only stages that differ between zebra and quagga mussels are the veliconcha and pediveliger, and plantigrade stages; one valve is larger than the other in the quagga mussel (see larval identification key, Section 1.4.1.5.1). The umbonal veligers of quagga mussels are also asymmetrical with one shoulder more square than the other. The phases and stages of the life cycle are similar for the three species, but the rates of development may be somewhat different. The rates of development shown in Figure 1.20 are based on numerous studies (outlined below) on the life cycle of the zebra mussel.

The larval life cycle of *Dreissena* usually takes about 18–30 days to complete (Stanczykowska, 1964, 1977; Morton, 1969a; Wiktor, 1969; Mackie et al., 1989; Hopkins, 1990; Neumann et al., 1993). Three larval events and two states or phases are recognized (Mackie, 1991; Ackerman et al., 1994): a veliger event in the planktonic phase and a postveliger event and a settling event in the benthic state (Figure 1.20). Each event can be identified by larval stages. Veliger larvae are identifiable by the presence of a tuft of cilia called the apical tuft and a "D"-shaped, bivalve shell that is secreted shortly after the trochophore emerges from the embryo.

Development begins with the external fertilization of gametes (Figure 1.20). Fertilized eggs (zygotes) pass through an embryonic stage where the development is nourished by materials stored in the egg rather than by direct feeding. The embryo develops into a free-swimming trochophore larvae in 6–20 h, depending on the ambient temperature. With the rapid development of the velum, a ciliated swimming organelle, the trochophore larva becomes a veliger larva (Figure 1.20). The "veliger stage" includes any larva that possesses a velum. Several days after fertilization, the veliger secretes the first larval shell (Prodissoconch I) from its shell gland. Prodissoconch I shells are unornamented and D shaped in profile, hence the terms "D-form" or "straight-hinged veliger." The second larval shell (Prodissoconch II) is secreted later by the mantle tissue and can be ornamented in various ways. It is round or clamlike in profile with the typical umbonal (beak) region near the hinges. Because of the obvious umbone, the larval stage is often referred to as the "umbonal veliger stage" but more commonly a "veliconcha larva" (Figure 1.20). The veliconcha is the last stage of the veliger event that is free swimming and is found in the plankton.

As the veliconcha grows the velum slowly develops into the siphons, the foot lengthens, and the blood and some organ systems begin development. The acquisition of a foot leads to a change in behavior and the larva is now known as the pediveliger. With the loss of the velum the larvae enter the settling stage. The pediveliger marks the beginning of the postveliger stage. The foot can be used for swimming (by means of cilia on the base of the foot) near the bottom as well as for crawling on surfaces. This is typical behavior in the postveliger event. Once a pediveliger encounters an appropriate surface, it secretes a byssal thread and undergoes metamorphosis to become a plantigrade larva. Without an appropriate surface to settle upon, such as filamentous algae, they, like marine bivalves, may delay byssal attachment and metamorphosis. Metamorphosis, which follows primary attachment, transforms the larval bivalve into a "juvenile" bivalve. The principal changes during metamorphosis are associated with the loss of the velum, the development of the gills, and the secretion of the dissoconch, or adult, shell. Growth of the foot and mouth is facilitated by the new growth axis of the dissoconch shell, which results in the familiar triangular mussel shape. The completion of these developments transforms the plantigrade larva into a juvenile, which, with further growth and sexually maturity, becomes an adult. With the spawning of adults the life cycle begins again. Meisenheimer (1901) was one of the first to describe the early life history of *Dreissena polymorpha*, which appears to follow the general pattern described above. A more detailed review is available in Ackerman et al. (1994).

Stanczykowska (1977), Walz (1978a), Lewandowski (1982), and Sprung (1989) report up to 99% mortality of larvae during planktonic and settlement phases in European lakes. Veliger and postveliger mortality have been attributed to predation by fish or invertebrates, a lack of suitable settling substrate, and unfavorable oxygen or thermal conditions, but their respective intensities have not been determined (Stanczykowska, 1977; Lewandowski, 1982; Lewandowski and Ejsmont-Karabin, 1983). MacIsaac et al. (1991) tried to model settlement rates of larvae in western Lake Erie using the 99% mortality figure but could not account for the adult densities using such a high mortality rate. In order to account for the observed densities of adults, a mortality rate of 70% would be required between the pelagic and benthic phases. The 70% mortality rate includes predation of the pelagic larvae by adult *Dreissena*. Settlement is defined here as the transfer from the pelagic phase to the benthic phase, while recruitment shows settlement combined with postsettlement mortality.

Every aquatic ecosystem behaves differently with respect to development rates and settlement and recruitment patterns. This was nicely demonstrated by Lucy (2006) who found in the early years of invasion (1998–2000) increasing levels of larval density and settlement from 1998 to 2000 in Lough Key, Ireland. An exponential increase in the size of the adult zebra mussel population resulted in a high level of successful recruitment in those years. In subsequent years, the variations in development and settlement rates were likely related to environmental variations in summer water temperatures and food availability for veligers. In 2001, settlement estimates were extremely low relative to the larval densities present. Prolonged warm water temperatures in summer 2003 resulted in a long reproductive season and also high settlement rates. Because of the variation between annual datasets, Lucy (2006) concluded that in order to monitor the dynamics of dreissenids within lake ecosystems long-term studies of the early life stages of *Dreissena polymorpha* (and likely all invasive species) are required.

1.4.1.2 Mytilidae—Golden Mussel

1.4.1.2.1 Reproduction

The golden mussel is a dioecious species with external fertilization, the larvae requiring 30–70 days for development into plantigrade larvae to settle and attach to a hard substrate. Of this, *Limnoperna fortunei* could spend 15–20 days as planktonic larvae (Choi and Kim, 1985), leaving the rest of the time to settle out. The time required for settlement depends on temperature, currents created by winds, gravity (density of larvae relative to the water), and light (Crisp, 1974).

Spawning appears to be continuous for at least nine months for most populations (Iwasaki and Uriu, 1998; Boltovskoy and Cataldo, 1999; Darrigran et al., 1999).

For example, in an Argentina river reproduction was continuous between September and April in 1998 and again between August 1998 and January 1999; the threshold temperature level associated with the onset and termination of reproduction was 16–17°C (Cataldo and Boltovskoy, 2000). In Uruguay, veliger and postveliger larvae are present when the water temperature is 19–20°C and are absent between June and July or July and August when the temperature drops below 19°C (Brugnoli et al., 2005). Darrigran et al. (1999) found that gametogensesis and spermiogenesis were almost continuous but spawning periods could still be recognized. Reproduction of *Limnoperna fortunei* in subtropical climates has received little attention to date.

In some cases peaks in gonad maturation occur with peak temperatures [e.g., Bagliardi Beach, Hong Kong (Morton, 1977; Darrigran et al., 1999)], but in others no relationship has been observed between peak maturation and temperature (Boltovskoy and Cataldo, 1999). However, initiation of spawning appears to depend on temperature. The interruption of the reproductive activity of *Limnoperna fortunei* occurred when the water temperature dropped to 16–17°C (Cataldo and Boltovskoy, 2000).

Fecundity in *Limnoperna* has been poorly studied, but its very high rates of colonization indicate that fecundity must be very high (Karatayev et al., 2007a). In the laboratory fresh female mussels induced to spawn produced about 11,000 eggs [$N = 10$, range 133–29,800 per female (Cataldo and Boltovskoy, unpublished data, in Karatayev et al., 2007a)], which is similar to zebra mussels induced to spawn in the laboratory (Stoeckel et al., 2004). Egg size and larval duration in *Dreissena* and *Limnoperna* are almost identical. The time to sexual maturity is three to four months (Boltovskoy and Cataldo, 1999; Darrigran and Damborenea, 2005). The sex ratio is about 1:1 (Darrigran et al., 1998b).

1.4.1.2.2 Life Cycle

The life cycle is similar to dreissenids but with some differences, especially in sizes and shapes of pediveliger and plantigrade larvae (Figure 1.21). Dos Santos et al. (2005) studied the life cycle of the golden mussel in a lake in southern Brazil and much of the size variations in each stage shown in Figure 1.21 is based on their study. The trochophore is the first planktonic stage, measures 80–125 μm in length and lasts for only a few hours; D-form or straight-hinge larvae range in length from 120 to 190 μm and lasts for about seven days; the lengths of veliconcha larvae range between 190 and 220 μm and for pediveliger or umbonal forms between 220 and 250 μm. The larvae settle as plantigrade mussels (>300 μm in length) and then attach to substrate as juveniles. Peak numbers of larvae occurred in October to December in southern Brazilian lakes, but in many lakes two to three cohorts per year are produced (Iwasaki and Uriu, 1998; Boltovskoy and Cataldo, 1999; Darrigran et al., 1999). Juveniles remain immature for about one year and sexual maturity is reached between 5.5 and 8.0 mm in shell length.

Depending on climate, the life span of golden mussels has been reported to be about two years [e.g., Uji River, Japan (Iwasaki and Uriu, 1998)], three years [e.g., Cuenca del Plata locality, Argentina (Boltovskoy and Cataldo, 1999)], four or five years (e.g., Korea) to 10 years [central China (Iwasaki and Uriu, 1998)].

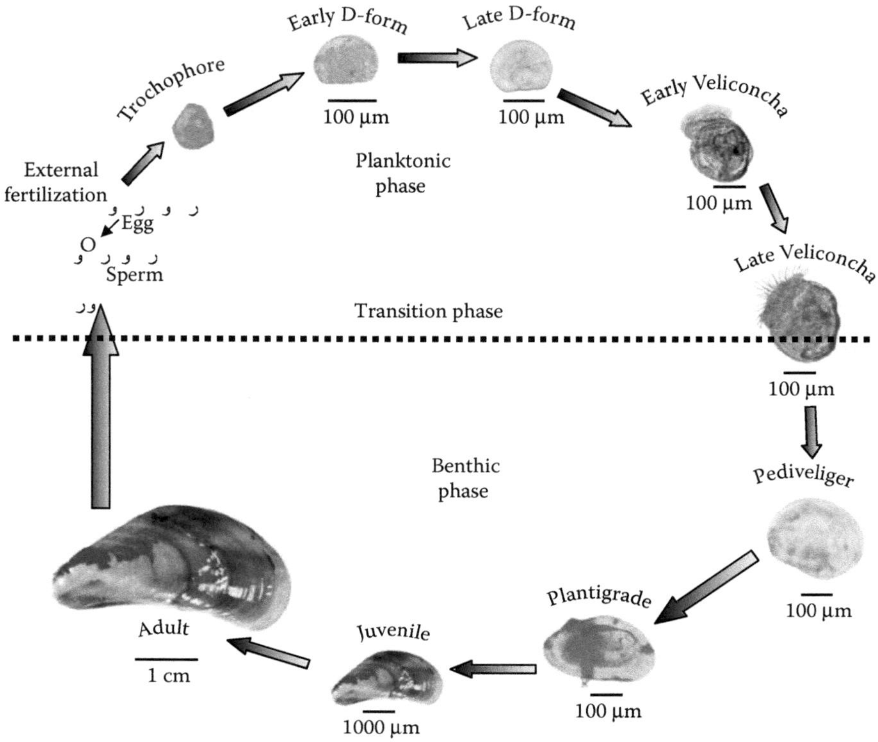

FIGURE 1.21 Life cycle of *Limnoperna fortunei* showing approximate sizes of each planktonic and benthic stage. (Modified from Dos Santos et al., *Revista Brasileira de Zoologica* 22(3): 702–708, 2005.)

1.4.1.3 Corbiculidae

1.4.1.3.1 Reproduction

In contrast to other hermaphroditic bivalve mollusks (e.g., Sphaeriidae), *Corbicula fluminea* is proto-oogamous and initiates oogenesis before spermatogenesis. During seasonal reproductive cycles of *Corbicula fluminea*, oogenesis is nearly continuous, but spermatogenesis is episodic (Kraemer, 1983). Kraemer and Galloway (1986) found that spermatogenesis is synchronously stimulated by a temperature rise in the spring and asynchronously stimulated by a temperature decline in the fall. Spermatogenesis, in turn, "times" the process of fertilization and ontogeny. In Corbiculidae, embryological development follows both self-fertilization and cross-fertilization, and include blastula, trochophore, veliger, and pediveliger stages all of which occur within the parent's inner demibranchs (Kraemer, 1983).

The genus *Corbicula* is distinguished by a wide spectrum of reproductive strategies, ranging from internal fertilization with the release of planktonic, free-swimming larvae to internal fertilization (e.g., *Corbicula fluminalis*, *Corbicula japonicus*) and incubation of larvae in the gills to the straight-hinge stage (e.g., *Corbicula fluminea*) (Kennedy and Huekelem, 1985; Morton, 1986; Kennedy et al., 1991; Byrne et al., 2000; Rajagopal et al., 2000). Species with planktonic larvae are tolerant to brackish water, while species with incubated larvae are restricted to freshwater (Morton, 1986; Rajagopal et al., 2000; Siripattrawan et al., 2000). The incubated larvae of *Corbicula* shed the velum (dreissenids reabsorb the velum) shortly after being released to the outside so the planktonic stage is very brief. A third reproductive strategy was observed in *Corbicula sandai* whereby the incubated larvae shed the velum early in development and released "creeping larvae" (Kornuishin, 2006). Byrne et al. (2000) suggested that the suite of life history traits exhibited by *Corbicula australis* (i.e., hermaphroditism), potential for self-fertilization/androgenesis (see below), brooding progeny to the crawl-away juvenile stage, and a high reproductive output provide for rapid colonization and population growth.

While self-fertilization is relatively common in hermaphroditic mollusks, androgenesis, a rare form of asexual male reproduction, was thought to be rare. However, Hedtke et al. (2008) found that through phylogenetic analyses of mitochondrial genes androgenesis has arisen repeatedly in the Asian clam genus *Corbicula*, including *Corbicula fluminea* and *Corbicula fluminalis*. Multiple lines of genetic evidence suggest that although nuclear recombination between these two species is rare, mitochondrial genome capture is a frequent consequence of androgenetic parasitism of heterospecific eggs (Hedtke et al., 2008). Development of larvae from an exclusively paternal genome has also been reported in East Asian incubating taxa (Okamoto and Arimoto, 1986; Komaru et al., 1997, 2000; Komaru and Konishi, 1999; Qiu et al., 2001) and Australian *Corbicula* (Byrne et al., 2000). Siripattrawan et al. (2000) suggested that most freshwater nominal taxa within the genus *Corbicula* are actually clonal lineages.

Corbicula fluminea does not display as great a fecundity as *Dreissena*, with only 25,000–75,000 veligers produced in the lifetime of a single clam (Aldridge and McMahon, 1978). The pelagic or planktonic development of larvae is common to all the invasive bivalves (zebra mussels, quagga mussels, Conrad's false mussel, golden mussel, and Asian clams), but the planktonic stage of *Corbicula fluminea* is very abbreviated, lasting for only two to four days.

1.4.1.3.2 Life Cycle

The larval stages of the Asian clam differ from the larval stages of dreissenids and mytilids in several ways: (1) most of the larval development in *Corbicula* occurs inside gill chambers of the female clam; (2) the development of internal body structures, including the foot and siphon, occurs at different times than seen in the free-swimming veliger larvae; (3) the size of the larvae at each stage differs from those of free-swimming veligers.

Within the genus *Corbicula* are two contrasting modes of reproduction, planktonic development via a free-swimming larva and intrabranchial incubation, or brooding, of shelled larvae. Kornuishin and Glaubrecht (2003) report three modes of brooding in *Corbicula*: (i) tetragenous brooding (i.e., in inner and outer demibranchs of both gills) in *Corbicula possoensis* (Sarasin and Sarasin, 1898) from Lake Poso, Sulawesi; (ii) prolonged incubation in the maternal gills, with juvenile shells reaching up to 1.3 mm in length and with a well-developed hinge in *Corbicula linduensis* (Bollinger, 1914) from the Lindu River system, Sulawesi; (iii) incubation of young in the inner demibranchs only until the stage of young with straight-hinged shells (D shaped) in *Corbicula fluminea* in North America and Asia, and *Corbicula matannensis* (Sarasin and Sarasin, 1898) from Lake Matano and Lake Mahalona, *Corbicula loehensis* (Kruimel, 1913) from Lake Masapi (all on Sulawesi), and *Corbicula moltkiana* (Prime, 1878) from Lakes Maninjau and Singkarak (on Sumatra). Only the third mode is discussed here.

The life cycle of *Corbicula fluminea* is shown in Figure 1.22. Gametogenesis occurs throughout the year but female tissue tends to be more common than the male tissue (Heinsohn, 1958; Kraemer and Lott, 1977; Britton et al., 1979; Kennedy and van Huekelem, 1985). The smallest clams carrying gametes are between 7 and 10 mm (10 mm, Aldridge and McMahon, 1978; ≥8 mm, Kraemer, 1978; 9.6 mm, Kennedy and van Huekelem, 1985). The embryos are incubated within the inner demibranchs and pass through a trochophore stage that lacks cilia. The larvae are released as D-form (straight-hinge) larvae (>195 μm in shell length) that already have a foot and a velum but it is shed shortly after emergence (~100 h) from the

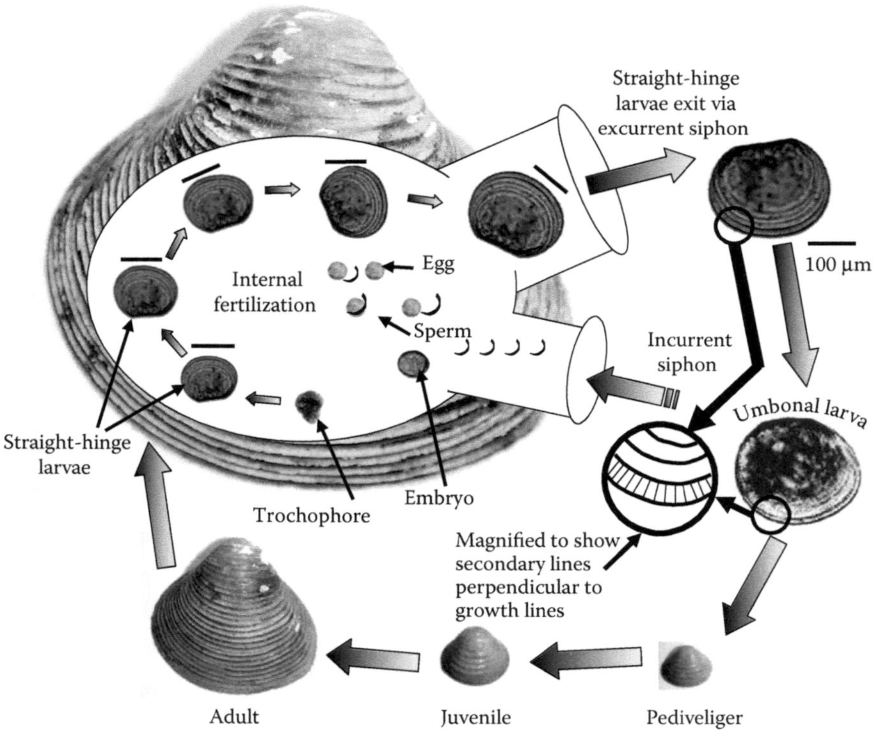

FIGURE 1.22 Life cycle of *Corbicula fluminea*, a hermaphrodite with internal fertilization of eggs by sperm of another individual. The different life stages are incubated within inner two demibranchs of the adult. All scale bars over shelled larvae are 100 μm.

parent, so its ability to swim is short lived (the velum is reabsorbed in dreissenids, not shed). The umbonal larvae (>280 μm long) look like miniature clams; they have siphons and the foot is the sole means of locomotion. A single byssal thread is secreted during the umbonal stage but it is soon lost. The byssal gland becomes nonfunctional by the time the clam is a young adult (Kraemer and Galloway, 1986). A plantigrade stage is lacking in *Corbicula fluminea*. Pediveligers are >500 μm in shell length, their shell is opaque and resemble miniature adults with heavy striae. No further development occurs except for continued growth and sexual maturity. The minimum size of adults that carry larvae is between 13 and 15 mm in shell length (Britton et al., 1979; Kennedy and van Huelekem, 1985). In North America latitude and longitude appear to have no effect on the larval production period, with larvae being produced from spring through fall, with peaks in spring and fall across the continent (Heinsohn, 1958; Aldridge and McMahon, 1978; Britton et al., 1979; Eng, 1979; Dreier and Tranquilli, 1981; Kennedy and van Huelekem, 1985).

In Hong Kong, *Corbicula fluminalis* and *Corbicula fluminea* (Morton, 1986) showed different spawning periods: the brooder, *Corbicula fluminea* released its straight-hinge larvae from May to September, when mean water temperature exceeded 15°C, while *Corbicula fluminalis* with planktonic development released its gametes during October–December and March–April, when mean water temperature was between 6°C and 15°C. Both species showed two spawning peaks per year, namely, October/November and March for *Corbicula fluminalis* and May/June and September for *Corbicula fluminea*. In both species, the second spawning peak was shorter than the first, and the percentage of spawning individuals was much lower during the second period. Spawning frequency of *Corbicula fluminea* correlated positively with the chlorophyll-*a* content of the water. Body mass development also showed a close correspondence with spawning, with two peaks per annum. In contrast to *Corbicula fluminea*, the body mass of *Corbicula fluminalis* increased from December to March, when chlorophyll-*a* concentrations were very low, indicating alternative food sources for this species other than algae (bacterioplankton, detritus). *Corbicula fluminea* allocated more energy resources (51% in May and 21% in September) to reproduction before spawning than *Corbicula fluminalis* (33% in October and 20% in March). The coexistence of *Corbicula fluminalis* and *Corbicula fluminea* in the Rhine branches is probably explained by their different reproductive strategies, reproductive periods, and possibly different food preferences.

1.4.1.4 Implications of Reproductive Behavior on Monitoring and Control

At temperate latitudes spawning in dreissenids, mytilids, and corbiculids may last for three to five months. In general, longer spawning periods are being recorded at southerly latitudes and in thermally enriched waters. Therefore, depending on the location of a facility, the monitoring program should be in place whenever nearby waters reach 12°C (55°F). This will allow one to determine the first appearance of larvae and when they reach settling size in order to optimize a control program.

It is important to realize that the most important sources of veligers, as far as impacts to a facility is concerned, are those well upstream of an intake structure;

since the planktonic stage of corbiculids is very abbreviated, the distance upstream will be much less than those relative to dreissenids and mytilids. Dreissenid and mytilid larvae require about three to five weeks to develop to the settling stage and only those that remain in the water for that period (e.g., those produced well upstream) will be of concern. The adults attached to the intake structure will contribute little or no pediveliger larvae that will settle inside a facility. Any veligers produced will be distributed and passed through a facility before they are even a few minutes (e.g., 30 min) old. The greatest percentage of settled juveniles within a facility originate from adult mussels that are far enough away that veligers will have three to five weeks to develop and settle downstream of prevailing currents (Claudi and Mackie, 1994).

The warmer waters of many of the southern states may allow the zebra and quagga mussels to be even more prolific than in the Great Lakes. The potential infestation level will depend on chemical variables, especially levels of calcium, pH, and alkalinity; physical variables, particularly thermal threshold levels for reproduction and growth, as well as upper and lower lethal thresholds; biological variables, especially primary productivity, which in turn is affected by levels of biologically available phosphorous and/or nitrogen. Knowing the levels of some variables will allow one to estimate the levels of others, for example, knowing conductivity will allow one to estimate levels of calcium and knowing the Secchi depth will allow one to estimate chlorophyll-*a* and total phosphorous levels. The roles of the chemical, physical, and biological variables are discussed later under physiology.

The viability of eggs and sperm of zebra mussels appears to be an inverse function of water temperature. At 12°C, eggs are viable for about 5 h and sperm for about 22 h. At 24°C, eggs and sperm are viable for only about 2 h. Gametes (eggs and sperms) will begin to form soon after the gonads are spent, usually in the fall, and will continue throughout the winter months. By early spring the gonads (ovaries and testes) will be massive and ripe for release of gametes. Spawning in zebra mussels will begin when water temperatures reach 12–16°C. Shallow bays may reach these temperatures sooner and spawning may be initiated sooner than in the main body of water. The resulting larvae may be transported out of such bays into colder water. Hence, it will not be unusual to find veligers in intake water that is well below 12°C. Warm surface (e.g., epilimnial) waters will be more enriched with gametes than lower, colder (e.g., hypolimnial) waters.

Since there is an inverse relationship between growth/development rate of veligers and temperature, the warmer waters of the southern states (relative to the Great Lakes) will probably shorten the development time (i.e., increase the rate of development) required for individuals to reach maturity. For example, peak densities of veliger larvae in the Great Lakes occur when water temperatures are between 18°C and 22°C, usually late June to late August, and produces one or two settlement events, each one lasting for two to three weeks. These water temperatures occur in some of the southern states (e.g., Tennessee) from early May to mid-October. Therefore, it is highly probable that there will be either a larger number of settlement events or the settlement events will be of much longer duration, if not continuous, especially in surface waters. Moreover, some recent studies

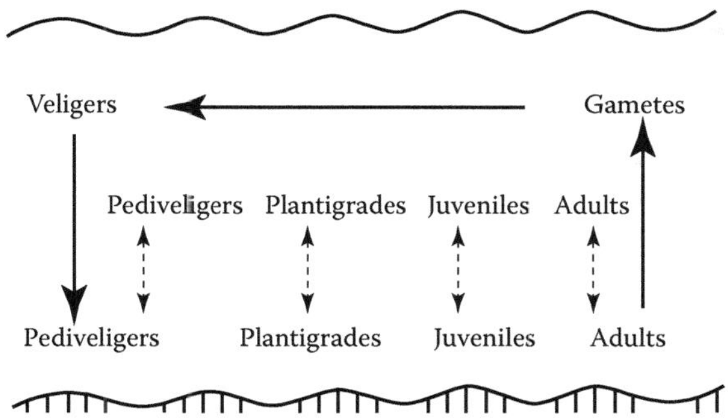

FIGURE 1.23 Movement of different life stages of dreissenids from the substrate back into the water to become translocators and then resettle again.

have shown that there is significant translocation of small adults (<3–5 mm) during the winter and spring. It appears that not only pediveligers readily translocate but plantigrade forms, juveniles, and adults readily move back into the water column after settling (Figure 1.23). The prevalence of translocation may require some facilities to monitor and treat year around. Because zebra mussels, Asian clams, and Conrad's false mussel may be found together in some locations, it is important that the planktonic larval stages of the three species be distinguished from each other (see next section). There is a high probability that *Corbicula* and *Mytilopsis* veligers will be misidentified as zebra or quagga mussel veligers. Although the general aim is to eliminate all veligers, the type of control to use will depend, in part, on the species present. For example, there is no point considering the use of coatings to control Asian clams. Qualified taxonomists should be consulted for final confirmation of identification.

1.4.1.5 Identification of Shelled Larvae of Invasive Species of Bivalves

The shelled larvae (embryos, trochophores) of dreissenids, mytilids, and corbiculids are difficult to distinguish by their form but Asian clam larvae can be readily separated from dreissenid larvae on their location of embryological development. In Asian clams, fertilization and embryo development through shell formation occur inside the female in gill chambers; trochophore larvae will die if removed from the gill chamber (Kraemer et al., 1986; Kraemer and Galloway, 1986). In dreissenids and mytilids, fertilization and all larval development occur outside the female, and gametes and preshell larvae are commonly collected in the water column.

Even the shelled forms are difficult to tell the species apart but the sizes at each stage differ and help in identifying the stage of each species. Nichols and Black (1994) provided a key to separate the different larval forms of *Dreissena polymorpha*, *Dreissena bugensis*, and *Corbicula fluminea*. Their key was used as the basis for the key below but incorporates two other species, *Mytilopsis leucophaeata* and *Limnoperna fortunei*. Characters provided by Conn et al. (1993) were useful for separating larvae of Conrad's false mussel from those of zebra and quagga mussels. As in Nichols and Black's (1994) key, the primary characteristics used here for preserved

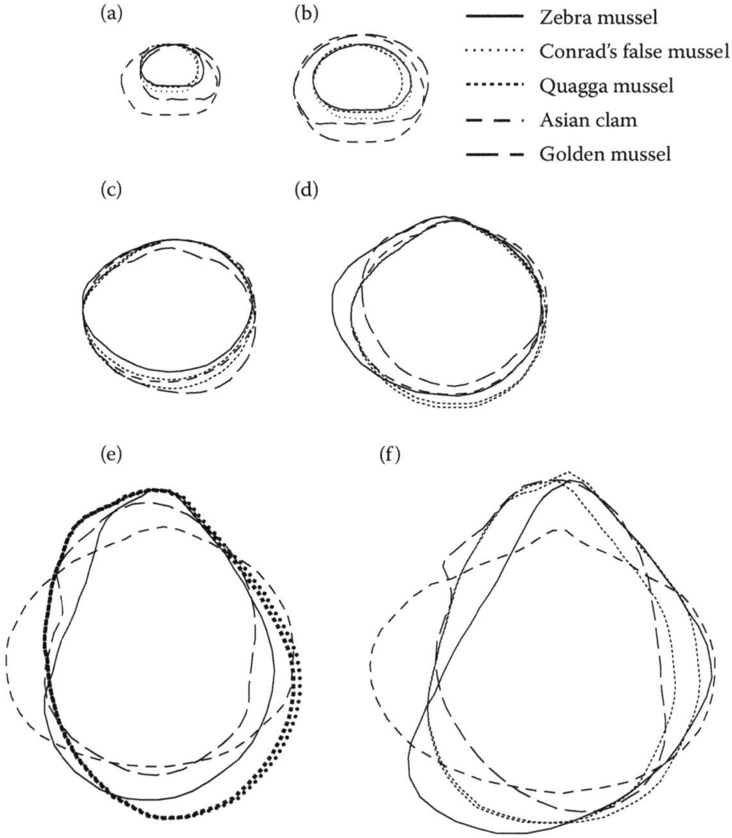

FIGURE 1.24 Shapes of larvae at six different stages for five species of invasive bivalves; (a) Early D-form; (b) Late D-form; (c) Umbonal or veliconcha stage; (d) Pediveliger stage; (e) Early plantigrade stage (Asian clam is juvenile); (f) Late plantigrade stage (Asian clam is juvenile). *Note*: The shapes of larvae for the zebra mussel and Conrad's false mussel are indistinguishable and shown as one line. The right valve extends beyond the left valve of the quagga mussel but valves are completely isometric in zebra and Conrad's false mussels. *Corbicula* lacks a plantigrade stage. See Figure 1.24b for comparisons within species.

specimens are shape, shell length, shell height, hinge length, and shell features, like presence or absence of striae. The shell shapes are given as line drawings for each of six stages in each of the five species (Figures 1.24 and 1.25). Photos of many of the larval stages are given in Figures 1.20 (dreissenid life cycle), 1.21 (mytilid life cycle), 1.22 (*Corbicula* life cycle), and 1.26 (quagga/zebra mussel comparisons). The shell dimension data are presented in Table 1.1. Secondary characteristics, such as the presence or absence of a foot and velum, are provided when living specimens are available. The development of internal organs and hinge development can be found in Section 1.4.1 and its subsections.

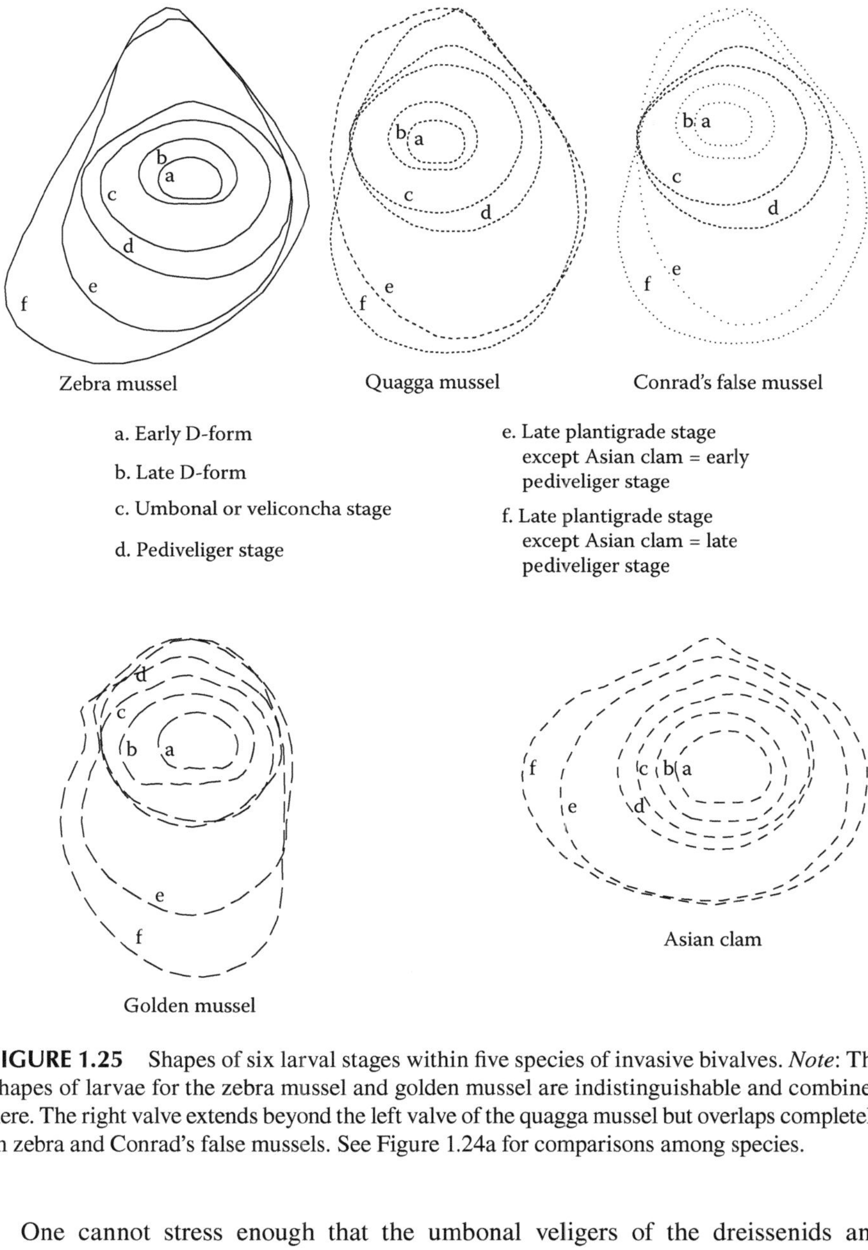

FIGURE 1.25 Shapes of six larval stages within five species of invasive bivalves. *Note*: The shapes of larvae for the zebra mussel and golden mussel are indistinguishable and combined here. The right valve extends beyond the left valve of the quagga mussel but overlaps completely in zebra and Conrad's false mussels. See Figure 1.24a for comparisons among species.

One cannot stress enough that the umbonal veligers of the dreissenids and mytilids are difficult to separate. This difficulty arises because the characteristics used to distinguish early umbonal larvae are not consistent throughout umbonal larval development (Nichols and Black, 1994). Note that as the veligers mature the shell proportions approach those of the zebra mussel. Apply both early and older characteristics to each veliger to assure an accurate identification. This is important, since it is extremely difficult to age veligers (Nichols and Black, 1994).

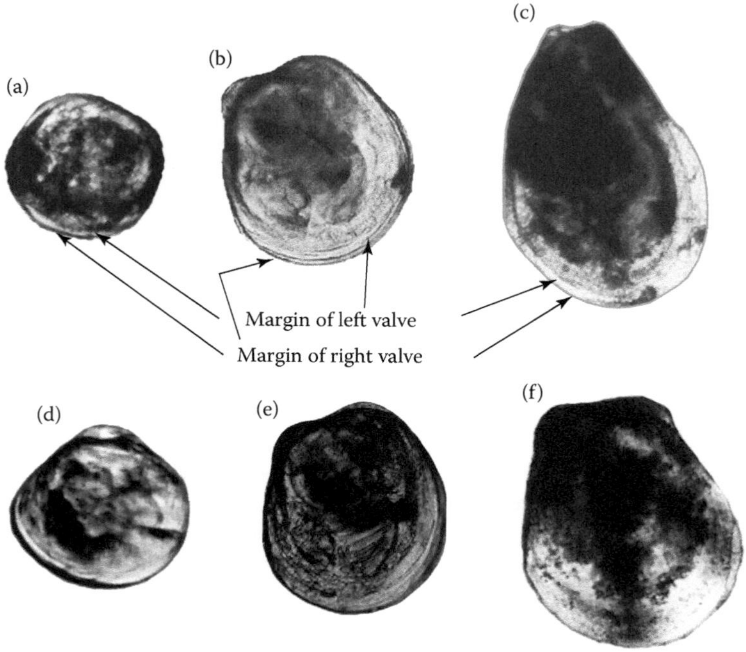

FIGURE 1.26 Valves showing increasing degree of overlap from late pediveliger (a) to early plantigrade (b) and to late plantigrade (c) of *Dreissena bugensis* relative to isometric shell shapes in late pediveliger (d) early plantigrade (e) and late plantigrade (f) of *Dreissena polymorpha*. (Modified from Nichols, S.J. and Black, M.G., *Canadian Journal of Zoology* 72: 406–417, 1994.)

1.4.1.5.1 Key to Identification of Shelled Larvae of Invasive Species of Bivalves

1a.	Hinge line straight (shell is D shaped) (Figures 1.24a, b and 1.25a, b)	2
1b.	Hinge line rounded, some with a protruding bump (umbo) in the center (Figures 1.24c, d and 1.25c, d)	3
2a.	D-forms: shell length 80–190 μm. Velum present, foot absent. Growth lines if any are weak and no secondary lines running perpendicular to the shell edge	4
2b.	D-forms: shell length 190–300 μm. Foot present, velum present or absent. Hinge line straight or sway-backed (sagged). Well-defined growth lines on valves. Numerous secondary lines perpendicular to growth rings, giving the growth ring region a pleated appearance (Figure 1.22)	*Corbicula fluminea*

TABLE 1.1
Mean Shell Dimension and Range (in Parentheses) for each of Five Larval Stages of Five Species of Invasive Bivalves

Species	Shell Length (μm)	Shell Height (μm)	Shell Height/ Length Ratio	Hinge Length (μm)	Hinge Length to Shell Length Ratio
		D-Form Larvae			
Dreissena bugensis	109 (89–129)	93 (83–118)	89 (85–93)	59 (48–54)	50 (48–54)
Dreissena polymorpha	108 (97–133)	72 (64–105)	71 (66–79)	60 (54–67)	46 (33–56)
Mytilopsis leucophaeata	90 (78–110)	81 (59–95)	84 (76–90)	54 (46–66)	60 (59–61)
Limnoperna fortunei	113 (80–150)	82 (58–108)	73 (72–74)	78 (53–99)	67 (66–70)
Corbicula fluminea	240 (190–265)	199 (165–246)	88 (83–93)	110 (74–129)	45 (39–48)
		Early Veliconcha Larvae			
Dreissena bugensis	166 (120–221)	150 (99–202)	88 (83–91)	56 (35–60)	30 (27–34)
Dreissena polymorpha	148 (106–209)	128 (92–191)	88 (86–91)	69 (60–75)	47 (36–57)
Mytilopsis leucophaeata	200 (177–232)	198 (175–230)	99 (98–100)	56 (50–65)	28 (25–30)
Limnoperna fortunei	185 (150–220)	126 (102–150)	68 (67–69)	118 (92–134)	64 (61–66)
Corbicula fluminea	280 (212–331)	247 (187–293)	88 (87–89)	90 (68–106)	32 (30–34)
		Late Veliconcha Larvae			
Dreissena bugensis	184 (150–228)	158 (133–177)	84 (78–89)	60 (51–76)	33 (30–35)
Dreissena polymorpha	214 (140–347)	213 (120–297)	90 (86–90)	91 (74–101)	42 (29–53)
Mytilopsis leucophaeata	270 (250–290)	251 (233–279)	93 (90–98)	81 (75–87)	30 (27–33)
Limnoperna fortunei	235 (220–250)	215 (201–238)	93 (91–95)	150 (141–160)	64 (60–68)
Corbicula fluminea	421 (380–502)	398 (340–472)	93 (85–94)	135 (120–176)	32 (28–35)
		Pediveliger Larvae			
Dreissena bugensis	251 (222–410)	220 (178–390)	88 (80–95)	NA	
Dreissena polymorpha	291 (231–462)	233 (189–367)	80 (79–82)	NA	
Mytilopsis leucophaeata	420 (350–490)	391 (323–460)	93 (92–94)	NA	
Limnoperna fortunei	450 (350–500)	328 (265–370)	75 (73–76)	NA	
Corbicula fluminea	>500	>375	75	NA	
		Plantigrade Larvae			
Dreissena bugensis	>365	>325		NA	
Dreissena polymorpha	>365	>290		NA	
Mytilopsis leucophaeata	>500	>375		NA	
Limnoperna fortunei	>500	>375		NA	
Corbicula fluminea	Plantigrade lacking	Plantigrade lacking		Plantigrade lacking	

Source: Data calculated from Conn, D.B. et al., 1993. *New York Sea Grant Institute*, Stonybrook. Nichols, S.J. and Black, M.G., *Canadian Journal of Zoology* 72: 406–417, 1994; Dos Santos et al., *Revista Brasileira de Zoologica* 22(3): 702–708, 2005.

3a.	Umbonate forms with symmetrical shells and oval outline	4
3b.	Shells may be umbonate but shell outline distinctly asymmetrical	11
4a.	Shell length 300–500 μm. Well-defined growth lines on valves. Numerous secondary lines perpendicular to growth rings, giving the growth ring region a pleated appearance (Figure 1.22)	*Corbicula fluminea*
4b.	Shell length 150–350 μm. Growth lines on valves weak or absent, no secondary lines perpendicular to growth rings	7
5a.	Hinge length to shell length ratio <0.62	Dreissenidae and Mytilidae 6
5b.	Hinge length to shell length ratio >0.62	*Limnoperna fortunei*
6a.	Hinge length to shell length ratio <0.58	8
6b.	Hinge length to shell length ratio >0.58	*Mytilopsis leucophaeata*
7a.	Shell height >85% of shell length, giving shell a rounded appearance. Margin of one valve extends past the margin of the other, particularly just below shoulder line (Figure 1.26a vs. b). One umbo slightly higher or more convex than the other, but often difficult to see	*Dreissena bugensis*
7b.	Shell height >85% of shell length, giving shell a rounded appearance. Shell margins completely isometric	*Dreissena polymorpha*
8a.	Hinge length to shell length ratio <0.58	Dreissenidae 9
8b.	Hinge length to shell length ratio >0.60	*Limnoperna fortunei*
9a.	Hinge length to shell length ratio <0.35	*Mytilopsis leucophaeata*
9b.	Hinge length to shell length ratio 0.35–0.58	10
10a.	Shell height >85% of shell length, giving shell a rounded appearance. Margin of one valve extends past the margin of the other, particularly just below shoulder line. One umbo slightly higher or more convex than the other, but often difficult to see	*Dreissena bugensis*
10b.	Shell height >85% of shell length, giving shell a rounded appearance. Shell margins completely isometric	*Dreissena polymorpha*
11a.	Margin of one valve extends past the margin of the other, particularly just below shoulder line. One umbo slightly higher or more convex than the other, but often difficult to see	*Dreissena bugensis*

11b. Margins of isometric and valves do not overlap. Umbos of similar shape and height	12
12a. Shell height usually <78% of shell length.	*Limnoperna fortunei*
12b. Shell height usually >78% of shell length	13
13a. Shell height <85% of shell length	*Dreissena polymorpha*
13b. Shell height >85% of shell length	*Mytilopsis leucophaeata*

1.4.2 Gastropoda

Freshwater gastropods have two types of reproduction. Oviparous forms lay eggs in water (or above the water surface for some amphibious species), embryo development occurs within the egg shell, the eggs hatch and the snails grow, become mature, copulate, eggs are laid (oviposition), and the life cycle begins again (Figure 1.27). In ovoviviparous forms, embryonic development occurs within the parent and the eggs hatch within the parent as miniature adults, which are then released to the outside where growth and maturity occurs (Mackie, 1978a) (Figure 1.28). The life spans are usually very short, from months to one or two years. Because reproduction and the life cycle are much abbreviated in invasive gastropods, relative to invasive bivalves at least, reproduction and life cycle are discussed together for each family and species.

1.4.2.1 Ampullariidae

Most ampullariids are dioecious; a few like *Marisa cornuarietis* are hermaphroditic. Still, others, like *Pomacea canaliculata*, have been reported to be capable of

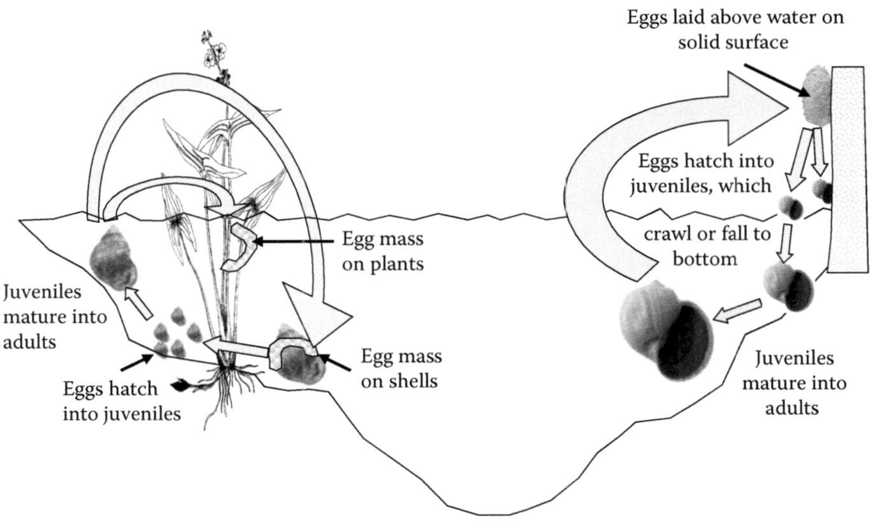

FIGURE 1.27 Oviparous reproduction in which some species (e.g., *Bithynia tentaculata*) lay eggs under the water surface (left side) and others (e.g., *Pomacea haustrum*) lay eggs above the water surface. Both use sexual reproduction with both males and females present.

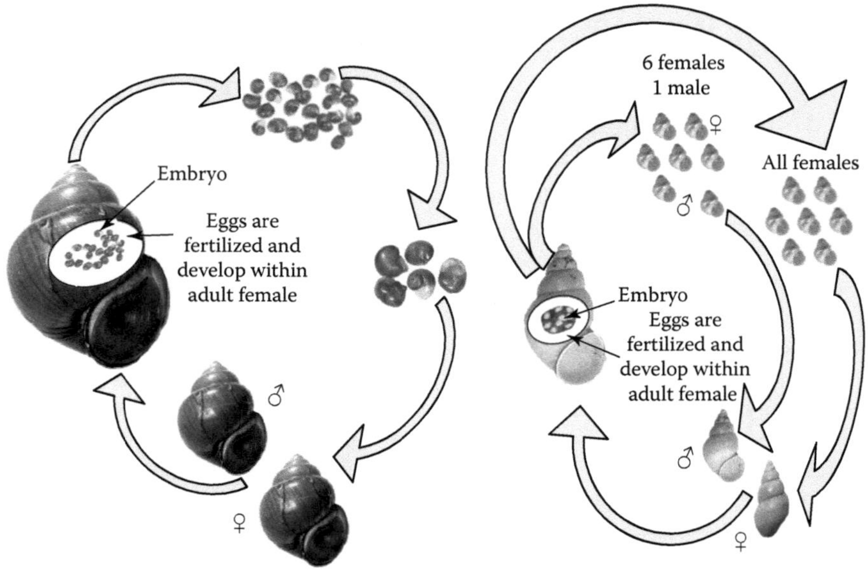

FIGURE 1.28 Ovoviviparous reproduction. Left shows sexual reproduction with both male and female (*Bellamya chinensis* shown). Right shows both parthenogenetic (all females, right side) and sexual reproduction (both male and females) in *Potamopyrgus antipodarum*; parthenogenesis prevails in the species, see text.

protandry and change from male to female at any time of the year (Keawjan, 1987; Keawjan and Upatham, 1990). Breeding tends to be seasonal and related to air and water temperature and rainfall, both influenced by latitude. In equatorial regions, most ampullariids breed in the rainy season and aestivate during the dry season as habitats lose water, but in subtropical regions, ampullariids may breed only during the summer as temperatures reach a threshold level (Andrews, 1964; Cowie, 2002), which differ among species.

All ampullariids are oviparous, laying gelatinous egg masses on vegetation or hard surfaces like rocks either above or below the water surface. The color of the egg masses and clutch sizes differ among species, as described below. The eggs of most ampullariids are individually encased in a calcium carbonate shell. The eggs are fertilized internally with mating occurring anytime during the day or night (Andrews, 1964; Albrecht et al., 1996; Naylor, 1996). On each oviposition event, a single clutch is laid of variable egg numbers depending on species, temperature, and locality (Cowie, 2002). Oviposition usually occurs shortly after copulation, usually in early evening, at night or in early morning (Andrews, 1964; Halwart, 1994a,b; Albrecht et al., 1996). The interval between successive ovipositions is 5–14 days (Cowie, 2002). Embryo development occurs within the egg shell, with some of their calcium requirements being derived from the calcareous shell (Andrews, 1964; Tompa, 1980; Turner and McCabe, 1990). The interval of time required for many life cycle events is highly variable, depending on species and location. For example, hatching occurs about two weeks after oviposition but period varies greatly from 7 to 30 days; the time from hatch to maturity varies from about two months to two years; longevity varies from 119 days to 3 years (Cowie, 2002).

1.4.2.1.1 Giant Ramshorn Snail, Marisa cornuarietis

Marisa cornuarietis is a hermaphrodite and lays its eggs in gelatinous clusters and adhering to a variety of substrates below the water surface, but mostly on macrophytes (Hunt, 1958; Robins, 1971; Thompson, 1984). Each clutch is about 10 cm long and contains 100 to 200 eggs. The sizes of the eggs are about 2–3 mm. During the first three days of development the eggs swell to double or triple their original volume, become more transparent as small white snails inside the enveloping transparent gelatinous mass (Caspers et al., 2008). The eggs take eight days to hatch at 25–30°C and 20 days at 15–20°C (Demian and Yousif, 1973). At a constant water temperature of 26°C the young snails hatch within 10 days (Behrendt, 2007). The eggs are orange when first laid but soon lose this color (Michelson, 1961) and become grayish-white (Robins, 1971). Hunt (1958) reported observing egg masses and newly hatched young in December and January in southern Florida. The Giant ramshorn snail lives for three to five years.

1.4.2.1.2 Golden Apple Snail, Pomacea canaliculata

The golden apple snail is dioecious (Andrews, 1964; Kaneshima et al., 1986; Ito, 2003) but sex determination in *Pomacea canaliculata* appears to be oligogenic, that is, a phenotypic trait produced by two or more genes working together. In this case, the gender is controlled by a small number of additive sex-determining genes, apparently scattered through the genome, inherited from both parents (Yusa, 2004, 2006). Females in North and South American populations lay large eggs in pink clusters on emergent vegetation and other solid surfaces above the water line during the summer (Figure 1.27) (Scott, 1957). In ponds in Japan, females deposit reddish egg masses above the water from June to November (Suzuki and Fukuda, 1988). The brood sex ratio is determined genetically and not environmentally and can vary significantly from 1:1 (Yusa, 2004).

The eggs are fertilized internally with mating occurring during the day or night and lasting for 10–18 h and as often as three times per week (Andrews, 1964; Albrecht et al., 1996; Naylor, 1996). Females in North and South American populations lay a single clutch of 25–1000 eggs, which is laid on each oviposition event (Cowie, 2002). In Japanese ponds, egg production per female per breeding season varies from 400 to 3500 depending on the snail density and food availability in the paddy fields (Albrecht et al., 1999; Tanaka et al., 1999; Estoy et al., 2002). Oviposition generally occurs about 24 h after copulation, usually in early evening, at night or in early morning (Andrews, 1964; Halwart, 1994a,b; Albrecht et al., 1996) but Chang (1985) reports oviposition occurring as much as two weeks after mating. The interval between successive ovipositions is 12–24 days (Chang, 1985). Hatching occurs one to two weeks after oviposition, the newly hatched snails either falling or crawling to water.

Mochida (1998a,b, 1999) estimated that on average a mature female can produce 4375 (maximum 8680) eggs each year. Assuming an average clutch size of about 200 eggs, then about 22 clutches are produced by each female every year. However, development rates are dependent on temperature and locality (Robins, 1971; Demian and Yousif, 1973).

The time from hatch to maturity varies from 50 to 70 days to 2 years (Kiyota and Sogawa, 1996; Cowie, 2002). Maturity is highly variable from as little as three or four months to two years, depending on the environment and temperature, with iteroparous reproduction occurring thereafter (Andrews, 1964; Albrecht et al., 1999; Martin and Estebenet, 2002). Snails maintained at 25°C bred continuously for a single season of four months and then died; when maintained under seasonally fluctuating temperatures (7–25°C), snails took two years to reach sexual maturity, then bred for two distinct annual breeding seasons and died after four years (Estebenet and Cazzaniga, 1992). In Japanese paddy fields, the snail reaches sexual maturity when the shell is about 30 mm high (Ito, 2003). *Pomacea canaliculata* generally has one or two reproductive seasons in the course of its lifetime (Estebenet and Cazzaniga, 1992). However, most of the snails in paddy fields live for only up to two years, and have only one reproductive season but most newly hatched juveniles do not seem to reach their mature size within a single year (Ito, 2003).

Growth of the snails is highly variable and influenced by water and air temperatures, population density, habitat size, and food availability (Johnson, 1958; Keawjan, 1986; Estebenet and Cazzaniga, 1992; Donnay and Beisinger, 1993; Tanaka et al., 1999; Estoy et al., 2002; Ito, 2003). Growth and reproduction of the snail strongly depends on snail density. Hence, a low-density population shows a high population growth rate (Tanaka et al., 1999). This implies that reductions in snail density by artificial methods cannot suppress the density of the snail in future generations.

The habitat and perhaps locality appear to play a role in determining shell size and life span. Maximum size varies among localities, with snails reaching 30 mm in Hawaii, 65 mm in Asia (Schnorbach, 1995), and 90 mm in Germany (Heidenreich, 1997). Snails inhabiting lakes seem to live longer than those from streams in semi-arid regions, probably due to the highly variable hydrological regime of the latter (Martín and Estebenet, 2002). Males and females from lakes, streams, and reservoirs with hard bottoms showed lighter shells than those from shallow lakes with sandy bottoms or streams with sand-muddy bottoms (Estebenet et al., 2006).

1.4.2.1.3 Titan Applesnail, Pomacea haustrum

Pomacea haustrum is dioecious. After successful insemination the female lays green eggs (Winner, 1989, 1996), deposited in dense honeycomb-like clutches on emergent vegetation or hard surfaces above the waterline; Snyder and Snyder (1971) and Guimaraes (1981) reported both green and red/pink eggs but this may represent taxonomic confusion (Cowie et al., 2006). Depending on the female's size and the environment the clutch can be 4–8 cm long. The average clutch develops in high humidity and at warm temperatures within about 7–30 days, after this time 100 to 236 young snails hatch (Cowie, 2002). The time from hatch to maturity varies from 374 to 529 days if time to hatch was 15–23 days and less than one year if time to hatch was 9–30 days (Cowie, 2002). *Pomacea haustrum* can live to be about 2–4 years old in the conditions of its habitat.

1.4.2.1.4 *Island Applesnail,* Pomacea insularum

Pomacea insularum is dioecious and lays light pink eggs in a loose cluster on plants and rocky or cement walls at or below about 1 m above the waterline (Figure 1.27). The egg clutches are large and consist of a mean of 2064 eggs, with clutch size increasing over the reproductive season (Barnes et al., 2008). Not all eggs survive to adulthood, an average clutch yielding 14 to 144 new apple snails (Barnes et al., 2008). The high fecundity of *Pomacea insularum* translates into considerable ecological impact because adult females can contribute >1 clutch/week over an extended growing season in the southeastern USA (Barnes et al., 2008). The island applesnail has a life span of about two years (Cattau, 2008).

1.4.2.1.5 *Spiketop Applesnail,* Pomacea diffusa

The spiketop applesnail is dioecious and somewhat sexually dimorphic. The shell opening of the male tends to be larger and rounder than females, due to the penial complex. The penial complex lies to the right of the mantle in front of the gills; females lack the penial complex (Perrera and Hall, 1996). Females deposit pale pink to reddish eggs above the water surface on vegetation or hard surfaces like rocks (Figure 1.27). The eggs, 2.20–3.5 mm (0.5–0.9 inch) diameter, are closely attached to each other and average 200 to 600 eggs to a clutch. The average size of a clutch is 3.8 cm (1.5 inches). Eggs hatch out after two to three weeks but high temperature causes earlier hatching. In aquaria, the life span appears to be temperature dependent; at higher temperatures they live shorter, but grow, reproduce, and move faster; at 20–21°C, they live for three years or more; at 25°C or above, the life span is shorter, about two years (Perrera and Halls, 1996).

1.4.2.1.6 *Florida Applesnail,* Pomacea paludosa

This native snail of Florida is dioecious. Females lay between 10 and 80 eggs in pinkish-white calcareous clusters well above the water surface on broad-leaved vegetation, especially Sawgrass (*Cladium jamaicense*) (Hanning, 1977; Turner, 1996; Darby et al., 2001) and hard surfaces including rocks, boats, and logs (Figure 1.27). Peak egg cluster production by the apple snails in central and south Florida consistently occurs between March and July; the majority of eggs are laid over a 4–12-week period. Dry downs that encompass the time period of peak reproduction may reduce or eliminate recruitment in the effected area (Darby et al., 1997). Each egg is relatively large, measuring between 3 and 6 mm (Turner, 1996), the hatchlings also being 3–5 mm in shell length (Kushlan, 1975). The young snails hatch and drop into the water below. Maturity is reached at about 30–40 mm, a size that typically requires a year to achieve (Dillon, 2000). Thereafter, adult snails reproduce iteroparously and live for one to two years (Dillon, 2000).

1.4.2.2 Thiaridae

Sexes are separate in the thiarids and most are ovoviviparous, oviparity occurring only rarely (e.g., *Melanatria*; Brown, 1994). Parthenogenesis occurs in some, like the quilted melania. The eggs are brooded inside the female after being inseminated by males; there is no intromittent organ in males. The young hatch inside the female and are released as free-swimming larvae or as small crawling snails (Brown, 1994).

1.4.2.2.1 Quilted Melania, Tarebia granifera

The quilted melania is ovoviviparous with a well-developed brood pouch on the top of the head and neck (Chaniotis et al., 1980c). The species is parthenogenetic although males have been reported (Morrison, 1954; Chaniotis et al., 1980c; Prentice, 1983). Chaniotis et al. (1980c) reported that the sperm ducts of any males present have a blind termination, suggesting that the males are nonfunctional (Figure 1.28). The young are released through a birth pore on the right side of the head at the level of the mantle edge (Chaniotis et al., 1980c) when they are approximately 2 mm in length (Prentice, 1983). Maturity can be reached 97 days after birth when specimens are as small as 6 mm in shell length (Chaniotis et al., 1980a).

1.4.2.2.2 Red-rimmed Melania, Melanoides tuberculata

Red-rimmed melania are dioecious and reproduce sexually and also by parthenogenesis (Figure 1.28) (Morrison, 1954; Berry and Haji Kadri, 1974; Livshits and Fishelson, 1983; Livshits et al., 1984). In Guadeloupe, French West Indies reproduction peaks during the rainy season between June and November (Pointier et al., 1993). As in the quilted melania the brood pouch is located in the head and the number of young within the brood pouch ranges from 1 to 42 (Subda Rao and Mitra, 1982) and from 1 to 71 (Livshits and Fishelson, 1983). The number of young within the pouch appears to vary with the size of the adult (Subda Rao and Mitra, 1982; Livshits and Fishelson, 1983). The young are released when the shells are 1.5–2 mm high. When the young first appears their shells consisted of a large protoconch and the first two whorls. Maturity is reached when individuals reach 10 mm. Pointier et al. (1993) found that the red-rimmed melania needed 4.5 months to reach a size of 10 mm, 14 months to reach 15 mm, and more than 30 months to reach 20 mm; they calculated life expectancies of 0.74 year for the first year of age, 0.9 year for the second year, and 0.5 year for the third year.

1.4.2.3 Bithyniidae

1.4.2.3.1 Faucet Snail, Bithynia tuberculata

Bithynia tentaculata is dioecious but the shells lack sexual dimorphism. It is oviparous and a prolific and opportunistic species that often produces two broods per year. Females become sexually mature at about 7 mm shell height (Mattice, 1970; Vincent et al., 1981). Adult females born the previous year lay eggs in the spring (May–July) when the water temperature is 20°C or higher (Jokinen, 1992). The eggs are laid in organized aggregates arranged in double rows, in clumps of 1–77 mostly on shells of other animals or other species (Kipp and Benson, 2007) or on rocks and broad-leaved submersed macrophytes (Figure 1.27) (Mackie, personal observation). Growth is rather slow with increments of 3.5–4.0 mm in shell height each year (Mackie, 2000b). However, females born in early spring may reach sexual maturity in October and November and lay a second brood of eggs if water temperature is still above 20°C (Jokinen, 1992). Animals take up to two years to reach maximum size (~1.5 cm shell height) and fecundity may reach up to 347 eggs and is greatest for the 2nd year class (Jokinen, 1992). The density of eggs on the substrate can sometimes reach 155 clumps/m^2. Eggs hatch in three weeks to three months, depending on water temperature. The upper lethal threshold temperature for reproduction is about 30°C since oocytes develop poorly at temperatures of 30–34°C (Korotneva and Dregol'skaya, 1992). The lifespan varies regionally and can be anywhere from 17 to 39 months (Kipp and Benson, 2007) or 2 (Pinel-Alloul and Magnin, 1971) to three years (Vincent et al., 1981).

1.4.2.4 Hydrobiidae

1.4.2.4.1 New Zealand Mud Snail, Potamopyrgus antipodarum

Potamopyrgus antipodarum is ovoviviparous, but populations may be entirely parthenogenetic or contain varying proportions of sexually functional males (Figure 1.27) (Mackie, 2000b). Sexual reproduction probably occurs to some extent in many populations because males constitute up to 20% in some European populations (Wallace, 1985), 9% in some Australian populations (Wallace, 1978), but usually 0 to <3% in most populations (Ponder, 1988). Zaranko et al. (1997) reported only females in a Lake Ontario population. Native populations in New Zealand consist of diploid sexual and triploid parthenogenetically cloned females, as well as sexually functional males (less than 5% of the total population). Of the dioecious group (i.e., sexually functional males and females), females bear live young (Winterbourn, 1970a,b; Wallace, 1978). In the asexual female group, females develop eggs that can grow without fertilization and produce genetically identical offspring (clones). Therefore, one female is sufficient to initiate a new population. So far, introduced populations to North America appear to be entirely clonal, consisting of genetically identical females (Zaranko et al., 1997).

Ova develop within the female's brood pouch inside the first whorl of the shell of most adult snails. The young emerge into the environment as fully functional snails, most or all of which are clonal in North American populations. Each female can produce between 20 and 120 embryos per brood (Winterbourne, 1970b). In New Zealand, the life cycle is annual and young can be born every three months (Winterbourne, 1970b). Each snail can produce approximately 230 young per year (Schreiber et al., 1998; Lively and Jokela, 2002). Reproduction occurs in spring and summer, and the life cycle is annual. The frequent and rapid reproduction rate of the snail can result in enormous population densities, up to 800,000 m^{-2} in Lake Zurich, Switzerland, 300,000 m^{-2} in parts of the Greater Yellowstone Ecosystem (Kerans et al., 2005) and as high as 500,000 snails/m^2 in others (Hall et al., 2003).

1.5 POPULATION DYNAMICS

Population dynamics is a function of a species' reproductive potential, its life cycle or life history, and biotic and abiotic modifying factors. According to Mackie (2004) several aspects of reproductive potential need to be considered in order to predict its population dynamics and, hence, the rarity or invasiveness of a species: (i) the species' sexual state (e.g., monoecious or dioecious); (ii) its egg-laying habit (e.g., oviparous, ovoviviparous, viviparous); (iii) its fecundity (number of eggs produced);

(iv) its natality (number of eggs surviving); (v) its annual frequency of egg-laying habits (e.g., univoltine, bivoltine, multivoltine); (vi) its life-time frequency of egg-laying habits (e.g., semelparity, iteroparity).

Hermaphroditism reduces the risk of a species being eliminated during periods when it is difficult to find a mate (Mackie, 2004). Hermaphrodites, like sphaeriids and corbiculid clams, tend to have shorter life spans and higher frequencies of reproductive events per year (e.g., bivoltinism, trivoltinism, or multivoltinism) but many (the Asian clam being an exception) have few reproductive events in their life time (i.e., are semelparous, see below) than do most dioecious species. Indeed, it is uncommon to find an invasive, nuisance species that is hermaphroditic, the Asian clam being an exception. Most (at least of the species discussed here) are dioecious.

Dioecious animals can be oviparous, ovoviviparous, or viviparous. Viviparity is mostly absent in freshwater animals. Oviparity occurs in two forms, laying of internally fertilized eggs with young hatching from egg attached to the substrate (e.g., *Pomacea* spp.) and spawning of eggs that are then fertilized in the water (e.g., dreissenids). Oviparity is more common than ovoviviparity (brooding of eggs and young, with birth of juveniles that look like miniature adults) in freshwater animals. Oviparity often results in enormous numbers of eggs (e.g., >1,000,000/female/yr in dreissenids), many of which perish during development (e.g., 99% in dreissenids), but those that do survive seem to mature and complete their life span (e.g., dreissenids). Hence, oviparity is a trait represented by many invasive nuisance species like dreissenids, mytilids, and ampullariids.

Snails of the families Thiaridae and some of Hydrobiidae are ovoviviparous and dioecious, but many species are also capable of parthenogenetic reproduction. Like hermaphroditism, parthenogenesis would also allow a species to reproduce when mates are difficult to find. However, the brooding behavior is usually accompanied by a few, small-sized young and although they tend to have a high survival rate (i.e., high natality rate), the population sizes rarely attain the levels seen in oviparous species. Nevertheless, the high survival rate is conducive to forming large enough populations to be a nuisance.

In summary, species that are dioecious, oviparous or ovoviviparous and parthenogenetic, and iteroparous and multivoltine with high fecundities and/or natality rates are more likely to become an invasive nuisance than species that are monoecious, ovoviviparous (without parthenogenesis), semelparous, and univoltine with low fecundities and/or natality rates. Key to the invasive traits, however, is dispersal ability, the greater the variety of dispersal mechanisms, the more potential a species will be invasive and a nuisance (Mackie, 2004). Hence, based on reproductive traits alone, the potential for a species' dispersal ability can be ranked from high (e.g., ovoviviparity, high fecundity, high natality, multivoltine) to moderate (e.g., ovoviviparity, low fecundity, high natality, bivoltine) or low (e.g., oviparity, low fecundity, low natality, univoltine) (Mackie, 2004). The dispersal mechanisms of the invasive mollusks are discussed in the next section.

The life cycles described above incorporate the type of life history events that may occur annually or in a species' life span. A semelparous animal reproduces only once in its life time; an iteroparous animal reproduces two or more times in its life time. Voltinism describes the *annual* number of reproductive events (generations): univoltine is one generation per year, bivoltine is two generations per year, and multivoltine is many generations per year. A semelparous species normally dies after releasing its young and if the young aestivate for the remainder of the year the population is univoltine. If the young begin to grow immediately after birth and achieve adulthood and produce another brood (generation) of young within the same year as the first brood, the population is bivoltine.

Mackie (2004) suggests that species are thought to adopt one of three life history strategies in order to thrive in stable and unstable environments: (i) *r*-selection, (ii) *K*-selection, or (iii) bet-hedging. Stable environments are those in climates that are relatively constant and/or predictable, as in tropical climates or in large, deep cold lakes. Unstable and/or unpredictable environments are characteristic of variable climates, like many in the temperate zone, and also include vernal ponds, artificial impoundments, and rivers with widely fluctuating physical and chemical variables.

Stearn (1976) states that stable environments are characterized by species with a *K*-strategy, while fluctuating environments are characterized by species with an *r*-strategy. Advocates of *r*- and *K*-selection deal with models in which fecundity and mortality schedules fluctuate. Table 1.2 summarizes Stearn's (1976) correlates of *r*- and *K*-selection. In fluctuating environments, such as temporary ponds, species must develop quickly and reproduce before the pond dries. The species has a high resource threshold that permits the organisms to utilize the resources that suddenly become abundant and available when the pond fills in the spring. The parents die but the small body size of the young enables the organism to aestivate in the mud. Because adult mortality is usually high and life spans are short, semelparity, or "big bang" reproduction (a bunch of eggs or young are released and then the parent dies), is favored in unstable environments. However, or, in some cases (e.g., *Pomacea* species), adult mortality is low because the species have developed adaptations like lungs, an operculum, and/or producing shelled eggs to help increase their survival rates.

Organisms in stable environments, on the other hand, tend to grow slowly and put more energy into growth and development early and then apportion energy to reproduction. With environmental stability, species have evolved long life spans (e.g., 10 to >100 years in unionids), which leads to larger body sizes, reproduction once a year (after a long period of reaching sexual maturity), and low adult mortality until they reach senescence (a case where being big helps) (Mackie, 2004).

Bet hedging occurs when fluctuations in these life history traits occur or there is a mix of traits. For example, the life history traits for *r*-selection favor invasive species while those of *K*-selection favor rare species or species at risk. Using dreissenids as an example of an invasive, nuisance species, *r*-selection in Table 1.2 predicts high adult mortality (but larval mortality greatly exceeds adult mortality), semelparity (iteroparity predominates), life span <1 year (but life span is typically >2 years). Stearn (1976) advocates a "bet-hedging" strategy for such species and results in reversals of some *r*- and *K*-selection tactics (Table 1.3). It appears that *r*-selection traits prevail for aquatic invasive nuisance mollusks but the mix of traits that are revealed differ from one population to the next. Hence, population dynamics can be expected to vary accordingly from one ecosystem to the next.

TABLE 1.2
Some Strategies of *r*- and *K*-Selection Traits of Stearn (1976)

Traits	*r*-Selection Tactic	*K*-Selection Tactic
Population size	Variable in time, no equilibrium; recolonization frequently needed; usually below carrying capacity	Constant in time and equilibrium; recolonization rarely needed; at or near carrying capacity
Selection favors:	1. Rapid development	1. Slow development
	2. High intrinsic rate of increase	2. Low intrinsic rate of increase
	3. Early reproduction	3. Delayed reproduction
	4. High resource thresholds	4. Low resource threshold
	5. Small body size	5. Large body size
	6. Semelparity	6. Iteroparity
	7. Increased birth rate	7. Decreased death rate
Competition	Often slack due to exploitation of poorly used habitat	Usually intense due to sharing of habitat
Mortality	Density-independent; high juvenile survival; high adult mortality	Density-dependent; high juvenile mortality; low adult mortality
Life span	Short, <1 year	Long, ≫1 year
Relative allotment of energy to reproduction:	Relatively large	Relatively small
1. Mass of young/parent/brood	Larger	Smaller
2. Mass of young/parent/lifetime	Larger	Smaller
3. Size of young	Smaller	Larger
4. Parental care	Less	More
Leads to:	Productivity	Efficiency

Source: Modified from Mackie, G.L., 2004. *Applied Aquatic Ecosystem Concepts*, 2nd ed., Dubuque: Kendall/Hunt Publishing Company.

Population dynamics is not only a function of a species' reproduction potential and of its life cycle or life history but of changes in population size resulting from biotic and abiotic forces that directly or indirectly control and regulate populations over time. Biotic forces include food supply and availability, predation and parasitism, and generally fall under the category of "density-dependent" factors. Abiotic forces include all the physical and chemical variables that affect the physiology of a species. Mollusks require sufficient levels of anions like calcium and magnesium and cations like bicarbonates, as well as dissolved oxygen and temperature; if these are relatively constant in magnitude, the population remains healthy and its size tends to vary in a predictable manner. If the levels of these variables vary significantly above or below the physiological thresholds of a species, the population size falls due to direct mortality and/or impaired reproductive effort and/or growth; these generally fall under the category of perturbations or "density-independent" factors.

Population size is usually expressed as either density/unit area or biomass/unit area (e.g., kg/m^2). The change in biomass per unit area over time is called

TABLE 1.3
Contrasting Predictions of *r*- and *K*-Selection and Bet-Hedging According to Stearss (1976)

Stable Environments	Fluctuating Environments
Bet-Hedging Predictions when Adult Mortality is Variable	
Slow development, late maturity	Rapid development, early maturity
Iteroparity	Semelparity
Smaller reproductive effort	Larger reproductive effort
Fewer young	More young
Long life span	Short life span
Bet-Hedging Predictions when Juvenile Mortality is Variable	
Early maturity	Late maturity
Iteroparity	Semelparity
Larger reproductive effort	Smaller reproductive effort
Fewer broods but more young per brood	More broods but fewer young per brood
Shorter life span	Longer life span

Source: Modified from Mackie, G.L., 2004. Applied Aquatic Ecosystem Concepts, 2nd ed., Dubuque: Kendall/Hunt Publishing Company.

production, expressed as biomass/unit area/unit time (e.g., $kg/m^2/yr$). Most control programs follow changes in population densities because it is easier to measure than production.

1.5.1 Measuring and Interpreting Variations in Population Sizes

Chapter 4 gives methods for measuring population sizes of all the invasive species of mollusks. When a nuisance species first invades an aquatic habitat, the population density is low but grows quickly in an exponential fashion (Griffiths et al., 1991) until carrying capacity of the habitat is reached or even exceeded. For a thorough discussion of population growth and density-dependent and independent factors that affect growth in mollusks, see Dillon (2000). Chapter 4 also describes how to interpret variations in population sizes in the nuisance invasive species of mollusks. Typically, populations of invasive species grow quickly, often in orders of magnitude in the first three to five years (Figure 1.29), such as reported for zebra mussels in Lake St. Clair by Claudi and Mackie (1994). This is known as "boom or bust" pattern of population behavior by introduced species, whereby a population in attaining its peak density overexploits its food and habitat resources and then declines (Williamson, 1996). The time at which carrying capacity is reached varies from one aquatic habitat to the next but can be expected to be quicker in warmer climates than in cooler ones.

If a major perturbation occurs, such as an increase in temperature above the specie's lethal survival threshold, an introduction of a parasite or a predator, or a decline

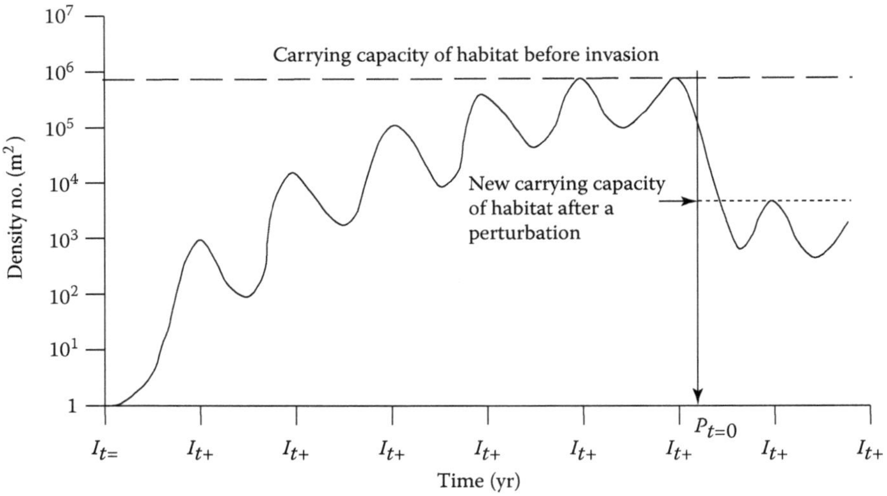

FIGURE 1.29 Characteristic population growth pattern of an introduced species and a population crash after a major perturbation. I_t is time of introduction with 0–8 years. $P_{t=0}$ is initial time of perturbation. Carrying capacities before and after the perturbation are indicated by dashed lines.

in food supply and availability, the population will suddenly decline to a new level (Figure 1.29). If the decline is due to its own exploitation of its food resources, a new carrying capacity will be established and the population will carry on in its cyclic pattern of peaks and valleys in abundance. Graney et al. (1980) took benthic samples from six stations monthly to bimonthly for *Corbicula fluminea* from a thermally influenced river in Virginia and found that densities peaked at about 11,500 m^{-2} and then declined and stabilized at about 2000 m^{-2} from 1976 to 1978. Similar trends for the Asian clam were reported by Foe and Knight (1985) and Sickel (1986) in other rivers and all appear to be influenced in part by density-dependent factors, such as self-mediated reductions in phytoplankton food resources. Population crashes in densities of zebra mussels of 10 years or more are common in lakes that develop unusually high densities over several years, especially in eutrophic lakes in Poland (Stanczykowska and Lewandowski, 1993). The crashes were attributed to mortality of planktonic veligers in the settlement stage and in the postveliger stage. If the decline is due to an increase in temperature above the lethal thresholds for growth and reproduction, the population will crash and continue to crash if the high temperatures prevail. However, if the increase in temperature is of short term the population will gradually return to its original carrying capacity, assuming no other perturbations (e.g., declines in dissolved oxygen and essential elements) occur in the mean time.

2 Distribution, Dispersal Potential, and Dispersal Mechanisms

2.1 INTRODUCTION

This chapter examines the native range, the current distribution in North America, and the different dispersal mechanisms employed by each of the invasive mollusk species discussed in Chapter 1. The chapter also offers suggestions to limit the rate of spread among different bodies of water.

Dreissenids are unique from other freshwater bivalves and gastropods in having two dispersal stages, one as planktonic larvae and the other as byssally attached adults. Asian clams also have a planktonic larval stage but it is much shorter in duration than those of dreissenids. Corbiculids are infaunal (buried in the sediments), whereas all four dreissenids are epibenthic or typically attached to the surface of solid objects. The gastropods are represented by oviparous and ovoviviparous forms. Oviparous forms can be dispersed in both the egg and adult stages. The ovoviviparous forms are dispersed only as the adult (or juvenile) stage. The bivalves (dreissenids, mytilids, and corbiculids) are examined first, then the oviparous forms of gastropods (Ampullariidae and Bithyniidae), and finally the ovoviviparous forms (i.e., Thiaridae and Potamopyrgus antipodarum).

2.2 DISTRIBUTION

2.2.1 Bivalvia

2.2.1.1 Dreissenidae

2.2.1.1.1 Zebra Mussel, Dreissena polymorpha *(Pallas, 1771)*

Native range: The zebra mussel is commonly reported as native to the Caspian and Black Seas. Son (2007) re-examined the distribution of the different species in Europe and found that the native range of the zebra mussel is limited by the Northern Black Sea Region and the Caspian Basin. The native range of the zebra mussel includes Danube, Dniestr, Berezan, Southern Bug, Dniepr, Molochnaya, Don, Kuban, Kamchia, and Veleca river basins; isolated and semi-isolated relic estuarine reservoirs along the Bulgarian, Romanian, Ukrainian, and Russian Black and Azov sea coasts and in zones of these seas influenced by freshwater (Son, 2007). It is also indigenous to Turkey and is still expanding its range in that country (Bobat et al., 2004). One empty valve of *Dreissena polymorpha andrusovi* was found in the

Taganron Bay, Turkey (Bobat et al., 2006). It has been expanding its range in Europe since Pallas first described in 1771.

Current range: The zebra mussel has been expanding its range to most of western Europe and the United Kingdom. Invasion of the Iberian peninsula, starting in the Ebro River, was a recent range expansion. Son (2007) also found new locations of the zebra mussel within the Ponto-Caspian region (Lake Sasyk, Baraboj River) and Sukhoj Liman Basin.

The species has also been expanding its range in North America since it was first discovered in Lake St. Clair and the Lake Erie in 1988 by Hebert et al. (1989). In 1989, it had spread to all the Great Lakes with one record in Lake Superior, two in each of Lake Huron and Lake Michigan, numerous records on the U.S. shores of Lake Erie and Niagara River west end of Lake Ontario and one record in the St. Lawrence River (Benson, 2008a). A year later the zebra mussel had expanded its range throughout the Great Lakes and it was not until 1991 that it had appeared in the Ohio, Mississippi, and Hudson rivers, and several inland lakes in southern Ontario and New York (Lake George). By 1992, the mussel had moved into several tributaries of the Mississippi River, including Tennessee, Kansas, Kentucky, Arkansas, and Kanawha rivers, and other river systems like the Susquehanna River. From 1993 to 1998 the zebra mussel had colonized most of the major river systems in the United States. It was first found in the Missouri River in 1999, but it took four years to expand its range in the Missouri and Delaware rivers. The zebra mussel expanded its range through Oklahoma and Kansas from 2003 to 2007. The next major range expansion was in 2008 when it was found in Colorado (Grand Lake and Pueblo Reservoir), Utah (Huntington Creek), and California (San Justo Reservoir) (Benson, 2008a). The current distribution (as of October 2008) of the zebra mussel is shown in Figure 2.1, but Kraft et al. (2002) suggest that zebra mussels have not yet saturated European and North American lake landscapes, including many suitable lakes; similar distribution patterns of invaded lakes in Belarus and North America suggest that similar processes have influenced zebra mussel spread in both landscapes.

2.2.1.1.2 Quagga Mussel, Dreissena bugensis *(Andrusov, 1897)*

Native range: The native range of the quagga mussel has commonly been reported to be in lower stretches of Bug River (the Southern Bug), where Andrusov first found and described it, and the Dnieper River (Mills et al., 1996). However, Son (2007) found that this designation of its native range was due to vagueness of the term "Dniepr-Bug Liman" in some English-written literature; in his opinion, the native range is the estuarine Southern Bug–Bug Liman (expanded river mouth). Son (2007) gives reasons to believe that the Dniepr Delta may be considered as the beginning of the quagga mussel's expansion into the Dniepr Basin. See also Section 10.1 where Orlova adds further to the distribution of zebra mussels in Europe and adjacent countries.

Current range: The range expansion of the quagga mussel in Europe has been attributed to the extensive network of canals in most countries and it now occurs in almost all Dneiper reservoirs in the eastern and southern regions of Ukraine and deltas of the Dnieper River tributaries (Mills et al., 1996).

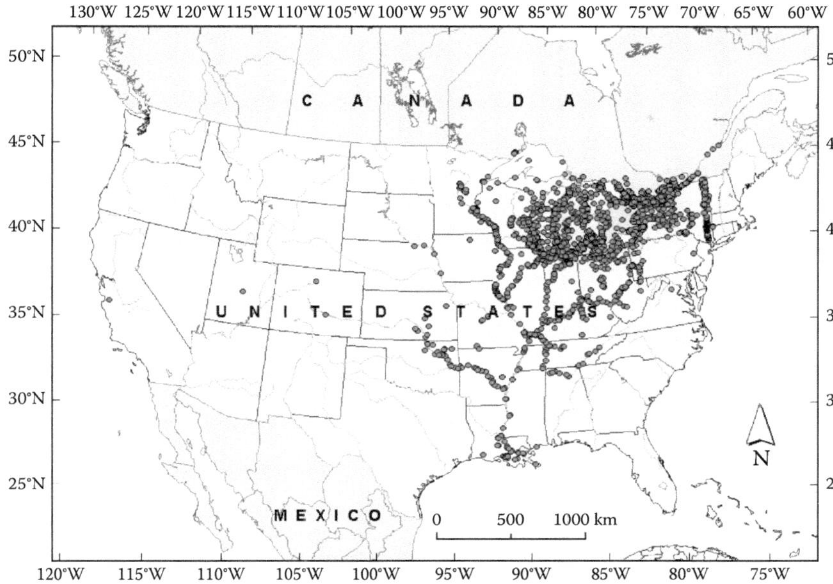

FIGURE 2.1 Distribution of the zebra mussel, *Dreissena polymorpha*, in North America as of May 2009. (Courtesy of Amy Benson, National Biological Services, Southeastern Biological Sciences Center, Gainsville, FL.)

In North America, the quagga mussel was first found in Lake Erie in September 1989 near Port Colborne (Mills et al., 1996). However, it was first considered a variant of the zebra mussel and was not recognized as a quagga-type mussel until 1991 (Mills et al., 1996). A mussel with a different genotype was found in Erie Canal near Palmyra, New York, in August 1991 and was named by May and Marsden (1992) as the "quagga mussel" after the "quagga," an extinct African relative of the zebra. The following timelines are taken from Benson (2008b). The quagga mussel slowly spread to the other Great Lakes, through Lake Erie and Lake St. Clair from 1992 to 1994, to Lake Huron and Lake Michigan in 1997, and to Lake Superior in 2005. The first report of the quagga outside the Great Lakes watershed was in 2000 from Dutch Springs Reservoir, Pennsylvania. It was found in the Ohio River near Cincinnati in 1993 and Mississippi River (River Mile 256) in 1995. The quagga mussel expanded its range in these two rivers from 1995 to 2006. It was first reported beyond the 100th meridian in 2007 in California (e.g., Murray Reservoir, Colorado River Aqueduct, Lower Otay Lake, Lake Miramar, Lake Skinner, Copper Basin Resevoir), Nevada (e.g., Lake Mead, Colorado River), and Arizona (e.g., Lake Pleasant, Central Arizona Project canal) (Figure 2.2). The species appeared in Colorado in 2008 in Pueblo Reservoir, Tarryall Reservoir, Granby Reservoir, and Julesburg Reservoir (Benson, 2008b).

2.2.1.1.3 Conrad's False Mussel, Mytilopsis leucophaeata *(Conrad, 1831)*

Native range: Conrad's false mussel, *Mytilopsis leucophaeata*, is a brackish water species with an original distribution in the subtropical and temperate area of the Gulf of Mexico (Marelli and Gray, 1983).

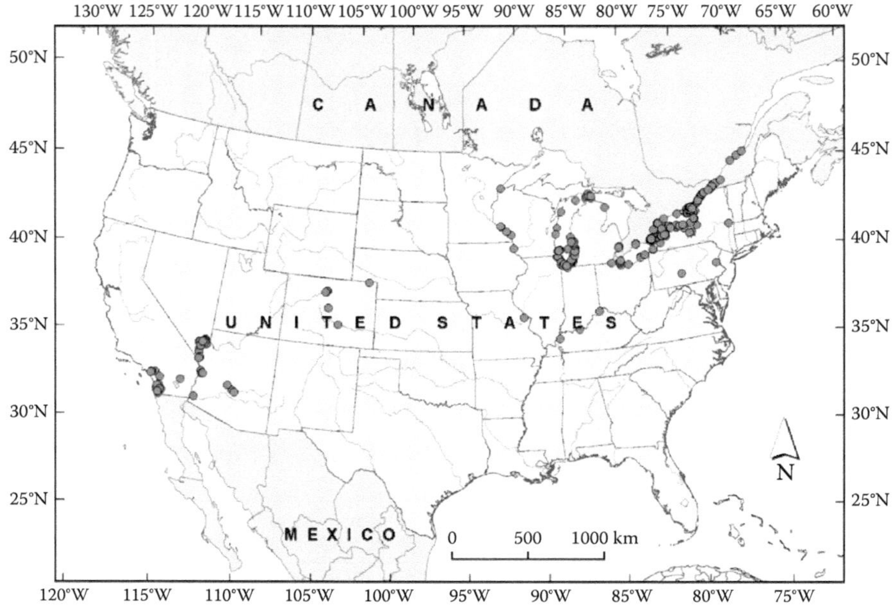

FIGURE 2.2 Distribution of the quagga mussel, *Dreissena bugensis* as of May 2009. (Courtesy of Amy Benson.)

Current range: Mytilopsis leucophaeata has been reported from European waters as early as 1835 (Wolff, 1999; Verween et al., 2005). Gittenberger and Janssen (1998) suggest that the species is dispersed as larvae in ballast water. This, however, is unlikely since ballast water was not used before the 1870s (Carlton, 1985). Marelli and Gray (1983) suggest that the most probable vector was adults on solid ballast "in somewhat less than watertight wooden vessels." However, Wolff (2005) thinks this vector is unlikely because one would expect anaerobic conditions in the hold of such a vessel. Wolff (2005) therefore suggests that *Mytilopsis leucophaeata* has been introduced into Europe and into the Netherlands on the outer hulls of ships. Laine et al. (2006) state that the introduction of the species into the Baltic Sea "has obviously taken place via ballast water transport."

It is found in North Western Europe in estuaries along North Sea coasts from Germany to France (Marelli and Gray, 1983; Oliver et al., 1998; Verween et al., 2005) and the River Thames estuary, England (Bamber and Taylor, 2002). The mussel was reported in the Netherlands in the western margin of the Zuiderzee area and in the coastal areas of the Wadden Sea, Friesland, Noord-Holland and Zuid-Holland (Van Benthem Jutting, 1936, 1943), as well as in Noordzeekanaal and Kanaal van Gent naar Terneuzen (Kuijper, 2000). The species has long been present in the Kiel Canal connecting the North and Baltic Seas (Boettger, 1933; Schlesch, 1937), as well as in Kaliningrad area (Brohmer, 1962) but the latter population has probably become extinct (Laine et al., 2006). More recently, the species has also been reported in the river Warnow estuary, northern Germany

(Darr and Zettler, 2000), in Spain (Escot et al., 2003), and the Black Sea (Therriault et al., 2004). The most recent record is in the North Baltic Sea (Laine et al., 2006). European populations occupy both freshwater and brackish estuary habitats (Reise et al., 1999).

In North America *Mytilopsis leucophaeata* is found on the east coast from New York south to Florida, and in Texas and Mexico (Abbott, 1974; Emerson and Jacobson, 1976; Mondadori, 1980; Andrews, 1981; Marelli and Gray, 1985). Conrad's false mussel colonized the Hudson River at Piedmont, New York, in the 1930s as a result of the discharge of a ship's solid ballast (Rehder, 1937; Jacobson, 1953). Since it is an estuarine (oligohaline) species, it is probably endemic to most of the reported occurrences in coastal states, including Charles River, Massachusetts (Rehder, 1937; Jacobson, 1953; Carlton, 1992), St. James River, Virginia (Jordan and Sutton, 1984), St. John's River (Williams et al., 2008) and Doctor's Lake, Florida (Cooksey et al., 2001), Perry River Bass Marine Research Station, Calhoun County, Texas (Howells et al., 1996), and Mobile Basin, Alabama (Williams et al., 2008) (Figure 2.3). The species was also reported in southern New England (Smith and Boss, 1996) and most recently at several locations (e.g., South, Severn, Magothy, and Patapsco rivers) in Chesapeake Bay (Milke and Kennedy, 2001). The species has been reported in inland freshwaters in the Upper Mississippi River in Illinois (Koch, 1989) and Kentucky Lake, Kentucky/Tennessee (USACE, 1993) (Figure 2.3), but it apparently cannot reproduce in freshwaters (Marelli and Gray, 1985) so the populations may not persist.

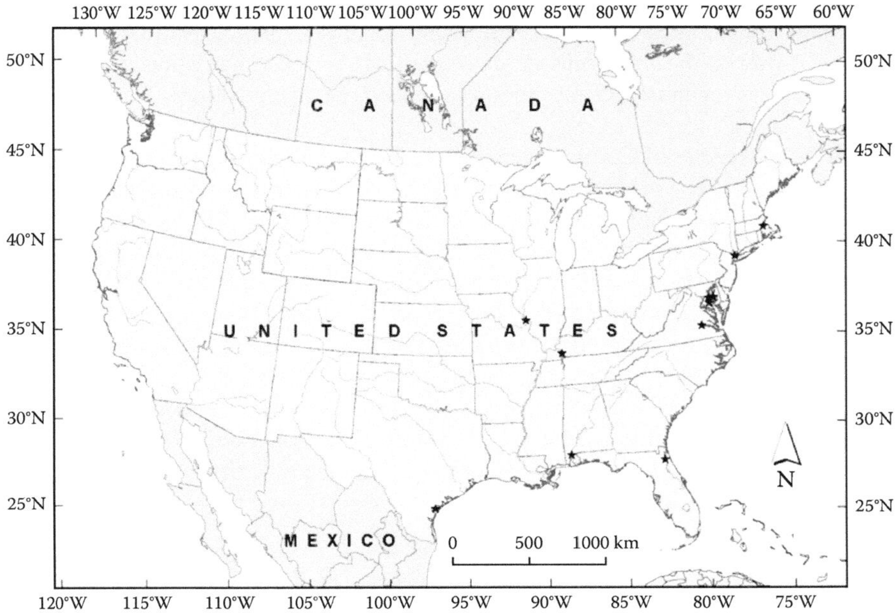

FIGURE 2.3 Distribution of *Mytilopsis leucophaeata* (solid stars) in North America as of May 2009. Most occurrences are on east coast and Gulf of Mexico coast.

2.2.1.2 Mytilidae

2.2.1.2.1 Golden Mussel, Limnoperna fortunei *(Dunker, 1857)*

Native range: The golden mussel, *Limnoperna fortunei*, is an epifaunal mytilid, native to rivers and creeks of China and southeastern Asia. It became established in Hong Kong in 1965 and in Japan and Taiwan in the 1990s. In 1991 it invaded America through the Plata Basin in South America.

Current range: The golden mussel was introduced to South America in 1991 through ballast water discharge into La Plata River basin at Argentina (Pastorino et al., 1993; Darrigran and Pastorino, 1995). It has since expanded its range to the Paraná, Paraguay, Pilcomayo, and Uruguay rivers in Argentina, Brazil, Paraguay, and Uruguay (Boltovskoy and Cataldo, 1999; Oliveira, 2003; Brugnoli et al., 2005). In 1998, the golden mussel was recorded at the northern reaches of the Patos Lagoon drainage basin (Mansur et al., 1999, 2003). The species invaded the southern portion of the Patos Lagoon drainage basin between 1998 and 2002 (Mansur et al., 2003; Capítoli and Bemvenuti, 2004). In 2005, *Limnoperna* was captured at São Gonçalo channel, which connects Patos and Mirim lagoons and then at Taim reserve (Burns et al., 2006). The populations had steadily increased at these sites from 2004 to 2006 (Magara et al., 2001; Garcia et al., 2006).

The golden mussel is not yet present in North America.

2.2.1.3 Corbiculidae

The term "Corbiculacea" was coined to include the two families, Corbiculidae and Sphaeriidae (Mackie, 2007). The most comprehensive zoogeographic study of any corbiculacean species is that of *Corbicula fluminea*, an introduced species that has become one of the most important freshwater pest species in the United States. However, the zebra mussel, *Dreissena polymorpha*, has since outranked the Asian clam as the most nuisance mollusk, indeed organism in North American fresh waters.

Of the 39 species of Corbiculacea in Nearctic freshwaters, five appear to be introduced. These are *Corbicula fluminea*, *Sphaerium corneum*, *Pisidium amnicum*, *Pisidium punctiferum,* and perhaps *Pisidium henslowanum.* The systematics of all species, except *Corbicula fluminea*, has been relatively stable since Herrington's revision in 1962. However, the taxonomy of *Corbicula* in North America became confusing when Hills and Patton (1982) reported in south Texas a smaller and thinner variant or morph with a dark purple porcelaneous layer. However, allozyme studies show that the species present in North America is *Corbicula fluminea* (Kijviriya et al., 1991). Readers are referred to McMahon (1999) and references therein for recent information on dispersal (and on biology, ecology, and physiology) of the Asian clam in North America. Details of the systematics of *Corbicula* can be found in Section 6.1 of Mackie (2007).

2.2.1.3.1 Asian Clam, Corbicula fluminea *(Prashad, 1929)*

Native range: Asian clams are native to southeast China, Korea, and in the Ussuri Basin, southeastern Russia (Lachner et al., 1970; McMahon, 1982).

Current range: The introduction of *Corbicula fluminea* into North America has been attributed to immigrants from China, where it is very common to the west coast between the late 1840s (Fox, 1969, 1970a,b, 1971, 1972) and the mid-1930s (Britton and Morton, 1979; Mackie, 2007). Some (e.g., Fox, 1970a) believe that the introduction occurred during the Gold Rush in the late 1840s but, as Counts (1986) pointed out, it is not likely that *Corbicula* could survive the 180-day ocean voyage that the Chinese immigrants required to arrive in North America. Others (e.g., Abbot, 1975) suggested that *Corbicula* may have been imported from China with the Giant Pacific Oyster, *Crassostrea gigas*. Counts (1986) reported shells found on Vancouver Island, British Columbia, in 1924 and on this basis considered the introduction of *Corbicula fluminea* to occur in the early to mid-1920s. However, living specimens have never been recorded from this site since its original discovery. The most popular theory is that *Corbicula* arrived in North America with the Chinese escaping the political upheaval in China during the mid- to late-1930s. This period probably permitted more favorable shipboard conditions for the successful transport of *Corbicula fluminea* (Counts, 1986).

Cherry et al. (1980) have recorded the distribution of *Corbicula* in North America, but Britton and Morton (1982), McMahon (1982), and Counts (1986) have documented the pattern and pace of its spread. The point of original introduction is probably near the mouth of the Columbia River separating Washington and Oregon, since the first living specimens were found here in 1938 (Burch, 1944). Successive invasion of adjacent West Coast drainage systems brought *Corbicula* southward, and by the end of the 1950s it had invaded most of Southern California (Figure 2.4). Apparently, the Rocky Mountains formed an effective barrier for the eastward dispersal of *Corbicula* into the northwestern states at this time. Within 10 years,

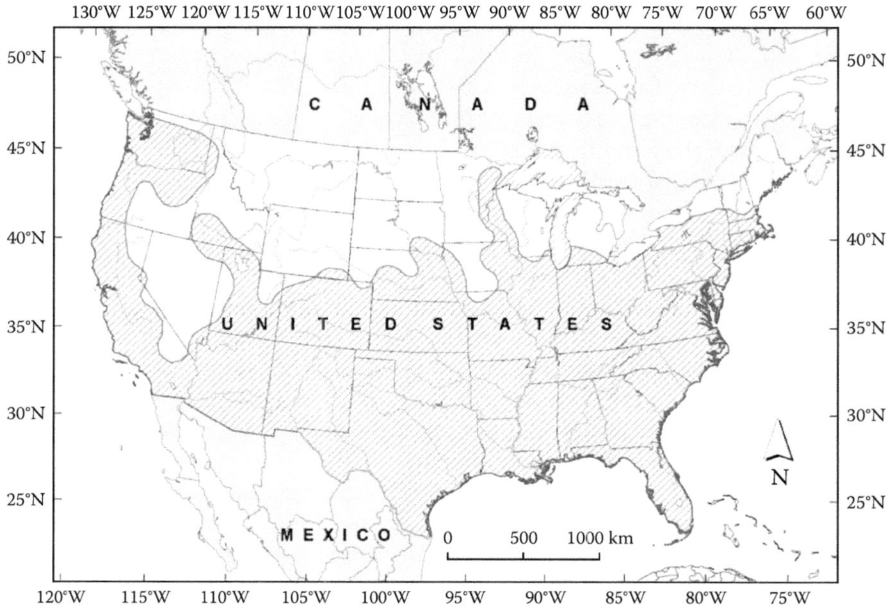

FIGURE 2.4 Distribution of *Corbicula fluminea* in North America as of May 2009.

Corbicula had made its way eastward to the gulf coast of Texas, and by 1972 it had already begun its northward invasion through several Texas river drainage basins (Counts, 1986).

Meanwhile, in 1957, a second, unexpected infestation of *Corbicula* was recorded in the Ohio River drainage system. This anomalous population is thought to represent an unnatural extension (i.e., cannot be explained by dispersal through drainage patterns) that probably originated by a human vector (Mackie, 2007). The artificially introduced population formed the epicenter of a second major expansion into the eastern half of the United States (McMahon, 1982). Using rivers of the Mississippi River drainage system, *Corbicula* spread eastward to the Appalachian Mountains and southward into Texas and Florida (Figure 2.4). The rapid southward extension of the range of *Corbicula* may have been accelerated by an infestation from another epicenter formed in the late 1960s or early 1970s, but this time in Oklahoma (McMahon, 1982). The species spread through Oklahoma, Texas, and New Mexico in the 1960s and 1970s (Metcalf, 1966; Metcalf and Smart, 1972). Cordeiro et al. (2007) describe its recent expansion through Colorado.

From the drainage systems of the eastern Gulf Coastal Plain, *Corbicula* quickly invaded the adjacent fresh waters of the Atlantic Coastal States, and by the early 1970s had reached Virginia and New Jersey (Counts, 1986). By 1982, *Corbicula* had been reported in 35 of the continental states (Mackie, 2007) (Figure 2.4). Using a network of numerous drainage systems, *Corbicula* eventually spread into most of the suitable habitats in the eastern United States south of 40° latitude, the waters north of this apparently being below the 2°C absolute lower thermal limit (Mattice and Dye, 1976) for three or four months of the year. However, recent surveys on the east coast suggest that *Corbicula fluminea* may be able to tolerate low temperatures under certain conditions. The species was discovered in the Connecticut Yankee Nuclear Power Station on the Connecticut River, New England, in 1990 (Morgan et al., 1992). This northern most range extension of the Asian clam in eastern United States was a surprise because the water temperature frequently drops below 2°C. The population dynamics was studied for 10 years by Morgan et al. (2003) and they found the species can survive low-temperature periods (winter) only because of the influence of thermal discharges from upstream facilities.

Indeed, because of its low-temperature tolerance, the Asian clam was never expected to reach Canadian waters and, in fact, until recently, reproducing populations had never been found in Canada. Clarke (1981) recorded specimens from the western end of Lake Erie (Figure 2.4) in warm-water plume areas, but none in cold-water areas. However, Mr. Dave Zanatta (personnel communication) found a small population in Mitchell's Bay of Lake St. Clair in 1999, but none were found living in 2000. It was generally agreed that *Corbicula* had probably reached the extent of its northern distribution because of its intolerance of low winter temperatures for extended periods of time (Horning and Keup, 1964; Mattice and Dye, 1976). Indeed, so far the occurrences of *Corbicula* in rivers and lakes in northern states like Michigan, Minnesota, and Wisconsin are in areas receiving thermal effluents from power stations (e.g., Thomas and MacKenthum, 1964; Eckblad, 1975; Rodgers et al., 1979; Dreier and Tranquilli, 1981; French and Schloesser, 1991, 1996;

Kreiser and Mitton, 1995). Such thermally protected populations may serve as stepping stones in further northern expansions (Graney et al., 1980).

However, Cordeiro and MacWilliams (1999) reported the species in western high altitudes in Colorado, all in areas that experience temperatures below 2°C and suggested that it can survive and reproduce in colder climates. The presence of thermal plumes in these areas was not reported, but Werner and Rothhaupt (2008) found that lethal effects of low water temperatures on *Corbicula fluminea* may become effective only after a time lag. French and Schloesser (1991, 1996) suggest that low water temperature in the St. Clair River may limit the success of *Corbicula* in the river and in the Great Lakes by reducing growth, delaying the onset of sexual maturity and reproduction, and by causing heavy overwinter mortality in the first year of life.

The species has also been introduced into Europe (Mouthon, 1981), South America (Ituarte, 1981), and Mexico (Hillis and Mayden, 1985). Range expansions are occurring in these countries as well. Range expansions for Europe are described for Britain by Elliott and zu Ermgassen (2008), France by Mouthon (2000, 2001), Germany by Hasloop (1992), and the Netherlands by Bij de Vaate (1991). In the western world, range expansions of the Asian clam into southern Mexico are described by Torres-Orozco and Revueltas-Valle (1996). Further range expansions in South America since its introduction in 1975 are described for the Brazilian Amazon Basin (Beasley et al., 2003) and the Guadiana River Basin (Pérez-Bote and Fernández, 2008).

Before leaving the discussion of the distribution of *Corbicula fluminea* as depicted in Figure 2.4, there is a caveat; the current range is assumed based on the fact that many or all populations may have declined but have not been extirpated. The "boom or bust" pattern of population behavior by introduced species in which its population grows quickly to a very high level (e.g., see Figure 1.29, Chapter 1) and then declines as its food and habitat resources are overexploited (Williamson, 1996). Such an event appears to have occurred for the Asian clam population in the Potomac River estuary (Phelps, 1994). In some cases, populations are so large that after the population peaks and reaches its maximum age, massive die-offs occur. The consequence of these massive die-offs is accumulations of total ammonia and unionized ammonia (NH_3-N) in the pore water and overlying water. The reductions in dissolved oxygen levels result in accumulations of NH_3-N that are acutely toxic not only to Asian clams, young more sensitive than old (Cherry et al., 2005; Cooper et al., 2005) but also native mussels (Scheller, 1997). Some populations may even be extirpated over time, such as occurred over 10 years in the Clear Fork of the Trinity River population of *Corbicula fluminea* in North Central Texas (McMahon, personnel communication).

2.2.2 Gastropoda

2.2.2.1 Ampullariidae

2.2.2.1.1 *Giant Ramshorn Snail,* Marisa cornuarietis *(Linnaeus, 1758)*

Native range: Northern South America (Honduras, Costa Rica, Panama, Venezuela, Columbia, Brazil) (Hunt, 1958).

Current range: The species was first discovered in Coral Gables, Miami-Dade County, Florida, in 1957 (Hunt, 1958). It has invaded many other counties in southeastern states, including southern Florida (Thompson, 1984; Rawlings et al., 2007) and Texas (Neck, 1984) (Figure 2.5). The most remote invasions appear to be California and Idaho (Frest and Johannes, 2000; Howells et al., 2006). Locations for Idaho are given by Frest and Johannes (2000) as "scattered over the state, but mostly southern" and by Bowler and Frest (1992) as "middle Snake River." Figure 2.5 shows the distribution of the species based on information above and in USGS (2009a), which lists 16 records, 2 in Texas, 4 in Florida, and 4 in Puerto Rico.

The snail has spread to other tropical regions due to the aquarium trade and its use for controlling aquatic vegetation. *Marisa cornuarietis* appeared in Cuba in the late 1940s (Penalver, 1950) and in Puerto Rico in 1952 (Oliver-Gonzales et al., 1956).

Rawlings et al. (2007) examined the phylogenetic relationships of several populations of five species of *Pomacea* from the United States and South America and found that the Arizona sequence of *Pomacea canaliculata* differed from the Argentinean one by a single nucleotide, suggesting that the source of the introduced populations of *Pomacea canaliculata* is likely Buenos Aires, Argentina. They also found a match of haplotypes to Hawaiian samples and suggested that the most possible pathway for its introduction to the western United States was via the food trade. Rawlings et al. (2007) also believe that some *Pomacea haustrum* specimens in Florida and some *Pomacea insularum* specimens from Texas are *Pomacea canaliculata*. The golden apple snail entered California in 1997 in Lake Miramar, San Diego County (Cerutti, 1998). Other

FIGURE 2.5 Distribution of *Marisa cornuarietis* in North America as of May 2009 as indicated by diamonds in locations in counties of Florida, Texas, Idaho, and California.

counties in California where the species was introduced are Riverside County and Alameda County (Howells and Smith, 2002). The Arizona population is in a pond near Yuma (USGS, 2008). There have been several reports of the golden apple snail in Texas but some have failed to survive. Howells and Smith (2002) list over a dozen records from at least five counties in Texas (Figure 2.6) and USGS (2009b) lists 17 records, 11 of them from Hawaii.

2.2.2.1.2 Golden Apple Snail, Pomacea canaliculata *(Lamarck, 1828)*

Native range: The golden apple snail is a neotropical species that is apparently native to Argentina, Bolivia, Paraguay, Uruguay, and Brazil (Cowie and Thiengo 2003), but the species is so variable in appearance that it may in fact be endemic to most of tropical and warm-temperate South America (Cazzaniga, 2002). The type locality of *Pomacea canaliculata* is purported to be "rivières de la Guadeloupe," but Cazzaniga (2002) suspects it may be "Lago Guadeloupe, Argentina."

Current range: Florida, Texas, Arizona, California, and Hawaii (Figure 2.6), as deduced from below and from Howells and Smith (2002) and USGS (2009b). North

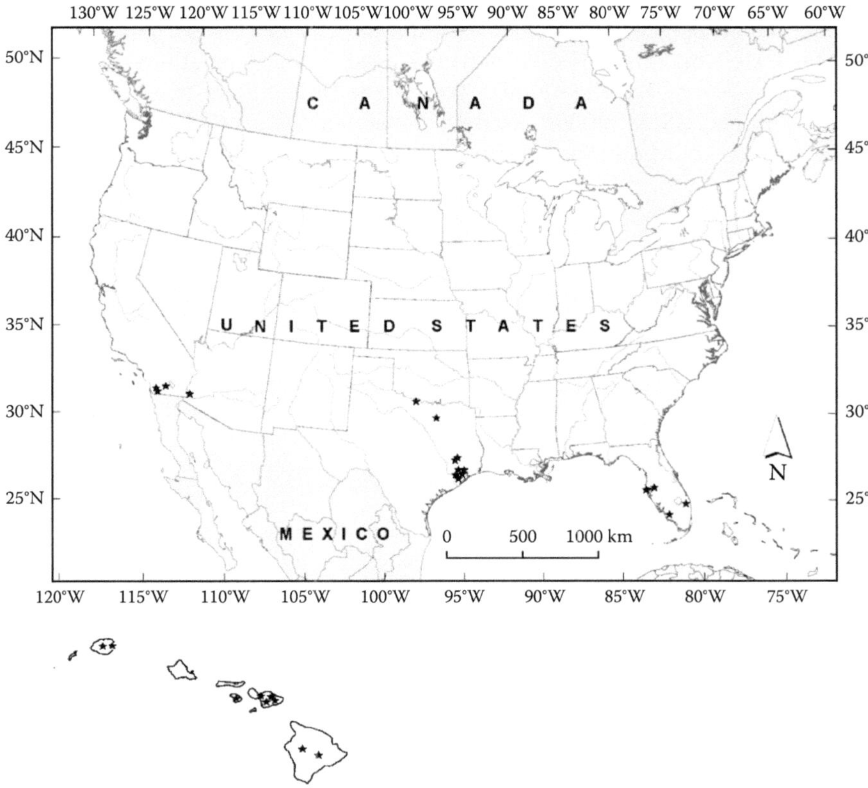

FIGURE 2.6 Distribution of *Pomacea canaliculata* in North America as of May 2009. Distribution shown as stars in locations in counties of Florida, Texas, Arizona, California, and Hawaii islands (bottom left).

Carolina has also been reported as a state with the species but the record was based on specimens collected in 1992 from a reservoir in Stoneville, Rockingham County; the population apparently has not become established (Howells and Smith, 2002).

Pomacea canaliculata invaded Florida in 1978, first in Palm Beach County, but by the late 1990s other populations were found in Collier, Hillsborough, and Pinellas counties (Figure 2.6), no populations of which apparently threaten major agricultural areas (Howells and Smith, 2002).

Outside of North America, the golden apple snail has also extended its range to East Asia (Philippines, Vietnam, Malaysia, Indonesia, China, Taiwan, Korea, Cambodia, Japan, and Papua New Guinea) (Cowie et al., 2006).

2.2.2.1.3 Titan Applesnail, Pomacea haustrum

Native range: Brazil, Peru, and Bolivia (Pain, 1960). Confusion in its taxonomy and current status are discussed in Chapter 1. Rawlings et al. (2007) also found problems with its type locality (Río Marañón, Peru, not Brazil).

Current range: Rawlings et al. (2007) extracted DNA from a specimen identified as *Pomacea haustrum* (in the Florida Museum of Natural History) from the Rio Ucayali in Peru, above its confluence with the Río Marañón and found that its sequence grouped closely with DNA from *Pomacea haustrum* specimens from Florida. Based on this evidence the only established populations of *Pomacea haustrum* known by Rawlings et al. (2007) in the United States are in Palm Beach County, Florida; USGS (2009c) lists five collections, four of them in Palm Beach County and one in Miami-Dade County (Figure 2.7).

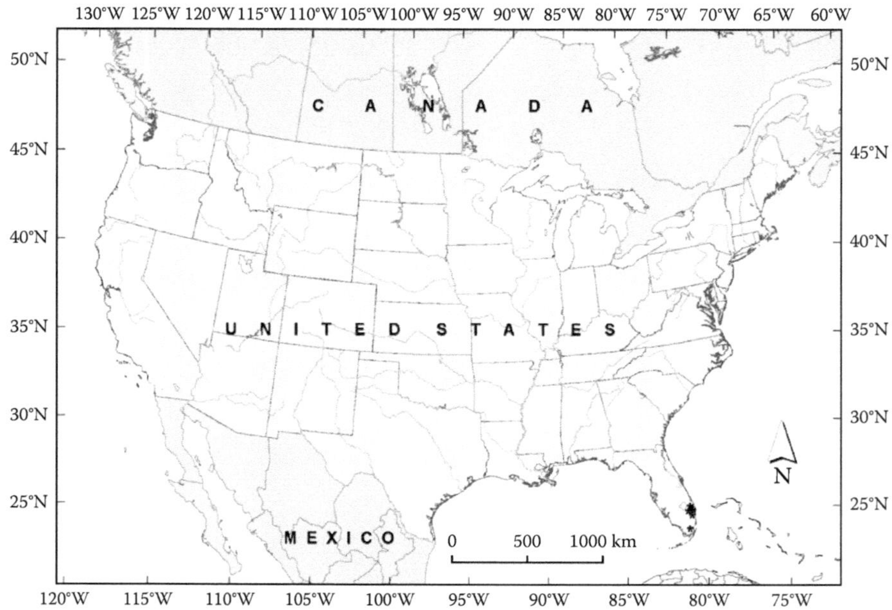

FIGURE 2.7 Distribution of *Pomacea haustrum* in North America as of May 2009. Only locations are in counties in Florida (see asterisks).

2.2.2.1.4 *Island Applesnail,* Pomacea insularum *(D'Orbigny, 1839)*

Native range: Argentina, Brazil (Cowie and Thiengo, 2003), Bolivia, Uruguay, and Paraguay (Rawlings et al., 2007). The type locality is the Río Paraná (Rawlings et al., 2007).

Current range: Based on Rawlings et al.'s (2007) analyses, *Pomacea insularum* is restricted to Texas, Florida, and Georgia; USGS (2009d) lists the species as collected from Alabama, Mississippi, and South Carolina, and collected and established in Florida, Georgia, Louisiana, and Texas (Figure 2.8). The majority of the localities in Florida are in the center of the state between Tampa and Orlando, other populations near Jacksonville and Tallahassee and some populations in ecologically sensitive areas including Everglades National Park and Loxahatchee National Wildlife Refuge (Rawlings et al., 2007).

The year of invasion of the species into North America is ambiguous. The first established population of the island applesnail was apparently in 1989 in Harris County, Texas (Neck and Schultz, 1992). The first record of *Pomacea insularum* in Florida was in 1991 from Hillsborough County (Winner but because of the taxonomic confusion, genetic analyses of material collected from Florida to date may reveal that its period of introduction into Florida may have been very recent) (Rawlings et al., 2007). In Georgia, *Pomacea insularum* was first collected from the Alabaha River, Pierce County, in February 2005, but other populations, some established, have since been discovered in ponds and streams in Glynn, Camden, Pierce, Bacon, and Brooks

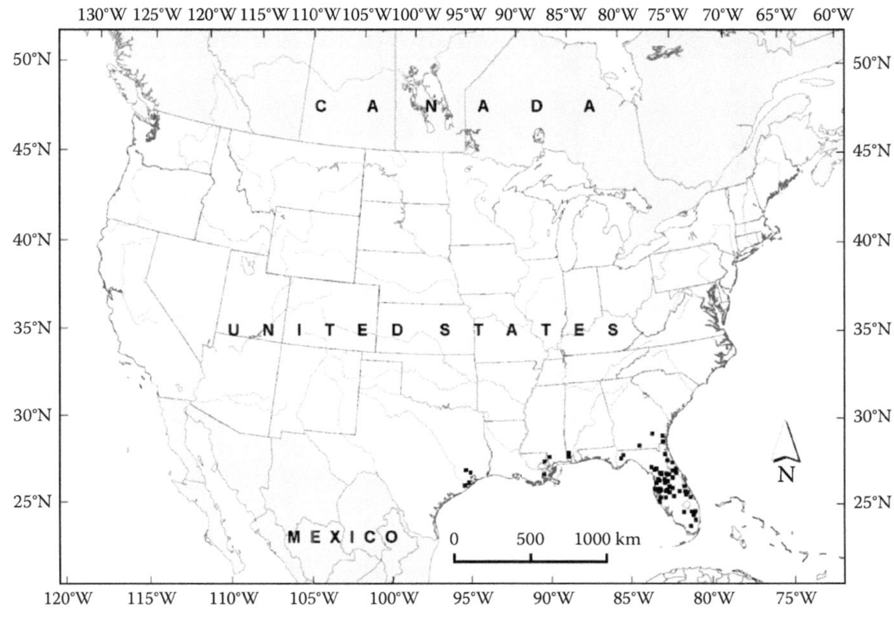

FIGURE 2.8 Distribution of *Pomacea insularum* in North America as of May 2009. Solid squares represent localities in counties of Georgia, Florida, Alabama, Mississippi, Louisiana, and Texas.

counties (Rawlings et al., 2007; USGS, 2009d). On the basis of genetic analyses, the Georgia, Texas, and the majority of the Florida samples evidently originated from Río Uruguay, Argentina (Rawlings et al., 2007).

2.2.2.1.5 *Spiketop Applesnail,* Pomacea diffusa *(Blume, 1957)*

Native range: The taxonomic status of this species was discussed in Chapter 1 and was formerly called *Pomacea bridgesii* (Rawlings et al., 2007). The spiketop applesnail is native to Brazil, Peru, and Bolivia (Cowie et al., 2006).

Current range: Figure 2.9 shows the distribution of *Pomacea diffusa* in North America. The distribution is based on records of collection and/or established populations in Florida as cited by Rawlings et al. (2007) and an established population in Mobile County, Alabama, cited by USGS (2009e). The first record in Florida (as *Pomacea bridgesii*) was in 1966 by Clench (1966) and the first record for Alabama was in 2003 by Howells et al. (2006).

2.2.2.1.6 *Florida Applesnail,* Pomacea paludosa *(Say, 1829)*

Native range: Central and southern Florida where the species has existed continuously since the Pliocene (Thompson, 1994).

Current range: Widely distributed throughout the Florida peninsula, occurring sporadically west of the Suwannee River to the Choctawhatchee River, in Georgia in isolated springs along the Flint and Ocmulgee rivers and in Covington County, Alabama, in a small reservoir artificially heated by industrial wastewater

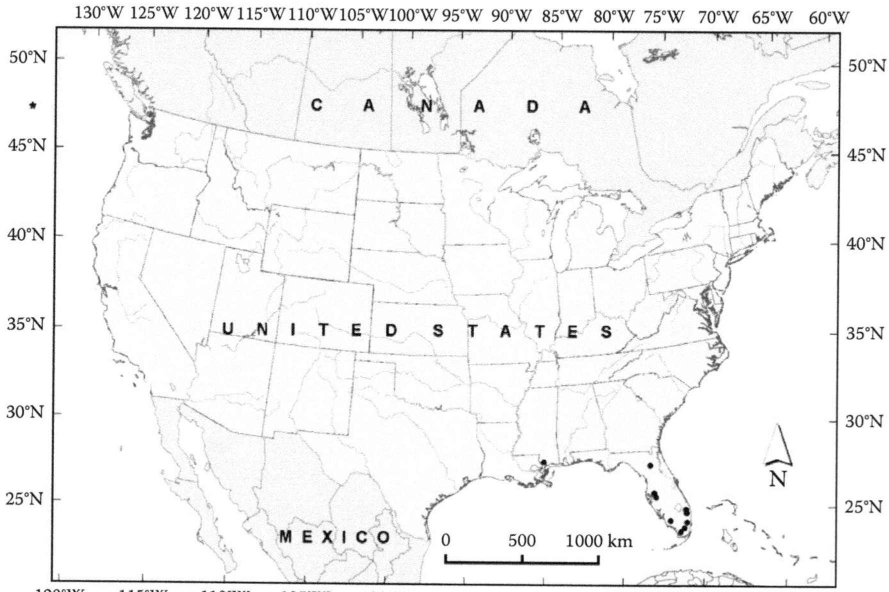

FIGURE 2.9 Distribution of *Pomacea diffusa* in North America as of May 2009. Solid circles represent localities in counties of Florida and Alabama.

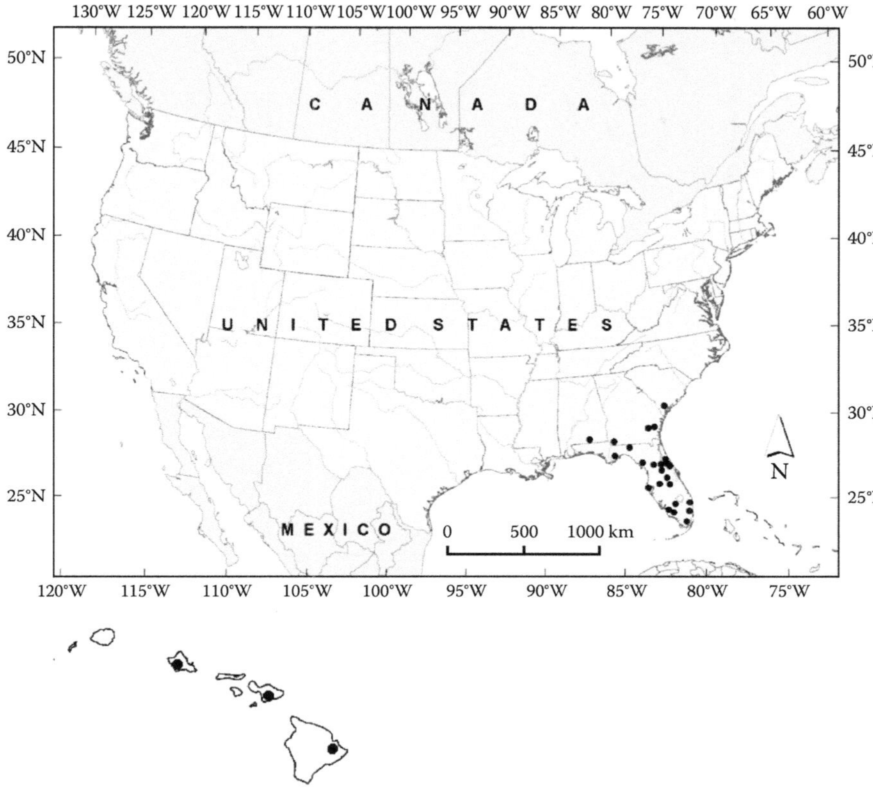

FIGURE 2.10 Distribution of *Pomacea paludosa* in North America as of May 2009. Distribution shown as solid circles in locations in counties of Florida, Georgia, South Carolina, Alabama, and Hawaii islands (bottom left).

(Thompson, 1994). According to 22 records from GSMFC (2009) and 31 records from USGS (2009f), the Florida applesnail also occurs in Georgia, South Carolina, and Oklahoma, and three records for Hawaii islands (Figure 2.10).

The Florida Applesnail now occurs in Cuba and has been introduced extensively into Pacific islands and southeast Asia, where it is raised for human consumption (Thompson, 1994).

2.2.2.2 Thiaridae

2.2.2.2.1 Quilted Melania, Tarebia granifera *(Lamarck, 1822)*

Native range: India and Ceylon east to the Philippines and Hawaii, north to southern Japan and south to the Society Islands (Abbot, 1952).

Current range: The species is established in Florida in Lithia Springs, Hillsborough County, Florida (Lachner et al., 1970; Dundee, 1974) and in Miami and Coral Gables in Dade County (Chaniotis et al., 1980a) (Figure 2.11). It has also been collected in Twin Falls County, Idaho (USGS, 2009g). For Texas, it has been reported from the

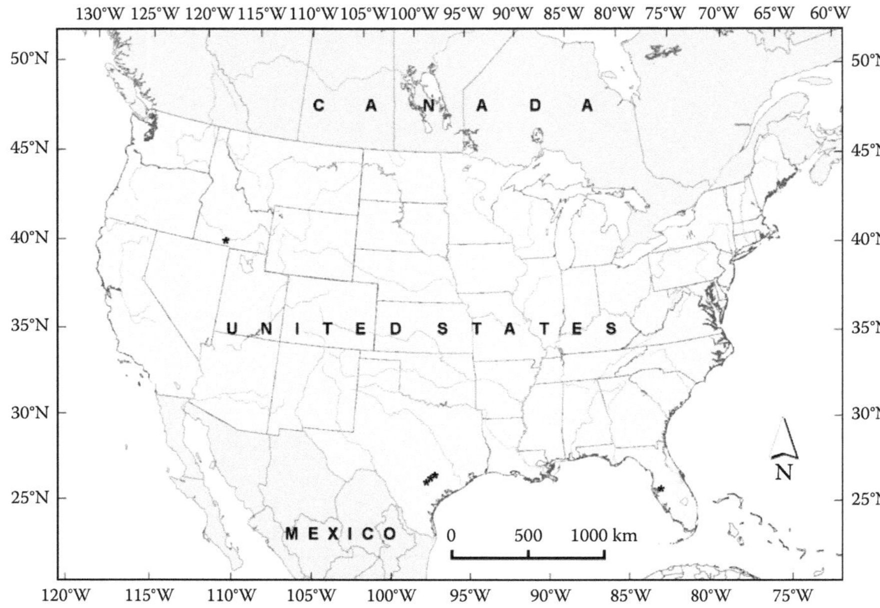

FIGURE 2.11 Distribution of *Tarebia granifera* in North America as of May 2009. Solid stars represent localities in counties of Florida, Texas, and Idaho.

San Antonio River in Bexar County, Landa Park in New Braunfels, Comal County (Murray, 1964; Dundee, 1974; USGS, 2009g), and the San Marcos River, Hays County (Howells, 1992; USGS, 2009g). USGS (2009g) also has records for established populations in Puerto Rico and collections from St. John, U.S. Virgin Islands.

The quilted melania has also been introduced to Cuba (Jacobson, 1978) and possibly South America (Chaniotis et al., 1980a,c).

2.2.2.2.2 Red-Rimmed Melania, Melanoides tuberculata *(Müller, 1774)*

Native range: Native to subtropical and tropical areas of northern and eastern Africa and southern Asia from Morocco and Madagascar to Saudi Arabia, Iran, Pakistan, India, southern China, and Indonesia east to Java and Celebes (Clench, 1969; Neck, 1985).

Current range: This species is widely distributed throughout the United States (Figure 2.12). As with the quilted melania, the species has been recorded from the Gulf of Mexico ecosystem in Lithia Springs and Hillsborough State Park, Hillsborough County, Florida (Abbot, 1952; Clench, 1969; Dundee, 1974; USGS, 2009h); numerous localities in Miami-Dade County and from Collier County (Clench, 1969; Russo, 1973; Roessler et al., 1977; Taxson and Taxson, 1980; USGS, 2009h). It is in Louisiana in several canals of New Orleans (Dundee and Paine, 1977) and in Orleans/Jefferson counties (USGS, 2009h); in Texas from San Antonio, New Braunfels (Clench, 1969; Kotrla, 1975) and San Antonio River, both in Bexar County, San Marcos River, Hays County (Howells, 1992), Las Moras Creek, Kinney County, (Murray, 1975; Howells, 1992), Brownsville, Cameron County (Neck, 1985), as well as in Hays, Comal, Hildago, Pecos, Reeves, Val Verde, Travis and Goliad counties (USGS, 2009h). USGS (2009h)

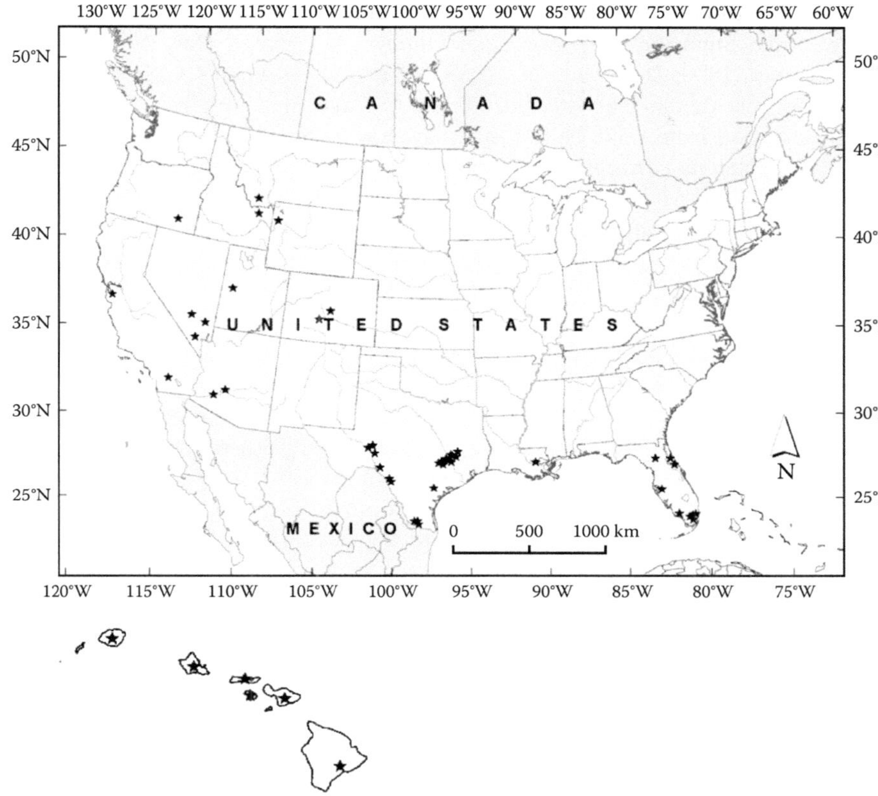

FIGURE 2.12 Distribution of *Melanoides tuberculata* in North America as of May 2009. Solid stars represent localities in counties of Florida, Louisiana, Texas, Colorado, Arizona, Utah, California, Oregon, Montana, Idaho, Wyoming, and the Hawaii islands (below left).

lists several other states with collected or established records: Arizona (Maricopa County), California (Fremont, Saguache counties), Montana (Madison County), Nevada (Clarke, Lincoln, and Nye counties), Oregon (Harney County), Utah (Juab County), Wyoming (Teton County), North Carolina (Brunswick County), all Hawaii islands. It is also recorded in Lajas County, Puerto Rico, and on St. John, U.S. Virgin Islands (USGS, 2009h). It is probably more widely spread than the collections indicate.

The species has invaded other countries as well, including Veracruz, Mexico, Panama (Abbot, 1973), Brazil (Dos Santos et al., 2007), and New Zealand (Duggan, 2002; Duggan et al., 2007).

2.2.2.3 Bithyniidae

2.2.2.3.1 Faucet Snail, Bithynia tuberculata *(Linnaeus, 1758)*

Native range: Europe to west Siberia and Kashmir (Kerney, 1999).

Current range: The earliest record of the faucet snail in North America is in Lake Michigan in 1871 (Robertson and Blakeslee, 1948), although Berry (1943) believes it invaded the lake in 1870. It spread quickly to the other Great Lakes, except Lake Superior, reaching Lake Ontario at Oswego New York by 1879 (Beauchamp, 1880; Duggan et al., 2003). Carr and Hiltunen (1965) and Krieger (1985) suggest the species was introduced to Lake Erie sometime before 1930. It was recorded from Lake Huron, Saginaw Bay in benthic samples in the 1980s and 1990s (Nalepa et al., 2002). It was not recorded in Lake Superior in a 2002 survey by Grigorovich et al. (2003) when five other invasive species, including *Potamopyrgus antipodarum*, were found. It is now widespread throughout most of the states bordering the Great Lakes and from southern Ontario to southern Quebec (Mackie et al., 1980; Clarke, 1981; Mackie, 2000b; Carter et al., 2006) (Figure 2.13).

The faucet snail was discovered outside the Great Lakes watershed for the first time in 1918 in the Hudson River (Baker, 1928) and then in the Potomac River at Alexandria, Virginia, in 1927 (Pilsbry, 1932; Marshall, 1933). It spread through New York during the 1980s (Jokinen, 1992; Mills et al., 1993a) and probably through Michigan, Ohio, Pennsylvania, and Wisconsin in the 1980s and 1990s. The species was recorded from Chesapeake Bay, Maryland, in 2000 (Ruiz et al., 2000; USGS, 2009i). The records from Montana in Figure 2.13 date from 1991 to 2006 (USGS, 2009i).

Bithynia tuberculata is present in the northern parts of Ireland and Britain (Kerney, 1999; Anderson, 2005), southwestern Finland (Carlsson, 2000), central Europe North

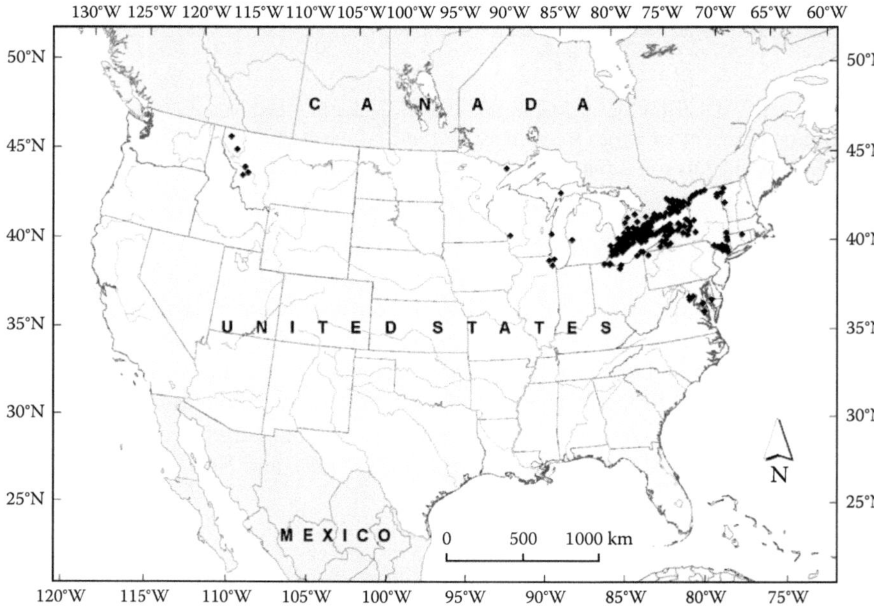

FIGURE 2.13 Distribution of *Bithynia tuberculata* in North America as of May 2009. Solid diamonds represent localities in counties of Ontario and Quebec, Canada, and in the United States.

of the Alps, and the northwestern parts of the Balkans (Gloer et al., 2007). It is one of the more common species and has existed in the Curonian Lagoon of the Baltic Sea for more than 50 years (Zettler and Daunys, 2007). It has recently been recorded in the Czech Republic along with another alien snail, *Ferrissia fragilis* (Beran and Horsák, 2007). The species has been recorded from early Holocene (8800–11,500 cal yr BP) deposits in Sweden (Hammarlund et al., 2003) and Denmark (Bennike et al., 2008); Bennike (personnel communication, January 26, 2009) says *Bithynia tentaculata* is common in Denmark.

2.2.2.4 Hydrobiidae

2.2.2.4.1 New Zealand Mudsnail, Potamopyrgus antipodarum *(Gray, 1843)*

Native range: New Zealand (Winterbourn, 1970; Zaranko et al., 1997).

Current range: The New Zealand mudsnail was first reported in North America in the mid-1980s, from the Snake River, Idaho (Bowler, 1991; Langenstein and Bowler, 1991; Bowler and Frest, 1992). The species appeared in Gallatin County, Montana, shortly after in 1995 and in Teton County, Wyoming, in 1997 (Kerans et al., 2005; USGS, 2009h).

The first report of the species in the Great Lakes was in 1991 in Lake Ontario by Zaranko et al. (1997). Since then the species has been reported from Lake Superior in 2001 (Grigorovich et al., 2003; USGS, 2009h), Lake Erie in 2005 (Levri et al., 2007), and Lake Michigan in 2006 (USGS, 2009j). In the mean time, the mudsnail was spreading beyond the 100th Meridian and appeared in California and Utah in 2001, and Washington, Arizona, and Oregon in 2002 (USGS, 2009h). Sarina et al. (2007) predicted the invasion range of the New Zealand mudsnail in Australia and North America based on modeling its native and invaded ranges; their predictions for westward invasion are substantiated by the current range shown in Figure 2.14 but it is apparent that the species are slower to invade eastern states and provinces of Canada. The most recent record is from Port Alberni, British Columbia (Davidson et al., 2008) (Figure 2.14).

The New Zealand mudsnail is native to New Zealand and, not surprisingly, is now naturalized and an invasive species in neighboring Australia and Tasmania. It is spreading quickly in Europe and Asia. The New Zealand mudsnail came to Europe in ballast water and was first observed in the Thames estuary in 1859 (Lassen, 1978). It was reported from the Baltic Sea area for the first time in 1887 in Wismar Bight and in 1898 in the Kiel Canal (Lassen, 1978) and is now throughout the Baltic Sea area (Gruszka, 1999; Leppäkoski and Olenin, 2000; Ezhova et al., 2005). According to Carlsson (2000) the species invaded Finland in 1926 in the Åland archipelago and then Sweden shortly after. It is now in Norway (Okland, 1990; Okland and Okland, 1992), Poland (Strzelec and Krodkiewska, 1994), Slovak Republic (Čejka, 1994; Elexova, 1998; Čejka et al., 2008), France (Costil et al., 2001), and has appeared more recently in Ukraine (Alexandrov et al., 2007), Greece (Radea et al., 2008), Azov-Black Sea region (Son, 2008), and Russia (Dmitry et al., 2008; Son et al., 2008). In Asia the species has invaded Japan (Davidson et al., 2008).

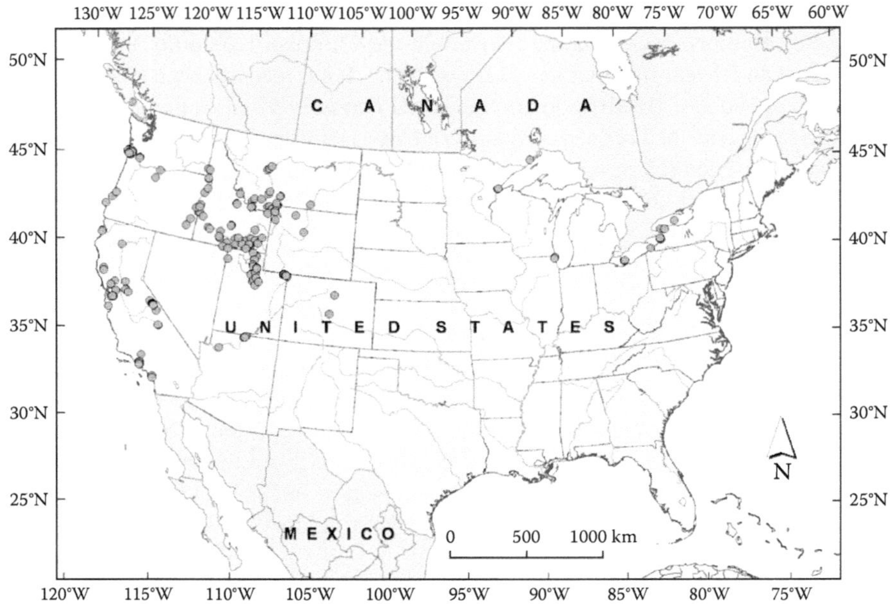

FIGURE 2.14 Distribution of *Potamopyrgus antipodarum* in North America as of May 2009. Solid circles represent localities in counties of Ontario, Canada, and in the United States. (Courtesy of Amy Benson, with most recent finding in Port Alberni, BC [Davidson et al., 2008] added.)

2.3 DISPERSAL POTENTIAL AND MECHANISMS

2.3.1 Dispersal Potential

2.3.1.1 Anthropogenic Vectors

There are two main vectors to consider in assessing dispersal potential: one is the human or anthropogenic vector, and the other is nonanthropogenic or natural vector. The dispersal potential based on natural vectors is a bit easier to predict (see next section) than that based on anthropogenic vectors because the human mind can often be devious, unintentional, and unpredictable.

Intentional introductions: Some introductions are intentional and, it can be argued, required, such as releasing the beetles, *Galerucella calmariensis* and *Galerucella pusilla*, in 1992 to control the spread of purple loosestrife, *Lythrum salicaria*; however, this was done only after rigorous screening to ensure the beetles were host specific on purple loosestrife (Kok et al., 1992; Malecki et al., 1993; Manguin et al., 1993; Blossey et al., 1994). There are molluskan examples too. Kew (1893) attributed the introduction of species of *Campeloma*, *Goniobasis*, *Somatogyrus*, *Bithynia*, and *Viviparus* into the Erie Canal, Mohawk River, and Schuyler's Lake by naturalists who saw a perceived need to improve diversity in the systems. Some of the largest snails, such as *Bellamya* (*Cipangopaludina*) *chinensis*, were imported for their food value (Mackie, 2000c). Ahlstedt (1979) transplanted mussels from the Clinch River into the rejuvenated North Fork of Holston River, southwest Virginia, to help recover 8 of the 42 species of mussel fauna that had been extirpated by chemical pollution. Transplantation of mussels was considered an option to save mussel communities from infestations of the zebra mussel, but there are several concerns related to genetic alterations, survival rates, and other issues that must be addressed before endorsing it as a national strategy (National Native Mussel Conservation Committee, 1998). Hallac and Marsden (2001) recommend cleaning of mussels from unionids and transplant may be used as the first step to conserve small populations of fouled unionids living in environments where food is not limiting and where collection and cleaning are logistically feasible.

Unintentional introductions: Some anthropogenic introductions of mollusks are unintentional. Schmeck (1942) attributes the introduction of *Viviparus malleatus* to dumping of aquarium water containing the species into the Niagara River, not realizing that the species could establish a breeding population. Aquarium releases are attributed to the introduction of *T. granifera* and *Melanoides tuberculata* (Abbott, 1952; Hunt, 1958; Mackie, 2000a) and applesnails (Howells and Smith, 2002).

Malicious introductions: Finally, some anthropogenic introductions are intentional or unintentional but not for the reasons given above. Often referred to as "biological terrorism" there are foreign animal disease threats, such as avian influenza, foot and mouth disease, hog cholera, and West Nile virus, that have been attributed to accidental or malicious introduction of the disease (Renner, 2005). The zebra mussel is a good example of malicious introductions. Zebra mussels were either accidentally introduced into some water bodies in the former Soviet Union with fish transplants or were deliberately planted in new waters to provide food for fish or to improve water quality (Karatayev et al., 2007b). In the Netherlands, zebra mussels were introduced into two small ponds as an experiment to control eutrophication (Reeders et al., 1993). The major pathway seems to be the transport of boats, personal watercraft, and related equipment from zebra mussel infested to uninfested waters without washing the craft and equipment before immersing them in the uninfested water. It is well known that zebra mussels attach to hulls, trailers, and other exposed locations on boats, boating equipment, and personal watercraft, and the adult zebra and quagga mussels can survive for three to five days (Ricciardi et al., 1995) or up to 10 days (Hincks, 1994) of aerial exposure on hulls of boats; the larvae can also survive in motors, live wells, or other moist areas for eight days (Johnson and Padilla, 1996). It is the propensity of boaters to move their boats from one body of water to another that has facilitated the rapid spread of zebra mussels to throughout their current range (Johnson and Carlton, 1996; Johnson and Padilla, 1996; Schneider et al., 1998a). The 100th meridian initiative (Mangin, 1998) was developed to prevent the spread of zebra mussels and other aquatic nuisance species in the 100th meridian jurisdictions and west and to monitor and control aquatic nuisance species if detected in these areas. It should be noted that geographic proximity to a source population and overall boater use are poor predictors of probability of invasion (Padilla et al., 1996). Padilla et al. (1996) do advise that attention to human activity patterns helps to predict the spread of invading taxa.

Havel et al. (2005) provide arguments for reservoirs acting as stepping stones for the dispersal of exotic species across landscapes. A variety of passively dispersing species have invaded reservoirs, spread through interconnected waterways, and transported to nearby disconnected habitats. Havel et al. (2005) hypothesize that reservoirs are more readily invaded than natural lakes, because of their physiochemical properties, greater connectivity, and higher levels of disturbance. They argue that reservoirs typically have a high level of physical disturbance, solute loading, and unstable food webs that would enhance the invasibility of reservoirs over that of natural lakes. Reservoirs may be located in such a way as to bridge gaps between standing-water systems, thus facilitating movement across landscapes, by homogenizing habitats and serving as stepping stones for invasive species (Havel et al., 2005).

2.3.1.2 Natural Vectors

Mackie (2002) discusses how a species' reproductive potential, its life span and size, its tolerances and requirements, and, most of all, its dispersal mechanisms determine a species' potential to become widespread, abundant, or to decline in a changing environment. He examined the distribution of these traits among invasive species of mollusks and compared them to molluskan species at risk in North America.

Reproductive potential: Several aspects of reproductive potential were considered by Mackie (2002): (1) its sexual state (e.g., separate sex, hermaphrodite, or parthenogenetic); hermaphroditism and parthenogenesis reduce the risk of a species being eliminated during periods when it is difficult to find a mate and therefore increase its invasive potential. (2) Its egg-laying habit (e.g., oviparous, ovoviviparous, viviparous); oviparity generally results in larger numbers of eggs than ovoviviparity or viviparity. (3) Its fecundity (number of eggs produced); the numbers of eggs produced by ovoviviparous and viviparous forms are limited by the internal volume of the female. (4) Its natality (number of eggs surviving); eggs can be fertilized either internally in the female and then laid or externally in the water. In freshwater mollusks, external fertilization occurs only in bivalves (Dreissenidae and Mytilidae) and millions of eggs are produced, or at least orders of magnitude larger numbers than by oviparous forms with internal fertilization or ovoviviparous and viviparous forms, regardless of whether they are monoecious, dioecious, or parthenogenetic. (5) Its annual frequency of egg-laying habits (e.g., univoltine, bivoltine, multivoltine); clearly the more times one reproduces every year, the greater the potential for increasing population size. (6) Its lifetime frequency of egg-laying habits (e.g., semelparity, iteroparity); again, the more times one reproduces in its life time, the greater the potential for increasing population size.

Life span and size: A species' life span and size affects dispersal potential because one would think that the longer-lived species have a greater opportunity to disperse, but this is limited by its size, because smaller individuals are easier to transport than larger ones. As it turns out, however, smaller animals have shorter life spans than larger ones, at least in the Mollusca (Mackie, 2002). Hence, small, short-lived mollusks have a greater dispersal potential than large, long-lived mollusks. On the other hand, large, long-lived mollusks are preferred foods and aquarium pets in the anthropogenic vector.

Physiological and ecological tolerances and requirements: A species physiological and ecological tolerances and requirements dictate how successful the potential invasion will be. Each species has its own physiological requirements for variables such as calcium, pH, alkalinity, dissolved oxygen, temperature, turbidity, and so on. Similarly, each species has its own ecological requirements for food and habitat, and yet be relatively free of predation and parasitism. But the physiological and ecological variables vary within and from one environment to the next, and the ability of the species to tolerate the extremes in these variations will determine its population dynamics and its nuisance potential. For example, a species will more likely be an invasive nuisance if it can avoid desiccation or survive prolonged periods of exposure; exploit an unused or poorly used habitat, tolerate high turbidities, have a high upper lethal thermal limit, and/or be able to deal with short periods of anoxia or low oxygen tensions. Chapter 3 is devoted to the use of physiological and ecological tolerances and requirements for risk assessments.

2.3.2 Dispersal Mechanisms

There are two basic types of dispersal mechanisms, passive and active dispersal. The type used has important consequences for processes of colonization, gene flow, and evolutionary divergence (Bilton et al., 2001). Passive dispersal is hitchhiking a ride using abiotic (e.g., water currents, wind, ships, boats, etc.) or biotic (e.g., birds, insects, mammals) vectors. However, most of the active dispersal mechanisms are natural and include the swimming (e.g., fish) or flying (e.g., adult insects with aquatic larval stages) abilities of the species, and most of the passive dispersal mechanisms are anthropogenic. As discussed above, dispersal by anthropogenic means can be intentional or nonintentional and are typical for mollusks valued as food and those in the aquarium trade. The dispersal ability of a species introduced intentionally by humans can often be ascertained by the rapid spread of the species throughout the continent, as in the case of the Asian clam in North America and other continents and zebra and quagga mussels west of the 100th meridian.

There are numerous mechanisms available to organisms for dispersal over short or long distances (Table 2.1); each mechanism has a potential for dispersal within a region (e.g., by moving from lake to lake to eventually disperse throughout a province or state), a continent (e.g., intracontinental, such as within North America or within Eurasia), or intercontinentally (e.g., from Eurasia to North America). The list is based mainly on dispersal mechanisms used by zebra mussels (Carlton, 1993), but it is not necessarily restricted to them or other dreissenids.

Of the natural mechanisms, external transport (e.g., feet and feathers) is generally a more effective dispersal mechanism than internal transport via the digestive tract. Sphaeriids can survive passage through the intestinal tract of waterfowl as extramarsupial larvae but internal transport is probably not prevalent (Mackie, 1979). Large insects are able to disperse only small organisms, like *Potamopyrgus antipodarum* or young *Bithynia tentaculata*. Lake currents disperse primarily plankton or planktonic stages, but translocation of adult dreissenids is well known (see Chapter 1 and next section). River currents disperse organisms mainly downstream of their introduction. Water spouts are probably of little dispersal value unless they are large and

TABLE 2.1
Dispersal Mechanisms Available to Dreissenids and Mytilids for Short-Range (e.g., from Lakes within a State), Mid-Range (e.g., from Lakes within all States in North America), and Long-Range (e.g., from Ships Arriving to Canada or USA from Europe, South America, etc.) Transport

Dispersal Mechanisms	Potential
Natural Mechanisms	
By insects, waterfowl, or mammals	Possibly waterfowl but unlikely. See text
By fish or semiaquatic vertebrates	Not likely
Currents	Not likely, but see text
Water spouts (planktonic stages only)	Not likely
Wind	Not likely
Unintentional Anthropogenic Mechanisms	
Ocean vessels	No
Interiors (e.g., fish wells) or exteriors (e.g., hulls) of boats and crafts of rivers and lakes	High, see text
Canals (irrigation and vessels)	Not likely
Navigation and marker buoys and floats	Not likely
Marina and boatyard equipment	Not likely
Fisheries equipment (e.g., cages, nets, bait, buckets), stocking, fishing activities	High, but angling and stocking protocols. Can be monitored and controlled. See text
Amphibious and fire-fighting planes	Possible, see text
Fire-truck water	Not likely if water is from lakes in the same watershed
Commercial products (e.g., aquatic plants)	Not likely
Aquarium releases	Not likely
Recreational equipment (e.g., floating docks)	Not likely
Litter (e.g., tires)	Not likely
Scientific research	Not likely
Intentional Anthropogenic Mechanisms	
Food	Not likely
Sport	Possible

spill over into nearby and adjacent water bodies, and then mostly planktonic larvae (e.g., veligers) can be dispersed this way.

The anthropogenic mechanisms are split between intentional (or deliberate) and nonintentional (or accidental) releases because the former usually involves dispersal of large organisms for their food or sport value. Most alien fish species have both food and sport value, but mollusks such as the Chinese mysterysnail, *Cipangopaludina chinensis malleata* (Reeve), and the Japanese mysterysnail, *Cipangopaludina japonica* (von Martens), have no sport value and were introduced by Orientals purely for their food value as escargots. Mills et al. (1993a) have attributed deliberate releases to 11 of 139 alien species in the Great Lakes, most of these being fish. Of the 144 species introduced into the Great Lakes (at least five more have been introduced since 1993), 81 have originated in Eurasia and were introduced by ballast water exchange, the main intercontinental release mechanism. About 32 species originated from somewhere in North America (e.g., southern United States, Mississippi drainage, and Atlantic and Pacific sources). Mills et al. (1993a) attribute aquarium releases to some species in the Great Lakes, claiming that many people released their pets without any intention of establishing self-sustaining populations.

Many of the most widely distributed species have selected a variety of dispersal mechanisms as part of the natural selection process. Others have evolved efficient dispersal stages. For example, the veliger larva can be considered an efficient dispersal stage of zebra mussels. Because humans seem to be continually altering aquatic habitats globally, the species most destined to extinction are those with poor dispersal mechanisms and/or very narrow ranges of physiological and ecological tolerances and requirements.

Karatayev et al. (2007a) presented arguments that the spread of exotic species has not been a continuous process, but rather punctuated by periods of rapid long-distance spread, or jumps, during which species greatly expanded their geographic ranges. They provide evidence that "each jump has been associated with changes in the tempo of some human activity, such as the construction of shipping canals for trade, building of reservoirs for water storage and power production, political boundary changes or changes in political systems, which affected the position or permeability of national borders, human migration, changes in the mode and volume of international trade, or recent industrial practices and environmental laws." Karatayev et al. (2007a) hypothesized that the rate of spread of exotic species depends on the spatial scale of spread and is related to rates of human activities. In general, aquatic exotic species may quickly spread along connected waterways in a new continent they invade and soon reach their maximum range, which is at the continental scale (Karatayev et al., 2007a). However, it will take much longer to colonize all isolated regions, which is at a regional scale, and longer still to spread to all isolated lakes and river systems at a more local scale; the difference in the rate of colonization across scales may be several orders of magnitude (Karatayev et al., 2007c).

The size of a species is important because it partly determines the potential to spread great distances. Generally, small species disperse greater distances than large species (within a taxon). For example, fingernail clams (family Sphaeriidae) have a greater global distribution than freshwater pearly mussels (family Unionidae). There are apparently two reasons for this trend. First, big is more noticeable than small and the dispersal agent is more likely to unload a large hitchhiker sooner than a small one. Second, big also means heavy and a vector would have to spend more energy, and therefore risk its own life, to transport itself and its "baggage" (i.e., large mollusks) great distances. Perhaps the only exception to the size rule is intentional introductions by humans, as discussed above. In this instance, large is an advantage if the introduction is for food (or sport). The mysterysnails, *Cipangopaludina chinensis malleata* and *Cipangopaludina japonica*, are large mollusks and, as stated earlier, were intentionally introduced for escargots.

2.3.2.1 Potential of Dispersal Mechanisms in Bivalvia

2.3.2.1.1 Dreissenidae and Mytilidae

The juvenile and adult dreissenids and mytilids are epibenthic and attach to hard surfaces by a set of byssal threads. The surfaces can be as small as fingernail clams and hydrobiid snails, or as soft as macrophytes leaves. After several settlement events, the mussels clump and when the clump becomes large, small clumps, called druses, made of living and dead mussels break off and come to lay free on the bottom, even in mud, and act as a new substrate for new recruits. Quagga mussels are known to live alone in soft sediments, especially in the hypolimnial waters (Claudi and Mackie, 1994).

Both juveniles and adults can release their threads and if not restrained by the threads of other mussels can relocate and reattach in a new location. This behavior, called translocation (see Chapter 1), appears to be more common in younger mussels than older ones (Yonge and Campbell, 1968; Koch, 1989; Iwasaki, 1997). Translocation can occur seasonally, vertically from the water to the sediments or from sediments into the water column. Translocators can be carried by water currents to other locations.

Dispersal of mussels in streams and rivers is predominantly downstream, any upstream dispersal occurring mainly by animals, particularly humans. Ackerman et al. (1994) report that adult mussels can move as much as 7 cm/night, but long-range dispersal by this method is highly unlikely. The distance that mussels are carried downstream is directly related to the velocity of flow and can be estimated by multiplying the velocity by the time to settlement (e.g., m/day × 21 days to settlement = m). The younger the stage of development the further the mussels are carried downstream in the water currents. Juveniles and adults, being heavier, will settle out faster, unless they are attached to floating vegetation or debris (logs, plastic, detached floats or buoys, etc.) (Bobeldyk et al., 2005). Juveniles and adults can be resuspended by turbulence, or can use their byssal threads as drag lines (Martel, 1993; Carlton, 1993).

In general, mussels that produce free-swimming veligers (dreissenids and mytilids) are not destined for a long span in streams because as soon as the adult mussels die, there is no longer a source of veligers, unless there is a seed population in an upstream impoundment (Mackie, 1995). Even then, the impoundment must have a retention time long enough for mussels to develop from embryo to plantigrade larvae and settle out as juveniles. If there is no seed population upstream, all populations that have settled downstream eventually die until there is no source of veligers remaining. Large rivers are an exception because currents are not as unidirectional as in small streams, and small areas with eddies or bays with little or no flow can act as refuge sites for mussels to settle, grow, and reproduce.

Under natural conditions, significant upstream dispersal of mussels even within the watershed but especially overland between drainages is challenging and unlikely to contribute to long-range dispersal. Mussels may be transported upstream if they attach to aquatic animals such as crayfish or turtles, or to logs and sticks carried by beavers, or if trapped in fur of aquatic mammals (e.g., otter, beaver, muskrat, etc.) or in feathers of aquatic birds (Carlton, 1993; Johnson and Carlton, 1996). Consumption and then defecation or regurgitation of living mussels by birds or terrestrial mammals is highly unlikely (Johnson and Padilla, 1996).

The two most widely suspected overland dispersal mechanisms are recreational boats and, to a much lesser extent, waterfowl (Johnson and Carlton, 1996; Johnson and Padilla, 1996; Padilla et al., 1996; Bossenbroek et al., 2007). Johnson and Carlton (1996) concluded that on a per trip basis, recreational boats are capable of transporting far more zebra mussels than are waterfowl. They examined the ability of waterfowl to transport veligers from small pools with veliger-laden water to target pools without veligers; they found that zebra mussels were transported at very low numbers, usually <0.5 mussel per duck per trip. The enhanced zebra mussel densities, the extremely short distances involved, and the fact that the ducks walked rather than flew to the target pools suggest that these rates of transfer overestimate the ability of waterfowl to transport zebra mussels overland.

The potential for introduction by recreational boats depends on the frequency of movements between infested and uninfested waters and the survival of mussels during transit. Recreational boating activity and the associated movement of boats between bodies of water are usually greatest in the summer and early fall. It is during this period when veligers are present in the water and aquatic plants are most abundant. "However, movements by waterfowl are usually limited during this period due to nesting, molting, and pre-migration feeding. Thus, waterfowl appear much less likely to move between bodies of water than recreational boats. This preliminary comparison demonstrates that the transport of zebra mussels by waterfowl is relatively low and thus of minor concern" (Johnson and Carlton, 1996). Hence, wildlife, including waterfowl as vectors, is of low risk, even if adjacent lakes get infected.

Johnson and Carlton (1996) identified seven different mechanisms by which recreational boats could transport zebra mussels overland: "(1) adults attached to exterior boat or motor surfaces, (2) adults attached to anchors or material snagged by the anchor, (3) adults attached to aquatic macrophytes (weeds) entangled on the boat trailer during retrieval of the boat, (4) larvae in bilge water, (5) larvae in engine cooling water, (6) larvae in bait buckets, and (7) larvae in live wells. Of these mechanisms, transportation on entangled macrophytes and in live wells appears to be the most common means of transporting adult and larval zebra mussels, respectively." Padilla et al. (1996) concluded that geographic proximity to a source population and overall boater use are poor predictors of probability of invasion. Thus, if quagga mussels are detected in a central state (e.g., Lake Pleasant, Arizona) it is possible that mussels can be transported by both anglers and recreational boaters to other reservoirs in the same state or even more remote states. Padilla et al. (1996) advise that attention to human activity patterns would help predict the spread of invading taxa. The probability of introduction of larvae in bait buckets to a lake is lessened if no live bait is permitted at the lake. However, the risk potential of dispersal of dreissenids to lakes by recreational boaters and fishing/boating equipment increases if there has been a recent incident of illegal release of warm water fish into a lake. If the water quality of a lake is appropriate and likely to support infestations by dreissenids, it would do well to reduce the risk of introductions of dreissenids by posting signage and monitoring human activity patterns. The risk can be further reduced by eliminating recreational boating on lakes and reservoirs.

Dispersal of veligers to lakes and reservoirs in a state by fish stocking is of very low probability if hatchery protocols are in place. However, if the water quality of waters in a state is appropriate, the lake is highly likely to support dreissenid infestations and it would be wise not to stock different lakes with the same truck on the same trip since they could transport veligers on trucks between lakes. In developing a program to prevent introduction of dreissenids to lakes and reservoirs, the program should consider the high risk of dreissenid infestations in the lakes and consider staffing boat inspector(s); closing the ramp to boats when an inspector is not on duty; eliminating internal combustion engine use, allow electric motors only; establishing a boat rental vendor at the lakes; close lakes to all boating use; and even closing lakes to all uses since it is possible for other alien species, including fish and macrophytes, to invade the lakes.

Introduction of veligers to waters by fire-fighting helicopters is possible. The United States Forest Service (USFS) often use helicopters to dip into a large (e.g., 2-million gal) priming tanks. Although most priming tanks are open and available to waterfowl, it is not likely that waterfowl will introduce veligers to the priming tank because veligers are transported in very low numbers and rates of transfer (Johnson and Carlton, 1996). It is recommended that helicopters dip only in priming tanks that are filled with water from lakes that surround the area when fighting fires, especially if there is a high risk of survival.

The foregoing describes dispersal mechanisms and their potential mainly for zebra and quagga mussels but they most likely apply to *Mytilopsis leucophaeata* and *Limnoperna fortunei* as well. Both of these mussels have been transported across oceans or along sea coasts in ballast water and/or attached to hulls of oceanic vessels (Morton, 1975; Hebert et al., 1989; Mackie et al., 1989; Griffiths et al., 1991; Kevin et al., 1992; Mills et al., 1996; Oliver et al., 1998; Darrigran and Ezcurra de Drago, 2000; Oliveira, 2003; Therriault et al., 2004; Souza et al., 2005; Verween et al., 2005; Zhulidov et al., 2005; Laine et al., 2006; Popa and Popa, 2006; Karatayev et al., 2007a,b). Transport of mussels attached to anchors or in anchor vaults is possible but unlikely for *Dreissena polymorpha* and *Dreissena bugensis* (Mackie et al., 1989; Carlton, 1993; McMahon, 1996), but highly possible for *Mytilopsis leucophaeata* and *Limnoperna fortunei* because of their high salinity tolerances (Bamber and Taylor, 2002; Laine et al., 2006; Verween et al., 2006). *Limnoperna fortunei* may have been accidentally transported across seas or oceans attached to live freshwater clams (*Corbicula*) imported for consumption (Morton, 1980; Kimura and Tabe, 1997).

Within continents the dispersal mechanisms varied. Dispersal of *Limnoperna fortunei* within Asia and South America appears to be mainly via ballast water (Darrigran and Patorino, 1993, 1995; Darrigran, 2000; Darrigran and Excurra de Drago, 2000; Karatyev et al., 2007b). In Europe, Van der Velde et al. (2002) reported dispersal of dreissenids on logs that had been rafted down rivers. The canal systems in Europe are extensive and commercial barge and other vessel traffic has helped to spread attached mussels among watersheds in different countries (Mackie et al., 1989; Stanczykowska and Lewandowski, 1993; Johnson and Padilla, 1996; Mills et al., 1996; Karatayev et al., 1997). Commercial vessel traffic has probably also spread *Limnoperna fortunei* through the Rio Plata watershed in Argentina, Paraguay, and Brazil (Darrigran, 2002; Karatayev et al., 2007b).

In 2004, *Mytilopsis leucophaeata* was reported from the Port of Recife in Brazil (Souza et al., 2005). It apparently arrived in ballast water (Souza et al., 2005) or attached to the hulls of cargo ships, fishing boats, and other vessels operating out of the Port of Recife (Farrapeira et al., 2007).

Predicting the potential spread of zebra mussels (and other invasive species) has been an active area of research (Strayer, 1991; Neary and Leach, 1992; Ramcharan et al., 1992; Padilla et al., 1996; Schneider et al., 1998a; Buchan and Padilla, 1999; Reed-Andersen et al., 2000; Kraft and Johnson, 2000; Bossenbroek et al., 2001, 2007; Drake and Bossenbroek, 2004; Leung et al., 2004, 2006; MacIsaac et al., 2004). Most recently, Bossenbroek et al. (2007) used a gravity model of boater movements to forecast the westward spread (i.e., beyond 100th meridian) of zebra mussels. Both distance and lake size were important factors in determining the destination of recreational boaters and the probability of invasion. Bossenbroek et al. (2007) also relied on many of the models of recreational boater movement patterns cited above to forecast the distribution of zebra mussels in the Great Lakes region. They extended these earlier forecasting efforts to a national scale and represented the dispersal of zebra mussels as a function of boater movement. Bossenbroek et al. (2007) predicted that population densities could achieve levels similar to those in the Midwestern United States, where zebra mussels have caused considerable economic and ecological impacts. Their analyses suggest that the dispersal of zebra mussels to the western United States is an event of low probability but could have potentially high impact on native biodiversity and human infrastructure.

2.3.2.1.2 Corbiculidae

Hanna (1966), Fox (1971), and Morton (1973) attribute the initial introduction of Corbicula in North America to the Chinese immigrants, who in their native land use Corbicula as food (Miller and McClure, 1931). Further historical introductions in North America can be traced to passive dispersal on water currents, anthropogenic transport in boat ballast and bait buckets, or through the aquarium pet trade (Counts, 1986; Isom, 1986).

McMahon (2000) describes the dispersal mechanisms of both adults and larvae in detail. He attributes the extraordinary dispersal capacities of pediveliger and juvenile to their small size and mass (Aldridge and McMahon, 1978), which allows them to remain suspended for long distances in even minimally turbulent water, although they are incapable of active swimming. Juveniles can also be transported by currents using their single long byssal thread (Kraemer, 1979), which acts as a drag line. In addition, juvenile byssal attachment to floating objects (i.e., debris, macrophytes, algal mats, or boat/barge hulls) facilitates their dispersal (Counts, 1986). Thus, channelization of waterways for flood control or navigation increases flow velocities and turbulence, improving conditions for *Corbicula fluminea* colonization and dispersal, while simultaneously rendering them unsuitable for native bivalves (McMahon, 1982, 1991, 2000). McMahon (2000) attributes the large numbers of pediveligers/juveniles suspended in the water column as the main dispersal mode of Asian clams, allowing rapid downstream colonization of lotic habits and re-establishment of disturbance decimated populations by dispersal from upstream populations. Pediveligers and small clams can be transported upstream and across drainages on feet or feathers of aquatic birds (McMahon, 2000). However, Voelz et al. (1998) found that the Asian clam moved upstream at a rate of 1.2 km/yr in a southeastern U.S. blackwater stream unaided by humans. They suspect that fish may have been partly responsible for dispersing the clams.

Adults produce a long, mucus dragline from the exhalent siphon, which suspends them in turbulent water allowing downstream dispersal (Prezant and Chalermwat, 1984). However, this mode of transport appears to be restricted to clams 2.0 mm or smaller and is somewhat seasonal (Williams and McMahon, 1986).

2.3.2.2 Potential of Dispersal Mechanisms in Gastropoda

2.3.2.2.1 Ampullariidae

2.3.2.2.1.1 Giant Ramshorn Snail, Marisa cornuarietis *(Linnaeus, 1758)* The initial spread of *Marisa cornuarietis* was probably the aquarium trade since the species is a popular aquarium snail (Oliver-Gonzales et al., 1956; Robins, 1971; Perera and Walls, 1996). Hunt (1958) attributes the spread of *Marisa* in Puerto Rican freshwater systems to natural migration of individuals and as a result of stream flooding during storms. Birds and other animals do not seem to be important vectors in dispersion of *Marisa* because it is such a large species. The spread of the species today is probably through releases and escapes from aquarium and pet trade sources (Howells, 2001). Alexander et al. (2009) attributed the aquarium and ornamental trade among the more important unintentional vectors of introduction of all freshwater exotics, including *Marisa cornuarietis*, in Texas, resulting in different patterns of distribution, spread, and ecological impacts. The species was also stocked to control noxious growths of aquatic weeds (Seaman and Porterfield, 1964; Simberloff and Stiling, 1996a,b; Cowie, 2002) but the high density needed to control macrophytes and its limited cold tolerance served to discourage introductions for this reason (Howells et al., 2006).

Marisa cornuarietis has also been introduced in attempts to control the snail vectors of schistosomes, either by outcompeting them (Pointier et al., 1988, 1991b) and/or by predation on eggs and juveniles (Robins, 1971; Peebles et al., 1972).

2.3.2.2.1.2 Golden Apple Snail, Pomacea canaliculata *(Lamarck, 1822)* New introductions of the species have been attributed to the live food trade imported both legally and illegally by smuggling for development of aquaculture projects for human food, sale of the snails (especially the bright colored forms like golden snails) by pet stores for the domestic pet/aquarium trade, and the nursery trade where snails were probably introduced as eggs or small juveniles attached to aquatic plants (Cowie, 1995, 1998, 2002). Short-range dispersal appears to have been mediated by the aquaculture industry, most likely by escape or release from aquaculture facilities, but also to deliberate movement to establish new aquaculture facilities, accidental dispersal of eggs or small juveniles on agricultural products, infested soils during new paddy preparation, deliberate introductions to uncontained wetland systems for the purpose of harvesting the snails for food, release of domestic aquarium snails, and if within a watershed, water currents (Anderson, 1993; Cowie, 1998; Howells et al., 2006).

2.3.2.2.1.3 Titan Applesnail, Pomacea haustrum *(Reeve, 1856)* The origin of species in the *Pomacea canaliculata* complex, including *Pomacea haustrum*, throughout southeast Asia, the Indo-Pacific, and Oceania, probably resulted from failed attempts to culture snails for escargot (Howells et al., 2006). Snails from Argentina were subsequently distributed to Taiwan and then to locations in the Indo-Pacific and southeast Asia, and eventually made their way through the culture trade to Hawaii in 1989 where they significant crop pests on *Oryza sativa* (rice) and *Colocasia esculenta* (taro or elephantear) (Howells et al., 2006). In the continental United States and Canada, nearly all introductions are related to releases and escapes from aquarium and pet trade sources (Howells, 2001). Species involved include *Pomacea haustrum*, *Pomacea canaliculata*, and *Marisa cornuarietis*, as well as relocations of native *Pomacea paludosa* to sites outside its native range in peninsular Florida (Howells et al., 2006).

2.3.2.2.1.4 Island Applesnail, Pomacea insularum *(D'Orbigny, 1839)* Genetic analyses of Asian and South American populations of *Pomacea insularum* by Hayes et al. (2008) indicate that the species was probably introduced from Brazil and Argentina independently. The introductions of the species into Asia and Texas were probably by aquarium releases (Hayes et al., 2008). Alexander et al. (2009) attributed dominating aquarium and ornamental trade among the important unintentional vectors of introduction of all freshwater exotics, including *Pomacea insularum*, in Texas, resulting in different patterns of distribution, spread, and ecological impacts.

2.3.2.2.1.5 Spiketop Applesnail, Pomacea diffusa *(Blume, 1957)* The spiketop snail was probably released through the domestic pet/aquarium trade, especially the brightly colored varieties. The bright yellow form was selectively bred by the aquarium trade and subsequently transported to many parts of the world (Cowie, 2002). Bright yellow, orange, and other shell color variants are also found in other species of *Pomacea* as discussed in Chapter 1.

2.3.2.2.1.6 Florida Applesnail, Pomacea paludosa *(Say, 1829)* *Pomacea paludosa* in native to Florida but its introduction into other states is thought to have been through aquarium trade (Howells et al., 2006). The use of native *Pomacea paludosa* and other imported, exotic ampullariids in the aquarium trade increased dramatically in the 1950s and 1960s but they aggressively consumed aquarium plants and often failed to eat algae; their popularity eventually decreased because of their propensity for macrophytes over algae and ultimately, *Pomacea bridgesii*, which does not feed on macrophytes but does eat algae, became the primary snail in the American aquarium industry (Howells et al., 2006). There has been little apparent interest in using ampullariids for human consumption in the mainland United States. *Pomacea paludosa* has been observed among frozen sea foods in Asian food markets in Houston (Howells et al., 2006).

2.3.2.2.2 Thiaridae

2.3.2.2.2.1 Quilted Melania, T. granifera *(Lamarck, 1822)* Quilted melania have been introduced into many parts of their nonnative range through the pet trade industry (Lachner et al., 1970; Oglesby, 1977). Alexander et al. (2009) attributed dominating aquarium and ornamental trade among the more important unintentional vectors of introduction of all freshwater exotics, including *T. granifera*, in Texas, which has resulted in different patterns of distribution, spread, and ecological impacts.

Along with the red-rimmed melania, the quilted melania has been reported to compete with native mollusks (Murray, 1971; Jacobson, 1975; Oglesby, 1977; Chaniotis et al., 1980c). Jacobson (1975) expressed concern over the possible detrimental effects of this species on native melanids in Cuba, and Oglesby (1977) reported possible competition for trophic resources with native pupfishes in California. Because of its ability to displace snails where introduced, the quilted melania has been considered for use as a biological control agent of *Biomphalaria glabrata*, which is a major intermediate host of *Schistosoma mansoni* in St. Lucia (Prentice, 1983).

2.3.2.2.2.2 Red-Rimmed Melania, Melanoides tuberculata *(Müller, 1774)* This species was introduced to California by a San Francisco aquarium dealer (Abbott, 1952). A Tampa Bay aquarium dealer acquired specimens from California and began selling them as "the Philippine horn of plenty" in 1937 (Roessler et al., 1978). It has gradually spread throughout the southern United States probably by the pet trade industry (Murray, 1971; Neck, 1984; Howells, 1992). Alexander et al. (2009) attributes an active aquarium and ornamental trade as among the more important unintentional vectors of introduction of *Melanoides tuberculata*, in Texas.

The introduction of red-rimmed melania has resulted in displacement of native gastropods in Cuba, Texas, and Florida as a result of competition (Murray, 1971, 1975; Jacobson, 1975; Roessler et al., 1977).

2.3.2.2.3 Bithyniidae

2.3.2.2.3.1 Faucet Snail, Bithynia tentaculata *(Linnaeus, 1758)* Mills et al. (1993a) suggest that *Bithynia tentaculata* was introduced to the Great Lakes basin in packaging material for crockery, through solid ballast in timber ships arriving to Lake Michigan, or by deliberate release by amateur naturalists into the Erie Canal, Mohawk River, and Schuyler's Lake. Jokinen (1992) thinks that its original introduction is the solid ballast vector, although Mackie (2000b) suggests the species was introduced in ballast water that contained plants or solid ballast-like wood. Natural dispersal of this snail is known to occur by passive transport in birds (von Proschwitz, 1997).

2.3.2.2.4 Hydrobiidae

2.3.2.2.4.1 New Zealand Mudsnail, Potamopyrgus antipodarum *(Gray, 1843)* The mechanism for its original introduction to North America is unknown. It was first discovered in Idaho so the most likely vector is sport fishermen or recreational boaters. The mud snails can live for 24 h without water, surviving for up to 50 days on a damp surface (Winterbourne, 1970), giving them ample time to be transferred from one body of water to another on fishing gear such as waders, boots, angling equipment, and boats, or by translocation of bait (Bowler, 1991; Haynes et al., 1985; Hosea and Finlayson, 2005). Intrawatershed dispersal may occur by birds that carry the snails among their feathers or by fish that consume but are unable to digest snails (Bondesen and Kaiser, 1949; Haynes et al., 1985).

3 Using Physiological and Ecological Requirements and Environmental Tolerances for Assessing Risk of Invasion

3.1 INTRODUCTION

The success of an invasion by an aquatic nuisance species depends on the quality of water and the health of the receiving ecosystem. The quality of water in a lake depends in part on the chemistry of the water that supplies the lake. Every lake has its own watershed, and the chemistry of the soil and bedrock in the watershed will largely (but not entirely) dictate the chemistry of the water flowing through it and over it (Mackie, 2004). The chemistry may change over time. The rate and amount of change depends on factors such as; type and amount of bedrock, climate and the weathering processes which take place. The changes are somewhat predictable, as long as humans do not interfere. In essence, if the chemistry of the watershed is known, one should be able to predict the chemistry of a lake; or if the chemistry of a lake is known, one should be able to predict the chemistry of the watershed that supplies the lake with water (Mackie, 2004).

For mollusks, as explained below, a source of calcium ions is required. There are two major sources of ions for inland freshwaters. The first is the soil and weathered rocks, collectively known as the edaphic source. Edaphic ions are released into the watershed by several mechanisms, including simple solution of ions such as carbonates, oxidation and reduction reactions, complexing or chelation reactions, and the action of hydrogen ions, particularly in carbonic acid that constitutes atmospheric precipitation (Mackie, 2004). The atmosphere is the second source of ions. It receives ions from the oceans, dust, volcanic explosions, and industrial and domestic pollution (Mackie, 2004).

The success of an invasion by an aquatic nuisance species also depends on the physiological and ecological requirements and tolerances of the species. Three levels of risk can be assessed using a species' physiological and ecological tolerances and requirements: its presence/absence potential; if it has potential for presence, estimate the potential density (number/m^2) of the population; from its potential density, estimate the potential biomass (kg/m^2) of the population.

A species' potential for invasion into a body of water depends not only on its dispersal potential but also on the ability of the body of water to support a viable population. By ability we mean the quality of the water, the quality and quantity of the habitat and food resources, and the predators and parasites that could limit the population size. The potential for dispersal into a body of water of concern depends on the current proximity of the species and on the species' dispersal prowess, as discussed in Chapter 2, and nothing more needs to be stated here on dispersal.

3.2 BACKGROUND INFORMATION ON KEY VARIABLES

Predicting the presence of an aquatic invasive species (AIS) and its ability to merely survive is easier than predicting the potential density or infestation level. The potential of a species to survive once it has been introduced depends on the levels of physical, chemical, and biological variables. For mollusks, the chemical physiological variables most likely to determine their survival are levels of calcium, alkalinity, pH, and total hardness, referred to herein as "chalk variables." One other chemical, dissolved oxygen concentration, is especially important for freshwater aquatic species with gills. Important physical variables are temperature, turbidity (often measured as Secchi depth), and conductivity. Of the biological variables, levels of chlorophyll *a*, the primary food source is critical. Exceedingly high or exceedingly low levels can be limiting to mollusk survival and/or production. The levels of chlorophyll *a* can often be estimated by spring levels of total phosphorus (for phosphorus-limiting lakes), total nitrogen (in nitrogen-limiting lakes), or Secchi depths.

Natural lake systems tend to be more stable than impoundments because many of the processes (e.g., flow rates, water depth, hydraulic retention times) are subject to the whims of humans. Reservoirs or impoundments often retain many riverine characteristics. The rivers provide the initial materials for the new ecosystem, the water, the flora, and the fauna. The result is an artificial lake ecosystem that develops its own limnological and physical characteristics and its own diverse populations of phytoplankton, zooplankton, macrophytes, benthos, and fish (Odum, 1969; Baxter, 1977). The reservoir has many predictable changes in water quality and biota, but the rate of change is dictated by four indices of the water balance regime: (i) the retention time of the lake, the internal flow regimes, and degree of thermal resistance to mixing; (ii) the ratio of amount of precipitation on the surface to total inflow; (iii) the ratio of groundwater inflow to total inflow; and (iv) the ratio of evaporation and evapo-transpiration from the lake water to the total discharge fluxes (Mackie, 2004). Every reservoir has its own set of indices, hence every reservoir behaves differently.

The potential for massive infestations or infestations requiring monitoring and control programs is dealt with in the next section. The key parameter for mollusks is calcium content and its availability. Calcium availability becomes a question only at pHs > 8.2. A pH of 9.5 is lethal to most freshwater mollusks, especially dreissenids, as is a pH < 6.5 (Harman, 1974a,b). Organisms living in the pH range 9.5–10.3 are known as salt-tolerant haloalkaliphiles that typically live in soda or salt lakes.

3.2.1 Chalk Parameters

Calcium levels, alkalinity, pH, and/or total hardness have been used as criteria to determine the potential spread and abundance of dreissenids in Ontario (Neary and Leach, 1992; Hincks and Mackie, 1997) and North America (Strayer, 1991; Cohen and Weinstein, 2001; Drake and Bossenbroek, 2004; Whittier et al., 2008). Of these "chalk" parameters, the calcium level is used the most and is the most important for making risk assessments since calcium (and carbonates) is essential for the production of shell material. Without adequate shell protection, mollusks cannot survive. The shells of most heterodont bivalves (Kennedy et al., 1969; Mackie, 1978b) and gastropods (Bøggild, 1930) consist of calcium carbonate in the form of orthorhombic aragonite. For most mollusks the calcium is obtained from the water and the food (Kennedy et al., 1969; Mackie and Flippance, 1983a–c). Increases in shell length occur by secretions of calcium carbonate by the outer fold of the mantle edge and increases in shell thickness occur by secretions from the outer wall of the outer lamella of the mantle (Wilbur, 1964).

Before delving into the role of calcium as a variable limiting population densities, a primer on the relationships between calcium, alkalinity, and pH is in order. The alkalinity informs us of the availability of the calcium. Total alkalinity consists of bicarbonate alkalinity (HCO_3) and carbonate alkalinity (CO_3). The pH governs the forms of carbon dioxide and hence of the carbonates available, pH values below 8.2 having all the calcium available in bicarbonate form (HCO_3) and values above 8.2 having the calcium present in monocarbonate form (CO_3). Figure 3.1 shows the effect of pH on the concentrations of free CO_2, bicarbonates, and carbonates in aquatic ecosystems. Carbon dioxide exists in freshwater in four forms: (i) free CO_2, the dissolved state that is available to plants; (ii) hydrated CO_2, or undissociated carbonic acid; (iii) half-bound CO_2 or bicarbonates; and (iv) bound CO_2 or monocarbonates.

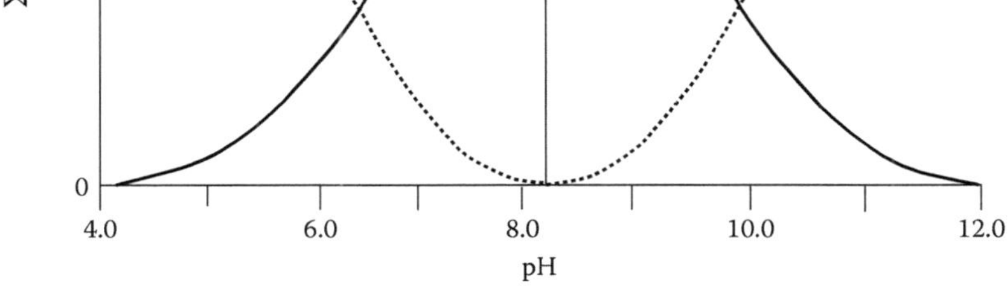

FIGURE 3.1 Relative proportions of three forms of carbon dioxide: free CO_2, bicarbonate, and monocarbonate.

The free dissolved form of CO_2 is required by photosynthetic organisms and, with the remaining forms, is important in the chemical equilibrium relationships of carbonates and bicarbonates that affect the pH, alkalinity, and buffering capacity of the water. Equation 3.1 shows that carbon dioxide combines readily with water in the atmosphere (i.e., precipitation) to form carbonic acid, which has a pH of about 5.6.

$$\underset{\text{Dissolved gas}}{CO_2} + \underset{\text{Water}}{H_2O} \leftrightarrow \underset{\text{Carbonic acid}}{H_2CO_3}. \tag{3.1}$$

Carbonic acid weakly dissociates into bicarbonate and monocarbonate ions, as shown in Equation 3.2:

$$\underset{\text{Undissociated carbonic acid}}{H_2CO_3} \leftrightarrow \underset{\text{Bicarbonate ion}}{HCO_3} + \underset{\text{Hydrogen ion}}{H^+} \leftrightarrow \underset{\text{Monocarbonate acid}}{CO_3^{2-}} + \underset{\text{Hydrogen ions}}{2H^+}. \tag{3.2}$$

Since most North American freshwater lakes lie within a pH range of 6–9, most of the CO_2 is present as bicarbonate (Figure 3.1). The monocarbonate ion can exist only in water with a pH > 8.2. Carbon dioxide in lakes with pH > 12 is present as monocarbonate ions. Such lakes will also have free hydroxide ions (OH^-), but they are rare in North America. Hence, plants in lakes with pH > 8.2 potentially have an inexhaustible supply of CO_2 for photosynthesis and mollusks potentially have an inexhaustible supply of bicarbonates.

The free CO_2 necessary to maintain bicarbonate in solution is called equilibrium CO_2. As carbonic acid flows through the soil and comes into contact with limestone, it reacts with the calcium carbonate to put calcium and bicarbonate ions into solution, as shown in Equation 3.3:

$$\underset{\text{Undissociated carbonic acid}}{H_2CO_3} + \underset{\text{Limestone}}{CaCO_3} \leftrightarrow \underset{\text{Calcium ion}}{Ca^{2+}} + \underset{\text{Bicarbonate ions}}{2HCO_3}. \tag{3.3}$$

Calcium and bicarbonate ions react to form soluble calcium bicarbonate, with equilibrium CO_2 maintaining the reaction (Equation 3.4) in the pH range 6–9. At pH levels >8.2, any CO_2 removed from the water by photosynthesis will cause calcium carbonate to precipitate out, producing a floc that causes lake whitening, which greatly reduces lake transparency. Calcium carbonate will also precipitate out as chalk, or marl, on leaves of submersed plants in high pH lakes:

$$\underset{\text{Calcium ion}}{Ca^{2+}} + \underset{\text{Bicarbonate ions}}{2HCO_3} \leftrightarrow \underset{\text{Calcium bicaronate}}{Ca(HCO_3)_2} \tag{3.4}$$

In hard water streams, attached algae deposit calcium carbonate on rocks as travertine. The travertine can also be produced by turbulence, where carbon dioxide is displaced by oxygen in the turbulent zone. The reaction that produces lake whitening, marl, and travertine is shown in Equation 3.5:

$$\underset{\text{Calcium bicarbonate}}{Ca(CO_3)_2} \leftrightarrow \underset{\text{Lime precipitates out}}{CaCO_3 \downarrow} + \underset{\text{Water}}{H_2O} + \underset{\text{Carbon dioxide}}{CO_2}. \qquad (3.5)$$

If any aggressive CO_2 is introduced, such as by dehydration of carbonic acid, dissolution of $CaCO_3$ can occur. However, the reaction is essentially irreversible since at pHs > 8.2, any additional CO_2 produced by dehydration of carbonic acid is quickly taken up by plants.

Reactions involving addition of CO_2 are similar to reactions involving addition of hydrogen ions. Any CO_2 that is added combines with hydrogen ions to form carbonic acid, which helps dissolve monocarbonates, such as calcium carbonate, to produce bicarbonates. The reactions (Equations 3.1 through 3.5) will continue to the left until all the carbonates and bicarbonates have been reduced to carbonic acid and, eventually, carbon dioxide. Lakes and rivers flowing over or through granitic bedrock have little or no calcium carbonate dissolved in the water and have a pH < 8.2 (see Figure 3.1). Even if the bicarbonate content is low, which is characteristic of lakes with a pH near 7, the pH level in the lake will drop very quickly if any hydrogen ions are introduced, such as those from acidic precipitation, because there is little or no bicarbonates to absorb, or buffer the hydrogen ions added. Such lakes acidify very quickly and generally lack mollusc populations.

Even above pH 8.2, some calcium is available as bicarbonates; calcium in monocarbonate form particularly as a precipitate, is unavailable for shell production. However, monocarbonates can be converted to bicarbonates by lowering the pH, for example by increasing the CO_2 concentration, which produces carbonic acid. Removal of carbon dioxide (e.g., by photosynthesis of plants and algae) results in precipitation of calcium carbonate, making it unavailable to mollusks. Hence, while calcium is the key variable, knowledge of the value of the total alkalinity is also important in predicting infestation potential of aquatic invasive mollusks. Total alkalinity is part of the total hardness and can be all of the total hardness. Rarely does total alkalinity exceed total hardness because most of the hardness is composed of cations of calcium and magnesium combined with carbonates, sulfates, and/or chlorides. Only when cations such as potassium and sodium combined with carbonates are prominent can alkalinity exceed the total hardness.

3.2.1.1 Using Calcium as a Limiting Chalk Variable

If calcium levels are too low, mollusks will not survive. If calcium levels of a lake are in the marginal area (i.e., may provide moderate to good growth), one may want to be certain and proceed as described below. Or one may want to play it safe and err on the side of safety and assume good growth when the data might indicate moderate to little growth. The following is especially useful if calcium measurements are lacking, but other measurements like conductivity, alkalinity, total hardness, and so on have been taken and can be used to estimate calcium levels.

Measure the calcium levels as mg Ca/L in the epilimnion and hypolimnion at several locations of the raw water supply. Estimate the annual means and the minima and maxima in the epilimnion and hypolimnion or of the raw water supply over a minimum of two years, but preferably five years to more accurately reflect variations in mean calcium levels that occur in the water. Variations in calcium content can occur with depth but they are generally much less than across the lake, especially if the watershed has outcrops of different kinds of bedrock. Compare the values to the threshold calcium levels for the species (e.g., if zebra mussels, use Table 3.7).

3.2.1.1.1 *Estimating [Calcium] from Alkalinity, Total Hardness, and Conductivity*

If calcium levels are unknown but the calcium hardness (as mg $CaCO_3$/L) or alkalinity (as mg $CaCO_3$/L) is known multiply each value by 0.40 since, in the equation

$$Ca + CO_3 = CaCO_3$$

100 parts of $CaCO_3$ is made up of 40 parts calcium (actually 40.08 is the atomic weight) and 60 parts CO_3 (i.e., 12 for C + 48 for O_3).

If calcium hardness is in meq or μeq (1 meq = 1000 μeq), multiply the calcium hardness by 0.02 in order to obtain the calcium level in mg/L.

If alkalinity as μeq/L is known, estimate calcium levels from the general relationship:

$$\text{Alkalinity}(\mu\text{eq/L}) = -0.013 + 1.396 \times \text{calcium}(\mu\text{eq/L}) \quad \text{(Neary et al., 1990).}$$

The equation is a linear regression of the type, $y = a + bx$, where $a(-0.013)$ is the y-intercept when $x = 0$ and $b(1.396)$ is the slope of the regression. Since the y-intercept is so small, the equation can be simplified to

$$\text{Ca } (\mu\text{eq/L}) = \frac{\text{Alkalinity } (\mu\text{eq/L})}{1.396}.$$

For example, if the alkalinity of the water is 2000 μeq/L, the calcium level is 1432.6 μeq/L (i.e., 2000/1.396). If mg/L is preferred and since alkalinity is often expressed in mg $CaCO_3$/L, and 50 mg $CaCO_3$/L alkalinity = 1000 μeq/L alkalinity and 20 mg Ca/L = 1000 μeq Ca/L, or a ratio of 2.5 parts of $CaCO_3$ to 1 part Ca, the equation can be rewritten as

$$\text{Ca(mg/L)} = \frac{\text{Alkalinity(mg CaCO}_3\text{/L)}}{3.49},$$

where 3.49 is simply the slope, 1.396, times 2.5. For example, if the alkalinity of the water is 100 mg $CaCO_3$/L, the calcium level is 28.65 mg Ca/L (i.e., 100/3.49). Using the calcium equivalent of 20 mg Ca = 1000 μeq above, the 28.65 mg Ca/L equates to 1432.6 μeq/L, which is the same value derived above in μeq/L.

If total hardness is known, estimate calcium levels from the general relationship:

$$\text{Calcium (mg/L)} = (1.08 + 0.272 \text{ total hardness}) (\text{mg } CaCO_3).$$

However, total hardness is composed of temporary and permanent hardness and the above relation works only for waters with temporary hardness, as discussed below. If the total hardness values exceed 400 mg $CaCO_3$/L, this relationship should be used with caution because some of the hardness may be due to anions other than carbonates (e.g., sulfates, chlorides).

If conductivity is known, estimate calcium levels from the general relationship:

$$\text{Calcium (mg/L)} = 0.141 \text{ Conductivity } (\mu S)^{1.175}.$$

However, this relationship assumes that sulfate levels are somewhat constant.

All the above are general relationships derived from a large number (2500–3950) of Ontario lakes (Dr. Peter Dillon, Ontario Ministry of Environment, personal communication) and should be used only as "guestimates" of mussel infestation levels based on "derived" calcium levels.

3.2.1.2 Using Total Alkalinity as a Limiting Chalk Variable

Alkalinity is a measure of the ability of water to resist a drop in pH. Water resists changes in pH due to the presence of anions of carbonates, bicarbonates (and occasionally hydroxides). The total alkalinity of water is the sum of the concentrations of bicarbonate (HCO_3) and carbonate (CO_3) anions (and hydroxide ions if present). This contrasts with total hardness, which is the total concentration of cations of calcium and magnesium (and iron if present at high levels). Since temporary hardness measures the amounts of anions of bicarbonates and carbonates, it is essentially the same as total alkalinity. If we use the same compounds in the water sample as used to discuss hardness above [i.e., $CaCO_3$, $Ca(HCO_3)_2$, $MgCO_3$, $Mg(HCO_3)_2$, $CaSO_4$, $MgSO_4$, $CaCl_2$, and $MgCl_2$], the compounds that contribute to alkalinity are $CaCO_3$, $Ca(HCO_3)_2$, $MgCO_3$, and $Mg(HCO_3)_2$ because only they have anions of carbonate and bicarbonate; the remaining compounds do not contribute to total alkalinity (Mackie, 2004).

Alkalinity is usually expressed in terms of equivalents of bicarbonate or carbonate or mg of $CaCO_3$ per liter of water. Total alkalinity is the sum of carbonate alkalinity and bicarbonate alkalinity. To measure total alkalinity, determine the carbonate alkalinity first and then the bicarbonate alkalinity. Carbonate alkalinity need not be measured if the pH is 8.2 or less because carbonate alkalinity can only exist in waters with pH > 8.2 (Figure 3.1). If the pH is 8.2 or greater, carbonate alkalinity is determined by adding an indicator, such as phenolphthalein, which changes color at pH 8.2. The sample is then titrated with a weak acid, such as 0.02 N H_2SO_4, until the color changes. Because phenolphthalein is the most commonly used indicator at this step, carbonate alkalinity is also known as phenolphthalein alkalinity. At this point (pH 8.2), all the carbon dioxide is present as bicarbonates (Figure 3.1). Bicarbonates can be reduced to carbonic acid by adding an acid until pH 4.4 (Figure 3.1). Hence, indicators that change color at pH 4.4, such as methyl orange and bromthymolgreen methyl red, can be used to detect the change.

Measure the total alkalinity levels in the epilimnion and hypolimnion as mg $CaCO_3$/L at several locations or of the raw water supply. Estimate the annual means and the minima and maxima in the epilimnion and hypolimnion or of the raw water supply over a minimum of two years, but preferably five years to more accurately reflect variations in mean alkalinity levels that occur in the water. Often the alkalinity is not clearly stated as total, bicarbonate, or carbonate alkalinity. In this event use the pH value to determine the kind of alkalinity. If pH is <8.2, all the alkalinity is present as bicarbonate alkalinity; if pH is >8.2, then one must assume that the alkalinity values are for total alkalinity. Variations in total alkalinity can occur with depth but they are generally much less (especially in high alkalinity lakes) than across the lake, especially if the watershed has outcrops of different kinds of bedrock. Compare the values to the threshold alkalinity levels for the species (e.g., if zebra mussels, use Table 3.7).

3.2.1.3 Using pH as a Limiting Chalk Variable

pH represents the hydrogen ion concentration in water on a negative base −10 logarithm scale, where each unit increase in pH represents a 10-fold reduction in hydrogen ion concentration. Waters with pH = 7 are neutral; values between 6.8 and 7.2 are referred to as circumneutral because they have poor buffering capacity and vary around pH 7.0. The pH of precipitation in equilibrium with CO_2 is about 5.6. The atmosphere in many areas also has significant amounts of sulfur dioxide and nitrous oxide, which when oxidized creates strong sulfuric and nitric acids that lower the pH of precipitation significantly below 5.6. Since normal precipitation has a pH of 5.6, it is not considered acidic until the pH falls below 5.6. pH values >7 are basic. As stated earlier, the pH of most natural waters is typically between 6 and 9. It is influenced by atmospheric precipitation and by normal biological processes in the water and soils, as well as by pollution, such as mine drainage. Of the biological processes, photosynthesis uses CO_2 causing an increase in pH. Respiration by aquatic plants and animals releases CO_2 to the water causing a decrease in pH. Decaying vegetation, particularly from wetlands that contain humic acids will also lower the pH (Shapiro, 1957; Kononova, 1966). Volcanic lakes can be acidic or basic, depending on the rock formations. If there are active sources of sulfur compounds, the lakes will be acidic, as low as 2 (Armitage, 1958). If there are sources of sodium carbonate (Na_2CO_3) or bicarbonate ($NaHCO_3$), the lakes will be basic and are called soda lakes with pH >9 (Satake, 1980). Any soils and bedrock within a watershed that contain carbonate minerals buffer fluctuations in pH by consuming hydrogen ions provided in by the precipitation and biological processes described above. If the soils and bedrock lack carbonates, and therefore buffering capacity, the pH varies widely. Any sulfide or chloride minerals in the watershed will lower the pH (Mackie, 2004).

Although pH and calcium, alkalinity and hardness levels are all correlated, the correlations are occasionally rather weak and it is better to plot pH directly and compare the variation to the threshold tolerances of mollusks to pH. However, for many species the threshold tolerance values are not as well known as calcium threshold values.

Most mollusks have an upper and lower lethal threshold level, the lower generally near 6.5, the upper near 9 (Harman, 1974a,b). Also, larval and juvenile forms are generally more sensitive than adults. For example, the threshold value for survival of adult zebra mussels is near pH 6.5 (McCauley and Kott, 1993) and for larvae pH 6.9 (Mackie and Kilgour, 1993). The incipient lethal level for larvae is near pH 7.4 (Mackie and Kilgour, 1993). Although no formal studies have been done, the pH needed for massive infestations to occur appears to be pH > 8.0, which is the average pH of Lake Erie.

Measure pH levels at one- to two-meter depth intervals at several locations in the lake, or of the raw water supply. Estimate the annual means and the minima and maxima in the epilimnion and hypolimnion or of the raw water supply over a minimum of two years, but preferably five years to more accurately reflect variations in mean pH that occur in the water. The pH can vary significantly with depth, depending on the buffering capacity (i.e., alkalinity) and the trophic status of the lake (Cole, 1994; Mackie, 2004). Eutrophic lakes with hypolimnetic oxygen deficits and therefore high concentrations of CO_2 will have much lower pH levels than oligotrophic lakes of the same alkalinity. Variations in pH can also occur across the lake, especially if the watershed has outcrops of different kinds of bedrock. Compare the values in the epilimnion and hypolimnion to the threshold pH levels for the species (e.g., if zebra mussels, use Table 3.7).

3.2.1.4 Using Total Hardness as a Limiting Chalk Variable

The concept of hardness originated in the early days to express the soap-consuming power of water (Mackie, 2004). Hard water (>120 mg $CaCO_3$/L) causes soap to precipitate, leaving a scum or scale on the walls of the vessel at the water surface. Producing lather with soap or shampoo is difficult when using hard water because the cations of calcium and magnesium precipitate the soap suds. Boiling removes the temporary hardness, leaving only permanent hardness in the water, the $CaCO_3$, or scale, settling on the walls of the boiling vessel (e.g., tubs, sinks, dishwashers, car radiators) when hard water is used. Durfor and Becker (1964) classified the hardness of water as soft (<60 mg $CaCO_3$/L), moderately hard (61–120 mg $CaCO_3$/L), hard (121–250 mg $CaCO_3$/L), very hard (≥251 mg $CaCO_3$/L) and this scheme is followed here. The total hardness of well water in limestone or dolomitic bedrock areas may have over 1000 mg $CaCO_3$/L (Mackie, 2004).

The hardness of water is contributed by the cations of calcium, magnesium, and, to some extent, iron. Total hardness is usually taken as the sum of the concentrations of calcium and magnesium and expressed as mg $CaCO_3$/L. Hardness is determined by titration with EDTA (ethylenediaminetetra-acetic acid) that chelates calcium and magnesium, the titrant being standardized to give results equivalent to mg of $CaCO_3$ per liter of water (Mackie, 2004). Hence, if water contains compounds of $CaCO_3$, $Ca(HCO_3)_2$, $MgCO_3$, $Mg(HCO_3)_2$, $CaSO_4$, $MgSO_4$, $CaCl_2$, and $MgCl_2$, the total hardness is the sum of the concentrations of the calcium combined with monocarbonates (CO_3), bicarbonates (HCO_3), sulfates (SO_4), and chlorides (Cl). Of the compounds that contribute to total hardness, only those combined with monocarbonates and bicarbonates contribute to total alkalinity, as discussed above. The total concentration of cations of Ca, Mg, and Fe combined with carbonates and bicarbonates is known as temporary hardness because the hardness can be removed by boiling the water and converting carbonates and bicarbonates to carbon dioxide, leaving calcium uncombined in solution or deposited on the walls of containers as "scale" (Mackie, 2004). The total concentration of cations of Ca, Mg, (and Fe) combined with noncarbonates (e.g., SO_4, Cl, PO_4, etc.) is known as permanent hardness because the hardness cannot be removed by boiling. If only the concentration of calcium combined with carbonates and noncarbonates is considered, the hardness is expressed as calcium hardness. Because 95–100% of the total hardness is usually due to cations of calcium and magnesium in surface waters (iron usually contributes <1% of the total hardness), an estimate of the magnesium hardness can be made by subtracting the calcium hardness from the total hardness. Hence, the total hardness can be estimated in two ways: (1) by summing the temporary and permanent hardness and (2) by summing the calcium and magnesium ions.

Measure total hardness levels in the epilimnion or raw water supply as mg $CaCO_3$/L. Estimate the annual means and the minima and maxima of the raw water supply over a minimum of two years, but preferably five years to more accurately reflect variations in mean hardness levels that occur in the lake or raw water supply. Variations in total hardness can occur with depth but they are generally much less than across the lake, especially if the watershed has outcrops of different kinds of bedrock. Compare the values obtained to the threshold hardness levels for the species (e.g., if zebra mussels, use Table 3.7).

3.2.2 Dissolved Oxygen

There are two primary sources of oxygen in aquatic systems, the atmosphere and photosynthetic production by algae and macrophytes. The rate of diffusion of oxygen from the atmosphere into water varies directly with the pressure difference across the air–water interface. The greater the difference in partial pressures, the faster the diffusion rate of oxygen from the atmosphere into water. The partial pressure and the solubility of oxygen at ambient temperatures determine how much of the gas can be dissolved in water. Air contains about 21% oxygen by volume, or approximately 300 mg/L of air. Oxygen is nearly twice as soluble as nitrogen at a given temperature but its partial pressure (21% in atmosphere) is only a quarter that of nitrogen (78% in atmosphere). Hence, the ratio of both gases dissolved in water is roughly one part oxygen to two parts nitrogen. Carbon dioxide is about 33 times more soluble than oxygen in water and has a partial pressure about 1/700 (0.0014% in atmosphere) that of oxygen. Because of the high solubility of carbon dioxide in water, carbon dioxide is relatively (i.e., compared to oxygen) more abundant in water than in air (Mackie, 2004).

Freshwater at 0°C and mean sea level contains about 14.6 mg/L. Oxygen solubility decreases with increasing temperatures (Table 3.1). Because of the effect of temperature on the solubility of oxygen, the concentration of oxygen in water is often expressed as percent saturation. To determine % saturation, divide the amount of oxygen (mg/L) dissolved in the water by the amount of oxygen that the water can hold at the observed temperature and multiply by 100. For example, suppose one observes 10 mg/L at 0°C. The water should have 14.62 mg/L (Table 3.1), so % saturation is $10/14.62 \times 100 = 68.4\%$ at mean sea level.

TABLE 3.1
Effects of Temperature and Altitude on the Solubility of Oxygen

	Dissolved Oxygen from Normal Atmosphere		Correction Factor for Oxygen Saturation at Various Altitudes			
			Altitude		Pressure	
Temperature (°C)	mg/L	ml/L	Ft	m	mm	Factor
0	14.62	10.23	0	0	760	1.00
1	14.23	9.96	330	100	750	1.01
2	13.84	9.69	655	200	741	1.03
3	13.48	9.44	980	300	732	1.04
4	13.13	9.19	1310	400	723	1.05
5	12.80	8.96	1640	500	714	1.06
6	12.48	8.74	1970	600	705	1.08
7	12.17	8.52	2300	700	696	1.09
8	11.87	8.31	2630	800	687	1.11
9	11.59	8.11	2950	900	679	1.12
10	11.33	7.93	3280	1000	671	1.13
11	11.08	7.75	3610	1100	663	1.15
12	10.83	7.58	3940	1200	655	1.16
13	10.60	7.42	4270	1300	647	1.17
14	10.37	7.26	4600	1400	639	1.19
15	10.15	7.11	4930	1500	631	1.20
16	9.95	6.96	5250	1600	623	1.22
17	9.74	6.82	5580	1700	615	1.24
18	9.54	6.68	5910	1800	608	1.25
19	9.35	6.55	6240	1900	601	1.26
20	9.17	6.42	6560	2000	594	1.28
21	8.99	6.29	6900	2100	587	1.30
22	8.83	6.18	7220	2200	580	1.31
23	8.68	6.08	7550	2300	573	1.33
24	8.53	5.97	7880	2400	566	1.34
25	8.38	5.86	8200	2500	560	1.36
26	8.22	5.75				
27	8.07	5.64				
28	7.92	5.54				
29	7.77	5.44				
30	7.63	5.34				

Altitude (and therefore partial pressure) also affects the solubility of oxygen. Since the solubility of oxygen increases with an increase in altitude, lakes at high altitudes, such as 1125 m for Lake Powell (Stanford and Ward, 1991), need to correct for the effect of altitude. Table 3.1 also gives correction factors for dissolved oxygen at different elevations. After accounting for temperature, multiply the % saturation value by the appropriate factor given in Table 3.1. For example, using the above example of 68.4%

for 10 mg/L dissolved oxygen at 0°C and Lake Powell, the adjusted % saturation value at elevation of 1125 m is $68.4 \times 1.1525 = 78.8\%$. The value 1.1525 was derived by taking the factor for 1100 m (1.15) and adding to it the value of 0.0025 for the additional 25 m since the factor increases by 0.0025 for every 25 m between 1100 and 1200 m.

Physical factors also affect the diffusion rate of oxygen across the air–water interface. Oxygen diffuses more rapidly into rough, or choppy, water than into a calm surface because of the increased surface area produced by the ripples or waves. Streams tend to be naturally aerated because of the turbulent flow of water.

The amount of oxygen produced by photosynthesis depends on the abundance of algae in the water. The reaction involved in photosynthesis is

$$nCO_2 + nH_2O \leftrightarrow (CH_2O)_n + nO_2,$$

where $n = 3$ (pyruvate), 6 (glucose), or 12 (sucrose) (Cole, 1994; Mackie, 2004). The relative concentrations of CO_2 and O_2 depend on whether the process moves to the left or to the right. It follows from this that when the concentration of carbon dioxide is high, the concentration of oxygen in the water will be low. Hence, an inverse relationship exists between the amount of O_2 and the amount of CO_2 dissolved in the water (Mackie, 2004).

Since photosynthesis requires light, diurnal fluctuations in dissolved oxygen concentration occur, with an increase during the day and a decrease during the evening. Conversely, the concentration of dissolved (free) CO_2 in the water decreases during the day as it is being used for photosynthesis and increases at night due to respiration by aquatic organisms, including plants. Lakes with high algal biomasses often have partial pressures greater than that provided by the atmosphere, resulting in supersaturation, or concentrations exceeding 100% (Mackie, 2004).

Lake productivity affects the dissolved oxygen depth profile. Oligotrophic lakes have close to 100% oxygen saturation at all depths (Figure 3.2a). The concentration of oxygen in the epilimnion is determined by the amount of photosynthesis occurring there as well as by the amount diffused from the atmosphere. The amount of oxygen in the hypolimnion remains high because of its high solubility at 4°C and because there are none to meagre amounts of organic material for bacteria to oxidize. Oxygen profiles that exhibit 75–100% saturation at all depths are called orthograde (Mackie, 2004).

Eutrophic lakes have an abundance of nutrients that leads to high primary productivity. While the epilimnion has 100% oxygen saturation, or even supersaturation, it also has large amounts of organic material composed largely of moribund algae that sinks and settles on the bottom of the lake. In the hypolimnion bacteria convert the organic material into carbon dioxide and in the process use up any oxygen present. The oxygen supply cannot be replenished if the water is thermally stratified. The stratified water column effectively isolates the hypolimnion from any renewable sources of oxygen. By the end of the summer the hypolimnion is anoxic to oxygen deprived. Lakes that exhibit high saturation values in the epilimnion and low values in the hypolimnion have clinograde oxygen profiles (Figure 3.2b and c). Both eutrophic (Figure 3.2b) and mesotrophic (Figure 3.2c) lakes have clinograde oxygen

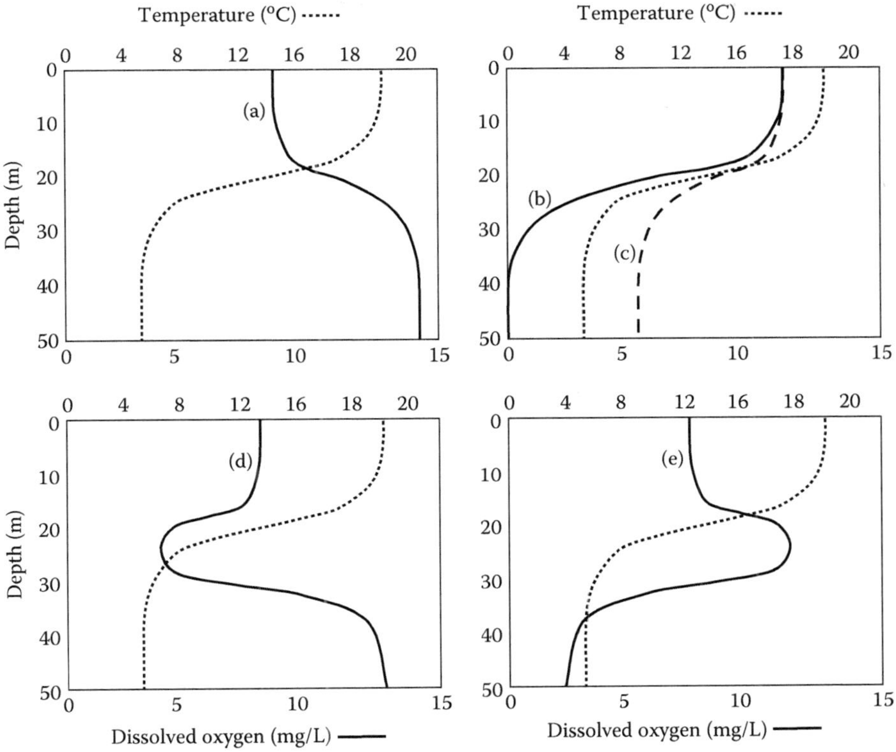

FIGURE 3.2 Variations in dissolved oxygen profiles in deep lakes: (a) orthograde; (b) clinograde with hypolimnial anoxia in eutrophic lakes; (c) clinograde with <25% oxygen saturation in hypolimnion; (d) negative heterograde curve; and (e) positive heterograde oxygen curve.

profiles, but eutrophic lakes have 0–25% and mesotrophic lakes have 25–75% oxygen saturation in the hypolimnion (Mackie, 1994).

At least two variations in oxygen profiles occur. One is the hypolimnetic oxygen minimum profile, also known as the negative heterograde oxygen curve, and exhibits a minimum concentration of oxygen in the metalimnion (Figure 3.2d). The other is the metalimnetic oxygen maximum or a positive heterograde oxygen curve and exhibits a maximum concentration of oxygen in the metalimnion. Mackie (2004) describes the curves and the reasons for their peculiar shapes.

Measure the dissolved oxygen content at 1- to 2-m depth intervals and at several locations across the lake. Estimate the annual means and the minima and maxima of the raw water supply over a minimum of two years, but preferably five years to more accurately reflect variations in mean dissolved oxygen levels that occur in the lake. Variations in dissolved oxygen usually occur with depth and will determine if the invasive species can survive in deep waters, even if the levels of chalk variables are appropriate for their survival. Compare the values obtained to the lower threshold dissolved oxygen levels for the species (e.g., if zebra mussels, use Table 3.7).

3.2.3 Chlorophyll *a*

Because phosphorus or nitrogen determines summer concentrations of algae, chlorophyll *a* levels can be used to assess the trophic status of a lake. The chlorophyll *a* level can either be measured if the appropriate equipment is available, or estimated from Secchi depth and/or total phosphorus relationships. To measure the level of chlorophyll *a*, a fluorometer (or spectrophotometer) and cuvettes with a 1-cm light path are required. A 1-L water sample is taken, preserved with 1 mL of 1% $MgCO_3$ solution and held at 4°C until it can be measured for [chlorophyll *a*]. First, the sample must be filtered through 0.8 μm maximum pore size; then the filter paper with the algae is folded and placed in a test tube to which 90% alkalized ($MgCO_3$) acetone solution (usually 20 mL) is added. This extracts the chlorophyll; the greater the chlorophyll content, the greener the solution. The sample is then transferred to a cuvette with a 1-cm light path, which is placed in a fluorometer or spectrophotometer. The absorbance (amount of light absorbed) is read at wavelengths that are optimal for chlorophylls *a*, *b*, and *c*. The greater the greenness, the more light is absorbed. For grass-green extracts, chlorophyll *a* dominates and absorbance need only be read at 663 nm. Use the following formulae to estimate the level of chlorophyll *a*:

$$[\text{Chl}\,a] \text{ in extract} = (\text{absorbance at 663 nm}) \times (13.4) = \text{mg Chl } a/\text{L of extract},$$

$$[\text{Chl}\,a]\ \text{mg/m}^3 \text{ of lake water}$$
$$= \frac{(\text{mg chlorophyll } a\,/\,\text{L of extract}) \times (\text{extract volume (mL)} \times 1000\,\text{L/m}^3)}{\text{filtered volume (mL)}}.$$

For example calculate the chlorophyll *a* level if 1 L of water is filtered, 20 mL of alkalized acetone is used to extract the chlorophyll, and the extract gives an absorbance of 0.20 at 663 nm:

$$[\text{Chl } a] \text{ in extract} = 0.20 \times 13.4 = 2.68\,\text{mg/L of extract},$$

$$[\text{Chl } a] \text{ in water mg/m}^3 = 2.68\,\text{mg/L} \times 20\,\text{mL} \times 1000\,\text{L/m}^3/1000\,\text{mL}$$
$$= 53.6\,\text{mg/m}^3, \text{ or } 53.6\,\mu\text{g/L}.$$

The concentration of chlorophyll *a* (in μg/L) can also be estimated from levels of total phosphorus (in μg/L). Dillon and Rigler (1974) developed the following model from data on numerous soft-water lakes in central Ontario:

$$\log_{10}[\text{Chl}\,a] = 1.45\log_{10} \text{P}_\text{T} - 1.14.$$

Brown et al. (2000) developed the following models from data on several lakes in Florida:

$$\log_{10}[\text{Chl}\,a] = -0.37 + 1.05\log_{10} \text{P}_\text{T},$$
$$\log_{10}[\text{Chl}\,a] = -2.42 + 1.21\log_{10} \text{N}_\text{T}.$$

Dillon and Rigler (1974) and Brown et al. (2000) cautioned that the models break down if the lakes are nitrogen limiting and should be used only when the total nitrogen (N_T) to total phosphorus (P_T) ratio is <10–17. This was confirmed by Smith (1982), who recommended the following model to reduce the error when N_T:P_T ratios were <10–17:

$$\log_{10}[\mathrm{Chl}\,a] = 0.653 \log_{10} \mathrm{P_T} + 0.548 \log_{10} \mathrm{N_T} - 1.517.$$

Considerable research has also been done relating Secchi depth to concentrations of chlorophyll *a*. Section 3.2.5 describes some regressions and precautions when using Secchi depth to estimate chlorophyll *a* levels.

There are instruments now that make a variety of water quality measurements, including dissolved oxygen, pH, and chlorophyll *a* with high levels of precision. Hydrolab® and YSI Environmental have probes which now give a direct reading of chlorophyll *a*.

3.2.4 Nutrients

The nutrient parameters, total phosphorus and total nitrogen, influence the chlorophyll *a* levels, which in turn affect the Secchi depth and the dissolved oxygen content. All five variables have criteria for assessing the trophic status of lakes and are known as trophic indicators (Mackie, 2004). The higher the values of the nutrient variables, the greater the biomasses of algae and hence of chlorophyll *a*, and dissolved oxygen (at the surface), and the lower the Secchi depth values (i.e., water is more turbid). Since mussels feed on planktonic algae, the values of the trophic indicators can be used as criteria for predicting dreissenid presence and abundance. Most gastropods are scraper feeders. Some like *Bithynia tentaculata* also filter feed. There are no criteria that relate periphytic chlorophyll *a* or macrophytes production to trophic status of lakes or marsh systems so the trophic indicators do not apply to the gastropods discussed here. Total phosphorus should be used when phosphorus is limiting and total nitrogen when nitrogen is limiting.

3.2.4.1 Using Total Phosphorus as a Limiting Nutrient Variable

Measurements for total phosphorus (P_T) need to be taken at the surface and in the spring since this is the period when [P_T] is maximum. Total phosphorus levels fall during the summer due to uptake by algae. Total phosphorus consists of particulate and soluble phosphorus. The soluble form is mostly orthophosphate. The particulate phosphorus is insoluble and consists of eroded forms from the rock and is supplied to the lake by parent streams and rivers. Some phosphorus is also supplied by aerial deposition. Nearly all (>90%) loadings end up on the bottom of the lake, largely because of clays that have a high phosphorus-holding capacity. Usually <10% enters lakes as soluble phosphate (Cole, 1994; Mackie, 2004).

All other sources of phosphorus are anthropogenic in origin, such as detergents in sewage, but most (>90%) of it is present in soluble form and very little (<10%) is insoluble. Because algae utilize soluble phosphate (e.g., orthophosphate) as a nutrient, anthropogenic sources are the main culprit causing algae blooms, and hence eutrophication (enrichment), in lakes. Most lakes are sinks for phosphorus and large amounts end up trapped in the lake's sediments. The mechanism for mobilizing phosphorus is complex but mobilization rates increase with decreasing dissolved oxygen levels. Hence, anoxic sediments of eutrophic lakes are more likely to release phosphorus and maintain eutrophication rates than are mesotrophic or oligotrophic lakes. This internal loading of phosphorus explains why many shallow lakes remain eutrophic long after all anthropogenic sources have been removed. In deep lakes, internal loading may be confined to the hypolimnion, with only small amounts of phosphorus being leaked to the epilimnion where primary production can occur (Cole, 1994; Mackie, 2004).

The solubility of phosphorus is also controlled by pH (and redox potential). Once phosphorus is mobilized, some or all of it, depending on the degree of thermal stratification and resistance to mixing, may reach the epilimnion where it can be utilized by phytoplankton for growth. The phosphorus available to algae is known as BAP, or biologically active phosphate (Mackie, 2004).

Estimate the annual means and the minima and maxima of the spring total phosphorus levels in the epilimnion in μg/L over a minimum of two years, but preferably five years to more accurately reflect variations in mean dissolved oxygen levels that occur in the lake. The measurements can be made using published standard methods or estimated using Secchi depth regressions (see Section 3.2.5.2). The species may have lower and upper threshold P_T levels for survival. A minimum amount of algae is needed for food and this amount will be determined by the spring $[P_T]$. There is upper limit of food, and hence of $[P_T]$ as well because too much algae may clog the gills of the mollusk. These lower and upper threshold values may limit the survival and abundance of invasive species of mollusks even if the chalk variables are optimal. If measured spring total phosphorus levels in the epilimnion are available as μg/L compare the values obtained to the upper and lower threshold P_T levels for the species (e.g., if zebra mussels, use Table 3.7). The assessment based on total phosphorus applies only if the phosphorus is the limiting nutrient for algal production (i.e., N_T:P_T ratios are <10–17). Otherwise, use the total nitrogen values as described in the next section.

3.2.4.2 Using Total Nitrogen as a Limiting Nutrient Variable

Nitrogen is the most abundant gas in the atmosphere, comprising a total of 78% of the volume. Approximately 50% of the dissolved gas in freshwater is nitrogen. The upper layers of water are usually saturated with nitrogen. The primary source of nitrogen is the atmosphere, with some nitrogen provided through the nitrogen cycle by algae and bacteria. Many organisms require nitrogen to build protein but they cannot use the free nitrogen in the atmosphere and water directly. The nitrogen must first be fixed into usable forms by specialized bacteria. The major forms of nitrogen available to autotrophs (bacteria, fungi, algae) are ammonium and nitrate. Nitrogen can be a growth-limiting factor to plants in some aquatic systems, especially those that are extremely oligotrophic or extremely eutrophic (Mackie, 2004).

The main inputs of nitrogen into lakes are of autochthonous and allochthonous origin. Losses occur by uptake by autotrophs (bacteria, fungi, algae, etc.), deposition to the sediments and losses of gaseous nitrogen to the atmosphere. The allochthonous sources are the atmosphere, edaphic supplies (includes soil elements in runoff and groundwater), and influents (parent streams). The autochthonous sources are everything produced within the lake, including remobilization of elements from the sediments, metabolic and excretory products from organisms, chemical transformations of soluble and insoluble compounds that occur in the water, the sediments, or at the sediment water interface, leaching of elements from living and moribund organisms, and so on. Allochthonous sources predominate for some elements, like phosphorus and sulfur, most of which ends up on the bottom of the lake, unavailable to autotrophs. Allochthonous sources contribute to the nutrient pools for carbon, nitrogen, and silica, but autochthonous sources are able to recycle and maintain the nutrient supply. Biologically active compounds are any elements or compounds that can be used and assimilated (digested) by autotrophs. For some elements (e.g., phosphorus and sulfur), the lake bottom is a sink and the elements accumulate in the sediments (Cole, 1994; Mackie, 2004).

The different processes that occur within the nitrogen cycle are described by Mackie (2004). Proteins are added to the autochthonous supply of nitrogen upon death and decay of organisms, and with the excretory products of organisms contribute to the ammonium pool, mainly as the dissociated ion, NH_4^+. The ammonium ion is useable by plants, but ammonia (NH_3) is very toxic. Most of the ammonia (NH_3) dissolves immediately in water to form ammonium hydroxide (NH_4OH), which dissociates to give ammonium (NH_4^+) and hydroxyl (OH^-) ions. Undissociated ammonium hydroxide is also toxic to organisms, the amount and toxicity of which increases with increasing pH and temperature and decreasing dissolved oxygen content.

Ammonium is oxidized to form nitrate by nitrifying bacteria, first by *Nitrosomonas* and then the by *Nitrobacter*. In eutrophic lakes, nitrate produced in the oxygenated epilimnion can move down to the anoxic hypolimnion to be denitrified in the sediments by denitrifying bacteria like *Escherichia*, *Bacillus*, *Pseudomonas*, and *Micrococcus*. Nitrogen (N_2) is produced in the denitrification process and is available for biological nitrogen fixation in the epilimnion by blue-green algae like *Anabaena, Nostoc*, and *Gloeotrichia*, and bacteria like *Clostridium* and *Azotobacter*. Since nitrogen fixation is an anoxic process and algae produce oxygen, the nitrogen fixation process occurs in specialized structures called heterocysts, which consumes the oxygen and prevents nitrogen from entering the cells. Nitrogen fixation can greatly accelerate eutrophication even when nitrate-nitrogen and ammonium-nitrogen are in low supply.

The concentrations and depth profiles for the different forms of nitrogen can be useful criteria for assessing trophic status of a lake. However, there are seasonal differences due to spring and fall turnovers, utilization of nutrient forms of nitrogen (nitrate and ammonium), and ice formation during the winter, which restricts photosynthesis and diffusion of oxygen into the water. Eutrophic lakes with anoxia in the hypolimnion will have either high levels of ammonia, methane, or hydrogen sulphide that is toxic to all mollusks.

Estimate the annual means and the minima and maxima of the spring total nitrogen levels in the epilimnion in μg/L over a minimum of two years, but preferably five years to more accurately reflect variations in mean dissolved oxygen levels that occur in the lake. The measurements can be made using published standard methods. The species may have lower and upper threshold N_T levels for survival. A minimum amount of algae is needed for food and this amount will be determined by the spring [N_T]. There is upper limit of food, and hence of [N_T] as well because too much algae may clog the gills of the mollusk. These lower and upper threshold values may limit the survival and abundance of invasive species of mollusks even if the chalk variables are optimal. If measured spring total nitrogen levels in the epilimnion are available as μg/L, compare the values obtained to the upper and lower threshold N_t levels for the species (e.g., if zebra mussels, use Table 3.7). If ammonia (unionized NH_3) measurements are available for the hypolimnion, compare them to the lethal threshold value for each species. There are tables (e.g., Table 12.1A in Mackie, 2004) that can be used to calculate the percentage of unionized ammonia in an aqueous ammonia solution at different temperature and pH conditions. For example, at 20°C and a pH of 8.0, a total ammonia concentration of 500 μg/L would give an unionized ammonia concentration of $500 \times 3.8/100 = 19$ μg/L, which is slightly less than the unionized ammonia objective of 20 μg/L that is toxic to most aquatic organisms (Mackie, 2004). The assessment based on total nitrogen applies only if the nitrogen is the limiting nutrient for algal production. Otherwise, use the total phosphorus values as described above.

3.2.5 Secchi Depth

The Secchi depth is a measure of the relative depth of light penetration, or its relative transparency. It is a relative measurement because the readings usually vary from one individual to the next, from one day to the next, and even from one hour to the next. Factors such as cloud cover, surface water conditions (smooth vs. rough), and whether you take the measurement from the shaded side or the sunlit side of the boat will affect the measurement. Normally, the measurement is taken at a similar time of day from the shaded side of the boat. Yet, in spite of all these variables, there is an excellent correlation between Secchi depth and trophic status variables of a lake.

The Secchi disk is a 20 cm in diameter disk, with black and white quadrats (Figure 3.3). The disks are easy to make out of 4-mm (1/4-in) thick sheet of either clear or opaque acrylic available from most hardware stores; the acrylic comes with paper masking on both sides. Using a 20-cm square piece of paper, fold it in half one way and then again in the other direction. Use a protractor to draw an arc on the flattened end. Cut the arc off with scissors and unfold the paper into a perfect circle that is 20 cm diameter, the center being at the point of the folded paper. Place the circle on the acrylic and draw a circle on the paper masking and mark the center. Drill an 8-mm (5/16-in) diameter hole in the center and cut the circle out using a jig saw. Roughen the surface of one side of the disk with sand paper. Paint the disk with marine flat black paint and let dry 24–48 h. Paint two opposing quadrats with marine white paint and let dry 24–48 h. Thread a nut on to a 7.6-cm (3-in) long eyebolt up to the beginning of the thread, place a lock washer and a flat washer under the nut. Place the eyebolt through the hole in the disk and then place another flat washer and

FIGURE 3.3 A 20-cm diameter Secchi disk.

lock washer and then tighten with another nut. Tie a 10-m (33-ft) long nylon line (not cotton because it shrinks when wet) firmly to the eyebolt. Keeping the line taught and starting from the surface of the disk, mark the line off in 1-m intervals with a black indelible ink pen.

The use of the Secchi disc is quite simple. Merely lower the disc into the water until the white quadrats disappear from view. Note the depth on the rope. Lower the disk a bit further and then raise the disk until it reappears. Note the depth again. The depth at which it disappeared will probably be different from the depth at which it reappeared. Repeat the procedure until the depth at which the disk disappears is the same as the depth at which it reappears. This is the Secchi depth. Readings should be taken weekly until relative transparency (Secchi depth) is minimal. The minimum Secchi depth should be used in the regressions given in Table 3.2.

TABLE 3.2
Some Relationships between Secchi Depth and Chlorophyll *a*, Total Phosphorus, and Color

	Reference
Secchi Depth (Z, m) and Chlorophyll *a* (Chl *a*, µg/L) Relationships	
$\ln Z = 2.04 - 0.68 \ln[\text{Chl } a]$	Carlson (1977)
$Z = -\ln(0.20)$/nonalgae extinction coefficient + (algae extinction coefficient × [Chl *a*])	Lorenzen (1980)
$\ln Z = 2.01 - 0.370 \ln[\text{Chl } a] - 0.278 \ln$ (Organic color mg/L)	Canfield and Hodgson (1983)
$\log_{10} Z = 0.65 - 0.43 \log_{10}[\text{Chl } a]$	Hoyer et al. (2002)
Secchi Depth (Z, m) and Total Phosphorus (P_T, µg/L) Relationships	
$\log_{10} Z = 0.88 - 0.48 \log_{10} P_T$ (Note: use [spring P_T])	Hoyer et al. (2002)
$\ln Z = 3.876 - 0.98 \ln P_T$ (Note: use [average summer P_T])	Carlson (1977)
Secchi Depth (Z, m) and Color (Pl-Co, mg/L) Relationships	
$\log_{10} Z = 0.76 - 0.43 \log_{10}$ Color	Hoyer et al. (2002)

The Secchi depth is often used to estimate several physical, chemical, and biological characteristics of a lake. The following describes its use for estimating levels of chlorophyll *a*, total phosphorus and total nitrogen, suspended solids, and light extinction.

3.2.5.1 Secchi Depth versus Chlorophyll *a*

Secchi depths have been related to summer [Chlorophyll *a*] by Dillon and Rigler (1975) based on data from numerous soft-water lakes in central Ontario; Oglesby and Schaffner (1975) based on several lakes in New York; Carlson (1977) based on data from some north mid-west lakes; and Canfield et al. (1983), Brown et al. (2000), and Hoyer et al. (2002) based on data from several lakes in Florida (Table 3.2). All studies found that a large decrease in Secchi depth was related to a small increase in chlorophyll *a* levels (e.g., 0–<4 μg Chl *a*/L from 10 down to 4–5 m Secchi depth) but a small decrease in Secchi depths was related to large increases in chlorophyll *a* levels (e.g., 5–30 μg Chl *a*/L from 2 down to 1 m Secchi depth, Figure 3.4). The curve in Figure 3.4 was derived from Dillon and Rigler's (1975) regression:

$$\log[\text{Chl } a] = 1.47 \log Z - 1.30,$$

where Chl *a* is in (μg/L) and *Z*, the Secchi depth, is in m. As Table 3.2 shows, there are numerous other regressions that relate chlorophyll *a* levels to Secchi depth and each results in the same kind of curve shown in Figure 3.4 but the regressions differ, some significantly.

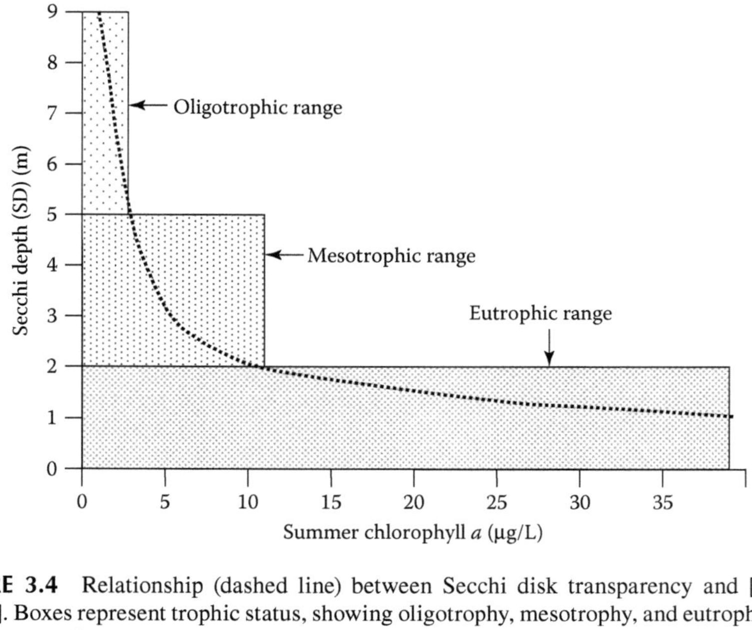

FIGURE 3.4 Relationship (dashed line) between Secchi disk transparency and [chlorophyll *a*]. Boxes represent trophic status, showing oligotrophy, mesotrophy, and eutrophy using the corresponding ranges for [chlorophyll *a*]. (Redrawn from Mackie, G.L., 2004. *Applied Aquatic Ecosystem Concepts*, 2nd edn, Dubuque: Kendall/Hunt Publishing Company.)

3.2.5.2 Secchi Depth versus Nutrient Levels, Color, and Light Extinction

The N_T:P_T ratio is not the only factor that affects the accuracy of predictions from Secchi depth and chlorophyll *a* regressions. In freshwater lakes, if the chlorophyll response per unit of phosphorus is below predicted averages, nitrogen may be the nutrient that most limits chlorophyll concentrations and especially when the total nitrogen to total phosphorus ratio (by weight) falls below 10 (Sakamoto, 1966), 12 (Dillon and Rigler, 1974), or 17 (Smith, 1982). High concentrations of inorganic suspended solids and color limit light transmission and have been shown to suppress chlorophyll concentrations per unit of phosphorus in freshwater lakes (Hoyer and Jones, 1983; Canfield and Hodgson, 1983; Hoyer et al., 2002—see Table 3.2). In fact, Armengol et al. (2003) found a weak or no correlation between chlorophyll *a* levels and Secchi depth, either when the whole studied period was analyzed or, for most years, when they were considered individually. Armengol et al. (2003) attributed the cause of the low contribution of chlorophyll *a* to water transparency to the large amounts of detrital matter present, due both to the inputs of allochthonous particulate organic matter and sediment mobilization from the bottom or the shores during the dry periods. They did find highly significant relationships between Secchi depth (Z_{SD}) and the light extinction coefficient (k_T) and total suspended solids (SS_T):

$$k_T = 1.36 Z_{SD}^{-0.806},$$
$$SS_T = 9.61 Z_{SD}.$$

In lakes where light transmission is affected more by silt or other nonalgal particles, Lorenzen (1980) found that Secchi depth–chlorophyll *a* relationships collapsed and incorporated extinction coefficients to account for both algal and nonalgal particles (see Table 3.2). He developed the regression

$$Z = \frac{-\ln(0.20)}{\alpha} + \beta C,$$

where α is the extinction coefficient from factors other than algae, β is the incremental extinction coefficient from algae, and C is the algal concentration. If $\beta C \ll \alpha$, the chlorophyll *a* level is low and Secchi depth is largely controlled by other suspended particles. Lorenzen (1980) examined the effect of α on Secchi depth using α values ranging from 0.05 to 1.0 m^{-1}, with very clear water having a value of 0.05 and turbid water ~40 μg Chl *a*/L, and a value of 0.02/m/μgChl *a*/L for β.

Cecala et al. (2008) examined the impact of zebra mussels on benthic (epipelic) primary production and water clarity using the relationship:

$$\log_{10} \text{Secchi} = -0.5203 \times \log_{10} k + 0.0487,$$

where *k* is the light attenuation coefficient. Section 3.3.1.1.2.4 (see also 6.2.1.2) discusses the impact of zebra mussels on benthic primary production.

Scavia and Chapra (1977) in comparing different ecological models recommend that the use of a particular type of model is governed by the nature of the problems being addressed rather than in the inherent superiority of one model over another. This emphasizes the need to understand the N_T:P_T ratios and which nutrient is limiting the primary production and the relative contributions of color, suspended solids, and algal cells to light transmission in a lake before selecting the appropriate model(s) to predict levels of chlorophyll *a*. In the final analysis, Secchi depth should be used to estimate nutrient levels only if they have not been measured empirically. Risk assessments based on Secchi depth alone is not recommended because (1) of the inherent errors in its measurement, (2) there is no correlation between Secchi depth and calcium levels (e.g., both low- and high-calcium lakes can have shallow Secchi depths), and (3) Secchi depth is a better predictor of a lake's trophic status than of population densities of aquatic invasive mollusks.

3.2.5.3 Trophic State Index

While the trophic state index (TSI) is not often used in risk assessments it has been examined as a variable influenced by the filtering activities of dreissenids. Stoermer (2007) used Carlson's TSI to examine the effect of zebra mussels on trophic state by comparing lake TSI values pre- and post-zebra mussel invasion and the influence lake morphometry had on the zebra mussels' effect on trophic state. Her results suggest that the zebra mussel's influence on trophic state variables were minor at this early stage in the invasion of these inland lakes. In contrast, Welling et al. (2001) did find that zebra mussel influenced the TSI, especially chlorophyll *a*, and Secchi depth. The TSI is introduced here because it may have potential in helping to explain levels of mussel infestations.

Carlson (1977) developed a TSI using Secchi depth, the average summer chlorophyll *a* level, or the average summer total phosphorus level (note: the index uses average summer values, not peak summer or spring values as used for risk assessments). The traditional concepts of lake enrichment are a slow progression from oligotrophy to eutrophy in a continuum over eons of time (Mackie, 2004). However, there is a tendency to classify lakes into three categories—oligotrophy, mesotrophy, and eutrophy—or even further into ultraoligotrophy to hypereutrophy. The determinations of trophic state are made from examination of several criteria, such as depth profile of the oxygen curve, concentrations of nutrients, various measures of biomass or production, and species composition of the bottom fauna or of the phytoplankton. However, Carlson (1977) pointed out that the changes do not occur at sharply defined places, nor do they all occur at the same place or at the same rate; some lakes may be considered oligotrophic by one criterion and mesotrophic by another. To resolve this dilemma, Carlson (1977) developed the TSI whose units increased by 10 from 0 up to 100 with each unit-10 increase correlating with Secchi depth, average summer chlorophyll *a* content, and average summer total phosphorus content (Table 3.3). Note that there is a halving of the Secchi depth and a doubling of the chlorophyll *a*

TABLE 3.3
Trophic State Index Values and the Associated Average Summer Surface Values for Secchi Depth, Chlorophyll *a*, and Total Phosphorus

TSI	Secchi Depth (m)	Surface Chlorophyll *a* (μg/L)	Surface Total Phosphorus (μg/L)
0	64	0.04	0.75
10	32	0.12	1.5
20	16	0.34	3
30	8	0.94	6
40	4	2.60	12
50	2	6.40	24
60	1	20	48
70	0.5	56	96
80	0.25	154	192
90	0.12	427	384
100	0.062	1183	768

and total phosphorus concentrations for every unit TSI increase of 10. The TSI relationships used for estimating the values of each variable are

$$\text{TSI} = 60 - 14.41 \ln Z(\text{m}),$$
$$\text{TSI} = 9.81 \ln \text{Chl}\,a\ (\mu\text{g/L}) + 30.6,$$
$$\text{TSI} = 14.42 \ln \text{P}_\text{T}\ (\mu\text{g/L}) + 4.15.$$

Note: The values for chlorophyll *a* and total phosphorus are average summer values. In order to equate the TSI with the traditional classifications into oligotrophy, mesotrophy, and eutrophy, Carlson (1977) used $[\text{Chl } a]_{\text{ave. summer}} < 1$ μg/L for oligotrophy, $[\text{Chl } a]_{\text{ave. summer}} = 1$–6.4 μg/L for mesotrophy, and $[\text{Chl } a]_{\text{ave. summer}} > 6.4$ μg/L for eutrophy.

3.2.6 Temperature

The annual temperature cycle of lakes in the temperate zone is described in many texts, including Wetzel (1975), Cole (1994), Horne and Goldman (1994), and Mackie (2004) and is only briefly summarized here. Not only lakes in the temperate zone stratify, but also lakes in southern latitudes and high altitudes. Many of these lakes turn over only once a year whereas lakes in northern latitudes turn over twice a year.

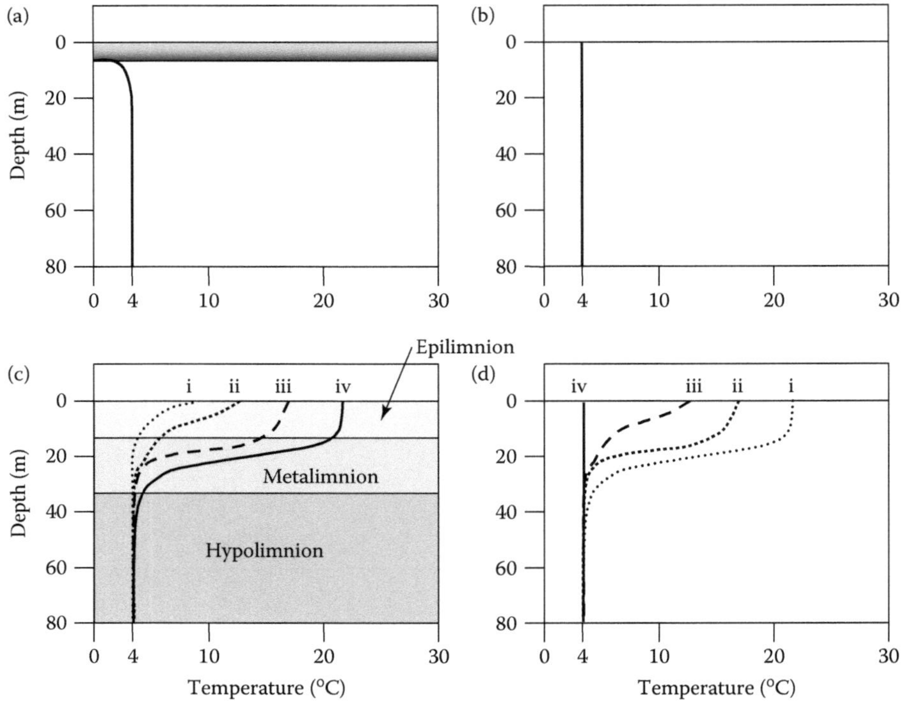

FIGURE 3.5 Annual temperature cycle. (a) Thermal inversion under ice. (b) Spring turnover. (c) Gradual formation of summer stratification beginning in early spring (i), late spring (ii), early summer (iii), and late summer (iv). (d) Gradual erosion of stratified layers beginning late summer (i), early fall (ii), late fall (iii), and fall turnover (iv).

The most important relationship in the annual temperature cycle is between density and temperature where water gets heavier as it is cooled, until 4°C at which point freshwater is most dense. Water gets lighter with further cooling below 4°C, until 0°C when ice forms and floats; the low-density surface layer just under the ice is at 0°C and sits on top of a dense layer near 4°C (Figure 3.5a). Normally, warm water sits on top of cold water but during winter under the ice, cold water (0°C) sits on top of warm(er) water (4°C) and the condition is known as "thermal inversion" (i.e., inverted from the summer norm).

During the spring, daylight hours increase and the amount of solar radiation and heating at the earth's surface increases causing the ice to melt from the surface down, the water temperature remaining close to 0°C. There is a brief period when all the ice has melted, the surface layer of water warms to 4°C, becoming more dense and sinks to mix with the water below that is already at 4°C. At this point the entire water column from top to bottom is at 4°C and the lake is in its "spring turnover" state (Figure 3.5b). With sufficient wind forces, water is capable of mixing at all depths.

As summer approaches the surface layers of the lake are rapidly heated and become much warmer, and lighter, than the water below. Because water is heated from the surface down, only the upper one meter or so can be heated by the red and infrared light waves. The clearer the water, the deeper the red and infrared light can penetrate and the deeper the heating process can occur. But the red light is absorbed in the first meter of water winds are needed to mix the warmer, lighter water of the surface layers with the cooler, heavier water below. Strong, frequent winds will mix the warm surface water into the deeper, colder water than weak and infrequent winds. The heating and mixing processes continue and eventually result in the formation of three thermal layers in the lake: an upper, warm layer called the "epilimnion," a lower, cold layer known as the "hypolimnion," and an intermediate layer known as the "metalimnion," often referred to as the "thermocline" (Figure 3.5c). The metalimnion is the zone of rapid change in temperature, more than 1°C per meter advance in the water column. In some lakes, the metalimnion may be extremely thin, as thin as 8–10 mm and the thermal gradient is very steep.

The temperature of the hypolimnion in deep northern temperate lakes is usually close to 4°C, but in shallow lakes or deep lakes in the southern and western temperate regions where wind can mix warmer water of the epilimnion with the deeper water in hypolimnion, the hypolimnion may be heated to 7–10°C, or whatever the temperature was before the waters began to stratify. Thermal stratification during the summer is referred to as "direct stratification."

As fall approaches and the air gets colder, the water begins to cool from the surface down. The cold water becomes heavier and sinks, gradually eroding the epilimnion and the thermocline [Figure 3.5d(i)–(iv)]. Eventually, the stratified thermal layers disappear, the entire destratification process occurring within the epilimnion and the metalimnion. The hypolimnial waters do not become part of the destratification process until the water above it approaches 4°C. At this point, the lake is in its "fall turnover" period and water mixes at all depths [Figure 3.5d(iv)].

Lakes with both a spring and fall turnover are called dimictic (two mixes). Some lakes do no freeze and the fall turnover period extends right through to the spring. In these lakes the water temperatures do not fall below 4°C and are called warm monomictic (one mix) lakes. Lake Superior is usually a warm monomictic lake. Lake Ontario may be dimictic one year and warm monomictic another. Most lakes in tropical and temperate climates are warm monomictic lakes. Lakes in polar regions that are permanently frozen and the water never mixes are amictic (no mix). The ice on some northern lakes may thaw but water temperatures never exceed 4°C and are called cold monomictic lakes. Many lakes at high altitudes near the equator are poorly stratified because they are constantly mixing; such lakes are polymictic.

Wilbur and Yonge (1983a,b,c) describe the importance of temperature to the physiology and development of mollusks. All freshwater mollusks have upper and lower lethal thermal limits and the physiological processes of many species are triggered by a specific temperature. Metabolic rates generally follow the Q_{10} principle, which states that rates double for every increase in 10°C (Prosser and Brown, 1962). In general, development and growth rates are directly related to temperature. Initiation of spawning generally requires warmer water than initiation of growth but gametogenesis may occur all year, even in winter.

Water temperature should be measured at 1-m intervals to the bottom, or if a deep lake, at least every 5–10 m in the hypolimnion. Temperature should be recorded every two weeks at several locations over a minimum of two years, but preferably five years to more accurately reflect variations in mean temperature that occur in the lake. The effects of extremes in upper and lower temperatures on survival, growth, and reproduction of a species should be examined. Lower and upper thermal threshold values may limit the survival and abundance of invasive species of mollusks even if the chalk variables are optimal.

3.2.7 Conductivity, Dissolved Solids, and Salinity

3.2.7.1 Conductivity

Intuitively, there should be upper and lower threshold levels of conductivity. Waters with too few ions, such as mountainous lakes filled with snowmelt runoff, will have too little calcium to support growth and reproduction. The upper threshold limit, however, would be in the salinity range, as discussed below.

The terms conductivity, specific conductance, and electrical conductance are often used interchangeably. Parker (1994) defines conductivity as "The ratio of the electrical current density to the electric field in a material. Also known as electrical conductivity; specific conductance." Aquatic biologists tend to use both conductivity and specific conductance. Specific conductance of a solution is calculated as the product of the measured conductivity and the electrode cell constant *d*/*A*, where *A* is the area of two electrodes (1 cm^2 each) and *d* is the distance between them, usually 1 cm. It is the inverse of resistance, the unit of which is ohms. Being the inverse of resistance, ohms was reversed to mhos and the unit of specific conductance was once reported as μmhos/cm. It is now most commonly reported for freshwater as μSiemens/cm (μS/cm), where μ1 S = 1 μmho [or 1 millisiemen (mS) = 1/1000 S, or 1 microsiemen (μS) = 1/1,000,000 S]. Temperature affects the flow of electrons, and therefore the specific conductance, the conductivity of a solution increasing with its temperature. Most meters with fixed temperature compensation (TC) that are used for freshwater measurements assume about a 2% increase in conductivity with each 1°C increase in temperature. Conductivity meters with a fixed TC should be calibrated to the conductivity value at the meters normalization temperature, usually 25°C. Calibration to values other than the normalization temperature would only be appropriate if the meter has no capability for temperature correction, or if the TC was adjusted to 0.0%. As a general rule, it is always best to calibrate and measure as close to 25°C as possible when a TC is applied. It is good practice to record the temperature during calibration.

3.2.7.2 Total Dissolved Solids

The conductivity of a freshwater solution is a measure of all the cations (e.g., calcium, iron, magnesium, and sodium) and anions (e.g., chloride, nitrate, sulfate, and phosphate) dissolved in the water. Hard waters of calcareous watersheds have higher conductivities than soft waters of granitic watersheds. The conductivity of the water can also be raised by mine effluents that often contain cations such as iron, sulfate, copper, cadmium, or arsenic; agricultural runoff that is often high in anion concentrations of phosphate and nitrate; and road runoff that can contain calcium, sodium, and magnesium salts used for deicing or keeping dust down, automobile fluids, and brake linings (Mackie, 2004).

The total concentration of cations and anions dissolved in water is expressed in mg/L as total dissolved solids (TDS). TDS is determined by taking a known quantity of water (usually 100 mL), filtering it to remove all the suspended solids and then evaporating the filtrate at 105°C in a tared (preweighed) container until a constant weight is obtained. The materials left on the bottom of the beaker are solutes that precipitated out as the water evaporated. The amount of water needed for a TDS measurement depends on the concentration of dissolved solids; the lower the TDS level, the more the water needed to detect the small amounts present (Mackie, 2004).

The suspended solids retained by the filter paper can be dried, at 60°C, and weighed to provide a total suspended solids (TSS) estimate, in mg/L. Both the TDS and TSS fractions contain inorganic and organic residues. The organic fraction can be removed simply by burning or ashing in a furnace at 500–550°C. The material remaining after ashing at 500–550°C is the inorganic fraction (e.g., Ca, Mg, Fe), and the loss in weight is the organic fraction (Mackie, 2004). Glass-fibre filter papers must be used when ashing for organic and inorganic weights of TDS and TSS.

Since conductivity is related to TDS, which is the total amount of cations and anions dissolved in the solution, a much easier method for measuring TDS is to determine the specific conductance, or conductivity, of the water sample and multiply by 0.64, as defined by

$$\text{TDS(mg/L)} = \text{conductivity}(\mu\text{S/cm}) \times 0.64.$$

The relationship is reasonably accurate for freshwaters in the normal range of conductivity (10–500 μS/cm), but the relationship begins to break down with the error increasing above 1000 μS/cm (Table 3.4). Chandhurti et al. (1965) considered the utility of using the factor 0.64 may be in order if the object is only classification of water for irrigation purposes but the exact value of TDS cannot always be obtained by it. Even their regression

$$\log_{10}\text{TDS}\,\text{mg/L} = -0.1601 + 0.9922\log_{10}\text{conductivity}\ (\mu\text{S/cm}),$$

which was derived from analyses of 263 groundwater samples, did not predict the exact TDS value, but the error was similar to that derived from multiplying by 0.64.

TABLE 3.4
Effect of Increasing Conductivity on Factor Value in the Relationship TDS = Conductivity × Factor

Class	Conductivity (μS/cm)	Number of Samples	Percentage Samples	Chaudhurti et al. (1965)	Darra and Mehta (1963)
I	Up to 250	1	0.40	0.60	0.668
II	250–750	15	6.1	0.50–0.89	0.670
III	750–2250	51	21.0	0.50–0.90	0.722
IV	2250–5000	176	72.5	0.40–0.95	0.722

3.2.7.3 Salinity

The levels of TDS of waters in the low salinity (<1% or <10 ppt) range should be determined by titration for chlorinity instead of measurement by conductivity. Chlorinity is the amount of chloride ions in water. It is the standard unit of measurement for salinity in seawater, where

$$\text{Salinity} = 0.03 + 1.805\ (\text{Chlorinity}).$$

The chloride content of seawater is directly proportional to the salinity (Sverdrup et al., 1942). However, in freshwater, the chloride content is not proportional to the salinity (or conductivity) because of the variations in dilutions of various ions. Although the amount of salt may differ among samples of seawater, the ratio of major dissolved components in the seawater is constant. The constancy of proportions of ions in relation to others in the oceans (known as Forchhammer's Principle) (Sverdrup et al., 1942) contrasts greatly with the inconstancy of ion proportions in freshwaters.

In 1978, oceanographers redefined salinity in the Practical Salinity Scale (PSS) as the conductivity ratio of a seawater sample to a standard KCl solution (Lewis, 1980; Unesco, 1981a,b). Commonly referred to as PSS78, its redefinition was based on the ratio of the specific conductance of a sample to the specific conductance of a standard potassium chloride solution at 15°C and standard atmospheric pressure, with a ratio of 1 yielding a salinity of 35 practical salinity units (psu). The ratios have no units, so a psu of 35 does not necessarily equal 35 g of salt per liter of solution.

Salinity is also defined in terms of conductivity and can be derived from specific conductance values using various formulae (e.g., Schemel, 2001) or internet calculators (e.g., http://gaea.es.flinders.edu.au/~mattom/Utilities/salcon.html).

Table 3.5 shows the specific conductance values at four temperatures for a range of salinity values from 0.5 to 10 ppt.

Freshwater is generally considered to be ≤0.5 ppt with brackish water ranging from 0.5 to 30 ppt (Remane and Schlieper, 1971). They classified freshwater species into two groups with regard to salinity tolerance: Stenohaline limnobionts are restricted to 0–0.5 ppt; euryhaline limnobionts to 0.5–>8 ppt. They further divided euryhaline limnobionts into three groups: those of the "first degree," tolerating 0.5–3 ppt; those of the "second degree," tolerating 3–8 ppt; and those of the "third degree," tolerating >8.

3.2.8 Turbidity, TSS

Turbidity is included as a variable for assessing infestation potential because silt-laden water affects the respiration and filtering ability of mussels.

Turbidity is the cloudiness or haziness of water caused by solid particles in suspension. The instrument used for measuring it is called a nephelometer or turbidimeter, which measures the intensity of light scattered at 90° as a beam of light passes through a water sample. Kits such as the 2100P Hach Turbidimeter are used to measure turbidity in nephelometric turbidity units (NTU). NTU has replaced the older unit of Jackson turbidity units (JTU) that used a Jackson Candle Turbidimeter. To convert JTU to NTU, use the relationship of Ross et al. (1982):

$$\text{NTU} = \text{JTU}^{b},$$

where $b = 0.860$ with a 95% confidence interval of ±0.012.

TABLE 3.5
Relationships between Salinity and Specific Conductance

Conductivity (μS/cm)	Salinity (ppt) 10°C	15°C	20°C	25°C
100	0.067	0.061	0.056	0.052
200	0.134	0.119	0.107	0.098
300	0.203	0.179	0.160	0.145
400	0.273	0.240	0.214	0.193
500	0.343	0.302	0.269	0.242
600	0.415	0.365	0.324	0.291
700	0.487	0.428	0.380	0.341
800	0.560	0.492	0.437	0.392
900	0.633	0.556	0.493	0.442
1000	0.706	0.620	0.551	0.493
	Specific Conductance (μS/cm)			
Salinity (ppt)	**10°C**	**15°C**	**20°C**	**25°C**
1.0	1400	1600	1800	2000
2.0	2700	3000	3400	3800
3.0	3900	4500	5100	5600
4.0	5200	5900	6600	7300
5.0	6400	7200	8100	9000
6.0	7500	8500	9600	10,600
7.0	8700	9800	11,000	12,300
8.0	9900	11,100	12,600	13,800
9.0	11,000	12,400	14,000	15,500
10.0	12,200	13,700	15,400	17,000

Top half gives conversions from conductivity to salinity for 100–1000 μS/cm in 100 μS/cm increments; bottom half from salinity to conductivity for 1–10 ppt in 1 ppt increments.

The turbidity in NTU can be estimated from the TSS. Christensen et al. (2001) examined relationships between turbidity (NTU) and TSS at two sites in Little Arkansas River in south-central Kansas and obtained two similar regressions with $R^2 > 0.89$. Averaging the two gave

$$\log_{10} \text{TSS (mg/L)} = 0.899 \log_{10} \text{Turbidity (NTU)} + 0.272.$$

Packman et al. (1999) also examined the relationship between TSS and turbidity but of streams carrying urban runoff in Puget Lowlands of Washington. He obtained a log-linear model:

$$\ln \text{TSS(mg/L)} = 1.32 \ln \text{Turbidity (NTU)} + C,$$

with an $R^2 = 0.96$ and a value for C that was not significantly different from zero for eight of the nine sampled streams. Only one stream had a C value of 0.15 that was significant at $P < 0.05$. However, the regression tends to give much higher values for TSS than Christensen et al. (2001).

Holliday et al. (2003) established relationships between turbidity (NTU) and three fractions of suspended sediments—clay, silt plus clay, and whole soil—from soils in southeastern Piedmont and obtained the following regressions:

$$\text{Whole soil}: \text{Turbidity(NTU)} = 0.4833 \times \text{TSS}^{1.012},$$

$$\text{Silt} + \text{clay}: \text{Turbidity(NTU)} = 1.0283 \times \text{TSS}^{1.0282},$$

$$\text{Clay}: \text{Turbidity(NTU)} = 0.7733 \times \text{TSS}^{0.9336}.$$

If the water contains significant loadings of silt and/or clay from runoff, one of these three regressions can be used to estimate TSS. Note that the exponents are all close to 1.0 and only the intercept values are significantly different. The regression for silt plus clay gives similar TSS values as Christensen et al. (2001).

3.3 ASSESSING SURVIVAL POTENTIAL OF AQUATIC INVASIVE BIVALVES

The foregoing provides background information on variables that are used for assessing potential of a species to: (1) survive its introduction to a new body of water; (2) develop nuisance population densities; and (3) develop nuisance biomasses. Its ability to survive an introduction can be based on the mean levels of chemical, physical, and biological variables that meet the needs of a species' suite of physiological and ecological requirements. However, the temporal and seasonal variations in levels about the mean values can be substantial and may exceed the species' physiological and ecological tolerances for growth, reproduction, and even survival in parts of the lake or during portions of the year. The suites of physiological and ecological variables chosen to make the assessments are: (1) chalk variables (calcium, alkalinity, pH, total hardness); (2) chemical and nutrient variables (dissolved oxygen, total phosphorus, total nitrogen); and (3) physical variables (temperature, conductivity, Secchi depth).

The following discussion is organized according to species, first the bivalves (in the order of Dreissenidae, Mytilidae, and Corbiculidae), then the gastropods (in the order of Ampullariidae, Thiaridae, Bithyniidae, and Hydrobiidae). For each species, the mean levels of variables needed for survival of adults is discussed first, then different ranges in variations in levels of the variables are used to assess its potential to support larval growth and development, moderate levels of infestation or intense levels of infestation. Risk assessments are made with respect to depth and seasons for bivalves but only for seasons for gastropods since the aquatic invasive mollusk species tend to be shallow water species.

3.3.1 Dreissenidae

3.3.1.1 Zebra Mussel, *Dreissena polymorpha* (Pallas, 1771)

The threshold limits for growth, reproduction, and survival for each of the physiological and ecological variables was determined from averaging the levels reported by the numerous researchers listed in Table 3.6. The ranges in values for each of the risk levels are given in Table 3.7.

TABLE 3.6
Studies Using Various Parameters to Assess Risk Potential of Survival and Potential Density of Zebra Mussels

Variables	Reference
Alkalinity	Claudi and Mackie (1994), Hincks and Mackie (1997), Ashby et al. (1998), Mackie (2004)
Calcium	Mackie et al. (1989), Neary and Leach (1992), Baker et al. (1993), Murray et al. (1993), Claudi and Mackie (1994), Koutnik and Padilla (1994), Tammi et al. (1995a,b), Doll (1997), Hayward and Estevez (1997), Hincks and Mackie (1997), Janik (1997), Sorba and Williamson (1997), Cohen and Weinstein (1998, 2001), Cohen (2001, 2007), Mackie (2004), Jones and Ricciardi (2005), Whittier et al. (2008)
Chlorophyll *a*	Claudi and Mackie (1994), Hincks and Mackie (1997), Mackie (2004)
Conductivity, TDS	Claudi and Mackie (1994), Sorba and Williamson (1997), Mackie (2004)
Dissolved oxygen	Mathews and McMahon (1995), McMahon (1996), Doll (1997), Hayward and Estevez (1997), Sorba and Williamson (1997), Ashby et al. (1998), Johnson and McMahon (1998), Cohen and Weinstein (1998, 2001), Cohen (2001), McMahon and Bogan (2001), Mackie (2004)
Nitrogen, total	Claudi and Mackie (1994), Koutnik and Padilla (1994), Mackie (2004)
pH	Mackie et al. (1989), Neary and Leach (1992), Claudi and Mackie (1994), Koutnik and Padilla (1994), Doll (1997), Hayward and Estevez (1997), Hincks and Mackie (1997), Janik (1997), Sorba and Williamson (1997), Ashby et al. (1998), Cohen and Weinstein (1998, 2001), Cohen (2001, 2007), Mackie (2004),
Phosphorus, total	Mackie et al. (1989), Claudi and Mackie (1994), Koutnik and Padilla (1994), Mackie (2004),
Salinity	Karpevich (1947, 1955), Wolff (1969), Barber (1992), Carlton (1992), Mackie and Kilgour (1992a), Baker et al. (1993), Strayer and Smith (1993), Kilgour et al. (1995), Walton (1996), Wright et al. (1996), Doll (1997), Hayward and Estevez (1997), Sorba and Williamson (1997), Cohen and Weinstein (1998, 2001), Cohen (2001)
Secchi depth, TSS	Claudi and Mackie (1994), Hayward and Estevez (1997), Sorba and Williamson (1997), Armengol et al. (2003)
Temperature (°C)	Strayer (1991), Domm et al. (1993), Spidle (1994), Armistead (1995), Spidle et al. (1995), McMahon (1996), Doll (1997), Hayward and Estevez (1997), Janik (1997), Roe and MacIsaac (1997), Sorba and Williamson (1997), Ashby et al. (1998), Cohen and Weinstein (1998, 2001), Thorp et al. (1998), Ciereszko et al. (2001), Cohen (2001), Cohen (2007)
Total hardness	Baker et al. (1993), Claudi and Mackie (1994), Hincks and Mackie (1997), Sorba and Williamson (1997), Mackie (2004)
Turbidity (NTU), TSS	Ross et al. (1982), Alexander et al. (1994), Packman et al. (1999), Christensen et al. (2001), Holliday et al. (2003)

The variables are listed alphabetically.

TABLE 3.7
Criteria Used in Determining Levels of Infestation by Zebra Mussels in North America

1. Parameter	2. No Potential for Adult Survival	3. Little Potential for Larval Development	4. Moderate Potential for Nuisance Infestations	5. High Potential for Massive Infestations
Chalk Variables				
Calcium (mg Ca/L)	<8	8–15	15–30	30–80
pH	<7.0, >9.5	7.0–7.8, 9.0–9.5	7.8–8.2 or 8.8–9.0	8.2–8.8
Alkalinity, total (mg $CaCO_3$/L)	<30	30–55	55–100	100–280
Hardness, total (mg $CaCO_3$/L)	<30	30–55	55–100	100–280
Nutrient Variables				
Dissolved Oxygen (mg/L) (% saturation)	<3 (<25%)	3–7 (25–50%)	7–8 (50–75%)	≥8 (>75%)
Chlorophyll *a* (mg/L)	<2.5 or >25	2.0–2.5 or 20–25	8–20	2.5–8
Total phosphorus (mg/L)	<5 or >50	5–10 or 35–50	10–25	25–35
Total nitrogen (mg/L)	<75 or >750	75–150 or 525–750	150–375	375–525
Secchi depth (m)	<1, >8	1–2, 6–8	4–6	2.0–4.0
Physical Variables				
Temperature (°C)	<10, >32	26–32	10–20	20–26
Conductivity (μS/cm)	<30	<30–60	60–110	≥110
TDS (mg/L)	<20	20–40	40–70	>70
Salinity (mg/L) (ppt)	>10	8–10 (<0.01)	5–10 (0.005–0.01)	<5 (<0.005)
Turbidity (NTU)	>80	20–80	5–20	<5
TSS (mg/L)	>96	28–96	8–28	<8

Use column 3 for assessing potential for survival (but little potential for nuisance infestation). Columns 4 and 5 indicate good potential for long-term survival and the criteria are used to assess temporal and seasonal variations in infestation potential. TSS was estimated using $\log_{10}$ TSS (mg/L) = 0.899 $\log_{10}$ Turbidity (NTU) + 0.272.

TABLE 3.8
Surface Water Quality Data of a Lake in Arizona, with Parameters Listed in Alphabetical Order

Variable	Mean Value	*N*	Risk Potential
Alkalinity, total mg ($CaCO_3$/L)	104.9	2	High
Calcium (mg/L)	30.4	8	High
Chlorophyll *a* (μg/L)	23.0	4	Little
Conductivity (μS/cm)	104	5	High
Dissolved oxygen (% saturation)	81.7	8	High
Dissolved oxygen (mg/L)	7.8	8	High
Nitrogen, total (μg/L)	490	5	None
pH	8.2	2	High
Secchi depth (m)	1.2	3	Little
Temperature (°C)	17.9	16	High
TDS (mg/L)	72	5	High
Total hardness (mg $CaCO_3$/L)	134.6	3	High
Total phosphorus (μg/L)	55	4	None
TSS (mg/L)	110	5	NA
Turbidity (NTU)	92	6	None
Overall likelihood of water quality supporting a zebra mussel infestation			Little

Data are averaged from up to eight collection dates with some parameters not being measured on all dates. *N* represents number of dates that were averaged. The risk potential for supporting a zebra mussel population (last column) is derived from Table 3.7.

In order to illustrate the assessment of survival and infestation potentials by zebra mussels, a hypothetical lake with the characteristics given in Table 3.8 is provided. The spatial and temporal data in Figures 3.6–3.12 are taken from different lakes in Ontario and United States.

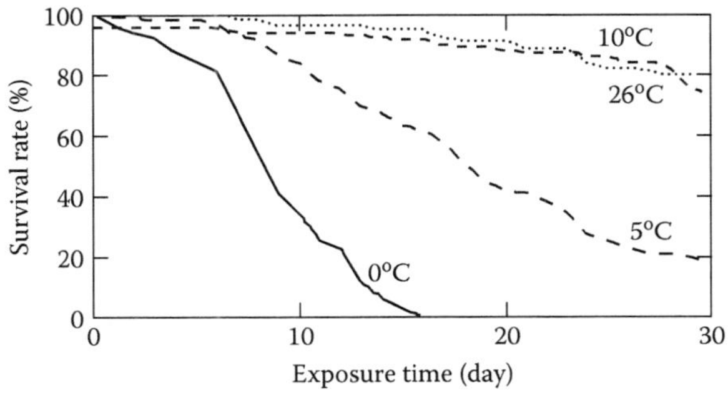

FIGURE 3.6 Survival rate of *L. fortunei* at temperatures of 0–1°C, 5–7°C, and 10–12°C. The lines are means of six replicates, except for 0°C which had nine replicates. Each replicate consisted of 20 mussels in a 600-mL beaker. Water in the beaker was replaced every week.

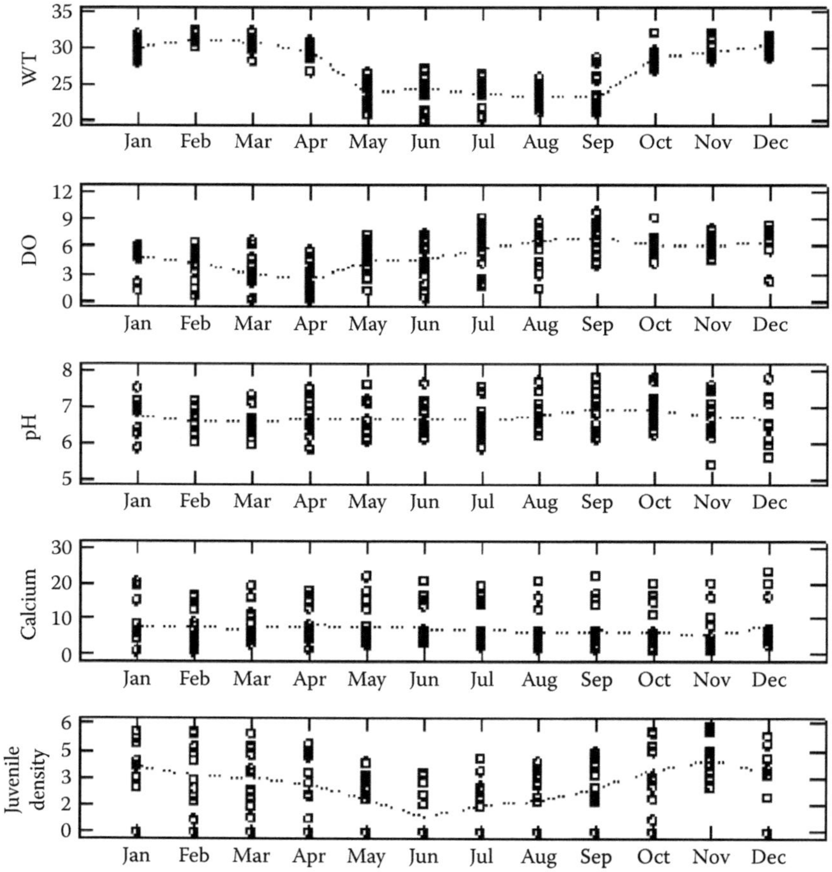

FIGURE 3.7 Seasonality of limnological variables and *L. fortunei* juvenile densities in the Paraguay River within the Brazilian Pantanal between February 2004 and November 2007, arranged by month of sampling.

3.3.1.1.1 Survival Potential of Zebra Mussels Based on Chalk Variables

3.3.1.1.1.1 Calcium Of the four variables, calcium concentration is considered by most researchers as the key factor for assessing the potential of zebra mussel colonization in North America, particularly based on the thorough reviews of Cohen and Weinstein (2001) and Cohen (2005). In North America the minimum level for survival of adults appears to be between 8 and 9 mg Ca/L (Hincks and Mackie, 1997; Kozlowski, 1997). Veligers appear to require 11–12 mg Ca/L to survive for at least a few days (Sprung, 1987; Hincks and Mackie, 1997) and 15–22 mg Ca/L to develop (Hincks and Mackie, 1997; Baldwin et al., 1997). There is considerable disagreement on the level of calcium required for establishment of a population of zebra mussels, particularly because of differences in reported values between the European and North American literature. Ramcharan et al. (1992), Karatayev (1995),

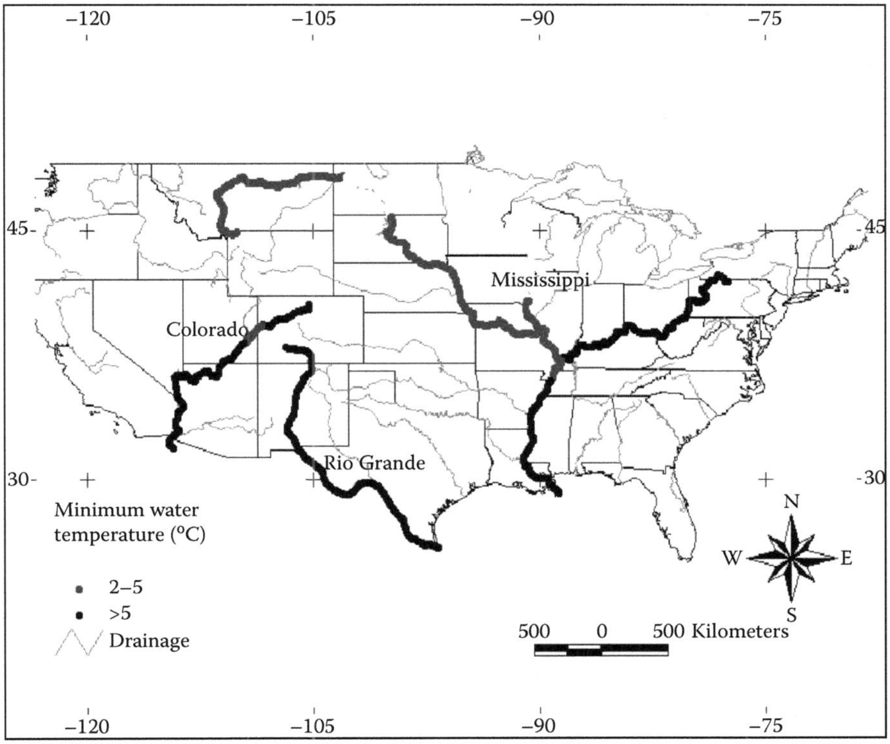

FIGURE 3.8 Spatial variation of water temperature (°C) in Mississippi, Ohio, and Missouri rivers, Rio Grande and Colorado rivers. Data were taken from U.S. Geological Survey's NASQAN program (description available at http://pubs.usgs.gov/wri/wri014255/). Map was prepared in Arc View 2.3 Software using geographic coordinates and mean water temperature at 23 sites, from December to February 1995 to 2000.

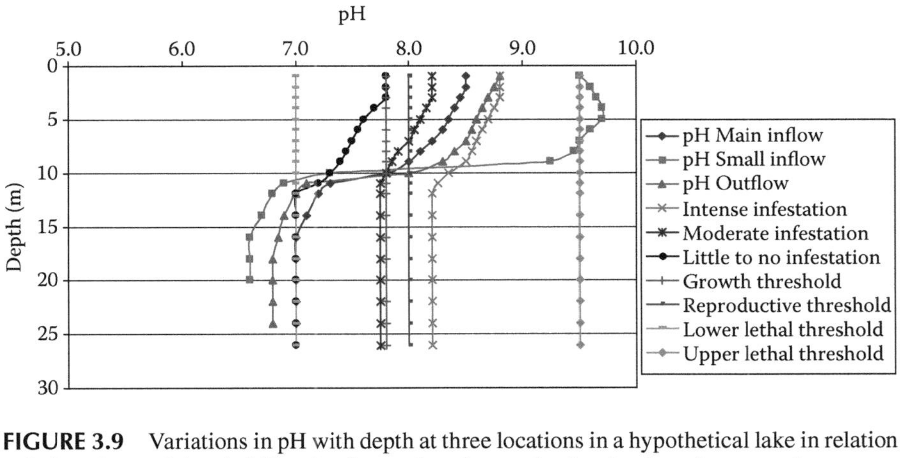

FIGURE 3.9 Variations in pH with depth at three locations in a hypothetical lake in relation to infestation and threshold levels of growth and reproduction for the zebra mussel.

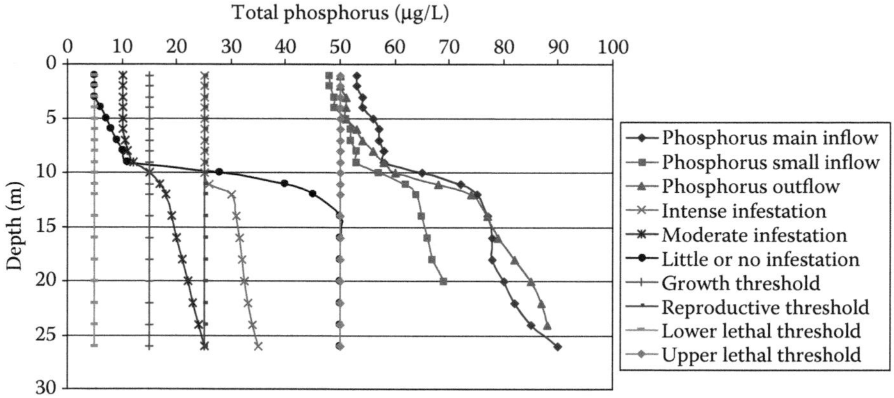

FIGURE 3.10 Variations in total phosphorus with depth at three locations in a hypothetical lake in relation to infestation and threshold levels of growth and reproduction for the zebra mussel.

and Padilla (1997) concluded that zebra mussel needed 25–28 mg/L of calcium to become established. However, there are exceptions, which may or may not be significant; Mikheev (1964) reported for European populations in excess of 500,000 m^{-2} in water with 12 mg Ca/L and a pH of 6.6. In North America, veligers of zebra mussels have been reported in as low as 4–7 mg Ca/L (Hincks and Mackie, 1993, 1997; Kraft, 1994; Baldwin et al., 1997; De Lafontaine and Cusson, 1997; Eliopoulos and Stangel, 1997, 1998, 1999, 2000, 2001, 2003; Stangel, 2004; Stangle and Shambaugh, 2005) but it is not likely that large adult populations, or even any adults, have become established. Most authorities agree that calcium levels above 28–30 mg/L will support intense infestations, and Lakes Ontario (Ca = 39 mg/L), Erie (Ca = 37 mg/L) (Hincks and Mackie, 1997), and Michigan (Ca = 32 mg/L) (Mackie, 2004), and in

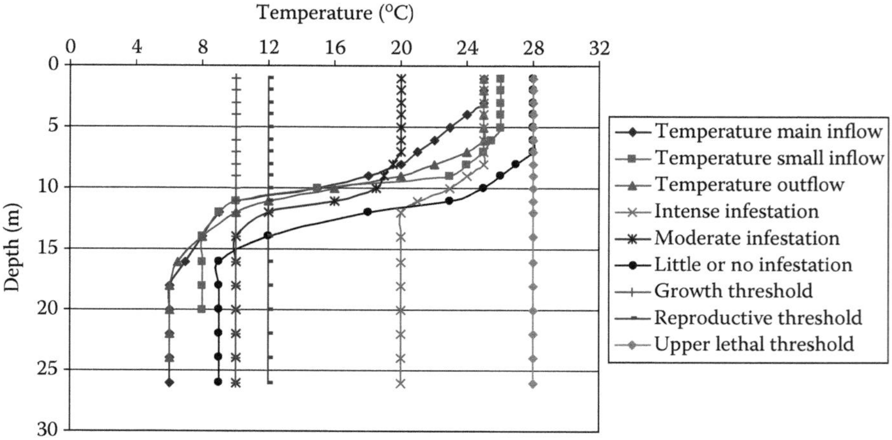

FIGURE 3.11 Variations in summer water temperature with depth at three locations in a hypothetical lake in relation to infestation and threshold levels of growth and reproduction for the zebra mussel.

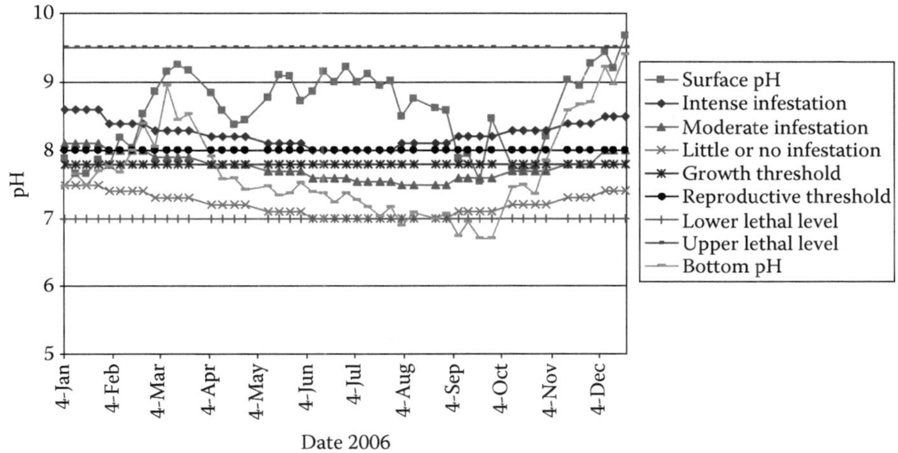

FIGURE 3.12 Average monthly pH at surface (1 m) and bottom in a hypothetical lake in relation to mussel infestation and threshold levels.

the Rideau River (40–45 mg Ca/L) (Martel, 1995) are testimonials to this criterion. Lake Huron (Ca = 23 mg/L) (Mackie, 2004) supports smaller densities and Lake Superior (Ca = 12–15 mg/L) (Beeton and Chandler, 1963; Goldman and Horne, 1983; Balcer, 1996a) supports depauperate populations of zebra mussels (Cohen and Weinstein, 2001). Occurrence and abundance of zebra mussels are high and unrelated to calcium levels above 30 mg/L (Stanczykowska, 1977).

This is not the case with veligers. Studies by Nierzwicki-Bauer et al. in 2000 documented that adult mussels are able to survive in Lake George water (Ca = 12 mg/L, pH = 7.15) but veligers fail unless both calcium and pH levels are raised. The study placed healthy veligers, up to two weeks of age in Lake George water. There was 100% mortality of the veligers within one week.

Experience from Lake George in New York State suggests what the borderline conditions for calcium and pH may be for zebra mussels. Hansen et al. (1998) document that calcium of 10.68 ± 1.17 mg/L and pH of 7.56 ± 0.18, both 10 year averages, have supported a very low level of zebra mussel presence in Lake George for the last 10 years. A confounding factor in Lake George is the presence of scattered limestone deposits, which may be creating microzones for dreissenid survival. This could mean that without these microzones both the calcium and the pH might not be adequate for long-term survival.

The above discussion centers around the minimum levels (>10–12 mg Ca/L) needed for survival and growth and the optimum levels (28–30 mg Ca/L) needed for intense infestations. While numerous authors report mussels in lakes with calcium levels between 12 and 28 mg/L, there is little information on the abundances or even life spans of mussels in this calcium range and presents a major enigma for predicting infestation potential of zebra mussel within this range. It is highly unlikely that an all-or-none principle applies (i.e., mussels will prevail only at 28–30 mg Ca/L or not at all), as implied by Ramcharan et al. (1992),

Karatayev (1995), and Padilla (1997) for European lakes. For example, there are three lakes in Ontario (Buckhorn, Opinicon, and Balsam) with calcium levels between 20 and 26 mg/L, showing that intermediate calcium levels do support reproducing populations in Ontario. Whittier et al. (2008) reported established populations at four locations in the Susquehanna River (17 mg Ca/L) and several locations in the Tennessee River (19 mg Ca/L). Higgins et al. (2008) report an estimated population size of 5.5×10^8 zebra mussels (4.9×10^5 kg) in Lower Doon Lake (Ireland) with an alkalinity of 90 mg $CaCO_3$/L (=25 mg Ca/L) and 2.2×10^4 zebra mussels (17 kg) in neighboring Upper Doon Lake with an alkalinity of 72 mg $CaCO_3$/L (=21 mg Ca/L), indicating that there are populations in European lakes with [Ca] < 28 mg/L. These results support Cohen and Weinstein (2001), who suggested 20 mg Ca/L as a functional lower calcium concentration needed for zebra mussels to establish reproducing colonies and agrees with others as well (Benson, 1998; Murray et al., 1993; Drake and Bossenbroek, 2004; Jones and Ricciardi, 2005). Table 3.7 uses a conservative approach and considers no likelihood of long-term survival in waters with Ca < 8.0 mg/L, little likelihood of larval development and hence nuisance infestations between 8 and 15 mg Ca/L and good potential for established populations between 15 and 30 mg Ca/L. *The survival potential of zebra mussels in the hypothetical lake based on calcium levels is high* (Table 3.8). It is worth stating at this point that if calcium level is too low, all other variables are of little consequence.

3.3.1.1.1.2 pH If calcium data are not available, use pH. Zebra mussels have both a lower and an upper lethal threshold to pH. For lower pH thresholds a pH of <7 is generally considered either lethal to zebra mussel veligers and adults, or is insufficient to support a reproducing population based on field or laboratory studies or a preponderance of evidence (Baker et al., 1993a,b; Baker and Baker, 1993; Claudi and Mackie, 1994; Hincks and Mackie, 1997), although some report survival at 6.3–6.8 (based on gill ciliary activity, McCauley and Kott, 1993), 6.5–6.8 (based on literature reviews, McMahon, 1996; Sorba and Williamson, 1997; Doll, 1996). Sprung (1993) found that veligers developed to settlement in the laboratory only at a pH between 7.4 and 9.4, with peak success at around pH 8.4 in 18–20°C. Based on a survey of 76 European lakes, Ramcharan et al. (1992) found that zebra mussels were present only at pH > 7.3. The criteria used in Table 7.6 use a conservative approach and consider no likelihood of long-term survival in waters with pH < 7.0, little likelihood of nuisance infestations between pH 7 and 7.8 and good potential for established populations between pH 7.8 and 8.2. The threshold levels of pH are also probably determined by the activity of spermatozoa, which are inactive at acid pH but become gradually motile when exposed to pH 6.0–9.0 (Ciereszko et al., 2001).

The upper lethal threshold of pH for zebra mussels is based largely on Sprung's (1993) research in which he found no survival of veligers in pH > 9.4. A pH of 9.5 or greater is considered lethal to larvae and adults (Doll, 1997; Hayward and Estevez, 1987) and pH between 9.0 and 9.5 is considered of little potential (Cohen and Weinstein, 1998), or low to moderate potential to support populations over the short term. Table 3.7 use a conservative approach and considers no likelihood of long-term survival in waters with pH > 9.5, little likelihood of nuisance infestations between pH 9.0 and 9.5 and good potential for established populations between pH 8.8 and 9.0. *The survival potential of zebra mussels in our hypothetical lake based on pH is high* (Table 3.8).

3.3.1.1.1.3 Total Alkalinity, Total Hardness Total alkalinity and hardness criteria are provided in the event that calcium and pH data are not available. The alkalinity criteria in Table 3.7 are simply the calcium levels times 3.49 and total hardness is estimated from the regression Calcium (mg/L) = (1.08 + 0.272 total hardness) (mg $CaCO_3$) (see Section 3.2.1.1.1). Criteria were rounded off to the nearest 10 mg/L (e.g., 30 instead of 28 mg/L). The total hardness assumes that all hardness is due to alkalinity (i.e., carbonates) and the criteria may overestimate the risk potential; for example, if total hardness has an additional 20% due to permanent hardness, the criteria should be <42, 42–76, 76–118, >118 mg $CaCO_3$/L for none, little, moderate, intense infestations, respectively. *The survival potential of zebra mussels in our hypothetical lake based on levels of total alkalinity and total hardness is high* (Table 3.8).

3.3.1.1.2 Survival Potential of Zebra Mussels Based on Trophic Variables

3.3.1.1.2.1 Dissolved Oxygen Dissolved oxygen levels can be reported as an absolute concentration (e.g., mg/L) or as percent saturation. The latter is preferred because it accounts for the effect of temperature on the solubility of oxygen (see Section 3.2.2). Therefore, if the absolute concentration is given, the temperature should also be provided. Also adjust for elevation, especially if at high altitudes.

Respiration rates of freshwater bivalves appear to vary tremendously from season to season. For example, oxygen consumption in summer months may be 230–3300% of that during the winter months, depending on species, body size, and temperature range (McMahon, 1991). Alexander et al. (1994) found that the increase in oxygen consumption with a 10°C increase in acclimation temperature (acclimation Q10 or $Q10_{(acc)}$) was 1.96 over the temperature range of 10–26°C in clear water, which agrees closely with data from Garton and Haag (1991) and Quigley et al. (1993). Alexander et al. (1994) suggested that the zebra mussel's seasonal response to temperature is intermediate compared with all bivalves; V_{O_2} at 26°C was 320% higher than V_{O_2} observed at 10°C in clear water. Quigley et al. (1993) reported declines in oxygen consumption rates with an increase in size of mussels and a significant correlation between dry weight and respiration rate in zebra mussels from Lake St. Clair. They also found that oxygen consumption progressively declined with corresponding decreases in dissolved oxygen, demonstrating the inability of zebra mussels to maintain a constant respiration rate when exposed to relatively low oxygen levels (4.8 mg/L).

There is a preponderance of evidence that European and North American zebra mussels are relatively intolerant of hypoxia or anoxia (McMahon, 1996). Mikheev (1964) held European zebra mussels in anoxic water for 37 h at 22°C and found 100% mortality among smaller mussels (1–4.9 mm shell length), survivorship increasing in larger size classes (0% mortality for 20–24.9 mm shell length). McMahon (1996) held mussels from Niagara River in water depleted of O_2 and also found that larger individuals were more anoxia tolerant than smaller individuals. There is a decrease in oxygen consumption rates with increasing size (Quigley et al., 1993) and an increase in survival times with increasing size. Mikheev (1964) and McMahon (1996) suggest that survival times of mussels can be increased by a lowering of their oxygen demand. McMahon (1996) also found that zebra mussel anoxia tolerance increased with decreasing acclimation temperature, survival times ranging from 907 to 1005, 228 to 428, and 53 to 83 h at 5°C, 15°C, and 25°C, respectively. In contrast, freshwater unionacean and sphaeriid bivalves are relatively anoxia/hypoxia tolerant (McMahon, 1991) and Asian clams, *Corbicula fluminea* (Miiller) are approximately twice as anoxia tolerant as zebra mussels (Matthews and McMahon, 1994), making zebra mussels among the least anoxia/hypoxia tolerant of all freshwater bivalves (McMahon, 1996).

The chronic, lower lethal P_{O_2} for North America adult zebra mussels is approximately 32–40 Torr (21–26% or 2.1–2.2 mg/L dissolved oxygen) at 25°C (McMahon, 1996), a value similar to the minimum of 32 Torr (21% or 2.1 mg/L dissolved oxygen) required for larval development at 18°–20°C (Sprung, 1987). Poor anoxia/hypoxia tolerance and O_2 regulatory ability probably restricts zebra mussels to well-oxygenated habitats (Stanczykowska, 1977), and explains why they are depauperate in eutrophic lakes (Stanczykowska, 1984; Mackie et al., 1989), and absent in hypoxic, hypolimnetic waters (Mackie et al., 1989; McMahon, 1990; Claudi and Mackie, 1994).

McMahon (1996) related oxygen consumption rates (V_{O_2}) to individual dry tissue mass (M) as

$$V_{O_2} = aM^b,$$

where "a" and the exponent "b" are constants relating dry tissue mass to V_{O_2}. Values of "b" range from 0.66 to 0.84 in European zebra mussel populations from February through October (Lyashenko and Karchenko, 1969). An average "b" of 0.63 was computed for zebra mussels based on combined data of several European authors (Alimov, 1975; McMahon, 1996). McMahon (1996) computed a value of 0.82 from the data of Quigley et al. (1992) for North American mussels. The average "b" value for zebra mussels is lower than those of other freshwater mollusks, with 0.721 (±0.026) cited by Alimov (1975) for freshwater bivalves in the former U.S.S.R. 0–1.24 cited by Hornbach (1985) for 28 sphaeriid species; and 0.25–1.45 cited by Burky (1983) for a wide variety of freshwater bivalves.

Several studies have recorded the temperature and dissolved oxygen concentrations at which zebra mussels were found living in nature, all of them at elevations <1830 m (6000′). Sprung (1987) concluded that Russian zebra mussel larvae can survive for a short time at 18°C with oxygen at 16–20% of saturation (1.5–1.9 mg/L). Smirnova and Vinogradov (1990) reported 80–85% oxygen saturation (7.1–9.1 mg/L at 10–25°C) as optimal for Russian zebra mussels. Boelman et al. (1997) reviewed the literature and found that zebra mussels are usually found where dissolved oxygen is over 90% of saturation, which is 7.5–10.2 mg/L at 10–25°C in surface waters at elevations, and become stressed at levels of 40–50% of saturation (4.2–5.7 mg/L at 10–25°C). Oxygen concentrations levels as low as 3.2 mg/L have been found in parts of the Illinois River, where zebra mussels are abundant (Kraft, 1994). Karatayev (1995) and Karatayev et al. (1998) reported adult mussels are oxyphylic and exist in waters no <50% saturation with optima above 80–85% the lowest tolerable being

25% (Shkorbatov et al., 1994). Under anoxic conditions, all zebra mussels die within "some" (Karatayev, 1995) to six days at 17–18°C and within three days at 23–24°C (Baker et al., 1993a; Mikheev, 1968). Most studies have used a limit of 3 mg/L (Claudi and Mackie, 1994) to 4 mg/L to assess zebra mussels' potential distribution (Doll, 1997; Sorba and Williamson, 1997; Cohen and Weinstein, 1998; Cohen, 2007).

Based on the above information, Table 3.7 uses a conservative approach and considers no likelihood of long-term survival in waters with dissolved oxygen levels <3 mg/L or <25% saturation, little likelihood of nuisance infestations between dissolved oxygen levels of 3–7 mg/L or 25–50% saturation, a moderate level of infestation between 7 and 8 mg/L or 50–75% saturation and a good potential for intensive infestations above 8 mg/L or >75% saturation. *The survival potential of zebra mussels in the hypothetical lake based on dissolved oxygen levels is high* (Table 3.8).

3.3.1.1.2.2 Chlorophyll a The European literature indicates that the trophic status of surface waters often dictates the presence and abundance of zebra mussels. Growth rates correlate well with chlorophyll *a* levels in the range 10–40 μg/L and not at all in the range 34–106 μg Chl *a*/L (Smit et al., 1993). Apparently the species composition of algal foods is as important as abundance in some lakes and rivers (Dorlego, 1993). Cyanobacteria in particular are selectively shunted into pseudofeces and thereby gain an advantage over other phytoplankton following invasions (e.g., Arnott and Vanni, 1996; Vanderploeg et al., 2001; Raikow et al., 2004; Bierman et al., 2005). Bastviken et al. (1998) and Vanderploeg et al. (2001) showed that chlorophytes are also a preferred food of zebra mussels.

In general, invasions by zebra mussels are quickly followed by increases in water clarity (e.g., Secchi depth) (Barbiero and Tuchman, 2004; Fahnenstiel et al., 1995; Idrisi et al., 2001) and a decrease in phytoplankton abundance (Lavrentyev et al., 1995; Caraco et al., 1997, 2006; Idrisi et al., 2001; Barbiero et al., 2004, 2006; Depew et al., 2006; Hall et al., 2007). In stark contrast De Stasio et al. (2008) found, after four years of postinvasion by zebra mussels in Green Bay, Lake Michigan, that (1) Secchi depth did not significantly change; (2) chlorophyll *a* levels were significantly greater in the postinvasion period; (3) phytoplankton biovolume was significantly higher following invasion; and (4) there was a significant shift to higher and more frequent dominance of the phytoplankton community by cyanobacteria (blue-green algae), also seen by Sarnelle et al. (2005). They attribute these changes to selective feeding and increased nutrient recycling by dense (e.g., >250,000 m^2) and scattered populations of zebra mussels (Reed et al., 2004) under high turbidity conditions, as has been observed elsewhere (e.g., Arnott and Vanni, 1996). A lack of correlation between Secchi depth and chlorophyll *a* is not unusual in highly turbid waters because high suspended sediment loads have been shown to affect feeding and energetics of zebra mussels (Alexander et al., 1994) (see the turbidity section). Carlson (1980) and Edmondson (1980) also noted that Secchi depth may not be as responsive to changes in chlorophyll *a* levels because of the known nonlinear relationship between these parameters, especially at high chlorophyll values and high levels of detritus (Armengol et al., 2003).

Jantz and Neumann (1992) found a significant correlation ($r^2 = 0.80$) between the rate of shell length growth and chloroplyll *a* concentration and between shell growth and temperature ($r^2 = 0.82$) but later Jantz and Neumann (1998) found that shell growth rate and the duration of the growing season were correlated with the quantity of available food. Sprung (1995a) found a strong correlation between zebra mussel shell growth and seston concentration. Apparently this correlation exists only when seston concentrations are below those at which the intestine is filled to capacity and when the animal filters at a maximum rate (Sprung, 1995b). Nichols (1993) and Schneider et al. (1998b) found that the scope for growth under laboratory conditions is strongly correlated with food quality.

There are few, if any, studies that relate chlorophyll *a* levels to abundance of zebra mussels, especially in North America. Stanczykowska and Lewandowski (1993) examined relationships between occurrence of mussels and trophic status of lakes in Poland and noted lowest densities in shallow, highly eutrophic, polymictic lakes; in general, densities were highest in mesotrophic lakes. Based on this information, Table 3.7 uses a conservative approach and considers no likelihood of long-term survival in either oligotrophic or highly eutrophic waters (i.e., [Chl *a*] < 2.5 or >25 μg/L) little likelihood of nuisance infestations between oligo-mesotrophic water (i.e., [Chl *a*] between 2.0 and 2.5 μg/L) or slightly eutrophic water (i.e., [Chl *a*] between 20 and 25 μg/L), a moderate level of infestation with [Chl *a*] 8 and 20 μg/L, and a good potential for intensive infestations between [Chl *a*] 2 and 8 μg/L. *The survival potential of zebra mussels in the hypothetical lake based on chlorophyll a levels is low* (Table 3.8). It is unlikely that a nuisance population will be established, and the high levels of algae will mitigate any accumulations due to smothering effects, even though the levels of chalk variables are high. If mussels are in hypolimnion, algae levels will be low, but mussels can survive on bacteria and other microbes at these depths.

3.3.1.1.2.3 Clearance Rates The clarity of water is related to the clearance rates of dreissenids. Collectively, zebra mussels are unrivaled in their capacity to clarify water (Morton, 1971; Stanczykowska, 1975; Piesik, 1983; Reeders et al., 1989; Claudi and Mackie, 1994). However, compared to other freshwater bivalves, the zebra mussel's filtration rate (10–100 mL/individual/h) is intermediate between Sphaeriidae (0.6 to 8.3 mL/individual/h) and Unionidae (60–490 mL/individual/h) (Stanczykowska et al., 1976). *Corbicula* has a filtration rate that far exceeds these on an individual basis (60–800 mL/individual/h). The well-known ability of *Dreissena* to clarify lakes and watercourses is a consequence of the enormous standing crops that usually prevail in aquatic systems.

Claudi and Mackie (1994) state that zebra mussels can efficiently filter food particles as small as 0.7 μm but they show a preference for the 15–40 μm range. Small particles, such as phytoplankton, small zooplankton and organic debris, form the bulk of the food (Claudi and Mackie, 1994). McMahon (1996) states that veligers can filter algae between 1 and 4 μm in size. They are also efficient at filtering particles out of sewage sludge and in the process reduce biochemical oxygen demand and total phosphorus (Mackie and Wright, 1994).

The pumping rates of zebra mussels depend on size classes of the animal at a particular location, their densities, and the substrate type. For example, pumping rates of zebra mussels in western basin of Lake Erie of 25–132 $m^3/m^2/d$ were recorded by MacIsaac et al. (1992) in nearshore rocky ares and 13.9 $m^3/m^2/d$ in the basin areas with mixed substrate; 1.7–2.3 $m^3/m^2/d$ by Pontius and Culver (2001) for nearshore shoal areas; 25 $m^3/m^2/d$ by Edwards et al. (2005) for nearshore cobble areas; 2.7–6.7 $m^3/m^2/d$ by Bunt et al. (1993) for nearshore rock areas; 16.2 $m^3/m^2/d$ by Ackerman et al. (2001) for offshore reefs; 3.08 $m^3/m^2/d$ by Noonburg et al. (2003) for basin area; 1.7 $m^3/m^2/d$ by Boegman et al. (2008) for basin area. These pumping rates can filter large percentage of the seston in the water column, between 8% and 67% in shallow areas (Edwards et al., 2005), 39% and 96% in the basin area (Bunt et al., 1993), 40% offshore (Ackerman et al., 2001) and 53% in the basin (Boegman et al., 2008). Much of the filtered material ends up on the bottom or in a boundary layer as pseudofeces, the boundary layer forming whenever pumping rates exceed 2 $m^3/m^2/d$ (Boegman et al., 2008). Associated with the clearance of seston in the western basin is increased deposition rates of organic matter, polychlorinated biphenyls, and cadmium at 8–10 times greater than natural sedimentation rates (Dobson and Mackie, 1997).

3.3.1.1.2.4 Nutrients (Total Phosphorus, Total Nitrogen) Stanczykowska and Lewandowski (1993) reported that zebra mussels are found in Polish waters with total phosphorus content ranging from 20 to 300 μg/L; they found mussels in 100% of dimictic lakes with 20–50 μg P_T/L (mesotrophic to eutrophic) and 51–100 μg P_T/L (eutrophic), 86% with 101–300 μg P_T/L (hyper-eutrophic), and none in lakes with 301–950 μg P_T/L (dystrophic). Karatayev et al. (2003a,b) also found that trophic type affects the probability of finding zebra mussels, most often in mesotrophic lakes, less often in oligotrophic and meso-oligotrophic lakes, but least often in eutrophic lakes and they do not inhabit dystrophic lakes. However, Karatayev and Burlakova (1995) pointed out that the highest densities tend to occur in eutrophic lakes. Dorgelo (1993) found that zebra mussel growth rates in two Dutch eutrophic lakes (P_T = 30–100 μg/L) were higher (0.54–0.59 mm/wk than those grown in a meso-oligotrophic lake (P_T = <10 μg/L) (0.35 mm/wk), both with similar mean summer temperatures.

Unfortunately, there are few, if any, studies that relate zebra mussel abundance to concentrations of total phosphorus (or total nitrogen), probably because zebra mussels play a major role in the cycling of phosphorus and, usually (Vanderploeg et al., 2002) but not always (e.g., Idrisi et al., 2001), result in declining phosphorus levels. The cycling process includes remineralization of filtered organic phosphorus into soluble forms that can be recycled back to the food web. Nutrient remineralization occurs directly via excretion of soluble nutrients (Heath et al., 1995; Mellina et al., 1995; Arnott and Vanni 1996; Effler et al., 1997; James et al., 1997, 2001) and indirectly through decomposition of feces and pseudofeces by the zebra mussels (Klerks et al., 1996). Their filtering activities can be beneficial in clarifying the water column (Reeders and de Vaate, 1990) and improving light transparency for submersed macrophytes (Skubinna et al., 1995; Zhu et al., 2006). The impacts of zebra mussels on the Great Lakes has not only been on high particle filtration rate with selective rejection of colonial cyanobacteria in pseudofeces and alteration of nutrient ratios, but also on ecosystem engineering (increased water clarity and reef building), fouling native mussels, and facilitation of the rapid spread of their Ponto-Caspian associates, the benthic amphipod *Echinogammarus ischnus* and the round goby, *Neogobius melanostomus*, which feeds on zebra mussels (Vanderploeg et al., 2002). The impact of selective grazing of zebra mussels is also being witnessed in Swedish freshwaters, where mussels showed a consistent preference for cryptophytes and avoidance of chlorophytes and cyanobacteria; diatoms, chrysophytes, and dinoflagellates responded differentially to zebra mussel grazing depending on their size; analysis of excreted products of the zebra mussels revealed that in addition to chlorophytes and cyanobacteria, phytoplankton >50 μm and very small phytoplankton (≤7 μm) were largely expelled in pseudofeces (Naddafi et al., 2006).

Studies of zebra mussel impacts on primary production and trophic variables in oligotrophic lakes are rare, probably because most oligotrophic lakes cannot support dreissenid populations. There is one study in ultraoligotrophic to oligotrophic Lake Leelanau, Michigan that shows the impact of a small population (about 3250 mussels/m^2 at north end, 400 m^{-2} central part, 1380 m^{-2} at south end) on nutrients (total phosphorus and nitrate nitrogen), Secchi depth, and pelagic algal diversity and density (Shankland and Lowe, 2005). Preinvasion (1996) P_T was 5.74 μg/L, NO_3 was 242.6, Secchi 5.15 m, Shannon algal diversity index was 1.71, and algal density 315.3 cells/mL; postinvasion values (2004) were $P_T = 4.70$, $NO_3 = 168.4$, Secchi depth = 5.55 m, algal diversity = 1.64, and algal density = 254 cells/mL.

Zebra mussels also impact primary production variables of benthic (epipelic and epithithic) algae. In mesotrophic Lake Oneida, New York, estimates of whole-lake epipelic and epilithic algal primary production showed a significant (4%) increase and exhibited significantly less interannual variability subsequent to the establishment of zebra mussels (Cecala et al., 2008). Also on Oneida Lake Mayer et al. (2000, 2002) and Idrisi et al. (2001) have shown that light attenuation (k) has decreased significantly since zebra mussel invasion; zebra mussel density was about 44,000 m^{-2} in 1992 but dropped to about 12,000 m^{-2} the following year and has been fluctuating between 4000 and 8000 m^{-2} since then. P_T has remained relatively stable at 27 μg/L (Cecala et al., 2008), although the lake has brief periods of nitrogen limitation (Idrisi et al., 2001). The area of the submersed macrophytes community has increased 23% and the maximum depth of macrophytes colonization has increased from 3.0 to 5.1 m (Zhu et al., 2006).

While population density appears to depend on trophic type, the time since initial colonization, relative abundance of substrate available for colonization, and lake morphometry are also important (Burlakova et al., 2006). In lakes where submerged macrophytes are the dominant substrate for zebra mussel attachment, populations may be less stable than in lakes with a variety of substrates, which will have a more balanced age distribution, and be less impacted by year to year variation in recruitment (Burlakova et al., 2006). Selecting trophic criteria for assessing vulnerability of a lake to *Dreissena polymorpha* invasion needs to be done before invasion or shortly after. Burlakova et al. (2006) found that most zebra mussel populations reach maximum density 7–12 years after initial introduction and both European and North American data suggest that zebra mussels reach a density large enough to be detected in about 2–3 years after their initial introduction. Population sizes may also fluctuate and follow 2- to 4-year cycles, like the zebra mussel population in the Hudson River estuary, which fluctuated approximately 11-fold across a 13-year period, following a cycle with a 2–4-year period (Strayer and Malcom, 2006).

Based on the above information, Table 3.7 uses a conservative approach and considers no likelihood of long-term survival in either oligotrophic or highly eutrophic waters (i.e., $[P_T] < 5$ or >50 μg/L; $[N_T] < 75$ or >500 μg/L), little likelihood of nuisance infestations between oligo-mesotrophic water (i.e., $[P_T] = 5–10$ μg/L; $[N_T] = 75–150$ μg/L) or eutrophic water (i.e., $[P_T] = 35–50$ μg/L; $[N_T] = 525–750$ μg/L), a moderate level of infestation meso-eutrophic water (i.e., $[P_T] = 10–25$ μg/L; $[N_T] = 150–375$ μg/L), and a good potential for intensive infestations in mesotrophic water (i.e., $[P_T] = 25–35$ μg/L; $[N_T] = 150–275$ μg/L). The N_T values are calculated from P_T, assuming an N_T:P_T of 15. *The survival potential of zebra mussels in the hypothetical lake based on nitrogen levels is low* (Table 3.8). It is unlikely that a nuisance population will be established, and the high levels nutrients will contribute to high levels of algae and mitigate any accumulations due to smothering effects, even though the levels of chalk variables are high.

3.3.1.1.2.5 Secchi Depth, Trophic Status The levels of chlorophyll *a* and total phosphorus and nitrogen given above are based on trophic status of the waters and since Secchi depth correlates well with each variable, waters with Secchi values of <1 and >8 m will not support long-term zebra mussel survival and waters with Secchi depth values of 2–4 m will support massive infestations. Secchi depths of 4–6 m will support moderate levels of zebra mussel densities and Secchi depths between 1–2 m and 6–8 m will support only low levels of zebra mussel abundances (Table 3.7).

3.3.1.1.3 Survival Potential of Zebra Mussels Based on Temperature

Zebra mussels have both a lower and upper lethal threshold limits. They cannot tolerate freezing and are rare in Europe, where mean annual air temperatures are below 3–6°C (Strayer, 1991; Karatayev, 1995; McMahon, 1996). In Europe zebra mussels were absent where mean annual air temperatures were above 18°C or where the highest mean monthly air temperatures were above 27°C (Strayer, 1991). Bij de Vaate (1989) reported adult growth down to 6°C and McMahon (1996) reported 6°C as the lower limit of occurrence in Europe, although it is not clear if it was mean annual air temperature. Mean summer temperatures in the range of 8–12°C (Claudi and Mackie, 1994), 9–15°C (Sorba and Williamson, 1997), or 10–15°C are considered the lower limiting values for potential distribution of zebra mussels (Strayer, 1991; Doll, 1997; Cohen and Weinstein, 1998; Cohen, 2007). Lower ranges would probably apply for large, moderately productive hard water lakes and rivers and upper ranges to smaller but deep and more productive lakes and rivers (Strayer, 1991).

The threshold temperatures for growth and reproduction apparently vary considerably among populations of zebra mussels. Claudi and Mackie (1994) reported preliminary analyses of data from Lake St. Clair populations that suggest threshold temperatures of at least 10°C for gametogenesis, 8–10°C for growth, and 14–16°C for appearance of veligers. These are similar to threshold temperatures of 11–12°C for growth (Walz, 1978) and 15–17°C for reproduction (spawning or first appearance of veligers) reported by Stanczykowska (1977) and Lewandowski and Ejsmont-Karabin (1983) for Polish populations and by Morton (1969b) for British populations. Bij de Vaate (1989) reported threshold temperatures of 6°C for growth and 12°C for reproduction for populations in the Netherlands. Other studies in Europe and North America report the lower thermal threshold for adult growth at around 10–12°C (Baker et al., 1993a; Morton, 1969; Stanczykowska, 1977). The threshold temperature for reproduction (spawning) was reported to be between 12°C (Borcherding,1991; Neumann et al., 1993; Baker et al., 1993a; McMahon, 1996; Cohen, 2007) and 15°C (Karatyaev, 1995; Doll, 1997; Claudi and Mackie, 1994). Spawning peaks between 12°C and 18°C, which probably envelopes the optimum temperature for larval development (Sprung, 1993). Limited spawning has been reported at <5°C (Mills et al., 1993b) and 10°C in the Great Lakes and Europe (Sprung, 1993; Nichols, 1996; McMahon, 1996). Claudi and Mackie (1994) also reported seeing larvae at <10°C, but they were residual larvae that had not yet developed to settlement stage in the fall in Lake St. Clair (October and November). The presence of larvae and adults in deep water at <5°C probably represent recently settled or translocated individuals that likely have not established large populations (McDermott, 1993; Claxton and Mackie, 1998; Claxton et al., 1998). Claudi and Mackie (1994) suggested that the threshold temperatures for gametogenesis, growth, and larval release are well within the ranges of temperate species and suggest that the zebra mussel will succeed in many lakes and rivers in most parts of North America.

The upper lethal threshold level for zebra mussels has been an issue of some debate largely because of discrepancies between observations in laboratory experiments and observations of occurrences and abundances in nature. In addition, most of the observations in the literature are based on lakes at low altitudes and/or high latitudes (e.g., states and provinces bordering the Great Lakes). Among the variables that affect the lethal threshold value are the rate of acclimation (in the lab) or acclimatization (in nature), the duration of the exposure, the frequency and extent of temperature variations, and the source or "previous experience" (i.e., genetics) of mussels to thermal extremes.

Acclimation rates and exposure times have been thoroughly reviewed by McMahon (1996). McMahon et al. (1993) derived acute upper lethal limits from models using North American adult zebra mussels exposed to constant warming rates ranging from 0.0167°C/min to 0.2°C/min. Values of LT_{50} (estimated temperature for 50% sample mortality) were affected by temperature increase rate (TIR) and prior acclimation temperature (AT) according to the regression:

$$\ln LT_{50}(°C) = 3.603 + 0.026 \ln TIR(°C/min) + 0.0036\ AT(°C).$$

McMahon et al. (1993) found that the LT_{50} increased exponentially with increasing warming rate and linearly with increasing acclimation temperature. Slower warming rates can induce mortality at lower temperatures, with warming rates of 0.05–0.0333°C/min inducing 50% kills at temperatures of 33.5–38°C depending on acclimation state.

In a later study McMahon et al. (1994) examined the chronic thermal tolerance of North American zebra mussels and found that tolerance depended on shell size according to the following regression:

$$\ln \text{Survivorship (h)} = 33.982 + 0.0579\ AT\ (°C) - 0.944 \ln ET\ (°C) \\ -0.0517\ SL\ (mm),$$

where AT is the acclimation temperature, ET is the exposure temperature, and SL is the shell length. Accordingly, tolerance times increase with increased acclimation temperature and decrease exponentially with increased exposure temperature. Temperatures ≥36°C were nearly instantaneously lethal regardless of acclimation state. The negative slope for SL indicates that larger specimens are less tolerant than smaller individuals, opposite to that reported for European mussels (Afanas'yev and Protasov, 1987).

The upper incipient lethal temperature was 31°C with mean tolerated exposure times of 52–292 h, depending on acclimatization (Armistead, 1995), but Domm et al. (1993) found <32°C to be the upper limit based on experiments without an acclimation period. McMahon and Ussery (1993) predicted 100% mortality at 35–36.5°C after a long-term exposure of mussels from the Niagara River acclimated to 5–15°C, which agrees with 100% mortality at 33–36°C for different zebra mussel populations in Volga River (Shkorbatov, 1986). Spidle et al. (1995) and Elderkin and Klerks (2005) reported 31–33°C as the upper limit for short-term survival (e.g., 100% mortality with one week exposure to 31°C, 100 h exposure to 32°C, or 24 h exposure to 35°C). Thorp et al. (1998) found the upper limit to be 34°C based on 85–95% mortality in bioassays with Lake Erie and Ohio River mussels.

Source and previous experience to thermal regimes were reported by Smirnova et al. (1993) and Smirnova and Vinogradov (1990); they noted that Volga River populations of zebra mussels varied in their heat tolerance (possibly due to genetic differences), with the southernmost population and a population living in waters heated by power plant discharges showing the greatest tolerance for high temperatures. Similarly, Thorp et al. (1998) found that zebra mussels from the Ohio River were more tolerant of high temperatures than zebra mussels from Lake Erie, and suggested that this may be an evolved, postinvasion difference that would increase over time. In the Zaporozhskoe Reservoir in the Ukraine, zebra mussels survive in water temperatures up to 33°C (Karatayev et al., 2007b). The fact is, all the variables can be tested separately or together in the laboratory but in nature the acclimatization rates, exposure periods, frequency of fluctuations in temperature, and source of mussels are difficult to impossible to predict.

Baker et al. (1993a) argue that in most lakes of temperate regions seasonal temperature fluctuations should result in some period each year with temperatures that allow successful reproduction. Countering this, however, is a period after which the duration and extent of a rise (or fall) in temperature could kill all larvae, with no recruitment as a net result. Baker et al. (1993a) suggest that water temperatures between 24°C and 30°C are unsuitable for reproduction, spawning, or larval growth and that temperatures above 26–36°C can be lethal. However, the duration of exposure to these temperatures is critical in effecting net recruitment. So, while in some respects selection of the appropriate threshold temperatures for growth, reproduction, and survival may seem irrelevant, it is a fact that the closer the mussel is to its upper lethal threshold temperatures for growth, reproduction, and survival, the lesser the extent of variation and duration of exposure needed to effect mortality of larvae and adults or cessation of growth. Also, instantaneous kills are not necessarily required. Once mussels are stressed they can "die a slow death." Walz (1978) reported that at high temperatures respiratory costs can exceed assimilation rates resulting in loss of body weight, which could prevent the establishment of zebra mussels without killing them outright. In laboratory experiments, Aldridge et al. (1994) found that an increase in water temperature between 20°C and 32°C results in an increase in zebra mussels' energetic costs and a decrease in their feeding rate, and concluded that energy losses start to exceed intake between 24°C and 28°C. Stanczykowska (1977) found that adult growth ceases above 26–33°C and below 11–12°C in Polish lakes. McMahon and Tsou (1990) noted that temperatures >26–32°C can kill larvae or adults. A temperature of 30°C is also commonly reported as the upper limit for feeding and growth of juveniles and adults (Smirnov and Vinagradov, 1990; Baker et al., 1993a; Iwanyzki and McCauley, 1993; Claudi and Mackie, 1994; McMahon, 1996).

As stated above, the upper lethal threshold limit is more vague but both zebra and quagga mussels can be killed instantly by short-time exposure to 40°C. Several factors affect the heat tolerance of dreissenids, including acclimation temperature, size, geographic location, rate of temperature increase, and genetics, much of which has been discussed above. McMahon et al. (2002), suggested "that there is a strong selection pressure for an increased heat-tolerance in the southern range of the zebra mussel but this has not (at least not yet) resulted in an adaptation. The lack of adaptation at this stage may be due to gene flow being too strong or due to insufficient genetic variation for selection to act on." Domm et al. (1993) reported the average survival time of zebra mussels tested at 32°C to be 275 min whereas that of quagga mussels was 75 min, showing that zebra mussels are more tolerant of high temperatures than quagga mussels. Iwanyzki and McCauley (1993), Spidle et al. (1995), McMahon (1996), and Thorp et al. (1998) consider the long-term lethal limit for zebra mussels to be approximately 30°C. However, growth and reproduction and byssal production of zebra and quagga mussels appear to begin to cease between 25°C and 28°C (Thorpe et al., 1998; Rajagopal et al., 1997; Elderkin and Klerks, 2005).

With such variances in thermal thresholds for each of growth, reproduction, and survival, what criteria do we use? First, as for most of the variables described above, it seems that our null hypotheses should use a conservative approach. That is, we hypothesize that no mussels will establish themselves if we set our criteria on the "safe side." For example, it is better to assume no survivorship at 32°C than 30°C because it is more likely that morality will occur at 32°C than at 30°C, but if mortality does occur at 30°C, we can accept our null hypothesis that mussels will not establish themselves at 32°C.

For potential distribution in southern U.S. waters, zebra mussels have been reported at temperatures up to about 30°C (Spidle, 1994), with die-offs occurring at 31–32°C (Karatayev, 1995; McMahon, 1996). Various studies have used mean summer temperatures in the range of 30–32°C and maximum temperatures of 31°C as the upper limiting values for potential distribution (Sorba and Williamson, 1997; Doll, 1997; Cohen and Weinstein, 1998). In conclusion, it seems that 15°C is the threshold limit for reproduction and 32°C is the upper threshold limit for survival, and hence potential distribution, of zebra mussels. The threshold values for growth and reproduction are discussed further in Section 3.3.3.3. *The survival potential of zebra mussels in the hypothetical lake based on mean annual summer temperature is high* (Table 3.8). It is unlikely that a nuisance population will be established, and the high levels of nutrients will contribute to high levels of algae and mitigate any accumulations due to smothering effects, even though the levels of chalk variables are high.

3.3.1.1.4 Survival Potential of Zebra Mussels Based on Conductivity, TDS, and Salinity

3.3.1.1.4.1 Conductivity Conductivity represents the total amount of cations and anions in solution of which calcium and magnesium are important cations and carbonates and bicarbonates are among the most important anions for zebra mussels. It stands to reason that the most important criterion for survival is a minimal conductivity, the upper limit not being critical until salinity becomes a factor. Sorba and Williamson (1997) are among the few that use conductivity as criterion for assessing survival potential of zebra mussels in Manitoba waters; they recommend 22 μS/cm for no likelihood for survival and 22–36, 36–40, and 40–200 μS/cm for low, moderate, and high likelihood of survival, respectively. Bartell et al. (2007) examined 14 physical–chemical factors to determine habitat suitability for the dreissenid mussels; they recommend a conductivity of 62.5 μS/cm as the threshold value for survival for zebra mussels.

Zhadin (1965) and Strayer and Smith (1993) suggest that zebra mussels may be able to tolerate higher salinity in waters that contain higher proportions of divalent ions (Ca^{2+} and Mg^{2+}) and sulfates relative to monovalent ions (Na^+ and Cl^-), and that chloride content rather than overall salinity may actually be the critical factor. Hence, in deriving criteria for assessing survival potential on the basis of conductivity (Table 3.7), regressions described in Sections 3.2.1.1.1 and 3.2.7 relating conductivity to calcium and alkalinity were used. Criteria recommended by Sorba and Williamsom (1997) represent very low levels of calcium and alkalinity, so a threshold level of 60 μS/cm (rounded off from 62.5 μS/cm as per Bartell et al., 2007) was recommended for moderate levels of infestation [≈15 mg Ca/L, the threshold level for reproduction (Hincks and Mackie, 1993; Mellina and Rasmussen, 1994; McMahon, 1996; Cohen and Weisnstein, 1998)] and then short-term survival values of 32 μS/cm [≈8 mg Ca/L, the mean lowest level reported for no to low survival (McCauley and Kott, 1993; Tammi et al., 1995; Doll, 1997; Hincks and Mackie, 1997; Sorba and Williamson, 1997)]. Most authorities on zebra mussel requirements agree that mussels do well in waters with >30 mg Ca/L (Ramcharan et al., 1992; Claudi and Mackie, 1994; Karatayev, 1995; Hayward and Estevez, 1997; Hincks and Mackie, 1997; Cohen and Weinstein, 1998, 2001; Bartell et al., 2007; Cohen, 2007; Karatayev et al., 2007a), which corresponds to 120 μS/cm. *The survival potential of zebra mussels in the hypothetical lake based on conductivity levels is high* (Table 3.8). This is not surprising because the conductivities are directly related to levels of chalk variables.

3.3.1.1.4.2 Total Dissolved Solids Levels of TDS can be determined by multiplying the conductivity by 0.64 (see Section 3.2.7.2). The values in Table 3.7 were rounded to the nearest 10 mg/L (e.g., 20 mg/L instead of 19 mg/L). *The survival potential of zebra mussels in the hypothetical lake based on levels of TDS is high, as it is for conductivity* (Table 3.8).

3.3.1.1.4.3 Salinity The lethal salinity based on laboratory experiments has produced as disparate results as occurrences of zebra mussels in estuaries of different salinities. Examining the literature of laboratory experiments in chronological order, Karpevich (1947) reported 10 ppt as the limit for long-term survival of gradually acclimated mussels. Later Karpevich (1955) reported the lethal concentration to be 5–7 ppt but 10–12 ppt when using stepwise acclimation. Smirnova (1973) reported 100% mortality after 168 days in 5 ppt. Mackie and Kilgour (1992) reported 85% survival of adult mussels in 4°C and 10°C water that were acclimated to 8 ppt salinity over 42 days. In contrast, Barber (1992) reported 100% mortality of adult mussels in 15°C water when salinity was raised from 0 to 2.7 ppt over 52 days. Wright et al. (1996) reared larval zebra mussels and quagga mussels through and beyond metamorphosis (settlement) at salinities of 0–8‰. Juvenile zebra mussels gradually acclimated to 8‰ and 10‰ survived over 8 mo (Wright et al., 1996). Wright et al. (1996) found that "tolerance to both higher temperatures and higher salinities increased with larval age in both species, but zebra mussel embryos and larvae had a greater degree of salinity tolerance than quagga mussel embryos and larvae. Only 6% of 3-day-old zebra mussel veligers survived exposure to 4‰ for eight additional days, whereas 22% of veligers placed survived in 4‰ at day 13 and grown to settlement 11 days later. Zebra mussel pediveligers, acclimated to increasing salinity in 2‰ increments beginning at day 23, continued to survive and grow in 8‰ after 5-mo exposure, though the growth rates of these juveniles were significantly less than those of juveniles reared in lower salinities."

Examining the literature reporting occurrences in estuaries of different and varying salinities, North American zebra mussels have been found in salinities in the Hudson River estuary at high densities at sites with maximum salinities up to 3 ppt, and at low densities at maximum salinities up to 5 ppt (Carlton, 1992; Walton, 1996). European zebra mussels have been found in up to <2 ppt in the extreme eastern Gulf of Finland and along the Black Sea (Wolff, 1969; Strayer and Smith, 1993); up to a mean of 0.6 ppt in the Netherlands; up to <1 ppt in the eastern Gulf of Riga; between 6 and 9 ppt in the northern Caspian Sea, but absent from the main body of the sea at 13 ppt (Strayer and Smith, 1993); up to 10 ppt in the Aral Sea but as water diversions raised the salinity, the populations began to decline at around 12 ppt and virtually disappeared by 14 ppt (Strayer and Smith, 1993). Populations of stunted mussels have been observed in the Vistula estuary and lagoon at up to 4.8 ppt, and in the Kiel Canal at 3.8–6.2 ppt (Strayer and Smith, 1993). 7.6 ppt was the maximum salinity at which zebra mussels have been found in the Dnieper-Bug estuary (Mills et al., 1996). Baker et al. (1993) believe that temperature (higher tolerance in colder water) and genetics could also affect salinity tolerance; they reported differences in salinity responses in different populations in the Volga River, zebra mussels at the most downstream locations tolerating exposures to high salinities better than the mussels from upstream locations.

While there is substantial variation in the estimate of the lethal salinity that will limit the distribution of zebra mussels [e.g., 2–15 ppt: (Doll, 1997; Hayward and Estevez, 1997; Cohen and Weinstein, 1998; Cohen, 2007)], most researchers on the subject agree that it is the rate of exposure or acclimation time, temperature, and stage of mussel development that determines the lethal level. If salinity is changing rapidly (e.g., in an estuary) zebra mussels appear to be able to tolerate only low levels of salinity, but can handle higher levels if allowed to acclimatize. Kilgour et al. (1994) found that zebra mussels are able to acclimate to slowly changing salinities (i.e., 1 ppt/d) such that the time to 50% mortality of a population should be >1 yr at temperatures near 20°C and salinities up to 8 ppt. They also found that the incipient lethal salinity of post-veligers is near 2 ppt, of larger adults (5–15 mm) between 2 and 4 ppt, and of veliger larvae near 4.5 ppt. At high temperatures (18–20°C), the condition of zebra mussels is reduced at salinities above 1 ppt, but at lower temperatures (3–12°C) the optimum salinity for zebra mussels is 2–4 ppt. Strayer and Smith (1993) also found that zebra mussels showed greater tolerance to higher salinity levels when the rate of increase was gradual. Karpevich (1947) found that respiration rates dropped when salinity was changed abruptly, but that there was no depression in respiration when salinity was changed at a slower rate of 2 ppt every two days up to a final salinity of 17 ppt.

3.3.1.1.5 Survival Potential of Zebra Mussels Based on Turbidity and TSS

3.3.1.1.5.1 Turbidity The effect of zebra mussel's filtering capacity on levels of TSS (also measured as turbidity and seston) consisting mostly of phytoplankton is well documented. Holland (1993) reported a decline in diatom abundance coincident with the arrival and establishment of the zebra mussel in western Lake Erie. Klerk et al. (1997) measured extremely high biodeposition rates averaging 28 mg/cm^2/d at a mussel density of 1180 individuals/m^2 at high-turbidity sites near the mouth of the Maumee River that resulted in the removal of 7–30% of the total suspended matter per day and a concomitant increase in biodeposition of sediment organic content. Similarly, establishment of the zebra mussel was associated with a 60% decline in chlorophyll concentrations and a 60% increase in water clarity in Saginaw Bay, Lake Huron (Fahnenstiel et al., 1995). Caraco et al. (1997) showed that the invasion of the zebra mussel into turbid rive Hudson River estuary reduced the standing crop of phytoplankton. There are also numerous studies on the effect of suspended solids on filter-feeding and growth in *D. polymorpha* (Morton, 1971; Sprung and Rose, 1988; Reeders and Bij de Vaate, 1990; Wisniewski, 1990).

While the impact of zebra mussels on TSS is well known, the reciprocal relationship has not been as well examined and little information exists on the effects of suspended solids and food concentrations on zebra mussel survival or even metabolic activities. Morton (1971) suggested but did not demonstrate that higher concentrations of suspended materials can depress growth rates by overloading the gut and gills with inorganic solids. Aldridge et al. (1987) and McMahon (1991) suggest, but did not show, that suspended solids may interfere with respiration and filter-feeding by fouling water currents passing over the gills or by physically impeding gas exchange across cell membranes of epithelial cells lining respiratory surfaces. However, there is a study by Alexander et al. (1994) that shows the impact of turbidity (and temperature, see the temperature section) on the oxygen consumption rate of zebra mussels. They examined the respiratory response (V_{O_2}, μL O_2/mg/h) of zebra mussels to acute exposure to various levels of turbidity (0–160 NTU). Zebra mussels were acclimated to three different temperatures and were tested at their acclimation temperature. Oxygen consumption was inversely related to turbidity at all three temperatures, with V_{O_2} of 1.25–1.75 μL O_2/mg/h from 0 to 5 NTU; V_{O_2} dropped from 1.25 to 0.75 μL O_2/mg/h between 5 and

20 NTU; V_{O_2} dropped from 0.75 to <0.50 μL O_2/mg/h between 20 and 80 NTU; V_{O_2} remained at <0.5 μL O_2/mg/h at NTU > 80. Respiration rates were significantly greater at 26°C than at 10°C and 18°C, which were similar to each other. Alexander et al. (1994) used their results to assess risk potential in the Ohio River where turbidity rises in fall and peaks in winter at about 37 NTU and then declines in spring and fluctuates at low levels of 3–4 NTU from May to October; they concluded that these turbidity levels may depress growth rates of adult *D. polymorpha* by increasing maintenance costs but probably will not by themselves prevent establishment of viable populations in this river. Based on these results, NTUs <5 are considered optimal for zebra mussel infestations; NTU of 5–20 will support moderate infestation levels; NTU of 20–80 will support little infestations; NTU >80 are lethal and will not support zebra mussels survival over a long term (Table 3.7). *The survival potential of zebra mussels in the hypothetical lake based on turbidity levels is low* (Table 3.8), the high levels mitigating any accumulations of mussels due to potential interference of suspended particles with filtering and feeding processes.

3.3.1.1.5.2 Total Suspended Solids TSS can be estimated from Christensen et al's. (2001) regression:

$$\log_{10}\text{TSS(mg/L)} = 0.899 \log_{10} \text{Turbidity (NTU)} + 0.272.$$

The criteria in Table 3.7 were derived using the regression and any inferences made for turbidity apply to TSS.

Madon et al. (1998) examined the effects of inorganic suspended sediment and food concentrations on clearance, ingestion, respiration, feces, and pseudofeces production rates and assimilation efficiencies of adult zebra mussels under laboratory conditions at 20 ± 1°C. They reported that suspended inorganic sediment above 1 mg/L greatly reduced clearance rates, ingestion rates, and assimilation efficiencies of zebra mussels, and ingestion of particulate organic matter (POM) was maintained only at high food concentrations. Copious amounts of pseudofeces were produced in the 100 mg/L sediment treatment, and there appeared to be an increased metabolic cost associated with pseudofeces production. Declining food quality (measured as IO = ratio of suspended inorganic to organic matter) caused exponential declines in clearance, ingestion and absorption rates, water processing potential, and assimilation efficiencies, with the sharpest declines between IO ratios of 0.304 and 5.0. The IO ratio often exceeded 1.71, and Madon et al. (1998) concluded that the IO experimental threshold for growth for zebra mussels was zero, and that zebra mussels in turbid rivers may exhibit low growth potential and may not stabilize at the high population densities seen in their lake counterparts.

Schneider et al. (1998b) also found that the scope for growth under laboratory conditions had a strong positive relationship with food quality.

Although there are no criteria for assessing impact potential based on 110 mg/L of TSS as this value is above the levels tested by Madon et al. (1998) and Schneider et al. (1998b) where clearance, ingestion, and absorption rates showed the greatest declines, it appears that *the survival potential of zebra mussel adults in the hypothetical lake based on TSS levels is low* (Table 3.8).

3.3.1.1.6 Summary of Survival Potential of Zebra Mussel in Hypothetical Lake

Table 3.8, last line, summarizes the potential for distribution into the hypothetical lake as "little." The reason for this assessment is that even though the levels of the chalk variables are sufficient to support a population of mussels, the high turbidity caused by blooms of algae and other kinds of suspended materials will mitigate heavy accumulations of mussels. However, there may be periods in the year (e.g., during spring) when algal biomass and the load of suspended solids is low enough to allow or even contribute to recruitment. An analysis of the seasonal variations in the variables would help show what periods, if any, will be conducive to zebra mussel recruitment, as discussed in Section 3.4.

3.3.3 Quagga Mussel, *Dreissena bugensis* (Conrad, 1831)

The threshold limits for growth, reproduction, and survival for each of the physiological and ecological variables for the quagga mussel were determined from averaging the levels reported by the numerous researchers listed in Table 3.9. The ranges in values for each of the risk levels are given in Table 3.10. Assessment of potential survival of quagga mussels in a hypothetical lake is pretty much the same process as described for zebra mussels, and is not repeated here.

TABLE 3.9
Studies Using Various Parameters to Assess Risk Potential of Survival and Potential Density of Quagga Mussels

Variables	Reference
Alkalinity	See Table 3.6
Calcium	See Table 3.6 and Ludyanskiy et al. (1993), Zhulidov et al. (2004, 2005, 2006), Grigorovich et al. (2008)
Chlorophyll *a*	See Table 3.6 and Bunt et al. (1993), Ackerman et al. (2001), Baldwin et al. (2002), Boegman et al. (2008), Zang et al. (2008)
Conductivity, TDS	Claudi and Mackie (1994), Sorba and Williamson (1997), Mackie (2004),
Dissolved oxygen	See Table 3.6 and Karatayev et al. (2007b), Son (2007)
Nitrogen, total	See Table 3.6 and Lohner et al. (2000), Conroy et al. (2005), Zhang (2006), Zhang et al. (2008)
pH	See Table 3.6
Phosphorus, total	See Table 3.6 and Lohner et al. (2000), Conroy et al. (2005), Zhang (2006), Zhang et al. (2008)
Salinity	See Table 3.6 and Spidle (1994), Spidle et al. (1995), Mills et al. (1996), Therriault et al. (2004), Karatayev et al. (2007b)
Secchi depth, TSS	Claudi and Mackie (1994), Hayward and Estevez (1997), Sorba and Williamson (1997), Armengol et al. (2003)
Temperature (°C)	See Table 3.6 and Roe and MacIsaac (1997), Claxton and Mackie (1998), Claxton et al. (1998), Karatayev et al. (2007b)
Total hardness	See Table 3.6
Turbidity (NTU), TSS	See Table 3.6 and Barbiero et al. (2006)

The variables are listed alphabetically.

TABLE 3.10
Criteria Used in Determining Levels of Infestation by Quagga Mussels in North America

1. Parameter	2. No Potential for Adult Survival	3. Little Potential for Larval Development	4. Moderate Potential for Nuisance Infestations	5. High Potential for Massive Infestations
Chalk Variables				
Calcium (mg Ca/L)	<10	10–12	12–30	30–120
pH	<7.0, >9.5	7.0–7.8, 9.0–9.5	7.8–8.2 or 8.8–9.0	8.2–8.8
Alkalinity, total (mg $CaCO_3$/L)	<35	35–42	42–100	100–420
Hardness, total (mg $CaCO_3$/L)	<35	36–42	42–100	100–420
Nutrient Variables				
Dissolved oxygen (mg/L) (% saturation)	<4 (<25%)	4–7 (25–50%)	7–8 (50–75%)	≥8 (>75%)
Chlorophyll *a* (µg/L)	<2.0 or >25	2.0–2.5 or 20–25	8–20	2.5–8
Total phosphorus (µg/L)	<5 or >50	5–10 or 35–50	10–25	25–35
Total nitrogen (µg/L)	<75 or >750	75–150 or 525–750	150–375	375–525
Secchi depth (m)	<1, >8	1–2, 6–8	4–6	2.0–4.0
Physical Variables				
Temperature (°C)	<2, >30	2–10 or >28	10–16 or 24–28	16–24
Conductivity (µS/cm)	<30	<30–60	60–110	≥110
TDS (mg/L)	<20	20–40	40–70	>70
Salinity (mg/L) (ppt)	>10	8–10 (<0.01)	5–10 (0.005–0.01)	<5 (<0.005)
Turbidity (NTU)	>80	20–80	5–20	<5
TSS (mg/L)	>96	28–96	8–28	<8

Use column 3 for assessing potential for survival (but little potential for nuisance infestation). Columns 4 and 5 indicate good potential for long-term survival and the criteria are used to assess temporal and seasonal variations in infestation potential. TSS was estimated using $\log_{10}$ TSS (mg/L) = 0.899 $\log_{10}$ Turbidity (NTU) + 0.272.

3.3.3.1 Survival Potential of Quagga Mussels Based on Chalk Variables

There is a paucity of experimental data and observational data correlating distribution and abundance to the ecological and physiological requirements and tolerances of quagga mussels. As a result, most assessments of the distribution potential of quagga mussel are based on zebra mussel criteria, with a few exceptions.

In 2003, Jones and Ricciardi (2005) found quagga mussels at 16 sites mussels in the St. Lawrence River down to near Montreal, where calcium concentrations measured 12.4–30.0 mg/L but absent from four sites below the Ottawa River where the calcium concentrations measured 7.6–10.0 mg/L. However, they did find zebra mussels at sites with calcium concentrations as low as 8.0 mg/L, suggesting that quagga mussels have a higher calcium threshold than zebra mussels, at least for settlement and growth (Jones and Ricciardi, 2005). Cohen (2007) suggested that the populations below the Ottawa River are almost certainly sink populations, that is, populations made up of mussels that spawned in higher calcium water in the upper St. Lawrence River or in Lake Ontario and that drifted downstream and settled in lower calcium sites where they cannot reproduce. In addition, the sites below the confluence with the Ottawa River may be subject to substantial fluctuations in calcium levels with changes in the relative flows from the two rivers, so conclusions based on correlations between calcium measurements taken on a single day and the presence or absence of mussels at these sites on that day should be suspect (Cohen, 2005).

Grigorovich et al. (2008) collected quagga mussels from Lake Superior, which has the lowest mean calcium and alkalinity levels of all the Great Lakes, 12–14 mg Ca/L and 51 mg $CaCO_3$/L, respectively (Horne and Goldman, 1994; Mackie, 2004). They found the mussels in 2005–2006 and based on size (2.65 cm long) suspected that the mussels had arrived in 2003. Zebra mussels were mixed among quagga mussels and outnumbered them 8:1 at the time. In Lakes Ontario and Erie with 39–42L and 32–38 mg Ca/L, respectively (Horne and Goldman, 1994; Mackie, 2004), quagga mussels far outnumber zebra mussels, their dominance occurring three years after their introduction (Mills et al., 1999), whereas in some European rivers, like the lower Dnepr River, it takes 4–9 years (Kharchenko, 1995; Mills et al., 1996). However, in some European rivers like the lower Don River in Russia, it has been more than 20 years that the quagga mussel was introduced and it still has not outnumbered zebra mussels (Zhulidov et al., 2004, 2005). In fact, the relative abundance peaked in 1999 at 30–50% (among 14 sites examined) and has since been declining, currently at 10–18% (Zhulidov et al., 2006). In upper parts of the Don River system quagga mussels dominated zebra mussels at sites with 119 mg Ca/L, while zebra mussels dominated at sites with lower calcium concentrations (45–78 mg/L), suggesting a higher calcium requirement for quagga mussels (Zhulidov et al., 2004). While the calcium levels are too high to determine a threshold for establishing a quagga mussel population, Zhulidov et al. (2006) suggest that 119 mg/L exceeds the optimum calcium level for zebra mussels (Ludyanskiy et al., 1993) but is apparently not above the optimum for quagga mussels.

The criteria given in Table 3.10 for quagga mussels is a minor variation of those given for zebra mussels in Table 3.7 due to the findings reported by Grigorovich et al. (2008) with quagga mussels surviving at 12–14 mg Ca/L in Lake Superior and up to about 120 mg Ca/L in the upper Don River (Zhulidov et al., 2006). The minimum value of 8 mg/L and an optimal value of at least 30 mg Ca/L (Table 3.10) are the same as for zebra mussels (Table 3.7). The criteria for total alkalinity and total hardness are based on the calcium levels (i.e., multiplying by 3.5 and rounding off to the nearest 10 units). The total hardness values assume that all the hardness is due to cations of calcium and magnesium carbonates and will be higher if cations of calcium and/or magnesium attached to noncarbonate anions are present.

Unfortunately, there is little or no information on pH as a variable limiting growth, reproduction, or survival of quagga mussels. The pH criteria for predicting survival and infestation levels of quagga mussels in Table 3.10 are the same as for the zebra mussel in Table 3.7.

3.3.3.2 Survival Potential of Quagga Mussels Based on Trophic Variables

3.3.3.2.1 Dissolved Oxygen

Compared with zebra mussels, quagga mussels have a lower respiration rate and devote a greater proportion of their energy input to body mass growth and less to reproduction (Stoeckmann, 2003). She found that quagga mussels respired at 6 μL O_2/h less at 16°C to 2 μL O_2/h less at 25°C than zebra mussels. While regressions had low r^2 values, the relationships between oxygen consumed and wet tissue weight were significant. For quagga mussels, Stoekmann (2003) found the following significant at $P < 0.0001$, $r^2 = 0.130$:

$$\log \mu\text{L O}_2/\text{mg soft tissue mass} + 1 = 0.0106\ (\text{temperature °C}) - 0.0073.$$

For zebra mussels, she found the following significant at $P = 0.001$, $r^2 = 0.116$:

$$\log \mu\text{L O}_2/\text{mg soft tissue mass} + 1 = 0.0138\ (\text{temperature °C}) + 0.0653.$$

Karatayev et al. (2007b) stated that quagga mussels require 1.5 mg/L of oxygen at 20°C. The quagga mussel extends to greater depths than zebra mussels in Lake Erie (Claxton and Mackie, 1998; Claxton et al., 1998) but is unlikely to penetrate the hypoxic hypolimnetic waters (McMahon and Bogan, 2001), especially the central basin which goes anoxic annually (Charlton, 1980). It is interesting to note that Wilhelm et al. (2006) found phycoerythrin-rich cyanobacteria to be abundant and active in a narrow region (~1.5 m) below the thermocline during hypoxic conditions, reaching abundances of greater than 10^5 mL^{-1} and being the primary agent releasing 1.5 mg O_2/L above the daytime demands in this region. If quagga mussels are present, this might be one or the only source of oxygen.

McMahon (1996) states that the profundal habitat of quagga mussels suggests that they are more hypoxia tolerant and better O_2 regulators than zebra mussels, although Son (2007) reported that both species of dreissenids occur in the Black Sea region in various habitats where hypoxia is absent. Birger et al. (1978) reported that low concentrations of *Microcystis aeruginosa* (10–25 g/L) were fatal to both *D. polymorpha* and *Dreissena bugensis* but the latter tolerated anoxic conditions associated with the algae better than *D. polymorpha*. It is assumed that the quagga mussel behaves much like the zebra mussel when deprived of oxygen; the partial pressure of oxygen and temperature influence survivorship dramatically with increasing survivorship at higher p_{O_2} and decreasing temperatures (Johnson and McMahon, 1998). The quagga mussel probably displays broad seasonal variation in hypoxia tolerance and its hypoxia limits may be used to assess infestation risk, as was suggested by Johnson and McMahon (1998). Since the temperature of the hypolimnion in most deep lakes is 4°C, and 25% saturation represents hypoxia typical of eutrophic lakes, the lower limit of hypoxia tolerance was chosen as 25% of 13.13 mg O_2/L [=100% saturation at 4°C (Mackie, 2004)], or about 3 mg O_2/L. This is slightly lower than Cohen's (2007) recommendation of 4 mg/L for both quagga and zebra mussels, based on a review of zebra mussels' requirements. Fifty percent saturation is selected as the criterion for the lower limit of moderate infestation levels (Table 3.10).

3.3.3.2.2 Chlorophyll a

Nicholls and Hopkins (1993) and Holland (1993) attributed rapid declines in zebra and quagga mussel populations in western Lake Erie and Long Point Bay to phytoplankton exploitation. However, Petrie and Knapton (1999) believe predation by waterfowl, especially lesser (*Aythya affinis*) and greater scaups (*A. marila*), buffleheads (*Bucephala albeola*), and redheads (*A. Americana*) also contributed to the declines in the two dreissenid species in Long Point Bay.

Quagga mussels probably have similar filtration rates as zebra mussels for filtering seston from the water. Unfortunately, published values are from Lake Erie where there is a mix of zebra and quagga mussels, although quagga mussels currently outnumber zebra mussels. For example, densities of quagga mussels in the shallow, warmer western basin of Lake Erie have increased from 20% of the dreissenid population in 1998 to 80% in 2001 (Stoekmann, 2003). Pontius (2000) reported a filtration rate (FR) that was related to dry weight (DW) in grams of $FR = 0.046DW^{0.88}$ for the western basin and Baldwin et al. (2002) reported a clearance rate that also depended on dry weight (DW) in grams of $FR = 0.005DW^{0.5419}$ for both central and eastern basins.

MacIsaac et al. (1992) suggest that the impact of settled mussel grazing activities on phytoplankton stocks may be up to 1162 times greater than that exerted by veliger populations in western Lake Erie. Based on 1990 size–frequency distributions and associated literature-derived clearance rates, reef-associated *Dreissena* populations (mix of zebra and quagga mussels) in western Lake Erie (mean depth similar to 7 m) possess a tremendous potential to filter the water column (up to 132 m^3/m^2/d) and redirect energy from pelagic to benthic foodwebs. Their preliminary analyses indicate that chlorophyll *a* concentration is strongly depleted (<1 mg/L) above *Dreissena* beds in western Lake Erie.

Bunt et al. (1993) found that small (2–11 mm long), settled zebra and quagga mussels comprise up to 90% of individuals inhabiting reefs in western Lake Erie and that pumping rates ranged between 0.20 and 4.45 mL/individual/h and increased in relation to mussel shell length. Based on 1990 size–frequency distributions for reefs in western Lake Erie and a pumping rate of 2.7–6.7 m^3/m^2/d, Bunt et al. (1993) estimated that dreissenids are theoretically capable of pumping between 39% and 96% of the water column daily, assuming the water column mixes thoroughly. Bunt et al. (1993)

agree with MacIsaac et al. (1992, 1999) that recent changes in water quality in western Lake Erie could be primarily related to dreissenid filtering activities, including those of small-bodied individuals.

Ackerman et al. (2001) conducted a field study including a series of cruises over an isolated offshore zebra mussel reef (7–11 m deep) in western Lake Erie to examine the effect of zebra and quagga mussels (*Dreissena* spp.) on the water column. The horizontal currents over the reef were found to be primarily due to the hydraulic flow and surface gravitational seiches. The turbulence generated by these currents was found to be too weak to fully mix the water column. Seasonal stratification is not normally observed in the western basin and Ackerman et al. (2001) believe solar heating during the day and intrusions of cold central basin water caused stable stratification of the water column 60% of the time. They also found a statistically significant mussel-feeding signature in chlorophyll *a* and organic seston concentrations measured within 2 m above the reef and at a mean pumping rate of 16.2 $m^3/m^2/d$ mussels could remove up to 40% of the total seston. In contrast to Bunt et al. (1993) and MacIsaac et al. (1992, 1999), Ackerman et al. (2001) concluded that offshore mussel colonies may have less of an effect on the water column than had been previously estimated by simple stirred reactor models, and the role of dreissenids in the clarification of Lake Erie needs to be investigated further.

Boegman et al. (2008) found that dreissenids (mostly quagga mussels) at a mean pumping rate of 1.7 $m^3/m^2/d$ grazed 53% of the western basin May through September. This grazing resulted in a 0.1-mg/L reduction in the pelagic algae concentration relative to a case without dreissenids. In comparison, dreissenids grazed 77% western basin net algal growth when the lake was modeled as a fully mixed water column. We found that the biomass grazed was governed by a balance between the timescales of vertical wind-induced mixing and benthic grazing. During calm conditions, weak diurnal stratification (~1°C between surface and bottom waters) was sufficient to suppress vertical mixing, when the mean daily wind speed of 4 m above the lake surface was >6 m/s. These conditions allowed a concentration boundary layer, 1 m thick to form, accounting for the reduced grazing effect relative to the fully mixed case proposed by Bunt et al. (1993) and MacIsaac et al. (1992, 1999).

Finally, Zhang (2006) and Zhang et al. (2008) used a two-dimensional model to investigate the impacts of zebra mussels and quagga mussels on phytoplankton in Lake Erie. In contrast to MacIsaac et al. (1992, 1999), their results show that dreissenid grazing impacts on nondiatom edible algae are weakened by a boundary layer above the basin bottom. However, dreissenid grazing impacts on diatoms are less affected by the boundary layer due to the higher sinking rates of diatoms. Zhang et al. (2008) found that dreissenid mussels increase nondiatom inedible algae rapidly with increasing mussel population size, because the dreissenid population excretes a large amount of ammonia and phosphate. In agreement with Ackerman et al. (2001) and Boegman et al. (2008), the results of Zhang (2006) and Zhang et al. (2008) indicate that dreissenid mussels have weak direct grazing impacts on algal biomass, but further suggest that indirect effects of their nutrient excretion have a greater impact on the system (see below).

There are no studies relating the presence and abundance of quagga mussels to chlorophyll *a* levels or even the trophic status of surface waters. However, Baldwin et al. (2002) reported that quagga mussel have a higher tolerance of starvation than zebra mussels. Assuming quagga mussels can tolerate lower levels of food than zebra mussels (Baldwin et al., 2002) but require an optimal amount of algae for growth and reproduction, and values above and below this amount result in negative growth, or even mortality if excessive amounts are present, the criteria given in Table 3.10 for limiting their survival are slightly lower than for zebra mussels (Table 3.7); levels of chlorophyll *a* for moderate and optimal growth and reproduction of quagga mussels (Table 3.10) are similar to those given for zebra mussels in Table 3.7.

3.3.3.2.3 Nutrients (Total Phosphorus, Total Nitrogen)

Recent increases in phytoplankton biomass and the recurrence of cyanobacterial blooms in western Lake Erie, concomitant with a shift from a community dominated by zebra mussels (*D. polymorpha*) to the one dominated by quagga mussels (*D. bugensis*), led Conroy et al. (2005) to test for differences in ammonia-nitrogen and phosphate-phosphorus excretion rates of the two species. They found significant differences in excretion rates both between size classes within a taxon and between taxa, with zebra mussels generally having greater nutrient excretion rates than quagga mussels. Comparing dreissenid ammonia and phosphate excretion with that of the crustacean zooplankton, Conroy et al. (2005) demonstrated that the mussels add to nitrogen and phosphorus remineralization, shortening nitrogen and phosphorus turnover times, and, importantly, modifying the nitrogen and phosphorus cycles in Lake Erie. Conroy et al. (2005) concluded that the increased nutrient flux from dreissenids may facilitate phytoplankton growth and cyanobacterial blooms in well-mixed and/or shallow areas of western Lake Erie.

Zhang (2006) examined the impact of dreissenid communities (mix of quagga and zebra mussels but dominated by quagga mussels) on phytoplankton and nutrients in Lake Erie and found that dreissenids excrete a large amount of ammonia and phosphorus. These amounts increase rapidly with increasing mussel population size and nitrification of ammonia might speed oxygen depletion in the hypolimnion of the central basin. Zhang (2006) and Zhang et al.'s (2006) results indicate that dreissenids had weak direct grazing impacts on algal biomass, but indirect effects of their nutrient excretion have a lot more impact on the system.

The structure and activities of the benthic bacterial community are also influenced by dreissenids. Lohner et al. (2007) examined the structure and activity of bacterial communities living in sediments (1) associated with *D. bugensis* clusters and (2) unassociated with established clusters (i.e., lake bottom sediments) from locations in Lake Erie, Lake Ontario, and several inland lakes. They found that overall bacterial activity and metabolic diversity were generally enhanced by the presence of clusters. Lohner et al. (2007) concluded that *Dreissena* appears to affect both structure and metabolic function of the benthic bacterial community and may have yet unexplored ecosystem and food web consequences.

Excretion rates of nitrogen and phosphorus by quagga mussels are probably similar to those by zebra mussels. Unfortunately, published values are from Lake Erie where there is a mix of the two species, although quagga mussels predominate.

As might be expected, excretion rates are related to body mass as DW in grams. Conroy et al. (2005b) estimated phosphorus (P) excretion rates (as mg P/DW/d) for dreissenids in the western basin of Lake Erie from the regression, $Log_{10} PO_4 = 0.505 Log_{10} DW - 1.172$, and for the central and eastern basins from the regression, $Log_{10} PO_4 = 0.297 Log_{10} DW - 1.195$. Nitrogen excretion rates (as mg N/DW/d) were estimated for all three basins by Conroy et al. (2005b) using one regression, $Log_{10} NH_4 = 0.37 Log_{10} DW + 0.021$.

It appears that the levels of nutrients that affect abundances of zebra mussels (Table 3.7) are the same as for quagga mussels and are therefore applied in Table 3.10.

3.3.3.2.4 Secchi Depth, Trophic Status

The levels of chlorophyll *a* and total phosphorus and nitrogen given in Table 3.7 are based on trophic status of the waters. Since Secchi depth correlates well with each nutrient, waters with Secchi values of <1 and >8 m will not support long-term quagga mussel survival and waters with Secchi depth values of 2–4 m will support massive infestations; Secchi depths of 4–6 m will support moderate levels of quagga mussel densities Secchi depths between 1–2 m and 6–8 m will support only low levels of quagga mussel abundances (Table 3.7).

3.3.3.3 Survival Potential of Quagga Mussels Based on Temperature

As with zebra mussels, quagga mussels have upper and lower thermal thresholds for survival, growth, and reproduction. Quagga mussels are well known to be able to inhabit the hypolimnion of lakes where the temperature seldom varies from 4°C all year, as in Lake Ontario (Hoyle et al., 1999; Dermott, 2001; Lozano et al., 2001; Owens and Dittman, 2003) and Lake Michigan (Nalepa et al., 1998; Pothoven et al., 2001), or spring and fall as in Lake Erie (Dermott and Kerec, 1997; Roe and MacIsaac, 1997; Claxton and Mackie, 1998; Claxton et al., 1998). Quagga mussels are also typically more abundant than zebra mussels at greater depths (Claxton et al., 1998; Mills et al., 1993b, 1996; Roe and MacIsaac, 1997; Ricciardi and Whoriskey, 2004). The quagga mussel has expanded its range in the Great Lakes from east to west (Mills et al., 1999; Jarvis et al., 2000). Initially found primarily in deeper, cooler waters (May and Marsden, 1992), quagga mussels can survive (MacIsaac, 1994) and are present in shallow, warmer waters (Mitchell et al., 1996), although their numbers appear to be declining (Holland, 1993; Petrie and Knapton, 1999).

Karatayev et al. (2007b) reported that quagga mussels' lower temperature limit for adult survival is 0°C. He also stated that quagga mussels are found in the upper Volga River, which freezes in the winter. It is unlikely that any mussel can survive freezing to the point that tissue cells rupture. The lower thermal threshold for growth and reproduction of quagga mussels appears to be near 4–5°C. Roe and MacIsaac (1997) found eggs in mussels collected from Lake Erie at 55 m depth maturing at 4.8°C; the mussels also showed evidence of spawning with 80% of the females having at least some mature eggs and 20% having spent gonads. Claxton and Mackie (1998) set cages in Lake Erie at 23 m; some cages had mussels from the 23-m depth, others had mussels transplanted from the epilimnion to the 23-m depth. After histological examination of the gonads of mussels pulled weekly between May 24 and July 20, 1994, they found that quagga mussels from the hypolimnion had spawned at 9°C and mussels transplanted from the epilimnion spawned at 9–10°C; all zebra mussels failed to spawn in the hypolimnion. Karatayev et al. (2007b) found spawned mussels at 6°C and reported that quagga mussels need water temperatures of 5–7°C to spawn in deep waters of Lake Michigan.

At 95% relative humitidy (RH), only 10% of *D. bugensis* survived three days exposure (none survived five days), contrasting with the high survivorship (70–100%) of *D. polymorpha* over the same period. Under cool, humid conditions (10°C, 95% RH), large *D. bugensis* (21–24 mm shell length) survived at least 10 days aerial exposure but with lower survivorships than large zebra mussels.

Comparisons of the upper thermal threshold limits of quagga and zebra mussels have been well studied. The acclimation temperature and rate of acclimation affects the upper thermal threshold. Domm et al. (1993), Spidle (1994), and Spidle et al. (1995) reported that the quagga mussel is more vulnerable than the zebra mussel under most combinations of acclimation temperature and rate of temperature increase. Spidle et al. (1995) also reported that mussels of both species acclimated to 5°C were less able to survive at high temperatures (30–39°C) than mussels acclimated to 15°C or 20°C. Quagga mussels acclimated to 20°C and then exposed to water temperatures rising at the rate of 0.3°C/min gaped open and did not respond to prodding at 36.4°C while zebra mussels only did so at 37.0°C (Domm et al., 1993). The upper thermal limit of the quagga mussel appears to be near 25°C while that of the zebra mussel is near 30°C (Domm et al., 1993; Spidle et al., 1995; McMahon, 1996). Metabolic functioning of the two species also probably differs, as indicated by lower tolerance of high temperatures in quagga mussels (Domm et al., 1993; Spidle et al., 1995). The reduced upper temperature limit of the quagga mussel implies that it will either not be able to expand as far south in North America as has the zebra mussel, or be limited to deeper parts of large lakes.

Thorp et al. (1998) examined the effects of small temperature increases and two turbidity levels on zebra and quagga mussels over two seasons (warm season: July 20 to October 4, 1994; cool season: October 17, 1994, to January 8, 1995); they exposed the mussels to three temperatures (ambient, ambient +2°C, and ambient +4°C) and two turbidities (ambient and twice ambient levels of suspended sediment) in a factorial design. The highest ambient (Ohio River) temperature was 26°C (August), the lowest was 4°C (January). Higher temperatures had little or no influence on growth in the warm season but had a positive effect in the cool season. For example, with a 4°C increase, the growth rates of quagga and zebra mussels tripled during the fall to early winter period. The ambient +2°C and ambient +4°C treatments caused substantial death in the warm season but had no detectable effect on survivorship in the cool season. Thorp et al. (1998) found 0% survival of quagga mussels at ambient +4°C (= ~25–32°C) in the warm season and 67–72% survival of quagga mussels at ambient +4°C (= ~5–10°C) in the cool season and the predicted maximum temperature for survival was 25°C. Turbidity of the Ohio River water varied between 2 and 44 NTU but there was no significant overall effects of turbidity on growth. There was some evidence that turbidity had a slightly negative influence on survivorship under some conditions.

While the above laboratory experiments using acclimated mussels suggest an upper thermal threshold for survival of 25°C, field observations and experiments using nonacclimated mussels suggest an upper limit for some populations of quagga mussels of at least 30°C, although the field observations are for European populations of quagga mussels. For example, in the Zaporozhskoe Reservoir in the Ukraine, quagga mussels survive in water temperatures up to 30.5°C (Karatayev et al., 2007b). On this basis, Karatayev et al. (2007b) concluded that the upper threshold limit for quagga mussels is 31°C. Domm et al. (1993) reported upper thermal limits of <32°C for mussels without acclimation and 36.4°C for mussels in rapidly rising temperatures. Spidle (1994) predicted 100% mortality at 34.3°C of mussels acclimated to 15°C with a temperature increase of 1°C/min and 35.3°C of mussels acclimated to 20°C with the same acclimation rate. Thorp et al. (1998) predicted >90% mortality of mussels at 34°C the maximum summer temperature of Lake Erie.

There is a preponderance of evidence above showing that both the lower and upper thermal threshold limits for quagga mussels are less than those for zebra mussels. Table 3.7 uses <15°C for the lower threshold limit for survival, which also applies to reproduction for zebra mussels. Field observations show that quagga mussels can spawn at 4.8°C (5°C), suggesting that this is the lower threshold limit for survival for quagga mussels and is used as the lower criterion in Table 3.10. Based on the above results, Cohen (2007) used 30°C as the upper thermal threshold limit in his analysis of potential distribution of quagga mussels in California. Again, there is a preponderance of evidence that the upper maximum thermal threshold limit is less for quagga mussels than for zebra mussels. In selecting the upper limit, one must consider the mixing effects of winds on water temperatures. For example, summer temperatures in the shallow margins of Lake Mead routinely reach 30°C, but the duration of exposure of quagga mussels to this temperature is probably not sufficient to kill all of them. The mean maximum temperature is more meaningful than a snapshot temperature, and assessing risk potential based on temporal and seasonal variations in temperature is better than assessments based on mean temperatures. In this context, the upper thermal threshold limit for zebra mussels is 30°C but the mean upper thermal threshold limit is 28°C (Table 3.7), which implies that mussels may survive but will be stressed sufficiently to negatively affect growth and reproduction. Based on the above evidence, the upper maximum thermal threshold limit for quagga mussels also appears to be near 30°C but the mean upper maximum thermal threshold limit is closer to 25°C (Table 3.10).

3.3.3.4 Survival Potential of Quagga Mussels Based on Conductivity, TDS, and Salinity

3.3.3.4.1 Conductivity

As with zebra mussels, calcium and magnesium are important cations and carbonates and bicarbonates are among the most important anions for quagga mussels, and the most important criterion for survival is a minimal conductivity, the upper limit not being critical until salinity becomes a factor. The criteria recommended for zebra mussels (Table 3.7) were derived from regressions described in Sections 3.2.1.1.1 and 3.2.7 relating conductivity to calcium and alkalinity and are used here for quagga mussels as well (Table 3.10). Also see discussion for zebra mussels under Section 3.3.1.1.4.

3.3.3.4.2 *Total Dissolved Solids*

Levels of TDS were determined by multiplying the conductivity by 0.64 (see Section 3.2.7.2) as was done for zebra mussels (Table 3.7). The values in Table 3.10 are rounded to the nearest 10 mg/L (e.g., 20 mg/L instead of 19 mg/L).

3.3.3.4.3 *Salinity*

There appear to be some differences in absolute salinity tolerances of dreissenids between Eurasian and North American studies. Mills et al. (1996) reviewed the Russian literature and found that salinity, which ranges from 0.5 to 10 ppt, is the main factor controlling the distribution of dreissenids in the Dnieper-Bug estuary with quagga mussels being less tolerant than zebra mussels. Mills et al. (1996) reported a maximum salinity of 4.0 ppt for quagga mussels compared to 7.6 ppt for zebra mussels in the Dnieper-Bug estuary. Forty-day laboratory exposures of Ukrainian quagga mussels also showed lower salinity tolerance, with high survival rates at up to 4 ppt at 18–21°C (compared to 6 ppt for zebra mussels), and high survival rates at up to 5 ppt at 7–15°C (compared to 8 ppt for zebra mussels).

Spidle (1994) and Spidle et al. (1995) found that Lake Erie populations of *D. polymorpha* and *D. bugensis* could not survive more than 18 days at salinity levels greater than 5 ppt. No interspecific difference in survival time when exposed to salinity was shown (Spidle, 1994; Spidle et al., 1995). They also showed that the negative effect of salinity on survival is enhanced at warmer temperatures (15–20°C), with mussels of both species having a much shorter survival time at all salinity levels. Mills et al. (1996) in their review of European and North American salinity tolerance studies concluded that in North America, there is no evidence that the salinity tolerance of the quagga mussel is any greater than that of the zebra mussel. However, Therriault et al. (2004) report a salinity range of up to 3 ppt in the quagga mussel's native geographic range and up to 2 ppt in its introduced range. Karatayev et al. (2007b) concluded, based on a literature review, that quagga mussels' upper salinity limit is 3.5 ppt. For an analysis of potential distribution in California, Cohen (2007) used an upper limit of 4 ppt in waters with relatively stable salinities and that value is used here as well (Table 3.10).

3.3.3.5 Survival Potential of Quagga Mussels Based on Turbidity and TSS

3.3.3.5.1 *Turbidity*

The impact of quagga mussels on water clarity is pretty much the same as that of the zebra mussel, that is reductions in chlorophyll *a* and plankton levels, as discussed in Section 3.3.1.1.2.4 (see also Section 6.2.1.2). A recent study by Barbiero et al. (2006), however, has revealed a new cause of increased water clarity, still caused by dreissenids; they showed that since the dreissenid (first zebra, then quagga mussels) invasion of Lake Erie, calcium concentrations in the downstream offshore waters of Lake Ontario have decreased by approximately 4–5 mg/L. This decline has coincided with a threefold reduction in August turbidity values and nearly a doubling of Secchi depths, presumably due to reduced summer calcite precipitation events (whitening) in the lake. Barbiero et al. (2006) reported declines in turbidity from 1.1 to 1.6 NTU between 1985 and 1991 down to 0.3 to 0.4 NTU between 1996 and 2004.

As with zebra mussels, the impact of zebra mussels on TSS is well known, but the reciprocal relationship has not been as well examined and little information exists on the effects of suspended solids and food concentrations on quagga mussel survival or even metabolic activities. It is assumed here that the criteria used for zebra mussels (Table 3.7) also apply to quagga mussels (Table 3.10). Hence, NTUs <5 are considered optimal for quagga mussel infestations; NTU of 5–20 will support moderate infestation levels; NTU of 20–80 will support little infestations; NTU >80 are lethal and will not support quagga mussel survival over a long term (Table 3.10).

3.3.3.5.2 Total Suspended Solids

The TSS criteria for quagga mussels (Table 3.10) are also assumed to be similar to those for zebra mussels (Table 3.7) and are estimated from Christensen et al's. (2001) regression:

$$\log_{10} \text{TSS (mg / L)} = 0.899 \log_{10} \text{Turbidity (NTU)} + 0.272.$$

3.3.4 Conrad's False Mussel, *Mytilopsis leucophaeata* (Andrusov, 1897)

The threshold limits for growth, reproduction, and survival for each of the physiological and ecological variables of *Mytilopsis leucophaeata* were determined from averaging the levels reported by the numerous researchers listed in Table 3.11. The ranges in values for each of the risk levels are given in Table 3.12. Assessment of potential survival of Conrad's false mussel in a hypothetical lake is pretty much the same process as described for zebra mussels, and is not repeated here.

TABLE 3.11
Studies Using Various Parameters to Assess Risk Potential of Survival and Potential Density of Conrad's False Mussel

Variables	References
Alkalinity	See Table 3.6 and Deaton et al. (1989)
Calcium	See Table 3.6 and Deaton et al. (1989)
Chlorophyll *a*	See Table 3.6 and Bergstrom (2004)
Conductivity, TDS	See Table 3.6 and Castagna and Chanley (1973), Marelli and Gray (1983), Deaton et al. (1989), MacNeill (1991), Pathy and Mackie (1993), Walton (1996)
Dissolved oxygen	See Table 3.6 and Bergstrom (2004)
Nitrogen, total	See Table 3.6 and Bergstrom (2004)
pH	See Table 3.6 and Deaton et al. (1989)
Phosphorus, total	See Table 3.6 and Bergstrom (2004)
Salinity	See Table 3.6 and Castagna and Chanley (1973), Marelli and Gray (1983), Deaton et al. (1989), MacNeill (1991), Pathy and Mackie (1993), Walton (1996)
Secchi depth, TSS	See Table 3.6 and Bergstrom (2004)
Temperature (°C)	See Table 3.6 and Siddall (1980), Rajagopal et al. (2005a), Verween et al. (2005, 2007b), Laine et al. (2006)
Total hardness	See Table 3.6 and Deaton et al. (1989)
Turbidity (NTU), TSS	See Table 3.6 and Barbiero et al. (2006)

The variables are listed alphabetically.

TABLE 3.12
Criteria Used in Determining Levels of Infestation by Conrad's False Mussel in North America

1. Parameter	2. No Potential for Adult Survival	3. Little Potential for Larval Development	4. Moderate Potential for Nuisance Infestations	5. High Potential for Massive Infestations
Chalk Variables				
Calcium (mg Ca/L)	<8	8–12	12–30	30–120
pH	<7.0, >9.5	7.0–7.8, 9.0–9.5	7.8–8.2 or 8.8–9.0	8.2–8.8
Alkalinity, total (mg $CaCO_3$/L)	<30	30–45	45–100	100–420
Hardness, total (mg $CaCO_3$/L)	<30	30–45	45–100	100–420
Nutrient Variables				
Dissolved oxygen (mg/L) (% saturation)	<2 (<15%)	2–4 (25–40%)	5–6 (40–50%)	≥6 (>50%)
Chlorophyll *a* (µg/L)	<5 or >25	5–10	10–20	20–25
Total phosphorus (µg/L)	<10 or >50	10–25	25–35	35–50
Total nitrogen (µg/L)	<150 or >750	150–375	375–525	525–750
Secchi depth (m)	<0.5, >6	4–6	3.0–4.0	0.5–3.0
Physical Variables				
Temperature (°C)	<10, >30	10–13 or 28–30	15–20 or 24–28	20–24
Conductivity (µS/cm), 25°C	<450 and see salinity	450–3400 and see salinity	3400–6600 and see salinity	5100–15,400
TDS (mg/L)	<300	300–2300 and see salinity	2300–4400 and see salinity	3400–10,300
Salinity (mg/L) (ppt)	<0.2 or >30	0.2–2, 25–30	2–4, 12–25	5–12
Turbidity (NTU)	>80	20–80	5–20	<5
TSS (mg/L)	>96	28–96	8–28	<8

Use column 3 for assessing potential for survival (but little potential for nuisance infestation). Columns 4 and 5 indicate good potential for long-term survival and the criteria are used to assess temporal and seasonal variations in infestation potential. TSS was estimated using $\log_{10}$ TSS (mg/L) = 0.899 $\log_{10}$ Turbidity (NTU) + 0.272.

3.3.4.1 Survival Potential of Conrad's False Mussel Based on Chalk Variables

There is a dearth of information on the calcium, pH, and total alkalinity and hardness thresholds of *M. leucophaeata* because water chemistry was not provided by researchers who reported the mussel from the different localities in North America, such as the Mississippi River (Koch, 1989), Hudson River (Rehder, 1937; Jacobson, 1953), Chesapeake Bay (Milke and Kennedy, 2001). The Mississippi River has calcium levels above 50 mg/L, a pH usually 7.6 and 8.0, and a total alkalinity above 170 mg $CaCO_3$/L (Foster et al., 1993; Claudi and Mackie, 1994). The Hudson River has low calcium (≈12–15 mg/L) and alkalinity (40–50 mg $CaCO_3$/L) levels (Simpson et al., 2006). Deaton et al. (1989) state that *M. leucophaeata* has been collected from waters with calcium levels as low as 12 mg/L and as high as 56 mg/L. It appears that the threshold limits for survival, growth, and reproduction of Conrad's false mussel based on chalk variables is not much different than that for the zebra mussel and criteria given in Table 3.7 are repeated in Table 3.12.

3.3.4.2 Survival Potential of Conrad's False Mussel Based on Trophic Variables

Virtually no studies have related threshold limits for survival, growth, and reproduction of Conrad's false mussel based on trophic variables. Bergstrom (2004) presented Secchi depth data to illustrate the impact of Conrad's false mussel on water clarity in Chesapeake Bay watershed. He examined several rivers with different trophic levels (based on Secchi depth), from very eutrophic waters as in Patapsco River (average Secchi depth range of 0.5–1.25 m; extremes 0.3–1.75 m) to eutrophic waters of Magothy River (average Secchi depth range of 1.0–1.8 m; extremes 0.5–3.0 m) and reported increases in Secchi depth at all rivers. Because these rivers had very high densities of mussels, it seems that Conrad's false mussel is able to tolerate high levels of nutrients, perhaps more than zebra and quagga mussels. The criteria given in Table 3.12 are based on Bergstrom's (2004) Secchi depth data and the criteria for chlorophyll *a*, dissolved oxygen, and nutrients were adjusted accordingly.

Dissolved oxygen: It appears that *M. leucophaeata* is able to tolerate very low levels of oxygen since Verween et al. (2007b) collected it in the Westerscheldt River near Antwerp where oxygen levels range from 0.8 to 12 mg/L. The threshold levels given in Table 3.12 for survival, growth, and reproduction are based on Bergstrom's (2004) Secchi depth data and the criteria for dissolved oxygen were adjusted accordingly.

Chlorophyll a: The threshold levels given in Table 3.12 for survival, growth, and reproduction are based on Bergstrom's (2004) Secchi depth data and the criteria for chlorophyll *a* were adjusted accordingly.

Nutrients (total phosphorus, total nitrogen): The threshold levels given in Table 3.12 for survival, growth, and reproduction are based on Bergstrom's (2004) Secchi depth data and the criteria for nutrients were adjusted accordingly.

3.3.4.3 Survival Potential of Conrad's False Mussel Based on Temperature

Verween et al. (2005) examined the seasonal variability of reproduction and growth of *M. leucophaeata* in an Antwerp harbour in 2000 and found the temperature at first detection ranged between 16.2°C and 19.5°C; no veligers were found later than

November 19 with average temperature 13°C ± S.E. 0.4°C. In 2001 the last veliger densities were found on November 20 (13.7°C, 4.2 psu). Based on these data the threshold temperature for gamete maturation in *M. leucophaeata* is near 13°C.

Verween et al. (2005) identified annual and geographic variation in temperature as the primary factor triggering reproduction of *D. polymorpha* and this may apply to Conrad's false mussel as well. Although temperatures at reproduction were not reported, the species has been collected in Florida where temperatures ranged from 13°C to –30°C (Siddall, 1980), in the harbor of Antwerp where temperatures ranged from 7°C to 26°C (Verween et al., 2007b), and in the Noordzeecanal in the Netherlands where temperatures ranged from 5°C to 20°C (Rajagopal et al., 2005a), and in a bay that is warmed by cooling waters from a power plant in Finland where temperatures range annually from around 5°C to over 25°C (Laine et al., 2006); they also reported its occurrence in areas outside of this bay where the water is ice-covered in the winter and near 0°C, but they were not sure if it reproduces in these waters.

In laboratory experiments, *M. leucophaeata* was more tolerant of exposure to high temperatures (36–41°C) than were zebra mussels (Rajagopal et al., 2005a,b). As noted above, Verween et al. (2007a) found good survival of embryos exposed to temperatures from 15°C to 24°C for 48 h at a range of salinities, and observed high rates of mortality only at the extreme test temperatures of 10°C and 30°C. They showed that 2-day-old larvae survived well over the entire test range from 5°C to 25°C, at a range of salinities.

3.3.4.4 Survival Potential of Conrad's False Mussel Based on Conductivity, TDS, and Salinity

Conductivity: Most of the literature cites salinity values rather than conductivity when examining tolerance thresholds for survival, growth, and reproduction. However, the conversion from salinity to conductivity can be determined from Table 3.5. Conrad's false mussel has more trouble-tolerating water with few ions than with a lot of ions. Deaton et al. (1989) reported 100% mortality after 80 days in 0 ppt (deionized water). It is unable to reproduce in freshwater (Miller et al., 1993) and the species is reported to be uncommon below 0.1 ppt (≈200 μS/cm, 25°C) (Deaton et al., 1989; MacNeill, 1991; Walton, 1996), 0.2 ppt (≈450 μS/cm, 25°C) (Rajagopal et al., 2005b), 0.3 ppt (≈620 μS/cm, 25°C) (Laine et al., 2006), and 0.4 ppt (≈800 μS/cm, 25°C) (MacNeill, 1991). In Belgium, *M. leucophaeata* has established large populations in conditions where salinity varies from 0.8 ppt (≈1600 μS/cm, 25°C) to 10.3 ppt (≈17,500 μS/cm, 25°C) total salinity during the reproductive period (Verween et al., 2005). It commonly occurs in the Netherlands between 2.0 and 2.5–3.0 ppt salinity (4700–5100 μS/cm, 25°C). In the Hudson River Conrad's false mussel is commonly found where the salinity range is 2–6 ppt (3400–9600 μS/cm, 25°C) (Deaton et al., 1989; MacNeill, 1991; Walton, 1996). Based on these data the lower threshold level for survival seems to be ≈450 μS/cm or 0.2 ppt salinity, and for growth and reproduction 3400 μS/cm or 2 ppt salinity. The upper thresholds are discussed under salinity.

Total dissolved solids: The TDS threshold levels were determined by multiplying the conductivity by 0.67 and rounding off to nearest 100. Conrad's false mussel cannot survive TDS levels below 400 mg/L (≈0.2 ppt) or above 30,800 mg/L

(≈30 ppt). The threshold level for growth and reproduction appears to be between about 2600 mg/L (≈2 ppt) and 6000 mg/L (≈5 ppt). The upper threshold levels are discussed under salinity.

Salinity: Both field studies and laboratory experiments demonstrate a remarkable ability of the species to survive sudden large changes in salinity (of up to 30 ppt in either direction), and to survive for many weeks at salinities from 0 to 30 ppt. Laine et al. (2006) give its salinity range as freshwater to "well over 20 ppt." Based on a literature search, MacNeil (1991) reported a maximum tolerated salinity for *M. leucophaeata* of 26 ppt and normal ranges with minimum salinities of 0.1–1.9 ppt and maximum salinities of 3–18 ppt. Verween et al. (2006a) stated that adult *M. leucophaeata* can survive in salinities from 0.1 to 31 ppt and that both fresh and seawater are outside its range for survival. Thresholds for survival, growth, and reproduction of *M. leucophaeata* have been reported for salinities up to 12 ppt in Chesapeake Bay (Castagna and Chanley, 1973), between 8 and 22 ppt in a Florida embayment (Siddall, 1980), between 6 and 15 ppt for a British population at Cliffe Fort Lagoon (Bamber and Taylor, 2002), between 1 and 15 ppt in the Noordzee canal in the Netherlands (Rajagopal et al., 2005a), between 0.3 and 8 ppt in the German Baltic Sea coast (Laine et al., 2006), and between 0.1 and 12 ppt in Belgium near Antwerp (Verween et al., 2006b). It occurs in the upper reaches of the Hudson River only to where the range is 0–3 ppt (Deaton et al., 1989; MacNeill, 1991; Walton, 1996). Verween et al. (2006b) report it as living within the range of 1–18 ppt. Verween et al. (2005) examined the seasonal variability of reproduction and growth of *M. leucophaeata* in an Antwerp Harbour and found the salinity at first detection ranged from 2.6 to 4.9 ppt. The intensity and duration of reproduction is believed to be controlled by an interaction of environmental factors, such as temperature, food availability, and salinity. The variability in environmental factors indicates that for *M. leucophaeata* a threshold condition for spawning, like for *D. polymorpha*, does not seem to exist. Verween et al. (2005) found no veligers later than November 19 when salinities were highly variable, ranging from 0.8 ppt in 2002 to 10.1 ppt in 2003. Siddall (1980) stated that in the natural population that he studied, spawning coincides with decreasing salinities occurring during the rainy season but varying salinity up to 32 ppt has no effect on larval development but the species is normally found at salinities <10 ppt. In North America the distribution of *M. leucophaeata* and *D. polymorpha* overlap when salinities are below 3.0 ppt total salinity (Pathy and Mackie, 1993).

Based on laboratory experiments, Siddall (1980) found that *M. leucophaeata* was able to complete larval development in salinities up to 32 ppt salinity. In another experiment, Siddall (1980) found all larvae developed and metamorphosed normally, with all treatments growing at the same rate and to the same size when reared at 10, 24, and 32 ppt. Verween et al. (2007a) exposed 4-h-old embryos and 2-day-old larvae to a series of salinity and temperature bioassays using 48 h exposures. They found that most larvae survived well over the entire salinity test range of 5–25 ppt and over the temperature range of 5–25°C, with only 14% mortality in the highest combination of temperature and salinity (10°C and 30°C, and 0 and 25 ppt); embryo survival was generally good over a temperature range of 15–24°C and a salinity range of 3–22 ppt. Based on their results, Verween et al. (2007) predict that this species could establish populations in many brackish water bodies worldwide, with only colder regions potentially limiting its invasion success. Marelli and Gray (1983) suggested that *M. leucophaeata*'s upstream limits could be controlled by the physiological tolerances of the larvae, or by a lack of salinity pulses that may be needed to trigger spawning.

3.3.4.5 Survival Potential of Conrad's False Mussel Based on Turbidity and TSS

There is no information on threshold levels required for survival, growth, and reproduction of Conrad's false mussel.

3.3.5 Golden Mussel, *Limnoperna fortunei* (Dunker, 1857)

The threshold limits for growth, reproduction, and survival for each of the physiological and ecological variables of *Limnoperna fortunei* were determined from averaging the levels reported by the numerous researchers listed in Table 3.13. The ranges in values for each of the risk levels are given in Table 3.14. Assessment of potential survival of golden mussel in a hypothetical lake is pretty much the same process as described for zebra mussels, and is not repeated here.

TABLE 3.13
Studies Using Various Parameters to Assess Risk Potential of Survival and Potential Density of the Golden Mussel

Variables	Reference
Alkalinity	Deaton et al. (1989), Hamilton et al. (1997), Ricciardi (1998)
Calcium	Morton (1975), Deaton et al. (1989), Ricciardi (1998), Pareschi et al. (2008)
Chlorophyll *a*	Sylvester et al. (2004), Devercelli and Peruchet (2008), Molina and de Paggi (2008)
Conductivity, TDS	Devercelli and Peruchet (2008), Pareschi et al. (2008), see also salinity and derive from Table 3.5
Dissolved oxygen	Ricciardi (1998), De Oliviera et al. (2004), Devercelli and Peruchet (2008), see also Table 3.5
Nitrogen, total	Molina and de Paggi (2008), Devercelli and Peruchet (2008)
pH	Hamilton et al. (1997), Ricciardi (1998), De Oliviera et al. (2004), Oliviera et al. (2006), Devercelli and Peruchet (2008), Pareschi et al. (2008)
Phosphorus, total	Devercelli and Peruchet (2008), Molina and de Paggi (2008)
Salinity	Ricciardi (1998), Darrigran (2002), Brugnoli et al. (2005), Karatayev et al. (2007a)
Secchi depth, TSS	Devercelli and Peruchet (2008), Pareschi et al. (2008)
Temperature (°C)	Morton (1977, 1982), Ricciardi (1998), Cataldo and Boltovskoy (1999), Cataldo et al. (2000), Magara et al. (2001), Goto (2002), Darrigran et al. (2003), Kim et al. (2003), Devercelli and Peruchet (2008), Pareschi et al. (2008)
Total hardness	See calcium and estimate by multiplying by 3.5
Turbidity (NTU), TSS	Pareschi et al. (2008)

The variables are listed alphabetically.

TABLE 3.14
Criteria Used in Determining Levels of Infestation by the Golden Mussel in North America

1. Parameter	2. No Potential for Adult Survival	3. Little Potential for Larval D	4. Moderate Potential for Nuisance Infestations	5. High Potential for Massive Infestations
Chalk Variables				
Calcium (mg Ca/L)	<1, >100	1–5	5–10, 80–100	10–80
pH	<3.0, >10	3.0–5.0	5.0–7.0	7.0–9.0
Alkalinity, total (mg $CaCO_3$/L)	<3, >350	3–18	18–35, 280–350	35–280
Hardness, total (mg $CaCO_3$/L)	<3, >350	3–18	18–35, 280–350	35–280
Nutrient Variables				
Dissolved oxygen (mg/L) (% saturation)	<1 (<10%)	1–3 (10–20%)	3–6 (20–50%)	≥7 (>50%)
Chlorophyll *a* (μg/L)	0–1, >25	1–2 or 20–25	8–20	2–8
Total phosphorus (μg/L)	?	?	?	?
Total nitrogen (μg/L)	?	?	?	?
Secchi depth (m)	<1, >8	1–2, 6–8	4–6	2.0–4.0
Physical Variables				
Temperature (°C)	<5, >40	5–15 or 32–40	16–26	26–32
Conductivity (μS/cm)	<30, >25,000	9000–25,000	3800–9000	30–3800
TDS (mg/L)	<0, >17,000	6000–17,000	2500–6000	20–2500
Salinity (mg/L) (ppt)	<0, >15	5–15	1–5	0–1
Turbidity (NTU)	>80	20–80	5–20	<3–65
TSS (mg/L)	>96	28–96	8–28	<8

Use column 3 for assessing potential for survival (but little potential for nuisance infestation). Columns 4 and 5 indicate good potential for long-term survival and the criteria are used to assess temporal and seasonal variations in infestation potential. TSS was estimated using $\log_{10}$ TSS (mg/L) = 0.899 $\log_{10}$ Turbidity (NTU) + 0.272.
? = No data available.

3.3.5.1 Survival Potential of Golden Mussel Based on Chalk Variables

In the most places where *L. fortunei* occurs naturally or as an invader, pH is above 6.4 and calcium concentrations are higher than 2.4 mg/L (Morton, 1975; Ricciardi, 1998). *L. fortunei* can establish itself in waters with low calcium concentrations (e.g., 1.0–6.0 mg/L in the Paraguay River) and a pH of around 6.0. Pareschi et al. (2008) examined densities of *L. fortunei* in two Brazilian reservoirs and found that the pH ranged between 7.27 and 9.96, so it appears to be tolerant of high and low (pH 6, Darrigran, 2002) pH; the calcium levels ranged between 4.66 and 16.9 mg/L. Mata and Campos (2005) conducted studies on the maximum levels of dissolved calcium tolerated by *L. fortunei* and found that any concentration higher than 100 mg/L produced 100% mortality. See Chapter 10 for additional information.

3.3.5.2 Survival Potential of Golden Mussel Based on Trophic Variables

Dissolved oxygen: Pareschi et al. (2008) examined densities of *L. fortunei* in two Brazilian reservoirs and found the dissolved oxygen levels ranged between 6.78 and 10 mg/L, which when related to the temperature range (discussed below) corresponds to 74% and 89% saturation. Devercelli and Peruchet (2008) in their examination of trends in nine water quality parameters in several man-used basins in Argentina, but the two that had *L. fortunei*, had dissolved oxygen levels range from 5.3 to 12.6 mg/L, or 67–116% saturation in one (Setúbal Lake) and 6.3 to 12.5 mg/L, or 82–91% saturation in the other (Santa Fe River). Oliveira (Section 3.3.6) has found survival of *Limnoperna* adults at dissolved oxygen levels near 1 mg/L.

Chlorophyll a: As with the dreissenids, most of the information on trophic status is on the impact of mussels on levels of chlorophyll *a* and nutrients. Sylvester et al. (2004) reported that the clearance rates (125–350 mL/individual/h) of *L. fortunei* are generally higher than those reported for other suspension feeding bivalves, including the invasive species *D. polymorpha*, *D. bugensis*, and *C. fluminea*, and with densities up to over 150,000 m^{-2}; they suggest that the levels of chlorophyll *a* will be significantly reduced. The threshold levels of chlorophyll *a* on survival, growth, and reproduction of *L. fortunei* are unknown. Devercelli and Peruchet (2008), however, note that *L. fortunei* can tolerate high levels of eutrophication and algal densities are reduced significantly after post-invasion. Molina and de Paggi (2008) examined changes in environmental variables in Santa Fe and Colastiné rivers of the Middle Paraná River floodplain during 2004–2005 and prior to invasion by *L. fortunei* during the early 1990s; they found a reduction in chlorophyll *a* from 9.6 to 1.9 μg/L in the former river and 4.8 to 1.9 μg/L in the latter. The initial levels in the two rivers suggest that the species spans the trophic range from oligotrophy to eutrophy. Devercelli and Peruchet (2008) examined trends in chlorophyll *a* levels during 2005 and 2006 in several man-used basins in Argentina, two of which (Setúbal Lake and Santa Fe River) have *L. fortunei*; compared to studies in the 1990s, they found a diminution of chlorophyll *a* that they attribute in part to the introduction in the last decade of the bivalve and its high clearance rate of algal cells. The current levels range from 0.14 to 6 μg/L (mean 2.0 μg/L) in Setúbal Lake and from 0.6 to 6.7 μg/L (mean 2.8 μg/L) in Santa Fe River.

Nutrients (total phosphorus, total nitrogen): The threshold levels of phosphorus and nitrogen on survival, growth, and reproduction of *L. fortunei* are unknown.

However, Molina and de Paggi (2008) reported golden mussels in the Santa Fe River of the Middle Paraná River floodplain with nutrient levels ranging from 190 to 1040 μg PO_4/L and 650 to 1400 μg NO_3/L, both of which are in the eutrophic range. The initial levels in the two rivers suggest that the species can tolerate higher nutrient levels than dreissenids. Devercelli and Peruchet (2008) in their examination of trends in nine water quality parameters in several man-used basins in Argentina, and the two that had *L. fortunei* had total phosphorus levels ranging from 10 to 2750 μg/L (mean 810 μg/L) in one (Setúbal Lake) and from 80 to 2750 μg/L (mean 1100 μg/L) in the other (Santa Fe River). With such high levels, it appears that nutrients are not limiting survival, growth, and reproduction of golden mussels.

Secchi depth, trophic status: Devercelli and Peruchet (2008) note that *L. fortunei* can tolerate high levels of eutrophication but they did not report any threshold levels for survival, growth, or reproduction. Similarly, Molina and de Paggi (2008) reported golden mussels in the Santa Fe River of the Middle Paraná River floodplain with levels of chlorophyll *a* and nutrients in the high eutrophic range. The initial levels of chlorophyll *a* in the two rivers suggest that the species spans the trophic range from oligotrophy to eutrophy. Devercelli and Peruchet (2008) in their examination of trends in nine water quality parameters in Setúbal Lake and Santa Fe River in Argentina, reported Secchi depths ranging from 20 to 81 cm (mean 29 cm) in the former and 15 to 38 cm (mean 23 cm) in the latter, again indicating its ability to live in oligotrophic to eutrophic waters.

3.3.5.3 Survival Potential of Golden Mussel Based on Temperature

Morton (1975, 1977, 1982) noted that this species has broad tolerance to chemical and physical variables and is able to live in temperatures ranging from 0°C to 31°C in a Hong Kong reservoir. The threshold temperature for reproduction seems to be different from one location to another. In southern China, the first spawn occurs when the 27–28°C is reached (June and July) and the second occurs during the minimum, 16–17°C (January–February) and appears to be triggered by falling temperature (Morton, 1982). In northern China, *L. fortunei* has only one reproductive cycle at the beginning of winter, when temperatures are between 16.0°C and 21.0°C (Morton, 1982). Generally, reproduction appears to begin when ambient temperatures reach 16°C (Morton, 1982; Cataldo et al., 2000). As with dreissenids, larval development rates increase with increasing temperatures, reaching an optimum around 30°C (Cataldo et al., 2005). See Chapter 10 for additional information. Silva and Campos (2004) recommend 19.5°C as the ideal survival and reproduction temperature for the golden mussel.

3.3.5.4 Survival Potential of Golden Mussel Based on Conductivity, TDS, and Salinity

Conductivity and TDS: Pareschi et al. (2008) examined densities of *L. fortunei* in two Brazilian reservoirs and found the conductivity ranged between 44 and 232 μS/cm and the TDS between 10 and 180 mg/L, which are levels in the same range as zebra and quagga mussels. Devercelli and Peruchet (2008) in their examination of trends in nine water quality parameters in 20 man-used basins in Argentina, and the two that had *L. fortunei* had conductivity levels range from 70 to 293 μS/cm (mean

126 μS/cm) in one (Setúbal Lake) and 76 to 259 μS/cm (mean 152 μS/cm) in the other (Santa Fe River), again in the same range as zebra and quagga mussels.

Salinity: Darrigran (2002) states that *L. fortunei* is a freshwater and brackish water species until salinity reaches 3 ppt. Brugnoli et al. (2005) report its presence in Uruguay Rivers with salinities up to 1.6 ppt. However, in the Río de la Plata estuary, where salinities can vary from 5 to 15 ppt within a matter of hours, the golden mussel is found at salinities as high as 13.7% (Karatayev et al., 2007). It may be that the species can tolerate short-term increases of salinity at least up to 15 ppt.

3.3.5.5 Survival Potential of Golden Mussel Based on Turbidity and TSS

Pareschi et al. (2008) examined densities of *L. fortunei* in two Brazilian reservoirs and found that the TSS ranged between 3 and 62.4 NTU, which is within the same ranges as *D. polymorpha*, *D. bugensis*, and *M. leucophaeata* for high and moderate infestation levels, respectively. See Chapter 10 for additional information.

3.3.6 Predicting Potential Spread of the Asian Golden Mussel (*L. fortunei*) in North America Based on Observations of Its Environmental Tolerance in South America

Márcia D. Oliveira, Stephen K. Hamilton, and Claudia M. Jacobi

Zebra and quagga mussels (*D. polymorpha* and *D. bugensis*) have spread widely through North American freshwaters since their introduction into the Laurentian Great Lakes in the 1980s. This invasion has resulted in costly industrial biofouling as well as inducing drastic ecological changes that rippled through aquatic food webs (O'Neill, 1997; Strayer et al., 1999; Knoll et al., 2008). One limiting factor for zebra and quagga mussels is their upper thermal tolerance, which effectively excludes them from colonizing shallow, warm waters. The golden mussel, *L. fortunei* (Dunker, 1857), although of a different family (Mytilidae rather than Dreissenidae), is a remarkably similar species native to China and Southeast Asia. It has recently been introduced into the Paraná River watershed in southern South America (Boltovskoy et al., 2006). Like the dreissenids, *L. fortunei* uses byssal threads to attach to surfaces, is a suspension feeder, can occur in very high population densities, grows rapidly within a short life span, and goes through a planktonic veliger larval stage (Morton, 1973). The ability of *L. fortunei* to colonize warm waters, combined with apparently lower calcium requirements, has raised concerns that this species might be able to colonize waters that are too warm or ionically dilute for dreissenid mussels. This would include the rivers and reservoirs of the southeastern United States and elsewhere in South America (Ricciardi, 1998; Kluza and McNyset, 2005; Karatayev et al., 2007). Furthermore, *L. fortunei* appears be more tolerant of salinity than the dreissenids, allowing it to live in brackish waters (Karatayev et al., 2007).

The successful establishment of *L. fortunei* in the upper Paraguay River of Brazil (a tributary of the Paraná River that is associated with extensive floodplains known as the Pantanal) offers an opportunity to explore the environmental limits of this species. These tropical waters span a broad range of temperature and chemical conditions. These water also experience periodic oxygen depletion events that can last for weeks. This environment represents inhospitable conditions for mussels survival, including *L. fortunei*. The objective of this chapter is to summarize our recent studies of the distribution and environmental limits of *L. fortunei* in rivers of the Pantanal. Further, we combine available information from previous studies in other parts of the world to predict the potential distribution of *L. fortunei* in river systems of North America. Successful invasion of these systems by *L. fortunei* would result in economic and environmental problems similar to those caused elsewhere by the dreissenids.

3.3.6.1 Water Temperature and Chemical Variables as Limiting Factors

Morton (1975, 1977, 1982) described the first recorded invasion by *L. fortunei* in Hong Kong around 1965. He noted that this species has broad tolerance to chemical and physical variables. *L. fortunei* is able to live in temperatures ranging from 0°C (reviewed in Ricciardi, 1998) up to 33°C in the Pantanal wetland (Oliveira et al., in review). Thermal tolerance involves duration of exposure as well as maximum or minimum temperatures. Observations in the literature indicate that *L. fortunei* can exist at high densities in waters that fall as low as 5–8°C (Japan: Magara et al., 2001; Goto, 2002) or near 0°C (South Korea: Kim et al., 2003), although in both of these cases the period of low water temperatures is short, lasting about two months. In laboratory assays we conducted, mussels from the tropical Paraguay River survived up to 15 days at water temperatures of 0–1°C, although they were inactive during that time (Figure 3.6; Oliveira et al. in prep.). Survival at 5–7°C was 50% in the first 20 days, with a maximum survival length of about 38 days. At temperatures above 10°C but up to 26–28°C, the mussels were active and maintained a survival rate of approximately 30% after more than two months (Figure 3.6).

Temperature appears to control the reproductive cycle of *L. fortunei* (Morton, 1977, 1982; Cataldo and Boltovskoy, 1999; Darrigran et al., 2003), even though the seasonal characteristics and lower and upper ranges can be different in Japan and Korea than those observed in Brazil. The reproductive cycle of *L. fortunei* in the Pantanal resembles that observed in Argentine rivers (Cataldo and Boltovskoy, 1999). There is almost continuous recruitment of larvae for about seven months at temperatures around 26°C in the Paraná River and 31°C in the Paraguay.

River: This is followed by relatively little reproductive activity at lower temperatures, which are typically around 13°C in the Paraná River and around 20°C in the Paraguay River. This pattern is quite different from that observed at Plover Cove Reservoir (Hong Kong), where spawning occurs for four months in two distinct periods. The first spawn is when the summer maximum temperature is reached (27–28°C) in June and July and the second occurs during a winter minimum (16–17°C, January–February) and appears to be triggered by falling temperature (Morton, 1982). In northern China, *L. fortunei* has only one reproductive cycle at the beginning of winter, when temperatures are between 16.0°C and 21.0°C (reviewed in Morton, 1982). The threshold temperature for reproduction is therefore difficult to determine as it seems to be different from one location to another. Generally, reproduction appears to begin when ambient temperatures reach 16°C (Morton, 1982; Cataldo et al., 2000). Larval development rates increase at higher temperatures, reaching an optimum around 30°C (Cataldo et al., 2005). Water temperature can reach 31°C in the Hong Kong reservoir, but peak spawning usually occurs at lower temperatures. In contrast, in the rivers of the Pantanal we have observed larval production at temperatures of up to 32°C.

Although *L. fortunei* might tolerate water temperatures of around 28–30°C in southern Asia, the duration of high temperatures may be shorter than found in Pantanal. In southern Asia, temperatures around 30°C generally last for about three months, whereas in the Pantanal water temperatures of 30–32°C commonly last for seven months (Figure 3.7). This longer warm season likely stimulates changes in the reproductive cycle of *L. fortunei*. The larval release lasts much longer in the Pantanal than in the Asian waters. Longer recruitment periods in warmer water could be one of the causes of high densities of *L. fortunei* observed in some South American reservoirs and hydropower systems. In spite of the high temperatures and prolonged spawning periods in the Pantanal, *L. fortunei* does not achieve population densities as high as have been observed in other South American locations. This seems likely to be caused by mitigating control factors other than temperature.

In the most places where *L. fortunei* occurs naturally or as an invader, pH is above 6.4 and calcium concentrations are higher than 2.4 mg/L (Morton, 1975). In general, waters of the Pantanal are low in dissolved ions, particularly calcium. The neutral to slightly acidic conditions might control the density of *L. fortunei* by interfering with its shell calcification (Hamilton et al., 1997; Figure 3.7). Nevertheless, we have observed that in the Pantanal, *L. fortunei* can establish itself in waters with low calcium concentrations (e.g., 1.0–6.0 mg/L in the Paraguay River) and a pH of around 6.0 (Figure 3.7). Perhaps pH and calcium concentrations in the Paraguay River are near the lower limit for *L. fortunei*, but more research is required to confirm this. Experiments have shown that *L. fortunei* can produce byssal attachments and begin filtering within one week of incubation in deionized water (Deaton et al., 1989), bringing the hypothesis of a lower threshold of calcium concentration into question. Yet in the field we have observed apparent thinning of their shells.

In addition to the relatively low calcium and pH in some rivers of the Pantanal where *L. fortunei* has been found, oxygen depletion events are common every year, particularly during the rising water phase (Hamilton et al., 1997; Calheiros and Hamilton, 1998). Oxygen depletion events in the Pantanal are associated with simultaneous stress factors including not only low or no O_2 but also higher temperature, lower pH, and higher free CO_2. During one of these events we observed that populations of *L. fortunei* were extirpated from a floodplain lake in the Pantanal connected to the Paraguay River (Oliveira et al., in review).

Fluctuating water levels present another limiting factor for *L. fortunei*. Perhaps a sequence of stressful factors, such as low water levels that killed mussels by exposing them to the air, low pH during the falling and rising water phases, and depletion of dissolved oxygen during rising water might explain the lower densities of *L. fortunei* in the Pantanal. However in spite of these limitations, *L. fortunei* has spread through the Pantanal and into the tributaries of the Paraguay River. Fluctuations in densities of *L. fortunei* are related to the intensity and duration of oxygen depletion events. The probability of recruitment was less than 10% when dissolved oxygen was close to 0.0 mg/L (Oliveira et al., in review).

L. fortunei can survive in turbid river waters with low food availability, and evidently it can use detrital organic carbon sources in addition to algae (Calheiros, 2003; Sylvester et al., 2005). In rivers of the Pantanal, TSS varied from 10 to 120 mg/L and chlorophyll *a* concentrations averaged <1.0 μg/L. High suspended sediment concentrations were observed mainly during the reproductive period of *L. fortunei* and may explain the low rate of larval survival, perhaps together with water velocities exceeding 0.9 m/s. Thus, the combination of high water velocity, low food quality and quantity, and high concentration of sediments, common in unimpeded lotic ecosystems, likely contributes to the low larval survival rate and low adult population densities in the Pantanal.

3.3.6.2 Predicting Invasion Potential of *L. fortunei* in North American Rivers

If given the opportunity, *L. fortunei* probably will colonize North American waters, especially the warmer waters of the southeast. The potential for introduction exists in the ports along the Gulf of Mexico (Ricciardi, 1998; Kluza and NcNyset, 2005; Karatayev et al., 2007). We can apply the knowledge of environmental tolerances of *L. fortunei* in the Pantanal and elsewhere to project its potential distribution in representative large rivers of North America. Calcium availability and pH are unlikely to be limiting in these rivers as the levels are generally much higher (Ramcharan et al., 1992) when compared to the native environment and Pantanal waters where *L. fortunei* is present.

Low water temperatures may be more likely to be the key environmental variable determining distribution patterns of *L. fortunei* in North American rivers (Ricciardi, 1998), although Karatayev et al. (2007) suggested that the lower thermal limits of *L. fortunei* may resemble those of the invasive dreissenids that flourish in the temperate climate of the Laurentian Great Lakes region (United States and Canada).

Figure 3.8 shows the spatial variation of minimum water temperature in three large U.S. rivers. Based on our water temperature tolerance assays, the probability of *L. fortunei* establishment will increase in water with temperature minimum above 5°C. In the Upper Mississippi the minimum temperature is about 0°C, with about five months of temperatures below 5°C. It is possible that this protracted cool period is too long for *Limnoperna* survival. The probability of establishment and survival is greater in the Mississippi River below its confluence with Ohio River (Mississippi River at Thebes, IL), as well as in the Colorado River and Rio Grande, where the minimum water temperature is above 5°C.

Other variables limit the abundance of *L. fortunei* rather than preclude its establishment. Suspended sediment levels in the three North American rivers are usually higher than those found in the rivers of the Pantanal. In the lower Mississippi River (where water temperature stays above 5°C), concentrations of TSS suspended range between 51 and 1360 mg/L, and dissolved organic carbon concentrations are low in comparison to the Paraguay River. This combination of high suspended sediment levels and low food availability may reduce the abundance of *L. fortunei*. High population densities are observed mainly in lakes or reservoirs with low suspended sediment levels and greater algal biomass than observed in these North American rivers. Lakes and reservoirs connected to these rivers, which have less suspended sediment and more phytoplankton, may act as a source of larvae for downstream rivers.

The North American rivers do not generally experience oxygen depletion events akin to those of the Pantanal waters that control the population densities of *L. fortunei*. Hypoxia is common during the summer in the Mississippi River estuary and when combined with high summer temperatures can provoke mortality of *D. polymorpha* (Mihuc et al., 1999), even when dissolved oxygen concentrations are higher than 1.0 mg/L. *L. fortunei* is more resistant to low dissolved oxygen levels and would be more likely to survive such episodes of oxygen depletion, mainly because the temperature is generally lower in the Mississippi estuary than in the Pantanal wetland. Above the river delta, the oxygen concentrations in the Mississippi River are normally higher than 1.0 mg/L.

Another indicator of the possible success of *L. fortunei* colonization in the North American rivers is the distribution of the Asian clam *C. fluminea* (Müller), an invasive species that has been studied in considerable detail (McMahon, 1999). This species is native to the same region as *L. fortunei* and is widely distributed in South America, including the waters of the Pantanal (Mansur et al., 2004). *C. fluminea* is also widely distributed in North America from the northern U.S. to Mexico (Foster et al., 2008). Cold winters can cause mortality of *C. fluminea* (French and Schloesser, 1991; Cordeiro et al., 2007; Werner and Rothhaupt, 2008); temperatures below 2°C for two months or more may reduce densities or completely eliminate populations. Temperature threshold for reproduction of *C. fluminea* is around 15–16°C (reviewed in McMahon, 1999) and for *L. fortunei* is around 16°C (Morton, 1982; Cataldo et al., 2000). With these thermal characteristic thresholds, *L. fortunei*'s invasionary range may overlap with those of *C. fluminea* and *D. polymorpha* in North America, especially in waters warmer than 2°C.

In conclusion, based on our observations in the waters of the Paraguay River and the Pantanal in Brazil, combined with information from studies in Hong Kong, South Korea, Japan, and Argentina, we agree with previous assessments that *L. fortunei* is capable of and is likely to eventually spread into rivers and lakes throughout the Americas. Its potential distribution overlaps with the current distribution of *Corbicula* and dreissenid mussels in North America but it will extend into waters that are warmer, lower in calcium, and subject to oxygen depletion events. Thus waters that are currently free of invasive bivalves may experience invasion by *Limnoperna*. The ecological and economic impacts of the spread of *L. fortunei* will add to those caused by the dreissenids, and efforts to prevent *Limnoperna* introduction and contain its spread are well justified.

3.3.6.3 Acknowledgments

The work was supported by The National Council for Scientific and Technological Development (CNPq), PELD and CTHIDRO programs, and Embrapa Pantanal. A visit by Márcia D Oliveira to Kellogg Biological Station was partially supported by the U.S. Environmental Protection Agency's Science to Achieve Results (STAR) Program (grant #RD83170801). We are grateful to R. Alexander and D. Dudgeon for assistance.

3.3.7 Asian Clam, *C. fluminea* (Prashad, 1929)

The threshold limits for growth, reproduction, and survival for each of the physiological and ecological variables of *C. fluminea* were determined from averaging the levels reported by the numerous researchers listed in Table 3.15. The ranges in values for each of the risk levels are given in Table 3.16. Assessment of potential survival of Asian clams in a hypothetical lake is pretty much the same process as described for zebra mussels, and is not repeated here.

3.3.7.1 Survival Potential of Asian Clams Based on Chalk Variables

Survival, growth, and reproduction of *C. fluminea* do not appear to be limited greatly by levels of chalk variables. It is found in a wide range of calcium levels, from a mean level of 5.7 mg Ca/L (range 2.1–8.5 mg Ca/L) up to at least a mean of 39.6 mg Ca/L (range 29.5–45.7 mg Ca/L) in lentic and lotic systems in South Carolina and New York (Voelz et al., 1998; Strayer and Malcom, 2007). The pH levels in these same systems ranged from a mean low of 6.6 (range 5.8–6.9) to a mean high of 8.03 (range 7.5–8.4) and alkalinity ranged from 3.3 to 27.8 mg $CaCO_3$/L in the South Carolina stream. The threshold levels for survival (columns 1 and 2), growth (column 3), and reproduction (column 4) in Table 3.14 are based on the low and high values reported above.

TABLE 3.15
Studies Using Various Parameters to Assess Risk Potential of Survival and Potential Density of Asian Clams

Variables	Reference
Alkalinity	Voelz et al. (1998), Strayer and Malcom (2007)
Calcium	Voelz et al. (1998), Strayer and Malcom (2007)
Chlorophyll *a*	See Table 3.6 and Sousa et al. (2008)
Conductivity, TDS	See Table 3.6 and Castagna and Chanley (1973), Marelli and Gray (1983), Deaton et al. (1989), MacNeill (1991), Pathy and Mackie 1993), Walton (1996)
Dissolved oxygen	See Table 3.6 and Fast (1971), Horne and McIntosh (1979), Sickel and Heyn (1980), McMahon (1983), Belanger (1991), Sousa et al. (2008)
Nitrogen, total	See Table 3.6 and Sousa et al. (2008)
pH	Voelz et al. (1998), Strayer and Malcom (2007)
Phosphorus, total	See Table 3.6 and Sousa et al. (2008)
Salinity	See Table 3.6 and Castagna and Chanley (1973), Marelli and Gray (1983), Deaton et al. (1989), MacNeill (1991), Pathy and Mackie (1993), Walton (1996)
Secchi depth, TSS	See Table 3.6 and Sousa et al. (2008)
Temperature (°C)	See Table 3.6 and Siddall (1980), Rajagopal et al. (2005a), Verween et al. (2005, 2007b), Laine et al. (2006)
Total hardness	Voelz et al. (1998), Strayer and Malcom (2007), Sousa et al. (2008)
Turbidity (NTU), TSS	See Table 3.6 and Barbiero et al. (2006)

The variables are listed alphabetically.

TABLE 3.16
Criteria Used in Determining Levels of Infestation by Asian Clams in North America

1. Parameter	2. No Potential for Adult Survival	3. Little Potential for Larval Development	4. Moderate Potential for Nuisance Infestations	5. High Potential for Massive Infestations
Chalk Variables				
Calcium (mg Ca/L)	<1	1–2	2–5	>5
pH	<5, >10?	5–6, 9.5–10?	6–7	7–9
Alkalinity, total (mg $CaCO_3$/L)	<3	3–7	7–17	>17
Hardness, total (mg $CaCO_3$/L)	<3	3–7	7–17	>17
Nutrient Variables				
Dissolved oxygen (mg/L) (% saturation)	<0.5 (<5%)	0.5–2.0 (5–10%)	2–6 (10–50%)	≥6 (>50%)
Chlorophyll *a* (µg/L)	<5, >25	5–10	10–20	20–25
Total phosphorus (µg/L)	<5, >300?	5–20	20–100	100–250
Total nitrogen (µg/L)	<?	?	?	?
Secchi depth (m)	<0.1, >8	0.1–0.3, 6–8	0.3–0.5, 3–6	0.5–3
Physical Variables				
Temperature (°C)	<2, >36	2–14, 30–36	15–18 or 25–30	18–25
Conductivity (µS/cm), 25°C	>12,600	11,000–12,600	8100–11,000	<8100
TDS (mg/L)	>8400	7400–8400	5400–7400	<5400
Salinity (mg/L) (ppt)	>8	7–8	5–7	<5
Turbidity (NTU)	?	?	?	?
TSS (mg/L)	?	?	?	?

Use column 3 for assessing potential for survival (but little potential for nuisance infestation). Columns 4 and 5 indicate good potential for long-term survival and the criteria are used to assess temporal and seasonal variations in infestation potential. Conductivity values were estimated from Table 3.5 and TDS by multiplying conductivity by 0.67 and rounding off to nearest 100. ? = No data available.

Sousa et al. (2008) performed multivariate analyses of 17 abiotic variables against biomass and abundance of *C. fluminea* at 16 stations separated by 1–7 km in a 42-km long Portuguese estuary and found that hardness had a significant relationship; they rationalized its importance in shell production and hence its value in predicting occurrence and distribution of the species. The hardness values at the 16 sites varied from 5 (average ~500 individuals/m^2) to 60 (average ~3000 individuals/m^2) mg $CaCO_3$/L. The total hardness criteria in Table 3.14 were estimated from the calcium and alkalinity levels (see Section 3.2.1.1.1) and tend to support the results reported by Sousa et al. (2008).

3.3.7.2 Survival Potential of Asian Clams Based on Trophic Variables

Dissolved oxygen: Dissolved oxygen levels have been well investigated as a variable limiting survival of Asian clams. Asian clams are able to survive dissolved oxygen concentrations of 0–0.5 mg/L for seven days with 89% survival rate (Horne and McIntosh, 1979). However, Fast (1971) and McMahon (1983) have been reported that Asian clams are intolerant of acute hypoxia. Mass mortality of *C. fluminea* were reported by Sickel and Heyn (1980) in near anoxic conditions in the lower Tennessee and Cumberland rivers below Kentucky and Barkley dams. McMahon (1979) found that *C. fluminea* increases its O_2 uptake as temperature increases up to 25°C, but the uptake rate is depressed at 30°C. Matthews and McMahon (1995) found specimens of *D. polymorpha* were much less tolerant of anoxia than those of *C. fluminea*. Mean tolerance times, LT_{50} values for zebra mussels were 2–7 times lower than those of Asian clams. Mathew and McMahon (1995) also reported that zebra mussels acclimated to 25°C showed the greatest anoxia tolerance at both test temperatures, surviving a mean of 3.45 days at 25°C and a mean of 17.8 days at 15°C; in contrast, Asian clam survival times did not appear to be influenced by temperature acclimation. The highest mean anoxia tolerance times for *C. fluminea* were 11.8 and 35.1 days among 15°C-acclimated clams at test temperatures of 25°C and 15°C, respectively. Asian clams show no capacity to regulate oxygen uptake rate with declining oxygen concentration (McMahon, 1979) and has been reported to be unable to tolerate reduced oxygen concentrations associated with discharge of treated sewage (Belanger, 1991) and hypolimnetic waters (Fast, 1971). In warm summers, a combination of factors, including low flow, decrease dissolved oxygen and the presence of large quantities of organic matter causes a decrease in the redox potential and an increase in ammonia and any one or a combination of these factors can cause massive mortalities of *C. fluminea*, native mussels, and other benthic species (Scheller, 1997; Johnson and McMahon, 1998; Strayer, 1999; Cherry et al., 2005; Cooper et al., 2005; Sousa et al., 2006a,b).

Sousa et al. (2008) performed multivariate analyses of 17 abiotic variables against biomass and abundance of *C. fluminea* at 16 stations separated by 1–7 km in a 42-km long Portuguese estuary and found that dissolved oxygen did not explain much of the variation in abundance and biomass of the clams. This is not surprising because the dissolved oxygen levels ranged from 5.5 mg/L (55% saturation) to 11.3 mg/L (116% saturation), all well above the threshold levels reported above. The threshold levels for survival, growth, and reproduction based on dissolved oxygen levels for Asian clams in Table 3.14 are based on the above results, especially Sousa et al. (2008).

Chlorophyll a*:* In the same study above, Sousa et al. (2008) found no relationship between chlorophyll *a* levels and abundance and biomass of Asian clams. However, the chlorophyll *a* values only ranged from 4.3 to 5.6 μg/L, which are surprisingly low considering the levels of phosphates available (see below).

Nutrients (total phosphorus, total nitrogen): Sousa et al. (2008) found that concentrations of nutrients were negatively correlated with *C. fluminea* biomasses; they concluded that this species seems to be sensitive to organic pollutants and are not well adapted to organic enrichment conditions. They found this situation to be particularly problematic for survivorship and/or development of juveniles since areas with high nutrient levels had adults but not juveniles. The phosphate levels ranged from 30 μg/L (average ~1200 individuals/m^2) to 230 μg/L (average ~500 individuals/m^2) among the 16 stations; the station with the greatest density (~3000 m^{-2}) had an average of 115 μg PO_4/L and the station with the fewest clams (~50 m^{-2}) had 220 μg PO_4/L. The threshold levels for survival, growth, and reproduction based on phosphorus levels for Asian clams in Table 3.14 are based on the above results, especially Sousa et al. (2008). There is no information on threshold levels for total nitrogen.

Secchi depth, trophic status: The Asian clam is normally a filter feeding species but it can invoke deposit feeding and use the organic matter available in the sediment as a food resource (Hakenkamp and Palmer, 1999; Hakenkamp et al., 2001; Vaughn and Hakenkamp, 2001; Mackie, 2007). This alternative food source can play an important role in habitats with low phytoplankton concentrations like some oligotrophic sites in the River Minho estuary of Portugal (Sousa et al., 2008). They also examined the relationship between percent organic matter in the sediments and abundance and biomass of clams. The percent organic matter ranged from a low of 0.4% to a high of 14% but they found no significant relationship. However, Vaughn and Hakenkamp (2001) believe that *C. fluminea* can achieve higher biomasses and growth rates in habitats with high organic matter content. The threshold levels for survival, growth, and reproduction based on Secchi depth for Asian clams in Table 3.14 are based on the above results, especially Sousa et al. (2008).

3.3.7.3 Survival Potential of Asian Clams Based on Temperature

As with all three dreissenids discussed above, Asian clams have lower and upper thermal thresholds for survival, growth, and reproduction. Being native to semi-tropical/tropical southeast Asia (Morton, 1979, 1986), *C. fluminea* was rarely exposed to temperature extremes (McMahon, 2000). Temperatures <2°C can kill *Corbicula* (French and Schloesser, 1991) and results in winter population kills in the northern edge of its range (McMahon, 1983a). However, healthy populations have been reported within thermal plumes in Lake Erie (Garton and Haag, 1993), in the 10–15 m depths of dimictic Lake Erie (Kreiser and Mitton, 1995), and in northern U.S. streams (Burky and White, 1989; Janech and Hunter, 1995). The explanations for its survival in these cold conditions are that the thermal discharges maintain temperatures above 2°C in the winter (French and Schloesser, 1991); water temperatures at the 10–15 m depths may be at 4°C, at which point water is most dense and cannot be cooled further, and if adequate supplies of oxygen are present, the clams will survive (Kreiser and Mitton, 1995); winter water temperature in small streams are unstable and may rise above 2°C long enough to prevent mortality (McMahon,

2000); and northern waters do not reach thermal thresholds for reproduction (Burky and White, 1989; Janech and Hunter, 1995).

The upper lethal threshold of the Asian clam is 36°C in the field (McMahon and Williams, 1986a) and 34°C in the laboratory (Mattice and Dye, 1976). McMahon (2000) attributes depressions of oxygen uptake and filtration rates above 30°C, which then triggers metabolic suppression and ultimately death of the clams. Williams and McMahon (1985) found that metabolic rates are suppressed above 33°C in field-collected clams and above 30°C in laboratory-acclimated clams.

McMahon and Williams (1986b) state that the Asian clam's temperature threshold for reproduction is 15–16°C. Based on a literature review, Claudi and Mackie (1994) report 16°C and 22°C as the threshold temperatures for growth and reproduction, respectively. Spawning can continue year around in water temperatures higher than 16°C. The water temperature must be above 16°C for the clams to release their larvae. Table 3.14 gives the thermal thresholds for survival (column 2), growth (column 3), and reproduction (column 4). Column 5 gives the optimal temperature range.

3.3.7.4 Survival Potential of Asian Clams Based on Conductivity, TDS, and Salinity

Conductivity: Asian clams are freshwater animals, although most of them can tolerate salinities between 3 and 8 ppt. Hence, the upper threshold of conductivity for survival is in the low range of salinity and the lower threshold level is probably more limiting than the upper threshold in freshwaters. The threshold levels of conductivity for growth and reproduction probably depend more on the contributions of calcium and carbonate ions to the conductivity than the absolute conductivity value. This is confirmed by Sousa et al. (2008), who found no significant relationship between conductivity and biomass and abundance of *C. fluminea* at 16 stations in a Portuguese estuary where the conductivity ranged between 50 and 1459 µS/cm, but did find a significant relationship between abundance and biomass of Asian clams and total hardness.

Total dissolved solids: Not surprisingly, Sousa et al. (2008) found no significant relationship between TDS and biomass and abundance of *C. fluminea* at the 16 stations in the Portuguese estuary where the TDS ranged between 0.04 and 1.32 mg/L. As with conductivity, the threshold levels of TDS for growth and reproduction probably depend more on the contributions of calcium and carbonate ions to the TDS than on the TDS value itself.

Salinity: Filice (1958), Heinsohn (1958), Copeland et al. (1971), Diaz (1974), and Balcom (1994) state that Asian clams are limited to salinities <5–8 ppt. King et al. (1985) raised clams in the laboratory with salinities between 0 to 8 ppt. At salinities below approximately 13.0 ppt, *C. fluminea* can osmoregulate, maintaining fairly constant tissue chloride levels of periods of up to seven days, with little mortality but prefer lower salinities (Morton and Tong, 1985).

3.3.7.5 Survival Potential of Asian Clam Based on Turbidity and TSS

While most (e.g., Rinne, 1974; Way et al., 1990; Bilos et al., 1998; McMahon, 2002) agree that turbidity affects biomass and densities of Asian clams, the threshold levels of turbidity and TSS for survival, growth, and reproduction are unknown. The

Secchi depth could probably be used as a surrogate measure until the turbidity/TSS and biomass and abundance of Asian clams have been modeled.

3.4 ASSESSING SURVIVAL POTENTIAL OF AQUATIC INVASIVE GASTROPODS

3.4.1 AMPULLARIIDAE

Being amphibious and grazers, the survival, growth, and reproduction of ampullariids depend on a different suite and/or magnitudes of variables than do aquatic bivalves. For example, their calcium requirements are met by eating plants while on land and by absorption of dissolved calcium and ingestion of plants while grazing in the water. Ampullariids have both a lung and a gill and can emerge from the water if it is anoxic to breathe fresh air. The lung is formed by a fold of the mantle skirt and is used for air storage (Thiengo, 1987). When submersed, oxygen is obtained by the gill, from a water current drawn through the mantle cavity (McClary, 1964). Metabolic rates depend on temperature and thermal thresholds depend on the rate of change of temperature. Extremes in temperature are greater in air and can vary from several degrees below freezing to 40–45°C, while water temperatures can vary from 0°C to 30–35°C, or if in deeper water, from 4°C to 10°C. The specific heat and latent heat of vaporization and of fusion of water moderates the rate of change of temperature from hours to days, while in air temperatures can change within minutes, so acclimatization rates are very different in air and water.

3.4.1.1 Giant Ramshorn Snail, *Marisa cornuarietis* (Linnaeus, 1758)

The threshold limits for survival, growth, and reproduction for each of the physiological and ecological variables of *Marisa cornuarietis* were determined from averaging the levels reported by the researchers listed in Table 3.17. The ranges in values for each of the risk levels are given in Table 3.18 when known.

3.4.1.1.1 Survival Potential of Giant Ramshorn Snail Based on Chalk Variables

The threshold levels of chalk variables needed for survival, growth, and reproduction of ramshorn snails rely on laboratory experiments by Meier-Brook (1978), who found calcium uptake peaked at 50–75 mg Ca/L and that 25 mg Ca/L was suboptimal for uptake of calcium. Most laboratory cultures use high levels of calcium, from 40 mg Ca/L to 130–160 $CaCO_3$/L and a pH of 7.5 to 8.5 (Selck et al., 2006; Aufderheide et al., 2006; Forbes et al., 2008). It is possible that calcium, alkalinity, pH, and total hardness may not be limiting above a certain value.

3.4.1.1.2 Survival Potential of Giant Ramshorn Snail Based on Trophic Variables

Dissolved oxygen: Dissolved oxygen levels are critical only when the species is submerged in water. Ramshorn snails can tolerate short periods (2 days) of anoxia before having to emerge from water for atmospheric oxygen (Horne, 1979). The oxygen uptake rates increase with an increase in water temperature between 10°C

TABLE 3.17
Studies Using Various Parameters to Assess Risk Potential of Survival and Potential Density of *Marisa cornuarietis*

Variables	Reference
Alkalinity	Infer from Meier-Brook (1978), Selck et al. (2006), Aufderheide et al. (2006), Forbes et al. (2008)
Calcium	Meier-Brook (1978), Selck et al. (2006), Aufderheide et al. (2006), Forbes et al. (2008)
Chlorophyll *a*	Perera and Walls (1996)
Conductivity, TDS	Infer from Hunt (1961), Robin (1971), Santos et al. (1987) and Table 3.5
Dissolved oxygen	Akelund (1969), Freiberg and Hazelwood (1977)
Nitrogen, total	No information
pH	Aufderheide et al. (2006), Forbes et al. (2008)
Phosphorus, total	No information
Salinity	Hunt (1961), Robin (1971), Santos et al. (1987)
Secchi depth, TSS	Ferguson and Palmer (1958)
Temperature (°C)	Blackburn et al. (1971), Demian and Ibrahim (1971), Robins (1971), Thomas (1975), Freiburg and Hazelwood (1977), Aufderheide et al. (2006), Selck et al. (2006)
Total hardness	Infer from Meier-Brook (1978), Selck et al. (2006), Aufderheide et al. (2006), Forbes et al. (2008)
Turbidity (NTU), TSS	No information

The variables are listed alphabetically.

and 35°C and decline at the lethal temperature of 40°C (Freiberg and Hazelwood, 1977). The oxygen uptake rates are lower during emersion than during submersion. *M. cornuarietis* showed an increase in V_{O_2} with increasing temperature between 10°C and 35°C and then a decrease at 35°C (Akelund, 1969; Freiberg and Hazelwood, 1977). Unfortunately, the threshold levels of dissolved oxygen required for survival, growth, and reproduction of ramshorn snails are unknown but laboratory cultures generally keep the water at ≥60% saturation (Forbes et al., 2008). Hunt (1958) reported this species as "an exceptionally hardy snail" and kept snails alive for months in aquaria where the dissolved oxygen regularly dropped to under 0.5 mg/L.

Chlorophyll a*:* *M. cornuarietis* feeds rapaciously on aquatic macrophytes without any selectivity (Perera and Walls, 1996). The threshold levels of chlorophyll *a* required for survival, growth, and reproduction of ramshorn snails are unknown.

Nutrients (total phosphorus, total nitrogen): The threshold levels of nutrients required for survival, growth, and reproduction of ramshorn snails are unknown. It is possible that nutrients may not be limiting above a certain value.

Secchi depth, trophic status: The species is a shallow water species, generally less than a meter (Ferguson and Palmer, 1958; Akerlund, 1973), and it is unlikely that that there are data for threshold levels of enrichments for survival, growth, and reproduction of ramshorn snails. However, *Marisa* has been found to thrive in both nonpolluted and heavily enriched waters, surviving in streams receiving effluent from hog sites, abatoirs, and privies (Ferguson and Palmer, 1958).

TABLE 3.18
Criteria Used in Determining Levels of Infestation by *Marisa cornuarietis* in North America

1. Parameter	2. No Potential for Adult Survival	3. Little Potential for Egg Development	4. Moderate Potential for Nuisance Infestations	5. High Potential for Massive Infestations
Chalk Variables				
Calcium (mg Ca/L)	<20	20–25	25–50	50–75
pH	<6, >10	MNBL	?	>7.5
Alkalinity, total (mg $CaCO_3$/L)	<70	70–90	90–175	175–270
Hardness, total (mg $CaCO_3$/L)	<70	70–90	90–175	MNBL
Nutrient Variables				
Dissolved oxygen (mg/L) (% saturation)	<0.5 (<25%)	?	?	≥8 (>60%)
Chlorophyll *a* (μg/L)	?	MNBL	MNBL	MNBL
Total phosphorus (μg/L)	?	MNBL	MNBL	MNBL
Total nitrogen (μg/L)	?	MNBL	MNBL	MNBL
Secchi depth (m)	NA	NA	NA	NA
Physical Variables				
Temperature (°C)	<12, >40	12–15 or 35–40	15–25	25–30
Conductivity (μS/cm)	>14,700	5600–14,700	2000–5600	<2000
TDS (mg/L)	9800	3700–9800	1300–3700	>70
Salinity (mg/L) (ppt)	>8.5	3–8.5	1–3	<1
Turbidity (NTU)	?	?	?	?
TSS (mg/L)	?	?	?	?

Use column 3 for assessing potential for survival (but little potential for nuisance infestation). Columns 4 and 5 indicate good potential for long-term survival and the criteria are used to assess temporal and seasonal variations in infestation potential. MNBL = May not be limiting. NA = Not applicable. ? = No data available.

3.4.1.1.3 Survival Potential of Giant Ramshorn Snail Based on Temperature

Apparently the species survive at temperatures below 12°C for prolonged periods of time. Blackburn et al. (1971) reported a 50% reduction in egg-hatching success in *M. cornuarietis* to when eggs were exposed to temperatures of 14–16°C for 8 h; even exposure to 16°C for 2°h produced a measurable reduction in hatching success. Demian and Ibrahim (1971) maintained *M. cornuarietis* of Puerto Rican origin at ambient room temperature in the laboratory and found them breeding from March to early December at temperatures between 20°C and 33°C. They reported egg production to increase gradually, with increasing temperatures from March through July, and then to decrease again until ceasing during January and February, when room temperatures fell within the range 14°C to 20°C. Aufderheide et al. (2006) concluded that "25–28°C lies centrally within the tolerance range of individuals of *M. cornuarietis*, but that at temperatures of 22°C, snails begin to show signs of reduced reproductive performance." Selck et al. (2006) obtained the same results when rearing juveniles and found 25–28°C is optimal for growth of *M. cornuarietis*.

The species has both lower and upper lethal thermal thresholds. Robins (1971) reported cessation of egg development at 11°C and succumb when exposed to 8°C. But Thomas (1975) was able to keep *M. cornuarietis* alive at 6°C. The short-term upper lethal thermal threshold has been reported by Robins (1971) to be 39°C for adults and 37°C for juveniles and both adults and juveniles fed normally between 33.5°C and 35.5°C; eggs did not develop at 35–37°C. Thomas (1975) stated that the species can tolerate temperatures up to 45°C but Freiburg and Hazelwood (1977) gave 40°C as the upper temperature.

3.4.1.1.4 Survival Potential of Giant Ramshorn Snail Based on Conductivity, TDS, and Salinity

They are known to tolerate relatively high salt concentrations and they are occasionally found in light brackish waters (Ferguson and Palmer, 1958). They are able to survive in salinities of up to 8.5 ppt, but they do not reproduce above these salinities (Hunt, 1961). Hunt (1961), Robin (1971), and Santos et al. (1987) report that *M. cornuarietis* can grow and reproduce in salinities up to about 30% (3 ppt). Although conductivity and TDS can be estimated from salinity (e.g., Table 3.5), the threshold levels of conductivity and TDS required for survival, growth, and reproduction of ramshorn snails are unknown.

Remane and Schlieper (1971) designated salinity limits for 19 northern, temperate European freshwater gastropod species between 0.2 and 13.7 ppt. The majority of species fell within the 3–8 ppt range, characteristic of second-degree eurybionts. Among nine species of freshwater prosobranch snails, maximum tolerated salinities ranged between 0.5 and 18 ppt with that of all but two species falling within the 3–8 ppt range of second-degree eurybionts.

3.4.1.1.5 Survival Potential of Giant Ramshorn Snail Based on Turbidity and TSS

The threshold levels of turbidity and TSS required for survival, growth, and reproduction of ramshorn snails are unknown.

TABLE 3.19
Studies Using Various Parameters to Assess Risk Potential of Survival and Potential Density of *Pomacea canaliculata*

Variables	Reference
Alkalinity	No information
Calcium	No information
Chlorophyll *a*	No information
Conductivity, TDS	Infer from Fujio et al. (1991), Albrecht et al. (1996), Marfurt and Burks (2005) and Table 3.5
Dissolved oxygen	Ito (2002), Chunlin et al. (2006)
Nitrogen, total	No information
pH	Kim (1996), Ito (2002), Chunlin et al. (2006)
Phosphorus, total	No information
Salinity	Fujio et al. (1991), Albrecht et al. (1996), Marfurt and Burks (2005)
Secchi depth, TSS	Martín et al. (2001), Cazzaniga (2006)
Temperature (°C)	Oya et al. (1987), Mochida (1988, 1991), Eversole (1992), Li (1995), Sugiura and Wada (1999), Yin et al. (2000), Syobu et al. (2001), Ito (2002), Zhou et al. (2003), Heiler et al. (2008)
Total hardness	No information
Turbidity (NTU), TSS	Martín et al. (2001), Cazzaniga (2006)

The variables are listed alphabetically.

3.4.1.2 Golden Apple Snail, *Pomacea canaliculata* (Lamarck, 1822)

The threshold limits for survival, growth, and reproduction for each of the physiological and ecological variables of *P. canaliculata* were determined from averaging the levels reported by the researchers listed in Table 3.19. The ranges in values for each of the risk levels are given in Table 3.20 when known. Unfortunately, very little is known about the levels of calcium, alkalinity, pH, and hardness required for survival, growth, and reproduction of not only golden apple snails, coined GASs (Joshi and Sebastian, 2006), but other *Pomacea* species as well. Even the authoritative tome of Joshi and Sebastian (2006) has no information on the minimum levels of calcium needed for shell formation. Unlike dreissenids and mytilids that have lower and upper threshold levels of food (measured as chlorophyll *a*) for survival, growth, and reproduction, ampullariids are voracious herbivores of submersed (e.g., *Ceratophyllum*, *Hydrilla*, *Potamogeton*), emergent (e.g., *Typha*, *Sagittaria*), and floating (e.g., *Pistia*, *Nymphaea*, *Eichhornia*) (Howells et al., 2006) macrophytes, but the levels of nutrients required for growth and reproduction are unknown. Presumably some or all of their calcium requirements may come from food. All apple snails seem to be able to tolerate enriched conditions, the levels of which can be determined by Secchi depth. However, most species live in depths <1 m, many of them shallow enough to dry up for a portion of the year (Ito, 2002; Cazzaniga, 2006). Temperature and salinity seem to be the main factors controlling survival, growth, and reproduction. Since many ampullariid snails tend to inhabit shallow freshwater habitats such as slow moving streams,

TABLE 3.20
Criteria Used in Determining Levels of Infestation by *Pomacea canaliculata* in North America

1. Parameter	2. No Potential for adult Survival	3. Little Potential for Egg Development	4. Moderate Potential for Nuisance Infestations	5. High Potential for Massive Infestations
Chalk Variables				
Calcium (mg Ca/L)	?	MNBL	MNBL	MNBL
pH	<3, >10	3–5	5–7.5	7.5–10
Alkalinity, total (mg $CaCO_3$/L)	?	MNBL	MNBL	MNBL
Hardness, total (mg $CaCO_3$/L)	?	MNBL	MNBL	MNBL
Nutrient Variables				
Dissolved oxygen (mg/L) (% saturation)	<0.5 (<1%)	0.5–2 (1–10%)	2–5 (10–50%)	≥5 (>50%)
Chlorophyll *a* (µg/L)	?	MNBL	MNBL	MNBL
Total phosphorus (µg/L)	?	MNBL	MNBL	MNBL
Total nitrogen (µg/L)	?	MNBL	MNBL	MNBL
Secchi depth (m)	NA	NA	NA	NA
Physical Variables				
Temperature (°C)	<2, >40	2–10, 38–40	10–25, 32–38	25–32
Conductivity (µS/cm)	>13,800	7300–13,800	3800–7300	<3800
TDS (mg/L)	9200	4900–9200	2500–4900	<2500
Salinity (mg/L) (ppt)	>8	4–8	2–4	<2
Turbidity (NTU)	MNBL	MNBL	MNBL	MNBL
TSS (mg/L)	MNBL	MNBL	MNBL	MNBL

Use column 3 for assessing potential for survival (but little potential for nuisance infestation). Columns 4 and 5 indicate good potential for long-term survival and the criteria are used to assess temporal and seasonal variations in infestation potential. MNBL = May not be limiting. NA = Not applicable. ? = No data available.

canals, bayous, and marshes (Miyahara et al., 1986; Perera and Walls, 1996), they can be subjected to wide diurnal ambient fluctuations in temperature and salinity (McMahon, 1983).

3.4.1.2.1 Survival Potential of GASs Based on Chalk Variables

The threshold levels of calcium required for survival, growth, and reproduction of GASs are unknown. It is assumed that a minimum amount of calcium is required for shell formation but it is possible that the chalk variables may not be limiting above the minimum level needed for survival and growth. Chunlin et al. (2006) report that the GAS has a great resistance to high pH by closing the operculum and secreting mucus, but prefers slightly alkaline water in a pH range of 7.0–8.5; at pHs between 10 and 11.5 the operculum closes and the snails eventually die. Ito (2002) examined survival of an overwintering population of the GASs in northern Japan and reported the pH to vary between 6.3 and 7.5 (mean 6.8). Since the pH in rice fields can fluctuate between pH 2.7 and 10.3 (Kim, 1996), one would expect the pH tolerance range to be the same for GASs. With a maximum salinity tolerance of 85 ppt, the island applesnail is considered a second-degree euryhaline limnobiont.

3.4.1.2.2 Survival Potential of GASs Based on Trophic Variables

Dissolved oxygen: Chunlin et al. (2006) states that the GASs can tolerate low concentrations of dissolved oxygen. At 25°C and 1–2 mg O_2/L, the snail stretches its body to increase the surface area for air exchange and extrudes the siphon; when the dissolved oxygen falls below 0.23–0.45 mg/L the operculum closes. At concentrations of 0.08–0.10 mg O_2/L, 80% of the snails die. Ito (2002) examined survival of an overwintering population of the GASs in northern Japan and reported the dissolved oxygen to vary between 1.0 and 6.4 (mean 3.2), which corresponds to 7.2% and 54% saturation at the prevailing temperatures of 2°C and 8°C, respectively, but snails were hibernating at the time.

Chlorophyll a*:* This species feeds on a wide variety of submersed, floating, and emergent macrophytes, with a preference for submersed and floating forms (Howell et al. 2006). No information on quantities required to initiate growth and reproduction is available.

Nutrients (total phosphorus, total nitrogen): The threshold levels of nutrients required for survival, growth, and reproduction of GASs are unknown. It is possible that nutrients may not be limiting above a certain value. However, it is well known that food availability influences growth rate in most pomacean snails (Estebenet and Martin, 2002). Individuals of *P. canaliculata* grow more slowly when reared on reduced food rations (Estoy et al., 2002).

Albrecht et al. (1999) found that water temperature may be the critical factor that causes the seasonal onset of copulatory and spawning behavior but food availability may also be a determining factor, mainly to trigger spawning activity.

Secchi depth, trophic status: The GAS generally lives in shallow, quiet, turbid water, with low $Na^+/(K^+ + Mg^{2+})$ ratios (Martín et al., 2001). Cazzaniga (2006) states "The snail's preference for sites with high content of suspended matter is probably supported neither by trophic needs nor by nature of the sediments; it seems to be more related to the turbidity, which reduces the probability of predation."

3.4.1.2.3 Survival Potential of GASs Based on Temperature

The GAS has both low and high thermal thresholds for survival, growth, and reproduction. Addressing the lower thermal threshold issue first, the GAS in China goes into hibernation at 6°C to 8°C and temperatures lower than 6°C leads to their death (Yin et al., 2000). Snails exposed to air temperatures of 1–2°C were killed in 24 h, but snails in the sediment at 1–2°C required seven days to die, but snails survived for 30 days at 3–5°C (Chen et al., 2003). Oya et al. (1987) and Mochida (1991) reported that GASs survived for 15–20 days at 0°C, for 2 days at −3°C, and for 6 h at −6°C. Size and age affect the survival times; younger and smaller (2–3 cm high) snails tolerated cold better than larger (>3 cm high) snails (Oya et al., 1987; Syobu et al., 2001). Zhou et al. (2003) consider the lower lethal threshold to be 0°C and the upper thermal threshold to be >40°C. Mochida (1988) reported high mortality of the GSA at water temperatures above 35°C but later (Mochida, 1991) stated 32°C, which was supported by Eversole (1992).

The optimum temperature for the GAS is reported to be between 27.7°C and 30.6°C, with activity significantly reduced below 12°C (Li, 1995). In the laboratory, a temperature of 25°C is considered optimal for oviposition and growth in artificially cultured pomacean snails (Estebenet and Cazzaniga, 1992). Sugiura and Wada (1999) reported snails withdrawing into their shells under 10°C. Zhou et al. (2003) determined the suitable temperature range to be between 8°C and 38°C. Heiler et al. (2008) found that the activity of a GAS increased threefold with increasing temperature between 20°C and 32°C and crawling velocity increased 3.6-fold at these temperatures.

Ito (2002) found that the northernmost population of Japan cannot overwinter in dried paddies, but can overwinter in limited areas of the water canals. Places of hibernation in these canals were characterized by shallowness, high dissolved oxygen, and low pH, not by water temperature variations.

3.4.1.2.4 Survival Potential of GASs Based on Conductivity, TDS, and Salinity

Marfurt and Burks (2005) found that both hatchling and adult GASs survived 24-h plunge tests in 8 ppt for 24 h, but none survived at 14, 20, or 35 ppt. They also found that adult specimens actually increased feeding at 8 ppt, while hatchling feeding did not differ between 0 and 8 ppt.

Howells et al. (2006) citing a personal communication that stated no snails or egg masses were seen from May 25 to June 16, 2005 in areas where salinity reached 6 ppt at Armand Bayou in southeastern Texas. However, one egg mass and no snails were seen at another site where salinity ranged from 0 to 4 ppt, and 49 egg masses and >20 snails were counted at a third location where salinity was <1 ppt during the same time period. Apparently, different "strains" of *P. canaliculata* exhibit different degrees of salinity tolerance (Fujio et al., 1991). Albrecht et al. (1996) reported that the species tolerated a broad range of salinity.

3.4.1.2.5 Survival Potential of GASs Based on Turbidity and TSS

See comments under Secchi depth.

3.4.1.3 Titan Applesnail, *Pomacea haustrum* (Reeve, 1856)

Very little is known about the ecology and physiology of the Titan applesnail. Presumably, the threshold levels for survival, growth, and reproduction based on chalk variables, trophic variables, and physical variables are similar to the GAS.

3.4.1.4 Island Applesnail, *Pomacea insularum* (D'Orbigny, 1839)

The threshold limits for survival, growth, and reproduction for each of the physiological and ecological variables of *Pomacea insularum* were determined from averaging the levels reported by the researchers listed in Table 3.21. The ranges in values for each of the risk levels are given in Table 3.22 when known.

3.4.1.4.1 Survival Potential of Island Applesnail Based on Chalk Variables

Ramakrishnan (2007) reported 100% cumulative mortality data for *P. insularum* exposed to acidic pH levels of <3.0; the 96-h LD_{50} pH value was between 2.9 and 3.4, with lower pH needed to kill bigger individuals. There was no significant difference in raw cumulative percent mortality for pH treatments of 4.0–7.0 within the 28-d exposure period.

Ramakrishnan (2007) showed that 100% sample mortality was obtained for basic pH treatments ranging from 10.5 to 12.5, and no significant difference in raw cumulative percent mortality for pH treatments of 7.0–10.0; after 28 days, 5% mortality was seen at pH 9, 30% at pH 10.0, and 100% at pH 10.5. This is similar to the pH tolerance range reported by Prescott and Curteanu (2004) in the Central Parkland subregion of Alberta, Canada, for several freshwater gastropods; *Valvata tricarinata* was found in aquatic habitats with pH ranging from 8.1 to 10.0 (mean 9.1);

Lymnaea stagnalis appressa was found in water bodies whose pH ranged from 7.0 to 10.7 (mean 8.8). The pH in rice fields can fluctuate between 2.7 and 10.3 (Kim, 1996). Since the pH tolerance range of *P. insularum* without mortality at 4.0–10.5 is similar to that of cultivated rice fields, Ramakrishnan (2007) predicted that this

TABLE 3.21
Studies Using Various Parameters to Assess Risk Potential of Survival and Potential Density of *Pomacea insularum*

Variables	Reference
Alkalinity	No information
Calcium	No information
Chlorophyll *a*	No information
Conductivity, TDS	Infer from Ramakrishnan (2007) and Table 3.5
Dissolved oxygen	Ramakrishnan (2007)
Nitrogen, total	No information
pH	Kim (1996), Berezina (2001), Ramakrishnan (2007)
Phosphorus, total	No information
Salinity	Ramakrishnan (2007)
Secchi depth, TSS	NA
Temperature (°C)	Ramakrishnan (2007)
Total hardness	No information
Turbidity (NTU), TSS	No information

The variables are listed alphabetically. NA = Not applicable, see the text.

TABLE 3.22
Criteria Used in Determining Levels of Infestation by *Pomacea insularum* in North America

1. Parameter	2. No Potential for Adult Survival	3. Little Potential for Egg Development	4. Moderate Potential for Nuisance Infestations	5. High Potential for Massive Infestations
Chalk Variables				
Calcium (mg Ca/L)	?	MNBL	MNBL	MNBL
pH	<4, >11	4–6, 9.5–11	6–7.5	7.5–9.5
Alkalinity, total (mg $CaCO_3$/L)	?	MNBL	MNBL	MNBL
Hardness, total (mg $CaCO_3$/L)	?	MNBL	MNBL	MNBL
Nutrient Variables				
Dissolved oxygen (mg/L) (% saturation)	<0.5 (<1%)	0.5–2 (1–10%)	2–5 (10–50%)	≥5 (>50%)
Chlorophyll *a* (μg/L)	?	MNBL	MNBL	MNBL
Total phosphorus (μg/L)	?	MNBL	MNBL	MNBL
Total nitrogen (μg/L)	?	MNBL	MNBL	MNBL
Secchi depth (m)	NA	NA	NA	NA
Physical Variables				
Temperature (°C)	<15, >40	15–16, 36–40	16–25, 32–38	25–36
Conductivity (μS/cm)	>23,200	17,000–23,200	10,600–17,000	<10,600
TDS (mg/L)	15,200	11,400–15,200	7100–11,400	<7100
Salinity (mg/L) (ppt)	>14	10–14	6–10	<6
Turbidity (NTU)	MNBL	MNBL	MNBL	MNBL
TSS (mg/L)	MNBL	MNBL	MNBL	MNBL

Use column 3 for assessing potential for survival (but little potential for nuisance infestation). Columns 4 and 5 indicate good potential for long-term survival and the criteria are used to assess temporal and seasonal variations in infestation potential. MNBL = May not be limiting. NA = Not applicable. ? = No data available.

species is likely to be capable of invading the majority of wetland rice cultivation systems of the southeastern USA, and cites its distribution in the southeastern counties of Texas (Howells and Smith, 2002) as support for her predictions.

3.4.1.4.2 Survival Potential of Island Applesnail Based on Trophic Variables

Dissolved oxygen: Ramakrishnan (2007) found a linear increase in V_{O_2} with increasing dry tissue weight in juvenile specimens of *P. insularum*, with almost a twofold increase in whole animal V_{O_2} at test temperatures 20–35°C with a twofold increase in dry tissue weight; V_{O_2} was essentially anaerobic at 15°C. These results indicated that V_{O_2} in juvenile *P.insularum* was almost directly proportional to body mass and seem to have the lowest capacity for O_2 regulation of V_{O_2} when compared to other pomacean species and freshwater prosobranch species (Ramakrishnan, 2007). The poor O_2 regulatory ability of juvenile *P. insularum* may reflect that of adults and may account for the apparent restriction of this species to lotic (i.e., flowing water) freshwater inhabits habitats such as irrigation canals, streams, bayous, and slow-flowing rivers, where populations are less likely to experience extended periods of severe environmental hypoxia (Ramakrishnan, 2007). Oxygen uptake generally ceased at P_{O_2} between 0 and 16 Torr (i.e., 0–10% of full air O_2 saturation).

Chlorophyll a*:* This species feeds on macrophytes (Ramakrishnan, 2007) but no information is available on plant quantities.

Nutrients (total phosphorus, total nitrogen): It is possible that nutrients may not be limiting above a certain value. It is well known that food availability influences growth rate and food availability may trigger spawning activity of many pomacean snails (Albrecht et al., 1999; Estroy et al., 2002; Estebenet and Martin, 2002). Hence, it is likely that there is a minimum amount of food and nutrients required to stimulate growth and reproduction, but nutrients and food above these levels may not be limiting growth and reproduction.

Secchi depth, trophic status: This species lives in depths too shallow for Secchi readings.

3.4.1.4.3 Survival Potential of Island Applesnail Based on Temperature

Juvenile and adult specimens of *P. insularum* have a very high incipient lower lethal temperature of approximately 15°C and is likely to be a key factor limiting its eventual distribution in Texas and North America (Ramakrishnan, 2007). With an incipient lower thermal tolerance limit of approximately 15°C, this species could extend its present range longitudinally beyond southeastern Texas, into other Gulf of Mexico coastal regions of Texas, where mean air temperatures do not fall below 16°C in winter (Ramakrishnan, 2007).

Ramakrishnan (2007) found that the V_{O_2} of juvenile *P. insularum* increased with an increase in temperature up to 35°C but was greatly reduced at a temperature of 40°C, which she found lethal to this species. Chronic exposures of specimens of *P. insularum* to different temperatures revealed a maximum incipient upper lethal temperature limit of 36°C (Ramakrishnan, 2007). The adjusted mean heat coma temperatures for *P. insularum* were 38°C for 20.1°C acclimated individuals, 40.3°C for 25°C acclimated individuals, and 40.6°C for 30°C acclimated individuals. Ramakrishnan (2007) reported high survivorship throughout a 28-d exposure period in individuals held at temperatures ≤36°C, and progressively declined with increasing temperature from 37°C to 41°C, regardless of acclimation temperature. The upper thermal tolerance limit of Texas specimens of *P. insularum* was between 36°C and 37°C. The results also clearly indicated that temperatures ≥37°C are lethal to this species regardless of previous temperature experience or individual size (Ramakrishnan, 2007).

3.4.1.4.4 Survival Potential of Island Applesnail Based on Conductivity, TDS, and Salinity

Laboratory bioassays by Ramakrishnan (2007) indicate that specimens of *P. insularum* cannot tolerate salinities ≥13.6 ppt at which point snails failed to consume lettuce and remained with their operculum tightly withdrawn throughout a 28-d exposure period; the majority of specimens at 10.2 ppt consumed only small quantities of lettuce. Ramakrishnan's (2007) model revealed that salinities of 0–6.8 ppt were nonlethal to specimens of *P. insularum* with an estimated probability of survival >0.97 at the end of the 28-d exposure period. Hence, with a maximum salinity tolerance of 6.8 ppt, the island applesnail must be considered a second-degree euryhaline limnobiont as are the majority of other freshwater prosobranch and pulmonate gastropod species (Remane and Schlieper, 1971).

3.4.1.4.5 Survival Potential of Island Applesnail Based on Turbidity and TSS

Turbidity may not be a limiting factor because turbidity is typically elevated in the lotic freshwater habitats such as irrigation canals, streams, bayous and slow-flowing rivers (Miyahara et al., 1986; Perera and Walls, 1996) that the Island applesnail is found in.

3.4.1.5 Spiketop Applesnail, *Pomacea diffusa* (Blume, 1957)

Very little is known about the ecology and physiology of the spiketop applesnail, particularly the threshold levels for survival, growth, and reproduction based on chalk variables, trophic variables, and temperature and turbidity. There is some information on salinity thresholds.

3.4.1.5.1 Survival Potential of Spiketop Applesnail Based on Conductivity, TDS, and Salinity

Jordan and Deaton (1999) found that specimens of *Pomacea bridgesi* exhibited >80% survival when exposed to salinities ranging from 0 to 6.8 ppt for 14 days, but experienced 100% mortality when exposed to 13.6 ppt for three days. These results are similar to the 100% sample mortality within 4–7 days recorded for specimens of *P. insularum* exposed to 13.6 ppt reported by Ramakrishnan (2007). With a maximum salinity tolerance of 6.8 ppt, the spiketop applesnail is considered a second-degree euryhaline limnobiont.

3.4.1.6 Florida Applesnail, *Pomacea paludosa* (Say, 1829)

The threshold limits for survival, growth, and reproduction for each of the physiological and ecological variables of *P. paludosa* were determined from averaging the levels reported by the researchers listed in Table 3.23. The ranges in values for each of the risk levels are given in Table 3.24 when known.

TABLE 3.23
Studies Using Various Parameters to Assess Risk Potential of Survival and Potential Density of *Pomacea paludosa*

Variables	Reference
Alkalinity	Infer from Darby et al. (1997), Carrao et al. (2006)
Calcium	Infer from Darby et al. (1997), Carrao et al. (2006)
Chlorophyll *a*	Steinman (2001), Sharfstein and Steinman (2001)
Conductivity, TDS	Infer from Ramakrishnan (2007) and Table 3.5
Dissolved oxygen	Freiberg and Hazelwood (1977), Hanley and Ultsch (1999)
Nitrogen, total	Sharfstein and Steinman (2001), Carrao et al. (2006)
pH	Darby et al. (1997), Carrao et al. (2006)
Phosphorus, total	Sharfstein and Steinman (2001), Carrao et al. (2006)
Salinity	NA
Secchi depth, TSS	NA
Temperature (°C)	Freiburg and Hazelwood (1977), Darby et al. (1997), Stevens et al. (2002)
Total hardness	Darby et al. (1997), Carrao et al. (2006)
Turbidity (NTU), TSS	No information

The variables are listed alphabetically. NA = Not applicable, see the text.

3.4.1.6.1 Survival Potential of Florida Applesnail Based on Chalk Variables

There is little information on the threshold levels for survival, growth, and reproduction of the Florida applesnail based on chalk variables. Laboratory bioassays use hard water (190–210 mg $CaCO_3$/L) with pH ranging from 7.5 to 8.5 (Darby et al., 1997; Carrao et al., 2006). Since *M. cornuarietis*, *P. canaliculata*, and *P. insularum* have similar pH threshold levels for survival, growth, and reproduction, the threshold levels are probably similar for *P. paludosa* and are averaged in Table 3.24.

3.4.1.6.2 Survival Potential of Florida Applesnail Based on Trophic Variables

Dissolved oxygen: Hanley and Ultsch (1999) classified species as either oxygen conformers or oxygen regulators based on their responses of V_{O_2} to decreasing partial pressure of oxygen (P_{O_2}): they labeled a species as an oxygen conformer when V_{O_2} declines in direct proportion to declines in P_{O_2}; an oxygen regulator is a species in which the V_{O_2} remains constant over a broad range of declining P_{O_2} until P_{O_2} approaches a critical low value below beyond which V_{O_2} declines directly with further decreases in ambient P_{O_2}.

Freiberg and Hazelwood (1977) were able to sustain an increasing V_{O_2} between 10°C and 35°C for *P. paludosa*, suggesting that the species is able to remain metabolically active over a broader temperature range than *P. insularum* (Ramakrishnan, 2007). Hanley and Ultsch (1999) examined respiration in *M. cornuarietis* and *P. palludosa* and reported P_{O_2} values of 90 Torr at 25°C and 95 Torr at 15°C, respectively. In *M. cornuarietis*, the oxygen uptake rates are lower during emersion than during immersion while they were not different in *Pomacea palludosa* (Freiberg and Hazelwood, 1977).

TABLE 3.24
Criteria Used in Determining Levels of Infestation by *Pomacea paludosa* in North America

1. Parameter	2. No Potential for Adult Survival	3. Little Potential for Egg Development	4. Moderate Potential for Nuisance Infestations	5. High Potential for Massive Infestations
Chalk Variables				
Calcium (mg Ca/L)	?	MNBL	MNBL	MNBL
pH	<3.5, >9.5	3.5–5.5	5.5–7.5	7.5–9.5
Alkalinity, total (mg $CaCO_3$/L)	?	MNBL	MNBL	MNBL
Hardness, total (mg $CaCO_3$/L)	?	MNBL	MNBL	MNBL
Nutrient Variables				
Dissolved oxygen (mg/L) (% saturation)	<0.5 (<1%)	0.5–2 (1–10%)	2–5 (10–50%)	≥5 (>50%)
Chlorophyll *a* (μg/L)	?	MNBL	MNBL	MNBL
Total phosphorus (μg/L)	?	MNBL	MNBL	MNBL
Total nitrogen (μg/L)	?	MNBL	MNBL	MNBL
Secchi depth (m)	NA	NA	NA	NA
Physical Variables				
Temperature (°C)	<10, >38	10–18, 35–38	18–25, 32–38	25–35
Conductivity (μS/cm)	?	?	?	?
TDS (mg/L)	?	?	?	?
Salinity (mg/L) (ppt)	?	?	?	?
Turbidity (NTU)	MNBL	MNBL	MNBL	MNBL
TSS (mg/L)	MNBL	MNBL	MNBL	MNBL

Use column 3 for assessing potential for survival (but little potential for nuisance infestation). Columns 4 and 5 indicate good potential for long-term survival and the criteria are used to assess temporal and seasonal variations in infestation potential. MNBL = May not be limiting. NA = Not applicable. ? = No data available.

Chlorophyll a, *nutrients (total phosphorus, total nitrogen):* Sharfstein and Steinman (2001) fed three assemblages of plants [*Utricularia* sp./periphyton complex, an *Eleocharis* sp./periphyton complex, a metaphyton (loosely attached benthic algae)] to the Florida applesnail and reported that *Utricularia* had the highest nutritive quality based on its relatively low C:N ratio and ash content, and high phosphorus and chlorophyll concentrations. *Eleocharis* had a high C:N ratio and low P content, but also had moderate chlorophyll and low ash concentrations. The metaphyton assemblage had a low C:N ratio, but high ash and low chlorophyll concentrations. Sharfstein and Steinman (2001) also reported positive correlations between snail biomass and length with treatment chlorophyll content and P content and an inverse correlation with treatment ash content. Snail excreta, collected at the end of the experiment, showed significant differences in N content among all treatments, with the excreta N content of metaphyton-reared snails being twice that of the other two treatments. Changes in snail biomass were inversely correlated with snail excreta N content (Sharfstein and Steinman, 2001). Unfortunately, the data set does not allow estimates of threshold levels for survival, growth, and reproduction in the species.

Corrao et al. (2006) found no correlation between snail density and nitrate concentration measured in six Florida springs. Laboratory bioassays that exposed snails for 96 h to nitrate levels ranging from 0 to 1308 mg NO_3/L found that survival ranged from 93% to 100%. In two 14-day growth experiments, Corrao et al. (2006) reported a decreasing percentage of shell growth in juveniles with increasing nitrate levels and EC_{50} values of 504 and 622 mg NO_3/L. They concluded that elevated nitrate levels (at least up to 1300 mg/L) is an unlikely cause of snail mortality in Florida springs with nitrate pollution. Therefore, it appears that nitrate levels probably do not affect survival, growth, and reproduction of the species in most areas where it occurs.

Secchi depth, trophic status: Species lives in depths too shallow for Secchi depth readings.

3.4.1.6.3 Survival Potential of Florida Applesnail Based on Temperature

Darby et al. (1997) performed field and laboratory experiments to determine the effect of temperature on the survival of Florida applesnails. They reported significant mortality of snails in the field when air temperatures fell below 0°C to –3.3°C. Eleven of 23 snails that were stranded, all survived one to three nights, and all but one were died after 8.3 weeks, and it finally succumbed after another 8 weeks. In the water, the greatest number of snail deaths occurred during the week when water temperatures reached 38°C. The laboratory experiments supported the field findings, with the upper lethal threshold occurring at 35°C. Freiburg and Hazelwood (1977) report 40°C as the upper lethal temperature for *P. paludosa* when exposed for 1–4 h.

Temperature data collected by Darby et al. (1997) during the 8-week session indicated that oviposition began once temperatures reached approximately 20°C in March and peaked around 28°C in April. Stevens et al. (2002) found that snails are inactive when temperature drops below 10°C. Cary (1985) reported average capture rates of 5.28 snails per hour above 30°C, 2.68 between 21°C and 30°C, 0.72 between 11°C and 20°C, and 0.0 below 10°C.

3.4.1.6.4 Survival Potential of Florida Applesnail Based on Conductivity, TDS, and Salinity

There is no information on the salinity tolerance of *P. paludosa.*

3.4.1.6.5 Survival Potential of Florida Applesnail Based on Turbidity and TSS

There is no information on the turbidity tolerance of *P. paludosa.*

3.4.1.7 Quilted Melania, *Tarebia granifera* (Lamarck, 1822)

The threshold limits for survival, growth, and reproduction for each of the physiological and ecological variables of *Tarebia granifera* were determined from averaging the levels reported by the researchers listed in Table 3.25. The ranges in values for each of the risk levels are given in Table 3.26 when known.

3.4.1.7.1 Survival Potential of Quilted Melania Based on Chalk Variables

Little is known about the pH tolerance of quilted melania except that they are able to survive at pH between 7.1 and 8.5 (Oglesby, 1977). Cañete et al. (2004) found the species in waters with pH ranging from 7 to 8.5 and that hardness was an important factor limiting its abundance but did not give values.

3.4.1.7.2 Survival Potential of Quilted Melania Based on Trophic Variables

Dissolved oxygen: Little is known about the threshold levels of dissolved oxygen required for survival, growth, and reproduction of the quilted melania. Since it

TABLE 3.25
Studies Using Various Parameters to Assess Risk Potential of Survival and Potential Density of *Tarebia granifera*

Variables	Reference
Alkalinity	Infer from Darby et al. (1997), Carrao et al. (2006)
Calcium	Infer from Darby et al. (1997), Carrao et al. (2006)
Chlorophyll *a*	No information
Conductivity, TDS	Infer from Oglesby (1977), Roessler et al. (1978), Chaniotis et al. (1980a) and Table 3.5
Dissolved oxygen	No information
Nitrogen, total	No information
pH	Oglesby (1977), Cañete et al. (2004)
Phosphorus, total	No information
Salinity	Oglesby (1977), Roessler et al. (1978), Chaniotis et al. (1980a)
Secchi depth, TSS	NA
Temperature (°C)	Abbot (1952), (Chaniotis et al., 1980a, b, c), Chen (2004)
Total hardness	No information
Turbidity (NTU), TSS	No information

The variables are listed alphabetically. NA = Not applicable, see the text.

TABLE 3.26
Criteria Used in Determining Levels of Infestation by *Tarebia granifera* in North America

1. Parameter	2. No Potential for Adult Survival	3. Little Potential for Egg Development	4. Moderate Potential for Nuisance Infestations	5. High Potential for Massive Infestations
Chalk Variables				
Calcium (mg Ca/L)	?	MNBL	MNBL	MNBL
pH	<6, >?	6.0–7.0	7.0–7.5	7.5–8.5
Alkalinity, total (mg $CaCO_3$/L)	?	MNBL	MNBL	MNBL
Hardness, total (mg $CaCO_3$/L)	?	MNBL	MNBL	MNBL
Nutrient Variables				
Dissolved oxygen (mg/L) (% saturation)	<0.5 (<1%)	0.5–2 (1–10%)	2–5 (10–50%)	≥5 (>50%)
Chlorophyll *a* (μg/L)	?	MNBL	MNBL	MNBL
Total phosphorus (μg/L)	?	MNBL	MNBL	MNBL
Total nitrogen (μg/L)	?	MNBL	MNBL	MNBL
Secchi depth (m)	NA	NA	NA	NA
Physical Variables				
Temperature (°C)	<7, >40	7–13, 36–40	13–22	22–36
Conductivity (μS/cm)	>5600	3800–5600	2000–3800	1000–2000
TDS (mg/L)	>4000	1300–4000	700–1300	700
Salinity (mg/L) (ppt)	>3.0	2.0–3.0	1.0–2.0	<0.5–1.0
Turbidity (NTU)	MNBL	MNBL	MNBL	MNBL
TSS (mg/L)	MNBL	MNBL	MNBL	MNBL

Use column 3 for assessing potential for survival (but little potential for nuisance infestation). Columns 4 and 5 indicate good potential for long-term survival and the criteria are used to assess temporal and seasonal variations in infestation potential. MNBL = May not be limiting. NA = Not applicable. ? = No data available.

prefers the riffles of shallow fast-flowing freshwater streams (Lachner et al., 1970), it may high dissolved oxygen requirements.

Chlorophyll a*:* This species primarily feeds on algae (e.g., diatoms), associated microorganisms, and small particles of organic matter (Ogesby, 1977; Chaniotis et al., 1980a; Prentice, 1983).

Nutrients (total phosphorus, total nitrogen): The threshold levels of nutrients required for survival, growth, and reproduction of the quilted melania are unknown.

Secchi depth, trophic status: Although red-rimmed melania have been reported from relatively deep portions of freshwater pools (3–3.7 m deep) (Murray, 1975), the species typically is found in rivers and pools too shallow for Secchi depth readings. It can probably withstand enriched conditions because the species hides in poorly oxygenated microhabitat beneath decaying plants and stones or buries it in the mud during the day.

3.4.1.7.3 Survival Potential of Quilted Melania Based on Temperature

Chaniotis et al. (1980b) reported a lower lethal temperature for specimens under experimental conditions of 7°C. Abbot (1952) stated that this species is rarely found in water under 10°C and generally not found in water <18°C. Chaniotis et al. (1980c) reported total kills at water temperatures higher than 40°C. Habitat temperatures typically range from 22°C to 36°C for the species (Chaniotis et al., 1980a). The minimum air temperature observed for a *T. granifera* population in Jinlun River, Southeastern Taiwan during the winter was 13°C and peaked at 20°C; during the summer embryos were first released when temperature reached 25°C and peaked at 28°C (Chen, 2004).

3.4.1.7.4 Survival Potential of Quilted Melania Based on Conductivity, TDS, and Salinity

This species typically occurs in freshwaters and does not appear to tolerate salt water well (Chaniotis et al., 1980a). Oglesby (1977) noted that the highest level of salinity that the snails were observed was about 1.0 ppt. Specimens tend to bury themselves in response to rising salinity (Chaniotis et al., 1980a). Roessler et al. (1978) found red-rimmed melania at concentrations of up to 37,500 specimens/m^2 in estuarine areas of Florida dominated by red (*Rhizophora mangle*) and white (*Laguncularia racemosa*) mangroves. The species appears to be a first-degree euryhaline limnobiont tolerating 0.5–3 ppt.

3.4.1.7.5 Survival Potential of Quilted Melania Based on Turbidity and TSS

The threshold levels of turbidity and TSS required for survival, growth, and reproduction of quilted melania are unknown.

3.4.1.8 Red-Rimmed Melania, *Melanoides tuberculata* (Müller, 1774)

The threshold limits for survival, growth, and reproduction for each of the physiological and ecological variables of *Melanoides tuberculata* were determined from averaging the levels reported by the researchers listed in Table 3.27. The ranges in values for each of the risk levels are given in Table 3.28 when known.

TABLE 3.27
Studies Using Various Parameters to Assess Risk Potential of Survival and Potential Density of *Melanoides tuberculata*

Variables	Reference
Alkalinity	Yacoubi et al. (2007) and infer from Duggan et al. (2007)
Calcium	Infer from Duggan et al. (2007) and Yacoubi et al. (2007)
Chlorophyll *a*	No information
Conductivity, TDS	Russo (1973), Roessler et al. (1977), Yacoubi et al. (2007)
Dissolved oxygen	Livshits and Fishelson (1983), (Neck 1984)
Nitrogen, total	No information but see Yacoubi et al. (2007)
pH	Duggan et al. (2007),Yacoubi et al. (2007)
Phosphorus, total	No information but see Yacoubi et al. (2007)
Salinity	Infer from Russo (1973), Roessler et al. (1977), Yacoubi et al. (2007) and Table 3.5.
Secchi depth, TSS	NA
Temperature (°C)	Murray (1971), Duggan et al. (2007), Mitchell and Brandt (2003, 2005)
Total hardness	Infer from Duggan et al. (2007) and Yacoubi et al. (2007)
Turbidity (NTU), TSS	No information

The variables are listed alphabetically. NA = Not applicable, see the text.

3.4.1.8.1 Survival Potential of Red-Rimmed Melania Based on Chalk Variables

Duggan et al. (2007) reported the species present in New Zealand sites with little or no geothermal influence typically had pH between 5 and 8. Four of the geothermally influenced streams had alkalinities that ranged from highly acidic (pH 3.0) to highly alkaline (pH 9.1).

Yacoubi et al. (2007) reported finding *M. tuberculata* highly associated with *Bulinus truncatus* in waters with pH ranging from 5.9 to 9.2 and alkalinity from 13.7 to 44.0 mg $CaCO_3$/L.

3.4.1.8.2 Survival Potential of Red-Rimmed Melania Based on Trophic Variables

Dissolved oxygen: Although the quilted melania is very resistant to low oxygen levels (Neck, 1984) and it is active mostly at night, and hides in poorly oxygenated microhabitat beneath decaying plants and stones or buries itself in the mud during the day (Livshits and Fishelson, 1983), little is known about the threshold levels of dissolved oxygen required for survival, growth, and reproduction of the red-rimmed melania. Yacoubi et al. (2007) reported finding the species associated with *Bulinus truncatus* in waters with dissolved oxygen levels ranging from 5.1 to 9.3 mg/L; temperature was not reported so percent saturation levels are unknown.

Chlorophyll a*:* This species primarily feeds on microalgae (Roessler et al., 1978; Subda Rao and Mitra, 1982) but the levels of chlorophyll *a* needed to initiate growth and reproduction are unknown.

TABLE 3.28
Criteria Used in Determining Levels of Infestation by *Melanoides tuberculata* in North America

1. Parameter	2. No Potential for Adult Survival	3. Little Potential for Egg Development	4. Moderate Potential for Nuisance Infestations	5. High Potential for Massive Infestations
Chalk Variables				
Calcium (mg Ca/L)	?	MNBL	MNBL	MNBL
pH	<5, >10?	5.0–6.0	6.0–7.5	7.5–9.5
Alkalinity, total (mg $CaCO_3$/L)	10	MNBL	MNBL	MNBL
Hardness, total (mg $CaCO_3$/L)	10	MNBL	MNBL	MNBL
Nutrient Variables				
Dissolved oxygen (mg/L) (% saturation)	<0.5 (<1%)	0.5–2 (1–10%)	2–5 (10–50%)	≥5 (>50%)
Chlorophyll *a* (µg/L)	?	MNBL	MNBL	MNBL
Total phosphorus (µg/L)	?	MNBL	MNBL	MNBL
Total nitrogen (µg/L)	?	MNBL	MNBL	MNBL
Secchi depth (m)	NA	NA	NA	NA
Physical Variables				
Temperature (°C)	<10, >50	10–15, 35–50	13–22, 30–35	18–30
Conductivity (µS/cm)	<100, >46,000	100–200, 25,000–46,000	200–1200, 14,000–25,000	1200–14,000
TDS (mg/L)	<70, >31,000	70–140, 17,000–31,000	140–800, 9400–17,000	800–9400
Salinity (mg/L) (ppt)	>30	15.0–30.0	0.1–2.0, 8–15	2.0–8.0
Turbidity (NTU)	MNBL	MNBL	MNBL	MNBL
TSS (mg/L)	MNBL	MNBL	MNBL	MNBL

Use column 3 for assessing potential for survival (but little potential for nuisance infestation). Columns 4 and 5 indicate good potential for long-term survival and the criteria are used to assess temporal and seasonal variations in infestation potential. MNBL = May not be limiting. NA = Not applicable. ? = No data available.

Nutrients (total phosphorus, total nitrogen): The threshold levels of nutrients required for survival, growth, and reproduction of red-rimmed melania are unknown but Yacoubi et al. (2007) reported finding the species associated with *Bulinus truncatus* in waters with phosphate levels ranging from 170 to 1520 μg/L and nitrates from 3000 to 5000 μg/L.

Secchi depth, trophic status: The species lives in water too shallow for Secchi depth readings, but it appears to be tolerant of enriched conditions because of its occurrence in highly enriched habitats.

3.4.1.8.3 Survival Potential of Red-Rimmed Melania Based on Temperature

Red-rimmed melania are tropical snails believed to be restricted in the United States to waters between 18°C and 25°C (Murray, 1971). Duggan et al. (2007) reported the species present in New Zealand sites with little or no geothermal influence typically had temperatures <15°C and four of the geothermally influenced streams had temperatures >25°C. Livshits and Fishelson (1983) reported specimens of *M. tuberculata* bury themselves in the mud and hibernate in the winter in Israel.

Mitchell and Brandt (2003) performed 96-h bioassays on 10–20 mm high snails and reported 100% mortality when exposed to 5°C for 24 h and 9°C for 96 h; and some mortality was observed when exposed to 15°C for more than 12 days. All snails died when exposed to 50°C for 1 min and 45°C for 10 min. In a later bioassay, Mitchell and Brandt (2005) reported that all red-rimmed melania would be expected to indefinitely survive temperatures ranging from about 18°C to 32°C. About 90% survived at 17°C and 32.5°C for 27 days and 15°C and 35°C for 14 days. The snails survived <24 h at 5°C and 40°C. At 50°C, all snails tested were dead in <3 min. Young snails 1–4 mm high survived <30 s. Mitchell and Brandt (2005) recommend a dip of fisheries equipment in hot water at 50°C for 5 min to kill any small red-rimmed melania trapped in nets or other fisheries equipment.

3.4.1.8.4 Survival Potential of Red-Rimmed Melania Based on Conductivity, TDS, and Salinity

Russo (1973) and Roessler et al. (1977) reported this species from both fresh and estuarine waters in south Florida in salinities ranging from 0 to 30 ppt. This is supported by the findings of Brown (1994), who reported the species in freshwater lakes with conductivities of 200 μS/cm and in saline lakes with conductivities up to 20,000 μS/cm. Yacoubi et al. (2007) reported finding the species associated with *Bulinus truncatus* in waters with conductivity levels ranging from 120 to 6020 μS/cm.

3.4.1.8.5 Survival Potential of Red-Rimmed Melania Based on Turbidity and TSS

The threshold levels of turbidity and TSS required for survival, growth, and reproduction of red-rimmed melania snails are unknown.

3.4.1.9 Faucet Snail, *Bithynia tentaculata* (Linnaeus, 1758)

The threshold limits for survival, growth, and reproduction for each of the physiological and ecological variables of *B. tentaculata* were determined from averaging

TABLE 3.29
Studies Using Various Parameters to Assess Risk Potential of Survival and Potential Density of *Bithynia tentaculata*

Variables	Reference
Alkalinity	Harman (1974), Jokinen (1992), Berezina (2001), Paavola et al. (2007)
Calcium	Harman (1974), Jokinen (1992), Berezina (2001), Paavola et al. (2007)
Chlorophyll *a*	No information
Conductivity, TDS	Russo (1973), Roessler et al. (1977); Yacoubi et al. (2007)
Dissolved oxygen	Harman (1974), Jokinen (1992), Hahn (2005)
Nitrogen, total	Jokinen (1992), Paavola et al. (2007)
pH	Harman (1974), Jokinen (1992), Berezina (2001), Paavola et al. (2007)
Phosphorus, total	Jokinen (1992), Paavola et al. (2007)
Salinity	Infer from Carlsson (2006), Berezina (2007), Ramakrishnan (2007) and Table 3.5.
Secchi depth, TSS	Jokinen (1992), Paavola et al. (2007)
Temperature (°C)	Jokinen (1992), Hahn (2005)
Total hardness	Infer from Harman (1974), Jokinen (1992), Berezina (2001), Paavola et al. (2007)
Turbidity (NTU), TSS	No information

The variables are listed alphabetically. NA = Not applicable, see the text.

the levels reported by the researchers listed in Table 3.29. The ranges in values for each of the risk levels are given in Table 3.30 when known.

3.4.1.9.1 Survival Potential of Faucet Snail Based on Chalk Variables

Boycott (1936) considers the faucet snail a calciphile, requiring >20 mg Ca/L. Briers (2003), however, found that calcium requirements alone did not explain the geographical range of the species in Britain and other niche variables were required.

Berezina (2001) states that the acidic pH tolerance of *B. tentaculata* is comparable to that of *P. insularum*, which under experimental conditions has been known to tolerate a pH as low as 4.0 for 96 h and can survive for more than 60 days in a pH range of 6.0–11.0. The ranges given by Harman (1974) for central New York populations for pH range from 7.2 to 8.2 and for alkalinity from 80 to 330 mg $CaCO_3$/L. Jokinen (1992) gives pH ranges of 6.6–8.4 and calcium levels of 5–89 mg Ca/L for Connecticut populations of *B. tentaculata*.

3.4.1.9.2 Survival Potential of Faucet Snail Based on Trophic Variables

Dissolved oxygen: The dissolved oxygen concentrations given by Harman (1974) for central New York populations range from 4.5 to 13 mg/L, which correspond approximately to 50–100% saturation. *B. tentaculata* is a strict oxygen conformer (Berg and Ockelmann, 1959).

Chlorophyll a*:* The threshold levels of chlorophyll *a* required for survival, growth, and reproduction of faucet snails are unknown.

TABLE 3.30
Criteria Used in Determining Levels of Infestation by *Bithynia tentaculata* in North America

1. Parameter	2. No Potential for Adult Survival	3. Little Potential for Egg Development	4. Moderate Potential for Nuisance Infestation	5. High Potential for Massive Infestations
Chalk Variables				
Calcium (mg Ca/L)	<5	5–15	15–25	25–90
pH	<6, >11	5.0–6.0, 9.5–11	6.0–7.0, 8.5–9.5	7.0–8.5
Alkalinity, total (mg $CaCO_3$/L)	<18	18–50	50–90	90–330
Hardness, total (mg $CaCO_3$/L)	<18	18–50	50–90	90–330
Nutrient Variables				
Dissolved oxygen (mg/L) (% saturation)	<3 (<25%)	3–5 (25–50%)	5–8 (50–75%)	≥8 (>75%)
Chlorophyll *a* (μg/L)	?	MNBL	MNBL	MNBL
Total phosphorus (μg/L)	?	MNBL	MNBL	MNBL
Total nitrogen (μg/L)	?	MNBL	MNBL	MNBL
Secchi depth (m)	<0.5	0.5–1	1–3	>3
Physical Variables				
Temperature (°C)	<0, >34	0–15, 30–34	15–22, 26–30	22–26
Conductivity (μS/cm)	<75, <13,800	75–100, 9000–13,800	4700–9000	100–4700
TDS (mg/L)	70, >9200	6,000–9200	3100–6000	800–9400
Salinity (mg/L) (ppt)	<0.03, >8	0.03–0.5, 5–8	2.5–5	0.05–2.5
Turbidity (NTU)	MNBL	MNBL	MNBL	MNBL
TSS (mg/L)	MNBL	MNBL	MNBL	MNBL

Use column 3 for assessing potential for survival (but little potential for nuisance infestation). Columns 4 and 5 indicate good potential for long-term survival and the criteria are used to assess temporal and seasonal variations in infestation potential. MNBL = May not be limiting. NA = Not applicable. ? = No data available.

Nutrients (total phosphorus, total nitrogen): Jokinen (1992) suggested that the ability of the faucet snail to filter feed may play a role in allowing populations of the faucet snail to survive at high densities in relatively eutrophic, anthropogenically influenced water bodies. However, the threshold levels of nutrients required for survival, growth, and reproduction of faucet snails are unknown.

Secchi depth, trophic status: Paavola et al. (2007) reported seasonal variations in Secchi depths of 4–6 m in one and 0.8–1.1 m in another brackish water harbour in Finland, suggesting the species can tolerate all levels of trophic status.

3.4.1.9.3 Survival Potential of Faucet Snail Mussel Based on Temperature

Hahn (2005) reported a lack of thermal acclimation of V_{O_2} in this temperate freshwater species. Jokinen (1992) states that egg-laying occurs from May to July when the water temperature is 20°C or higher. Jokinen (1992) also reported that oocytes develop poorly at temperatures between 30ºC and 34ºC.

3.4.1.9.4 Survival Potential of Faucet Snail Based on Conductivity, TDS, and Salinity

Carlsson (2006) found that the salinity limiting most freshwater species in southwestern Finland was 2–4 ppt, and for *B. tentaculata* it was found in conductivities ranging from 100 to 1050 μS/cm, or a maximum salinity of about 0.5 ppt. Berezina (2007) found *B. tentaculata* co-occurring with an amphipod, *Gammarus tigrinus*, in Neva estauary of the Baltic Sea where salinities range from 0.2 to 2.5 ppt. Klekowski (1961) reported a survivorship of 36% when exposed to a salinity of 8‰ for four days. Ramakrishnan (2007) considered *B. tentaculata* a second-degree euryhaline limnobiont, but a salinity tolerance somewhat lower than that of *P. insularum*, which had >65% survival at 10.2 ppt. Paavola et al. (2007) noted the occurrence of both *B. tentaculata* and *P. antipodarum* in brackish water harbours of Finland in which salinities range from 5.5 to 7 ppt. Jokinen gives 87–2320 μS/cm as its range for Connecticut waters.

3.4.1.9.5 Survival Potential of Faucet Snail Based on Turbidity, and TSS

The threshold limits of turbidity for survival, growth, and reproduction in *B. tentaculata* are unknown, but they generally occur in relatively clear, unpolluted, near-shore areas (Jokinen, 1992).

3.4.1.10 New Zealand Mudsnail, *Potamopyrgus antipodarum* (Gray, 1843)

The threshold limits for survival, growth, and reproduction for each of the physiological and ecological variables of *P. antipodarum* were determined from averaging the levels reported by the researchers listed in Table 3.31. The ranges in values for each of the risk levels are given in Table 3.32 when known.

3.4.1.10.1 Survival Potential of New Zealand Mudsnail Based on Chalk Variables

There is very little information on the levels of chalk variables to derive threshold levels of survival, growth, and reproduction of the New Zealand Mudsnail (NZMS). However, Kerans et al. (2005) found the NZMS in 23 of 32 locations in the Greater Yellowstone Ecosystem, and recorded levels of pH, calcium, and bicarbonates in

TABLE 3.31
Studies Using Various Parameters to Assess Risk Potential of Survival and Potential Density of *Potamopyrgus antipodarum*

Variables	Reference
Alkalinity	Kerans et al. (2005)
Calcium	Kerans et al. (2005)
Chlorophyll *a*	No information
Conductivity, TDS	Infer from Todd (1964), Duncan and Klekowski (1967), Jacobsen and Forbes (1997), Costil et al. (2001), Carlsson (2006), Herbst et al. (2008) and Table 3.5.
Dissolved oxygen	Paavola et al. (2007)
Nitrogen, total	No information
pH	Kerans et al. (2005)
Phosphorus, total	No information
Salinity	Infer from Todd (1964), Jacobsen and Forbes (1997), Costil et al. (2001), Carlsson (2006), Paavola et al. (2007), Herbst et al. (2008) and Table 3.5.
Secchi depth, TSS	Paavola et al. (2007)
Temperature (°C)	Winterbourn (1968, 1969, 1970), Hylleberg and Siegismund (1987), Quinn et al. (1994), Cox and Rutherford (2000), Richards et al. (2004), Paavola et al. (2007)
Total hardness	No information
Turbidity (NTU), TSS	No information

The variables are listed alphabetically. NA = Not applicable, see the text.

each location but reported levels for 12 locations. The pH ranged from means of 6.9 to 8.0; calcium from 3.3 to 6.9 mg Ca/L; bicarbonate from 24.5 to 126.0 mg HCO_3/L. The threshold levels in Table 3.32 are based on these values but substantially more data are required to verify their validity.

3.4.1.10.2 Survival Potential of NZMS Based on Trophic Variables

Paavola et al. (2007) reported seasonal variations in Secchi depths of 4–6 m in one harbor and 0.8–1.1 m in another brackish water harbor in Finland, suggesting the species can tolerate all levels of trophic status. Based on the Secchi depth ranges, the NZMS can probably tolerate low levels of dissolved oxygen, and high levels of phosphates and nitrates. Levri et al. (2007) found two small specimens in the hypoxic zone of Lake Erie, which suggests that the NZMS can tolerate hypoxia but probably cannot grow and reproduce under low oxygen conditions. Unfortunately, there is insufficient information on the levels of dissolved oxygen, chlorophyll *a*, and nutrients required to derive threshold levels of survival, growth, and reproduction of the NZMS.

3.4.1.10.3 Survival Potential of NZMS Based on Temperature

Winterbourn (1969) and Hylleberg and Siegismund (1987) state that the NZMS has a wide thermal tolerance of 0–30°C. Winterbourn (1970) reported that *P. antipodarum* from New Zealand was able to remain dormant and survive desiccation for up to

TABLE 3.32
Criteria Used in Determining Levels of Infestation by *Potamopyrgus antipodarum* in North America

1. Parameter	2. No Potential for Adult Survival	3. Little Potential for Egg Development	4. Moderate Potential for Nuisance Infestation	5. High Potential for Massive Infestations
Chalk Variables				
Calcium (mg Ca/L)	<2	2–4	4–7	7–?
pH	<6.0?	6.0–6.8	6.9–7.5	7.5–8.5?
Alkalinity, total (mg $CaCO_3$/L)	<10	10–50	50–100	100–350
Hardness, total (mg $CaCO_3$/L)	10	10–50	50–100	100–350
Nutrient Variables				
Dissolved oxygen (mg/L) (% saturation)	?	?	?	?
Chlorophyll *a* (µg/L)	?	MNBL	MNBL	MNBL
Total phosphorus (µg/L)	?	MNBL	MNBL	MNBL
Total nitrogen (µg/L)	?	MNBL	MNBL	MNBL
Secchi depth (m)	<1	1–2	2–4	>4
Physical Variables				
Temperature (°C)	<0, >35	0–15, 30–35	15–20, 26–30	20–26
Conductivity (µS/cm)	<25, >46,000	25–200, 25,000–46,000	200–1200, 9000–25000	1200–9000
TDS (mg/L)	<17, >30,800	17–130, 16800–30,800	130–800, 6000–16,800	800–6000
Salinity (mg/L) (ppt)	<0.01, >30	15.0–30.0	0.1–2.0, 5–15	2.0–5.0
Turbidity (NTU)	MNBL	MNBL	MNBL	MNBL
TSS (mg/L)	MNBL	MNBL	MNBL	MNBL

Use column 3 for assessing potential for survival (but little potential for nuisance infestation). Columns 4 and 5 indicate good potential for long-term survival and the criteria are used to assess temporal and seasonal variations in infestation potential. MNBL = May not be limiting. NA = Not applicable. ? = No data available.

50 days on a wet substratum at 20–25°C. Richards et al. (2004) performed desiccation bioassays and showed that *P. antipodarum* in the western USA can survive desiccation on damp substratum at 9°C for at least 48 h.

The NZMS is capable of tolerating a wide range of temperatures with upper thermal limits of 28°C and lower thermal limits near freezing (Winterbourn, 1968). Cox and Rutherford (2000) acclimated snails at 16°C and then exposed them to several constant temperatures and to diurnal temperature ranges of 10°C over 24 h periods for 96 h. At constant temperature they observed 50% mortality in 96 h at 31.0 ± 0.6°C. For diurnally varying temperatures 50% mortality occurred when the daily mean temperature was 28.6 ± 0.4°C (snails), which is about 10% (2.5 ± 1.3°C) lower than the constant LT_{50} value. Conversely, 50% mortality occurred when the daily maximum temperature was 33.6 ± 0.4°C, which is about 10% (2.5 ± 1.3°C) higher than the constant LT_{50} value.

Quinn et al. (1994) found the upper thermal limit of *P. antipodarum* to be 32.4 ± 2.5°C. They also found that LT_{50} values declined with exposure time, *P. antipodarum.* Winterbourn (1968) and James (1985) found that the upper limit of the natural distribution of *P. antipodarum* in the New Zealand was 28°C, which agrees with diurnal upper limit reported by Cox and Rutherford (2000) but approximately 4°C lower than the upper lethal limit observed by Winterbourn (1969) and Quninn et al. (1994) when temperatures were held constant. Winterbourn (1969) attributed the 4°C difference to the decline in the snail's activity above 28°C. Quinn et al. (1994) suggests that if this response is typical of other species, the 96 h LT_{50} temperatures may be a few degrees higher than the temperatures limiting distribution in the field.

3.4.1.10.4 Survival Potential of NZMS Based on Conductivity, TDS, and Salinity

The species has a wide range of salinity tolerance, from saline and brackish to freshwater, with populations in saline conditions producing fewer offspring. *Potamopyrgus antipodarum* sustains feeding, growth, and reproduction at salinities of 0–15‰ with 5‰ being optimal (Jacobsen and Forbes, 1997). Todd (1964) and Costil et al. (2001) consider this species to be a third-degree euryhaline limnobiont, being found in natural habitats at 0–34‰, with juveniles surviving up to 64‰ (Duncan and Klekowski, 1967). These data explain why this species is often found in brackish water habitats (Costil et al., 2001; Jacobsen and Forbes, 1997). Carlsson (2006) found that the salinity limiting most freshwater species in southwestern Finland was 2–4 ppt, and for *P. antipodarum* it was found in conductivities ranging from 200 to 1050 μS/cm, or a maximum salinity of about 0.5 ppt.

Herbst et al. (2008) performed laboratory bioassays that showed significant reductions in survival and growth among treatments diluting river water from 300 to 50 μS/cm. They found: no growth at or below 25 μS/cm; growth was inhibited in calcium-free artificial water compared to natural river water with the same conductivity, showing that lack of this mineral impedes development; many streams in the range of 25–200 μS/cm cannot support productive NZMS populations and that nuisance invasions may be most prevalent in waters above 200 μS/cm where sufficient dissolved mineral content is present for growth. Paavola et al. (2007) noted the occurrence of both *P. antipodarum* and *B. tentaculata* in brackish water harbors of Finland in which salinities range from 5.5 to 7 ppt.

3.4.1.10.5 Survival Potential of NZMS Based on Turbidity and TSS

The species is able to tolerate turbidity, clear water, and degraded conditions (including sewage) and may pass through the digestive tracts of many fish species.

3.5 ASSESSING POTENTIAL INFESTATION LEVELS

Modeling of historical data sets could be helpful in predicting spatial and temporal impacts on environmental variables in freshwater ecosystems. For example, Reed-Anderson et al. (2000) modeled chlorophyll and temperature data collected biweekly from Lake Mendota during 1977–1993 to predict the potential impact of zebra mussels on water clarity. Their model suggests that zebra mussel invasion would lead to increased water clarity in Lake Mendota and cyanobacterial blooms would be reduced but not eliminated. Ricciardi (2003) suggested that the impact history of a species, particularly a well-studied invader like the zebra mussel, is a valuable guide for predicting the consequences of its introduction into a new environment. Ricciardi (2003) recommends a synthesis of data from multiple invaded sites to reveal predictable patterns at various spatial scales. Although precise predictions are often elusive, lack of precision should not be viewed as a deterrent to developing predictive models where none exist. Even crude models (constructed from reliable data) have potential value and can be refined as additional data become available (Ricciardi, 2003). Where possible, models that relate an invader's abundance to physical environmental variables should be linked with models that relate the invader's impact to its abundance. The resulting composite models would allow us to predict, prior to invasion, which habitats would be vulnerable to high impacts and thus should be a priority for management (Ricciardi, 2003).

Lee and Bell (1999) believe that three important traits needed to facilitate transition from salt to freshwater include phylogenetic bias of recent invaders, physiological traits and habitat of origin, and life history traits. The phylogenetic bias of recent invaders implies that not all invaders succeed, but those that do, invade and spread quickly. A species' physiological traits and habitat of origin implies that populations within species often vary in their propensity to invade freshwater because of differences in physiological traits among populations. This variation might result from variation in acclimation or selection regime among habitats (Lee and Bell, 1999). For example, freshwater invaders might originate more readily from habitats that have wide temporal or spatial ranges in salinity, where broad salinity tolerances can form. Temperature might pose a barrier to freshwater invasion because freshwater habitats tend to fluctuate more widely in temperature than do brackish or marine habitats. Given that temperature fluctuates more at higher latitudes, it has been proposed that the propensity to invade should increase at lower latitudes (Hutchinson, 1957). However, salinity and temperature interact in their effects on tolerance, with temperature affecting metabolic rate, ion uptake rate, and membrane permeability (Lee and Bell, 1999). Ion proportions in freshwaters vary more than in seawater, necessitating freshwater invaders to either tolerate the mix of ions or find habitats with an appropriate mix of ions. Life history traits may predispose certain invading species to persist in lakes, by countering problems such as drought and fluctuations in temperature and food availability (Lee and Bell, 1999). Zebra mussels are free spawners with earlier maturity, more rapid growth, and higher fecundity and possess r-selected traits that will drive those, like unionids, with K-selection traits to extinction (Mackie, 2004).

Koutnik and Padilla (1994) used a geographical information system to test for associations between predicted lake population density classes and three landscape-scale characteristics (surficial deposits, bedrock type, ecoregions) to predict (i) absence or presence, (ii) categorical population density, and (iii) numerical abundance of zebra mussels for inland Wisconsin lakes. Although the models used differed in their predictions of specific lakes that would support *Dreissena*, they found a significant association between each landscape-scale characteristic and *Dreissena* density classes. Numerous researchers (e.g., see Table 3.6) have used available lake monitoring data to predict *Dreissena* density for inland lakes. It is clear that the more the information available, the less the uncertainty in the predictions. However, there are some parameters that better correlate with mussel survival and density than others. The most common parameters used (and listed in order of their predictive value from most predictive to less predictive) are: (1) calcium content, (2) alkalinity, (3) pH, (4) total hardness, (5) nutrient (total phosphorus, total nitrogen) levels, (6) chlorophyll *a* levels, (7) Secchi depth, (8) mean annual temperature, (9) dissolved oxygen content, and (10) conductivity (and/or salinity, TDS). Although mean annual values of each of the parameters can be used, as above, temporal (e.g., seasonal) and spatial (e.g., depth, horizontal) variations lend more certainty to the predictions of mussel survival and potential densities and are dealt with here.

If the species has a good potential for surviving in a body of water, it is possible to determine the potential level of infestation that it could develop because every species has a suite of optimal physiological requirements. However, the level of a required element can be extremely variable spatially (depth and across a lake) and temporally (hourly to seasonally), so the density predictions must account for any variations that exceed the upper and lower physiological thresholds. The potential infestation levels used here are: little to no potential for infestation, a moderate level of infestation, a high level of infestation, or an intense level of infestation. The definition of infestation varies from species to species (e.g., zebra mussel vs. Asian clam) and from one structure's function (e.g., mechanical filter screen would have different criteria than a navigation buoy) to the next. For the purposes of this book, for bivalves (dreissenids, mytilids, corbiculids) a high level of infestation is one where one square meter of surface is completely occupied by one layer of adult mussels, or 60–100% coverage and an intense level of infestation is more than one layer of mussels (>100% coverage). No infestation is <5% coverage, little infestation is 5–30% coverage, and moderate infestation is 30–60% coverage. This corresponds to coverage values given for zebra mussels in Chapter 1 (Section 4.2.5), except here even <5% coverage is considered a "no infestation level."

To demonstrate the use of the data in the Tables that provide criteria for delimiting thresholds for the different infestation levels (e.g., Table 3.7), only the zebra mussels will be used here as an example. The procedure for assessing infestation potential is the same for each species. When assessing risk potential it is assumed that risk potential varies with depth and with season because the levels of the different chalk, trophic, and physical variables vary spatially and temporally. For example, pH levels are lower in the bottom of a eutrophic lake where organic decomposition occurs, releasing carbon dioxide that promotes formation of carbonic acid and a lowering of the pH; in the epilimnion, photosynthesis by algae uses carbon dioxide, which raises the pH. The pH will vary seasonally as well, being lowest in spring after snowmelt, which releases hydrogen ions that had migrated down the snow pack and are released into the watershed and the lake during runoff.

Predictions on the long-term infestation levels can be made by analyses of depth and time series graphs for four chalk variables (calcium, alkalinity, pH, hardness), one nutrient variable (orthophosphate), and two physical variables (temperature and conductivity). Each graph shows the potential for little to no infestations up to massive infestations. Values that are above the high infestation curves indicate potential for massive infestations for the watershed/station; values that fall between the moderate and high infestation levels indicate potential for high infestation levels for the watershed/station; values that fall between the moderate and little to no infestation indicate potential for moderate infestations for the watershed/station; values that fall below the little to no infestation levels indicate no potential for infestation for the watershed/station being analyzed. Once again, it is important to stress that if calcium levels are not adequate, the other variables are of no consequence.

For demonstration purposes, pH, temperature, and total phosphorus will be used to assess infestation potential in the lake of Table 3.8. The variations in depth at three stations on the hypothetical lake will be examined first and then the temporal (seasonal) variations will be examined. One station is in the main inflow with high calcium and pH, one is at a smaller inflow with low calcium and pH, and one is at the outflow with moderate calcium and pH levels.

3.5.1 Assessing Depth Variations in pH, Total Phosphorus, and Temperature for Risk Potential at Three Locations in a Hypothetical Lake

3.5.1.1 pH

Figure 3.6 shows the variations in pH with depth in relation to infestation and threshold levels for growth and reproduction of zebra mussels at three locations in the hypothetical lake. Note that the small inflow must be carrying high levels of hydroxides or other bases because the pH is elevated above the upper lethal levels but drops quickly due to buffering action of the lake water. The epilimnial waters would support moderate levels of infestation but the hypolimnial waters would not support growth or reproduction of zebra mussels at any of the three locations.

3.5.1.2 Total Phosphorus

Figure 3.7 shows the variations in total phosphorus with depth in relation to infestation and threshold levels for growth and reproduction of zebra mussels at three locations in the hypothetical lake. The lake is highly enriched with phosphorus levels exceeding the upper lethal threshold of the zebra mussel. While phosphorus itself is not lethal, the amount of algal growth would probably be high and clog the gills of the mussels.

3.5.1.3 Temperature

Figure 3.8 shows the variations in water temperature with depth in relation to infestation and threshold levels for growth and reproduction of zebra mussels at three locations in the hypothetical lake. The epilimnial temperatures are sufficient to support massive infestations of mussels but the hypolimnion is too cold to support growth and reproduction of the mussels at all three locations. While temperature is appropriate to support massive mussel infestations, the pH and total phosphorus levels would mitigate accumulations during the summer months.

3.5.2 Assessing Seasonal Variations in pH, Total Phosphorus, and Temperature for Risk Potential in a Hypothetical Lake

3.5.2.1 pH

Figure 3.9 shows monthly variations in pH at the surface and bottom of a hypothetical lake in relation to infestation and threshold levels for growth and reproduction of zebra mussels. The surface pH is very high and approaches the upper lethal threshold during the winter and early spring; during May–July the surface pH fluctuates at or above 9.0, then drops to support heavy infestations in August and September; the surface pH exceeds the upper lethal threshold in late December. The bottom pH is below growth and reproductive thresholds from mid-spring through fall and rises above the growth and reproductive thresholds in early spring and late fall would support moderate to intense infestations during those periods.

3.5.2.2 Total Phosphorus

Figure 3.10 shows monthly variations in total phosphorus at the surface of a hypothetical lake in relation to infestation and threshold levels for growth and reproduction of zebra mussels. The total phosphorus levels are sufficient to support moderate infestation levels of zebra mussels throughout early summer and fall but greatly exceed upper threshold levels from late fall through early spring and will mitigate accumulations of mussels (Figure 3.13).

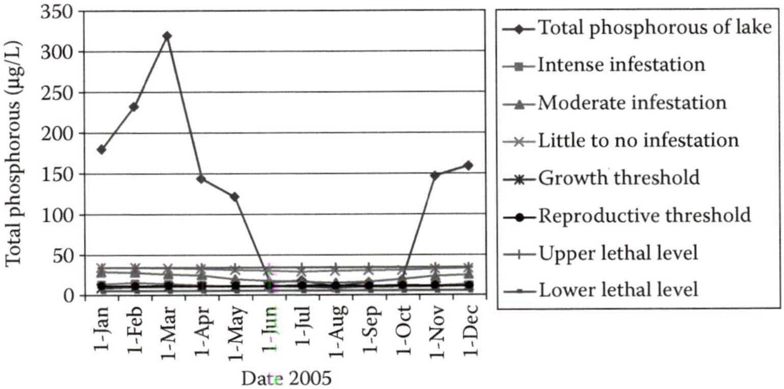

FIGURE 3.13 Variation in monthly mean surface (1 m) total phosphorus of a hypothetical lake in relation to zebra mussel infestation and threshold levels.

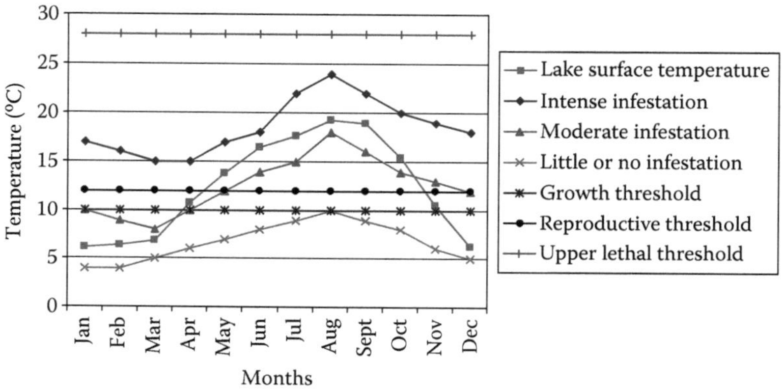

FIGURE 3.14 Variation in monthly mean surface (1 m) temperature of a hypothetical lake in relation to zebra mussel infestation and threshold levels.

3.5.2.3 Temperature

Figure 3.11 shows monthly variations in mean surface temperature of a hypothetical lake in relation to infestation and threshold levels for growth and reproduction of zebra mussels. The temperature is sufficient to support moderate levels of zebra mussel infestations from May to October. The surface water temperature is below the reproductive threshold from November to April and will not contribute to accumulations during this period (Figure 3.14).

3.6 RECOMMENDATIONS FOR FURTHER RESEARCH

The amount of information available for assessing risk potential of invasive bivalves is considerably greater than that for invasive gastropods. Comparisons of the numbers of researchers and amounts of data in Tables 3.6 through 3.16 for bivalves with those in Tables 3.17 through 3.32 for aquatic invasive gastropods is a prerequisite for more information on levels of calcium, pH, alkalinity, total hardness, dissolved oxygen, nutrients, air and water temperatures, conductivity and/or salinity, and turbidity for gastropods, particularly the amphibious ampullariids. There is a wealth of information on distribution, impact, and control of ampullariids and thiarids but very little on the quality of water they live in. Perhaps calcium, pH, and alkalinity are not limiting factors but there is no literature stating this fact. There are very few studies that examine the thermal and salinity tolerances for survival, growth, and reproduction in aquatic invasive gastropods, at least compared to those of bivalves. Indeed, even for bivalves more often than not, when a range extension is reported, no water chemistry is provided. It would be desirable to at least report the levels of calcium, pH, alkalinity, dissolved oxygen, temperature, and conductivity or salinity. In many cases there are data available on the seasonal variations in these variables from departments of natural resources or environment that could be included with any report on the occurrence of an aquatic invasive species. After about 50 or so reports of occurrences, the upper and lower limits of the variables can be derived and would help in assessing risk potential to yet-to-be invaded water bodies.

4 Monitoring: Searching, Finding, and Interpreting

4.1 INTRODUCTION

There are three levels of inspection for finding invasive mollusks, and each has its own purpose. (1) One level is for detecting the species in waters that do not yet have the species but the water quality and dispersal attributes are sufficient to warrant a monitoring program for early detection. (2) Once the species has invaded a system, then a postinvasion monitoring program is needed. The postinvasion monitoring program is largely a temporal one and is used to determine the level of infestations at different times of the year, the periods of reproduction, of settlement, and periods of most and least intense infestation levels. Growth rates and longevities are also part of the analyses at this level. Postinvasion monitoring is especially important to industries and utilities in order to determine when larvae are arriving in the cooling systems, follow the changes in numbers of veligers and attached adults within different parts of a facility; determine the effectiveness of a control program; and minimize the costs of a control program. (3) The third level is for long-term monitoring in large bodies of water, and involves spatial studies to determine variations in infestation levels in different parts (e.g., bays and depths) of the lake/reservoir. Due to horizontal and vertical currents, and differences in water quality among inlets, not all parts of a lake will be affected at the same level. Hence, some industries and utilities may experience low infestation levels while others will suffer from high infestation levels.

Of the 15 species of invasive mollusks, five have a planktonic stage followed by a benthic stage and 10 lay eggs on substrates either above or below the water and then develop into juveniles and adults, which are either amphibious or totally aquatic. Because of these differences, monitoring programs for the two groups are very different and are discussed separately. The five species with planktonic stages happen to be bivalves and are discussed first. Information about the spread of larvae and adults into new areas is critical for industries and public utilities concerned about water intake systems, and for fisheries agencies concerned about managing the ecosystem. Water users need to have an early warning of the arrival of invasive bivalves to their area, and especially into their facility, in order to properly prepare for biofouling problems. Once the mussels arrive, water users can benefit from a monitoring program that will establish the densities of mussels that have settled in their immediate area, when veligers are present in their water supply and how well their mitigation programs are working.

There are two basic types of monitoring, mainstream monitoring and sidestream monitoring. The sidestream monitor can be considered as a parallel stream of the system, mimicking conditions of the system, while being easily examined. Several different designs have been implemented both in Europe and in North America, and two of the most common are discussed in Section 4.4. In mainstream monitoring, devices are placed directly into the raw water supply or source. It usually entails taking samples directly from the lake or river or from a wet well or pump well that is supplied with water from the lake or river. In industries, remote areas, or areas too dangerous for divers to access, can be inspected with video cameras attached to remotely operated vehicles. All these alternatives are discussed below. Mainstream monitoring in the source water (i.e., lake or river supply) is recommended if mussels are not yet in the area and includes polymerise chain reaction (PCR) techniques described below.

4.2 SAMPLING FOR LARVAL AND ADULT BIVALVES

4.2.1 Measuring Populations of Planktonic Larvae of Dreissenids, Mytilids, and Corbiculids

This section provides enough information to enable one to interpret population dynamics of the planktonic veliger stages by explaining the units of density and how samples are taken at discrete depths.

Typically, the sampling devices are "grab samplers" [samplers that take a known quantity, or grab, of water for analysis, e.g., Schindler–Patalas trap (Figure 4.1c)] or composite samplers (e.g., plankton tow nets). Composite samplers are used when veliger abundance is low and a large volume of water is needed to detect the few larvae that are present. Grab samplers are used when veliger abundance is high and a small volume, usually less than 30 L, is sufficient to detect them. Plankton nets (Figure 4.1a and b) are most commonly used to sample a column of water for veligers. If the column sampled is vertical, it is called a vertical tow; if the column of water is horizontal, it is called a horizontal tow. The volume of water sampled is determined by the diameter of the opening and the distance through which the net was pulled, the volume sampled being calculated from the formula:

$$\text{Volume (m}^3) = \pi r^2 \times h,$$

where $\pi = 3.1416$, r = radius of plankton net opening in meters, and h = distance through which the net was pulled through the water in meters. Most nets have a funnel with a hose and clamp to empty the filtered contents of the net into a vial. Other nets (e.g., Wisconsin nets, Figure 4.1a) have a removable bucket with a plunger that has to be removed in order to pour the contents of the bucket into a vial. In either case, a relatively large volume of water (usually greater than 500 L) is reduced to about 20–50 mL. Usually, at least three samples are taken each time, each sample being kept separate from the other.

The recommended diameter for the plankton net is 50 cm. Most veligers have shell lengths greater than 90 μm but their shell height is only about 75 μm. Therefore, the maximum mesh opening size should be 75 μm. Mesh with 63–75 μm openings is recommended for routine sampling. If one wishes to retain embryos, a net with mesh openings smaller than 50 μm is required.

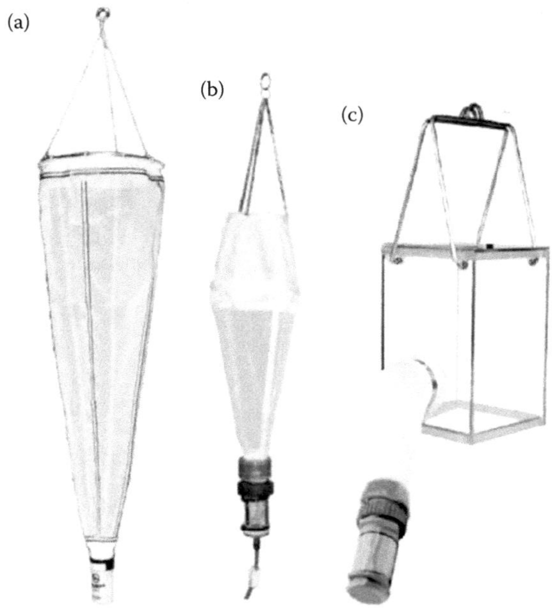

FIGURE 4.1 Plankton samplers commonly used for collecting dreissenid larvae. (a) A 50-cm diameter conical plankton net; (b) Wisconsin net; (c) 13-L Schindler–Patalas trap.

In areas where larvae are appearing for the first time, large volumes of water must be sampled in order to detect them. Sampling for veliger larvae with a plankton net can be done by towing the net through the water, either vertically or horizontally. It is easier to calculate volumes for a vertical tow than for a horizontal or oblique tow because the vertical distance is easier to measure than a horizontal or oblique distance. However, for detecting the first appearance of veligers, accurate estimates of volume of water sampled is not critical. For vertical hauls, start about 0.5 m above the bottom and bring it to the surface. If the net rests on the bottom, or is too close to the bottom, debris from the sediments will clog the net and also make sample analyses difficult. For horizontal or oblique tows, again keep the net at least 0.5 m off the bottom. Slowly drag the net through the different water depths while being towed forward. At least 500 L should be sampled with the plankton net. Therefore, If a 0.5-m diameter net is used, the net must be pulled through 2.55 m of water, either vertically or horizontally [$h = 500{,}000\ \text{cm}^3/(3.14)(25\ \text{cm})^2$, where $1\ \text{L} = 1000\ \text{cm}^3$ and radius = 25 cm]. The collected samples should be preserved in either 5% formalin or 70% ethanol. The 5% formalin solution is made by diluting concentrated (40%) formaldehyde. That is, add 80 mL of 40% formaldehyde to 920 mL of water.

A convenient way to take a plankton sample of known volume is to use a small 1/6th horsepower submersible sump pump. Connect the pump to a portable battery power pack of the type that is used for your car. Submerge the sump pump about 6 ft into the water and start the pump. The discharge of the pump is connected to a garden hose. Use a graduated pail and stop watch to determine the flow and then pass the flow through a plankton net for the required amount of time to collect your sample. A plankton pump work particularly well in shallow canals and fast flowing water.

It is recommended that samples are taken from at least three sites in open water, near shore, about 1 km apart, if possible. A minimum of two replicate samples should be taken at each location once per week while water temperatures are above 12°C.

At the industrial facilities plankton sampling can be done in the incoming water using either a plankton net or a plankton pump, provided there is access to the raw water source, for example, a wet well or a raw water supply line. Once again, replicate samples should be taken during each sampling event and the frequency of sampling if not daily should be at least twice a week.

4.2.1.1 Microscopic Analyses of Veliger Samples

Once plankton samples have been collected, there are two basic techniques for analysis. The first method is a fast way to establish the presence or absence of veligers in a plankton sample. Schaner (1991) recommends using a sugar solution to increase the probability of finding veligers. The sugar solution is made by dissolving 130 g of table sugar in 400 mL of distilled water. Either a 25-mL pipette or 25-mL burette can be used. If a pipette is used, first withdraw 10 mL of plankton sample into the pipette, and then 15 mL of sugar solution. If a burette is used, pour the 15 mL of sugar solution into the burette first, and then add 10 mL of plankton sample. The sugar increases the density of the solution so that lighter organisms, like filamentous algae, sink at a slower rate than heavier organisms, like veligers. Therefore, veligers and sand will arrive to the bottom of the column first. After 20 min of settling time several drops are withdrawn from the bottom of the burette or pipette. Examination of the sample is best done at 60 × to 80 × magnification under a dissecting microscope equipped with polarized light. If veligers are present in the sample used on top of the sugar column, they will be present in these first few drops and will show up as glowing shapes bearing a black cross *when viewed under polarized light.* The method works well in water with little algae (e.g., oligotrophic and mesotrophic lakes). If large amounts of algae are present, as in eutrophic lakes, they may form a blanket that interferes with the settlement of the veligers.

Cross-polarization is accomplished by using a microscope equipped with a polarizing filter, while a second polarizing filter is placed at the bottom of the microscope stage underneath the plankton sample (illumination is from below). By rotating the bottom filter, the two can be lined up to provide dark background in which the veligers glow brightly. The only other organisms that show similar characteristics as zebra mussel veligers under cross-polarized light are small ostracods. Details of the method are described by Johnson (1995). Figure 4.2 shows size comparison of zebra mussel veligers with other common plankton species.

The use of cross-polarized light is also useful for quantitative estimates. Once the presence or absence of zebra mussel veligers has been established, it may be desirable to establish their absolute densities in the water column. As stated above, it is easier to calculate volume of water sampled from a vertical tow than from a horizontal or oblique tow because the vertical distance is easier to measure than a horizontal or oblique distance. Because it is difficult to measure the distance of horizontal and oblique tows, you can use either time to pull the net over a given distance

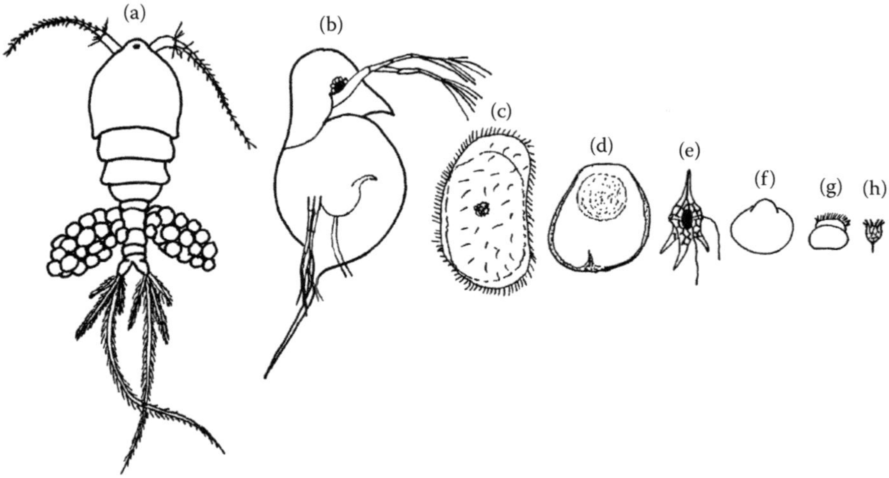

FIGURE 4.2 Relative sizes of veligers in relation to some common plankton in fresh water. (a) *Cyclops* copepod; (b) Daphnia cladoceran; (c) *Candona* ostracod; (d) *Anodonta* glochidium; *(e) Ceratium dinoflagellate; (f) Dreissena polymorpha pediveliger; (g) Dreissena polymorpha D-form; and (h) Keratella rotifer.*

or a flow meter to measure the actual volume sampled. If time is used and you are in boat, you must know how far the boat travels in a given time. If you are on shore, you can simply measure the average distance towed, the tow being a combination of horizontal and oblique hauls. Obviously, there is bound to be considerable error in either estimate, so it is better to use a metering device. Flow meters can be purchased and attached to the mouth of the net to measure the volume of water actually passing through the mouth opening.

If veliger abundance is high (e.g., >50,000 m^{-3}), the sample size will have to be reduced. Ideally, you should end up with between 100 and 200 veligers in a sample. Therefore, if the raw water supply has about 100,000 m^{-3}, all you need is about 10 L (i.e., 100 L^{-1}). Under these circumstances, a 12-L Schindler–Patalas trap (Figure 4.1c) is ideal for taking samples. They are available in usually two sizes, 12 and 30 L. The 12-L trap is usually preferred because it weighs much less than the 30-L trap when it is filled with water. If 20 L of water needs to be sampled (see rationale below), it will be necessary to take 3, 12-L samples and combine them into one sample. This should be repeated two more times so that you end up with three replicate samples for analyses. As with the plankton nets, relatively large volumes of water are filtered through a fine mesh (usually with mesh openings less than 75 μm) and reduced to about 20–50 mL.

Alternatively, if available, a plankton pump can be used to pump water through a net slowly. Again a metering device should be attached to the pump to measure the volume of water pumped and filtered or an estimate can be made using a pail of known volume and a stop watch to estimate the rate of flow.

For quantitative estimates, the volume of water to be filtered should take into account the abundance of veligers and other plankton present in the water. If veligers and plankton are abundant, a large sample will contain too many organisms to count and may have to be diluted. This is likely to introduce a statistical bias. If veliger densities are expected to be high, a relatively small sample of water will be sufficient. A good rule of thumb to follow is to try to have about 100 veligers/L of sample. Hence, if your initial estimates show about 5000 veligers/m^3 (=10 L^{-1}), a 20-L sample is sufficient; or if there are 100,000 m^{-3} (=100 L^{-1}), a 1-L sample will suffice.

Once the samples are taken, they can be examined fresh for living veligers, or preserved with 5% formalin or 70% ethanol for analyses later. Samples should be kept fresh only if you wish to determine the mortality of veligers. It is best to preserve samples whenever possible because it is difficult to count and measure specimens that are continually moving. Counting is particularly difficult because you can never be certain whether you are counting a specimen more than once as it appears in out of your field of view several times.

If the plankton sample is of ideal size, the entire sample can be examined (either fresh or preserved) and the total number of veligers can be counted. The density (abundance) will then be a simple matter of equating the numbers found to volume filtered and then converting this to either number of veligers/m^3 or number of veligers/L.

More often than not, however, the sample is like "pea soup" and either has to be diluted or subsampled. The volume of the preserved sample is taken and is then thoroughly mixed by gentle shaking. Immediately (i.e., before any plankton settle), take 1 mL of sample from the middle of the jar. Count all veligers in the 1 mL sample under a stereomicroscope at 80 × magnification. Repeat for at least five subsamples for each jar and then determine the average number in the five subsamples. In a well mixed sample, the numbers counted in each subsample should be within 5% of each other. Then convert the number/mL to number/jar by multiplying the number/mL by the total volume (in mL) of the sample in the jar. The result will be abundance per volume of water sampled. Now convert this to number/L by dividing the total number/jar by volume of water sampled. Multiply by 1000 to convert number/L to number/m^3. Densities as high as 1,000,000 m^{-3} (1000 L^{-1}) have been reported in Lake Erie. Table 4.1 is an example of how to record the numbers of veligers in a water sample. If an excel file is used, formulae can be inserted to calculate the number per cubic meter (last column).

4.2.1.2 Settling Plates and Calculating the Number of Settled Juveniles per Square Meter

Settlement substrates are basically plates of solid (natural or artificial) substrate, attached to a rope and submerged in the water body that is being monitored. Periodically, the plates can be removed and examined visually or the surface can be lightly scraped into a dish and the collected material is then examined under a binocular stereomicroscope (binocular) with an ocular lens to provide up to 80 × magnification. Settling plates can be made out of any number of nontoxic materials such as carbon steel, stainless steel, polyvinyl chloride (PVC), bricks, clay, or even cement tile.

Settlement plates or coupons can be of any size but the larger the plate, the more area for juveniles to settle on. The advantage of using smaller plates in a rack is that one can obtain several replicate samples over the summer; for example, if removing two each week for 16 weeks, 32 coupons would be needed.

TABLE 4.1
An Example of a Table That Can Be Used to Record Numbers of Larvae in Different Size Classes

Location		Stage, with Size Class in μm, Estimated Time to Settlement in Approximate Days					
Sample #, M/d/08	Volume Filtered (L)	Pre-D form <80 μm	D-form 80–150 μm	Veliconcha 150–250 μm	Pediveliger 250–300 μm	Plantigrade >300 μm	Total Number (m^{-3})
1. May 27	1000	0	0	0	0	0	0
2. May 31	850	0	0	0	0	0	0
3. June 8	900	120	244	0	0	0	404
4. June 14	1000	100	300	0	0	0	400
5. June 23	1100	88	440	180	0	0	644
6. June 28	1000	244	1152	216	0	0	1612
7. July 5	850	244	1260	360	180	90	2511
8. July 13	850	122	1512	1060	360	180	3805
9. July 22	1000	72	2080	1600	360	180	4292
10. July 27	1000	102	4480	2200	124	360	7266
11. August 3	1000	36	5520	2580	244	440	8820
12. August 10	1000	0	7560	4840	440	440	13,280
13. August 15	600	0	6760	6010	360	360	22,483
14. August 25	600	0	5140	3536	476	720	16,453
15. August 26	800	0	4047	3180	952	576	10,944
16. September 2	850	0	3600	3240	1800	1440	11,859
17. September 8	900	0	6540	4480	3180	1800	17,778
18. September 16	1000	0	6480	3240	3240	1008	13,968
19. September 22	1000	0	4480	3672	1800	576	10,528
20. October 1	900	0	3240	2580	2088	360	9187
21. October 8	800	144	448	1550	1224	180	4433
22. October 16	750	288	224	840	750	144	2995
23. October 25	1000	72	122	640	440	120	1394
24. November 4	1000	0	36	36	0	0	72
25. November 11	1000	0	0	0	0	0	0

Note: Numbers of zebra mussel larvae in different size classes in samples. Last column converts #/sample into #m^{-3}. Estimated time to settlement is given as a range because of the effect of temperature, for example, more time is needed for development in cooler water than in warmer water, inherent differences in rates of development among individuals.

The area of the settlement plates (or coupons) needs to be calculated. If two sides were scraped, double the calculated area. For example, suppose the plates are 10 cm by 10 cm and two sides were scraped, the total area settled upon is $2(10 \times 10) = 200\ cm^2$. Hence, if you counted 300 juveniles, the number per square meter is $300 \times 10{,}000\ cm^2/m^2/200\ cm^2 = 15{,}000$.

Care should be taken not to let the sampling assembly touch any other submerged surfaces that are likely to be colonized. For example, a sampling plate that touches the face of a concrete dam already colonized will show large number of adults. This phenomenon is called translocation and is regularly observed in dreissenids. Adult mussels crowded onto a colonized surface will move to a new surface where the competition for space is low (Figure 4.3).

Periodic visual observation of the sampling substrates will determine if macrofouling organisms are settling at this location. Once initial settlement has been detected, the settling plates are used to determine the density of settlement.

Determination of settlement density is best done by having at least three identical replicates of settlement plates deployed at the same time and depth. The sampling frequency of the settling plates is flexible. The sampling can be weekly, biweekly, or monthly. After the settlement plates have been exposed for a predetermined time period, each string is lifted from the water. Both sides of each plate are scraped off and preserved in collection jars. The material collected can be either placed into prelabeled sampling jars and preserved in 70% ethyl alcohol or 3–5% formalin for later examination or examined immediately. When diluting the alcohol or formalin, consider the water retained within the mantle cavities of the mussels.

Depending on the macrofouling organism involved and the size of the population on the sampling plate, actual number of individuals can be recorded and reported on a square meter basis or square foot basis if the plate size is known. Alternatively, the material collected can be recorded as wet weight or dry weight. Data from plates occupying the same depth on each of the three strings should be added together and averaged for a more statistically robust result. Once the population data are collected, as suggested above, the collected material can be examined for the presence of discrete size classes. Usually maximum shell length (tip of umbone to posterior edge of the mussel) is taken using either an ocular micrometer on a microscope for small shells, or digital calipers for large shells.

As the plates are scraped clean after each sampling period, the largest shells found on the plates during the next sampling period will have been settled for the

FIGURE 4.3 **(See color insert following page 286.)** Sampling substrates, the one on the far right was touching a heavily colonized surface.

duration of the interval between sampling. For example, if maximum growth is occurring, 1-mm long mussels will be present after one week, 2-mm long after two weeks, and up to 4-mm long mussel shells after one month. An occasional large mussel outside the expected size range could be found on the sampling plates. These mussels are usually called translocators and may have moved onto the plate from another location. They should not be included in the weekly growth calculations. The presence of more than one size class of shells in the material collected from the plates suggests multiple settling events have taken place while the settling plates have been in the water. This is also valuable information for the industrial facility as is the relative growth rate of the mussels in their area. It will give the facility the information the maximum shell size that will be attained by the population in between periodic or end of season treatments. After all the plates have been scraped, the sampling string is replaced in the water column for the next sampling period.

It is strongly recommended to measure and record ambient water temperature at each plate depth when collecting samples from the settlement plates. Having a record of the temperature at various depths could greatly help interpret settlement and growth data from the settling plates.

It may also be useful to expose additional strings of sampling plates at the same location and leave these additional strings undisturbed for a 6–12 month period. At the end of this time follow the same procedure as above. These additional plates will then give you the cumulative densities of the macrofouler at each depth for the time period selected.

Table 4.2 is an excel sheet that shows settlement of zebra mussels in five bioboxes of a steel factory drawing water from Hamilton Harbour, Ontario, Canada. The plates used in the example here were 15.24 cm × 22.85 cm = 384 cm^2 or 696.6 cm^2, including both sides. Hence, the numbers of juveniles counted are multiplied by 10,000 $cm^2/m^2/696.6\ cm^2 = 14.4$. Note that there are two bioboxes in Pumphouse #2, one is chlorinated and the other is not (i.e., a control). If the chlorine is effective, there should be no juveniles settling in the bioboxes receiving chlorinated water, which is the case in Table 4.2, except on September 22, 2005. The operators took immediate action and remedied the problem.

4.2.2 Interpreting Variations in Population Sizes of Planktonic Dreissenids, Mytilids, and Corbiculids

Plotting the length frequency distribution of larvae every two weeks allows one to determine when reproduction (spawning of eggs and sperm) and settlement occurs, the frequency of reproduction and settlement events, and when spawning and settlement ceases. For example, Figure 4.4 shows the variations in numbers of different size classes of zebra mussel larvae in Hamilton Harbour in 2008. Samples were taken about every two weeks using plankton net (with 75 μm mesh openings) hauls through the top 5 m of water beginning May 1 (no larvae were found). Larvae first appeared between May 27 and June 10. Spawning occurred throughout the summer and early fall as indicated by the presence of D-forms until November 4.

TABLE 4.2
Settlement Intensity on Different Materials After 4 Months in Lake St. Clair

Material	Mean Number m^{-2}
Copper	0[a]
Galvanized iron	548[b]
Aluminum	2324[b,c]
Acrylic	6896[c,d]
PVC	7471[c,d]
Teflon	8593[c,d]
Vinyl	12,068[d]
Pressure-treated wood	15,255[d]
Black steel	15,420[d]
Pine	16,117[d]
Polypropylene	17,554[d]
Asbestos	21,333[d]
Stainless steel	21,812[d]

Source: Modified from Kilgour, B.W. and Mackie, G.L., *Zebra Mussels: Biology, Impact and Control*, eds. T. Nalepa and D. W. Schloesser, 167–174, CRC Publications, Boca Raton, 1993.

[a–d] Abundances with the same letters are not significantly different ($P > 0.05$).

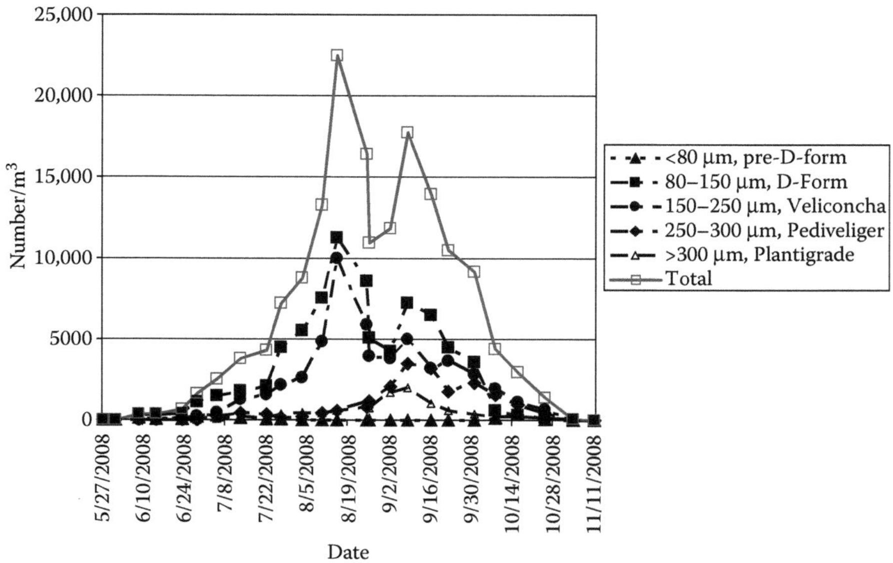

FIGURE 4.4 Seasonal variations in different length classes of zebra mussel veligers in Hamilton Harbour from May 27 to November 11, 2008.

It is wise to include a plot of the densities of larvae and the water temperature with the bar graphs in the length frequency distributions because they aid greatly in interpreting the dynamics of the larval population. For example, on May 15 there was only one length class present (Figure 4.4). A total of 100% of the larvae at 90 μm would mean little if only 20 larvae/m^3 were found, but if 200,000 larvae/m^3 were found, one would be concerned and probably prepare for a major infestation. Determining the number of peaks in densities of larvae helps to predict numbers of cohorts (generations) that could potentially settle. Generally the onset and cessation of spawning and the development rate of larvae are related to water temperature, so monitoring water temperature helps one to predict the period when infestations are likely to occur.

The growth rate of adults can be exceedingly fast, as much as 0.5 mm/day, as measurements of adults on artificial substrates (cement blocks) in Lake St. Clair have shown (Mackie, 1991). Typically, however, adults grow 1.5–2.0 cm/yr, with average daily growth rates in the summer of about 0.10–0.15 mm/day. Most adults attain 1.5 cm shell length by the fall (Figure 4.4). In the second year, a few (less than 5%) grow to exceed 3 cm in shell length. In the Great Lakes region most adults die after 1.5–2 years of age.

Understanding the population dynamics of mussels entering a facility is critical for cost effective control of the mussels. Mussel infestation intensity will decline over time and then rise again, but to levels below the initial infestation rate (Claudi and Mackie, 1994). The rate of growth of individual mussels and of the population will vary from one year to the next, depending mostly on temperature. The easiest way to measure growth rates of individuals and of the population is to use side-stream monitors or bioboxes that continuously sample water from the intake structure. Details of the procedures are given later.

The average growth rate of zebra mussels in the southern states will probably be greater than the growth rate of zebra mussels in the Great Lakes. In Lake St. Clair, where the most extensive historical data on population growth rates have been collected, population densities increased in logarithmic fashion for the first three years and then geometrically in the next two when peak densities were observed (Claudi and Mackie, 1994). Densities exceeding 3 million/m^2 were recorded in the summer of 1990 immediately after a major recruitment (settlement) event. The turbidity of Lake St. Clair is highly variable, ranging from about 5 NTU to about 200 NTU; pH ranges from 7.9 to 8.6; calcium levels range from 35 to 40 mg/L; chlorophyll *a* levels range from 0.6 to 4.2 μg/L; temperatures range from 0°C to 30°C (32°F to 86°F). These physical–chemical variables, all known to greatly influence survival, reproduction, and growth rates of zebra mussels, are very similar in magnitude in the Mississippi River, Illinois River, and Missouri River, suggesting that population growth rates in many southern states will emulate, or even exceed, those in Lake St. Clair. Populations can be expected to increase by an order of magnitude for each of the first three years and peak after about five years at which time average densities can be expected to be near 100,000 m^{-2}, but may range up to 1,000,000 m^{-2}. These densities are particularly common in pipelines and tunnels where a continuous flow of water supplies the mussels with a constant source of food and oxygen.

For *Mytilopsis leucophaeata*, a threshold condition for spawning, like for *Dreissena polymorpha*, does not seem to exist (Verween et al., 2005); the temperature at first detection ranged between 16.2°C and 19.5°C and the salinity ranged from 2.6 to 4.9 PSU in Antwerp Harbour, Belgium. Schütz (1969) indicates that reproduction in *Mytilopsis leucophaeata* usually starts at a temperature higher than 15°C or even higher than 20°C (Rajagopal et al., 1995).

Mytilopsis leucophaeata reproduces once a year, with spawning between end of May and September–October in Europe, somewhat later in the United States (Verween et al., 2005, 2007). During this period, embryos are the most vulnerable phase in larval development and need a relatively high temperature (Verween et al., 2007); they believe that the need for high temperature will limit the potential locations for future invasion.

The seasonal flexibility in larval production patterns suggested that adults carry ripe gametes for a very long time in the harbour. After initial spawning, the exposure to ripe eggs and sperm in the water column often triggers gamete release by other ripe mussels, as such creating variability in recruitment (Verween et al., 2005).

In Mexico, *Mytilopsis leucophaeata* is restricted to warmer temperate waters (Marelli and Gray, 1983), but in Europe, it survives in much lower temperatures, ranging from 5°C in Finland (Laine et al., 2006) to up to 30°C in Miami (Siddall, 1980). In contrast, adult *Dreissena polymorpha* can survive temperatures up to 29°C (Karatayev, 1995) but reproduction and body functions are impaired (McMahon, 1996).

Verween et al. (2007) agree with de Vooys (1999) that temperature remains an important species-specific factor for spawning initiation. The threshold temperature for gamete maturation may be 13 ± 1°C (Verween et al., 2005) but the optimal temperature for embryonic survival in *Mytilopsis leucophaeata* appears to be 22 ± 2°C, although salinity rather than temperature seemed to be the limiting factor for survival (Verween et al., 2007).

4.2.3 Measuring Populations of Juvenile and Adult Dreissenids and Mytilids

This section describes how to sample juveniles and adults of epibenthic species of invasive bivalves for studies of their population dynamics in lakes and rivers. The procedures are different from those described earlier and are appropriate for water bodies that are infested and where no artificial substrate has been deployed. In both cases, quantitative estimates of abundance are desired but in the former (lakes and rivers) the desired unit is $\#m^{-2}$, whereas in the latter (facilities) the desired unit is $\#coupon^{-1}$ (of unit area). Some plants still want $\#m^{-2}$ or weight/m^2 and will scrape smaller areas, such as 1 ft^2 of wall than do a rock estimate but will still look for the number of settling events that took place.

In lakes and rivers the mussels generally attach to irregular surfaces, such as rocks, for which surface areas are difficult (but possible) to measure, and less likely on soft or sandy substrates. This leads to a very patchy distribution with mussels congregated on rocks and none or few in the fine sediments between the rocks. Since the objective is to estimate the average density per unit area of substrate (e.g., $\#m^{-2}$), one must determine the number of mussels per unit area of rock as well as the number of rocks per square meter.

The number of rocks per square meter can be determined by lifting all the substrates (e.g., rocks, tires, tin cans) within a square-meter quadrat. The rocks from three to five quadrats need to be taken as replicates in order to derive mean densities with an error estimate (e.g., standard error of the mean). The area on each substrate to be determined is the "exposed" area on the substrate (e.g., rocks) since zebra mussels do not attach to that part of the substrate that is submerged and buried in the sediments. If the substrate has a normal geometric shape, such as a rectangle, square, circle, oval, cylinder, and so on there is a formula for calculating its area. However, for rocks and other objects with ill-defined shapes, the "aluminum foil method" is recommended. It can also be used on the normal shapes listed above. Follow these procedures:

1. Scrape all the mussels off the rock surface noting especially the line that marks the part of the rock that was exposed. Usually it is a different color because chemical reactions within the substrate differ from those above. Also, algae and byssal threads often delimit the area of attachment.
2. In order to determine numbers of generations in the population, the sizes (shell lengths) of mussels (or a portion thereof) on each rock need to be measured. Measure either all the mussels or at least 200 if there are more than 200 mussels on the rock. The choice of the 200 mussels must be random. This needs to be done for each sampling event throughout the year to determine voltinism. The length–frequency distribution plots (see Figure 4.5) will also allow one to estimate mean generation times, longevity, and maximum lengths attained for a population.
3. Take a piece of aluminum foil and wrap it around the rock, pressing the foil into grooves and holes, if the grooves and holes are large enough for mussels to attach. In many places excess foil will gather. Squeeze the excess together to make flaps that can be easily cut off. Use scissors to cut off all the excess foil, including any foil that falls below the line where mussels were not found.
4. In order to estimate the area of the foil, one must first make a standard "curve" that relates size to weight of foil. This is done by cutting squares of 1, 4, 16, 64, 256, 900, 2500, and 3600 cm^2 (i.e., pieces 1×1, 2×2, 4×4, 8×8, 16×16, 30×30, 50×50, and 90×90 cm) and weighing each square piece. Plot the values on graph paper to produce a straight line through the origin. Now weigh your unknown (from step 3) and interpolate the area from the standard curve. Standard curves should be drawn for each roll of foil because the foil on different rolls may vary in thickness.
5. Once the area of each substrate and the number of substrates in each square meter are known, the $\#m^{-2}$ can be calculated. For example, if there are five substrates in quadrat 1 (=1 m^2, or 10,000 cm^2), and the total exposed area of the five substrates is 2575 cm^2 and the total number of mussels on all five substrates is 8550, the mean density in quadrat 1 is $10{,}000/2575 \times 8550 = 33{,}204$ mussels/m^2. Repeat for all quadrats and derive an overall mean density/m^2.

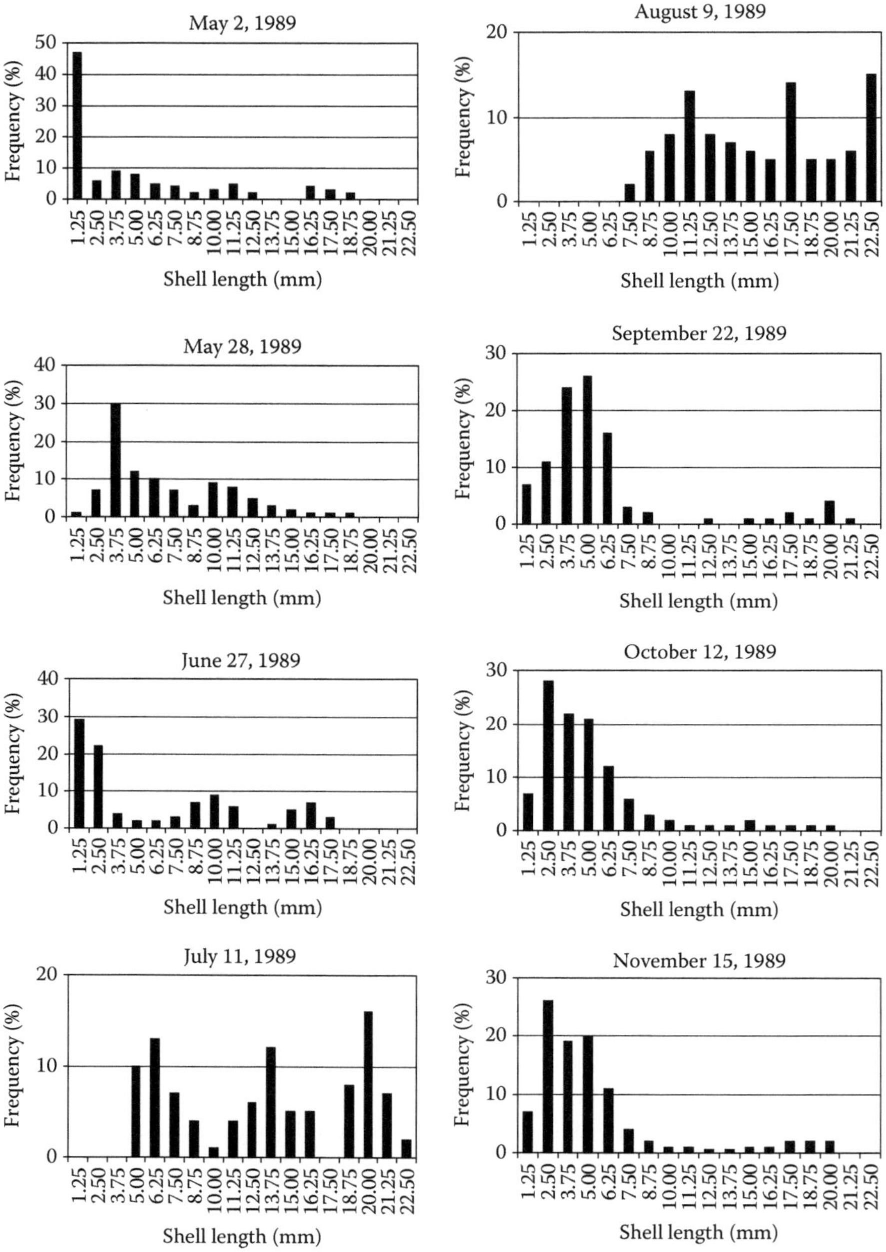

FIGURE 4.5 Length–frequency histograms showing growth of up to three cohorts of adult zebra mussels in Lake St. Clair from May 2 to November 15, 1989, one year after the mussels were first found.

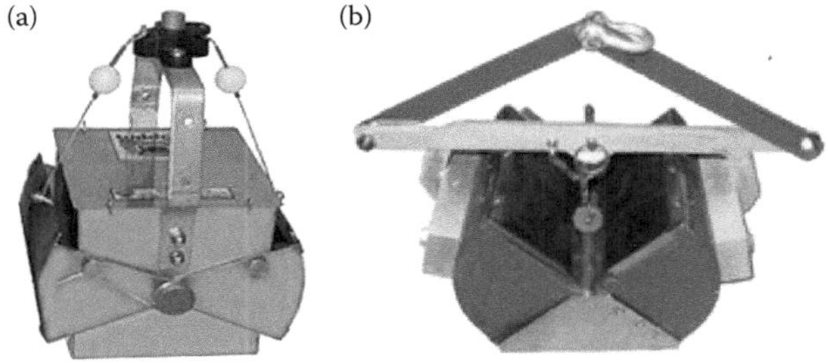

FIGURE 4.6 Benthic sampling devices. (a) Ekman grab used in soft sediments; (b) Ponar grab used in gravel substrates.

4.2.4 Measuring Populations of Juvenile and Adult Corbiculids

Corbiculids are infaunal organisms and their densities need to be determined with common quantitative benthic sampling devices like Ekman grabs or Ponar grabs (Figure 4.6a and b). Each grab takes a "bite" of the sediments, typically 15 cm × 15 cm (=225 cm^2). The sediments are washed through a sieve with 0.50 mm openings, and the material retained on the mesh is sorted and searched for clams larger than 50 mm in shell length. Usually five grab samples are taken and the mean number of clams in the five samples is multiplied by 44.4 to obtain the mean density/m^2. As for dreissenids, a representative number of clams is measured for shell length to determine the length–frequency distribution in each sampling event and hence the number of generations in the population. The length–frequency plots will also allow one to estimate mean generation times, longevity, and maximum lengths attained for a population.

4.2.5 Interpreting Variations in Population Sizes of Adult Dreissenids, Mytilids, and Corbiculids

Plotting the length–frequency distribution of juvenile and adult mussels not only allows one to verify the settlement period based on the presence of pediveligers in plankton samples but also to determine the number of cohorts (generations), growth rates, life span, and mean and maximum shell lengths attained in a population. Weighing and measuring the shell length, width, and height of a few small to large shells also allows one to determine the length–weight relationship and allometry (length–width, length–height, width–height relationships) of shell growth. Knowing the length–weight relationship allows one to determine the total weight that a mass of mussels adds to the weight of a floating structure and to its decrease in buoyancy. Knowing the allometry of shell growth permits calculations of pore sizes needed in meshes of filters to keep mussels out of a facility.

Figure 4.5 shows the length–frequency distribution of juveniles (<8 mm) and adults in Lake St. Clair in 1989, one year after they were first discovered. Densities were determined as described above in lists 1–5 in Section 4.2.3. The first sample was taken on May 2 when the water temperature was 14°C, too cool for spawning and too soon to explain the presence of the large numbers (~46%) of new recruits <1.25 mm; these new recruits probably settled the previous fall (1988) and did not grow over the winter months. There appears to have been three cohorts, the oldest being >15-mm long and probably contributed to the mass of fall 1988 recruits. Adults become sexually mature at 8 mm shell length (Mackie, 1991) and it is unlikely that the P2 cohort (Figure 4.5) contributed to the biomass of the fall 1988 cohort. The fall 1988 recruits continued growth with a rise in water temperature, so the peak in shell size of the fall 1988 cohort had reached 3.75 mm and the oldest cohort began disappearing by May 28, 1989. A mass of new recruits occurred between May 28 and June 27, 1989, probably due to reproduction of P2 because the oldest cohort had disappeared by June 27, 1989. There were no new recruits produced between June 27 and August 9, 1989 (Figure 4.5). Another cohort of new recruits appeared between August 9 and September 22, while most adults of P2 and P3 cohorts had died. There was little growth of mussels between September 22 and November 15, 1989 (Figure 4.5).

Both tissue and total (shell + tissue) dry weight were taken of a range of sizes of zebra mussels to derive the regressions shown in Figure 4.7. Dry weight was measured because it is weight added by a mass of mussels; wet weight includes water in the tissues and does not add to the weight of a mass of mussels in water.

An example of the utility of the length–weight relationship in Figure 4.7 is realized when one wishes to know the effect of a predicted density of mussels and its added weight on a buoyancy interceptor sewer system (BISS). Assuming a predicted density of 1200 mussel/m^2 and an average shell length of 20 mm, the additional weight would be about 1200 mussels/m^2 × 403 mg (from Figure 4.7) = 483,768 mg

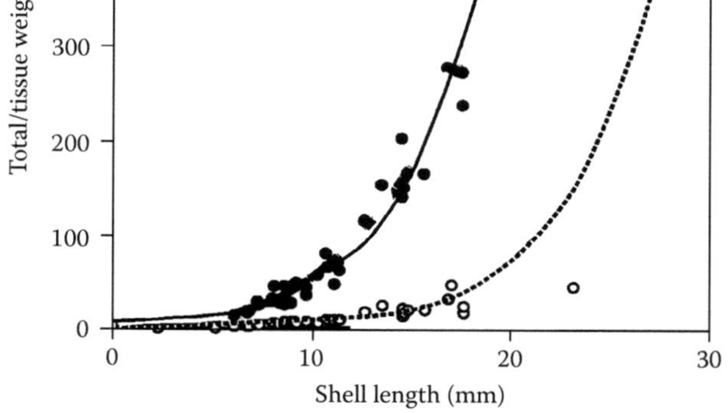

FIGURE 4.7 Relationships between shell length and (a) total dry weight (closed circles) and (b) tissue dry weight (open circles) of *Dreissena polymorpha*.

or 0.484 kg/m^2. The total length of BISS that is suspended in the water is 60,960 m, with a total surface area of 18,580 m^2. The estimated total weight to the BISS is 8993 kg or 9.9 ton. Engineers would use the added weight figure to determine how much buoyancy would be needed to add to the system to prevent it from sinking.

The length–weight relationships differ among species, among water bodies, and among populations within water bodies. Tables 4.3 through 4.5 give length–weight relationships (all are of the form $y = ax^b$) for *Dreissena polymorpha*, *Dreissena bugensis*, and *Corbicula fluminea* for different kinds of masses [dry tissue mass, wet tissue mass, total wet mass (shell + tissue), and total dry mass (shell + tissue)]. Differences among each of the *a* and *b* values have been attributed to depths, pH, calcium levels, latitude, lake nutrient levels, age, predation pressure, reproductive state, and ranges in shell sizes used to develop regressions [Shevtsova (1971); Joy and McCoy (1975); Zolotareva (1976); Sickel (1976); Stanczykowska (1977); Mackie et al. (1989); Mackie (1991); Dermott et al. (1993); Roe and MacIsaac (1997)].

Roe and MacIsaac (1997) discovered that the shells of *Dreissena bugensis* from deep water were very fragile, with thin and brittle shells, as reflected in the allometry of shells (Table 4.4) and may reflect reduced calcium carbonate availability in deepwater environments. They suggested that deepwater *Dreissena bugensis* allocate very little energy to shell formation, possibly reflecting the low degree of

TABLE 4.3
Allometric Relationships (Shell Weight = $a \times \text{length}^b$) between Mussel Mass (mg) and Shell Length (mm) of *Dreissena polymorpha* from Different Depths in Several Water Bodies

Water Body	Depth (m)	Mass Type	*a*	*b*	Source
Lake Erie (east)	3	DTiM	0.040	2.198	Dermott et al. (1993)+
Lake St. Clair	<2	DTiM	0.007	2.982	Mackie (1991)
Slesinskie Lake	NS	DTiM	0.015	2.406	Stanczykowska (1977)
Lichenskie Lake	NS	DTiM	0.012	2.576	Stanczykowska (1977)
Patnowskie Lake	NS	DTiM	0.014	2.520	Stanczykowska (1977)
Neusiedler Lake	NS	DTiM	0.022	2.355	Stanczykowska (1977)
Mikolajskic Lake	NS	DTiM	0.005	2.931	Stanczykowska (1977)
Taltowisko Lake	NS	DTiM	0.061	1.937	Stanczykowska (1977)
Mean DTiM			*0.022*	*2.488*	
Lake St. Clair	<2	ToDM	0.051	2.996	Mackie (1991)
Dneiper-Krivoy Rog Canal	NS	ToDM	0.278	2.701	ShevtSova (1971)
Mean ToDM			*0.165*	*2.849*	
Lake Erie (east)	3	ToWM	0.244	2.663	Dermott et al. (1993)
Lake Erie (east)	3	WTiM	0.135	2.695	Dermott et al. (1993)

DTiM = dry tissue mass; WTiM = wet tissue mass; ToWM = total wet mass (shell + tissue); ToDM = total dry mass (shell + tissue). + indicates equations have been modified from original form in Dermott et al. (1993). NS = Not stated.
Means of *a* and *b* are calculated when two or more values are provided.

TABLE 4.4
Allometric Relationships (Shell Weight = $a \times \text{length}^b$) between Mussel Mass (mg) and Shell Length (mm) of *Dreissena bugensis* from Different Depths in Several Water Bodies

Water Body	Depth (m)	Mass Type	*a*	*b*	Source
Lake Erie (east)	37	DTiM	0.005	2.725	Roe and MacIsaac (1997)
Lake Erie (east)	55	DTiM	0.004	2.964	Roe and MacIsaac (1997)
Mean DTiM			*0.0045*	*2.845*	
Lake Erie (east)	37	WTiM	0.044	2.831	Roe and MacIsaac (1997)
Lake Erie (east)	55	WTiM	0.034	2.966	Roe and MacIsaac (1997)
Mean WTiM			*0.039*	*2.899*	
Lake Erie (east)	37	ToWM	0.064	2.911	Roe and MacIsaac (1997)
Lake Erie (east)	55	ToWM	0.054	3.002	Roe and MacIsaac (1997)
Zaporozh'ye Reservoir	NS	ToWM	0.064	3.32	Zolotareva (1976)
Mean ToWM			*0.061*	*3.078*	
Lake Erie (east)	55	DToM	0.035	2.877	Roe and MacIsaac (1997)
Dneiper-Krivoy Rog Canal	NS	ToDM	0.244	2.687	Shevtsova (1971)

DTiM = dry tissue mass; WTiM = wet tissue mass; ToWM = total wet mass (shell + tissue); ToDM = total dry mass (shell + tissue). NS = Not stated.
Means of *a* and *b* are calculated when two or more values are provided.

turbulence encountered in deep water in eastern Lake Erie. The thinner, lighter shells would also be less likely to sink into the soft sediments typical of hypolimnia than would thicker shells. Appleton and Palmer (1988), Miller et al. (1994), and MacIsaac (1994a,b) pointed out the importance of the strength of shells in relation to types and sizes of predators. Roe and MacIsaac (1997) suggested that the thin, weak shells of deepwater *Dreissena bugensis* may be more vulnerable to predators than other *Dreissena* populations or, alternatively, thinner shells of deepwater *Dreissena* may be a result of lower predation pressure (Appleton and Palmer, 1988).

Corbicula fluminea has a heavy shell as reflected in the slope values of the length–weight relationships (Table 4.5). The shell length–weight relationships appear to vary significantly among populations, which Sickel (1979) attributes to differences in environments.

Figure 4.8 gives the allometric relationships between shell length and shell height and shell width of zebra mussels. The relationship between shell length and shell width is linear (i.e., $y = a + bx$), whereas the relationship between shell length and shell height is curvilinear (i.e., $Y = a + bx + cx^2$).

Figure 4.9 relates the ventral area of the shell to length. The area was calculated as the area of an ellipse ($A = ab\pi$). The values of a and b are half the long axis x (i.e., $a = x/2$) and half of the short axis of y (i.e., $b = y/2$). The values of x and y were determined from Figure 4.8.

TABLE 4.5
Allometric Relationships (Shell Weight = $a \times \text{length}^b$) between Mussel Mass (mg) and Shell Length (mm) of *Corbicula fluminea* from Two Water Bodies in United States

Water Body	Mass Type	*a*	*b*	Source
Mud R., West Virginia	DTiM	0.00333	3.558	Joy and McCoy (1975)
Altamaha R., Georgia	DTiM	0.00693	3.127	Sickel (1979)
Mean DTiM		*0.00513*	*3.343*	

DTiM = dry tissue mass.
Means of *a* and *b* are also provided.

The utility of the shell length–surface area graph is realized when one wishes to determine the level of infestation on different structures. For example, if a filter screen is to be protected, there will be a certain level of tolerance of coverage by zebra mussels before flow rates through the screen affect performance of pumps. If the screen is 1 m^2 and performance begins to be affected when 10% of the screen is blocked, then coverage of 0.10 m^2 can be tolerated. If the average annual density of mussels around the facility is 1000 m^{-2} and the average shell length of mussels

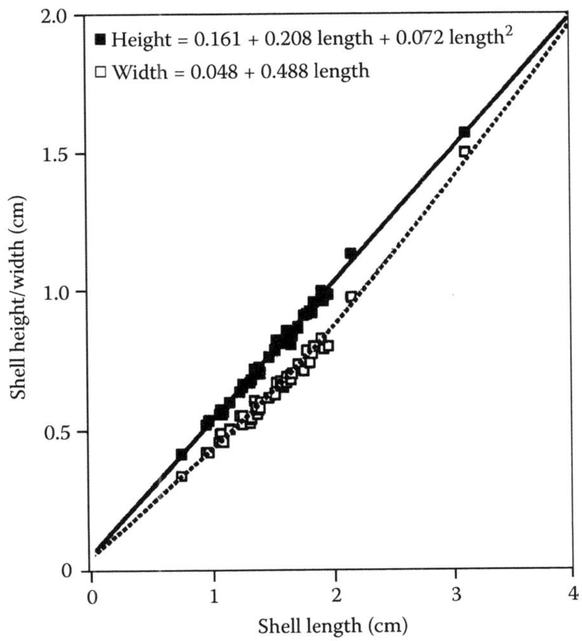

FIGURE 4.8 Relationships between shell length and shell height (solid squares) and shell length and shell width (open squares) of *Dreissena polymorpha*.

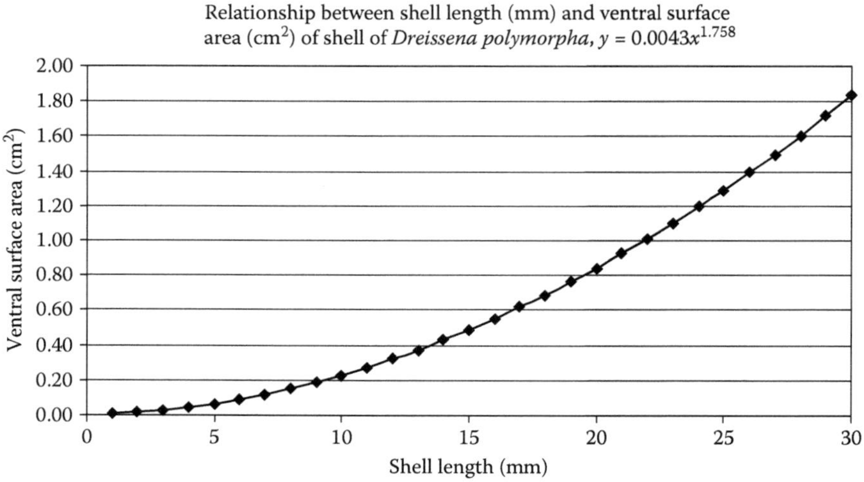

FIGURE 4.9 Relationship between shell length and ventral surface area of shell of the zebra mussel.

is 20 mm, then the average surface area covered by a single layer of mussels would be 1000 mussels × 0.833 cm^2/mussel (from Figure 4.8) = 833 cm^2 or 0.0833 m^2. This represents 83% of the 0.10 m^2, indicating that flow rates through the screen should still be sufficient to maintain effective performance of the pumps.

Perhaps this is the point where one should ask "what are low, moderate and high densities of mussels?" Ramcharan et al. (1992) consider low densities between 1000 and 3000 m^2 and high densities >~4800 m^2 (Figure 1A in Ramcharan et al., 1992), but no rational was given. A high density is considered one in which a square meter of surface is completely covered by mussels. If we assume an adult mussel (~3-cm long) has a ventral surface area of 2 cm^2 (see Figure 4.9), it would take 5000 mussels to cover 10,000 cm^2 (i.e., 1 m^2), so >5000 m^{-2} is considered the criterion for high density and supports Ramcharan's et al. (1992) estimate. Ramcharan et al. (1992) considered <3000 m^2 as the upper threshold of low densities because their data (Figure 1A) "appeared to represent a natural break in the density–frequency distribution." However 3,000 m^2 represents 60% coverage of a square meter of surface. A 30% coverage seems more appropriate, so for this book a "little to no infestation level" is considered one in which the density is less than 1500 m^2 and moderate infestation is considered for densities between 1500 and 3000 m^2. Infestation levels of 100,000 m^{-2} have been observed in the Great Lakes and the levels in Lake Havasu are probably not far off.

Growth of *Mytilopsis leucophaeata* has not been studied as thoroughly as the zebra mussel. One study by Verween et al. (2006) examined growth of a population in an Antwerp harbor, Belgium. They used growth cages to determine shell growth of three cohorts with initial shell lengths of ≤5, 10, and 15 mm, and found that the mussel had an oscillatory growth pattern with a single summer growing period per year (May–August). Growth decreased during wintertime, but never ceased completely. It had an average growth rate of 3–6 mm/yr. They found temperature to be the main environmental factor affecting growth. The von Bertalanffy growth function was used to model growth of individuals ≤5 mm, resulting in a maximum length of 16.7 mm and a growth constant of 0.56. Based on a combination of growth of all three cohorts, the hypothetical growth of an average individual was modeled over a 5-year period, resulting in a maximum length of 19 mm with a growth rate of 0.41 and a longevity of more than five years. Verween et al. (2006) found that the predicted maximum length corresponded well with most of those reported in the literature [10–20 mm in the United States (Abbott, 1974; Emerson and Jacobson, 1976); 22 mm in Miami (Siddall, 1980)]; 14 mm in North Sea Channel, The Netherlands (Rajagopal et al., 1995); 27 mm, the Netherlands (Gittenberger et al., 1998); 15.2 mm in Thames River, UK (Bamber and Taylor, 2002); 26.5 mm, Grigorovich et al. (2005); 17.2 in Schelde, Belgium (Verween et al., 2006).

4.3 SAMPLING FOR GASTROPOD EGGS AND ADULTS

4.3.1 Measuring Population Sizes of Eggs of Gastropods

Of the gastropods dealt with here, the ampullariids and bithyniids are oviparous and the thiarids and hydrobiids are ovoviviparous. Because most gastropods live among dense stands of macrophytes, measuring and interpreting variations in populations of eggs and adults can be a challenge.

4.3.1.1 Measuring Populations Sizes of Eggs of Ampullariids and Bithyniids

Ampullariids lay 25–250 eggs per clutch one to four times a week throughout the year, at least in some tropical areas (Estebenet and Martin, 2002). The eggs of *Pomacea* species are laid above water on any solid surface including emergent vegetation, concrete walls, and boulders, making it a challenge to find the eggs and then to estimate density (of eggs or adults) per unit area. Hence, fecundity (number of eggs per female) is usually recorded. The eggs have a calcareous shell and are quite firm. Look for clutches of eggs, carefully scrape them from the surface, preferably in a single clump, and then count the number of eggs in each clutch. If the eggs are on flat surfaces such as concrete walls, the density per unit area can be more easily estimated than if on plant stems.

Eggs of *Marisa cornuarietis* are laid in gelatinous clusters under the water surface (<1 m deep) typically among thick stands of aquatic vegetation (Hunt, 1958; Thompson, 1984) and are difficult to find. Population densities are rarely reported because the eggs are usually attached to plant leaves and/or stems, often on the underside of broad floating leaves. Hence, as with *Pomacea* species, only fecundity is usually recorded. They are prolific reproducing often even in December and January in southern Florida (Hunt, 1958). The eggs are not enclosed in a calcareous shell and are quite soft, so greater care must be taken when scraping them from substrate surfaces than eggs of *Pomacea* species.

Bithynia tentaculata is also dioecious but lays its eggs in double rows in clutches of 1–77 eggs on hard surfaces such as rocks, logs, and shells of its own species or of other snails and bivalves. The clutches are about 1-cm long, 6–7 mm in diameter, and semitransparent. Look for eggs from May to August when water temperature is 20°C or higher. Any females born early in the year may lay a second mass of eggs in October and November. The egg masses are firmly attached, soft, and fragile, and great care needs to be taken to remove them intact.

4.3.1.2 Interpreting Variations in Populations Sizes of Eggs of Ampullariids and Bithyniids

Mating, oviposition, eggs, and fecundity of ampullariids and bithyniids have already been discussed under reproduction. Variations in population dynamics is discussed here. Breeding tends to be seasonal and related to latitude, longitude, temperature, and rainfall (Andrews, 1964). Growth and reproduction ceases during the dry seasons, and breeding occurs during the wet season. Local variations in reproductive regimes appear to be related to local climatic variation in availability of water (Bourne and Berlin, 1982).

Of the species discussed here, all usually lay their eggs above water on the exposed parts of vegetation, rocks, and so on. Snyder and Snyder (1971) suggest that this behavior may be to avoid aquatic predators or perhaps is in response to anoxia in their aquatic habitats. Since the eggs are enclosed in a calcareous shell, perhaps there is physiological explanation, for example, during warmer temperatures air heats up faster than water so eggs develop faster. Heiler et al. (2008) showed that activity and vertical crawling velocity of golden apple snails increased with temperature and 1–2 h after the onset of darkness. The observed activity rhythms were hypothesized to represent an adaptation to avoid specialized predators (Heiler et al., 2008). The influence of temperature may explain the large variability in hatching period after oviposition and why the eggs (that are submerged and lack a shell) of *Marisa cornuarietis* take eight days to hatch at 25–30°C and 20 days at 15–20°C. Temperature strongly influences the seasonal pattern of reproduction in *Marisa* (Jobin, 1970), so it may influence oviposition in *Pomacea* species. It is clear that population dynamics of eggs and young of ampullariids is poorly understood; variations in fecundity and recruitment periods appear to vary from one species to the next and from one population to the next, and if populations are to be controlled, the population dynamics of all life stages need to be determined for species in each water body.

The egg masses of *Bithynia tentaculata* are small, each one being about 1-cm long and 6–7 mm in diameter and semitransparent. The density of egg masses sometimes reaches 155 clumps/m^2. Fecundity may reach up to 347 eggs and is greatest for the second year class. Eggs hatch in 3 weeks to 3 months, depending on water temperature. Oocytes develop poorly at temperatures of 30–34°C. Egg masses are produced when water temperature reaches 20°C, which in the northern temperate zone is May or June to August. Growth usually does not occur from September to May. The lifespan varies regionally and can be anywhere from 17 to 39 months (Jokinen, 1992; Korotneva and Dregol'skaya, 1992). As with ampullariids, fecundity and recruitment periods vary from one population to the next and if populations are to be controlled, the population dynamics and the biotic and abiotic variables that accompany the changes need to be determined for the species in each water body.

4.3.1.3 Measuring Populations of Juvenile and Adult Ampullariids and Bithyniids

As with studies on fecundity, there are challenges in sampling adult ampullariids because they are associated with heterogeneous and often densely vegetated habitats (Hann, 1999). Traditional macroinvertebrate sampling techniques such as seines and dredge or grab samplers are usually ineffective among dense vegetation and shallow, intertwined rhizomes. Even though apple snails are large, they generally have cryptic coloration and are difficult to find and collect by visual searches. Heavy 1-m^2 throw traps sufficiently heavy to penetrate plant stems and root-bound substrates have been used to collect wetland macroinvertebrates and fish with some success (Chick et al., 1992; Jordan et al., 1994, 1996) but the most efficient for collecting ampullariids are funnel traps (Figure 4.10); Darby et al., (2001) found that the funnel trap decreased collection time of *Pomacea paludosa* 10-fold and required considerably less labor than throw traps.

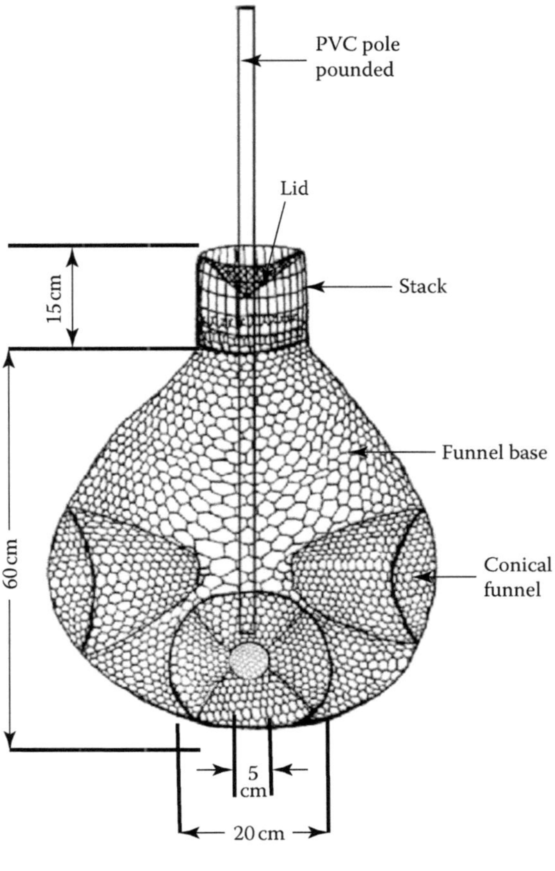

FIGURE 4.10 Funnel trap for collecting *Pomacea* spp. (Modified from Darby et al. *Wetlands* 21, 308–311, 2001. With permission.)

Darby et al. (2001) modified a commercially available crayfish trap that consisted of a funnel base, a stack, and a lid (Figure 4.10). The base of the funnel is constructed of vinyl-coated wire with 19 × 24 mm hexagonal mesh and includes three conical funnel entrances that taper from 20 cm on the outside to 5 cm on the inside. The size of mesh and funnel openings dictate the sizes of snails captured (e.g., 20–50 mm here). The stack is made from 25 × 13 mm mesh vinyl-coated wire. The lid is made from extruded polyethylene plastic with 13 mm mesh and is attached to the stack with nylon cable ties. The trap is firmly set adjacent to submersed plants by inserting a nylon cable tie through the stack wire and around a 2.5-cm diameter PVC pole driven into the substrate (Figure 4.8).

Darby et al. (2001) placed 50–100 traps 2–5 m apart in Florida wetlands each month for 3–5 days from February through August. Snail escape rates were determined by placing one to three marked snails (plastic numbered tags affixed to the shell with marine epoxy) in funnel traps and checking traps after one day (14 snails in 13 traps), two days (30 snails in 20 traps), and eight days (21 snails in 16 traps). Escape rates were recorded as the percentage of marked snails missing from each trap. Snail densities ranged from 0.25 to 0.70 snails/m^2 at a capture rate of 0.5–2 snails per trap over a 3–5 day period during spring–summer months. Escape studies indicated that trap intervals should be limited to 3–5 days, as nearly 35% of snails escaped after 8 days. In another study Valentine-Darby et al. (2008) used a mark-recapture approach and calculated capture probabilities (p^); they found that p^ varied significantly between sampling grids and it can be misleading to compare counts of animals without adjusting for p^. Although the funnel trap approach proved especially advantageous over throw traps in densely vegetated habitats, there was no clear advantage in areas with relatively sparse vegetation. Both approaches have their limitations, but the mark-recapture approach using funnel traps appears to accommodate a wider range of habitat types.

To collect *Marisa cornuarietis* look in lakes, ponds, irrigation systems, and swamps close to the surface, they prefer shallow waters with a lot of vegetation. They can tolerate the relatively high salt concentrations of brackish waters (Hunt, 1961; Fujio et al., 1991; Santos et al., 1987) but apparently they do not reproduce under these conditions. In general, they cannot tolerate temperatures below 12°C for prolonged periods. Jobin (1970) used an Emery dredge to estimate the total number of snails in three farm ponds in Puerto Rico as well as the amount of vegetation.

Take random samples from the population after recruitment has occurred and measure the shell height/diameter of several snails, and then plot their length–frequency distributions for each collection date, as for dreissenids (Figure 4.4). This will allow one to determine the size of adults at recruitment, recruitment dates, numbers of cohorts, growth rates, maximum size, and life span.

The faucet snail, *Bithynia tentaculata*, is generally found on the bottom (rocks, gravel, sand, clay, mud) in slow rivers and in ponds, shallow lakes, and canals or on submersed macrophytes (e.g., *Potamogeton* spp., *Elodea canadensis*, *Myriophyllum* spp., *Chara*) (Mackie et al., 1980; Clarke, 1981; Vincent et al., 1981; Jokinen, 1992). For specimens living on the bottom use Ekman (sand or mud) or Ponar grab (gravel, rocks). For specimens on macrophytes use sweep nets to determine size class variations or pull vegetation within 1-m^2 quadrat or use plant samplers (Gillespie and Brown, 1966; Gerking, 1957; Mackie and Qadri, 1971) for quantitative estimates of density.

4.3.1.4 Interpreting Variations in Population Sizes of Juvenile and Adult Ampullariids and Bithyniids

Very few studies have examined ampullariid population dynamics directly. Seasonal changes in water availability make such studies difficult, especially since many species aestivate and bury themselves in mud for variable lengths of time. Growth and reproduction ceases during aestivation periods and mortality may be high, with larger snails apparently more able to withstand dry periods (Kushlan, 1975).

Densities of apple snails appear to be highly variable, perhaps because there are numerous biotic and abiotic factors involved, such as food availability, temperature, water hardness (Perera and Yong, 1990), amount, quality and kind of vegetation (Bryan, 1990), and possibly predators such as the Everglade snail kite, *Rostrhamus sociabilis* (Pain, 1950; van Dinther, 1956; Snyder and Snyder, 1971; Kushlan, 1975; Beissinger et al., 1994), the limpkin, *Aramus guarauna* (Peterson, 1980), Caiman lizard, *Dracaena guianensis*, and several aquatic animals like fish, turtle, crayfish, and insects (Robins, 1971; Snyder and Snyder, 1971; Kushlan, 1975; Donnay and Beissinger, 1993). However, Donnay and Beissinger (1993) suggest that more research is needed to evaluate whether any of these predators impact snail population dynamics significantly. In general, habitats with more permanent standing water (e.g., irrigated rice fields as opposed to natural wetlands) have larger and higher population densities of ampullariids because the period spent in aestivation (no growth) is shorter. For example, densities of *Pomacea dolioides* ranged from 33/100 m^2 in Venezuelan rice fields to 3/100 m^2 in natural wetlands (Donnay and Beissinger, 1993). But densities of *Pomacea canaliculata* also vary among rice paddies, probably due in part to latitude and longitude; in Hawaii, densities of *Pomaceacanaliculata* in taro patches have been recorded at over 130 m^{-2} (Tamaru, 1996; Tamaru and Hun, 1996) while in Philippine rice paddies, densities vary from 1–5 m^{-2} up to 150 m^{-2} (Halwart,1994a,b; Schnorbach, 1995) or even an incredible 1000 m^{-2} (Anderson, 1993), and in Japanese rice fields 3–7 m^{-2} (Okuma et al., 1994a,b) or 12–19 m^{-2} (Litsinger and Estano, 1993). Likewise, for *Pomacea haustrum* fluctuations in population densities up to 215 m^{-2} have been reported by Freitas et al. (1987). Over the reproductive season in Florida, *Pomacea paludosa* populations were estimated to produce 1.2–1.5 million snails per ha (=120–150 m^2) (Hanning, 1979).

Marisa cornuarietis is a voracious herbivore and can attain high population densities in habitats with thick stands of vegetation, particularly elephant-ear-type, *Xanthosoma atrovirens*, plants, water lily, *Castalia ampla*, both common in Puerto Rico (Ferguson and Palmer, 1958), and water hyacinth, *Eichhornia crassipes*, in Florida (Emerson and Jacobson, 1976). *Marisa cornuarietis*, introduced to Florida, has been reported at densities over 200 m^{-2} (Robins, 1971). Population crashes can occur when it depletes the available vegetative food supply in a body of water. Taxson and Taxson (1982) reported that birds such as grackles are major predators in south Florida but did not state the impact on population dynamics.

Marisa is also known to devour the egg sacks and immature stages and even eradicating populations of *Biomphalaria glabratus*, *Bulinus truncatus*, and *Lymnaea cailliaudi*, all known carriers of *Schistosoma mansoni* (Oliver-Gonzales et al., 1956; Chernin et al., 1956; Ferguson et al., 1958; Radke et al., 1961; Jobin, 1970). Jobin (1970) reported a life span of over 500 days for *Marisa cornuarietis* and that water temperature was an important environmental parameter influencing the number of snails in a habitat by controlling the seasonal pattern of reproduction.

4.3.1.5 Measuring Population Sizes of Juvenile and Adult Thiarids and Hydrobiids

Most hydrobiids are oviparous but *Potamopyrgus antipodarum* being ovoviviparous is an exception. It and the two thiarids, *Melanoides tuberculata* and *Terbia granifera*, brood their young and produce miniature, adult-like juveniles. All three are mostly benthic, living in the mud, under stones, and leaf litter in running or standing water (Dundee and Paine, 1977; Roessler et al., 1978; Livshits and Fishelson, 1983). The red-rimmed melania is most active at night, hiding beneath decaying plants and stones or burying themselves in the mud during the day (Livshits and Fishelson, 1983). The quilted melania and the mudsnail in particular are extremely resilient, thriving in disturbed waters, enriched habitats in silt, mud, sand, and gravel. Being small the mudsnail also commonly occurs on submersed macrophytes in lakes or slow streams. Being mostly benthic in nature, the three species, especially the mudsnail, are best sampled with grab devices. The melania are larger and can be quantitatively sampled by scraping sediments from within 1-m^2 quadrats and washing the sediment through sieve boxes with mesh size to retain the juveniles. For snails on submersed macrophytes, use sweep nets to determine size class variations or pull vegetation by their roots from within 1-m^2 quadrat or use plant samplers (Gillespie and Brown, 1966; Gerking, 1957; Mackie and Qadri, 1971) for quantitative estimates of density.

Capturing the adults before they release the young provides a good estimate of recruitment. Alternatively, taking random samples from the population after recruitment has occurred and measuring the shell height of several snails, then plotting their length–frequency distributions for each collection date, as for dreissenids (Figure 4.5), will allow one to determine the size of adults at recruitment, recruitment dates, numbers of cohorts, growth rates, age and maximum size.

4.3.1.6 Interpreting Variations in Population Sizes of Juvenile and Adult Thiarids and Hydrobiids

Population sizes of the two species of adult melania and adult mudsnails vary by orders of magnitude. Dundee and Paine (1977) reported up to 2700 red-rimmed melania/m^2 in prime habitat in Louisiana. Roessler et al. (1978) reported densities of red-rimmed melania up to 37,500 specimens/m^2 in estuarine mangrove areas of Florida. Individuals as small as 10 mm may begin to reproduce. The newborn measured between 1.5 and 2 mm. Maturity of the quilted melania can be reached 97 days after birth when specimens are as small as 6 mm in shell length (Chaniotis et al., 1980a); their young are released into the environment when they are approximately 2 mm in length (Prentice, 1983). While there is active research on many aspects of the New Zealand mudsnail, the population dynamics of the species is not well described. As with other gastropods discussed herein, control of the species can be accomplished best when the dynamics of each population is known.

Little is known about shell allometry of growth in hydrobiids and thiarids. Chen (2004) reported a shell width/height regression of shell width = 0.3888 (shell height) + 0.49 ($r^2 = 0.914$, $P < 0.0001$) for a river population of *T. granifera* in Taiwan.

4.4 EARLY DETECTION OF *DREISSENA* SPP. VELIGERS USING MOLECULAR METHODS

Paul A. Rochelle, Anne M. Johnson, and Ricardo De Leon

4.4.1 Introduction

Due to their relatively small size and lack of easily identified morphological characteristics, detection and identification of quagga and zebra mussel veligers can be difficult using microscopy. Although cross-polarization makes the task easier, microscopy is time and labor intensive. Consequently, alternative molecular-based methods have been investigated for detection and monitoring, primarily the polymerase chain reaction (PCR). Over the past 20 years, molecular methods such as PCR have revolutionized the clinical diagnosis of human infections, criminal and environmental forensics, and the detection of pathogens in food. Although not as widely adopted as in clinical and food microbiology laboratories, they have also been used within the water industry for detecting pathogens and other microbes in source and drinking water. They have radically altered and improved our understanding of the composition, phylogeny, and function of microbial communities in diverse environments. Molecular methods have also been used to assess breeding and management practices in shellfish aquaculture (Dias et al., 2008; Selvamani et al., 2001), to monitor the distribution and spread of various terrestrial and aquatic invasive species, including oysters (Patil et al., 2005), and to identify bivalve larvae (reviewed by Garland and Zimmer, 2002).

Molecular methods such as PCR enable rapid detection of target nucleic acids with a sensitivity and specificity that are difficult to achieve with culture- or microscopy-based methods. A recent review of molecular technologies concluded that DNA-based methods offer a wealth of benefits for researchers and managers involved in the detection, identification, and monitoring of invasive species (Darling and Blum, 2007). These include confirming the identity of invading species, identifying unknown species, screening for the presence of target species while still in the early, juvenile, or larval stages of development, comparing geographically distinct populations within a species, and assessing the diversity and abundance of invading species. The evolutionary relationships within the genus *Dreissena* were studied using PCR-based analysis targeting multiple genes and it was concluded that the Great Lakes invasion by both species involved a large number of individuals from multiple sources, rather than a single introduction (Stepien et al., 2003). The goal of this section is to provide sufficient information to allow utility operators and managers to decide whether to incorporate PCR into their early warning *Dreissena* spp. monitoring program. While there are certainly many advantages to molecular approaches, some of the limitations of PCR and necessary precautions are also discussed.

4.4.2 PCR-Based Veliger Detection

PCR is an *in vitro* enzymatic method for rapidly amplifying specific DNA sequences so that they can be detected by optical, chemical, or physical techniques. It has revolutionized approaches to clinical diagnostics and life sciences research. It is based on the recognition of a specific sequence of target DNA by two short DNA primers typically located 100–1500 nucleotides apart and a heat stable polymerase enzyme that synthesizes multiple copies of double-stranded DNA between the two primers. The sequences of the primers, along with annealing temperature and $MgCl_2$ concentration, control the specificity of the reaction. The DNA sequence between the two primers is amplified exponentially, yielding high concentrations of double-stranded DNA amplification product from small amounts of template material.

The process involves multiple cycles of three temperature steps to amplify target-specific DNA sequences using a programmable thermocycler: (1) denaturation of target double-stranded DNA into single strands at 94°C, (2) annealing of target-specific primers to the single strands of DNA template at a temperature determined by the sequence of the primer-target hybrids (typically 50–65°C), and (3) synthesis of a copy of the target sequence by the thermostable polymerase at 72°C. Under ideal conditions, each cycle doubles the amount of target DNA, so after 40 cycles the initial target DNA has been amplified 2^{40}-fold. While actual amplification reactions do not usually achieve the maximum theoretical yield, billions of copies of a specific target sequence can be made in a short period; this is the basis of the technique's extreme sensitivity.

The advantages of PCR compared to traditional organism detection methods such as culturing and microscopy are its sensitivity, with the potential to detect a single copy of the target gene, and its high degree of specificity. In addition, assays can target multiple genes within an organism, thereby increasing confidence in positive results. In addition, many samples can be processed simultaneously for high throughput applications, which reduces the per sample cost. Detailed reviews of PCR are provided by Sambrook and Russell (2001) and McPherson and Moller (2006). Although other amplification methods such as transcription-mediated amplification and strand displacement amplification have been used in clinical diagnostics, they have not been applied to environmental samples as extensively as PCR.

PCR-based methods have been used to identify bivalve larvae, study species composition within bivalve larval assemblages, and detect ostreid herpes virus in commercial shellfish (Larsen et al., 2005; Livi et al., 2006; Sawada et al., 2008; Webb et al., 2007). PCR has also been applied for the detection of *Dreissena* spp. (Baldwin et al., 1996; Claxton and Boulding, 1998; Claxton et al., 1997; Frischer et al., 2002) and interest in its use has increased as the invasion of the western United States has continued. Although PCR for *Dreissena* spp. veliger detection is still a research-grade tool, the availability of commercial PCR kits for clinical diagnostics and detection of bioterrorism agents demonstrates that the method can be converted to routine applications using off-the-shelf kits and reagents. With well-designed primers and optimization of reaction conditions, PCR assays may be specific to most species within a genus (e.g., all *Dreissena* spp.), a single species (e.g., just *Dreissena polymorpha*), or strains within a particular species. Primers that have been used for *Dreissena* spp. detection assays are described in Table 4.6.

TABLE 4.6
PCR Primers for Detecting *Dreissena* spp. Larvae and Veligers

Target Gene	Primer Sequences[a] (5′–3′)	Reaction Conditions[b]	Specificity and Amplicon Size	Reference
Mitochondrial cytochrome C oxidase	F: GGTCAACAAATCATAAAGATATTGG R: TAAACTTCGGGTGACCAAAAAATCA	2.5 mM, 54°C	Bivalve species, 710 bp	Baldwin et al. (1996)
Mitochondrial cytochrome C oxidase	F: SCTTGTKGGMACRGGTTTTAGTG R: GGATCTCCTAACCCTGTWGGATCAA	4 mM, 50°C	Zebra and quagga mussels, 608 bp	Claxton and Boulding (1998)
18S ribosomal RNA	F: CTGCCAGTAGTCATATGC R: AGGGGACGGTGACGC	1.5 mM, 55°C	Zebra and quagga mussels, 701 bp	Frischer et al. (2002)
Mitochondrial cytochrome C oxidase	F: GAAACTGGTTGGTCCCGATA R: TAAGGCACCGGCTAAAACAG	2 mM, 59°C	Quagga mussels, 383 bp	Rochelle (this chapter)
Mitochondrial cytochrome C oxidase	F: TGTCACCACTCATGGGCTTGTT R: TGCAGAACAAAGGGACCCGGTAAA	1.5 mM, 61°C	Zebra mussels, 386 bp[c]	Rochelle (this chapter)
28S ribosomal RNA	F: TCAGTAACGGCCAGTGAA R: CATAGTTCACCATCTTTCGG	na	Zebra and quagga mussels, 878 bp	Hoy and Rodriguez (2008) (unpublished[d])

[a] F, forward primer; R, reverse primer.
[b] Concentration of $MgCl_2$ and PCR primer annealing temperature (primary specificity controllers).
[c] Reaction conditions can be manipulated so that these primers also amplify a 491 bp fragment from quagga mussels.
[d] Information obtained from GenBank sequence submission FJ4554436. na, not available.

Standard PCR is usually applied as a presence/absence tool with agarose gel electrophoresis as the end-point analysis method (Figure 4.11a). However, real-time PCR using fluorogenic probes and fluorescence detectors to monitor the accumulation of amplification products as they are formed during each cycle (Figure 4.11b) provides automatic probe-based confirmation of the PCR product and quantitative PCR (QPCR). These methods do not allow real-time monitoring of organisms in a sample matrix since sample concentration and DNA extraction are still required prior to the detection assay. In QPCR, the production of amplified DNA is monitored after each amplification cycle and the amount of amplicon is calculated based on the early exponential phase of the amplification process. While QPCR is useful for relative quantification of target sequences or organisms, absolute quantification leading to direct enumeration of target organisms is more difficult.

4.4.2.1 Equipment and Supplies

All equipment and supplies necessary for PCR are available from a wide variety of biochemical and biotechnological suppliers (e.g., Roche®, ABI®, Qiagen®, Invitrogen®).

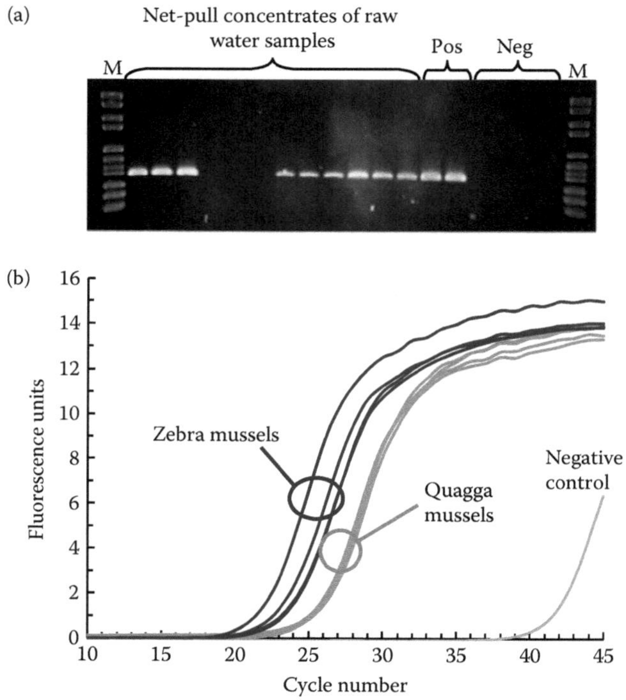

FIGURE 4.11 Detection of *Dreissena* spp. veliger DNA by PCR. The results of a PCR detection assay targeting quagga veligers in a plankton net concentrate were visualized by postamplification agarose gel electrophoresis and staining with ethidium bromide (a); 383 bp amplification product): Pos, positive control; Neg, negative control; M, molecular size standard. Alternatively, amplification products can be detected as they are formed using real-time fluorescence quantitative PCR methods (b).

A thermal cycler is the only dedicated apparatus that is not readily available in most microbiology laboratories that have not yet made the move to molecular methods. The capacity of thermal cyclers ranges from 24 to 384 simultaneous reactions, which can be increased with slave units operating from a central controller. The equipment and supplies that are essential for PCR are the following:

- Centrifuge capable of spinning 225 mL bottles at approximately 3200× *g*
- DNA extraction kits
- Standard laboratory supplies
    - Pipettes and tips
    - Microfuge tubes
    - Centrifuge bottles
- Microfuge capable of spinning 1.5–2 mL tubes at 10,000× *g*
- Programmable thermal cycler
- PCR primers (available from many commercial custom synthesis companies)
- PCR reagents: thermostable DNA polymerase, deoxynucleotide triphosphates (dNTPs: adenine, guanine, cytosine, thymine/uracil), reaction buffer, $MgCl_2$
- Gel electrophoresis boxes and power supply
- Agarose
- Ethidium bromide or other DNA-binding fluorescent dye
- Cloning kit if amplification products will be sequenced (recommended).

4.4.2.2 Protocol

Veligers are typically recovered from large volumes of water by passing through 63 μm plankton nets to yield a concentrate of approximately 20–50 mL (see Section 4.2.1). The plankton net concentrate is further concentrated by centrifuging at 3260× *g* for 20 min and the resulting supernatant completely aspirated. Final pellet volumes are typically <0.5–20 mL, depending on the type of sample and turbidity. Many methods have been published for extracting nucleic acids from environmental samples, but there is no single technique that can be used for all types of sample (Tsai and Rochelle, 2001). The simplest approach, particularly for laboratories new to molecular methods, is to use a commercially available extraction kit; suitable kits are available from MoBio®, Qiagen®, and Invitrogen®, among others. These kits are available in a variety of sizes and formats and typically include a mechanism for lysis of target organisms (enzymatic, detergent, physical disruption by bead beating, or a combination), purification of DNA away from sample and cellular debris, and proprietary reagents or processes for removal of inhibitory compounds. The MoBio UltraClean Soil DNA Isolation kit involves cell lysis by a combination of heat, detergent, and bead beating followed by binding of the released DNA to a silica spin filter. It usually yields high-quality amplifiable DNA from net concentrates, which rarely contains inhibitory material. However, a thorough side-by-side comparison of extraction kits and methods has not yet been conducted on plankton net collected veliger concentrates. The kit is designed for small samples, so it can only accommodate relatively small pellet volumes (≤0.3 mL). If the pellet is larger than 0.3 mL, it must be split into as many samples as necessary to process the entire pellet. Larger capacity extraction kits are available but have not been thoroughly evaluated with veliger concentrates. At the end of the extraction process, purified DNA is eluted in 50 μL of molecular grade water or Tris buffer and used as the template for amplification reactions.

The amplification reaction typically contains 20 mM Tris–HCl, pH 7.4 and 50 mM KCl (provided as a 10× reaction buffer with the polymerase), 1.5–2 mM $MgCl_2$, 0.2 mM of each dNTP (dATP, dCTP, dGTP, dUTP), 0.25 μM each of the forward and reverse primers, 0.01 μg/μL T4 gene 32 protein, 0.01 U/μL uracil DNA glycosylase (UDG), 0.025 U/μL Platinum Taq DNA polymerase (Invitrogen), and 1–15 μL of template DNA, in a total volume of 50 μL. T4 gene 32 protein is added to increase stability of the polymerase-template complex as an aid to overcoming sample inhibition.

A less expensive alternative is bovine serum albumin. There are many different versions of native and cloned DNA polymerases, with varying properties and performance characteristics. These include high fidelity polymerases with proof-reading capability, which is important if looking for minor DNA sequence variation between different target organisms, enzymes designed to maximize amplification yield, and to amplify long DNA fragments. Thermal cycling conditions are 25°C for 10 min to allow UDG to digest any amplification products carried over from previous reactions, 94°C for 10 min to simultaneously inactivate UDG and activate the 'hot-start' Taq polymerase, followed by 40 cycles of 94°C for 20 s, 59–61°C for 20 s, and 72°C for 40 s. A final incubation at 72°C for 10 min ensures that all amplification products are extended to full length. Reaction products can either be analyzed immediately by gel electrophoresis (Figure 4.1) or stored at 4°C for later analysis. Although PCR can be very specific, depending on primer design and reaction conditions, confirmation of amplification products by DNA sequencing is recommended. Sequencing is also necessary if PCR is being used to study the population genetics of invading mussels and to compare them with geographically and temporally disparate populations. Amplification products can either be sequenced directly following a relatively simple purification step or they can be cloned into a suitable *Escherichia coli* vector and then sequenced. Contract sequencing services are readily available and most offer 48 h turnaround.

4.4.2.3 Data Interpretation and Method Limitations

The widespread use of PCR for environmental applications may be seen to imply that the procedures are routine and the results definitive. However, optimization and objective evaluation are necessary to ensure reliable and consistent results. Without a thorough understanding of the limitations of the technique, end users and clients may be disappointed by the outcome of molecular assays. Both false-positive and false-negative results can make interpretation of PCR-based monitoring difficult if sufficient controls and Quality assurance and control (QA/QC) measures are not incorporated into monitoring protocols. Inhibition of the amplification reaction by compounds coextracted with DNA from the sample can be particularly problematic when applying PCR to DNA extracted from environmental concentrates. Both inorganic and organic compounds in water (particularly humic acids) can be concentrated along with target organisms and can carry through the DNA extraction process. These materials can reduce the specificity of reactions or completely inhibit PCR, leading to a negative result even though the target organism was present. Many extraction kits contain proprietary inhibitor removal solutions and procedures. Also, there are many published methods for removing inhibitory material from extracted DNA, including gel filtration through spin columns, immunomagnetic capture of DNA, and various DNA binding columns and materials. Although high-quality DNA can be extracted from environmental samples, there is still a need to improve DNA extraction methods for plankton net concentrates so that a greater proportion of extracted DNA can be included in a single PCR reaction.

Ideally, positive results would only arise from DNA extracted from the target organism that was concentrated by the plankton net. However, while extreme sensitivity is the greatest strength of PCR, it can also be its weakness and PCR positives must also be interpreted with caution. Because it is so sensitive, even trace amounts of contaminating DNA may be amplified as well as, or instead of, the intended target. False positives may result from nonspecific amplification of nontarget organisms, contamination with amplification products from previous reactions (carryover contamination), or from target organisms that are maintained, manipulated, and grown in the same laboratory or facility. Nonspecific amplification is controlled by carefully designing primers so that they only recognize target sequences, optimizing reaction conditions, and removing contaminating salts and other materials from the extracted DNA. Design of efficient PCR primers is dependent on the availability of sufficient high-quality DNA sequence information for the target genes and organisms. Even after an assay has been optimized, it may be necessary to revisit the design of primers as new DNA sequence information becomes available. Laboratory and carryover contamination are prevented by stringent quality assurance and quality control procedures, as described in the following section. Even when a positive result is due to amplification of the correct target sequence, it does not guarantee that the target organism was actually present in the collected sample. Naked DNA released by live or dead organisms upstream of the sample site may be concentrated by the plankton net if it is particle associated, and carry through the extraction and amplification processes. However, a positive in this instance indicates that the organism is or has been present in waters connected to the specific sample location at some recent time. Samples can be treated with DNA digesting enzymes prior to extraction to remove naked DNA but there is no guarantee that such treatment will not damage DNA inside target organisms and thereby reduce sensitivity. However, despite the potential for false positives, a positive PCR result generated by a well–designed assay with all of the appropriate controls and QA/QC measures in place should be considered as a reliable true positive.

Standard PCR assays are presence/absence tests but provide no measure of the viability of the target organism. Reverse transcriptase-PCR (RT-PCR) assays that target RNA instead of DNA have been used to assess viability based on the instability of RNA in dead cells compared to DNA. Incorporation of propidium monoazide (PMA) into amplification reactions allows standard PCR to amplify DNA only from viable cells (Nocker et al., 2006). However, most of these molecular viability assays were developed for bacteria and single celled eukaryotes in environmental samples (Keer and Birch, 2003) but have not been evaluated with more complex, multicelled organisms such as mussel larvae.

4.4.2.4 Quality Assurance and Quality Control

One of the biggest challenges in applying PCR to environmental samples is maintaining the integrity of the sample prior to and during DNA extraction. Cross contamination between samples, carryover contamination from previous analyses, and contamination of test samples with positive control material can all compromise the integrity of new samples and invalidate the results, or at the very least lead to confusion in data interpretation. Ideally, dedicated laboratories should be used for sample storage, sample preparation, DNA extraction, set up of molecular assays, and end point analysis. However, few facilities have the luxury of large amounts of dedicated space so a high degree of laboratory organization is necessary to minimize contamination where multiple functions are performed in a single laboratory. These measures include coordination of activities and workflow within the laboratory, spatial and temporal separation of different activities, and rigorous QA/QC procedures. Reference to the U.S. Environmental Protection Agency guidance manual on QA/QC procedures for PCR on environmental samples is recommended for anyone contemplating the use of PCR as a monitoring or screening tool (USEPA, 2004).

Stringent measures must be used at every step of the process to prevent contamination of the sample and to ensure the validity of the results. All work surfaces are decontaminated before and after each use to eliminate bacteria, DNA, and DNA degrading enzymes by wiping with solutions of Lysol, ethanol, and various proprietary DNA and DNase eradication solutions. While nucleic acid extractions can be performed on the open workbench, it is preferable to use biological safety cabinets that are equipped with germicidal UV lights. Dedicated UV hoods used for PCR set up should be decontaminated by UV illumination for 60 min before and 30 min after reaction set-up and micropipetting devices should be dedicated to PCR assays. Inhibition controls, hot-start PCR assays, and addition of UDG should all be routine parts of the assay. Inhibition controls (also called amplification controls) are used to determine whether a negative result is due to inhibitors in the sample or if the sample is actually negative. These controls can either be external or internal to the test sample. An external inhibition control involves adding a known concentration of unrelated DNA to an aliquot of the test sample and amplifying it using primers specific to the control DNA. Internal inhibition controls involve adding DNA to the test sample with primer binding sites that are identical to those of the target sequence, an unrelated internal sequence that is similar to the target sequence in length and base composition, and a unique region of sequence that differentiates the internal control from the target amplification product. If the sample contains PCR inhibitors, the inhibition control sequences will not be amplified leading to a false negative. The control sequence should always be amplified regardless of whether the sample contains the target organism (refer to Hoorfar et al., 2004 for a detailed discussion of amplification controls). UDG is an enzyme that degrades uracil-containing PCR products prior to amplification. Contaminating PCR products from previous reactions, during which dTTP was either partially or entirely replaced by dUTP, are thereby digested and no longer available as amplification templates. Hot-start methods reduce background amplification and increase yield by preventing nonspecific primer annealing and extension during the lower temperatures of reaction set-up. Hot-start DNA polymerases are activated by a brief incubation at 95°C immediately prior to initiating temperature cycling and reduce the likelihood of false priming.

For each batch of samples that are to be screened by PCR, an extraction negative control should be included. This consists of sterile water or buffer going through the entire extraction and amplification process. While at least one extraction control should be included with each batch of samples, it is advisable to periodically test multiple extraction controls since contamination may be transient and sporadic. In addition, each PCR run should include a positive control, a reagent negative control (DNA template replaced by water), and a nontarget negative control. Positive controls are DNA from individual confirmed quagga or zebra mussels. The nontarget control should be DNA extracted from a nontarget but a similar organism that is known or expected to occur in the test samples. In the case of dreissenid-specific PCR, *Corbicula* spp. is a suitable control that should not be amplified by the dreissenid primers.

4.4.3 Conclusion

A carefully designed PCR assay that incorporates all of the appropriate controls and QA/QC precautions can be reliably used as part of a routine monitoring program to detect the early stages of a quagga and/or zebra mussel invasion. Monitoring by PCR allows detection of both species and discrimination between them if they are both present in a body of water. Once veligers are routinely detected by microscopy or adult mussel infestation is observed in a body of water, the utility of PCR as an early detection or monitoring tool is greatly diminished. However, once mussels are established in a water body, molecular-based assays can be modified for studying other aspects of an infestation. By sequencing the amplification products, it is possible to determine whether veligers and adult mussels are genetically similar to mussels in other bodies of water and different geographic regions. There is also the potential to adapt PCR for studying veliger viability and stress-induced changes in adults that may result from an array of possible mussel control strategies.

4.5 MONITORING FOR MACROFOULING CONTROL IN INDUSTRY

The monitoring strategies described here are primarily used for attached macrofoulers that possess a free swimming life stage as larvae. In fresh water, the most important species in this category includes the *Dreissenids* and the mytillids.

The primary objectives of a monitoring program should be to

- Determine when larval stages of macrofouling organisms are present in the raw water
- Determine how fast the macrofouler grows once settled
- Determine the density of settled population of macrofoulers at the end of a discreet time period
- Determine the susceptibility of the macrofouler to various control methods

To achieve these monitoring objectives, a combination of settling plates, plankton net sampling, and in-plant monitoring is generally recommended. Plankton sampling and settlement plates have been already covered in depth earlier in the chapter. This portion of the text will focus on the use of sidestream monitors that are uniquely designed for industrial applications.

4.5.1 Sidestream Monitoring

To monitor the settlement of mussels within the power plant, we recommend the use of sidestream samplers. The most frequently used device in North America is an aquarium-like structure, commonly known as a biobox. The biobox is designed to mimic the flow in the industrial piping that is in danger of fouling. The flow rates through the biobox should provide a retention time of at least 20 min to mimic most industrial cooling systems.

A number of North American facilities use various forms of the "biobox" (Figure 4.12). Several settling plates, made up of different materials, are installed in the biobox. The outflow (which goes to a drain) balances the inflow to maintain continuous circulation in the biobox. A standpipe assures constant level in case of supply interruption. While mussels are lucifugous (avoid light) the clear plexiglass does not inhibit settlement, as one would expect since in nature they are exposed to light most of the day.

When connecting the biobox to the system to be monitored, it is best to tap off from the lower portion of the system pipe. Water from a location on the lower portion of the system pipe will be likely to contain more planktonic life stages than supply from the upper portion of a pipe. This is because most of the plankton sediment out by gravity after they pass through a pump. The flow will become less turbulent, depending on bends in the conduits and after ca. 30–50 times the diameter of the pipe the flow will become more or less laminar. The sinking rate of juvenile mussel shells is about 10 mm/s at a length of 0.5 mm and 40 mm/s at a length of 2 mm with the valves closed. Do we have a reference for this?

FIGURE 4.12 Typical biobox. (Reprinted from Green Peak Services, Picton, Ontario, Canada. With permission.)

General recommendation is to place one biobox as close as possible to the intake of the water for the facility. The water supplying the biobox should be raw water that is never going to be treated. Some plants inject the chemical immediately before the intake pumps. If this is the case, *a separate pump must be installed to pump water from the source directly to the biobox.* A second biobox should be placed further in the piping system, but prior to any heat exchangers. If macrofouling control measures are deployed in the piping systems, the absence of macrofoulers in the bioboxes placed prior to the heat exchangers or at the discharge end of the piping system (if heat exchangers are not present) would verify that the control measures are working, provided that a control biobox placed at the front of the system prior to application of the control strategy is colonized. Conversely, if settled, live macrofoulers are found in both bioboxes, the control measures are not working.

The data that are collected by the monitoring program described can assist industrial facilities in their decision-making process on how to best deal with macrofoulers.

4.5.2 Timing of Treatment for Elimination of Macrofoulers

If prevention of settlement is the aim, knowing when the settlement starts and ends will determine the start and the end of the period for a mitigation program. As soon as settlement is observed on the settlement plates outside the plant or in the bioboxes in the plant, treatment to eliminate settlement in the cooling pipes should commence. Once new settlement is no longer observed, and the previous settled organisms have disappeared, the treatment has been successful.

Typically in the Northeast bioboxes are used in May to June to determine the initial appearance of veliger larvae; in June and July to determine the development rate of veliger larvae; from July through October, or even December, to determine the settling rate of postveliger larvae, to assess the effectiveness of a control program (typically chlorination). In the southern states it is conceivable that monitoring may be required throughout the year.

4.5.3 The Length of Periodic or End-of-Season Treatment Required Using BioMonitors

If the treatment method chosen is to periodically eliminate any settlement in the piping, the growth rate of individual shells combined with the rate of accumulation of biomass will determine when and how often such periodic treatments may have to be administered. The size of individual shells prior to treatment should be small enough that a single shell could not block a vital part of the equipment when the macrofouler dies. The allowed accumulation of biomass should be such that the posttreatment sloughing off would not block strainers or heat exchangers and overwhelm the mechanical maintenance staff. Therefore, the data on biomass accumulation will aid in predicting the required timetable for maintenance of various systems in the plant. This would include the need for maintenance of external structures such as trash racks, rotating sieves, and at hydro power stations spill gates or penstock gates. If heavy biomass accumulation is observed on the settling plates, these external structures may have to be cleaned more frequently.

How long a periodic treatment has to last to remove all settlement from inside of the piping is best judged by observing macrofoulers that have settled in the bio-boxes at the end of the system. If macrofoulers have settled in the bioboxes in sufficient numbers, it is important to observe the biobox during the periodic treatment. Depending on the treatment method chosen, the macrofoulers may disappear from the biobox or die *in situ*.

Once the last individual has either exited the biobox or is observed gaping and unresponsive, the treatment can be terminated. If there is a low number of mussels settled in the biobox prior to the start of the treatment, it is a good idea to "seed" or introduce additional macrofoulers from another location into the monitor. Having a large number of individuals to observe during the treatment will give greater assurance that the treatment method is working. To seed coupons for the biobox, place coupons on the bottom of a large container (plastic garbage can perhaps). Remove clumps of adults from natural substrate (walls of the intake, rocks) infested with adult mussels (of the size class required or larger) and place them on the coupons. *When removing mussels from the rock, the least damage or stress is created by removing mussels by the byssal apparatus rather than by the shell.* Grab the byssal apparatus firmly with a pair of watchmaker's forceps and tear it from the rock surface. Leave the mussels in the vessel for about a week, or long enough for the mussels to attach onto the coupons. Seed the monitor with large adults, or at least with adults of approximately the same length class that are present in the distribution system. If you are not sure what the largest length class is, opt for the highest length class found.

4.5.4 Efficacy of the Preventative Treatment

Some facilities wish to prevent the settlement of macrofoulers in their system rather than eliminating an existing population of adults. The use of bioboxes gives some assurance that the strategy chosen is working. If settlement is observed in the biobox at the beginning of the system (which is not receiving treatment) and no settlement is observed in the biobox that is exposed to treatment, the treatment is working and there is reason to believe that the piping is staying free of settlement. To verify that a chemically based preventative treatment is eliminating adults that may have settled prior to the start of the treatment, seed the biobox with adult mussels (as described above). If all of the seeded adults are eliminated and no new settlement is found, the preventative treatment is serving dual purpose. If the adults are not eliminated, but no new settlement is found, adult macrofoulers may still be present in your piping either from prior settlement or as adult translocators that have gained access to the system.

Table 4.7 is an excel sheet that shows settlement of zebra mussels in five bio-boxes of a steel factory drawing water from Hamilton Harbour, Ontario, Canada. The plates used in the example here were 15.24 cm × 22.85 cm = 384 or 696.6 cm^2, including both sides. Hence, the numbers of juveniles counted are multiplied by 10,000 $cm^2/m^2/696.6\ cm^2 = 14.4$. Note that there are two bioboxes in Pumphouse #2, one is chlorinated and the other is not (i.e., a control). If the chlorine is effective, there should be no juveniles settling in the bioboxes receiving chlorinated water,

TABLE 4.7
Sample Data Sheet Showing Numbers of Zebra Mussels Settling on Coupons in Bioboxes Placed in Different Locations of an Industry

Week No.	Pump House #2, Unchlorinated			#1 Caster			Pump House #2, Chlorinated			Annealing Line			Boiler House		
& Date/05	Plate	No.	No. m^{-2}	Plate	No.	No. m^{-2}	Plate	No.	No. m^{-2}	Plate	No.	No. m^{-2}	Plate	No.	No. m^{-2}
1. June 23	2	0	0.0	2	0	0.0	2	0	0.0	2	0	0.0	2	0	0.0
2. June 28	3	0	0.0	3	0	0.0	3	0	0.0	3	0	0.0	3	0	0.0
3. July 5L	1	4	57.4	1	0	0.0	1	0	0.0	1	0	0.0	1	0	0.0
4. July 13	2	0	0.0	2	0	0.0	2	0	0.0	2	0	0.0	2	0	0.0
5. July 22	3	0	0.0	3	0	0.0	3	0	0.0	3	0	0.0	3	0	0.0
6. July 27	1	0	0.0	1	0	0.0	1	0	0.0	1	0	0.0	1	0	0.0
7. Aug 3	2	0	0.0	2	0	0.0	2	0	0.0	2	0	0.0	2	0	0.0
8. Aug 10	3	0	0.0	3	0	0.0	3	0	0.0	3	0	0.0	3	0	0.0
9. Aug 15	1	0	0.0	1	0	0.0	1	0	0.0	1	0	0.0	1	0	0.0
10. Aug 25	2	12	173	2	0	0.0	2	0	0.0	2	0	0.0	2	0	0.0
11. Aug 26	3	21	301	3	0	0.0	3	0	0.0	3	0	0.0	3	0	0.0
12. Sep 2	1	32	459	1	0	0.0	1	0	0.0	1	0	0.0	1	0	0.0
13. Sep 8	1	24	344	1	0	0.0	1	0	0.0	1	0	0.0	1	0	0.0
14. Sep 16	2	28	401	2	0	0.0	2	0	0.0	2	0	0.0	2	0	0.0
15. Sep 22	3	25	358	3	0	0.0	3	0	0.0	3	0	0.0	3	6	86
16. Oct 1	1	42	602	1	0	0.0	1	0	0.0	1	0	0.0	1	0	0.0
17. Oct 8	2	26	373	2	0	0.0	2	0	0.0	2	0	0.0	2	0	0.0
18. Oct 16	3	22	315	3	0	0.0	3	0	0.0	3	0	0.0	3	0	0.0
19. Oct 25	1	13	186	1	0	0.0	1	0	0.0	1	0	0.0	1	0	0.0
20. Nov 4	2	2	28	2	0	0.0	2	0	0.0	2	0	0.0	2	0	0.0
21. Nov 11	3	0	0	0	0	0.0	3	0	0.0	3	0	0.0	3	0	0.0

Numbers of zebra mussel juveniles found on plates in bioboxes at five locations in an industry. Each plate is 15.24 cm × 22.85 cm (= 6″ × 9″), with surface area of 384 or 696.6 cm^2 (= 0.0697 m^2) including both sides. Only larvae larger than 250 μm (i.e., settlement size) are counted.

which is the case in Table 4.7, except on September 22, 2005. The operators took immediate action and remedied the problem.

4.6 NAVIGATIONAL BUOYS (NAVAIDS) AS MONITORING PLATFORMS

David Bruce Conn

The use of navigational buoys and channel markers (navaids) as platforms for monitoring hard-bottom benthic organisms is among the most cost-effective ways of obtaining accurate and robust data, especially over large geographical areas and over long time periods. The cost effectiveness derives from the fact that the buoys are deployed and maintained by agencies that are responsible for marine transport, so there is no initial cost to the scientific investigating team. In the St. Lawrence Great Lakes Region, buoys are maintained by the Canadian Coast Guard and the United States Department of Transportation in their respective national waters. Comparable government agencies maintain the navigational aids in their waterways throughout the world, and no major navigable waterway lacks such devices. These agencies are generally willing to cooperate with legitimate scientific studies to make the buoys available to scientists, as long as it does not unreasonably detract from the agency's primary mission of maintaining safe and efficient shipping lanes.

Examples of buoys used for monitoring invasive dreissenid mussels and native benthic macroinvertebrates are shown in Figures 4.13 and 4.14. In our studies on the St. Lawrence River and lower Great Lakes, we use both quantitative and semiquantitative sampling methods. Figure 4.13 shows buoys retrieved from different sites in Lake Ontario and the St. Lawrence River, demonstrating how semiquantitative ranking can give general but clear distribution trends for biofouling dreissenid mussels. This method works well in this situation, in which buoys are removed from the water for the winter period each year when the seaway is closed to navigation. The buoys are withdrawn from the water during the coldest part of the year, so detailed quantitative sampling is difficult, especially when 200–300 buoys, dry-docked at several locations, must be sampled in the few days following removal from the water. However, ranking density of mussels provides good overall distributional data. In our work the ranks are 0 (no mussels), 1 (1–25% area coverage by mussels), 2 (25–50%), 3 (50–75%), or 4 (75–100%). The coverage percent is estimated by eye and recorded on a digital voice recorder that functions at extreme environmental temperatures encountered year round in the field.

Figure 4.14 shows use of buoys for quantitative sampling of populations of dreissenid mussels and other benthic organisms, using actual counts of organisms from digital photographs taken with a measured quadrat of known area. This method works well in situations where a smaller number of buoys are to be sampled, and in which the species to be quantified are regularly distributed on the substrate.

A major advantage of using navaids for biomonitoring is that the buoys are deployed at precise geographical points year after year. The buoy location must be exact and consistent each year to meet the demands of shipping navigators. Another

FIGURE 4.13 Navigational buoys retrieved and examined in January 2009 following 2008 deployment. From left to right, Buoy "P" is from Seaway mile 491.8, in Lake Ontario. Dreissenid mussels have attached at a density of Rank 2 (25–50% area coverage; dark patches on base column). Buoy "T8" is from Seaway mile 509.5, in Toronto Harbour, and dreissenids are at a density of Rank 4 (75–100% area coverage; dark areas on base column). Buoy "D91" is from Seaway mile 167.5, near Cornwall, Ontario, and has no mussels attached (Rank 0; dark areas on D91 are filamentous green algae).

FIGURE 4.14 **(See color insert following page 286.)** Fixed-area sampling quadrat used for quantification of dreissenid mussels and other benthic organisms attached to navigational buoys. This buoy was retrieved and examined in January 2008 following 2007 deployment in the St. Lawrence River near Cornwall, Ontario, Canada. Attached mussels can be counted, along with adhering byssal threads of mussels that have been recently dislodged.

advantage is that the water depth is known and consistent, and the substrate remains uniform from year to year and from location to location. Thus, these factors can be eliminated as variables in the experimental design.

There are some drawbacks to using navaids. The buoy itself only covers the top meter or so of the water column. However, when considered together with the chain and anchor that attach the buoy to the bottom, the entire water column is invariably covered by artificial hard substrate. By coupling use of the buoys with methods such as Scuba-based collections or retrieval of the entire chain and anchor, the samples can be taken at all depths.

In recent years, mussels and other aquatic filter feeders have been employed in biomonitoring for pathogens and chemical pollutants that are removed from the water and sequestered during feeding. By sampling mussels from navaids, the advantages of these techniques can be combined for collection of robust data of epidemiological value.

5 Impacts of Macrofoulers on Man-Made Structures

5.1 INTRODUCTION

When discussing macrofouling impacts, the macrofoulers can be grouped into two basic groups: those that attach to surfaces using byssal threads and those that do not.

The impacts caused by all dreissenids and mytillids are very similar. They have free living larval stages that will penetrate most strainers and filters and an adult stage that is firmly attached to the substrate.

Corbiculids and aquatic snails have no means of attachment but they are capable of locomotion. The presence of the corbiculid larval stage in the plankton is generally very brief compared to the dreissenids and mytilids. Some of the aquatic snails are capable of trapping an air bubble in their mantle cavity and thus are being carried with the water stream. For example, juvenile mud snails (*Hydrobia ulvae*) disperse by floating at the water surface in summer (Armonies and Hartke, 1995) and it is not uncommon for apple snails to float when they have sufficient air in their lung (Perera and Walls, 1996).

Most facilities that pump raw water have common components, albeit sometimes arranged in unique ways. It is beyond the scope of this publication to describe in detail the many different facilities that must deal with macrofouling. Rather, we feel it is more helpful to discuss the impact each macrofouler group has on generic components of man-made structures that are in contact with raw water. At the end of the chapter, we will discuss some of the unique challenges faced by some of the facilities.

5.2 MARINE STRUCTURES

These structures are primarily at risk from attached macrofoulers. Buoys and navigational markers, especially near shore, can become heavily infested. The concern is that such an infestation can partially sink these markers and affect shipping safety.

Bridges, docks, and access ladders are all subject to fouling below the waterline. In some geographic locations fouling will be visible right at the water/air interface. Elsewhere, either due to ice scouring or extreme summer temperatures, fouling may start deeper.

It is common to place floating surface barriers in the area approaching an industrial intake. These barriers are there to prevent large floating objects from reaching the intake or they may be there to prevent access by boaters. Sometimes the floats suspend nets to prevent fish from entering the facility. The floats and the net can become weighed down with mussels and no longer able to perform their duty.

5.3 WATER INTAKES

Intake systems provide the means by which water enters a facility. Intake systems encompass large static structures such as a concrete-lined forebays, intake pipes, intake towers, trash racks, and floating barriers. Some structures are active, such as gates, valves, and trash rack cleaning equipment. Once again, attached macrofoulers are the group of concern for all external and intake structures. Unattached macrofoulers have not been identified as a concern unless the intake flows only periodically. In intakes with long idle flow periods, Asian clams are able to build up large colonies and either maintain their position when the water is flowing or end up downstream on fixed screens or strainers if the flow lifts them up. An example of such intakes is the irrigation systems of golf courses, which periodically withdraw water from surface water bodies.

5.3.1 Deep Water Intakes

Large diameter intakes (Figure 5.1), which draw from depths greater than approximately 30 m (100 ft), will generally experience the least amount of fouling. This is due primarily to three factors. In most cases, depths greater than 30 m (100 ft) contain water that is permanently cooler than shallower depths. This area is described as being below the thermocline. The water entering the intake will be much cooler than that above the thermocline during the reproductive season. Low ambient temperatures translate into slower growth of mollusks. Secondly, food availability may be limited. Light at 30 m (100 ft) depth or greater will be very low in most cases, limiting the growth of green algae (the primary food source for most mollusks). The third moderating factor is that the number of viable larvae reaching the intake may be limited depending on the bottom profile of the lake or reservoir, and on the circulation patterns that exist. Furthermore, the ambient water temperature may be below the minimum temperature required by free swimming larvae to settle.

In some reservoirs in the southern United States, where water is withdrawn from the bottom of the reservoir, dissolved oxygen may plummet at certain times of the year, precluding survival of dreissenids in these intakes. However, the low oxygen levels may not be sufficient to discourage *Limnoperna*, as this species has much lower dissolved oxygen requirements than the dreissenids (see Chapter 3).

5.3.2 Mid-Water Intakes

Water intakes that are of relatively small diameter (60–180 cm; 2–6 ft), situated in the upper 30 m (100 ft) of water and which are very long (several hundred meters), are particularly vulnerable to macrofouling. The lake end of the intake usually terminates in a fixed grate designed to exclude large debris. Water flows through the grate and into the intake most often under gravity feed, and subsequently into the facility. The intake grate is likely to be the most visible point of fouling. Attached mussels can easily close gaps in the grate and decrease the water intake flow rate.

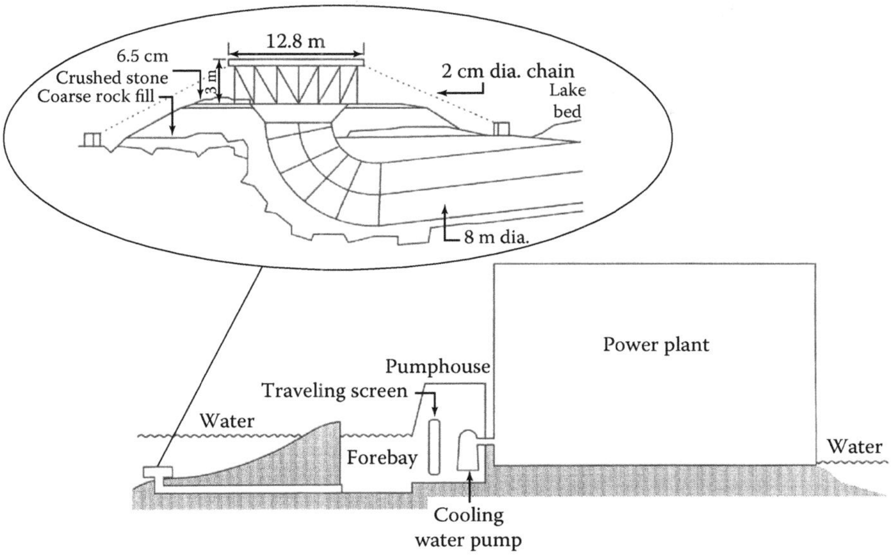

FIGURE 5.1 Submerged water intake of a power generating station.

In the intake itself, even before the mussels settling in the intake pipes reduce the amount of water the pipe can carry, they increase the friction and turbulence of flow by increasing the roughness of the pipe's surface. During the intake design, the material roughness is incorporated into required diameter calculations (using the Hazen and Williams formula) as the "*c*" factor (or coefficient of friction) to determine the size of pipe required to satisfy the projected flow rate. When the roughness increases due to mussel settlement, the flow rate drops and plants pumping water near their design capacity will have trouble meeting demand. Van Cott et al. (1993), while working on the Great Lakes, has shown that increased roughness and decreased capacity were recorded within four weeks from the onset of settlement, when zebra mussel densities rose from 0 to 6000 m^{-2}.

Drinking water plants, which prefer to draw water from as far offshore as possible (Figure 5.2), generally have very long intakes that need to be kept clear of fouling.

5.3.3 Shallow and Shore Water Intakes

Facilities that have very large, shallow, or onshore intakes generally do not have macrofouling problems until the raw water reaches the pumphouse. Before the entrance into the pumphouse, fixed trash racks are frequently used to trap large floating debris. Trash racks are often the first visible structure fouled when mussels arrive. At the Monroe, Michigan, power plant, trash bars that are set on 7.6 cm (3 in.) centers became badly fouled within one season. More than 75% of the straining surface was occluded. Anywhere below the first 2 m (6.5 ft), the mussel layer spanned the 7.6-cm (3-in. gap) between vertical slats and also extended as much as 15.2 cm (6 in.) past the downstream side of the rack (Kovalak, 1993). Such observations have

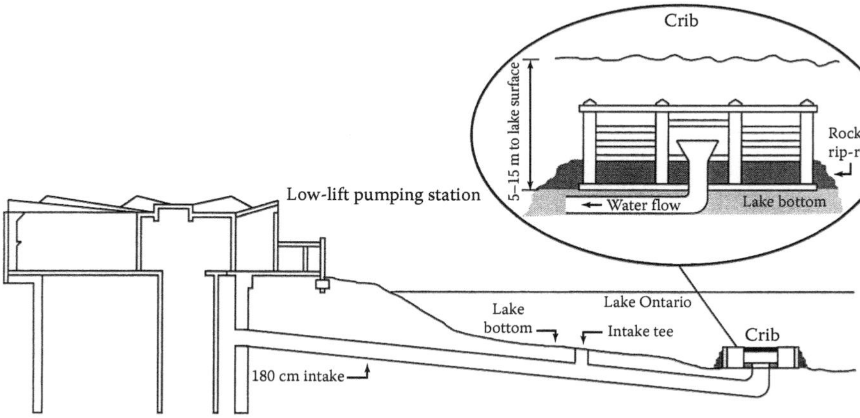

FIGURE 5.2 Typical drinking water plant intake on the Great Lakes.

since been reported from a number of different industries in a variety of geographic locations for both dreissenids and *Limnoperna*.

In the northern or cooler climes, shallow intakes or grates on surface intakes may develop a secondary problem of frazzle ice due to the presence of macrofoulers on the intake grate. At the Monroe, Michigan drinking water plant, zebra mussel populations in and around the inlet to the raw water intake contributed heavily to the frazzle ice formation, which in turn interrupted raw water flow for more than 56 h (Lepage, 1993). It is likely that the velocity changes and turbulent flows created by the zebra mussel population enhanced the transformation of supercooled water into frazzle ice.

5.3.4 Penstocks

Penstocks are the specialized intake structures found at many hydraulic power plants. They channel the incoming water through a vertical drop toward the turbines (Figure 5.3). Under full operation, the water flow in the penstock is generally too fast for mussels to settle; however, during short outages, settlement is possible if the penstock is not dewatered. Typically, one penstock will feed one turbine and therefore penstock flow can be stopped on a regular basis depending on the grid demand. Penstocks of several large hydraulic facilities have developed 2–3 in. of dreissenid fouling in one season due to their intermittent operation (Claudi, unpublished data).

During longer outages, especially in warm waters, the drop in dissolved oxygen in the water within the penstock may be sufficient to prevent settlement from taking place.

Infestation of the penstock by mussels will result in increased hydraulic roughness, which then translates into loss of power production. In cold climates, the penstock may be subject to frazzle ice formation as described in the previous section.

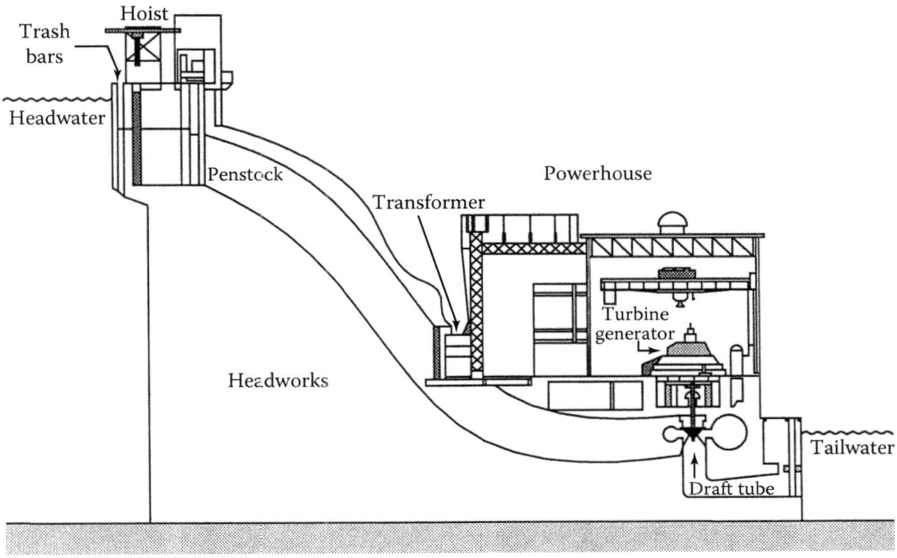

FIGURE 5.3 Cross-section of a typical hydroelectric generating facility.

If penstock vents are present, the wetted portions of the vent piping can be mussel settlement locations as there is little or no flow in the vent branch. A blocked or sufficiently reduced penstock vent could present a risk of pipe collapse during penstock draining.

5.4 PUMPHOUSE STRUCTURES

Once the water intake brings the raw water to the facility, it flows through a number of different structures and is pumped into various systems.

5.4.1 Trash Racks, Grates, and Screens

Most trash racks are at risk of mussel settlement (Figure 5.4). The flow velocity is generally low and permits easy attachment but is high enough to bring a continuous supply of nutrients. These areas are prime real estate for mussels. Past the trash bars of the intakes, there are frequently other screens designed to exclude fish and smaller floating debris. These are usually travelling screens that rotate at frequent intervals and the impinged debris is washed off by a screen wash system. These screens may have mesh sizes as fine as 1 mm (0.06 in.) and as coarse as 13 mm (0.5 in.). Although the screens themselves are not colonized due to their rotation in and out of water, the supporting structures are usually heavily infested. Severe colonization of supporting structures may interfere with the operation of the screen. Further, when the intakes, intake channels, or forebays have high densities of adult mussels, the mussels tend to form clusters. These clusters are subject to sloughing by current and gravity.

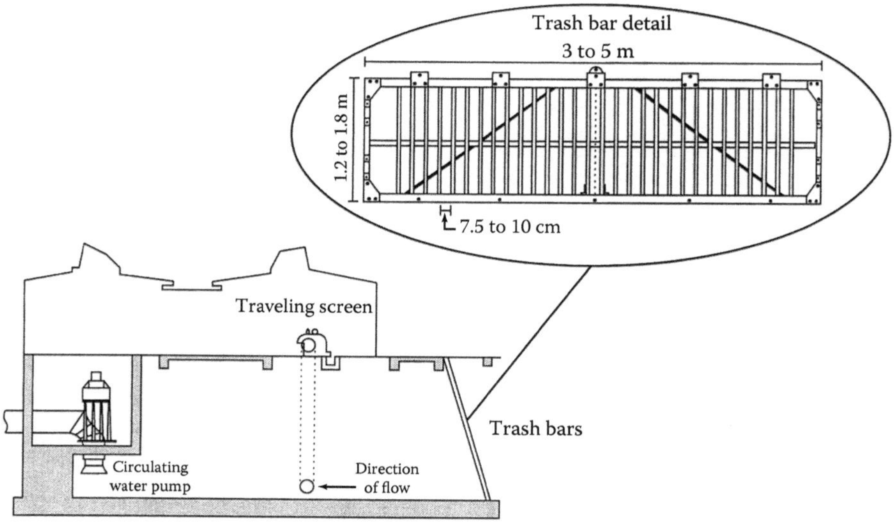

FIGURE 5.4 Possible pumphouse components.

They tend to roll along the bottom with the current and frequently impinge on the travelling screens. These clusters are usually removed from the travelling screen by the screen-wash system, but they may be carried over and fall from the travelling screen into the pump well that is being protected, thereby forming a readymade adult colony, providing shell debris to be ingested by the pumps.

Additional grates covering small raw water intakes or draft tube drains may be present in the system, and all are susceptible to fouling.

5.4.2 Gates and Stop Logs

Isolation gates are often suspended in the water above an opening. Fouling of the gate could interfere with its proper insertion. Many gate panels have structural stiffening members such as channel sections or I-beam sections oriented horizontally. These sections have drain holes drilled in the web portion of the section. The drain holes can be plugged by mussels, adding a significant amount of water that must be lifted when the gate is raised. If the gates are particularly heavily fouled, the motors used for lifting and lowering of the gates may not be up to the task.

Gates and stop logs all fit into vertical guide channels in the supporting side structures. The slots or channels are usually carbon steel C-channel shapes. Mussel fouling in the channel could make it difficult to properly seal the gate or stop logs.

5.4.3 Pump Wells

Within the pumphouse, the pump wells are generally made of concrete and subject to heavy colonization by dreissenids and mytilids. The walls and pump bells, which are immersed in water and through which water is withdrawn, can be the source of adult mussel clumps and individual shells for downstream systems, if those systems are not protected by strainers downstream of the pump discharge.

Fire protection pumps, frequently located in the pumphouse, are of particular concern. These pumps operate infrequently and shell or clump accumulation may build up near the pump intake during the period when the pumps are not operating. A large ingestion of individual shells or clumps can occur on pump start-up. Such an ingestion of debris could impair the performance of nozzles or sprinklers in an emergency situation.

The floor of the pump wells frequently collects lots of loose sediment and is a good area for colonization by Asian clams.

5.4.4 Pumps and Turbines

All pumps have air-cooled or water-cooled motors. If motors are air cooled, macrofoulers will have no impact on their operation. If the motors are water cooled, as is the case with many of the larger pumps, the motor windings, motor thrust and guide bearings, pump guide bearing, and pump shaft seal all may require cooling water. Most cooling water for such applications is passed through strainers.

Similarly, large generators have stator coolers and lube oil coolers that will require cooling water to maintain proper operating temperatures. The cooling water for these systems can carry settling sized mussels or shell debris into the cooler heat exchangers, if strainers are not present.

Strainers are effective at preventing adult macrofoulers or shells large enough to cause plugging problems from entering into the piping system. However, larvae and small juveniles will readily pass through the strainers and can settle in areas of the piping that are not made of copper or copper alloy.

The pump/turbine shaft seal water system, if not supplied with filtered water, may be exposed to small shell fragments and some increased wear on mechanical seal faces may be experienced. The seal cavity or stuffing box lantern ring cavity may become a settlement area for macrofoulers and will need to be inspected and cleaned if cavity temperatures rise above permitted levels.

Pumps and turbines usually have wear rings on the rim of the rotating parts. These wear rings are of a sturdy material. However, if there is a large amount of shell debris in the incoming water, it could increase the wear rate on the rings.

Seals on slurry pumps used for ash disposal failed on ten occasions at the Monroe plant during 1989. Water supply to the seals was blocked by shells of small mussels, causing the failure (Kovalak, 1993).

5.5 RAW WATER SYSTEMS

Once water is pumped into the facility, it will ultimately pass through one of the piping systems in the plant. The largest volume of water (up to 90% of the total raw water) drawn into industrial and power generating plants is used for cooling and heat transfer. The rest of the volume is used for plant processes other than cooling, such as makeup in steam systems; service systems such as cleaning; air conditioning, fire protection; and human consumption. As water is drawn into the various systems, so are the free swimming larvae during the reproductive season and small juveniles all year round. Although the free swimming stage of the corbiculids is much shorter than that of dreissenids and mytilids, if adult corbiculids are present in the pump wells, their larva can be drawn in and settle in the piping. Copper piping or alloys with high copper content are at much lower risk of fouling but can still become fouled if a biofilm builds up on the copper material.

Most of the systems of concern are composed of small diameter piping where the water flow is chronically less than 2 m/s (6.5 ft/s). Under such conditions, larvae may settle and start to grow, reducing laminar flow at the pipe surface. This creates ideal conditions for further settlement. The subsequent development of dense colonies in turn decreases the amount of water available to the plant system. In some instances, areas, that (on paper) have design flow rates high enough to preclude attachment, have been found to be full of juvenile mussels. This may happen during partial or short-term outages when the flows are reduced for a period of time. During the outage or reduced flow period mussels may attach. Once the byssal threads are firmly attached, the mussels will not be dislodged by flow rates that would typically not permit them to settle. Embedded, small diameter piping is of particular concern, as in most cases the piping is not easily accessible for cleaning.

Surge tanks are special cases of pipe with a large diameter. The wetted portions of the tank are at risk of mussel settlement. If the inlet to the tank becomes partially plugged, then the piping is at risk of water hammer. If the tank also serves as a vent, then increased risk of pipe collapse will exist unless the entrance to the tank is kept clear.

5.5.1 Strainers and Filters

Many raw water systems are protected from small debris by strainers placed at the discharge of the pumps. Generally these strainers are self-cleaning and will not be impacted by macrofoulers unless they do not operate continuously. Figure 5.5

FIGURE 5.5 **(See color insert following page 286.)** Heavily colonized strainer.

illustrates what happens when a self-cleaning strainer is left filled between intermittent bouts of operation.

Many systems have small, fixed, in-line strainers designed to protect critical areas. Adult macrofoulers (attached and unattached), which may be carried into the system or are detached from upstream locations, will continue to move downstream until an in-line strainer prevents further movement. At such locations, large aggregates of macrofoulers may form, even when the overall infestation is light. The strainers may limit flow to downstream portions of the system and cause unexpected system upsets. Fixed strainers are often plugged during end-of-season or periodic treatments. During these treatments, the strainers should be inspected on a daily basis once the macrofouling die-off begins.

When carrying out a field inspection, it is usually a good idea to inspect several strainers for the presence of shells.

5.5.2 Instrument Tubing and Instrumentation

Raw water systems generally have a number of different instruments associated with them. The instrumentation is there to monitor the performance of various components and relay the information to the control room. Any instrument using raw water could attract either type of macrofouler, which in turn could affect the readings obtained by the instrument (Figure 5.6). In lines that have no flow, such as pressure sensing lines, the settlement is likely to occur near the tap end of the line, where oxygen and nutrients are available.

Water sampling stations that have a regular, low flow of water through them for analytical purposes are at particular risk of fouling by dreissenids or mytillids.

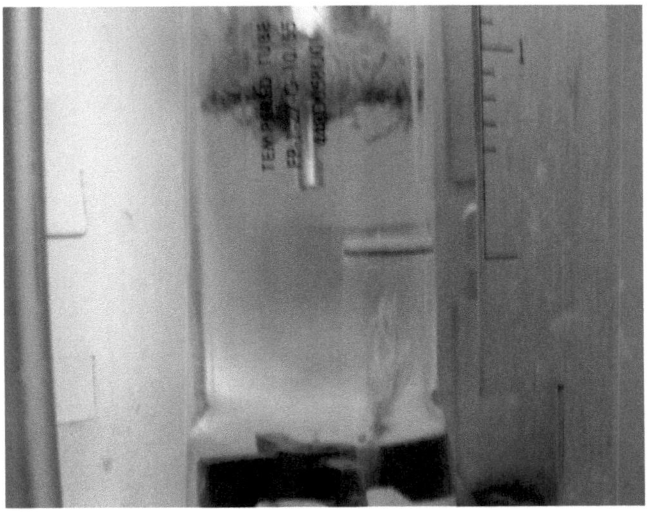

FIGURE 5.6 Fouled level gauge.

5.5.3 Heat Exchangers

The largest volume of water (up to 90% of the total raw water) drawn into industrial and power generating plants is for cooling and heat transfer. To cool various components and flow streams, heat exchangers are used. Heat exchangers are of two basic types: shell and tube heat exchangers and plate heat exchangers. The plate-type heat exchangers generally have smaller passages for water to pass through and therefore are more vulnerable to macrofouling. Several plants on the St. Lawrence River experienced severe plugging of such heat exchangers by aquatic snails. The heat exchangers had to be dismantled to remove the shells (Claudi, unpublished data).

Shell and tube heat exchangers can carry the cooling water either on the tube side or on the shell side. If cooling water is carried on the tube side, the inlet plenum is typically at risk of fouling by both types of macrofoulers. The inlet plenum of a cooler can be a catchment area for any shells that manage to find their way along the upstream piping, and shell material can gradually accumulate in the plenum, ultimately blocking tubes and causing poor performance of the coolers.

Shell and tube heat exchangers that carry cooling water on the shell side can accumulate shell debris and possibly primary growth by all types of macrofoulers in the body of the heat exchanger. Performance may deteriorate and mechanical removal of the debris is the only remedy once this occurs (Figure 5.7).

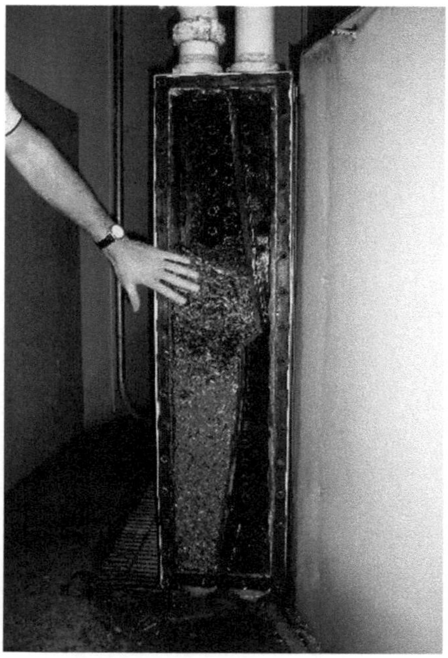

FIGURE 5.7 **(See color insert following page 286.)** Shell debris accumulated in heat exchanger plenum.

5.5.4 Condensers

The main condenser cooling systems of thermal power plants usually flow at rates greater than 2 m/s (6.5 ft/s). This generally prevents settling of free swimming larvae in most areas of the system. However, there is a second mechanism of fouling that comes into play in the main cooling systems. The travelling screens are generally housed in a pumphouse. Downstream of the screens are the pump bells of the circulating water pumps hanging into the pump well. The walls of the pump well are subject to similar or greater (more water passes through them) fouling as are all other external structures. Clusters, that may break off from the walls, are entrained by the pumps as these generally have no screening to exclude debris. In power plants, these clusters may be carried along to the main steam condenser, which is the first place the conduit narrows substantially. The face of the steam condenser is composed of a number of small diameter tubes that may become blocked by incoming clusters. The clusters, or even individual large shells, can block flow and allow other debris to accumulate behind them (Figure 5.8). This reduces heat exchanger efficiency and aggravates tube wall corrosion rates in low-flow areas. Both types of macrofoulers have been found blocking small diameter tubes.

Even partial blocking of a tube by a shell may be detrimental, as it will prevent the entry of the balls used for automatic tube cleaning.

5.5.5 Valves

The development of dense colonies of attached macrofoulers places proper operation of valves at risk, particularly valves that do not operate frequently. For example, at one facility a 10 cm butterfly valve failed to close because the disc was completely covered by mussels (Claudi, unpublished data).

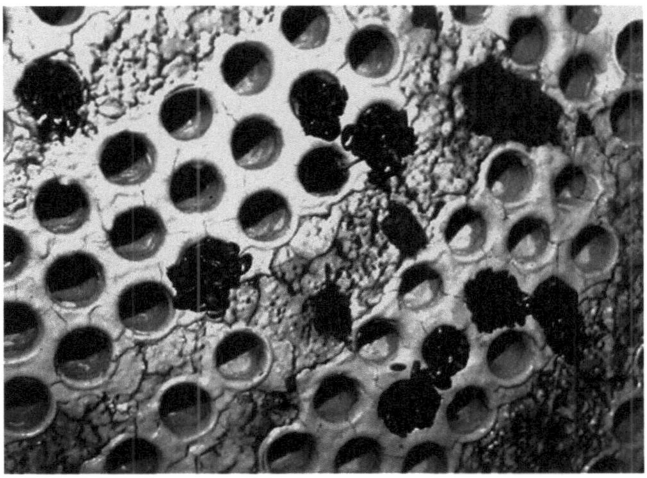

FIGURE 5.8 Heat exchanger tubes plugged by clumps of mussels.

Some hydraulic facilities use a Johnson valve to regulate the flow of penstock water to the turbine. Embedded small diameter piping, associated with the operation of this large valve, is at great risk of fouling. In most cases, these systems are not easily accessible for maintenance or repair on line. If a shut down is necessary to affect repairs, the lost opportunity cost to the facility can be large.

Valves are often placed in series pairs. The interchamber cavity may have a vent line and pressure balancing line. Either or both of these could be disabled by plugging with shells.

5.6 PIPING AND COMPONENTS OF MATERIALS OF CONSTRUCTION

Mussels will attach to almost any material of construction. Typical materials such as concrete, carbon steel, and stainless steel are all good settlement substrates for mussels. Less common materials such as aluminum, wood, and plastics will also serve as settlement substrates, but generally the densities and the strength of attachment may be less.

Copper, copper alloys, and tin are toxic to mussels. Copper pipe and brass components are generally at much smaller risk of fouling than other materials. Mussels can occasionally be found to be covering a copper material. This typically occurs in areas of low flow where a protective biofilm has accumulated on the substrate. In such situations, insufficient copper ions are present at the surface to discourage mussel attachment. Mussels attached to copper-based material coated with a biofilm are typically easy to remove by light scraping or washing.

When mussels are removed from steel and aluminum, the byssal threads are often left behind. The byssal thread/metal interface can serve as a site for corrosion pitting.

5.7 FIRE PROTECTION SYSTEMS

Safety systems, such as fire protection, frequently use treated city water. In such cases they are safe from macrofoulers. If, however, they draw water directly from a raw water source, they are as vulnerable as other service water systems. Fire protection systems are designed to be filled with water and then maintained in a static, ready-to-go state. In real life, the hose stations are often a convenient source of water for washing equipment or flushing debris from spills on floors. This means that a supply of fresh raw water is needed to replace the volume of water used. This makeup stream is the main point of entry for macrofoulers. It also provides food and oxygen necessary for their survival. Macrofoulers may enter these systems both as larvae and as adult translocators if there are no strainers present at the discharge of the pump. In addition to makeup water, which is drawn in to compensate for leaks, fire systems tend to be tested on a regular basis, sometimes monthly. During these tests, raw water is frequently drawn through the entire piping system. Such tests, if conducted during the breeding season, may introduce larvae or adult macrofoulers into the fire protection system. Adult macrofoulers may be eliminated by up-front strainers and larvae may fail to survive if the system is truly static and experiences rapid decay of oxygen.

Sometimes, portions of fire suppression systems such as deluge systems are kept in a partially drained state, which protects that portion of the piping system from growth in place.

5.8 DRAINAGE AND SUMPS

Drain piping and sumps are often overlooked as being at risk from mussels. However, they can easily become colonized by macrofoulers. Typically, the drain piping and sumps do not threaten the operability of a facility but they can be an operational nuisance if infested. Open drains and sumps can often be good places to observe for the presence of macrofoulers.

Both types of macrofoulers will settle in these structures. Dreissenids and mytilids can settle on structures such as sump floats used for water level detection and cause unreliable level readings. Mussels will also settle on the external portions of submerged sump pump casings and on the walls of the sump. The floor of the sump can be colonized by both types of foulers; large colonies of Asian clams have been found in such locations. The water in sumps can very often be several degrees warmer than the raw water. Mussels may grow more rapidly in the warmer sump water than in other areas. Conversely, when the ambient raw water is very warm, the slightly elevated temperature of the sump may be warm enough to prevent mussel growth.

5.9 HEATING, VENTILATING, AND AIR-CONDITIONING SYSTEMS

Heating, ventilating, and air-conditioning systems (HVAC) employ a variety of technologies. Some systems use air as the medium to exchange heat and will not be at risk of mussel problems. Other systems use raw water to chill a coolant and in very dry climates the raw water may be used in an evaporative cooler.

HVAC systems are normally associated with personnel comfort and therefore impairment of the system is not an operational threat. However, sometimes areas of a facility contain rooms of essential control equipment, usually electrical in nature, that generate heat and require cooling. These areas will need to be examined to make sure that any raw water used is protected from impairment by macrofoulers.

5.10 COMPRESSED AIR

Compressed air can be an essential operational system in a facility to operate valve actuators and maintain water level suppression for rapid-start turbines. Compressed air systems commonly rely on cooling water for intercoolers and after-coolers. More rarely, the compressor motors may have water-cooled bearings or even water-cooled motors, making these susceptible to macrofouling.

5.11 FACILITY-SPECIFIC IMPACTS

As we have mentioned earlier, all facilities that draw in raw water will experience some level of macrofouling if the macrofoulers are present and no control strategy is used.

5.11.1 Dams, Reservoirs, and Aqueducts

There are many structures in North America which deliver water, provide inland waterways, regulate water levels, control flood flows, or provide water to industrial users. When evaluating the potential impact a macrofouler may have in such a system, you need to consider all the elements associated with dams, impoundments, holding ponds, recharge ponds, siphons, dikes, turnouts, and canals.

All dams incorporate dam wall drain tubes. The drains will collect dam water that manages to seep past the dam seals and percolate through the dam wall. Dreissenid, mytilids, and corbiculids may be able to travel with the normal dam seepage into the drain tubes where they could settle and grow. The occurrence of such infestation is likely to be rare but has been documented at some dam facilities (Claudi, unpublished data).

Dams incorporate foundation uplift drain pipes. Water in these pipes is generally expected to be ground water seepage and is not likely to transport macrofoulers.

Dam structures usually have a rigorous inspection program. In the unlikely event that sufficient mussels should accumulate to restrict the drain flow, the reduced drainage should be picked up during the frequent routine inspections. Sometimes the seepage through the dam drains to a sump. The sumps could become an ideal inspection area to check for the presence of macrofoulers.

All dams have outlet works. The outlet works will have an upstream (dam pressure) side and a downstream side. Macrofouling impact on the trashracks and upstream tunnel need to be considered, especially as it is usually very difficult to drain these areas. The downstream tunnel may also be at risk of macrofouling but usually this area has greater ease of access for cleaning. Impacts on pressure gates need to be considered as well as any interchamber piping and vent lines.

Structures that rely on movable gates to direct or moderate water flow need to consider impact of macrofoulers. Attached mussels can colonize the outside and inside of submersed gates, thereby increasing their weight. If the design parameters of the lifting equipment cannot cope with much additional weight, the gates could become inoperable. Improper sealing is also a concern if the seals around the doors are colonized. Small drains associated with submersible gates can become clogged as can weep holes in tunnels.

Reservoirs often have large annual level fluctuations. The level changes can sometimes be used to assist in controlling the amount of mussel infestation through desiccation during the normal level changes. Floating structures within a reservoir need to be identified as they will change level with the reservoir and attached mussels can inoculate the reservoir when the levels increase.

Aqueducts may have concrete-lined sides that can provide an ideal settling substrate. These water conveyances have many associated structures such as moving gates, pumping plants, agricultural turnouts protected by fixed screens, and instrumentation in direct contact with the raw water. All these components must be considered.

Aqueducts often pass under roadways, rail lines, or other man-made and natural obstacles. The water moves through an inverted siphon then re-emerges to continue on its path. Mussel attachment to the surface of the siphon wall will increase hydraulic friction and reduce the capacity of the siphon. Siphons are difficult and costly to drain for cleaning, so mussel fouling can be an expensive problem on these structures.

Gauging stations and float wells are used to track the water levels in the various water bodies and the movement of water between the water bodies via aqueducts. The gauging stations usually include a tap line, stilling well, and float. The inlet pipe could become colonized with attached mussels and could impair the accuracy of the gauging station where there are frequent or sudden changes in flow and levels. For locations where flow rates and water levels change very slowly, there may not be a reduction in accuracy as a plugged pipe would still allow water to percolate to the stilling well. The floats could have sufficient mussels attached such that the level readings become unreliable.

5.11.2 Drinking Water Plants

Depending on actual configuration, the greatest macrofouling problem most drinking water plants have is with attached macrofoulers in their intake lines as described in Section 5.1.2.

Once the water is drawn into the facility, various chemicals are added to the water in preparation for making potable water. Both flocculants and acids, if used, are capable of eliminating larval-stage bivalves. Filtration beds, if present, will eliminate all macrofoulers. The exception is the raw water sampling system used by many drinking water plants to monitor the quality of the incoming raw water. By definition, no chemicals are added to the water and the system is usually of small diameter, low flow, and is therefore prone to fouling. A weekly flush with hot water may be the best available strategy.

Drinking water plants that use the technology of direct filtration may suffer from all types of macrofoulers in their filtration chambers. Invasion of snails into this area has caused a great deal of grief to at least one drinking water facility (Claudi, unpublished data).

5.11.3 Fish Diversion Structures

Many facilities draw water from water bodies that contain plentiful numbers of desirable fish species. A means of protecting the fish from injury or mortality is a common regulatory requirement.

Barrier structures such as nets suspended from floating booms can become coated with mussels causing the nets to submerge and fish to pass over top of the net.

Submerged intakes designed to have very low flow (to prevent fish entrainment) by nature have very large surface areas. The low speed of flow permits easy mussel attachment and the flow provides a continuous supply of fresh food for the mussels.

FIGURE 5.9 Portion of a fish diversion capture structure.

This creates an ideal mussel habitat which leads to rapid colonization and loss of flow.

Structures to divert fish around a facility or to capture and transport desirable species can be very elaborate and complex. Fish diversion and capture systems are themselves facilities with different structures that allow the fish to be managed (Figure 5.9).

Diversion facilities will typically include trash racks, screens, and turbulence louvers (to guide fish into bypass pipes) and holding tanks if the fish are to be transported.

All these structures are at risk of mussel settlement. Control methods chosen to mitigate the effect of the mussels must be benign to the fish.

6 Ecological, Socio-Economic, and Beneficial Impacts of Aquatic Invasive Mollusks

6.1 INTRODUCTION

This chapter describes only the ecological, socio-economic, and beneficial impacts for the aquatic invasive mollusks described in Chapters 1 through 3. Sometimes the impacts and benefits are similar among species within families, and descriptions are provided at the family level.

Impacts to utilities and industries, other than agriculture, are described in Chapter 5. Agricultural impacts are described here because most are ecologically related. Socio-economic impacts include health-related impacts, and vectors of parasites and diseases of humans and other animals like birds and fish. Beneficial impacts can be direct, like predation of snails that are intermediate hosts of trematode parasites, or indirect, like improving water clarity while respiring and filter feeding.

6.2 DREISSENIDAE

6.2.1 Ecological Impacts

6.2.1.1 Impacts on Native Mussels

6.2.1.1.1 *Zebra Mussel,* Dreissena polymorpha *and Quagga Mussel,* Dreissena bugensis

One of the earliest impacts of zebra mussels was on native species of unionid clams in the Great Lakes (Nalepa and Gauvin, 1988; Mackie, 1991b; Nalepa et al., 1991, 1996; Gillis and Mackie, 1992a,b; Schloesser and Nalepa, 1994; Schloesser et al., 1996, 2006; Metcalfe-Smith et al., 1998, 1999, 2000, 2003, 2004). Zebra mussels are not selective and have colonized all species of unionids. The rare species were eliminated first, the most common were eliminated last, or at the very least, were significantly reduced in numbers. Losses of native species of unionids have also been reported for other water bodies in North America, including the Illinois, Ohio, Mississippi, and Missouri watersheds (Baker and Hornbach, 1997; Richter et al., 1997; Ricciardi et al., 1998), Hudson River (Strayer and Smith, 1996), and interconnecting rivers of the Laurentian Great Lakes (Schloesser et al., 2006).

Zebra mussels have contributed to the losses of populations of many clam species in the Great Lakes and to their listing in Canada as either endangered, including the Rayed Bean, *Villosa fabalis* (West et al., 2000); Snuffbox, *Epioblasma triquetra* (Watson et al., 2000a); Mudpuppy mussel, *Simpsonaias ambigua* (Watson et al., 2000b); Kidneyshell, *Ptychobranchus fasciolaris* (COSEWIC, 2003a); Round Hickorynut, *Obovaria subrotunda* (COSEWIC, 2003b); Round Pigtoe, *Pleurobema sintoxia* (COSEWIC, 2004); Rainbow mussel, *Villosa iris* (COSEWIC, 2006a), or threatened, such as the Mapleleaf, *Quadrula quadrula* (COSEWIC, 2006b). Ricciardi et al. (1998) reported that in North American lakes and rivers that support densities >3000 m^{-2} of *Dreissena polymorpha*, native mussel populations are extirpated within four to eight years following invasion. They compared species losses at various sites before and after invasion and found that *Dreissena polymorpha* has accelerated regional extinction rates of North American freshwater mussels by 10-fold. Ricciardi et al. (1998) predicted that if the current trend persists, the regional extinction rate for Mississippi basin species will be 12% per decade and will result in extinction of over 60 endemic mussels in the Mississippi River basin.

The density of zebra mussels required to cause mortality of native unionids depends on the sizes of the attached mussels. Schloesser et al. (1996) suggested that unionids die if they are infested with more than 100–200 zebra mussels. Based on the lowest mean infestation at which an impact has been documented, Ricciardi et al. (1997, 1998) suggested that mortality may occur at mean infestations as low as 10 zebra mussels/clam if the zebra mussels are large, but 100 individuals/clams were lethal to most native clams. Their analyses suggested that mass mortality will occur if the mean infestation "load" (the ratio of the wet mass of zebra mussels to the wet mass of the unionid on which they live) approaches 1. Strayer and Smith (1996) found low densities of zebra mussels (<5 per clam) and only 30% of the unionid populations were infested. Nevertheless, after the arrival of the zebra mussel, densities of unionid clams fell by 56%, recruitment of young-of-the-year unionids fell by 90%, and condition of unionids fell by 20–50%. These changes occurred in all three of the common unionid species in the river (*Elliptio complanata*, *Anodonta implicate*, and *Leptodea ochracea*).

While the forgoing implies total destruction of unionids populations in lakes, McGoldrick et al. (2009) found that some parts of lakes can be a refuge for unionids. In Lake St. Clair, where the first zebra mussels in North America were found (Hebert et al., 1988), infestations on unionids are being mitigated by dominant offshore currents, which limit densities of zebra mussel veligers in nearshore compared to offshore waters (13,600 vs. 28,000 m^{-3}, respectively). Glycogen concentrations in the tissues of a common and widespread species in the delta (*Lampsilis siliquoidea*) suggest that zebra mussels may be adversely affecting physiological condition of unionids in a portion of the Lake St. Clair delta. Physiological condition and community structure of unionids within the delta may also be influenced by differences in food quantity and quality resulting from the uneven distribution of water flowing from the St. Clair River. The delta likely supports the largest living unionid community in the lower Great Lakes and includes several species that have been listed as endangered or threatened in Canada and/or the state of Michigan, making it an important refuge for the conservation of native unionids (McGoldrick et al., 2009).

The mechanisms for mortality of the unionids by zebra mussels are varied. Mackie (1991) reported the following mechanisms for zebra mussel kills of clams in Lake St. Clair: (i) infestations were so great on the posterior end that clams could not fully open their valves and impaired normal metabolic functions for feeding, growth, reproduction, respiration, and excretion; (ii) infestations were invasive and clams could not close their valves, and lost their ability to protect themselves from predators, parasites, disease, and noxious water quality; (iii) zebra mussels engulfed the posterior end and had first access to food and oxygen, essentially "robbing" the clams of food and oxygen; (iv) weights of mussels were too great for clams to upright themselves and impaired their balance and equilibrium during locomotion. Indeed, Hebert et al. (1991), Schloesser and Kovalak (1991), and Nalepa (1994) showed that the weight of infesting zebra mussels may exceed host unionid weight by a factor of four, or as high as 8.5; (v) clams were not able to rebury themselves because of the weight of the mussels when on the surface of the substrate; (vi) infestations of the mantle margin of the shell caused shell deformities that ultimately result in the death of the clam; (viii) complete occlusion of the gape at the posterior end and of siphons which probably caused death by smothering. Strayer and Smith (1996) suggested that unionids in the Hudson River may be affected by zebra mussels by competition for food as well as by fouling. Baker and Hornbach (1997) used laboratory studies to examine the physiological effects of zebra mussel infestation on *Actinonaias ligamentina* and *Amblema plicata* and showed that (i) zebra mussel infestation caused stress and symptoms of starvation in unionid mussels, (ii) the two species were affected unequally, and (iii) symptoms of starvation are greater when initial condition is low. Hebert et al. (1991) and Haag et al. (1993) showed lipid reserves of infested unionids were 50% lower than that in noninfested unionids. Similarly, Haag et al. (1993) reported lower glycogen and cellulase activity of infested unionids than noninfested unionids. Finally, Tucker (1994) observed infested unionids lying on their sides in windrows along the shores in Pool 26 of the Mississippi River and were thus stranded and exposed to fluctuating water levels. Tucker (1994) described the substrate as "a pavement"-like bottom covered by zebra mussels and prevented unionids from burrowing into sediments once they were dislodged.

6.2.1.2 Impacts on Plant and Animal Benthos

Zebra mussels also have altered the structure and functioning of benthic animal and plant communities. The increased water clarity resulting from zebra mussel feeding has caused increased biomass and spread of benthic algae and macrophytes (Griffiths, 1993; Lowe and Pillsbury, 1995; Skubinna et al., 1995; Stuckey and Moore, 1995; Caraco et al., 2006). The changes in the plant community were accompanied by changes in local zoobenthic communities (Strayer et al., 1998). Studies by Stewart and Haynes (1994), Botts et al. (1996), Howell et al. (1996), Dermott and Kerec (1997), and Ricciardi et al. (1997) have shown that zebra mussels dramatically alter the structure and density of the macrobenthos in and around zebra mussel beds themselves, presumably by locally changing both sedimentation rates and the physical structure of the habitat. Some taxa are affected more than others. Strayer and Smith (1996) and Strayer et al. (1998) showed that after the arrival of the zebra mussel in the Hudson River, densities of native bivalves declined sharply, 67% in the case of sphaeriids and 56% in the case of unionids. In addition, recruitment and body condition [body mass at a given shell length (SL)] of unionid bivalves dropped sharply between 1991, 1992, 1993, and 1995. They argued that declines in native bivalves were probably caused by competition for food, as zebra mussels caused a decline of 81% in phytoplankton biomass (Caraco et al., 1997), as well as benthic algae (Caraco et al., 2006).

Idrisi et al. (2001) analyzed a data series on zooplankton and young-of-the-year fish from Oneida Lake, New York, and found that although algal standing crop declined, *Daphnia* spp. biomass and production did not, and dominance shifted from *Daphnia galeata mendotae* to *Daphnia pulicaria*. They found no evidence that zebra mussels had a negative impact on young yellow perch (*Perca flavescens*) growth, biomass, or production. Idrisi et al. (2001) concluded that, despite the order of magnitude increase in grazing rates and associated decrease in algal biomass, pelagic production at primary, secondary, and tertiary levels did not decline in association with zebra mussels.

6.2.1.3 Impacts on Water Quality

Large populations of zebra mussels have an enormous capacity to remove seston and clarify water. The filtration and clearance of seston has generally reduced levels of phytoplankton, which results in reduced levels of chlorophyll *a* and total phosphorous, and a concomitant increase in Secchi depth (MacIsaac et al., 1992; Leach, 1993; Lavrentyev et al., 1995; Madenjian, 1995; Mellina et al., 1995; Roditi et al., 1996; Horgan and Mills, 1997; Karatayev et al., 1997; Baker et al., 1998). The suspended materials and nutrients are biodeposited on the bottom as pseudofeces. One of the more dramatic changes in water clarity occurred in Lake Erie. Within three years of invasion, Leach (1993) reported that chlorophyll *a* levels declined by 43% in the western basin and 27% in the west-central basin; mean Secchi depths increased by 85% in the western basin and 52% in the west-central basin; however, there was no observed effect on walleye spawning. The accumulation of pseudofeces on the bottom of Lake Erie has altered habitat structure and resulted in a change of the benthic community, with increased abundances of amphipods, flatworms, and worms (Griffiths, 1993).

Idrisi et al. (2001) analyzed a data series on nutrients and phytoplankton from Oneida Lake, New York, to test several hypotheses relating the response of the pelagic food web to grazing by zebra mussels. They found that the most dramatic change was increased water clarity and overall decrease in algal biovolume and chlorophyll *a*. Unlike other studies, primary production did not decline. They attributed the lack of decline in primary productivity to the compensating effect of increased water clarity resulting in deeper penetration of light. They also did not observe any change in total or dissolved phosphorus concentrations.

A recent study by Barbiero et al. (2006), however, has revealed a new cause of increased water clarity, still caused by dreissenids; they showed that since the dreissenid (first zebra, then quagga mussels) invasion of Lake Erie, calcium concentrations in the downstream offshore waters of Lake Ontario have decreased by approximately 4–5 mg/L. This decline has coincided with a threefold reduction in August turbidity values and nearly a doubling of Secchi depths, presumably due to reduced summer calcite precipitation events (whitening) in the lake. Barbiero et al. (2006) reported declines in turbidity from 1.1 to 1.6 NTU between 1985 and 1991 down to 0.3–0.4 NTU between 1996 and 2006.

6.2.1.4 Impacts to Fisheries and Wildlife

Fouling of fishing gear such as gill nets, trap nets, and pound nets that are left in the water for extended periods of time were commonly reported in Lake Erie; accumulations of mussels are often so large that nets are rendered useless and are difficult to retrieve (Claudi and Mackie, 1994). Some *Dreissena* have elevated levels of organic and inorganic contaminants that are passed up the food chain and biomagnified, resulting in reproductive problems, such as reduced clutch sizes and high embryo mortality in ducks. Zebra mussels and gobies are implicated in the life cycle of Type E Botulism (*Clostridium botulinum*). Major outbreaks of Type E Botulism have occurred in fish-eating birds on the Great Lakes and have involved loons, mergansers, grebes, scaup, and gulls in particular. An outbreak of Type E Botulism among common loons (*Gavia immer*) occurred in Michigan's upper peninsula in 1988. In the fall of 1999, there were major outbreaks when at least 700 common loons died on Lake Huron and several thousand birds, mostly red-breasted mergansers, died on Lake Erie. Large numbers of horned grebes, various diving ducks, and three species of gull also died in these incidents. It is believed that round gobies infected with the virus are being consumed by the birds. Gaygusuz et al. (2007) showed that gobiid fishes are an important consumer of zebra mussels and their predation would affect the mussel populations, especially in ecosystems where the zebra mussel is abundant. Zebra mussels may be filtering out spores of *Clostridium botulinum*; the spores are a resting or dormant stage that is resistant to changes in physical conditions and allows persistence in the environment. The vegetative form grows and replicates and produces the toxin that kills birds and fish. In order for the vegetative form to grow, it requires a nutrient-rich substrate, a complete lack of oxygen, and an optimum temperature. These spores, themselves, are harmless and can be filtered out by zebra mussels without killing them. It is not until the bacterium begins to grow and produce toxin that disease can occur. The death of fish, mudpuppies, or birds that have spores in their bodies will allow the production of toxin during decomposition of their carcasses.

6.2.2 Impacts on Industries and Utilities

Impacts of zebra mussels to industries and utilities were discussed in Chapter 5. To quickly summarize, the main impacts of accumulations are (i) reductions in the bore of pipes; (ii) reduced flow through the pipe due to friction loss (turbulent flow instead of laminar); (iii) electro-corrosion of steel or cast iron pipes; (iv) deposition of empty mussel shells at the pipe outlet; (v) tainting and possible contamination of the water upon death (especially when killed as part of a massive control program). Underground irrigation systems of golf courses that draw water from the Great Lakes have reported plugged irrigation systems (Claudi and Mackie, 1994). Bobat et al. (2004) describes clogging problems in irrigation systems in Euphrates Basin in Turkey.

6.2.3 Impacts to Navigational and Vessel Impacts

Encrusting the hulls of boating and sailing vessels is a common sight in the Great Lakes and smaller, more inland lakes. Navigational aids, such as fishing buoys and markers, have such heavy infestations that they sink below the surface. Hulls of boats and ships are so laden with zebra mussels that their sailing efficiency is impaired (Claudi and Mackie, 1994). Cooling systems of vessels may also be impacted.

6.2.4 Private Property Impacts

Formation of shoals of shell debris on beaches detracts from the beach's recreational and aesthetic value. It is not an uncommon sight to see signs posted warning bathers to wear sandals or running shoes before walking on the beach or swimming. Fouling of cottage plumbing and intake structures is common and has forced cottagers to develop annual opening and closing strategies. Mackie and Taylor (2007) recommend maintenance and protection of three main components: the intake structure (e.g., foot valve), the delivery system (piping and plumbing), and the cottage itself.

6.2.5 Socio-Economic Impacts

In 1991 economists predicted that the total economic costs to users of Great Lakes water and the ecological costs caused by the zebra mussel would amount to about $5 billion over a 10-year period. O'Neill (1997) showed that the total zebra mussel economic impacts on water users in the United States and Canada (Ontario and Quebec only) from 1989 to 1995 were approximately $113,000,000 just to combat the zebra mussel. These costs include research, training monitoring, mechanical repairs, planning, prevention, chemical and no-chemical mitigation, retrofitting and lost revenue due to shutdowns. However, the total annual zebra mussel-related costs have risen from an average of $8 million in 1991 to $17.5 million in 1995, and these costs were predicted to continue to increase as the zebra mussel extends its range into the United States. Pimentel (2002) cites an article that predicted annual costs of $5 billion for damage and associated control costs of zebra and quagga mussels by the year 2000. Pimentel et al. (2005) reported $1 billion/yr for damage and associated control costs of zebra and quagga mussels as of 2005. Lovell et al. (2005) review the literature on economic costs associated with zebra mussel damage and control, mostlyz in the Great Lakes; using the literature citations and extrapolating to all facilities in the Great Lakes area, total monitoring and control costs were $120 million from 1989 to 1994, and averaged $30 million annually from 1992 to 1996.

6.2.6 Impacts on Submerged Archaeological Sites

Nancy Binnie, Jonathan Moore, and Darren Keyes

With the introduction of zebra and quagga mussels to North America, thousands of underwater archaeological sites have been adversely affected by mussel biofouling. The effects are multifaceted and include biological, chemical, and physical changes that alter the rates of decay and deterioration in the submerged environment. Some sites such as historic shipwrecks have been underwater for hundreds of years and their construction materials, typically wood and iron, were already highly altered and weakened before mussel colonization. Over the last 15 years there have been efforts to understand the negative effects that mussel colonization and environmental changes have imposed upon underwater heritage sites, most notably historic shipwrecks (Cohn et al., 1996; Binnie et al., 2000; Watzin et al., 2000, 2001).

Mussel colonization immediately complicates archaeological investigation as both discontinuous and continuous layers of shells obscure structural features and artifacts from view, thus impeding survey and recording, inventory, condition assessment and monitoring activities. Indiscriminate and repeated mussel removal to allow direct observation causes damage to the fragile and already degraded surfaces. Industrial mitigation methods developed for robust modern infrastructure are inappropriate for underwater archaeological sites or for open-water use where nontarget biological organisms will also be affected.

Nondestructive mussel removal for the purposes of one-time archaeological recording can be carried out by divers, but is time consuming, and is not always possible due to constraints such as water depth; such mussel removal is only temporary as surfaces will quickly become resettled during the next spawning season or populated through translocation of mussels. The searching for and identification of small-scale artifacts lying on lake and river beds alike can be nearly impossible in places where mussel occupation is particularly dense. In these cases, mussel-covered artifacts are indistinguishable from natural features such as rocks, tree branches, and other materials that are similarly covered. One positive result from the mussel occupation of archaeological sites, many of which are sport diving attractions, is the accompanying dramatic improvement in underwater visibility that has ensued since the late 1980s. This in turn has some archaeological benefit in that general site photographic and video recording can often capture more panoramic views, versus the limited views that could be achieved before the arrival of the filter-feeding mussels.

Recent site condition inspections of the heritage shipwrecks USS *Hamilton* and USS *Scourge*, sunk in 1813 in western Lake Ontario near Port Dalhousie, Ontario, (Figure 6.1) provide a vivid example of how impacts to even deepwater sites of archaeological significance can be extensive (Figures 6.2 and 6.3). Both exterior and interior surfaces of almost all material types are now colonized by mussels, including wood (e.g., hull, rigging, deck equipment), metal (e.g., fasteners, ordnance, anchors, artifacts), and even delicate artifacts such as rope, leather, and human remains (Binnie, 2009; Keyes and Moore, 2009; Moore and Keyes, 2009).

The environment at the 90 m depth of the USS *Hamilton* and USS *Scourge* is a flat, relatively featureless bottom with low current (4–10 cm/s). It has low water clarity, temperatures generally ranging from 2°C to 4°C with occasional incursions of warmer water during summer months, always aerobic and oxidizing conditions, fine surface sediment that is easily disturbed, and biological life composed mostly of amphipods, sculpins, bryozoans, zebra, and quagga mussels. Images obtained during expeditions in 1980, 1982, and 1990 using a submersible and remotely operated vehicles showed the wrecks before mussel coverage. As of 2008 almost all surfaces suffer from mussel fouling with variable coverage. With the blanketing effect

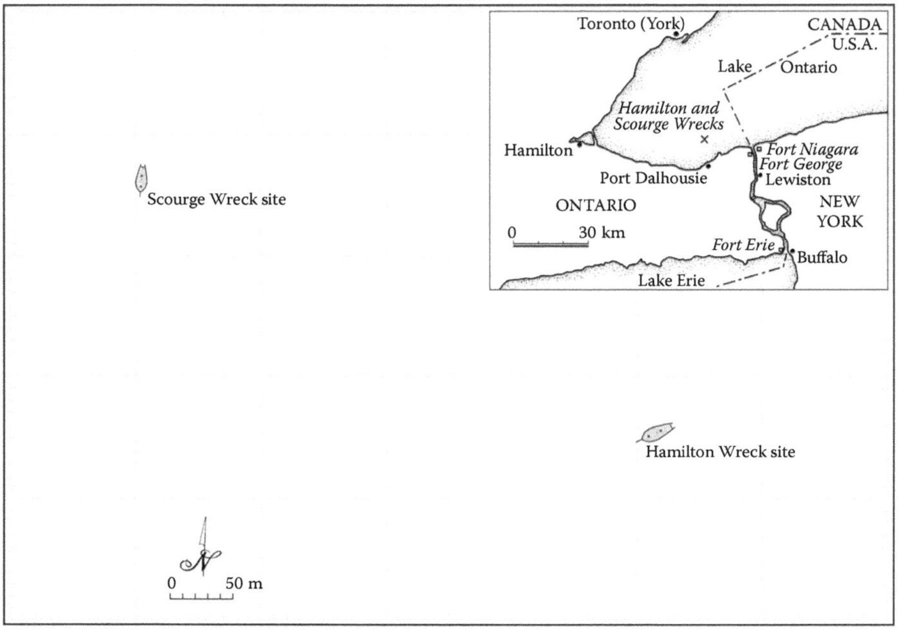

FIGURE 6.1 Location of the USS *Hamilton* and USS *Scourge* Shipwreck Site, Lake Ontario. (Map by Parks Canada, Underwater Archaeology Service, Ottawa, Ontario.)

of the mussels, it is now impossible to obtain photographs or video, or other types of normally precise acoustic-generated "images" of many structural features such as seams or timber junctures. Although the obscuring effect of mussel attachment is offset by existing premussel photograph and video collections, future condition

FIGURE 6.2 **(See color insert following page 286.)** Iron gun and wooden carriage projecting through a gunport on the port side of the USS *Scourge*, prior to mussel occupation in 1982. (Photograph by Emory Kristof, courtesy of City of Hamilton, Ontario, Canada.)

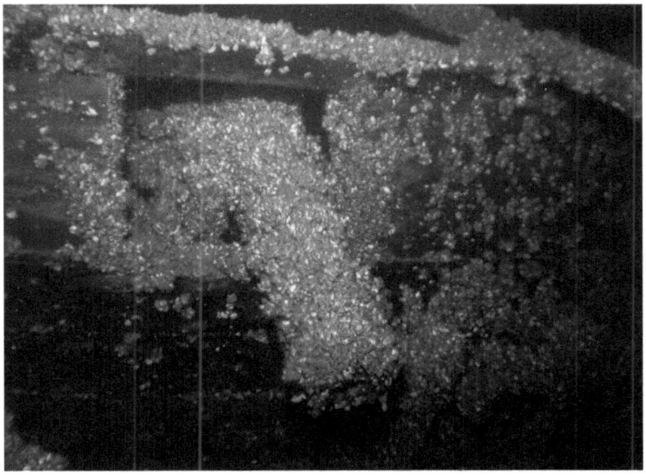

FIGURE 6.3 **(See color insert following page 286.)** Iron gun on the USS *Scourge* (shown in Figure 6.1) in 2008 after a minimum of approximately 10 years of mussel occupation. (Photograph by ASI Group Ltd., courtesy of City of Hamilton, Ontario, Canada.)

monitoring tasks will become increasingly difficult with increased attachment. Presently the lakebed immediately surrounding each wreck is covered by a continuous layer of mussels estimated to be 20–30 cm deep that in some cases has buried artifacts (Moore and Keyes, 2009).

Examination of site photographs taken before and after the dreissenid invasions show two settlement stages (Binnie, 2009). *Primary settlement* included bacterial colonization and secretion of slime film onto surfaces, accumulations of organic and inorganic detritus, and slime film invasion by other organisms. *Secondary settlement* followed primary settlement with the bacterial film providing a site for attachment by macrofouling organisms.

Binnie (2009) describes biological changes to a localized shipwreck environment as being initiated by the formation of a biofilm, then subsequent attachment of other species; microsystem alterations also occur with changes in dissolved oxygen, pH, and redox potential; physical changes include accelerated decay rate of wood and accelerated corrosion of metals. Other physical impacts on shipwrecks include increased mass loading, localized turbulence changes, development of anaerobic conditions below thick layers of mussels and feces/pseudofeces, and changes in species diversity of bacteria and microorganisms colonizing the surface of the shipwreck. The amount of loading has been calculated by the U.S. Army Corps of Engineers (1993) as 6800 g/m^2 in water (24 kg in air) for a 2-cm thick layer of mussels using an exponential relationship between SL and loading mass (Y), $Y = 0.002307 \times e^{0.24966x}$, ($r^2 = 0.96$).

The magnitude of impact of dreissenids on shipwrecks, particularly the amount of loading, will depend on the location of the wreck and fragility of structures. Custer and Custer (1997) in their survey of eight nearshore sites in Lake Erie reported percent coverage of lake bottom by zebra mussels ranging from 0% to 70%; average coverage in the west end of Lake Erie was 17% ± 4.0% ($n = 27$). The percent coverage of zebra mussels was relatively high in the northern portion (28–70% coverage) and in the southwestern portion (18–40%), but relatively low (<5%) in the southeastern portion of the study area. Custer and Custer (1997) also found that percent coverage by zebra mussels, as determined from underwater videography, was highly correlated ($r^2 = 0.96$) with zebra mussel biomass.

Binnie (2009) observed what are believed to be both zebra and quagga mussels on the USS *Hamilton* and USS *Scourge* shipwrecks, but quagga mussels seem to far outnumber the zebra mussels (Figure 6.4). The few zebra mussels and apparent lack of a distinct range in length classes suggest that the survival rate for larvae or young mussels is low, or that zebra mussels may have translocated from other sites. However, the wide range in length classes of quagga mussels (Figure 6.4) suggests that quagga mussels are able to more successfully grow and tolerate the environmental conditions. It is not known whether adaption to the 2–4°C temperatures has allowed reproduction for either species. Notice also the greatly extended incurrent siphons (top right, Figure 6.4), indicating active filtration at these temperatures.

Evidence that accelerated rates of corrosion are caused by mussels include (1) iron concentration is higher above wrecks (Cohn, 1996; Watzin et al., 2000, 2001); (2) iron concentration correlates to fecal production and mussel density (Marsh et al., 1999); (3) pitting corrosion is rapid in areas with mussel coverage, with 2.5–3 mm in three years on sheet pile, and pit depths are double under mussels compared to uninfested areas (MIE, 1994); (4) corrosion rate under mussels is 0.30–0.40 mm/yr.

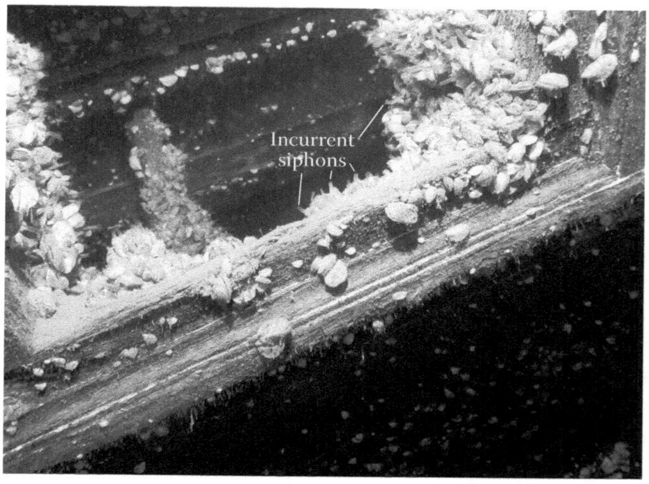

FIGURE 6.4 (See color insert following page 286.) Detail of mussels attachment at a stern gallery window of the USS *Hamilton*. Notice the mussels actively filtering and their elongated siphons. Mussel occupation in the stern cabin is also apparent, with attachment to the underside of the deck. (Photograph by ASI Group Ltd., courtesy of City of Hamilton, Ontario, Canada.)

6.2.7 Beneficial Impacts

Entrepreneurs have developed numerous chemical and physical methods for controlling zebra mussels in a variety of industrial and domestic facilities. Most molluscicides were developed specifically for control of nuisance bivalves like the Asian clam and the zebra and quagga mussels.

A few entrepreneurs in Florida have used the more colorful zebra mussel shells to make earrings, broaches, and paper weights (when embedded in plastic). Zebra mussels are not eatable, as many Ontario Ministry of Natural Resource personnel can attest to after a challenge for one of their annual picnics; apparently more than a dozen "dishes" were prepared, including entrees, soups, and desserts, and not one was eaten.

Since the appearance of the zebra and quagga mussels, their filtering activities have almost tripled water clarity in Lake Ct. Clair and Lake Erie to the pleasure of SCUBA divers who are able to see objects (e.g., sunken ships) deeper in the water than ever before. Unfortunately, the clearer water permits deeper penetration of light that has extended the areas of weed beds (Lowe and Pillsbury, 1995; Skubinna et al., 1995; Stuckey and Moore, 1995).

Orlova, in Chapter 10 of this book, describes various beneficial uses of dreissenids in the former USSR where they have been used for feeding of various livestock and as fertilizer for agriculture.

Ontario Power Generation has carried out numerous studies on the benefits of using of dreissenids as soil amendment and as mulch for both fruit trees and ornamental trees.

6.2.7.1 Conrad's False Mussel, *Mytilopsis leucophaeata*

Rajagopal et al. (2002a) describe *Mytilopsis leucophaeata* as a biofouling and nuisance organism that causes problems in industrial cooling water systems. Verween et al. (2002) considers the cooling water conduits of a power station an ideal habitat for *Mytilopsis leucophaeata*. The conduits provide perfect conditions for settlement and rapid growth, which ultimately interferes with the system operations and their failure. Bergstrom (2004) reports that *Mytilopsis leucophaeata* causes fouling of cages, boats, ropes, and so on, and that the species also competes with barnacles and other filter feeders.

6.3 MYTILIDAE

6.3.1 Ecological Impacts

6.3.1.1 Golden Mussel, *Limnoperna fortunei*

Like dreissenids, golden mussels settle in large numbers on native South American bivalves (Hyriidae and Mycetopodidae), causing their starvation and suffocation, and ultimately their death. Since its invasion of the Plata Basin, dense colonization of hard substrates has modified the presence and abundance of several species of native macroinvertebrates, homogenized the habitat, and altered the diet of fish. The introduction of the golden mussel produces a rapid change in benthic communities and threatens native biodiversity (Darrigan and Ezcurra de Drago, 2000). Garcia and Protogino (2005) reported that several fish species had altered their diets to consume large amounts of these mollusks; Penchaszadeh et al. (2000) reported that the golden mussel is now the main food source for *Leporinus obtusidens*.

Sylvester et al. (2004) found similarities between the zebra mussel and *Limnoperna fortunei*, suggesting that the latter may have an even stronger environmental impact than *Dreissena*. They performed filtration experiments and concluded that *Limnoperna's* filtering activity will strongly affect trophic interactions and the availability of food for both pelagic and benthic species, and the rates of other ecosystem processes including mineralization of nutrients, oxygen availability, and sedimentation rates. High filtration rates indicate that suspension feeding may (i) reduce phytoplankton standing stocks and biomass; (ii) suppress zooplankton populations; (iii) outcompete native species for available food; (iv) increase sedimentation rates; and (v) alter contaminant and nutrient cycling.

6.3.2 Industrial Impacts

The golden mussel produces macrofouling problems in the water systems of numerous facilities. De Oliveira et al. (2004) reported that despite low densities in Paraguay River, Brazil, *Limnoperna fortunei* can colonize water cooling systems of boats, obstructing water circulation and causing motor overheating. They also observed accumulations in water supply equipment, such as pumps and pipes in the river. Shortly after arrival of *Limnoperna fortunei* in Rio de la Plata estuary (Argentina), byssally attached populations reach densities in excess of 150,000 m^{-2} clogging pipes, filters, heat exchangers, condensers, and so on in industrial and power plants that use raw water, chiefly for cooling purposes (Sylvester et al., 2004). These and other impacts are described in Chapter 10 in Sections by Darrigan and by Rolla.

6.4 CORBICULIDAE

6.4.1 Ecological Impacts

The impact of Asian clams on the native clam populations has not been as dramatic as seen with zebra mussels, and impacts occur mainly when major Asian clam die-offs occur. In most cases, Asian clams and native unionids coexist (Beaver et al., 1991; Miller and Payne, 1994), but some negative impacts have been reported. Clarke (1984) reported adverse impacts of *Corbicula* on native unionids in the Tar River system, Virginia. Buchanan (1987) reported high mortality of native mussels after die-offs of *Corbicula* in Bourbeuse and Meramec rivers, Missouri. Scheller (1997) attributed die-offs of native clams population crashes of *Corbicula*, which during putrefaction release ammonia and reduce dissolved oxygen in the sediment. In a two-year ecological study in the Altamaha River, Georgia, Sickel (1976) concluded that the Asian clam did not appear to be affecting adult native clams directly; however, they may be excluding the juvenile unionids, which may result in the loss of many endemic species.

The Asian clam has had similar other types of ecological impacts as the dreissenids. *Corbicula's* high filtration rates and densities make it a major consumer of phytoplankton (McMahon, 2000). The Asian clam reduced phytoplankton and chlorophyll *a* concentrations in the Potomac River by 20–75% (Cohen et al., 1984). Similar seston reductions by *Corbicula fluminea* occurred in the Chowan River (Lauritsen, 1986), Savannah River (Leff et al., 1990), and other lotic systems (McMahon and Williams, 1986b). The removal of seston by Asian clams has altered the zooplankton community by increasing the abundance of copepods and reducing the abundance of rotifers in some systems (Beaver et al., 1991). Hass et al. (2002) reported that the mass invasion by *Corophium curvispinum*, *Dreissena polymorpha*, and *Corbicula* spp. strongly influenced ecosystem functions by linking pelagic and benthic processes by their intense filter feeding activity and thereby enhance the capacity for self-purification of the river system. Hakenkamp and Palmer (1999) found that *Corbicula* consumed significant quantities of organic material in the streambed when conditions favored pedal feeding but increased buried organic matter stores when filter feeding promoted deposition of organic matter as feces and pseudofeces. *Corbicula* contributed significantly to total benthic community respiration (and thus carbon dioxide production), and used pedal feeding on benthic organic material to grow at a faster rate than that possible by filter feeding alone (Hakenkamp et al., 2001).

6.4.2 Economic Impacts

Like dreissenids, the Asian clam's greatest economic impact in North America has been macrofouling of raw water systems in fossil-fueled or nuclear power stations (McMahon, 1983; McMahon and Bogan, 2001). Even though the planktonic stage is short lived, the nonswimming and settling pediveligers (~200–250 μm) and juveniles (<5 mm) (McMahon, 1983a; Page et al., 1986; Williams and McMahon, 1986) enter the facility through the intakes, then through traveling screen and strainers and settle in areas where flow rates are <1.2–1.5 m/s (4–5 ft/s) (Eng, 1979; Neitzel et al., 1984; Page et al., 1986). Settlement occurs where sediments accumulate, typically in wet wells at the intake structure, eddies created at ends of piping or at tees and bends, emergency systems with reduced water flows or at partially closed valves. Accumulations of adult shells also reduce flow velocities below the threshold for settlement, promoting accumulations of silt and sand in which clams burrow (Smithson, 1986). The filter-feeding adults also greatly increase sedimentation rates and enhance accumulations of sediments. In the Delta Mendota Canal, California, sediment accumulation induced by Asian clams required dewatering and removal every two years (Prokopovich and Herbert, 1965). Once settled, juveniles grow quickly to SLs >2 cm (0.75 in.) in 6 to 12 months and begin to foul small-diameter openings, such as condenser tubes, coarse strainers, and fire protection lines (McMahon, 1977; Aldridge and McMahon, 1978; McMahon, 1983a; Williams and McMahon, 1986b; McMahon, 2001).

Isom (1986) estimated annual U.S. costs in 1986 at $1 billion, which included costs for control, replacement, and repair. Pimentel (2002) and Pimentel et al. (2005) also report $1 billion/yr for damage and associated control costs for the Asian clam up to 2005.

6.5 AMPULLARIIDAE

6.5.1 Ecological and Agricultural Impacts

6.5.1.1 Giant Ramshorn Snail, *Marisa cornuarietis*

Pointier et al. (1988, 1991b) stated that *Marisa cornuarietis* competed with and caused the decline of *Biomphalaria glabrata*, the intermediate snail host of schistosomiasis. Similar impacts through competition and/or predation on *Biomphalaria* and several species of native snails (e.g., *Lymnaea*, *Pila*) have been reported in Puerto Rico (Peebles et al., 1972), Dominican Republic (Perera and Walls, 1996), Egypt (Cedeño-León and Thomas, 1983), and South-east Asia (Acosta and Pullin, 1991; Halwart, 1994a).

6.5.1.2 Golden Applesnail, *Pomacea canaliculata*

In Argentina the golden applesnail (GAS) is "harmless and useless" (Cazzaniga, 2006). But in the United States (Levin et al., 2006; Rawlings et al., 2007), southeast Asia (Carlsson, 2006), Philippines (Adalia and Magsino, 2006; Halwart, 2006), China (Yin et al., 2006; Ranamukhaarachchi and Wickramasinghe, 2006), Taiwan (Cheng and Kao, 2006; Yang et al., 2006), Japan (Wada, 2006), Malaysia (Yahaya et al., 2006), Indonesia (Hendarsih-Suharto et al., 2006), Vietnam (Cuong, 2006; Huynh, 2006), Cambodia (Preap et al., 2006), and Sri Lanka (Epa, 2006), *Pomacea canaliculata* (and *Pomacea insularum*) poses the greatest threat to rice and taro crops, native wetland ecosystems, and in biodiversity and ecosystem functioning. Currently, the occurrence of *Pomacea canaliculata* has been confirmed in California and Arizona in the continental United States. However, it has the potential to spread into other areas, including the rice-growing parts of California, where it could cause serious damage (Rawlings et al., 2007).

GAS are also disease vectors, causing at least three diseases in humans: cercarial dermatitis ("swimmers itch") caused by trematode cercariae (Mackie and Taylor, 2007); intestinal problems caused by *Echinostoma* flukes; eosinophilic meningitis caused by *Angiostrongylus cantonensis*, a nematode (Hollingsworth and Cowie, 2006).

6.5.1.3 Titan Applesnail, *Pomacea haustrum*

Rawlings et al. (2007) state that *Pomacea haustrum* is currently of relatively minor concern in the United States, given its failure to spread beyond Palm Beach County after 30 years or more in Florida. They agree with Cox (2004), however, that many species have maintained limited distributions, sometimes for decades, before becoming invasive.

6.5.1.4 Island Applesnail, *Pomacea insularum*

Some of the ecological and agricultural impacts in Asia associated with *Pomacea canaliculata* are almost certainly attributable to *Pomacea insularum* (Rawlings et al., 2007). This species has not been explicitly acknowledged as a serious pest because of the confusion in identification of these two species, with most of the literature referring to *Pomacea canaliculata* (Rawlings et al., 2007). *Pomacea insularum* may therefore be likely to have a significant impact on aquatic ecosystems and pose a threat to crops in the southeastern United States, particularly given the potential for it to spread through parts of Alabama, Mississippi, Louisiana, and Texas (Rawlings et al., 2007). *Pomacea insularum* also poses threats for the Florida apple snail, *Pomacea paludosa*, and the species that rely on it for food (Rawlings et al., 2007).

To predict the potential impact of this snail on aquatic ecosystems, Burlakova et al. (2008) tested the feeding rate of *Pomacea insularum*, under laboratory non-choice experiments, for three species of invasive macrophytes and 13 species of native aquatic plants that are important for wetland restoration and health. They observed high levels of consumption for four native submersed species (*Ceratophyllum demersum*, *Hymenocallis liriosme*, *Ruppia maritima*, and *Sagittaria lancifolia*) and three invasive species (*Colocasia esculenta*, *Alternanthera philoxeroides*, and *Eichhornia crassipes*). In contrast, less than 10% of the biomass of emergent species *Spartina alterniflora*, *Scirpus californicus*, *Thalia dealbata*, and *Typha latifolia* was consumed by *Pomacea insularum* over the test period (Burlakova et al., 2008). They also found the palatability of macrophytes to be negatively correlated with dry matter content, and suggested that their results could be applicable to all regions where this invader may be present. They recommended: wetland restoration in areas invaded by *Pomacea insularum* should focus on emergent structural species with low palatability; apple snails should not be considered as agents of biocontrol for invasive plants; although apple snails fed on invasive plants at a high rate, their consumption of many native species was even greater.

6.5.1.5 Spiketop Applesnail, *Pomacea diffusa*

Cowie (2002) states that *Pomacea diffusa* has generally been assumed to pose little threat in the United States. It is the only apple snail for which interstate transport is permitted, but Cowie (2002) suggests that this lack of concern may be unwarranted. Howell (2002) found in a feeding study that the spiketop applesnail feeds primarily on aufwuchs, not macrophytes. Aditya and Rau (2001) reported that it will feed quite readily on animal carcasses, live worms, and the eggs of planorbid snails. Cowie (2002) suggests that it may have direct effects on both aquatic vegetation and native snails and compete for food with native scavengers such as crayfish, shrimp, and fish.

6.5.1.6 Florida Applesnail, *Pomacea paludosa*

There appear to be no ecological or agricultural threats by the Florida applesnail. In fact, this species is of great ecological importance in Florida because it serves as a major prey item for the endangered Everglades kite, *Rostrhamus sociabilis* (Harper, 1936; Kushlan, 1975, 1978; Emerson and Jacobson, 1976). It is also eaten by the white ibis, *Eudocimus albus* (Kushlan and Kushlan, 1975), the limpkin, *Aramus guarauna*, and the boat-tailed grackle, *Cassidix mexicanus* (Snyder and Snyder, 1969), among others (Turner, 1996).

The Florida applesnail may serve as a vector for parasites and diseases, since Hanning and Leedom (1978) reported dermatitis-producing schistosomes in apple snails from Lake Okeechobee, Florida.

6.5.2 Socio-Economic Impacts

Halwart (1994a) gives details on the extent of snail damage in irrigated rice and states that the amount of damage depends on the age of the crop, the snail density, and the age structure of the snail population; snails with a shell height of less than 1.5–1.6 cm are too small to feed on rice seedlings; adult snails larger than 5 cm can ingest 7 to 24 seedlings per day; the number of seedlings consumed correlates positively with shell height. Damage also increases with snail density (Halwart, 1994a); 0.5 snails/m^2 cause 6.5% of missing rice hills, 8.0 snails/m^2 cause 93% of missing rice hills; at 30 days after transplanting, medium-sized snails (2–3 cm shell height) at a density of 1 snail per m^2 reduced the number of tillers by 19%, 8 snails/m^2 reduced the number of tillers by 98%. Under continuous flooding, snail damage to transplanted rice was severe: 52% of 16-day-old seedlings were consumed after 6 h, 75% after 24 h, and 100% after four days (Halwart, 1994a). Snails do not normally feed on rice plants when there was no standing water at the base and damage was limited to newly transplanted seedlings (Watanabe and Ventura, 1990). The fast growth and reproduction of snails in irrigation systems and rice fields lead to population levels that can destroy entire crops.

Halwart (1994a) estimated $2.4 million for pesticide expenditures in 1988 in Asian rice farming areas. In Taiwan, in 1982 loss of rice was estimated as $2.7 million but in 1986 the costs increased to $30.9 million because huge areas (103,350 ha) were treated with pesticides (Mochida, 1991). In Japan, control in just 176 ha cost $64,385 (Mochida, 1991).

The most detailed published economic analysis so far is for the Philippines where farmers spent $10 million on pesticides between 1987 and 1990 (Anderson, 1993). In the analyses of Naylor (1996), he included not only the cost of loss of rice, but also the costs of replanting, application of pesticides, and handpicking snails; total costs in 1990 due to *Pomacea* infestation were estimated as $28–45 million, which was 25–40% of what the Philippines spent on rice imports in 1990 (Vitousek et al., 1996).

Levin and Kalo (2006) reported impacts on human health due to the rat lung worm, *Angiostrongylus cantonensis*, a nematode that causes eosinophilic meningitis in humans. The disease is contracted by eating raw or insufficiently cooked apple snails, playing with infected snails, drinking raw vegetable juice, or eating transport hosts such as shrimp, crayfish, crabs, frogs, or fish (Tsai et al., 2004).

6.5.3 Beneficial Impacts

In spite of the nuisance aspects of the GAS, some have found positive attributes.

Human consumption: Tamaru et al. (2006) examined the cost-effectiveness of raising wild-caught snails on commercially available diets. They showed that a high-end product could be made but only 25% of the snail's biomass could be used; finding a use for the remaining 75%, including the shell, needs further investigation. Joshi (2006) provides numerous recipes for GAS from Philippines, Suriname, China, Hawaii, and Cambodia.

Aquarium trade: Pomacea species, especially *Pomacea diffusa*, is a popular aquarium snail in North America (Howells et al., 2006). The golden color is especially sought after (Perrera and Walls, 1996; Mackie, 2000a).

Weed control: The channeled applesnail has been used in the Philippines (Castillo and Casal, 2006; Joshi et al., 2006; Porte et al., 2006), Japan (Wada, 2006), and Korea (Porte et al., 2006).

Food for agricultural animals: The GAS is high in protein and calcium and has been used as food for fish, ducks, and tiger shrimp and freshwater prawn cultures (Castillo and Casal, 2006). Apparently the GAS is rich in carotenoids that Koi need for coloration and Japanese use GAS meal diets to promote the coloration.

6.6 THIARIDAE

6.6.1 Ecological Impacts

6.6.1.1 Quilted Melania, *Tarebia granifera*

The quilted melania has been reported to compete with native mollusks in Texas (Murray, 1971), Cuba (Jacobson, 1975; Cañete et al., 2004), Salton Sea Basin (Oglesby, 1977), West Indies (Prentice, 1983) where it displaced *Biomphalaria glabrata*, is a major intermediate host of *Schistosoma mansoni*, and Puerto Rico (Chaniotis et al., 1980a–c). The snail was introduced to Puerto Rico in 1954 (Harry and Aldrich, 1958) and it is the most common species there, outcompeting all native snail species (Chaniotis et al., 1980a). Oglesby (1977) reported possible competition for trophic resources with native pupfishes in California and suggested that similar negative interactions with native species may occur within the Gulf ecosystem. Jacobson (1975) predicted possible detrimental effects of this species on native melanids of the genus *Hemisinus* in Cuba. Because of its ability to displace *Biomphalaria glabrata* in St. Lucia, West Indies, the quilted melania has been considered for use as a biological control agent (Prentice, 1983).

6.6.1.2 Red-Rimmed Melania, *Melanoides tuberculata*

Red-rimmed melania have been reported to displace native gastropods in Texas (Murray, 1971), Cuba (Jacobson, 1975), and Florida (Roessler et al., 1977). Lee (1973) reported this species as an agricultural pest in Hong Kong, primarily in Chinese white cabbage plantations.

6.6.2 Socio-Economic Impacts

6.6.2.1 Quilted Melania, *Tarebia granifera*

Quilted melania serve as intermediate hosts to several trematodes, which as adults are known to parasitize man, including the oriental lung fluke, *Paragonimus westermani* (Abbott, 1952; Chaniotis et al., 1980a–c), *Philophthalmus* sp. (Murray and Haines, 1969), *Philophthalmus megalurus*, which affects the nictitating membrane of aquatic birds (Jacobson, 1975), and *Metagonimus yokogawai* (Murray and Stewart, 1968). A second intermediate host, such as a freshwater crayfish or crab, is a necessary part of the life cycle of *Paragonimus westermani*, and for human infection to be successful, the infected crustacean must be eaten raw (Lachner et al., 1970).

6.6.2.2 Red-Rimmed Melania, *Melanoides tuberculata*

This species serves as an intermediate host for several dangerous human parasites, including the Chinese liver fluke, *Clonorchis sinensis* (Murray, 1964, 1971; Kotrla, 1975), and the Oriental lung fluke, *Paragonimus westermani* (Murray, 1971; Roessler et al., 1978). A second intermediate host, such as a freshwater crayfish or crab is a necessary part of the life cycle of *Paragonimus westermani*, and for human infection to be successful, the infected crustacean must be eaten raw (Lachner et al., 1970). Other parasites of this species include *Diorchitrema formosanum* (Kotrla, 1975), *Opisthorchis sinensis* (Dundee and Paine, 1977), and *Philophthalmus* sp., an ocular parasite of waterfowl which occasionally infects mammals (Murray and Haines, 1969; Russo, 1974; Kotrla, 1975).

6.6.3 Beneficial Impacts

6.6.3.1 Quilted Melania, *Tarebia granifera*

The quilted melania has been considered for use as a biological control agent because of its ability to displace snails where introduced. For example, Prentice (1983) reported displacement of the snail *Biomphalaria glabrata*, which is a major intermediate host of *Schistosoma mansoni* in St. Lucia.

6.6.3.2 Red-Rimmed Melania, *Melanoides tuberculata*

Pointier et al. (1992, 1993a) and Pointier and Guyard (1992) state that *Melanoides tuberculata* has been employed in the Caribbean islands and Pointier et al. (1991a, 1994) in Venezuela, as an agent for the biological control of *Biomphalaria* mollusks, which are intermediate hosts of *Schistosoma mansoni.* This thiarid has a great longevity, ability to maintain high population densities for long periods, is a strong competitor with other snails for space and food, has a rapid growth rate, and a low mortality rate (Bedë, 1992). However, in China, Taiwan, Japan, and Korea, it is the first intermediate host of *Clonorchis sinensis* and *Paragonimus westermani* (Malek and Chen, 1974), two trematodes that are responsible for serious parasitic diseases in humans.

6.7 BITHYNIIDAE

6.7.1 Ecological Impacts

6.7.1.1 Faucet Snail, *Bithynia tentaculata*

The common name, faucet snail, is derived from its propensity to infest municipal water supplies from intake pipes to household faucets in the early 1900s in Britain (Baker, 1902). Sterki (1911) described the intense infestations in Erie, Pennsylvania water supplies as "wagon loads" of snails having to be removed from municipal water pumping stations (Sterki, 1911).

The faucet snail has higher growth rates per unit respiration than most native gastropods due to its ability to filter feed, as well as suspension feed (Tashiro and Colman, 1982). Brendelberger and Jürgens (1993) reported an increase in suspension feeding with increasing food concentrations up to a certain level above which feeding rate was kept constant by lowering the filtering rates. Suspension feeding also increased with increasing size of snail. All size classes of *Bithynia* showed a circannual fluctuation of their filtration rates.

Jokinen (1992) attributes the decrease in abundance of the two pleurocerid species, *Elimia virginica* and *Elimia livescens* in the Erie Canal to the invasion of the faucet snail. Harman (2000) reported a 15% reduction in the species richness of mollusks in Oneida Lake between 1917 and 1968 after the arrival of *Bithynia tentaculata*. It is very probable that impacts on abundance and diversity of native snail species is one of the consequences of *Bithynia's* ability to switch between two feeding modes, grazing and suspension feeding. Harman (2000) reported a decrease in the density of the faucet snail after colonization by the zebra mussel, *Dreissena polymorpha*, in Oneida Lake, although the overall mollusk abundance also decreased. Similar effects occurred in Lake Ontario between 1983 and 2000 due to competition with invasive dreissenid mussels (Haynes et al., 2005).

6.7.2 Socio-Economic Impacts

6.7.2.1 Faucet Snail, *Bithynia tentaculata*

In its native Eurasian habitat, the faucet snail is host to many different species of digeneans, rediae, cercariae, metacercariae, cysticercoids, and other parasites (Mattison et al., 1995; Toledo et al., 1998; Morley et al., 2004). The faucet snail is first intermediate host for *Prosthogonimus ovatus* (Hromada et al., 2000), second intermediate host for *Echinostoma revolutum* (Morley et al., 2004), first and second intermediate host for *Cyanthocotyle bushiensis* (Cole and Friend, 1999), and intermediate host for *Sphaeridiotrema globulus*, *Leyogonimus polyoon*, and *Syngamus trachea* (Cole and Friend, 1999).

Major waterbird die-offs every spring and fall since 2002 in Lake Onalaska (Navigation Pool 7 of the Upper Mississippi River) located near La Crosse, Wisconsin, have been attributed to infection by two trematode parasites, *Cyathocotyle bushiensis* and *Sphaeridiotrema globulus* carried by the faucet snail (Sauer et al., 2007). The area lies within the Mississippi Flyway, through which an estimated 40% of the continent's waterfowl migrate. Through the 2006 spring migration, an estimated 22,000 to 26,000 birds died, primarily lesser scaup (*Aythya affinis*) and American coots (*Fulica americana*); other waterbirds affected included northern pintail (*Anas acuta*), American wigeon (*Anas americana*), northern shoveler (*Anas clypeata*), blue-winged teal (*Anas discors*), mallard (*Anas platyrhynchos*), American black duck (*Anas rubripes*), gadwall (*Anas strepera*), redhead (*Aytha americana*), ring-necked duck (*A. collaris*), bufflehead (*Bucephala albeola*), tundra swan (*Cygnus columbianus*), herring gull (*Larus argentatus*), and ruddy duck (*Oxyura jamaicensis*) (Sauer et al., 2007).

6.7.3 Beneficial Impacts

6.7.3.1 Faucet Snail, *Bithynia tentaculata*

Desy et al. (2000) and Flessas et al. (2000) found good correlations between environmental concentrations and snail tissue concentrations of cadmium, zinc, and methylmercury and recommend the faucet snail as a biomonitor for these contaminants. *Bithynia tentaculata* serves as a food item for many species of fish, especially introduced common carp *Cyprinus carpio* (Ricciardi, 2001), and waterfowl. Vincent et al. (1981) and Ricciardi (2001) reported a commensal relationship between the faucet snail and the introduced milfoil, *Myriophyllum spicatum*, with the snail presumably deriving benefit grazing on algae attached to the milfoil. Ricciardi (2001) and Ricciardi et al. (1997) reported a mutual relationship between the faucet snail and zebra mussels, the snail presumably providing its shell as substrate for settling juveniles and the mussel providing food in its pseudofeces for the snail.

6.8 HYDROBIIDAE

6.8.1 Ecological Impacts

6.8.1.1 New Zealand Mudsnail (NZMS), *Potamopyrgus antipodarum*

NZMS can achieve extremely high densities (>500,000 m^2, Hall et al., 2003) which have the potential to manifest large negative ecological impacts. Hawkins and Furnish (1987), Feminella and Hawkins (1995) showed that grazing herbivores like NZMSs are extremely effective primary consumers in aquatic systems and can directly or indirectly affect native biota in the following ways.

- Densities of NZMS can be extremely high, from 100,000 to 800,000 individuals/m^2 in Europe (Dorgelo, 1987), to densities over 500,000 individuals/m^2 in the Greater Yellowstone Ecosystem (Hall et al., 2003). At such high densities, NZMSs may simply physically exclude other grazing organisms by occupying attachment space.
- Consume large quantities of periphyton, which is the main source of primary production in streams (Riley, 2003; Hall et al., 2003).
- Outcompete and displace native gastropods, especially threatened and endangered species, which are rare to begin with (Richards, 2004; Riley, 2003; Brown et al., 2008). For example, Bowler (1991) and Bowler and Frest (1992) speculated that NZMSs could have an impact on the diversity of Snake River (Idaho) gastropods. In the Snake River NZMS have invaded areas occupied by five threatened or endangered species of native aquatic snails (Richards et al., 2001).
- Compete with other grazing and detritivorous invertebrates that are at the base of aquatic food webs (Cada, 2004; Kerans et al., 2005).
- Negatively impact both invertebrates and vertebrates at higher trophic levels in aquatic food webs that depend on the aquatic invertebrate food base (Cada, 2004; Hall et al., 2006). For example, NZMSs negatively affected the survivorship but not the growth of mayfly species in experiments conducted in a tributary of the Madison River in Montana (Cada, 2004). Similarly, Cuker (1983) found that at high gastropod density, the densities of attached invertebrates such as Chironomidae were lower. Cuker (1983) suggested that the high snail densities resulted in the dislodgement of these organisms and their fixed benthic feeding retreats.

- Hall et al. (2006) reported that in the Gibbon River and Polecat Creek, Idaho, NZMSs constitute 88–93% of total secondary production. The displacement of native biota could cause an alteration in the energy flow pathways among trophic levels and damage or weaken aquatic ecosystems. Indeed, this degree of dominance by a single species is comparable to highly degraded communities.
- Hall et al. (2003) suggest that when densities and biomass are so high mudsnails excrete large amounts of ammonia, which dominates the nitrogen cycle and alters nutrient cycling and ecosystem processes in rivers. Chavaud et al. (2003) believe that high densities of NZMS make large sources of CO_2 by precipitating calcium bicarbonate to calcium carbonate to make their shells, which could alter the carbon cycle as well.
- The majority of mudsnails apparently pass through fish intestinal tracts undigested, and have low energetic value for these secondary consumers (Ryan, 1982; McCarter, 1986). A feeding study by Vinson and Baker (2008) showed that rainbow trout fed an exclusive and unlimited amount of NZMS lost 0.14–0.48% of their initial body weight per day, trout feces showed that only 8.5% of the NZMS shells were empty and assumed to have been digested, and 53.8% of NZMSs passed through the digestive system alive. Bioenergetics modeling showed that starved rainbow trout would have lost between 0.52% and 0.54% of their initial body weight per day, depending on their initial weight. Hence, community structure might be directly and indirectly affected at higher trophic levels in the food web.
- Interference competition has been demonstrated in several studies. Brown et al. (1994) showed that aggressive interactions between *Physella* and NZMSs increased significantly as densities increased. Interference competition for space may also result from extremely high densities known from invasive NZMS populations. Field experiments by Kerans et al. (2005) in Madison River in Yellowstone National Park showed that densities of macroinvertebrates from many different functional feeding groups were lower on tiles with high abundances of NZMSs, suggesting that interference was the mode of competition.
- Exploitative competition for periphyton also may occur between NZMSs and other benthic invertebrate grazers by depressing periphyton biomass in several New Zealand streams (Winterbourn and Fegley, 1989; Death, 1991). Carpenter et al. (1985) suggested that reduction of periphyton biomass may negatively affect other invertebrates and have wide-ranging effects on ecosystem processes in streams dominated by bottom-up interactions.

6.8.2 Socio-Economic Impacts

6.8.2.1 New Zealand Mudsnail (NZMS), *Potamopyrgus antipodarum*

In their native habitat, NZMSs pose no problem because of parasites, which sterilizes many snails keeping the populations to a manageable size. Winterbourn (1974) and Dybdahl and Krist (2004) described larval forms of 11 species of Trematoda parasitizing the New Zealand species of *Potamopyrgus*. Fourteen trematodes are now known to use the NZMS as an intermediate host and two of these species have also been found parasitizing the brackish water species, of *Potamopyrgus*, *Potamopyrgus estuarinus*, and *Potamopyrgus pupoides*. Metacercarial infections were more frequently found than infections by active larval stages and they also produced the highest intrapopulation infection rates. Highest cercarial infection rates were produced by monostome species.

There is also concern that the high densities of New Zealand mudsnail could produce biofouling in facilities that become infested (Zaranko et al., 1997). NZMSs have been documented to pass through water pipes and to emerge from domestic taps (Ponder, 1988) and block water pipes and meters (Cotton, 1942 in Zaranko et al., 1997). With population densities as high as 800,000 individuals/m^2 (Dorgelo, 1987), there is a potential for biofouling, particularly in irrigation systems in arid regions.

The NZMS Management and Control Plan Working Group (2007) identified several associated costs to combating NZMS invasions.

- A related to the vulnerability of threatened or endangered native fauna to mudsnail invasion. For example, in the Snake River NZMSs have invaded areas occupied by five threatened or endangered species of native aquatic snails (Richards et al., 2001). There will be costs associated with protection of these species.
- Costs for research and development expenses will be incurred by agency and university personnel to prevent further spread.
- Monitoring the distribution and spread of the snail to determine whether sensitive native species being placed at risk will be incurred.
- Extra monitoring that must occur for threatened and endangered species within the range of the NZMS invasion.
- Extra steps taken by agency personnel to ensure that facilities such as hatcheries do not act as vectors.
- Extra requirements placed in permits for activities such as dredging, canal maintenance, and so on.
- Extra steps and materials used by agency personnel, researchers, citizen monitors, and consultants to decontaminate gear.
- Extra costs incurred for materials, transport, and time of public outreach and information dissemination.
- Finally, contamination of private hatcheries and subsequent regulation and prohibition of their operations results in direct economic impact to these operators.

6.8.3 Beneficial Impacts

6.8.3.1 New Zealand Mudsnail (NZMS), *Potamopyrgus antipodarum*

There are some positive interactions among NZMSs and other species. For example, Schreiber et al. (2002) showed that NZMSs facilitated the colonization of macroinvertebrates in an experiment in an Australian stream.

FIGURE 1.1 (a) *Mytilus edulis*, outer view of left valve; (b) inner view of right valve. (c) *Quadrula quadrula*, outer view of left valve; and (d) inner view of right valve.

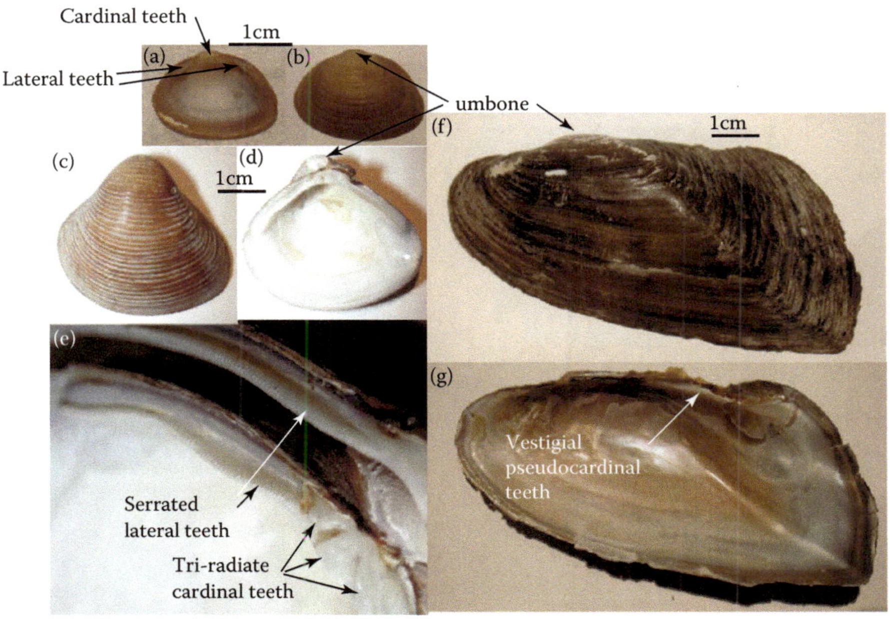

FIGURE 1.3 (a–c) Infaunal (burrowing) bivalves. All upper images are outer views of left valves; all lower images are of inner views of valves. (a, b) Sphaeriidae; (c, d, e) Corbiculidae [(e) shows serrated lateral teeth]; and (f, g) Unionidae.

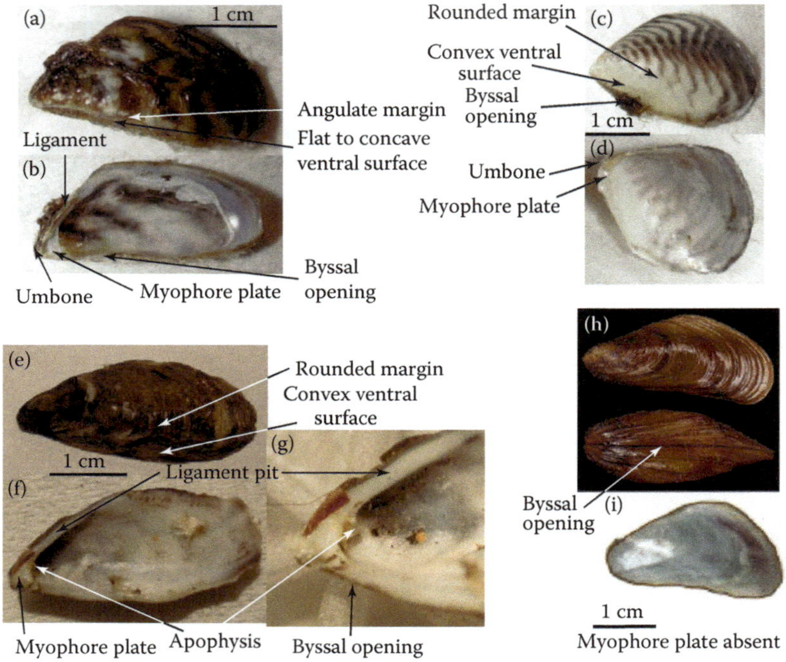

FIGURE 1.4 (a–i) Epibenthic invasive bivalves. All upper images are outer views of left valves; all lower images are inner views of right valves. (a–g) Dreissenidae; (h, i) Mytilidae. (a, b) *Dreissena polymorpha*; (c, d) *Dreissena bugensis*; (e, f, g) *Mytilopsis leucophaeata* [(g) = 3× magnification of umbonal end]; and (h, i) *Limnoperna fortunei* in side view (upper) and ventral view to show byssal opening.

FIGURE 1.9 Invasive ampullariids of North America: (a) *Marisa cornuarietis*; (b) *Pomacea canaliculata*; (c) *Pomacea haustrum*; (d) *Pomacea insularum* (Modified from http://www.nwrc.usgs.gov/invasive_species/applesnail_images2.htm. With permission.); (e) *Pomacea paludosa*; and (f) *Pomacea diffusa*.

FIGURE 1.10 *Tarebia granifera* showing aperture view.

FIGURE 1.11 *Melanoides tuberculata*, aperture view.

FIGURE 1.12 *Bithynia tentaculata*.

FIGURE 1.13 *Potamopyrgus antipodarum* in aperture view (left) and back view.

FIGURE 4.3 Sampling substrates, the one on the far right was touching a heavily colonized surface.

FIGURE 4.14 Fixed-area sampling quadrat used for quantification of dreissenid mussels and other benthic organisms attached to navigational buoys. This buoy was retrieved and examined in January 2008 following 2007 deployment in the St. Lawrence River near Cornwall, Ontario, Canada. Attached mussels can be counted, along with adhering byssal threads of mussels that have been recently dislodged.

FIGURE 5.5 Heavily colonized strainer.

FIGURE 5.7 Shell debris accumulated in heat exchanger plenum.

FIGURE 6.2 Iron gun and wooden carriage projecting through a gunport on the port side of the USS *Scourge*, prior to mussel occupation in 1982. (Photograph by Emory Kristof, courtesy of City of Hamilton, Ontario, Canada.)

FIGURE 6.3 Iron gun on the USS *Scourge* (shown in Figure 6.1) in 2008 after a minimum of approximately 10 years of mussel occupation. (Photograph by ASI Group Ltd., courtesy of City of Hamilton, Ontario, Canada.)

FIGURE 6.4 Detail of mussels attachment at a stern gallery window of the USS *Hamilton*. Notice the mussels actively filtering and their elongated siphons. Mussel occupation in the stern cabin is also apparent, with attachment to the underside of the deck. (Photograph by ASI Group Ltd., courtesy of City of Hamilton, Ontario, Canada.)

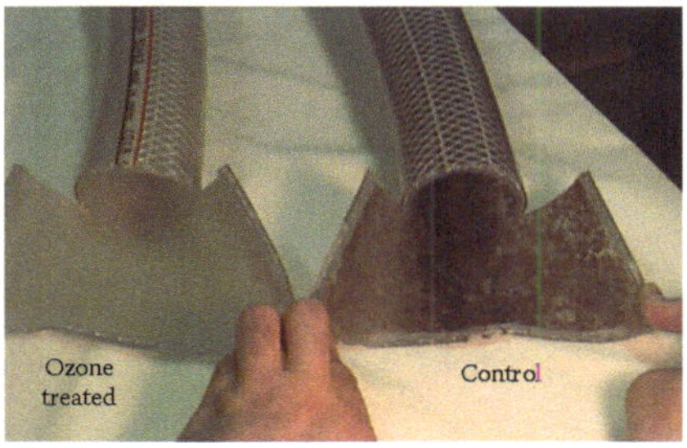

FIGURE 7.2 Biofouling prevention using intermittent ozone application.

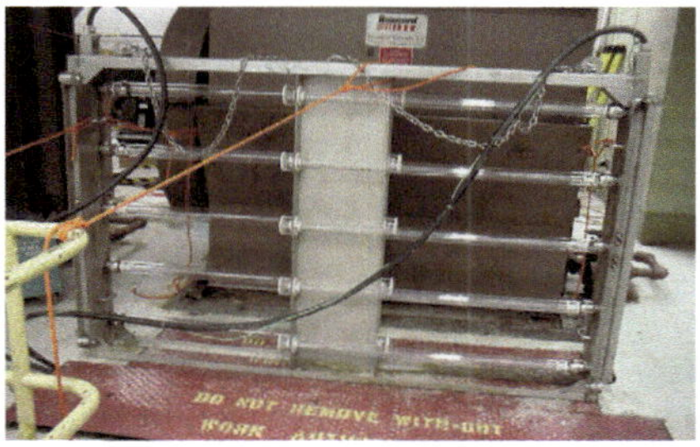

FIGURE 7.7 UV lamp rack used in an open channel system for dreissenid veliger control.

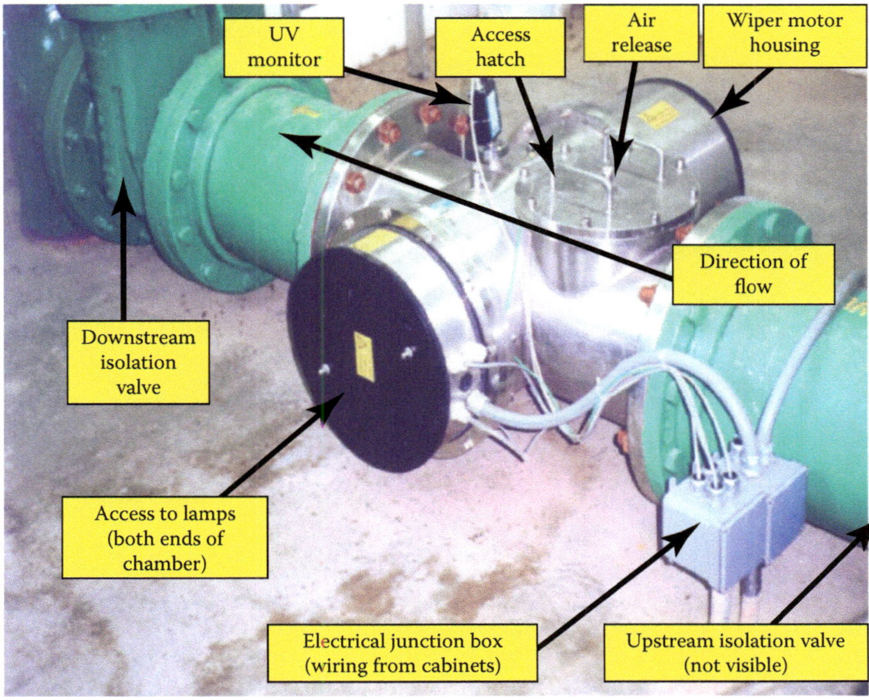

FIGURE 7.8 Within pipe UV system for ballast water control. (Reproduced from Hanovia Ltd. With permission.)

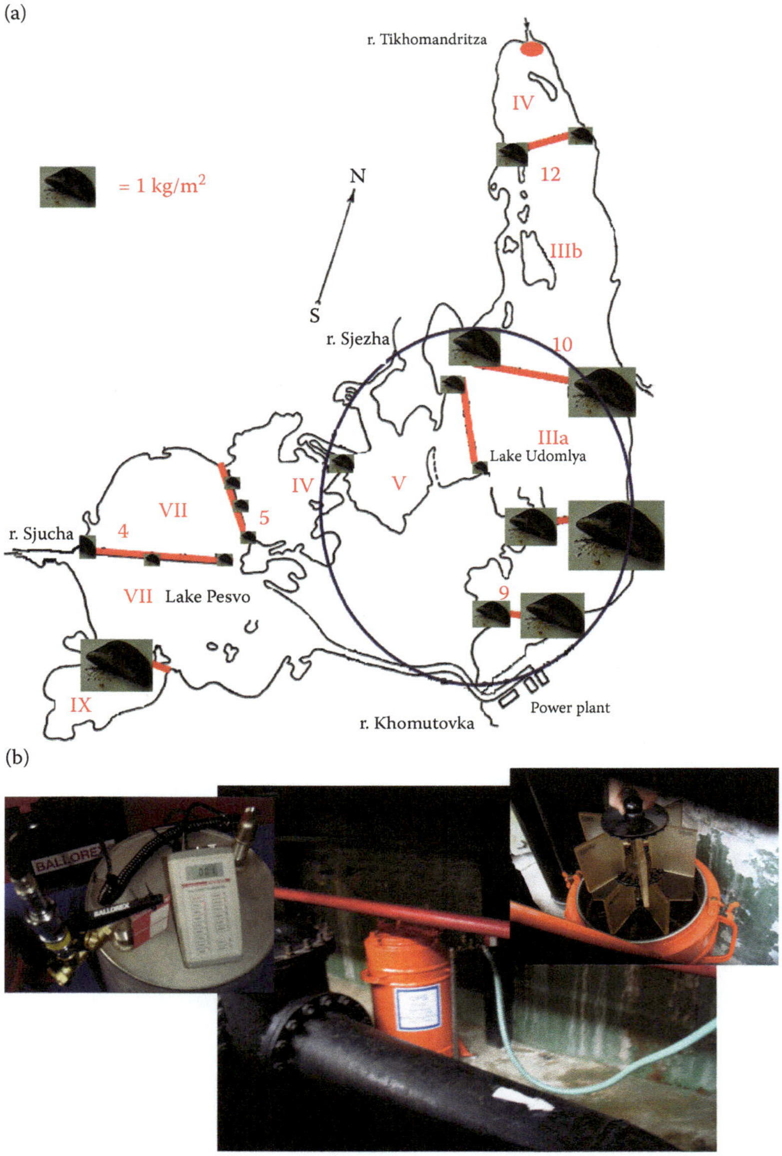

FIGURE 10.4 Zebra mussel distribution in lakes and side-stream monitoring at Kalininsk power plants. (a) Zebra mussel biomass distribution in lakes Udomlya and Pes'vo (gray area—the zone of water circulation, Arabic numbers—numbers of transects for diving survey). (b) Example of a monitor installed at a pump house of the fire protection system at Kalininsk power plant (monitors were constructed for both monitoring and experimentation; they are hermetic with regulated flow velocity and exchangeable interior equipment).

FIGURE 10.11 This specimen of *Anodontites trapesialis* is used as a settlement substrate by *Limnoperna*, preventing the specimen from opening or closing its valves, causing eventual mortality.

FIGURE 10.12 (a) Heat interchanger in a thermal power plant; (b) fish exclusion screens in a hydroelectric power plant; (c) cage for *Acipenser baeri* culture in Uruguay; and (d) detail of an aquaculture cage.

FIGURE 10.13 (a) Shell bank of the golden mussel in the littoral of the San Roque Dam, Cordoba, Argentina; (b) width of the shell bank; and (c) intake on the Tacuarí River of the rice watering system (Laguna Merin Basin).

7 Weapons of Mollusk Destruction

7.1 INTRODUCTION

Before deciding on a mitigation strategy, it is important to evaluate, in depth, both the environment from which the raw water comes, as well as the asset to be protected. When dealing with a newly introduced species, it is possible that the environment may support the invader for a short period of time after which the population, if not eliminated, will stabilize to a very low level. This has been the case with the introduction of *Corbicula fluminea* to the lower Colorado River. Following an explosive growth phase after introduction over 40 years ago, the population now remains at a very low level. The *Corbicula* present are small by North American standards (mostly around 1 cm in the Colorado River) and are not considered to be a problem in terms of macrofouling (McMahon, 1981, 1983; Strayer, 1999).

It is also possible that some areas will suffer from only seasonal colonization. Larvae will arrive from an upstream source into an area that can only support settlement and growth during certain times of the year. It is possible that parameters such as excessively high temperatures or low dissolved oxygen values may effectively eliminate settled macrofouling populations on regular basis. Mihuc et al. (1999) report seasonal colonization of a subtropical floodplain river by zebra mussels. The veligers coming in from upstream locations settle during the fall and winter. Resulting adults are eliminated in the following summer by a combination of high ambient temperature and low dissolved oxygen. A similar case involves the *Limnoperna fortunei* in the Pantanal region of Brazil. Annual floods in this area change the water chemistry in the region. Plummeting pH combined with very low levels of dissolved oxygen lead to massive kills of adult colonies of *Limnoperna* (see contribution by Oliviera in Chapter 3 of this book).

While it may be necessary to mitigate the effects of macrofoulers during the early part of the invasion, permanent retrofits should be delayed until it is clear that the local ecosystem will sustain problematic populations into the foreseeable future. Once it has been determined that the invasive species is here to stay, evaluation of individual assets should be carried out as described in Chapter 5. Following this evaluation, mitigation and treatment strategies can be selected based on the anticipated or observed level of infestation, the vulnerability of individual components, and systems to macrofouling and operational preference.

7.2 PREVENTATIVE VERSUS REACTIVE STRATEGIES

Some facilities prefer not to allow settlement of macrofoulers in any of their systems and therefore opt for preventative strategies. Other facilities feel their systems are adequately robust to cope with some level of fouling and select a reactive strategy that suits their needs. In both cases, treatment can be usually accomplished using several different options.

The following chapter reviews and examines both the preventative and reactive strategies using both chemical and nonchemical treatment options available.

7.3 MITIGATION OF MUSSELS THROUGH THE USE OF CHEMICALS

The treatment of choice for most facilities tends to be one of chemical control, as it has often proven to be convenient and effective. Chemical control has been the common practice in Europe and so far, it has been used extensively in North America. The major advantage offered by chemical treatments is that, by and large, they can be engineered to protect almost the entire facility, from intake to discharge. The difficulty in pursuing treatment through chemical means is in limiting the discharge of toxic materials to the environment while meeting local environmental regulations. As facilities continue to find themselves in an increasingly stringent regulatory environment, mitigation based on use of chemicals may become limited.

A wide variety of chemical treatment strategies is available for controlling mollusk populations. Even table salt, when applied in sufficient concentration for a long enough period of time will be effective. The aim, however, should always be to minimize local environmental impact whenever possible and practical. It is essential to work with the local regulators to ascertain that the chemical you wish to use is acceptable to them, especially when treated water is to be discharged into the local waterway. In closed systems, or under static conditions, a wider variety of chemical treatments can be used, provided there is no release to the environment.

7.3.1 Reactive Strategies Using Chemical Control in Piping Systems

These strategies are used after the adult form of the mollusk has become established in a raw water system. At this time there are no reactive treatments using chemicals for fouled external structures.

7.3.1.1 End-of-Season Treatment

The end-of-season treatment is usually performed after the breeding season of the mollusk is complete. Sufficient oxidizing or nonoxidizing chemical is applied for a long-enough period to kill all adults established in the system. The end-of-season treatment presupposes that the pipe system in question can tolerate one season's worth of macrofouling and that the accumulated biomass and shells can be removed from the system after the treatment. Adult mussels will release from the internal walls of systems during and also after the treatment. The system components must be able to tolerate the predicted mass of shell material which is released, and maintenance staff must be on hand to remove the debris. It is very hard to predict when the major release of shell material will take place as it depends on numerous variables.

When deciding on the timing for an end-of-season treatment, it would appear from operating experience in the continental northeast that most mollusks tend to be most vulnerable just after spawning. Adults may be most easily eliminated at the end of the spawning cycle. The explanation for this observation is not clear. Costa et al. (2008) concluded that while the susceptibility of zebra mussels does indeed increase significantly between spring and winter, it does not seem to be related to the body condition of the mussels. So, while we are not certain why in temperate regions the fall treatments are more effective than treatments in the spring, the fall treatments do eliminate most of the mussels that have settled during the summer in the least possible time.

7.3.1.2 Periodic Treatment

The periodic treatment is a variation on the end-of-season treatment. Once again, the adults are targeted but the chemical is applied more often—basically treating while the densities and physical size of the adult macrofoulers remain fairly low. The chemical concentration needed and the duration of application will be similar to that used for the end-of-season treatment. The biomass that has to be removed following the application is proportionally smaller but the system in question still must be capable of tolerating a high degree of fouling. If implemented frequently, periodic treatment will prevent the presence of large individuals in the system.

7.3.2 Proactive Chemical Strategies for Piping Systems

The following chemical treatments are designed to prevent the settlement of macrofoulers in raw water systems.

7.3.2.1 Intermittent Treatment

Chemical dosing at frequent intervals (every 12–24 h) is aimed at preventing infestations before they begin. It is generally accepted that the freshly settled postveligers are more susceptible than adults. Therefore, the concentration of chemical used and the duration of application to control this life stage will be significantly less than if adults are targeted. Most intermittent treatments will not eliminate adults already present in the system or translocators that gain access.

Chlorine as sodium hypochlorite has been used for 30 min, every 12 h, at 2 mg/L to prevent freshly settled postveligers from developing. This strategy was used at several plants belonging to Ontario Hydro for a number of years with good results (Claudi and Evans, 1993).

Ogawa and Prescott (2002) reported that an addition of 5–8 mg/L of ozone for 5 min every 12 h prevented the development of dreissenid mussel populations in the system treated.

In 2005, Rusznak et al., working with veligers and settled pediveligers observed a 95% reduction in settlement when chlorine dioxide was applied for 15 min every 6 h at a concentration of 0.25 ppm and an ambient temperature of 12.8°C.

An intermittent strategy, using the molluscicide Mexel has been developed in Europe. Giamberini et al. (1994) reported that an addition of 6 mg/L of the Mexel chemical for 3 h prevents infestation. Recently, the manufacturer has suggested that a once-per-day, 30 min addition of Mexel at a level of 4–5 ppm will control freshly settled mussels and avoid infestation of the system.

7.3.2.2 Semicontinuous Treatment

The semicontinuous treatment was developed after observing the response of zebra mussels to a chemical irritant. Upon exposure to a noxious substance (oxidizing chemical), mussels will stop filtering and quickly close their shell. It will then take 15–30 min before they will reopen the shell and attempt to resume filtering. This means that the treatment schedule can be adjusted to 15 min on and 15–45 min off. This is particularly advantageous to facilities that have multiple systems to be treated. The chemical addition system will work continuously, but the chemical can be directed sequentially to the different systems to be treated. This strategy has resulted in complete control of all stages of zebra mussels in the piping, while using significantly less chemical than in a truly continuous treatment (Evans and Sim, 1993).

This semicontinuous strategy has been further refined under the trademark Pulse-chlorination® by a European vendor. Using electrodes attached to shells of mussels within a specially constructed monitor, Pulse-chlorination determines the precise timing for semicontinuous chlorine treatment by monitoring if the mussels in the monitor have their shell open or closed. The system only applies chlorine when the mussels have opened their shells and discontinues the addition when the shells are closed. This technique significantly reduces the amount of chlorine required (up to 50%) for a treatment compared to a continuous application (Jenner and Polman, 1998).

7.3.2.3 Continuous Treatment

The continuous treatment strategy is designed to eliminate all settlement in all parts of the system. The incoming larvae do not necessarily suffer 100% mortality, but the presence of a noxious substance is enough to discourage settlement. Any adults present will either succumb to the toxin (if the low-level chemical addition is carried out for the entire reproductive season) or detach themselves and attempt to leave the system being treated. The concentrations of chemical needed can be quite low but the application must be continuous. Typically, continuous treatment is chosen for systems that cannot tolerate even the smallest degree of fouling such as fire protection and other safety-related systems. To date, this type of treatment has been attempted primarily with oxidizing chemicals such as chlorine or with a continuous feed of copper ions. Potentially, other chemical strategies, such as depression or elevation of pH or continuous addition of flocculants could be used as a continuous treatment.

7.3.3 Proactive Chemical Strategies for External Structures

Protection of external structures from settlement using chemical compounds is generally restricted to the use of antifouling or foul release coatings on these structures.

7.4 CHEMICAL OPTIONS AVAILABLE

If fouling organism mortality was the only consideration when designing a treatment, any number of chemicals could be used successfully to control fouling. Due to economic and environmental concerns, only a relatively small number of chemicals are ever likely to be used in practice. When approached by a vendor of either new or existing chemical treatment for macrofouling control, it is best to direct them to your local regulatory agency for an assessment of their product. In many cases, the manufacturer will not be willing or able to provide the money, data, or empirical studies required by regulators to gain approval for use. This reality unfortunately sometimes keeps promising new products from reaching the market.

7.4.1 Oxidizing Chemicals

Oxidizing chemicals have been utilized as disinfectant agents in water supply systems for over 100 years. In most cases their effect on the environment is understood and well documented. Treatment with oxidizing chemicals (primarily chlorine) has been the most frequently used proactive chemical treatment to date, while treatment with nonoxidizing chemicals, high temperature water flushing, and mechanical removal (scraping and water jetting) have also been used in periodic and end-of-season reactive treatments by a number of different industries.

7.4.1.1 Chlorine

One of the most effective and popular methods of macrofouling control is chlorination, where chlorine is added as diatomic chlorine gas, liquid sodium hypochlorite, or solid calcium hypochlorite. Chlorine has been used for over 100 years in the treatment of potable water. It is a well-known chemical, with well-documented use and by-products. One of the major concerns in using chlorination for surface water supplies is that it will combine with various organic compounds to form trihalomethanes (THMs), which are considered carcinogens. Most regulatory agencies permit the use of chlorine in flow-through systems but have stringent limitations on the level of total residual chlorine (TRC) in the discharge water. To meet these requirements, most facilities have to dechlorinate the treated stream, or else utilize a storage lagoon or a large volume of diluting water. Before deciding on a chemical treatment strategy, it is wise to contact your local regulatory agency and check on their policy for chemical treatment in general and chlorine in particular.

Chlorine-based mitigation can be used for all five strategies discussed above. Brady et al. (1996) confirmed that chlorine treatment strategies and concentrations used to control zebra mussels also successfully control quagga mussels. Details of chlorine chemistry and application levels for different strategies are discussed in Chapter 8, devoted solely to chlorination.

7.4.1.2 Chlorine Dioxide

Chlorine dioxide (ClO_2) has been implemented as a disinfectant in the water treatment industry for over 50 years. It acts effectively on both aerobic and anaerobic bacteria. Unlike chlorine-based treatments, it does not form THM by-products and is equally effective at all pH levels. Chlorine dioxide does not react with ammonia and therefore does not form chloramines. The by-products generated in the breakdown of ClO_2 in aqueous solution consist primarily of sodium chlorite, chlorate, and chloride, all considered to be acceptable at low levels to regulatory bodies. However, in the United States, the EPA did find that bromate and aldehyde by-products can be formed by chlorine dioxide addition to water. The by-products include the propanal to decanal series, benzaldehyde, methyl glyoxal, and glyoxal (Richardson, 1997), which may be of some interest depending on specific facility concerns.

Chlorine dioxide can be manufactured on-site from various precursors such as sodium chlorite and hydrochloric acid, sodium chlorite and chlorine gas, sodium chlorite and sodium hypochlorite and hydrochloric acid or sodium chlorate and hydrogen peroxide and sulfuric acid. The manufacture of chlorine dioxide requires specialized equipment and there have been concerns with worker safety in the past. In Germany, chlorine dioxide generators are permitted for drinking water applications only if they meet the technical requirements for automatic and airtight generation on-site.

Recently, manufacturers have started producing 3000 ppm solutions of chlorine dioxide off-site and delivering these solutions to the client. Although an aqueous solution of 3000 ppm chlorine dioxide is not classified as a hazardous substance, at room temperature it will sublimate to a gaseous phase, which is extremely poisonous. If the addition equipment is not airtight and carefully controlled, this can lead to health and safety concerns. For this reason, on-site generation with state-of-the-art equipment is recommended.

In terms of mitigation efficacy, differing opinions exist in the relevant research. Van Benschoten et al. (1993) reported chlorine dioxide to be less effective for veliger control than chlorine or chloramine. Fraleigh et al. (1993), working in a flow-through system, found that 0.125 mg/L of chlorine dioxide was sufficient to prevent settlement. Brooks and Matisoff (1993) reported chlorine dioxide as more effective in causing adult mortalities than chlorine. Khalanski (1993) observed adult zebra mussel mortality in eight days using a chlorine dioxide residual of 0.2 mg/L and an ambient temperature of >20°C (68°F). Belluati et al. (1997) found that all types of macrofouling and slime could be completely controlled using treatment concentrations varying between 0.07 mg/L in the winter and 0.22 mg/L in the summer months. Holt and Ryan (1997) achieved 100% mortality of adult mussels using 0.45 mg/L concentration of chlorine dioxide after 145.5 h. The ambient temperature was kept constant at 17°C (63°F). Additional data on dose versus mortality developed by Holt and Ryan are shown in Table 7.1.

Rusznak et al. (2005), working with veligers and settled pediveligers, observed 95% reduction in settlement when chlorine dioxide was applied for 15 min every 6 h at a concentration of 0.25 ppm and an ambient temperature of 12.8°C (55°F).

A number of facilities in Europe use chlorine dioxide for macrofouling control on both freshwater and seawater systems. European experience suggests that continuous treatment of seawater with ClO_2 injection at 0.1 ppm eliminated biofouling in cooling systems.

7.4.1.3 Chloramines

Chloramines are formed when free available chlorine (HOCl and OCl) reacts with nitrogen-containing compounds such as ammonia and amino acids. Chloramines are formed naturally when chlorine or sodium hypochlorite is added to lake water.

TABLE 7.1
Toxicity of Chlorine Dioxide to Zebra Mussel Adults

Concentration of Chlorine Dioxide (mg/L)	Hours to 100% Mortality at Average Temperature of 18.4°C (58°F)	Hours to 100% Mortality at Average Temperature of 25.8°C (78°F)
5.0	70	–
2.0	78	–
1.0	102	–
0.5	139	40
0.25	211	89

Source: After D. A. Holt and E. Ryan, 1997. *Proceedings of the Seventh International Zebra Mussel and Aquatic Nuiscance Species Conference*, New Orleans, USA, January 28.

The more ammonium that is present, the higher the level of chloramines formed. Chloramines can be generated in bulk by co-injection of ammonium as either ammonium gas or ammonium hydroxide and sodium hypochlorite. Although chloramines have generally been found to be a less powerful oxidant than hypochlorous acid, they have been used as disinfectants in various applications. Monochloramine (NH_2Cl) has been used as a disinfectant in drinking water and it was also found effective in controlling veligers of the Asian clam (Belanger et al., 1991). Van Benschoten et al. (1993) found over 90% veliger mortality in both static and flow-through tests at chloramine concentrations above 1.5 mg/L. At two French power plants, monochloramine is produced on-site by mixing sodium hypochlorite and ammonium chloride and is used to control zebra mussels as well as bryozoans, *Plumatella* sp. (Duvivier, 1993 in *Hydroecologie Appliquee 1998*). At this time, it would seem that the use of chloramines over chlorine may hold some advantages for some facilities. This is particularly true, once again, if the formation of THMs is a concern.

7.4.1.4 Ozone

Ozone is a well-known bactericidal agent in the sewage treatment and water treatment industry. The first experiments with ozone in drinking water treatment were conducted in France in the late nineteenth century. It was first used commercially as a drinking water disinfectant at Nice in 1908. Since that time, ozone has gained increasing popularity worldwide, particularly for its viral and bacterial inactivation properties. Viruses and bacteria are completely eliminated within 30 s by a dissolved ozone residual of <0.5 mg/L. Ozone improves taste, odor, and color of drinking water and can also be used to prevent various other forms of biofouling. Concentrations of 0.25–0.5 mg/L have been reported to eliminate the blue mussel (*Mytilus*) in some European studies. In terms of contact time at comparable residual levels, ozone outperforms chlorine (Lewis et al., 1993).

The most significant negative aspect to an ozonation-based mitigation strategy is the high initial capital cost of the equipment and the difficulties involved in maintaining it. Additionally, one of the main characteristics of ozone that make it attractive for use in once-through systems also turns out to be a major drawback. The relatively unstable ozone molecule dissipates quickly in water, depending on a number of factors including temperature, pH, and organic matter concentration.

$$O_3 + OH^- \rightarrow HO_2^- + O_2 \tag{7.1}$$

$$O_3 + HO_2^- \rightarrow {}^{\bullet}OH + O_2^{\bullet -} + O_2 \tag{7.2}$$

This ensures there will be little to no end of pipe residual and therefore no adverse environmental impact. This characteristic is slightly detrimental if control of downstream mussel settlement and growth is a consideration or in systems with a greater than 20 min retention time. It would be very difficult and expensive to maintain sufficient residual ozone in such a system. Early studies by Lewis et al. (1993) suggested that a minimum of 5 h of contact time was required at 0.5 mg/L at 15–20°C (59–68°F) for 100% mortality of veligers and postveligers in the water column. The same study also showed that adult mussels will suffer 100% mortality if they are exposed to an ozone concentration of 0.5 mg/L or more for 7–12 days (Figure 7.1). Time to death depends on both concentration and ambient temperature. The study did not consider the possibility that a shorter exposure to ozone would still result in the eventual mortality of the veligers.

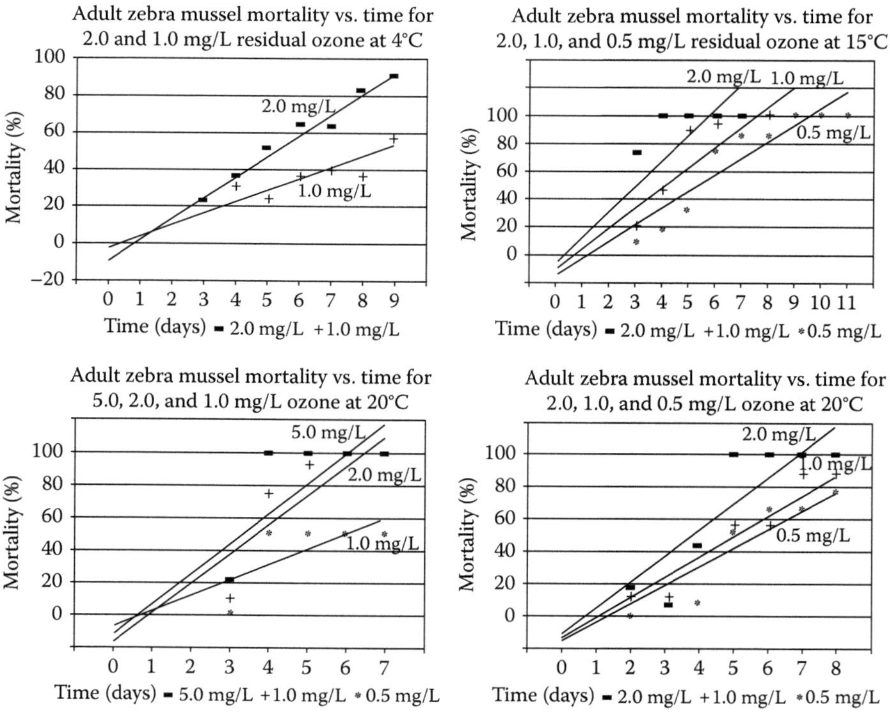

FIGURE 7.1 Adult zebra mussel response to various concentrations of ozone under different temperature regimes.

In practical applications ozone has performed remarkably well in controlling dreissenid infestations. In Canada, ozone treatment is currently being used in an electricity-generating station on Lake Ontario to prevent biofouling of once-through service water systems. The ozone-generating equipment was installed in the spring of 2000 and put into immediate use. Ozone is injected at the start of the system at 0.5–0.3 mg/L continuously throughout the mussel breeding season. At this location, ozone has provided excellent control of all biofouling, including the prevention of dreissenid settlement. The treatment has also eliminated any adults which had previously settled in the system prior to the installation of the ozone generator. This elimination took place within the first six months from the initiation of the treatment system. Caged fish fingerlings placed in the discharge stream of the power plant did not suffer any mortality, upholding ozone's reputation as an ecologically friendly treatment option (Brady, 2007).

The Lake Ontario results mirror the experience in Belgium (Verelst et al., 1999) at a power plant on the river Meuse. At this installation, ozone was injected at the level of 0.3–0.5 mg/L. This treatment achieved complete control of Asian clams and zebra mussels, as well as bacteria and algae in the secondary cooling system. The discharge into the river had <10 ppb of residual ozone. Significant mortality of adults was observed after 20 days of exposure, with complete mortality achieved in 48 days.

Ozone has performed equally well in the control of *Limnoperna fortunei* in Brazil. The addition of 0.3 ppm of ozone during the *Limnoperna* breeding season has prevented the macrofouling of heat exchangers at the Itaipu hydroelectric plant (Rothe, 2007).

Another successful application of ozone on the Great Lakes involved the use of an innovative ozone storage system from Mitsubishi Electric called MABOS (Karski, 2001; Ogawa and Prescott, 2002). The system operates in two cycles: (1) generation/adsorption and (2) desorption/injection. During the first cycle, ozone is generated from high-purity oxygen gas. The ozone/oxygen mix passes through an adsorption unit where the ozone is removed from the mixture and stored in chilled silica gel. This process continues for 12 h, at which time ozone is extracted from the silica gel (desorption) by increasing the temperature of the storage unit and is then injected into a carrier stream. In the test application, 1 kg of ozone was gradually released from storage every 12 h and injected into a service water carrier stream of 88 L/s via a venturi eductor. The desorption/injection process took about 5 min, resulting in an ozone carrier stream concentration of 38 mg/L. The carrier stream then delivered the ozone to the main service water flow (450–700 L/s or 7132–11,095 US gpm). Once mixed, the resulting ozone level in the service water was between 5 and 8 mg/L for approximately 5 min. This level of ozone, applied every 12 h was found to prevent all mussel settlement in the service water system during the three-month trial as well as controlling all other biofouling (Figure 7.2). The Mitsubishi system had numerous advantages over conventional ozone treatment technologies. The ozone generator was very small as it only had to produce 1 kg of ozone every 12 h, meaning smaller physical space requirements and lower energy requirements. It was also instrumented to allow for unattended operation and periodic treatment. The use of the system was discontinued for reasons other than performance.

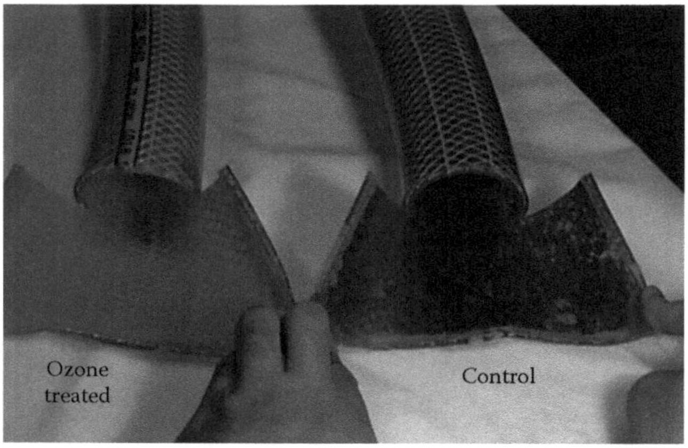

FIGURE 7.2 **(See color insert following page 286.)** Biofouling prevention using intermittent ozone application.

A fish hatchery in New England currently utilizes an ozonation treatment system for dreissenid control. The facility is relatively small, using approximately 11 MGD of raw water. The mussel-control regime involves continuous application of ozone at a concentration of 0.3 mg/L. The system is designed to treat a 2500-ft-long pipeline, inactivating the zebra mussel veligers in order to limit infestation in the hatchery raceways. The ozone injection at the intake is followed by removal of ozone at the hatchery using ultraviolet light.

7.4.1.5 Bromine

Bromine compounds can be used in several different forms as an antifouling measure. These include activated bromine, sodium bromide, bromine chloride, and proprietary mixtures of bromine and chlorine or bromine and other chemicals (e.g., Actibrome by Nalco Chemicals and BromiCide by Great Lakes Chemical Corporation). Bromine in all forms has been shown to be a more effective oxidizing agent when water pH levels are greater than 8.0 (Fellers et al., 1988). Previously bromine was thought to be less toxic to nontarget species than chlorine. However, research now suggests that the toxicity to nontarget species is in fact higher than for that of chlorine, in some cases by a factor of 2 (Burton et al., 1992; Howe et al., 1993). The breadth of research available for treatment using chlorine is not as readily available for bromine or bromine-based products. However, as a rough guide, the total amount of oxidant required would be the same whether using bromine or chlorine. Research conducted on zebra mussel veligers from Lake Erie found that bromine (0.5 mg/L) is not as nearly effective a settlement inhibitor as chlorine (Bidwell, 1993, 1999).

7.4.1.6 Hydrogen Peroxide

Hydrogen peroxide has the reputation of being a benign oxidizing agent, dissociating into hydrogen and oxygen, leaving no detrimental environmental by-products. While this may not be completely true, it is frequently used as an algaecide or biocide in small, contained systems such as spent fuel bays in nuclear-generating stations.

Several trials on both adult zebra mussels and veligers have shown that relatively high doses of hydrogen peroxide are required to induce mortality. A hydrogen peroxide concentration of 12 mg/L was required for adults and 6 mg/L was needed for veligers (Martin et al., 1992; Klerks et al., 1993). Limited studies on the efficacy of hydrogen peroxide in combination with ionic iron have demonstrated that concentrations of both peroxide and iron required for significant mortality are quite high, 5.0 and 1.25 mg/L hydrogen peroxide and iron, respectively. For adult mortality, the required exposure period was 56 days (Klerks et al., 1993).

Petrille and Miller (2000) recorded a 90% mortality rate of adult zebra mussels in 21 days of exposure at a concentration of 5.4 mg/L. The duration of the treatment decreased with increasing hydrogen peroxide concentration. Total mortality was achieved at 7.8, 8.8, and 3.0 days at concentrations of 10, 20, and 40 mg/L, respectively. In the same study, Asian clams were observed to be less sensitive to hydrogen peroxide than zebra mussels. Total mortality for the clams was observed after 13.5, 9.5, and 9 days at concentrations of 10, 20, and 40 mg/L, respectively.

As hydrogen peroxide is quite expensive compared to sodium hypochlorite, it would not seem economically practical to treat large volume, flow-through systems.

7.4.1.7 Potassium Permanganate

Potassium permanganate is another oxidizing chemical commonly used in municipal facilities for water purification. It is widely utilized to protect against oxidation of iron and manganese and for control of taste and odor problems. Some studies suggest that it can be used against adult zebra mussels although it is less effective than chlorine. In flow-through experiments, Fraleigh et al. (1993) found that 0.25 mg/L of potassium permanganate prevented settlement of zebra mussels in testing tanks. In static experiments, Klerks et al. (1993) saw less than 30% mortality in veligers exposed to 2.5 mg/L of permanganate for 3 h. Giacomo and Wymer (1997) established effective control of adults at a concentration of 2.0 mg/L and veliger settlement was prevented using a concentration of 1.0 mg/L. These results suggest that potassium permanganate may prevent the settlement of mussels, but that it is not acutely toxic to either veligers or adults. The Paul M. Neil Water Treatment Facility has employed the use of potassium permanganate at the intake structure on Lake Michigan, successfully preventing mussel settlement in the 2 miles of piping leading to the treatment facility for several years (Van Hemert, 2007).

Utilizing potassium permanganate as a mitigation strategy seems most applicable in potable water treatment facilities, especially considering that many already utilize this chemical for sanitation purposes or to eliminate THMs already in solution.

7.4.1.8 Ferrate

Ferrate (FeO_4^{2-}), which is also known as iron oxide with the iron atom having a valence state of +6, is considered a powerful and environmentally friendly oxidant and disinfectant. Fe(VI) is considered to have the highest reduction/oxidation potential of any common elemental oxidant. The oxidation and disinfection properties of Fe(VI) have been documented in the past. Treatment by Fe(VI) is thought not to yield any mutagenic/carcinogenic by-products and the final product of Fe(VI) reduction is nontoxic Fe^{3+}.

In the 1970s, ferrate was explored as a replacement chemical for chlorine, but synthesis methods at that time made its use cost prohibitive. Recently, Ferrate Treatment Technologies (FTT), LLC, has patented a new manufacturing process that may reduce ferrate production costs by up to 90%. The Ferrator® is a transportable on-site synthesis reactor which produces Fe(VI) in a liquid form that can be used to directly treat industrial water supplies.

To date, field trials with the Ferrator have focused on ballast water treatment, conducted at ports including Singapore and Rotterdam. Testing done to date suggests complete inactivation of microorganisms after an addition of 2–4 ppm of ferrate in the discharge stream of a 50 μm filter.

Ferrate(VI) has been suggested as a potential treatment for macrofouling. At this point there are no established protocols for the control of either veligers or adults; however, several studies are underway.

7.4.2 Nonoxidizing Chemicals

A number of nonoxidizing chemicals have been developed for bacterial disinfection, algae control, and have also been tested as molluscicides. Some of these chemicals have regulatory registration for use in once-through cooling systems. Most of these products are used for end-of-season or periodic treatments.

7.4.2.1 Proprietary Molluscicides

The term "molluscicide" is somewhat of a misnomer as generally these products are toxic to a wide variety of species and not just mollusks. These proprietary formulations are based either on quaternary amines (Betz–Clamtrol) or on isothiazolones (Buckman–Bulab 6002) or on other organic compounds (Bayer–Baylucide). The use of these products in closed systems is unrestricted. In once-through applications, most of the products have to be detoxified with the addition of bentonite clay slurry.

The major form of use for these chemicals is as an end-of-season or in some cases, periodic treatment. These chemicals are not detected by mussels as noxious compounds, allowing them to continue filtering normally without closing up until death. Thus mortality can be achieved quite quickly. Depending on the concentration used and the ambient temperature, significant mortality can occur in 4–24 h. Rolla and Motta in Chapter 10 of this book report that some proprietary molluscicides are not as effective on *Limnoperna* as they are on dreissenids.

Nonoxidizing chemicals have been used with good results in a number of U.S. facilities (Barton, 1993; Green et al., 1993) primarily as an end-of-season or periodic treatment. In some of these, the operators have been able to set up closed-loop systems where the same chemical was recirculated for the required period of time. This type of application significantly decreases the volume of chemical required, which is of both environmental and economic benefit.

There are several considerations when using nonoxidizing chemical treatments. When a facility uses bentonite clay for detoxification of periodic or seasonal treatments, the active product will adhere to the clay particles and be carried to the bottom. The amount of material which will accumulate in the discharge over the period of some years may be considerable. Possible further complication is the fate of the active ingredients in the sediments. The degradation process and the degradation products may not be well documented, but some of the active ingredients are believed to be quite persistent.

Another consideration is that most of these products require relatively warm ambient temperatures to work swiftly. In temperate zones this may mean a treatment well in advance of the end of spawning, leaving a fairly sizable population of macrofoulers in the system to grow over the winter.

One of the categories of nonoxidizing proprietary compounds are chemicals that act as filming agents and are based on fatty amines. The Mexel® line of products belong to this category. These products consist of filming aliphatic amines that are thought to inhibit corrosion, prevent slime, scaling, and various forms of macrofouling. The active ingredient in Mexel is an alkyl trimethylenediamine (ATMD). Mexel was found to inhibit the formation of byssal threads of dreissenid mussels at concentrations 2, 6, and 10 mg/L (Giamberini et al., 1994). Unlike other proprietary products, Mexel can be used as an intermittent treatment. When added at the water intake point, typically once a day for 30 min at a level of 4–5 ppm, it has been reported to keep all treated systems free of dreissenid mussels. Unlike other proprietary products, Mexel does not require detoxification on discharge. High concentrations of total suspended solids (TSSs) or organic matter content in the water will require larger concentrations of chemical to maintain an effective residual treatment level.

7.4.2.2 Bacteria-Based Molluscicide

Over the last decade, an enormous amount of work was undertaken by a team led by Dr. Dan Molloy from the New York State Museum studying the use of a common soil bacterium as a specific control agent for dreissenids. The team had found that the bacterial species, *Pseudomonas fluorescens*, strain CL145A, has the ability to cause mortality in adult mussels. When dreissenid mussels ingest artificially high densities of the bacteria (living or dead), a toxin within the bacterial cell destroys the mussels' digestive system. To date, no other aquatic species tested has demonstrated any susceptibility to this bacterium.

The product is currently in the commercialization stage and concurrently going through the regulatory approval process. Marrone Organic Innovations Inc. was selected as the commercial distribution partner by the research team. The first full plant treatment took place at Niagara Plant Group DeCew II hydraulic-generating station, Ontario, Canada in August 2009. This application was covered by a federal research permit. When this product becomes commercially available, it could offer a significant alternative to all forms of chemical control.

7.4.2.3 BioBullets®

Most mollusks are able to sense chlorine as well as a number of other oxidizing chemicals and toxins. When they detect a noxious substance, they respond by closing the shell and isolating themselves from the irritant. Depending on the ambient temperature, the shell can remain closed for up to three weeks. Aldridge et al. (2006) have developed a method to encapsulate the active ingredient for mussel control in an "edible" coating. The individual particles are of the size commonly fed on by dreissenids (~20 μm). The mussels filter these particles from the water and due to the coating, do not detect the active ingredient and they do not close their shell. They pass these particles on to the digestive system. Thus the BioBullet® overcomes the protective reaction that mussels have to noxious substances. Another perceived advantage is that much more of the active ingredient is filtered out by the mussels and therefore less of it ends up in the environment. Commercialization of this technology is proceeding rapidly in the United Kingdom.

7.4.2.4 Ammonium Nitrate

At concentrations of 400–500 mg/L, ammonium nitrate has been reported to cause 100% mortality of adults in five to six days. The ambient temperature was 61–67°F (16.1–19.4°C). Coon et al. (1993) report that ammonia concentrations exceeding 3 mg/L cause 100% mortality in veligers.

This treatment is not feasible for once-through systems, but could be used within a closed loop and also in agricultural circumstances where chemical fertilization is already being utilized. It does not appear to offer any advantages over the commercial molluscicides mentioned above, except perhaps economics.

7.4.2.5 Copper Ions

Dissolution of copper and aluminum anodes by electrolysis has been used to protect ship cooling system from macrofouling for at least 40 years. Based on the marine experience, Blume et al. (1994) conducted a series of experiments to determine if the same technology could be used against dreissenid mussels. They concluded that a continuous dose of 10 ppb of copper ion would limit veliger settlement in system protected. This technology was commercialized under the trademark of MacroTech® Copper Ion generator. Babinec (2003) reports that Wisconsin Energy Corporation We Energies utilizes this copper ion technology to control dreissenid mussel infestation in its Oak Creek Power Plant service water system. The system is still in use at the time of publication. While the copper ion generator does not eliminate all macrofouling in the service water system, the level of infestation is acceptable to the plant personnel.

As the toxicity of copper in freshwater systems is greatly influenced by the total hardness of water, one would anticipate that different bodies of water would require different levels of copper ion generation to control dreissenids. The discharge of copper ions into the aquatic ecosystem may not be permitted in all jurisdictions.

Kennedy et al. (2006) determined that 14–81 μgCu/L caused virtually complete mortality in zebra mussel veligers in 24 h. The copper was added both as copper sulfate and as copper-rich algaecide Cutrine-Ultra®. In October 2008, a collaborative effort involving many governmental and environmental agencies (including the EPA, the U.S. Army Corp of Engineers, and the Nebraska Game and Parks Commission) was undertaken to eradicate zebra mussels from Offut Base Lake in Nebraska. A strategy was devised utilizing copper sulfate pentahydrate as the control method and a total of 27,000 lbs of the substance was applied to the surface of the lake in aqueous format via vortex spreaders on pontoon boats. Although long-term studies will be required to assess the survival of the mussel population, at this time it is believed that a 100% mortality rate was achieved. The impact on native fish and wildlife species was not intended to be a significant component of the study; however, rough data obtained by the field teams showed direct impact on 14 species of fish, with approximately 29,000 lbs of dead fish washing up onshore (Morris, 2008).

7.4.2.6 Potassium Salts

Potassium compounds are toxic to most bivalves, including the dreissenids and corbiculids. Imlay (1973) found levels of K^+ as low as 11 mg/L to be lethal to 90% of the individuals of three unionid species in 52 days and 7 mg/L caused mortality in eight months. Imlay (1973) also postulated that the maximum safe K^+ level for long-term survival of most freshwater bivalves to be between 3.9 and 9.8 mg/L. In flow-through experiments, Fisher et al. (1993) found 50 mg/L of potassium chloride prevented settlement of zebra mussels in the test chamber while the chamber receiving concentrations of 25 mg/L had as much settlement as the untreated controls. In contrast, Lewis et al. (1996), also using flow-through experiments in mussel-infested waters, concluded that 20 mg/L caused 100% mortality in adults in 52 days while preventing any new settlement. Total mortality of small zebra mussels (7–11 mm) at a power plant on the Moselle River was achieved in 48 h using 600 mg/L of KCl (Khalanski, 1993). O'Donnel et al. (1996) found that KCl concentrations below 168 mg/L produced changes in gill physiology of the mussels but that these changes were reversible. However, at concentrations of 336 mg/L the potassium-induced changes became irreversible. In addition to causing mortality, potassium in various forms has been observed to prevent valve closure in mussels and reduced filtration rates (Wildridge et al., 1996). The ability to prevent valve closure suggests that combination treatments, using potassium to inhibit valve closure followed by a toxicant, could result in quicker elimination of adults and less chemical used. Study by Lewis et al. (1996) indicated that the addition of 100 mg/L of KCl to various control chemicals significantly decreased the time required to eliminate adult mussels.

The ASI group (St. Catharines, Ontario, Canada) has been using potassium chloride as an end-of-season treatment for a number of closed-loop systems such as fire-protection systems. Potassium chloride seems to be particularly effective in these types of semistatic systems due to mode of action and the fact that it does not possess the inherent demand characteristics typical of oxidants. They have found that 100 mg/L for two to four days at ambient temperature of 15°C resulted in 100% mortality of dreissenid adult mussels in the system treated. The length of treatment does increase with decreasing temperatures but the time required for treatment is still shorter than if the treatment was done with oxidizing chemicals. In 2006, ASI Group treated an isolated quarry in Virginia with potassium chloride in order to eradicate an infestation of zebra mussels. A total of 131,000 kg of muriate of potash (KCl) solution was dispersed into a 180 million USG body of water over an 18-day period. The temperature was <9°C. Complete mortality of all dreissenids occurred in 35 days (Virginia Department of Game and Inland Fisheries, 2005). As the quarry treated has no regular flow-through, the potassium levels will remain high for a number of years. Given that this quarry is used for recreational diving, the presence of high levels of potassium will prevent possible reintroduction of dreissenids.

Although potassium compounds are nontoxic to higher organisms such as fish, the toxicity to native bivalves makes the approval for use of potassium salts in once-through systems unlikely. In closed-loop systems however, the use of potassium salts is an attractive option.

7.4.2.7 Sodium Metabisulfite

Dreissenid mussels are relatively intolerant of low dissolved oxygen concentrations. Systems with less than 3 mg/L at 20°C (68°F) would have little chance of mussels surviving. Dissolved oxygen concentrations exceeding 50% saturation seem to be required for sustained, healthy populations. Depressing the dissolved oxygen level in a system infested by dreissenids could be a form of end-of-season treatment. Further details on the oxygen deprivation effects on mussels are given in Section 7.7.2.4.

Several water treatment plants in Wisconsin, which have dual intakes, have received permission to add either chlorine or sodium bisulfite to the intake they wish to take out of service. The chemical is added just before the intake is capped and kept static for 4–10 weeks. At the end of the lay-up period, all zebra mussels will be dead either from the effects of chlorine or from anoxia caused by sodium bisulfite removing all available oxygen.

The use of sodium metabisulfite as a dechlorinating and an oxygen-stripping agent prompted speculation if this chemical could be used effectively as a treatment for zebra mussels. Studies by Mackie and Kilgour (1992b) showed that sodium metabisulfite by itself is not very toxic to zebra mussels. A minimum concentration of about 177 mg/L of sodium metabisulfite would be required to kill all adult mussels in closed systems. Below this level, toxic effects are absent. Anoxia caused by the addition of sodium metabisulfite will contribute to the mortality on prolonged exposure as sodium metabisulfite is an effective oxygen stripper. Depending on facility-specific conditions, anoxic conditions combined with higher water temperatures will increase dreissenid mortality more than the effects of either alone (Matthews and McMahon, 1999; Yu and Culver, 1999).

Sodium metabisulfite is not considered to be a practically feasible chemical to use to kill dreissenid mussels in pipelines or conduits because enormous amounts of the compound which would be required for treatment where water is renewed continuously. However, it may be practical for use in closed systems, for example, fire-protection systems that hold water for long periods of time. The only caveat is that the lack of oxygen may result in uncontrolled growth of sulfate-reducing bacteria, which in turn could cause serious corrosion problems in the system.

7.4.2.8 Flocculation

There have been several reports concerning the effect coagulants/flocculants have on mussel veligers. Alum or aluminum sulfate is the flocculant most frequently used in the drinking water industry. Studies by Mackie and Kilgour (1995) investigated the effect of alum on the zebra mussel veligers. From this study it would appear that the concentrations of alum used in most water treatment plants (i.e., 10–30 mg/L) is not sufficient to kill the veliger larvae by direct toxicity. In the laboratory studies, 130 mg/L of alum were required to cause 50% mortality of zebra mussel veligers in 24 h. However, the data also indicate that a pH depression below 5, caused by the addition of alum, especially in the mixing zone, would cause an instantaneous kill of veligers. Further, the role of alum in removal of veligers under natural conditions (i.e., 10–30 mg/L alum) appears to be mainly a physical one, with the floc physically removing even living veligers. Most veligers remain alive for at least 24 h in the floc at concentrations below 100 mg/L. Prechlorination greatly improves the efficacy of alum in removing veligers from raw water supplies, probably by aiding sedimentation of veligers by causing them to close their shell and increasing their likelihood of becoming part of the floc. In short, alum will remove veliger larvae by chemical toxicity in the mixing zones of alum and chlorine and by flocculation of both living and dead veligers in other areas.

Several drinking water plants in Ohio have been experimenting with another flocculent both as means of minimizing the use of alum and as means of zebra mussel control (Blanck et al., 1996). Catfloc LS, which is a product of the Calgon Corporation, was introduced at the crib end of the intake at one drinking water plant on an experimental basis. The main function of this chemical is to act as a coagulant and thereby reduce the use of alum. At the end of the season, substantial mortality of mussels was observed in the intake structure and there was a substantial decrease in the amount of alum used. Another Calgon product, called VeliGON, is currently the only flocculent molluscicide available in the United States. It is permitted for use in potable water treatment facilities, assuming that the facility obtains a permit from and follows the guidelines of the National Pollutant Discharge Elimination System. This product is only approved for use in potable water treatment facilities and requires an end-of-pipe filter to remove the flocculent and mussel remnants (Sprecher and Getsinger, 2000).

This approach may be an appropriate treatment for some drinking water plant intakes, provided that flocculation does not cause problems with sediment formation in the intake pipe (i.e., that sludge can easily be removed).

7.4.2.9 Salinity

The potential for use of salinity as a control strategy exists mainly in industrial facilities situated on the freshwater and salt water interface (i.e., lower Hudson). Periodic flushing with water of high salinity is unlikely to eliminate adult dreissenids; however, for larval stages, which tend to be more susceptible, this strategy of control may work.

Kilgour and Kepple (1993) using zebra mussel veligers observed acute toxic effects at salinity levels above 8.5 ppt. Working with both zebra mussels and quagga mussels, Wright et al. (1996) found that, "tolerance to both higher temperatures and higher salinities increased with larval age in both species, but zebra mussel embryos and larvae had a greater degree of salinity tolerance than quagga mussel embryos and larvae."

The use of increased salinity levels as a possible control of *Mytilopsis leucophaeta* (a brackish water mussel) was tested by Verween et al. (2007). The authors subjected 4- and 48-h-old larvae to conditions of differing salinity and temperature. Both groups of embryos showed increasing mortality with increasing salinity levels. The earlier stage larvae were significantly more susceptible than the older group. It would appear that a salinity-based treatment would have to be applied as an intermittent treatment, in fairly short intervals, to target freshly settled juveniles.

7.4.2.10 pH Adjustment

As discussed in Chapter 3, there is debate as to what pH level can be tolerated by larval stages of the dreissenids and at what pH adults will survive for prolonged periods of time. Sprung (1987) found that successful development of zebra mussel veligers requires a pH range between 7.4 and 9.4, with optimal conditions at 8.5. The threshold value for survival of adult zebra mussels appears to be near pH 6.5 (McCauley and Kott, 1993) and for larvae, pH 6.9 (Mackie and Kilgour, 1993). The incipient lethal level for larvae is near pH 7.4 (Mackie and Kilgour, 1993). Lucy and Timpson (1998), working on Lough Allen (Ireland), observed a settlement of juvenile zebra mussels >5 mm on sampling substrate, in a marina, in mid-September. By the end of September the settlers were gone from the substrate. The pH level at this site ranged from 6.9 to 7.1, with an average total hardness of 24 mg/L $CaCO_3$ and a calcium level of 9 mg/L $CaCO_3$. It is assumed that the failure of Lough Allen juveniles to develop into adult zebra mussels was due to low pH and concomitant low water hardness and calcium levels in the lake.

Experience from Lake George suggests what the borderline conditions for calcium and pH may be for zebra mussels in North America. Hansen et al. (1998) document that calcium levels of 10.68 ± 1.17 mg/L and a pH of 7.56 ± 0.18 (both 10-year averages) have supported a very low level of zebra mussel presence in Lake George for the last 10 years. Studies by Nierzwicki-Bauer et al. in 2000 found that adult mussels are able to survive in Lake George water but veligers fail unless both calcium and pH levels are raised.

At the other extreme, Bowman and Bailey (1998) note that almost 100% adult zebra mussel mortality could be expected when pH reaches 9.3–9.5 and that the rate of pH change could likely affect the mortality rate observed.

In studying pH modification for reactive treatment, Smythe et al. (1998) concluded that there is a high probability that exposure to a pH of 5.0 or lower for one month would induce significant mortalities and reduce attached mussel density in piping.

Given the relatively narrow range of pH tolerance, water systems that draw water with pH near the limit for successful survival of dreissenid veligers may be protected if the pH is either further depressed below 7.1 or raised above 9.4 at the raw water intake. Adjusting the pH could create a hostile environment which would preclude dreissenid settlement. pH adjustment, if practical, would act as a continuous chemical treatment preventing all settlement and growth. This strategy would be particularly useful for drinking water facilities that adjust the pH of the incoming water before processing it. The point of pH adjustment, if it could be moved to the intake, would protect all subsequent structures and systems. At least one drinking water facility in Ireland is considering this strategy.

7.5 COMBINATION TREATMENTS

7.5.1 Elevated Temperature and Chemical Treatment

A number of chemicals have been observed to have greater efficacy at higher ambient temperatures. There is no doubt that both the time to death and the amount of chemical needed decreases as temperature increases. This would suggest that chemical treatments for adult populations will be most successful during the summer and fall, or at times when the temperature of the raw water can be artificially raised through some form of thermal enrichment. The achievable thermal enrichment may not be sufficient to cause mortality by itself, but when combined with oxidizing or non-oxidizing chemical treatment, it may result in swift elimination of adult zebra mussels. Studies by Harrington et al. (1993) indicate that the combined use of oxidants and increased temperature can be more effective than either treatment alone. For example, mussels acclimated to 0–5°C (32–41°F) and exposed to 32°C (90°F) required 11.4 h to obtain 95% mortality. Under the same circumstances, but with 0.5 mg/L of chlorine or ozone present, the 95% mortality was reached in 3.5 h.

7.5.2 Carbon Dioxide and Chemical Treatment

Many aquatic invertebrates, including bivalves, do not easily tolerate even relatively small increases in the ambient concentration of carbon dioxide (CO_2). Studies have indicated that zebra mussels are sensitive to elevated carbon dioxide. Elzinga and Butzlaff (1994), building on experiments with the Asian clam, *C. fluminea*, which demonstrated narcotizing effects of carbon dioxide at concentrations of 100 mg/L and toxic effects at concentrations of 500 mg/L, investigated the effect of CO_2 on zebra mussels. The study evaluated the effectiveness of CO_2 treatment using carbon dioxide by itself and as a pretreatment to chlorination. Results of these tests indicated that a narcotizing response that is identified by gaping of the mussel can be induced by exposure of adult zebra mussels to carbon dioxide at levels from 50 to 253 mg/L in less than 6 h. It should be noted that there was a sharp drop in pH, from 7.4 to 5.2 following the carbon dioxide addition.

Payne et al. (1998) documented that pretreatment with CO_2 greatly increased the mortality rate of dreissenid mussels on subsequent exposure to lethal levels of chlorine. The initial treatment with carbon dioxide induced valve gaping, which increased mussel sensitivity to chlorination. Jenner and Polman (2003) demonstrated a similar treatment for the marine mussel *Mytilis edulis*. Combined dosing with carbon dioxide (to achieve pH of 5.3) and 0.5 mg/L TRC resulted in 100% mortality of the marine mussels within six days of exposure. When exposed to the same TRC level only, the mussels required 63 days of exposure for 100% mortality.

7.6 CHEMICAL CLEANING

Chemical cleaning is an option to be used in the event that small diameter piping or perhaps heat exchangers become plugged by mussels and mechanical cleaning is difficult or impossible at that location. Several products exist on the market, mostly proprietary inorganic acid mixtures (e.g., phosphoric acid mixtures), which will rapidly dissolve mollusk shells as well as a number of corrosion products present without harming the piping itself. The vendor generally assists in delivering the product in a suitable container. The chemical product is then circulated through the piping in a closed-loop manner for 3–4 h and then removed for recycling by the vendor. The system is then flushed and, if found satisfactory, is returned to service. This is an expedient remedy for small, neglected systems (Gauthier and Nicolaides, 1991) but is not appropriate for large volume systems.

7.7 NONCHEMICAL CONTROL

Just as there are proactive and reactive techniques for macrofouling control using chemicals, there are proactive and reactive strategies using nonchemical options.

7.7.1 Proactive Techniques

7.7.1.1 Infiltration Galleries and Sand Filters

This technology is capable of removing all stages of all mollusk macrofoulers and protecting all downstream systems and components. An infiltration gallery can be defined as "a sub-surface water collection system, typically shallow in depth, constructed with open-jointed or perforated pipes that discharge collected water into a watertight chamber from which the water is pumped to treatment facilities and into the distribution system" (Figure 7.3). It can also be described as a "built-in-place" rapid or slow sand filter. Those designed as rapid sand filters have flow rates of 7–15 L/min (2–4 gpm) per 0.1 m^2 of filter area. Others slow filter at a rate of 0.15–0.3 L/min (0.04 to −0.08 gpm) per 0.1 m^2 (Keillor, 1991).

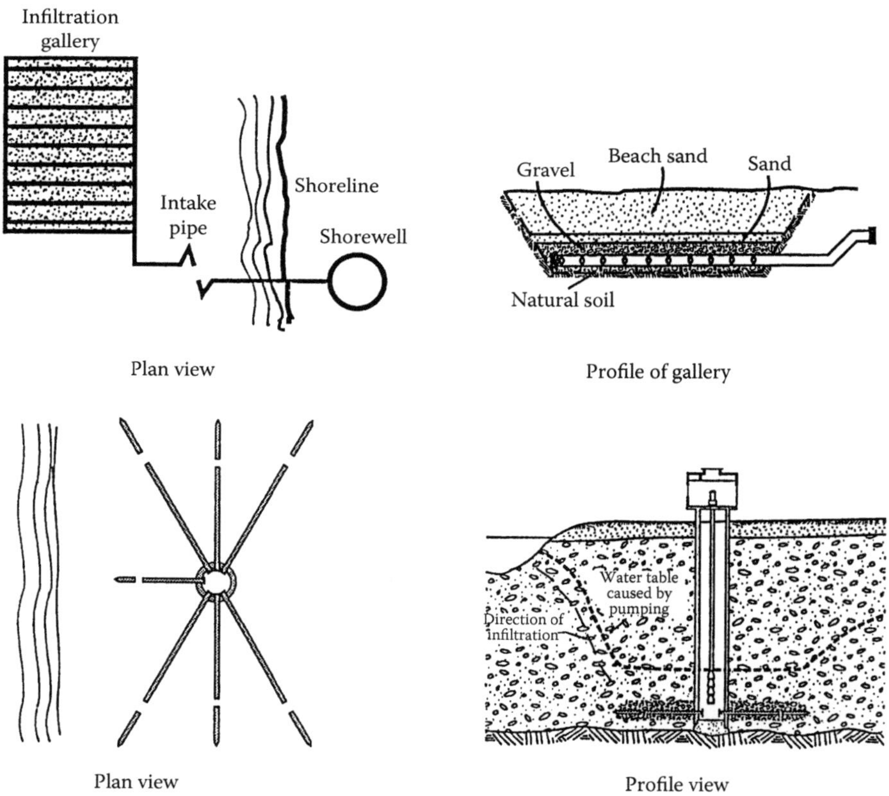

FIGURE 7.3 Typical infiltration gallery (top); typical infiltration Ranney Well (bottom).

Given these projected flow rates, obtaining large amounts of filtered water would require construction of an infiltration area of substantial dimensions. Such an undertaking undoubtedly requires regulatory approval as it takes place in or near a water body and generally involves shoreline alteration. Additional engineering factors that have to be incorporated into the design process include raw water quality, proximity to sources of high turbidity, and hydraulic considerations and cleaning method/frequency.

Although this technology offers full system protection, it is not appropriate for existing facilities using large volumes of water. The retrofit required is likely to be very expensive, difficult to implement, and may result in an unacceptable pressure drop to the system. For new intakes, an infiltration gallery could be a viable option. Currently, infiltration galleries are being considered for a number of water withdrawal applications. Jones (2008) proposes an infiltration gallery as an intake for a desalination plant as it "would eliminate undesirable elements from the feedwater such as garbage, debris, volatile organics, suspended solids and biologics such as toxic algae. Elimination of all stages of plankton greatly reduces impingement and eliminates entrainment from fouling the system."

7.7.1.2 Mechanical Filtration

This technology is capable of removing all stages of all mollusk macrofoulers if an appropriate screen size and configuration is used. Most conventional industrial strainers have strainer screen openings that will prevent some translocating macrofoulers and most shell debris from fouling the raw water systems. None however, will protect against the introduction of larval stage organisms. In most instances, it is not possible to retrofit existing strainers with finer screens and hope for successful mitigation. The performance of such modified strainer or filter tends to deteriorate, excessive clogging of the screen may result in stretching and tearing of the material, the backwash system may prove to be inadequate, and the pressure drop caused by the strainer may be unacceptable.

Different types of filters, designed primarily for the removal of small particles have been tested for dreissenid veliger control by a number of different organizations. Wedge wire slot filter/strainers have difficulty in excluding larval stages of macrofoulers. This is likely due to the fact that wedge wire-type screen filters are being designed to remove inorganic matter, such as quartz or metal shavings, but are not designed to stop organic matter from passing through the screen. Organic particles, due to their flexible nature, tend to "sneak through" the wedges of the screen.

Hydrocyclone or centrifugal separators filters were initially thought to be a mitigation option for facilities that already employ this technology for silt removal. A 1992 study by Acres (1992) tested two different centrifugal separators and concluded that the removal rate of larval mussel stages was no greater than 50%. A later study performed by Parsons and Harkins (2002), as part of a ballast water project, confirmed that a 100 μm hydrocyclone filter was 30% less effective in removing both biotic and abiotic materials than either the disk- or screen-type automatic backwash filter. The lack of success is probably due to the close to neutral buoyancy of dreissenid veligers. Recently, hydrocyclone filters have been used successfully for the removal of New Zealand mud snails in fish hatchery applications.

Excellent results were obtained by Ontario Hydro using a continuous backwash, pleated screen filter (Koopmans and Hughes, 1993) and by the New York Power Authority using a modified clean-in-place bag filter (Kahabka and Talgo, 1993) in eliminating dreissenid veligers from incoming water.

Many filters are very good at removing all or most particles from the water stream but most filters are not able to process large volumes of water efficiently. Filters which use stainless steel, square weave mesh, and automatic backwash seem to have the best balance between particle removal efficiency and volume of water filtered. A number of manufacturers produce such filters and the evaluation process of individual units can be confusing.

We offer the following points as aid in evaluation:

7.7.1.2.1 Filter Mesh

Some manufacturers do not distinguish between nominal and absolute size of the pores in the mesh they offer. It is important to understand the difference.

There are various test methods used to establish the absolute size of pores in woven wire cloths. One such widely used method is the Bubble Point Test. The absolute mesh-size rating done using this test corresponds to the diameter of the largest, hard spherical particle, which can pass through the filter medium under steady flow conditions.

Nominal value on the other hand is an arbitrary term, generally corresponding to the removal of 98% of all incident particles larger than the nominal value given. Various methods are used to determine the nominal rating and the reproducibility is poor. Therefore it is advisable to determine the quality and the absolute mesh-size rating the filter manufacturer is offering.

Smythe et al. (1993a) reports on the performance of the Kinney Strainer (equipped with mesh openings of 40, 95, or 142 μm mesh) and the Bromm Filter (nominal mesh size 60 and 100 μm). The Kinney Strainer (40-μm nominal mesh) and the Brom Filter (60-μm nominal mesh) reduced the densities of ready-to-settle veligers (>250 μm) in the filtrate by up to 97%. Given the size difference between the mesh opening and the size of veligers to be excluded, 100% elimination was anticipated. Examination of the mesh used in each case revealed that the nominal micron ratings were not reliable indicators of the largest opening found in the mesh.

Even great quality wire weave mesh will allow some organic particles greater than the absolute micron size. This is not surprising given the test protocol that uses hard spherical particles. Soft or flexible particles of a size greater than the absolute mesh pore rating but capable of distortion will be able to pass through a smaller than anticipated opening with even a small amount of backpressure. During recent filter trials, it was noted that a 120-μm absolute mesh allowed some veligers of up to 200 μm through (Claudi et al., 2009) (Figure 7.4a). A 57-μm absolute screen passed through some veligers of up to the 100 μm size (Figure 7.4b).

In order to have as much open area as possible in the mesh, the wires used to create the mesh tend to be very thin. This means that unless the mesh is properly supported, the individual pores may be distorted by pressure and the cloth may be torn by the backwash system. Strong support of the screen to prevent distortion and tearing is

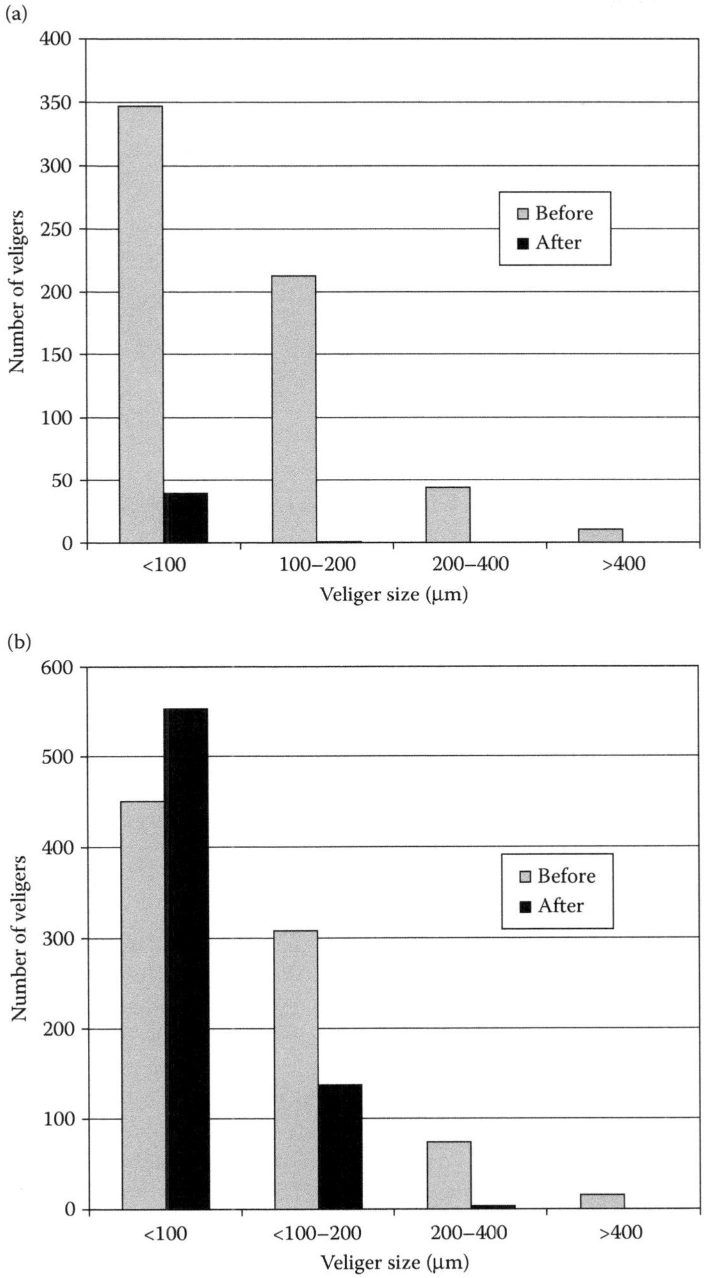

FIGURE 7.4 Results from filter evaluation tests carried out on the Lower Colorado River using (a) 40 μm and (b) 80 μm meshes.

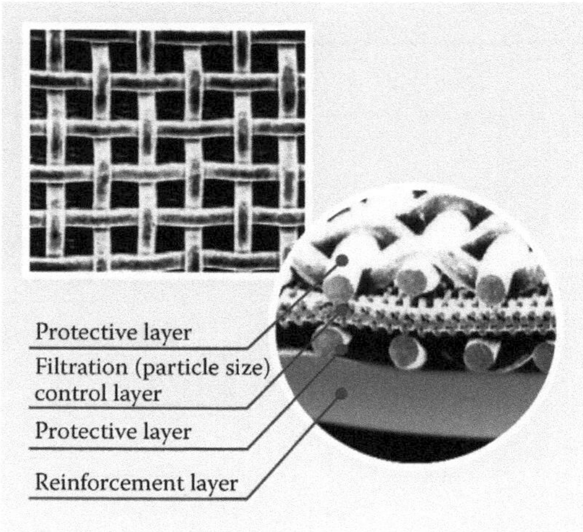

FIGURE 7.5 Filter mesh covered on both sides by protective mesh and a reinforcing layer beneath.

essential. Three to four layer "sandwich" construction is recommended (Figure 7.5). The various layers should be sintered together for best support and performance.

7.7.1.2.2 Construction

Sturdy materials of construction are a must. Plastic parts generally do not stand up to the rigors of industrial application. Excellent sealing between filter components is required to prevent water interchange between filtered and nonfiltered water in different chambers.

7.7.1.2.3 Backwash System

The more water a filter uses for its backwash, the less is available for application use. Under normal conditions, 1–3% of the total filtered flow is required for backwash. This percentage increases as the TSS load increases. The filter should be capable of backwash cycles that are based on both time elapsed and differential pressure across the screen. The greater the differential pressure across the screen, the more likely it is that soft organic material will be forced through. A differential of no more than 3–5 psi is generally recommended.

In addition to removing all larval stages of macrofoulers, filters remove substantial portions of particulate matter, such as sediment. How much they remove is a question of mesh size versus size distribution of the particulate matter in the water column. At one installation it was estimated that a self-cleaning filter, installed on a system carrying 4000 L/min of water, removed over 10,000 kg of silt each year (Rigden, 1993). Sediment removal necessarily results in improved performance and a decrease in maintenance for most industrial systems.

Filtration systems are not appropriate for water streams with continuously high sediment load. Under such conditions the backwash system may not be able to remove the sediment cake which builds up on the screen. Very efficient backwash systems are capable of coping with higher sediment loads. BallastSafe® company reported that a filter using a 40-μm screen continued to perform even when incoming water had 250 ppm of TSS. The filter flushed continuously, with the backwash consuming between 8% and 12% of the total water flow. As the amount of TSS a filter is able to cope with is somewhat related to the particle size distribution in the incoming water, site-specific trial with the filter you are considering is recommended.

7.7.1.2.4 Field Installation

Operating experiences suggest that in critical applications, a backup system (i.e., a parallel arrangement of multiple units) is a must if filtration is to be considered as a viable control measure for industrial systems. For example, two filters, each capable of filtering 100% of the flow would represent a guarantee that only filtered water reached downstream systems. If some ingress of unfiltered water can be tolerated, a system bypass may be installed to guarantee uninterrupted supply of water in case of filter upset.

To monitor the performance of the filter, at least one manufacturer suggests installing a small fixed filter screen downstream of the self-cleaning filter (Figure 7.6). The fixed screen should be the same or a slightly larger pore size than that the one used in the self-cleaning filter. If this fixed screen becomes plugged, it is an alert that the self-cleaning filter is passing particles larger than the absolute rating would suggest.

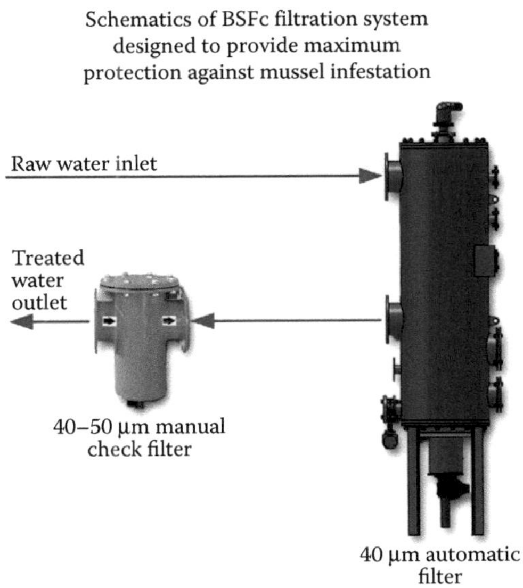

FIGURE 7.6 Schematic showing position of downstream fixed screen for filter performance monitoring. (Reprinted from the BallastSafe Filter Company. With permission.)

In the United States, the State of Vermont has successfully used a series of Amiad® automatic backwashing filters for zebra mussel control at the Edgar Weed Fish Culture Station facility since 1996 (Allhands, 2003). At this installation, due to the line pressure required by the filtration unit, the filter had to be located downstream from the pump. This requirement meant that the intake piping must be periodically mechanically cleaned using a "pigging" unit and that the pump itself must be allowed to dry out biannually, to allow for mussel desiccation and elimination.

In the fall of 1999, a full-scale filter experiment was set up at the Nanticoke TGS on Lake Erie by Ontario Power Generation. The self-cleaning filter, equipped with a 40-μm absolute woven mesh, was installed downstream of a self-cleaning wedge wire strainer with an approximately 3 mm gap. The designed filter capacity was 12,000 gpm but the system rarely required more than 6000 gpm. Typically, backwashing cycles were triggered about once every hour. During a backwash cycle, the water flow would drop down to about 4500 gpm. This represents a drop in flow of about 15–20%. The design specifications required that a minimum of 30 psi differential across the cleaning nozzles be available for the system to work effectively. This had to be increased to about 45–50 psi for the filter to cope with the incoming sediment load. The system generally ran well when total solids loading was mostly 15 mg/L or less. The filter was able to handle a loading of 60–80 mg/L, but it had to backwash almost continuously. It was concluded that the system would be unable to cope when the turbidity exceeded 80 mg/L.

In terms of efficacy, defined as the difference between veligers entering the system and those passing through the filter (dead or alive), the filter achieved between 95.9% and 100%. In all instances where veligers did pass through the filter screen, they were dead. The downstream biobox sampler was free of any adult settlement at the end of the experiment, while the biobox placed just ahead of the filter had hundreds of mussels of various sizes settled on the walls and sampling plates. Despite the impressive results, the pilot project was terminated as the site did not have the physical space to accommodate an adequate number of filters to protect the entire service water system flow. The physical dimensions of the filter could have been much less if the objective of the filter was to exclude only ready to settle veligers rather than all larval stages. A filter that uses an 80- or 100-μm absolute mesh would have a much smaller footprint than the filter tested.

Recent advances in filtration technology have allowed several manufacturers to design filters capable of removing all particles greater than 25 μm from relatively large streams of water. These filters are designed to minimize pressure drop in the system and they have corrected many of the problems encountered in the above described installation. Several of these filters are currently being tested as part of ballast water treatment systems.

7.7.1.3 Ultraviolet Light

The use of UV radiation is appropriate for larval stages of mollusk macrofoulers which have transparent shells.

The term "ultraviolet" is applied to that portion of the electromagnetic spectrum between visible light and x-rays, typically between 190 and 400 nm. This region is commonly subdivided into UVA, UVB, and UVC, where UVA corresponds to the longer wavelength (lower energy regime), through to UVC which corresponds to the shorter wavelength, higher energy end of the UV spectrum.

Based on work by numerous authors in the early 1990s, the UVB and UVC portions of the spectrum were found to be most effective for dreissenid veliger control. The basis for the control is thought to be the impact of the UV light on the essential functions of the veliger, thereby inactivating the organism and preventing attachment. The determination of the most effective wavelength and the radiation dose to achieve either immediate or latent mortality of the veligers or juveniles has been the subject of numerous studies.

In UV irradiation studies, the dose (=intensity × exposure time) and nature of the light source are important factors. Medium pressure mercury lamps have proven to be a convenient source of UV radiation since they provide reasonable intensity over the 240–310 nm range.

The original, small-scale tests at Ontario Hydro suggested that UV was effective in preventing primary settlement by veligers but it did not deter translocating juveniles. The trial evaluated the extent of mussel settlement following irradiation using a simple low-pressure germicidal Hg lamp (~7 W UVC). Veliger densities in the lake water ranged from 35,000 to 72,000 m^{-3} during the test period. The UV-treated chamber showed no primary settlement, while the control showed 2000 mussels in the 0.2–1 mm size range. The maximum estimated UV dose was 1000 mW cm^2 (=1 J cm^2) at a UV light flux of about 24 mW cm^2 (Evans et al., 1995).

Lewis and Whitby (1996) used medium pressure lamps in a flow-through system to determine both direct mortality of dreissenid veligers exposed to UV and to evaluate if settlement occurred downstream of the lamps. In ten individual experiments, the mortality of veligers averaged 87%. In five of the experiments, the veliger mortality reached 100%. The discrepancy in results was thought to be due to improper set up of the lamps during the initial tests. There was a 130-fold decrease, as compared to the control, in settlement past the UV lamps. The settlement that was found was likely due to veligers passing through during initial setup problems. The experiments supported the hypothesis that there is a threshold UV intensity or dose rate for mussel mortality, rather than a threshold based on integrated dose; that is, for a given dose, high dose rates of short duration are expected to be more effective than long, low-intensity exposures.

Wright et al. (1996) found that using 254, 280, and 297 nm wavelengths at doses of 12.5, 11.7, and 25.7 mW cm^2 s (note 1 mW cm^2 s = 0.001 J cm^2), respectively, produced mortality rates of 100%, 94%, and 87%, respectively, in quagga mussel veligers.

Wright et al. (1997), working with ready-to-settle veligers of the quagga mussel in a static setting, reported that an exposure of 7.5 s to broad-spectrum medium-pressure UV lamp resulted in 60% immediate mortality. When re-examined after 24 h, the veligers that survived the initial irradiation were all dead. Wright et al. (1997) also observed that a 60s exposure to wavelengths of 254 nm caused 100% mortality of veligers within eight days. They estimated the dose to which the veligers were exposed to as 0.1 W s cm^2 (i.e., 0.1 J cm^2).

Further studies performed by Lewis and Cairns (1998) using flow-through systems confirmed that a medium-pressure lamp with UV wavelengths between 200 and 400 nm would prohibit mussel settlement in pipes protected by a UV system.

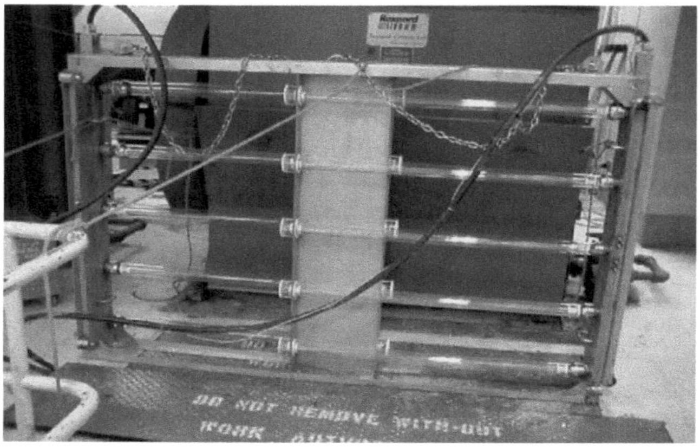

FIGURE 7.7 **(See color insert following page 286.)** UV lamp rack used in an open channel system for dreissenid veliger control.

A full-sized pilot UV system was installed at a power plant on Lake Huron in 1999. Twenty medium-pressure lamps were arranged in four frames each containing five lamps (Figure 7.7). The total volume treated was 760 L/s (12,000 US gpm). The system was sized to deliver a radiation dose of 70–100 mW s cm^2 to all particles passing through. This system was the only means of protection on this cooling system. During the operation of the UV system, lamps had to be serviced, the system experienced numerous upsets, and occasionally it was even taken out of service inadvertently. Despite these issues, there was an 85% reduction in settlement in the system, when compared to control (Pickles, 2000).

Additionally, Wight and Dawson (2002) attempted to determine an effective treatment regime for ballast water systems. They found that after the water was passed through a 50-μm filter, any remaining organisms could be eliminated by a dose of 200 mW cm^2 s applied at a flow rate of 1000 gpm.

The effectiveness of a UV system is dependent on the characteristics of the raw water being treated. Factors such as water transmittance, presence and size of suspended solids, iron, hardness, and temperature all affect the efficacy of the UV system. Treatment systems must be designed for the worst-case scenarios. This means designing for peak flows, end of lamp life intensity, minimum transmittance, and maximum suspended solids at the installation location. The aim of the system is to achieve 100% immediate or latent mortality in all ready-to-settle veligers that pass through. If an adequate dose is not delivered at the point treatment downstream settlement will occur as UV-based systems have no residual toxicity which could impact areas outside the influence of the lamps.

The UV systems currently being tested for control of various species in ballast water are rugged and dependable. They use UV arc tubes installed inside a stainless steel chamber, rather than the open array shown above and they have been engineered to deliver a maximum dose in the least amount of time (Figure 7.8). For shipboard use, UV systems are installed downstream of small-pore self-cleaning filters

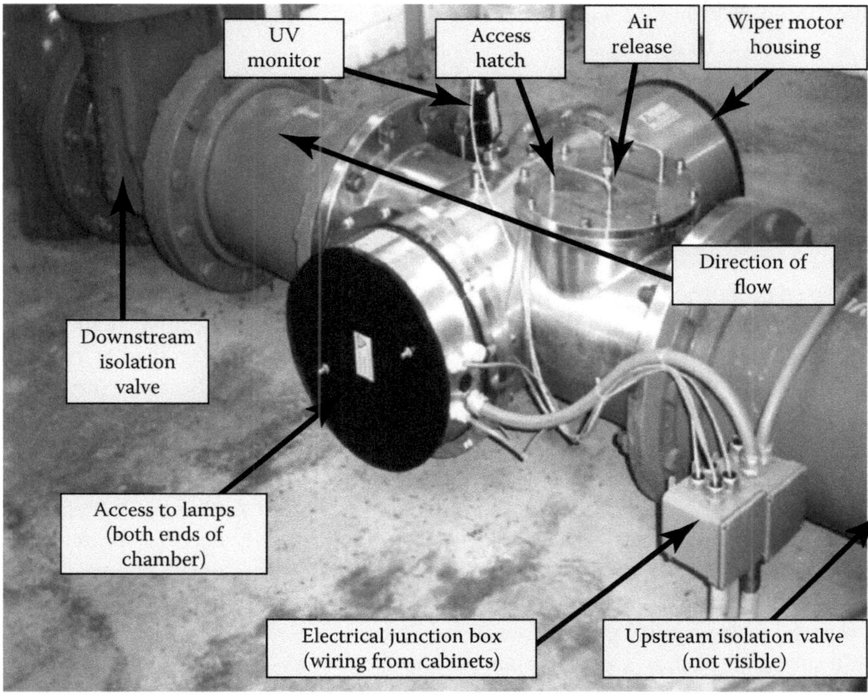

FIGURE 7.8 **(See color insert following page 286.)** Within pipe UV system for ballast water control. (Reproduced from Hanovia Ltd. With permission.)

to sterilize incoming ballast water. Such a configuration may well offer a robust, nonchemical control strategy for macrofouling control in industrial settings.

7.7.1.4 Electric Currents

Electric current can be used in a flowing stream of water to shock and disable incoming veligers, or it can be used to create an electrical field on or just above a substrate to prevent the settlement of macrofoulers. In experiments aimed at veliger destruction, veligers, postveligers, and juveniles were killed when passing through a strong electrostatic field. Exposure to 100 V/cm continuous alternating current (AC) field for 0.25 s seems to result in permanent physiological damage of veligers (McKay, 1991). The literature suggests that successful mitigation requires either a very high electrical intensity and short exposure or a longer exposure time at lower intensity to achieve the desired effect. Lange et al. (1993) tested the effects of low current fields, using both AC and DC (direct current). Using AC, at field strengths of up to 17 V/cm and veliger exposure of 0.1 s, there was no observed decrease in mussel settlement after they passed through the field. The DC voltages tested appeared to attract rather than discourage settlement.

Schoenbach et al. (1997) found that increased voltage at a reduced pulse rate was more effective than lower voltage at a higher pulse rate. Smythe and Lange (1999), using an electric pulse of 6000–8000 V/cm at 40 Hz, determined that increasing the pulse frequency showed a corresponding increase in veliger mortality and decrease in settlement activity.

Smythe and Dardeau (1999) , in reviewing the relevant literature, found that pulse rate, amplitude, duration, and amplitude shape were all important factors affecting the success of a system. Smythe and Miller (2003) found that when using a current of 100–110 A, approximately 45 pulses produced a maximum mortality rate of 40% in planktonic veligers. Studies performed on *Limnoperna* pediveligers found that pulsed electric fields of >40 V/cm at 25 pulses/s and >60 V/cm at 6 pulses/s were effective in stunning veligers and that the time required for larval recovery increased proportionately with increasing voltage (Satuito et al., 2000). The principle of increasing voltage in deterring veliger settlement has also been verified in studies by Schoenbach et al. (1997), which found that the increased voltage at a reduced pulse rate was more effective than lower voltage at a higher pulse rate. Smythe and Lange (1999) determined, using an electric pulse of 6000–8000 V/cm at 40 Hz, that increasing the pulse frequency showed a corresponding increase in veliger mortality and decrease in settlement activity.

When using electric currents to protect surfaces from settlement, a field study in 1993 suggested that electric current densities commonly used for protection of steel surfaces against corrosion in freshwater environments deterred zebra mussel settlement. Significant reduction in settlement was accomplished at 6–9 mA/ft^2 as compared to the control group (Lewis and Pawson, 1993).

Tests conducted at a power plant on the Great Lakes, using anode mesh panels in the pump wells showed a reduction of settled mussels versus control using a current density of <10 mA/ft^2 (Lewis et al., 1997). Unfortunately, the experiment was discontinued before long-term studies could be completed.

Electrodes installed at the Nanticoke Thermal Generating Station on both concrete and steel panels found that when using a voltage of 20 V and currents of 1.52 and 1.88 A, respectively, no mussels settled on either electrified surface, while the control panels contained mussel densities of 2877 m^{-2} (Fears and Mackie, 1997). In a similar study, complete prevention of settlement was achieved on both steel and wood surfaces utilizing a voltage of 3.15 V/cm (Fears and Mackie, 1995).

The overall conclusion at this time is that the voltages necessary, the required length of exposure and the amount of power needed, make this technique of control impractical for most industrial applications. We are not aware of any commercial systems currently using this technology.

7.7.1.5 Acoustics

Although the use of sound as a deterrent to mussel settlement is frequently mentioned, the research performed in this area often produces controversial results. The earliest recorded work is by Breitig (1965). He used high-frequency ultrasonics at 22 and 800 kHz, with sound intensities ranging from 2 to 5 W/cm. In this experiment, it was determined that mortality was achieved through the effects of cavitation alone and that it could cause complete veliger destruction (22 kHz) or perforation of the soft body (800 kHz). Differences in mortality rates between the two frequencies were negligible. At both frequencies 70% mortality was recorded after 3 s, 90% after 1 min, and 100% after 3 min.

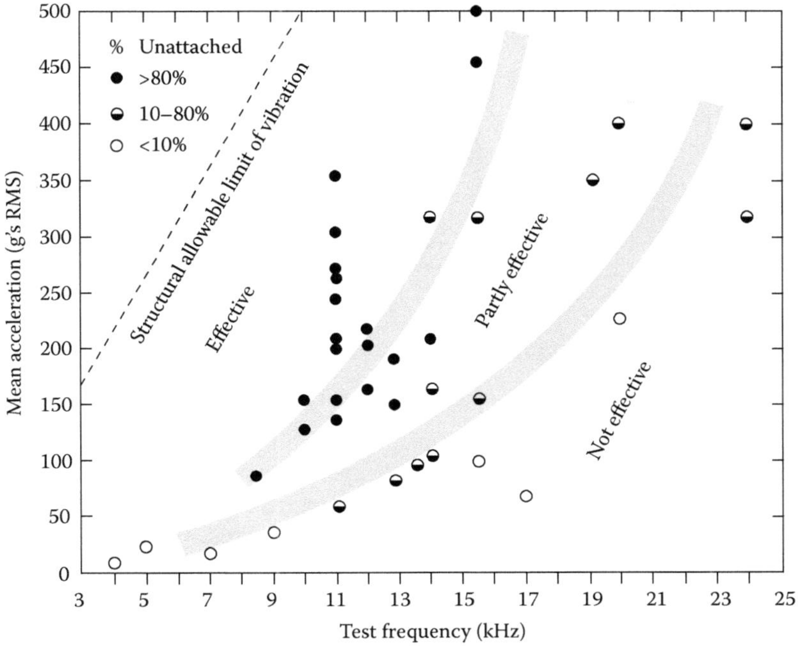

FIGURE 7.9 Attachment status of zebra mussels at various acoustic frequencies and acceleration.

Later macrofouling control tests have been performed using sound frequencies between 20 and 60 Hz and also at frequencies greater than 20 kHz. Results were often inconclusive or impractical, possibly because of the difficulty of effectively measuring, installing, and reliably operating some types of the sound-generating devices proposed. Kowalewski and Patrick (1992) subjected juvenile zebra mussels to a number of different sound frequencies and amplitudes. Figure 7.9 summarizes the results obtained.

The ESEERCO report (1992) describes how acoustic energy in the range of 39–41 kHz fragments veligers "within a few seconds" in flowing water. Attached adults were killed within 19–24 h. Donskoy and Ludyanskiy (1995) found that low-frequency sound waves below 200 Hz combined with sound waves of pressure intensity 315 Pa produced vibration which could prevent settlement. The combined vibratory effect was found to kill veligers, achieving a near 100% mortality rate using a sound frequency of <200 Hz and a sound level of 177 dB. This combination also prevented translocating juveniles and adults from settling on surfaces being bombarded with the treatment.

7.7.1.5.1 Sparker Technology

The fact that dreissenids, particularly larval stages, respond to shockwave of sound or pressure by closing the valves of their shell has been observed many times. Sharply tapping the glass of an aquarium full of dreissenids that have open shells

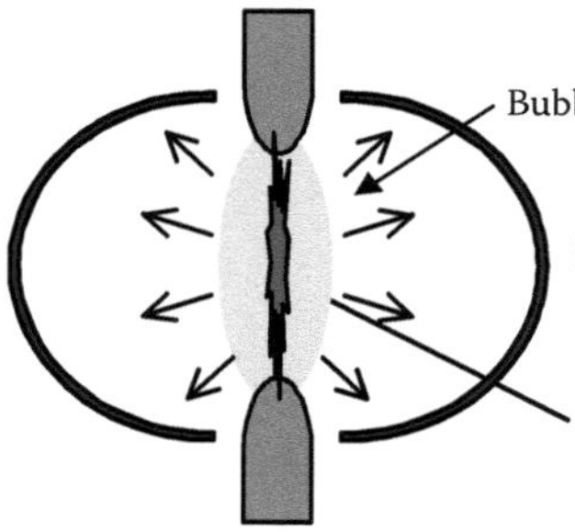

FIGURE 7.10 Sparker concept.

and are filtering will produce an immediate shell closure by most specimens. A Plasma Sparker device was used in several experiments and two field installations to produce an underwater pressure shock strong enough to cause incoming dreissenid veligers to close their shells and thereby prevent attachment.

Sparker technology is based on a high-voltage spark passing between two electrodes submerged in water. The spark arcs repeatedly between the electrodes at a predetermined rate of seconds or minutes. The electrical discharge quickly vaporizes the surrounding water producing a plasma bubble (high-temperature gas/water vapor), a pressure wave, and a light pulse. The vapor-filled "bubble" expands and contracts, producing additional pulses (Figure 7.10). The electric discharge creates a vacuum and produces UV light, breaking the molecular bonds of the water in the vicinity of the discharge. The process also produces some hydroxyl radicals ($OH^{\bullet}$).

Previous applications for this technology included water disinfection (inactivation of microorganisms), water remediation (destruction of organic contaminants), slime/algae control, and replacement for explosives for underwater profiling.

Several studies on the possible use of the Plasma Sparker technology for dreissenid mussel control have been conducted over the past decade with varying degrees of success. Mackie et al. (1999) performed a study on a sparker design by Sparktec Environmental Inc. This device utilized a high-voltage capacitive discharge of 5000 V released for 40 μs at 35 s intervals. When the device was used to treat groups of adult mussels, it produced a mortality rate of 54% versus 20% mortality recorded in control groups. Welch et al. (2000), working with the Sparker technology, found a wide range of variability in treatment effects depending on pipe radius, electrode gap spacing, and voltage potential utilized. Other, more recent Sparker designs tested incorporate a parabolic reflector to focus the shockwave and allow the pressure wave propagation along a pipeline. Such a Sparker was situated at the plant end of a 300 ft intake pipe drawing water from Lake Champlain. It was used to create a shockwave that traveled along the intake pipe right to the point of water withdrawal (Figure 7.11). The treatment concept was to expose incoming dreissenid veligers to repeated pressure shocks as they traveled through the pipeline, forcing them to close their shells and thereby prevent attachment.

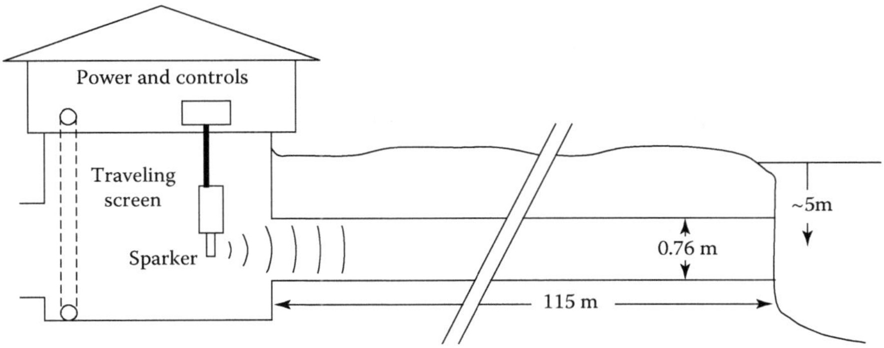

FIGURE 7.11 Schematic of Sparker installation.

Settlement of veligers was prevented up to a distance of 30 m (100 ft). At this location the pressure level was approximately 0.003 J/m^2 with a peak pressure of 7.5 psi (Schaefer and Claudi, 2004).

In a brief experiment performed on behalf of Transport Canada (ASI Group, 2005), juvenile fish, algae, and adult mussels were exposed to Sparker pulses in an enclosed vessel. Juvenile fish suffered internal injuries, but only adult mussels in the immediate vicinity of the Sparker were affected by the Sparker pulse, showing cracked and chipped shells. There were no obvious effects on the algal population.

The results to date are encouraging enough that this technology should be pursued for mussel settlement control in intake pipes. Locating the Sparker at the intake end of the pipe and increasing the number of "sparks" per minute could protect long intake pipes from dreissenid and possibly mytillid fouling.

7.7.1.6 Antifouling and Foul Release Coatings

Historically, the majority of developmental work in antifouling paints/coatings has been directed toward prevention of barnacle growth on ships. In freshwater, antifouling coatings' primary use has been for the prevention of mussel attachment to structures exposed to raw lake water. Coatings do not offer any protection to the rest of your facility, except perhaps by minimizing ingress of shell material from external structures into the piping systems.

For new facilities, appropriate selection of materials of construction may minimize the need for coatings.

Kilgour and Mackie (1993) established the following substrate preference by dreissenids by using settling plates: copper < galvanized iron < aluminum < acrylic < polyvinyl styrene (PVS) < teflon < vinyl < pressure treated wood < black steel < pine < polypropylene < asbestos < stainless steel. For pipes the preference was as follows: copper < brass < galvanized iron < aluminum < acrylic < black steel < polyethylene < polyvinyl chloride (PVC) < acrylonitrile butadiene styrene (ABS). Pipes oriented horizontally had significantly greater settlement than vertical pipes, and rough surfaces were more heavily colonized than smooth. Ackerman et al. (1996) tested the strength of attachment of zebra mussels to a variety of substrates. They found that the strength of attachment varied with substrate and within the substrate the strength of attachment increased with surface roughness. Strongest attachment was to limestone and mild steel of all roughness. Attachment was intermediate on marine concrete, polyvinyl chloride, stainless steel, and coal tar epoxy-coated mild steel. Smooth polytetrafluoroethylene, polymethylmethacrylate, and aluminum had the weakest attachment. However, most facilities have to deal with existing structures and therefore coatings present one of the best methods for minimizing the fouling on external surfaces.

Antifouling coatings containing the biocide TBTO (tributyl tin oxide, an organotin oxide) have been used successfully for the prevention of mussel attachment and growth in Europe. They are banned in Canada and in most of the United States. Copper has also been used successfully in antifouling coatings, but leaching of copper ions from such coatings may result in unacceptable copper concentrations in the discharge water. Johnson Screens Company has developed a copper alloy under the commercial name "Z-Alloy" for water intake structures which has been reported to be successful in mitigating biofouling settlement for over 10 years (Ekholm, 2000). Tests done by the U.S. Bureau of Reclamation (2008) on wedge wire screens made of Copper-Nickel (90-10) have shown quagga mussels settling heavily on this material within four months (USBR, 2009a).

The overall trend is toward the use of environmentally benign foul-release coatings that form a physical barrier to attachment. The most promising coatings at this time are nontoxic, silicone-based paints that prevent, or greatly decrease the strength of attachment (Leitch, 1993; Smithee and Kovalak, 1992). Silicone-based coatings applied to the pumpwell wall at a nuclear power plant were found to be effective at minimizing mussel settlement for almost 10 years. The silicone-based coatings usually require several different layers to be applied to a perfectly clean, white metal surface or very clean and almost dry (10% or less moisture level) concrete. This tends to make them very costly (\$80–\$100 m^2). In addition, the foul-release coatings tend to perform better in areas of high or moderate flow, rather than in quiescent areas.

There is renewed interest in developing environment-friendly coating alternatives. Diers et al. (2006) tested a product derived from sea sponges as an inhibitor of zebra mussel attachment. Sponges of the genus *Aaptos* have been observed to be free of biofouling. These sponges consistently produce aaptamine-type alkaloids. Aaptamine, isoaaptamine, and the demethylated aaptamine compounds prevented byssal attachment even when used at micromolar levels. No toxic response was observed in the nontarget organism, *Lemna pausicostata*, in a 7-day exposure. The authors propose the use of these aaptamine derivatives from *Aaptos* sp. as a potentially environmentally benign antifouling alternative to metal-based paints. Coincidentally, there have been many anecdotal observations of substrates, colonized by freshwater sponges, remaining clear of mussel fouling even in areas of high population density.

Angarano et al. (2007) investigated the plant-derived natural product capsaicin and 18 other capsaicin-like products to evaluate their potential to inhibit zebra mussel byssal attachment. The study determined that that capsaicin (8-methyl-*N*-vanillyl-*trans*-6-nonenamide), *N*-vanillylnonanamide, and *N*-benzoylmonoethanolamine benzoate all inhibited byssal attachment at levels in the micromolar range. At those levels, the active ingredients were not lethal to adult mussels or to the adults of the water flea, *Daphnia magna*.

When considering which coating to choose to protect some of the external structures of a facility, compare the cost of coatings and the problems associated with the application (i.e., sandblasting of surfaces, exposure of personnel to toxic fumes, problems of keeping large areas dewatered, etc.) with that of mechanical cleaning, and the disposal of the mussel debris on a regular basis. If after this evaluation, coating is still the appropriate strategy, carefully examine the data provided by the vendor. Make sure that the coating has been successfully used in an industrial environment for at least three years, preferably showing capability to prevent settlement of mussels. Further, make sure that no toxic substances are likely to either leach from this coating or be released into the environment when the "top-coat" of the coating is "reactivated," that is, abraded to expose fresh coating. Check with your local regulatory agency as to any possible constraints on the use of any particular product.

7.7.1.7 High-Speed Agitators

It was speculated that an in-line high-speed agitator would be able to kill or damage mussel veligers passing through. This would be done either by direct contact with the rapidly rotating blade or by hydraulic shear force. Smythe et al. (1993b) evaluated three off-the-shelf agitators. They found that none of the three agitators significantly affected the veligers present.

7.7.1.8 Speed of Flow

When the speed of flow in a raw water system exceeds 1.5 m/s (4.9 ft/s), there is minimal, if any, veliger settlement observed. Jenner et al. (1998) report that at velocities of 1.5 m/s, clusters of mussels detach in the cooling water circuits of some power plants. They also report "that zebra mussels are very abundant in areas where water velocity ranges from 10 to 50 cm/s. The species is essentially absent in areas where water velocity exceeds 1 m/s." Very few systems are designed for fast flow rates and it would involve a major expense to redesign them even if it was possible. In fact, intake structures are frequently designed to maintain as slow a rate of flow as possible to prevent entrainment of fish.

7.7.1.9 Magnetic Fields

Several studies have been performed to determine whether a standing magnetic field can perceptibly affect mussel settlement and populations. Although no solid conclusions about the mechanism responsible for any magnetic effects can be confirmed at this time, the dominant theories maintain that either the magnetism causes soft tissue damage directly, especially to the gills (Barnes et al., 1998) or else that elemental uptake and availability of certain minerals (e.g., magnesium and calcium) may be impaired. The studies performed in this area have yielded mixed results.

Smythe et al. (1997) found no significant mortality or settlement inhibition after treating mussels with water which passed through a 3000 gauss field. After 99 days, the treated mussels were not significantly different from the mussels in the control group in terms of mortality, shell length, or calcium-magnesium content in soft tissues. In 1998, experiments performed by Ryan exposed zebra mussels in a closed system to water that passed through an induced electromagnetic field. Mussels in this treatment died within 7–15 days. The loss of calcium by the mussels was thought to be the cause of mortality.

7.7.2 Nonchemical Reactive Techniques for Macrofouling Control

7.7.2.1 Thermal Shock

Hot water has proven very effective in killing macrofoulers. Thermal backwash appears feasible for some facilities and systems as an end-of-season or periodic treatment. The temperature and duration of the treatment can be combined in different ways. A temperature of 32°C (89.6°F) for 48 h has caused 100% mortality in dreissenids, so has 40°C (104°F) for 1 h. In between lies, a gray area where the exact temperature and time to death is dependent on several factors. One is the acclimation temperature of the mussels, that is, the ambient temperature of the water. The lower the acclimation temperature, the more susceptible the mussels are. A second factor is the rate of temperature increase. If the rate of increase is very gradual, the mussels may acclimate during the process and survive for a longer period than anticipated. The last factor is the possible genetic variation in local populations. It is possible that zebra mussels from a particular geographic area may be more temperature tolerant than mussels from another location.

Table 7.2 [from the study by Griffith (1992)] can be used as a guide to determine how long the mussels have to be maintained at a particular temperature to cause 100% mortality. In this study, the water temperature was raised gradually to a predetermined level and then maintained. The author observed that a very gradual increase in temperature allowed mussels to acclimate and survive longer than expected at the higher temperatures.

McMahon et al. (1993), on the other hand, developed the following multiple regression formulae for predicting the time to death for 50% of mussels (LT_{50}) and

TABLE 7.2
Summary of Resistance Times for 100% Mortality in Relation to Temperature

	Acclimation Temperature (°C)			
	5	10	20	25
Test Temperature (°C)	Resistance Time (Min)			
34	419	396	687	–
35	243	231	271	525
36	209	107	202	261
37	116	52	126	153
38	–	–	66	78

Source: After J. S. Griffiths, 1992. Ontario Hydro Report 92-65-K.

100% of all mussels (LT_{100}). In this experiment the temperature rise continued until either (LT_{50}) or (LT_{100}) was reached. Under these conditions, the slower rise in temperature actually results in death at lower overall temperature.

$$LT_{50} = 34.57 - 0.035\ (\text{min}/1°\text{C}) + 0.149\ (°\text{C Acclimation temperature})$$

$$LT_{100} = 36.10 - 0.040\ (\text{min}/1°\text{C}) + 0.147\ (°\text{C Acclimation temperature}).$$

The two studies provide useful tools when the thermal tolerance of a facility is close to that of the mussels. For example, if the temperature of acclimation is 20°C (in other words the temperature of the ambient water is 20°C) and only a rise of 1°C every 10 min can be achieved throughout the facility, then using the formula, mortality of all the mussels in the facility will be achieved when the temperature of the water in the system reaches 38.6°C. On the other hand, if the system can only be warmed to 34°C and the ambient temperature is 20°C, using Table 6.1 would predict that the treatment would have to last for 11 h.

Mytilopsis leucophaeta requires exposure to 37°C for at least 2 h to achieve 100% mortality (Jenner et al., 1998).

There are problems associated with using thermal shock. Regulations governing discharge of heated water have to be taken into account. Plants that do not already possess the capability to recirculate hot water are likely to be unable to retrofit the capability to do so and will only be able to apply thermal shock in small systems using an external heat source such as a steam generator. As plants have to be either taken off-line or production curtailed during thermal treatment, the cost of treatment tends to be quite high. In addition, manual cleaning of components may be required after a thermal treatment to clear away accumulated dead mussels and shells.

Nevertheless, a number of facilities have used this method of treatment and achieved very good results. Commonwealth Edison heat treated one of its plants by raising the water temperature from 31.6°C to 37.2°C (89° to 99°F) in 10 h. They then maintained this temperature for 6 h, resulting in 100% mussel mortality. Plants in Italy, France, and Spain have also used thermal treatments for macrofouling control. It is certainly worth considering a periodic or an end-of-season heat treatment as an alternative to chemical treatment. Most regulatory authorities regard heat treatment as a more environmentally safe and benign method than chemical treatment.

7.7.2.2 Desiccation

This process involves the draining of systems and subjecting the mussels to desiccation (drying). Unless the process is speeded up by the use of hot air circulating in the pipes, a prolonged shutdown may be required. Adult dreissenid mussels can survive considerably longer than 10 days in a cool (below 15°C), moist environment (McMahon et al., 1992). In the same study, the authors concluded that at 25°C, zebra mussels can survive for less than 150 h regardless of relative humidity. At 35°C, death occurred in less than 40 h, particularly at a high relative humidity (Figure 7.12). It appears that the inability to cool tissues through evaporation at a high relative humidity accelerates mortality rates rather than the actual loss of water from the tissues.

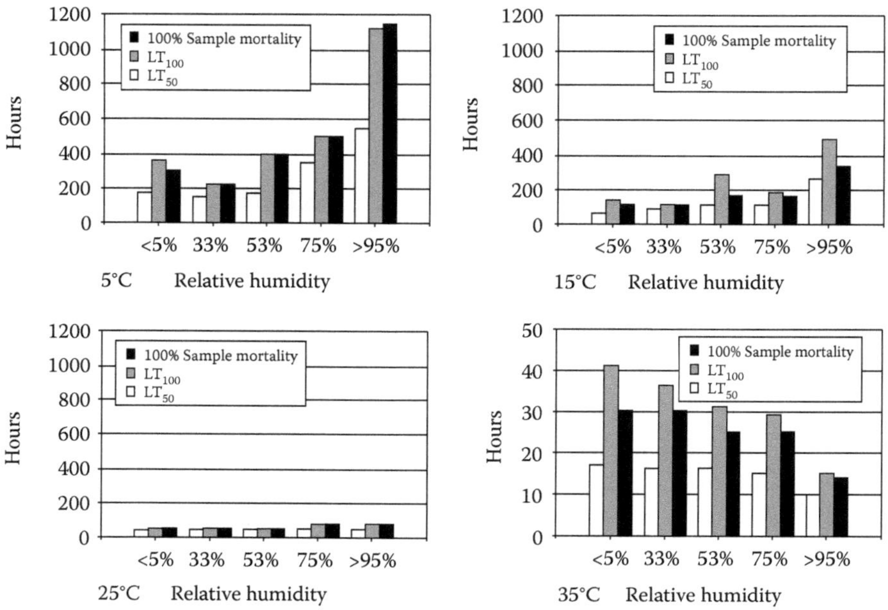

FIGURE 7.12 Effect of temperature and relative humidity on mortality of adult zebra mussels. (From R.F. McMahon, et al., 1993. *Proceedings of the Third International Zebra Mussel Conference,* Toronto, Canada, March. With permission.)

7.7.2.3 Freezing

When mussels are exposed to freezing temperatures after dewatering, mortality may occur more quickly than when they are exposed to high temperatures. Figure 7.13 from Clarke and McMahon (1993) illustrates the time to death of individual and clustered mussels exposed to several different subzero temperatures. It is interesting to note that clustered mussels survived freezing approximately twice as long as separated mussels and that time to death decreases exponentially with decreased exposure temperatures below –3°C. This would suggest that dewatering structures during low ambient temperatures is a viable and expedient way to control mussels on these structures.

7.7.2.4 Oxygen Deprivation

This strategy could be accomplished by adding an oxygen-scavenging chemical (see Chapter 4, Section 4.5.4) into a closed system or by keeping a system such as a pipeline static for a sufficient time period. Johnson and McMahon (1998) examined the role of temperature and low dissolved oxygen in causing mortality in zebra mussels and Asian clams. The following Tables 7.3 and 7.4 are an interpretation of their results.

For facilities that have two intakes but only need to use one, this could be an efficient method of control. One intake is capped until the other one is fouled and then the two are switched. The treatment would work best at high ambient water temperatures. A word of caution is given here. Lack of oxygen frequently results in a dramatic increase of sulfate-reducing bacteria which in turn are responsible for

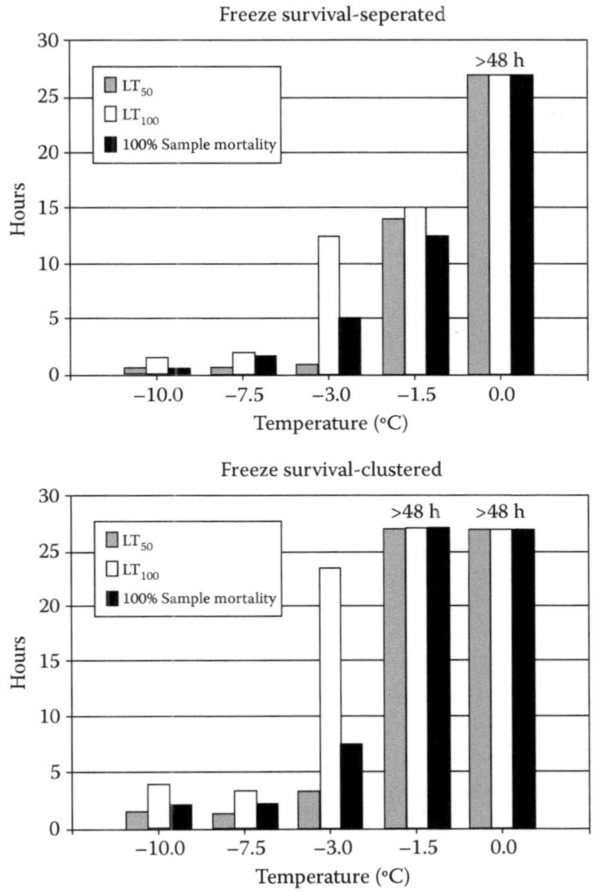

FIGURE 7.13 Freeze survival of separated and clustered adult zebra mussels. (From M. Clarke and R. F. McMahon, 1993. *The 3rd International Zebra Mussel Conference, Agenda and Abstracts,* Toronto, Canada, February. With permission.)

TABLE 7.3
Approximate Number of Days to 100% Mortality for Zebra Mussels

		Temperature (°C)		
O_2 Partial Pressure (Torr)	**O_2 % Saturation**	**5**	**15**	**25**
7.9	5.0	X	70	12
11.9	7.5		X	15
15.9	10.0		X	19
23.8	15.0	X	X	X

X = Mortality was observed but 100% mortality was not reached in the experimental time frame.

TABLE 7.4
Approximate Number of Days to 100% Mortality for Asian Clams

		Temperature (°C)		
O_2 Partial Pressure (Torr)	O_2 % Saturation	5	15	25
7.9	5.0	33	27	15
11.9	7.5	X	25	15
15.9	10.0	X	17	22

X = Mortality was observed but 100% mortality was not reached in the experimental time frame.

some microbially induced corrosion (MIC). By limiting the amount of oxygen in your system, you may exacerbate corrosion problems.

7.7.2.5 Mechanical Cleaning

Accumulated zebra mussel populations can be removed from all external structures and some large diameter piping by a variety of manual methods. This provides a short-term solution which must be repeated at regular intervals. New tools for mechanical cleaning are being introduced at such frequent intervals that it is impossible to mention them all. Further, everybody has a favourite method.

To date, all mussels removed by all types of mechanical cleaning have been disposed of in regular landfill sites or composted on site. Several tests done on these mussels did not uncover any high concentrations of toxic materials which would have forced disposal at a hazardous waste site (Evans, 1991; Kreis et al., 1991). This may not be true for locations that have high level of pollutants in the water or sediments.

7.7.2.5.1 Mechanical Cleaning of Large Diameter Pipes

Mechanical "pigs" or scrubbers have been used effectively to knock and scrape mussels and other debris from large-bore pipelines. Pigs are available in a wide variety of designs and they are manufactured to clean pipes up to 180 cm (70 in.) in diameter.

The pipeline is unavailable during cleaning, and the disposal of mussels dislodged can cause a problem. Drinking water plant intakes are particularly suitable for this method. However, several have expressed concern that their structures may not be able to withstand the pressure generated by the mechanical pig on the pipeline.

7.7.2.5.2 Underwater Cleaning Using Divers

Due to operational requirements, Ontario Hydro has concentrated on developing an efficient and economic strategy for underwater cleaning. A number of different diver operated tools and techniques were tested during the summer and fall of 1990 on a variety of infested surfaces (Wiancko and Lowther, 1991).

A continuous flow, 6-inch hydraulic pump reduced to 3-inch (600 gpm) equipped with two scraper assemblies (two diver operation) was found to be the best available option for the cleaning of vertical walls. Power wash was used on the pump bells with some success.

New diver-operated tools are being introduced all the time. This, as well as the development of remotely operated tools, could make mechanical cleaning a viable option for pipelines and external structures.

7.7.2.5.3 High- and Low-Pressure Water Cleaning

Hydroblasting or hydrolasing has been used to remove corrosion products, unwanted coatings, and biofouling. The area to be cleaned is generally dewatered and then cleaned with a jet of water. It is advisable to proceed with cleaning as soon as the structure is dewatered. If the mussels are allowed to die in place, the cleaning crew will be faced with very unpleasant working conditions. At Detroit Edison, a jet at 3000 psi was adequate to remove a thick build-up of mussels on the concrete wall of their pumpwell (Kowalak, 1992). A variety of nozzle and hose configurations is possible as is a combination of pressure and volume. The choice will depend on the personal preference of the user. However, the integrity of the surface being cleaned has to be preserved and it is desirable to remove as much of the byssal thread and byssal pad as possible. In 2008, the U.S. Bureau of Reclamation initiated a demonstration project on the use of water jetting for the removal of mussels from an underwater intake pipeline. A water jetting nozzle delivering water stream at 10,000 psi was inserted into a 25 cm (10 inch) diameter pipe that was over 100 ft in length. The pipe was heavily fouled by adult quagga mussels. The water jet was able to remove the majority of the fouling and restore the pipe to fully operational conditions (USBR, 2009b).

8 Domination by Chlorination

8.1 INTRODUCTION

Chlorination continues to be one of the most effective and popular method of microfouling and macrofouling control worldwide. For this reason, we devote an entire chapter to this control method. We will examine the behavior of chlorine in water, the different treatment strategies available, the principles of designing a problem-free chlorine delivery system, and the environmental impacts of chlorine. We will also examine the effect of chlorine on materials of construction, the dosage and consumption calculations, and the economics of chlorination.

8.2 HISTORY OF CHLORINE

Chlorine is a powerful oxidizing agent with many uses in water treatment. It is used for taste and odor control, maintenance of water quality, and disinfection. Chlorine has been used as a disinfectant for potable water for over 100 years. Its first continuous use for chlorination was in 1902 in Belgium for the dual objective of aiding coagulation and making water biologically safe (AWWA, 2006). Since its experimental introduction to the potable water supply in London in the late nineteenth century, the incidence of waterborne diseases such as typhoid fever, cholera, and dysentery has decreased dramatically. In North America the first continuous use of chlorine was in 1908 to disinfect water supply of the Jersey City water utility (AWWA, 2006). Chlorine is also effectively utilized for the control of macrofouling. Its mechanism of action occurs through:

1. Direct toxic effects on adult organisms
2. Inhibition of the settlement and growth during the larval stage
3. Weakening of the mechanisms by which the organism remains attached to the substrate

For many industrial users, chlorination using chlorine gas has been found to be a successful treatment option. Chlorine gas is kept under pressure in large tanks and added directly to the stream of water. As is the case with any gas kept under pressure, improper handling can lead to safety problems. In recent years, industrial use of gaseous chlorine has been discouraged in North America due to safety concerns surrounding accidental release in densely populated areas, as well as concerns over misuse and theft by the Department of Homeland Security in the United States. This mindset has gradually resulted in industries switching to liquid sodium hypochlorite as the chlorine source of choice.

8.3 SODIUM HYPOCHLORITE

Sodium hypochlorite, NaOCl, is considered a safe and versatile chlorinating liquid. Depending on the desired end use, various concentrations are available. For industrial use, 12% strength is generally recommended. This represents 12% available chlorine by weight. The sodium hypochlorite solution has a high pH due to the free caustic soda (sodium hydroxide) present in the liquid. This insures that the chlorine will stay in solution. Sodium hypochlorite solution is slightly heavier than water, with a specific gravity of 1.18.

Twelve percent hypochlorite solution freezes at −18°C to −21°C. Even before the freezing point is reached, some salt crystals may precipitate within this liquid. Freezing does not affect the product performance, but storage at warm temperatures does. Table 8.1 is a guide to the relative stability of sodium hypochlorite solution at different temperatures.

In some instances, dilution of the product may be required. Only deionized or distilled water should be used. If untreated water is used, the excess sodium hydroxide may react with free calcium and magnesium ions, resulting in an unstable product that will quickly decompose.

8.4 CHEMISTRY OF CHLORINATION

When hypochlorite is added to water, the following chemical reaction takes place:

$$NaOCl + H_2O \leftrightarrow HOCl + Na^+(OH)^+.$$

HOCl, hypochlorous acid, is the oxidizing agent formed in this reaction. Hypochlorous acid is a "weak" acid, which means that it tends to undergo partial dissociation:

$$HOCl \leftrightarrow H^+ + OCl^-$$

to produce a hypochlorite ion and a hydrogen ion. The amount of dissociated hypochlorite ion becomes appreciable above pH 6. When the pH value of the

TABLE 8.1
Effect of Storage on the Strength of Sodium Hypochlorite

Temperature (°C)	Month 0 (%)	Month 1 (%)	Month 2 (%)	Month 3 (%)	Month 4 (%)
21	12.3	10.7	9.5	8.5	8.5
27	12.3	10.0	8.0	6.8	8.8
32	12.3	8.3	6.2	4.6	3.0

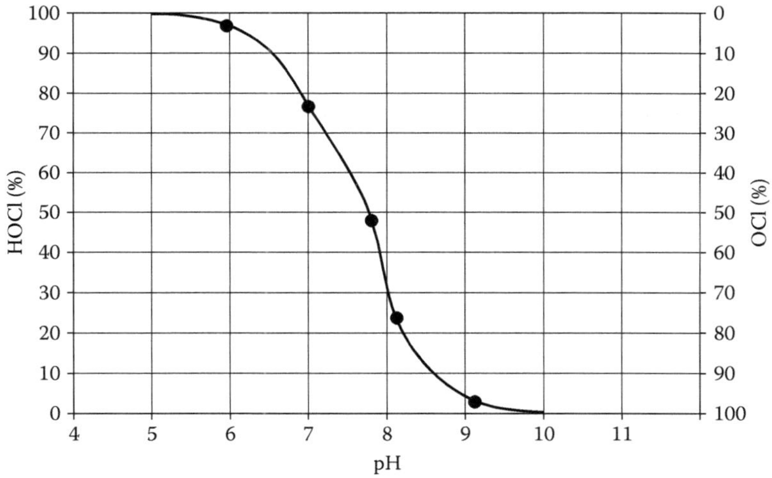

FIGURE 8.1 Dissociation of hypochlorous acid in relation to pH.

chlorinated water is 8.5, 50% of the chlorine concentration present will be undissociated hypochlorous acid and 50% will be the hypochlorite ion. Figure 8.1 shows the dissociation of hypochlorous acid versus pH. Hypochlorous acid (HOCl) is considered to have more biocidal activity than hypochlorite ion, making it a more efficient treatment agent. The effectiveness of HOCl is due to the relative ease with which it can penetrate cell walls. The hypochlorite ion (OCl^-) is less potent because its negative charge renders the ion incapable of diffusing through the cell wall of microorganisms (White, 1986). Thus, if a chemical treatment relies on the oxidative power of chlorination, more sodium hypochlorite will be needed at higher pH to obtain the required amount of HOCl. The concentration of hypochlorous acid and hypochlorite ions existing in chlorinated water represents free available chlorine (FAC).

In addition to forming hypochlorous acid and hypochlorite ions, chlorine reacts with various forms of nitrogen naturally occurring in water. Nitrogen appears in most surface waters in varying amounts as either inorganic (e.g., ammonia, nitrites, nitrates) or organic nitrogen (e.g., amino acids, proteins).

The reaction of chlorine with any compound containing the nitrogen atom with one or more hydrogen atoms attached (mostly inorganic nitrogen) will form compounds broadly classified as chloramines (Figure 8.2). These chlorine compounds include monochloramine (NH_2Cl), dichloramine ($NHCl_2$), and trichloramine (NCl_3), and represent what is called the total combined chlorine (TCC). Chloramine formation is also pH dependent. At pH 4.5, 100% of the chloramines formed will be in the form of dichloramine, while at pH 8.5, 100% of the total combined chlorine exists as monochloramines. Dichloramines are more potent disinfectants than monochloramines. At pH 4.5, using 0.5 mg/L chloramine (probably 100% dichloramine) disinfection is achieved in 20 min, whereas at pH 8.6 (probably 100% monochloramine) 60 min are required. Although chloramines are not considered to be as powerful a

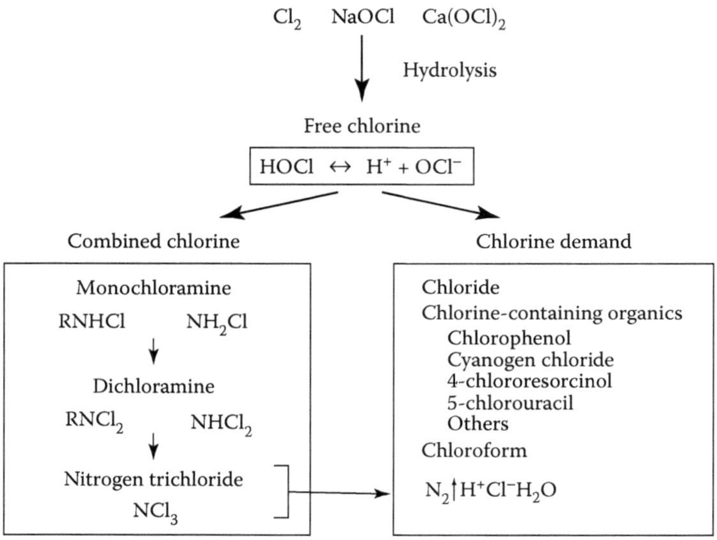

FIGURE 8.2 Chlorination by-products in freshwater.

disinfectant as hypochlorous acid, they have been shown capable of effective mollusk control. Belanger et al. (1991) observed 50% mortality in the pediveligers of Corbicula when they were exposed to 0.078 mg/L of monochloramine for 48 h at 20°C. They also found monochloramine to be completely effective when adults were exposed to 0.25 mg/L at 30°C for more than 10 days. It is therefore assumed that both the FAC and the combined chlorine, as chloramines, contribute to the overall effect chlorine as a macrofouling control agent. This combination of FAC and total combined chlorine is called total residual chlorine (TRC).

The reactions of chlorine atoms with organic nitrogen form complex chlorinated organic compounds that are referred to as the "chlorine demand" of the water being treated (Figure 8.2). The more that chlorine reacts with organic nitrogen, the less TRC will form and more hypochlorite has to be added to obtain the desired TRC level. Figure 8.3 illustrates the fact that the initial chlorine dosage produces no residual because of an immediate chlorine demand caused by fast reacting ions. Higher chlorine dosages develop chloramines. At still higher dosages, the FAC develops. The amount of chlorine demand tends to vary from one body of water to the next and it also varies seasonally.

8.5 FACTORS INFLUENCING THE EFFECTIVENESS OF CHLORINE

Many factors, some of them already mentioned (pH, organic and inorganic nitrogen content) and some listed below, can influence the effectiveness of chlorine for macrofouling control by determining the chlorine species present. All the influencing factors have to be examined when developing a treatment strategy for your facility.

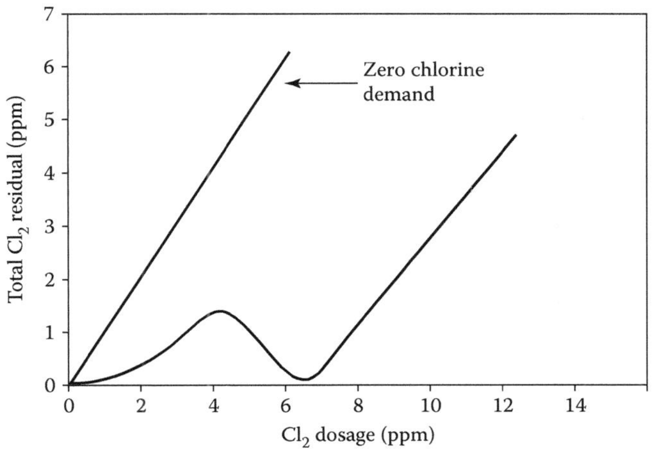

FIGURE 8.3 Theoretical chlorination curve.

8.5.1 Temperature

Temperature affects equilibrium reactions and thus the relative concentrations of the various chlorine fractions present. Similar to pH, temperature affects the relative proportions of hypochlorous acid versus hypochlorite ion. The higher the temperature, the lower the fraction of hypochlorous acid. Surprisingly, there is a synergistic effect of chlorine stress and elevated temperature, which is species specific. Overall, chlorine tolerance limits tend to decrease with increasing temperature. That may seem strange when we described how elevated temperature increases the amount of dissociation, thereby diminishing the level of HOCl in comparison to OCl^- ion.

We stated previously that OCl^- ion is less biocidally potent. Why then is there increased toxicity with elevated temperatures? Higher temperatures are thought to escalate the rate of diffusion or uptake of the chlorine compounds and further increase their chemical reactivity. Also, metabolic rates of cold blooded animals (such as mollusks) double with every 10°C rise in temperature. Thus, toxic effects occur more rapidly at higher temperatures. Increased filtration ability of *Dreissena polymorpha* at a temperature of 18.2–18.6°C supports this point (Zolotareva et al., 1978). Therefore, higher temperatures facilitate increased uptake of the chlorine compounds and, consequently, increase toxic effects.

The majority of studies have demonstrated this general trend of increasing toxicity with increasing temperature. Jenner (1978) observed that in the spring, with low water temperatures, it took more than three weeks for 100% mortality of zebra mussels at continuous chlorination of 0.4 mg/L total residual oxidants (TROs). But at the end of summer, with water temperatures of about 15°C, only two weeks were required. Seasonality may have been the confounding factor in this study. As mentioned previously, dreissenids appear more susceptible to chemical treatment in the fall than in the spring. However, van Benschoten et al. (1993) also reported that at TRC levels of 0.5 mg/L, adult zebra mussels reached 95% mortality in 42 days when the ambient water temperature was 10°C and it took only seven days at 25°C.

8.5.2 Biotic Factors

The physical state of the animal will generally have an effect on the amount of toxicant required for 100% mortality.

8.5.2.1 Age, Size, and Developmental Stage

The resistance to chlorine of a single individual within a test population can vary with age, size, and developmental stage. Most of the toxicity studies done on age- or size-related differences in chlorine tolerance tests indicate that older or larger organisms are slightly more resistant to chlorine than younger or smaller ones. Thus, the treatment of the most vulnerable stages of development does not require high levels of chlorine to produce significant mortality. In a number of different studies, the various veliger stages of the dreissenid have been shown to be vastly more susceptible to chlorine than any adults of the species.

8.5.2.2 Physiological Condition

Many biological processes may affect the sensitivity of an organism to chlorine. For example, changes in metabolic activity or levels of hormones or periodic changes in activity patterns might enhance or reduce the toxic effects of chlorine. Reproductive or disease states may also modify tolerance limits to chlorine.

In the Great Lakes region, peak reproduction for dreissenids normally occurs in July and August. At the end of a growing season, reports show that application of chlorine is more effective than when it is applied at other times. This may be due to the physiologically exhausted state of the animal following the reproductive effort or some other variable connected to seasonality. Costa et al. (2008) concluded that while the susceptibility of zebra mussels does indeed increase significantly between spring and winter, it does not seem to be related to the body condition of the mussels. General observation suggests that 30–50% of first-year spawners die following the reproductive season on the Great Lakes.

8.5.2.3 Population Biomass

It is possible that the degree of chlorine toxicity can be correlated with the ratio of population biomass relative to the concentration of chlorine oxidants applied. Thus, if natural biomass levels decrease, the magnitude of the toxic effects would increase without any change required in chlorine concentration. There are no reported studies done on dreissenids regarding this factor.

8.5.2.4 Other Water Quality Parameters

The decomposition rate of chlorine residual through oxidation–reduction (demand) reactions, volatilization as well as the extent of dilution, affects the toxicity exerted by chlorine. Waters rich in organic and inorganic compounds have high chlorine demand, consuming larger amounts of chlorine residuals. The rate at which residuals are lost is influenced by the type and amount of organic and inorganic compounds present. If reducing agents such as S^{2-}, Fe^{2+}, Mn^{2+}, and NO^{2-} are present in the water, the decomposition rate of chlorine residuals is much faster. If the chlorine dose at the intake does not take these parameters into account, then mussel mortality may be less than expected due to the reduction of the potential toxicity.

8.6 CHLORINATION TREATMENT STRATEGIES FOR ZEBRA MUSSEL CONTROL

Most toxicity studies that have been carried out involve the use of chlorine for controlling the settlement and growth of mussels on the walls of industrial and domestic intakes and raw water systems.

As mentioned previously, chlorine may be used in all five treatment strategies reviewed in Chapter 7. Toxicity of chlorine for dreissenid control is a function of chemical factors namely chlorine concentration, exposure time, and type and quantity of chlorine compounds formed. The exact dosing required for each strategy will differ based on the factors reviewed above, as well as configuration of the raw water system treated. Values given below are mostly the results of research and actual usage of chlorine by facilities on the Great Lakes.

8.6.1 END-OF-SEASON TREATMENT

The target of this treatment is to eliminate established mussels from infested systems at the end of the mussel breeding season. End-of-season treatment is acceptable if the mussel infestation accumulating in between treatments is not likely to be massive and if the affected system will tolerate a certain amount of macrofouling. Some means of dealing with the shell and soft tissue debris downstream must be in place, since there will be a large release of shells and soft body parts during the treatment. A route for disposing of this debris, such as sanitary landfill, should be secured prior to treatment. This has been a problem in some jurisdictions, mostly from the aesthetic standpoint.

Concerns about dreissenids accumulating enough toxic materials to require disposal at a hazardous waste site have so far been groundless. Chemical analysis of year-old mussels (10–15 mm in length) from several locations around the Great Lakes indicated no significant bioaccumulation of toxic substances such as heavy metals, pesticides, or herbicides (Evans, 1991; Kreis et al., 1991).

This may not be the case in locations where a significant amount of pollution is present. Roper et al. (1996, 1997) examined bioaccumulation of toxicants in the zebra mussel at the Times Beach Confined Disposal Facility, Buffalo, New York, and found that many contaminants were significantly accumulated in zebra mussel tissue. The contaminants that accumulated in wet wt. mg/kg were total polycyclic aromatic hydrocarbons (PAHs) (6.58), fluoranthene (1.23), pyrene (1.08), chrysene (0.98), benzo(a)anthracene (0.60), PCB Aroclor 1248 (1.64), As (0.97), Cr (2.87), and Ba (8.00). They suggested that accumulation of these contaminants in zebra mussel tissue represents a potentially realistic hazard to organisms (i.e., fish and birds) that feed on them. In addition, Luschützky (2005) examined bioaccumulation of heavy metals in the zebra mussel in the Danube and Drau rivers and found that the essential metals Zn and Cu were mainly accumulated in the gills, foot, and byssal gland tissue of the mussel, in contrast to the nonessential metals, Cd and Pb, that were found predominantly in the midgut gland.

In addition to disposing of shell debris at the end of the treatment, byssal threads tend to remain even after the adult mussel has died and sloughed off. The remaining byssal threads can promote under-deposit corrosion and can enhance attachment for incoming veligers during the next breeding season, possibly by slowing the flow in the boundary layer.

End-of-season treatment using sodium hypochlorite has been implemented by several hydraulic power facilities on the Great Lakes with demonstrated tolerance to macrofouling (Claudi and Van Oostrom, in print). The treatment is performed some time between mid-August and mid-September by continuous addition of 0.65 ppm TRC. It usually takes 10 days to two weeks to achieve 100% mortality of dreissenids in the system. This experience agrees with the European practice, which suggests that 0.5 mg/L TRC for two weeks at the end of the breeding season was sufficient to cause 95% mortality in zebra mussels (Jenner and Janssen-Mommen, 1993).

As the toxicity of chlorine is very much a function of the pH, ambient temperature, metabolic rate of the mussels treated, as well as their overall health, it is very difficult to define absolute levels applicable at all locations at all times. Table 8.2 summarizes research results of an end-of-season treatment experiment undertaken at the Nanticoke research site on Lake Erie in the late spring of 1992 (Evans and Sim, 1993). The lake pH was approximately 8.0–8.2, the temperature during the treatment fluctuated from 10°C to 12°C and the adult mussels used were assumed to be in peak, prespawn condition. These three factors probably resulted in the maximum amount of chlorine required for an end-of-season treatment. An inverse relationship of concentration versus exposure time was clearly present. However, if the treatment was done in the early fall, perhaps at the same temperature, the overall amount of chemical used and/or the time to death would likely be significantly less.

Another investigation of continuous chlorination impacts conducted on *Mytilus* species (EPRI, 1980), reported the results shown in Table 8.3. Once again, the inverse relationship of concentration and exposure time was noted.

Some investigators (Holmes, 1970; EPRI, 1980) stated that chlorine (chlorine-produced oxidants) caused the mussels to detach from its substrates not because of

TABLE 8.2
Summary of the Results Obtained by Evans and Sim (1993)

TRC Level (mg/L)	Amount Chlorine/Day (mg/d)	Number of Days to 80% Mortality	Total Amount of Chlorine Used (mg)	Number of Days to 100% Mortality	Total Amount of Chlorine Used (mg)
0.5	504.00	80	40,320	—	—
1.0	1008.00	42	42,336	64	64,512
1.5	1512.00	27	40,824	48	72,576
2.0	2016.00	21	42,336	42	84,672

TABLE 8.3
Time to 100% Mortality in Relation to TRO on *Mytilus*

[TRO] (mg/L)	100% Mortality (Days)
10	4
2.5	5
1.0	15

direct toxic action but because of an effect on the formation or strength of the byssal threads. The principal effect of chlorine was to depress the activity of the foot, leading to a reduction in the number of threads formed. Chlorinated mussels, with their weaker attachment systems, were swept from the walls when the speed of flow was increased. To achieve maximum effectiveness, flow rates had to be kept as high as possible, preferably above 1.5–2 m/s (James, 1967). Indeed, in direct observations of the behavior of caged dreissenid exposed to chlorinated water, declumping of mussel colonies into individual mussels was observed. If the mussel colonies were not caged, individual mussels cut or released their byssal threads and were swept from the system (Claudi, unpublished data).

8.6.2 Periodic Treatment

Periodic treatment during the breeding season is aimed at regular elimination of all adult mussels present in the system. This treatment is appropriate if the system can tolerate some degree of macrofouling but not fouling from 12 months of settlement and growth.

With periodic treatment, adults in the system will be smaller, generally less than 5 mm, so the resulting volume of debris will be less than in the end-of-season treatment. This represents the major advantage of this strategy over the previous one. Although 100% mortality is desirable, if these treatments are repeated periodically, it is not essential. The disadvantage is the increased amount of chlorine required and higher labour cost for administering the treatments. The amount of chlorine required will be roughly the same or a little less than used for the end-of-season treatment.

8.6.3 Intermittent Treatment

Intermittent treatment is aimed at the prevention of primary settlement in the system treated. It is recommended if the system cannot tolerate macrofouling caused by adult mussels, or if the operational preference is to exclude them. There are several advantages to this approach. With fragile shells, only about 350-µm long, the post-veligers will easily pass out of the system. This eliminates downstream plugging of system components and debris disposal problems. Under-deposit corrosion that can result from colonization by adults is avoided.

This strategy has been used extensively by Ontario Hydro (Claudi, 1992). A standard treatment regime consists of dosing with relatively high levels of chlorine:

2 mg/L (TRC) at the end of the system treated, for 1.5 h, at 12-h intervals, throughout the course of the zebra mussel breeding and settlement season. This treatment has proven effective on the six units of the Nanticoke Generating station on Lake Erie. These units were treated in this fashion in 1990. No new settlement was observed in these systems, but adult mussels that settled there in 1989 were still alive. Further, translocators that entered the system were able to survive the intermittent treatment. Barton (1993) also reports success of an intermittent treatment using 0.5 mg/L (TRC) at the end of the system treated, for 1.5 h, at 12-h intervals. This regime has prevented new settlement in all systems treated.

Unlike continuous treatment, the chemical addition and monitoring hardware does not need to be available 100% of the time, so it is more readily maintained. However, an outage of several days could result in newly settled veligers growing sufficiently to tolerate the intermittent treatment, thus leading to colonization of the pipework. In systems not protected by up-front strainers, the influx of translocators capable of tolerating intermittent chemical treatments may also be a problem.

8.6.4 Semicontinuous Treatment

Field observations as well as literature show that adverse physical and chemical stresses on zebra mussels result in shell closure. The duration of closure is dependent on the length and severity of the stress. The return of the mussels to normal filtering behavior (i.e., valves reopening), after removal of the stress, is slow compared to the rate of their closure response. In 1992, the semicontinuous regime was developed at Ontario Hydro to take advantage of this reaction. Frequent on/off cycling of the chlorine was postulated to be of comparable effectiveness to a truly continuous chlorination regime. In addition, this strategy was deemed to reduce the environmental impact (i.e., by-product formation and effect of residual oxidant on receiving waters) and the amount of chlorine required, by a factor of at least two.

The lower peak concentration tested, 0.3 mg/L TRC, corresponded to the lowest continuous chlorination concentration found to be completely effective in the 1991 field trials at the Nanticoke research site (Evans and Sim, 1993). The 0.5 mg/L peak concentration corresponded to the continuous chlorination level frequently used at Ontario Hydro facilities.

Figure 8.4 summarizes the results obtained with the various semicontinuous treatment regimes. It would appear that treatments consisting of 15 min on and 15 or 30 min off at the 0.5 mg/L level are as effective as a truly continuous treatment. The other treatments tested also inhibit mussel settlement, but not to the same degree. This treatment, applied for an entire mussel breeding season, not only inhibits new settlement but also causes mortality of any adult already present in the system.

This regime is ideal for facilities where several discrete systems need to be treated. The treatment can be done sequentially with a net saving in the amount of chemical used. Further, as only a portion of the facility is treated at any one time, more dilution water is available. This can assist facilities in meeting the discharge limit, and in some cases dechlorination can be avoided.

At a European power plant dealing with zebra mussels for the first time, a chlorine injection point was added to the service water system in 2008. The chlorine

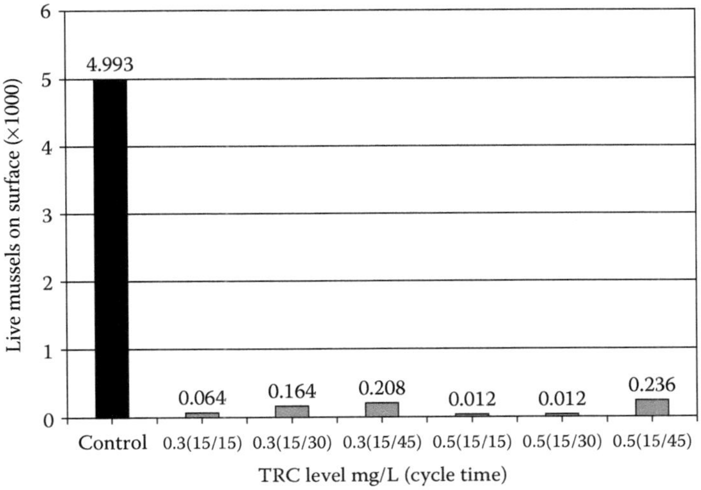

FIGURE 8.4 Mussel settlement under various semicontinuous regimes in a flow through test. (After D. W. Evans and B. F. Sim, 1993. *Third International Zebra Mussel Conference, Agenda and Abstracts*, Toronto, Canada, February.)

injection system was designed to deliver between 0.3 and 0.5 ppm TRC to the service water on a continuous basis. The chlorine injection pumps located at the start of the service water system are controlled by the output of a chlorine analyzer located at the end of the service water system just prior to discharge. When the chlorine analyzer detects levels above 0.3 ppm TRC, it sends a signal to the addition pumps to decrease injection. Chlorine injection is decreased, but this decrease is not registered by the chlorine analyzer until several minutes later when this water reaches the analyzer location. Once the analyzer sees levels below 0.3 ppm, it sends a signal to pumps to add more chlorine, but again the result of this command is not seen for several minutes; due to the lag time between chlorine addition and chlorine level detection, the system never reaches a steady state of chlorine concentration but rather cycles between a maximum and minimum value (Figure 8.5). Therefore, chlorine addition is in fact semicontinuous, rather than continuous, going from approximately 0.5 to 0.1 ppm and back up to 0.5 ppm in a 14-min cycle.

During the 2007 treatment season, this chlorine addition regime resulted in the extensive kill of zebra mussels that had settled in the service water system prior to the installation of the chlorine addition system. No new settlement was observed in the system piping during the summer breeding season (Cortes Cabanero et al., 2007).

8.6.5 Continuous Treatment

If chlorine is applied continuously at low level, the incoming veligers are able to sense the presence of noxious chemical when they enter the system. This causes them to keep their shell closed and pass through the system without any attempt

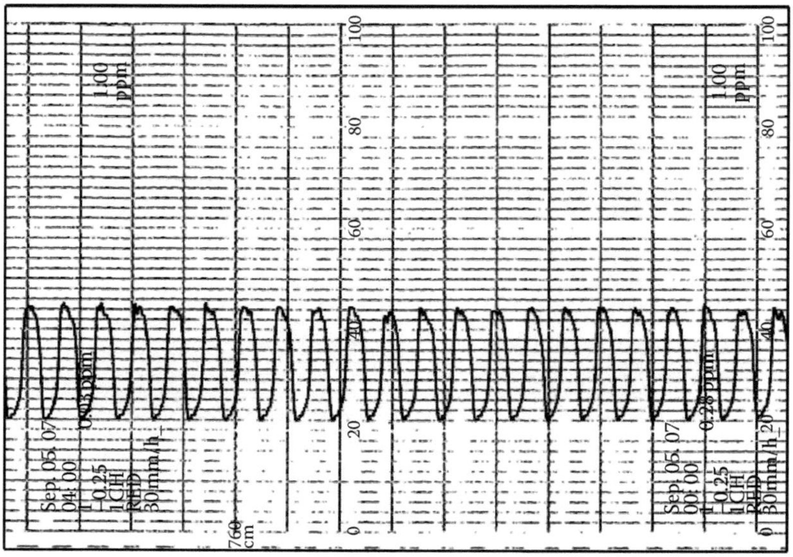

FIGURE 8.5 Tracing of chlorine concentration versus time.

to settle, provided the flow is swift. In areas of low flow, the closed valves of their shells will cause them to fall out of the water column. There they may survive for a short period, but eventually they are likely to succumb to the chronic toxicity of free chlorine or chlorine products (i.e., chloramines). Klerks et al. (1993) observed 100% mortality in veligers exposed to 0.5 mg/L TRC in static systems in 2 h.

Investigators have observed bivalve detachment from substrate due to the utilization of sublethal chlorine doses for the control of marine mussel fouling. Continuous chlorine application of 0.05 mg/L did not cause avoidance of toxicant by shell closure. Instead, continuous levels of 0.02–0.05 mg/L caused the organisms to detach after a few hours. This was followed by a flushing of mussels by the water current (EPRI, 1980). Field studies on *Mytilus* species indicated that a residual of 0.25 mg/L (analytical methods not specified and water quality not recorded) prevented the attachment and growth of mussel larvae even when the water velocities were as low as 0.4 m/s (EPRI, 1980). Similar data showed that under continuous laboratory exposures, a TRO level as low as 0.1 mg/L prevented the attachment of mussels to concrete test panels at water velocities as low as 0.76 m/s. These tests were conducted from May to September when water temperatures were 14.5–16.7°C (pH = 8.6–8.9; salinity = 31–32 ppt) (EPRI, 1980). Other studies (Holmes, 1970; EPRI, 1980) noted, however, that low-level continuous chlorination (0.05 mg/L) was not always effective in eliminating mussels from power plant cooling systems unless water velocities were greater than 1.5 m/s. Chlorination did not suppress mussel fouling in areas where the average speed was low. These reports noted that patterns of mussel fouling were related to water velocity patterns in the system. The greatest number of mussels (1000–3000 m^{-2}) occurred in areas where the speed was only 0.6 m/s. Mussel density was only 200 m^{-2} when the velocity in the system

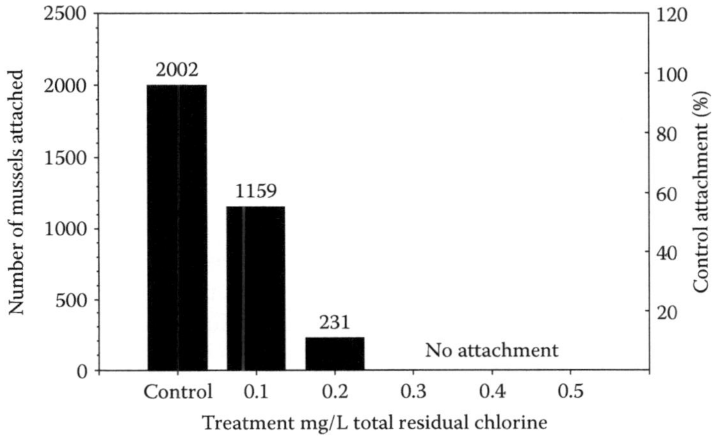

FIGURE 8.6 Continuous chlorination results from a flow through system. (Adapted from D. W. Evans and B. F. Sim, 1993. *Third International Zebra Mussel Conference, Agenda and Abstracts*, Toronto, Canada, February.)

was 1.2 m/s (Opresko and Hannon, 1979). As flows in the majority of systems tend to vary, it is difficult to rely on the speed of flow being sufficient to prevent zebra mussel infestation.

Continuous treatment, at 0.5 mg/L TRC, has been used successfully at several Ontario Hydro facilities (Claudi, 1992). This treatment not only prevented any new settlement, but a continuous application of 0.5 mg/L over 90 days also caused 100% mortality in the adult mussels that previously settled in the systems treated. This was the case in all systems, regardless of the speed of flow. In the field studies carried out in 1991, a level of 0.3 mg/L (TRC) had been observed to be as effective as 0.5 mg/L (Figure 8.6). At lower levels, some attachment of mussels was observed in the experimental chamber, but it is unclear if these animals would be capable of long-term survival (Evans and Sim, 1993). When this treatment (0.3 mg/L) was applied to a system previously colonized by adult zebra mussels, no new settlement and no surviving adults were observed at the end of the treatment period (July to November).

Continuous treatment, to our knowledge, has only been done with oxidizing chemicals. Concentrations needed to discourage settlement are lower than concentrations needed to actually kill freshly settled postveligers. This means lower peak levels of the residual oxidant will be observed in the receiving waters, compared with intermittent treatment. However, because it is a continuous addition, the total chemical loading to the receiving waters will be higher than with intermittent or semicontinuous treatment.

Continuous treatment is probably most suited to relatively small, critical subsystems (e.g., safety-related systems), where macrofouling cannot be tolerated and where a very high level of availability is required. As with all proactive treatments, under-deposit corrosion is avoided and microfouling is minimized. By limiting biofilm formed by the colonizing bacteria, the proactive treatments may provide additional control of settlement by the postveligers. Surface preconditioning by macromolecule sorption, followed by biofilm formation, is thought to be a prerequisite to freshwater and marine macrofouling. Disruption of this biofilm will inhibit subsequent postveliger attachment. Elimination of microfouling may also have performance benefits in systems containing heat exchangers.

8.6.6 Other Treatments

Many authors quote studies and field applications where chlorine levels vary greatly from those given here. It is important to remember that chlorine behaves differently in waters with different chemistry. The above values have been tried out on the Great Lakes and in some cases validated with rigorous research. The best strategy for facilities on different bodies of water is to make use of side-stream monitors (described in Chapter 4), to monitor the progress of the chemical treatment and then, if necessary, to adjust the levels of chemical used.

Further, it should be remembered that combined treatments using both heat and oxidants result in quicker mortalities than either treatment alone. If your facility lends itself to this type of combined treatment, savings in both time and amount of chemical used can be achieved.

8.7 MECHANISMS OF CHLORINE TOXICITY

It seems sensible that as chlorine is a simple compound its mechanism of toxicity to aquatic organisms should be equally straightforward and simple. This is far from reality. The literature documents at least three possible direct mechanisms of toxic action of chlorine and a host of other secondary mechanisms involving the formation of chlorine reaction products. The following are the direct mechanisms most often cited.

8.7.1 Membrane Oxidation

As mentioned previously, the form of chlorine that is considered the most effective biocide is hypochlorous acid (HOCl). This is a very strong oxidizing agent and subsequently it is very reactive. In a wide variety of organisms, the toxicity of chlorine has been reported to be due to the destruction of the respiratory membrane by oxidation, resulting in anoxia and suffocation. In fish, there have been studies to document this concept. Bass and Heath (1975, 1977) have determined histologically that the gill epithelium undergoes much damage during and after a chlorination event. They have also concluded that respiration rate, mucus production, and coughing frequency increases while blood p_{O_2} and pH decreases. They did, however, note no change in the oxygen-carrying capacity of the blood. From the data it was concluded that the primary mode of action of chlorine was gill damage, resulting in decreased oxygen exchange capability and ultimately asphyxiation. Acute chlorine exposure in people is also characterized by respiratory membrane damage (Menzel and Amdur, 1986). It seems obvious why this route would be the major route of damage since, in both human and aquatic species, this delicate membrane is in very intimate contact with the environment.

8.7.2 Action on Enzymes and Proteins

The following is the proposed mechanism by which chlorine acts to control bacterial growth. Lipids in the cell wall are attacked first, followed by the destruction of enzymes and cell structures. It has been generally concluded that the action of chlorine involves oxidation of any enzymes containing a sulfhydryl moiety. This enzyme inactivation is primarily due to the action of hypochlorous acid and occurs only at a considerably slower rate with chloramine compounds (Ingols et al., 1953; Green and Stumpf, 1946).

Many researchers attribute the toxic action of chlorine in higher organisms to its effects on blood parameters. Some observations suggest that hemoglobin is oxidized to methemoglobin with a resulting decreased oxygen-carrying capacity (Block, 1977; Block et al., 1977; Zeitoun, 1979). This is in conflict with the conclusions of Bass and Heath (1977). Other studies have found chlorine to cause a Heinz body hemolytic anemia in fish (Buckley, 1977; Eaton et al., 1973). This hemolytic anemia is ascribed to the effects of chloramines, which are secondary chlorine reaction products.

From all of the above studies it would appear that it is important to consider toxicities of TRC, rather than just FAC, when designing mussel treatments. Both mechanisms described result in the death of an organism from anoxia, and therefore either of the two above mechanisms are plausible and probably act in concert.

8.7.3 Effects on Osmoregulation

There is some indication in the literature that ion imbalance may also play a role in chlorine toxicity. Vreenegoor et al. (1977) determined that blue crabs lost the ability to excrete ammonia through their gills when exposed to chlorine. This in turn resulted in a strong ion afflux. In fact, this excretion of ammonia by the gills may mean that the gills will never come into direct contact with HOCl as the ammonium present will result in rapid formation of chloramines at the immediate vicinity of the gill.

This third mechanism is certainly feasible, although there is not much evidence to support these findings in detail. It is also questionable whether this mechanism of chlorine action is a major or minor one.

8.8 NONTARGET EFFECTS OF CHLORINE

When dealing with the chemical control of mussels, it is important to remember that other organisms could be affected by this treatment. The effects of chlorine have probably been documented better than the effects of any other chemical. However, *any chemical* entering the aquatic ecosystem is likely to have some effect on nontarget organisms.

A vast array of organisms from all trophic and taxonomic levels could be affected by chlorine. Since all species cannot possibly be included in this review, they will be divided into three classifications: fish, invertebrates, and phytoplankton. The objective of the following three sections is to provide some evidence of the impacts on the above three groups and is by no means an exhaustive review of the topic.

8.8.1 Fish

It has been suggested by several sources that when assessing chlorination toxicity, fish are the most sensitive species (Brungs, 1977). Diversity of fish populations have been seen to decrease downstream from chlorinated effluent, while no effect was observed in many other species in the same area (Tsai, 1968, 1970). In studies designed to set effluent guidelines, it was suggested that TRC levels of 0.01 mg/L from continuous treatment would affect salmon and trout. Truchan (1977) found that chlorinated effluent posed a serious hazard to fish species even at levels as low as 0.02 mg/L. In this study, a range of species, including brown trout, sunfish, and brown bullhead, were tested over two years using on-site caged fish bioassay systems at five Michigan power plants. The problem with this technique is that under normal circumstances, fish rarely remain in an area that has any traces of chlorine present. Therefore, in a lake environment, fish tend to avoid the chlorine plume. In a downstream, riverine environment, this could be more difficult.

It was noted throughout various studies that the younger fish were more sensitive to the effects of chlorine as well as temperature and dissolved oxygen (Hall et al., 1981; Heath, 1979).

In some locations, fish, particularly bass, have been observed to spawn in the discharge of power plants. As success of the hatch depends on the adult fish guarding and aerating the nest, it is suggested that chlorination for mussel control is delayed until after the hatch takes place and the young leave the area.

8.8.2 Invertebrates

The existing literature on toxicity of chlorine to invertebrates is somewhat sparse. However, an intensive study by Gregg (1974) found a great variation in the effects of chlorination depending on temperature and whether the organisms could withdraw into a shell during an intermittent chlorination event. Other investigators have noted the same wide variation in sensitivity of invertebrates, including conclusions that some invertebrate species are as sensitive as young salmon (Brungs, 1977; Gregg, 1974). There have also been observations regarding the dependence of toxicity on factors such as temperature and dissolved mineral concentration. Dickson et al. (1974) indicate that snails are particularly sensitive to the effects of chlorine. Seventy-two hour LC_{50} values for the snail were quoted at 0.04 mg/L, 477 ft downstream from the outfall of a power plant. They also suggested that toxicity of chlorine may also be a function of copper concentration. This observation should be viewed with caution since it is known that snails in particular are very sensitive to the effects of copper. The data may not be applicable to other species.

Seegert et al. (1977) have studied the effects of intermittent chlorine exposure on two species of invertebrates, *Cyclops biscuspidatus thomasi* and *Limnocalanus macrurs*. They found chlorine to be minimally toxic to both crustaceans with 30 min LC_{50} values at 10°C of 14.68 and 1.54 mg/L for *C. Biscuspidatus thomasi* and *L. Macrurs,* respectively. Other invertebrates that are affected by chlorine include *Cyclops* and *Daphnia* in which 100% mortality was seen using 0.5 mg/L for 1 h (Adams, 1927).

In summary, it is evident that the effects of chlorine on assorted invertebrates is quite varied and depends on many factors.

8.8.3 Phytoplankton

Any deleterious effect chlorine has on phytoplankton is very important since these species are the primary producers and subsequently the backbone of the aquatic community.

Algae are often the targets of many control programs. Studies have shown that free chlorine levels of 0.2–1.0 mg/L are sufficient to control algal growth (Courchene and Chapman, 1975). The actual mechanism of this inhibition was examined by Toetz et al. (1977). They found that nitrogen uptake, as nitrate and ammonia, was depressed by the addition of chlorine or chloramine. It was concluded that the toxicants act at the level of the cell membrane, to inactivate an enzyme involved in the nitrate or ammonia uptake process. Other investigators found chlorine to be extremely detrimental to the productivity of a community. Brook and Baker (1972) found both photosynthesis and respiration were depressed by 50% at chlorine levels of 0.32 mg/L, and reduced to 0 at 2.7 mg/L. Similar findings were also documented for marine systems (Hamilton et al., 1997).

Although effects on phytoplankton exist, it must be stressed that the effect of chlorination on phytoplankton is not as drastic as the effects on other organisms. Phytoplankton and aquatic plants can have generation times in the order of days. Thus, although a population could be decreased by half during one chlorination event, that population could be rapidly replaced. However, if a population is not allowed to recover, long-term deleterious effects on the community may be seen.

8.9 SUBLETHAL EFFECTS OF CHLORINE

One of the considerations when using chemical treatment is the possible sublethal effects it may have on the various aquatic species and parameters of life. These parameters include behavior, reproduction, growth, and mutagenesis. Again, these effects would have to be considered for *any* chemical entering the aquatic ecosystem.

8.9.1 Behavior

Probably the most important and most widely studied aspect of behavior with respect to chlorination is avoidance. This response is crucial to the survival of organisms in aquatic environments. Fish have been the most widely studied organisms for their avoidance response, probably because they are the easiest to handle and observe. For the measurement of avoidance various methods have been utilized. Most measure the time spent in a clean part of a tank versus time in a chlorinated part of a tank. Cherry et al. (1977) used a tank that was chlorinated at one end, not chlorinated at the other, and drained in the middle. In this study, the investigators found that fish tended to avoid lower concentrations of chlorine at lower temperatures. That is, avoidance occurred at 0.05 mg/L and at temperature of 6°C. At 30°C, avoidance occurred at 0.2 mg/L. Depending on species, avoidance was either continued throughout the 30-min exposure period, as was the case for the rosyface shiner, or avoidance declined with time, as was the case for spotted bass. It was also concluded that the HOCl fraction was the most stimulatory to the fish. Avoidance has also been observed in the blacknose dace at chlorine levels ranging from 0.51 to 1.52 mg/L TRC (Fava and Tsai, 1975). An interesting report put forth by Sprague and Drury (1969) implies that young salmon will avoid chlorine at levels of 0.001, 0.01, and 1.0 mg/L but not at 0.1 mg/L. In a follow-up study, Katz and Harder (1976) found that amines excreted by the fish, combines with the chlorine to produce chloramines, effectively "mopping" up the residual chlorine. This may explain lack of avoidance at higher concentrations but it does not explain the avoidance of the lower concentrations.

8.9.2 Reproduction

Reproductive success has always been a sensitive indicator of sublethal toxicity in mammals. Aquatic species are no exception. In one study, a level of chlorine of 0.01 mg/L was suggested to have deleterious effects on the reproduction of trout (Brungs, 1973). Similarly, Arthur and Eaton (1971) found that exposure of both the fathead minnow and the amphipod *Gammarus pseudolimnaeus* to chloramine resulted in decreased reproductive success. For the amphipod, reproduction was completely inhibited at levels of 0.054 mg/L and for the fathead minnow, spawning was almost completely inhibited at 0.085 mg/L. Various reasons have been discovered for decreased success of reproduction, from decreased sperm survivability (Muchmore and Epel, 1973) to decreased egg production by the female (Ward and DeGraeve, 1978). Khalanski and Bordet (1979) reported that continuous chlorination at the 0.2–0.5 mg/L level can cause degeneration of the reproductive tissue in *Mytilus edulis*.

8.9.3 Growth

As previously mentioned, chlorine has been used to keep the growth of algae in check. Chlorine effects on growth seem to be also substantial for other organisms. Arthur and Eaton (1971) found significant growth reduction at chlorine concentrations greater than 0.043 mg/L in the first generation of fish studied and growth reduction in the second generation only at 0.108 mg/L. Others have demonstrated growth inhibition in coho salmon. Exposure to 0.023 and 0.047 mg/L reduced growth to 80% of controls (Larson et al., 1977). Khalanski and Bordet (1979) reported growth reduction in *Mytilus edulis* when exposed to intermittent chlorination. The growth stopped completely under continuous chlorination (0.5–1 mg/L FAC).

8.9.4 Mutagenesis

The mutagenic effects of chlorine do not lie in its oxidizing form but in the many compounds that can be formed upon reaction with dissolved organic material in the water. Studies have indicated that chlorinated wastewater effluent can contain up to 103 chlorinated organics with 72 of these compounds being suspected mutagens.

8.10 DOSAGE AND CONSUMPTION CALCULATIONS

The following is a sample calculation that every facility should do to estimate the amount of sodium hypochlorite that will have to be injected for a particular treatment strategy. The outcome of this calculation will dictate the capacity of the metering system. Estimating the overall chlorine consumption will allow you to determine the size of the chlorine tanks you will require.

8.10.1 Chlorine Demand

As mentioned previously in Section 8.4, the amount of sodium hypochlorite required to maintain the desired TRC is dependent on the chlorine demand of the water as well as the system demand. The system demand is dependent on how much organic matter that will consume chlorine is in the piping. Operating experience with once through cooling water systems that have not been treated before, is that the demand is initially very high and then decreases as the system is cleaned up.

The lake water chlorine demand is site specific and it varies with the season and the runoff from either the watershed or municipal sewage treatment plants. Contact time, temperature, and mixing will influence the total chlorine demand of the water and consumption of sodium hypochlorite. As an example, a 0.5 mg/L chlorine demand is assumed by most water treatment plants on the Great Lakes in their calculations of how much sodium hypochlorite to inject. In other words, it is assumed that if you add 0.5 mg/L of chlorine to a sample of the raw water, you will not be able to measure any TRC in that sample. Most of the 0.5 mg/L will be consumed almost instantaneously by combing with organic nitrogen present in the form of amino acids, and so on.

The instantaneous chlorine demand averaged 0.2–0.3 mg/L over the chlorination season at the Nanticoke Generating station on Lake Erie in 1991, but it appeared to be only 0.02–0.04 mg/L at the Ontario Hydro Bruce facility (Lake Huron) in 1993. It is therefore imperative that you establish the chlorine demand of your incoming water. If it is unusually high, or unusually low, you will have to incorporate that fact into the design of your chlorination system. Some facilities periodically experience profound changes in water quality either due to lake turnover or due to switching from one water source to another. Under those circumstances the chlorination system has to be designed to handle the worst-case scenario.

8.10.2 Intermittent Chlorination Calculations

Basic Assumptions for Calculation

Sodium hypochlorite—12% active (10% min. to 16% max.)
Specific gravity—1.1725
1.0 L = 11,730 g
1.0 g/L = 1000 mg/L or 1 g/m^3 = 1 mg/L.

Calculation 1

No chlorine demand is present and the desired concentration is 2.0 mg/L TRC
Service water flow rate: 207 m^3/min

Hypochlorite required = 2.0 g/m^3/0.12 (12% active) = 16.7 g/m^3
Injection rate = 16.7 $g/m^3 \times 207\ m^3/min \times 1/1173$ g/L = 2.95 L/min.

Calculation 2

Chlorine demand is 0.5 mg/L and the desired chlorination level is 2.0 mg/L TRC
Dosage hypochlorite = 2.0 + 0.5 g/m^3/0.12 = 20.8 g/m^3
Injection rate =20.8 $g/m^3 \times 207\ m^3/min \times 1/1173$ g/L = 3.67 L/min.

8.10.3 Continuous Chlorination Calculations

Calculation 3

No chlorine demand is present and the desired chlorination level is 0.5 mg/L TRC
Dosage hypochlorite = 0.5 g/m^3/0.12 (12% active) = 4.17 g/m^3
Injection rate = 4.17 $g/m^3 \times 207\ m^3/min \times 1/1173$ g/L = 0.74 L/min.

Calculation 4

Chlorine demand is 0.5 mg/L and the desired chlorination level is 0.5 mg/L TRC
Dosage hypochlorite = 0.5 + 0.5 g/m^3/0.12 = 8.3 g/m^3
Injection rate = 8.3 $g/m^3 \times 207\ m^3/min \times 1/1173$ g/L = 1.46 L/min.

Table 8.4 provides summary of total hypochlorite consumption rates for intermittent and continuous treatment strategy.

8.11 MAIN COMPONENTS OF THE CHLORINATION SYSTEMS

This section by no means represents all the required components for a chlorination system as chlorination systems can be configured in many different ways. It is not the intention of this book to cover the topic in detail. Rather, the following represent some common components that have to be considered when evaluating proposed systems for the injection of liquid sodium hypochlorite.

TABLE 8.4
Daily Sodium Hypochlorite Consumption of a Hypothetical System

Chlorine demand (mg/L)	0.0	0.5	0.7	1.2
2.0 mg/L intermittent injection rate (L/min)	2.95	3.67	3.97	4.71
Volume (L/day) for 2 × 30 min doses	177	220	238	283
Total hypochlorite volume (L/day) for four systems	708	880	952	1132
0.5 mg/L continuous injection rate (L/min)	0.74	1.46	1.76	2.50
Volume (L/day)	1065	2102	2534	3600
Total hypochlorite volume (L/day) for four systems	4260	8408	10,136	14,400

Generally, a single point of chlorine application is most practical. Chlorination at the influent of the treatment process, in effective amounts for maintenance of the required TRC level throughout the entire treatment area, is the most popular control method for many industries. If this type of addition is not possible, then secondary points of chlorination may be necessary (i.e., at the discharge or intake of individual pumps).

8.11.1 Storage Tanks

The hypochlorite solution should be stored in a cool, clean, dark environment to minimize the rate of decomposition. High temperatures and exposure to sunlight will shorten product life. Storage in polyethylene chemical resistant fiberglass plastic (FRP) or polyvinyl chloride (PVC) tanks is common. The tanks have to be vented (to the outside if they are indoors) as the inevitable decomposition of the liquid results in the steady evolution of oxygen, which has to be released to the atmosphere. The vent size should be at least one and a half sizes larger than the larger of the tank inlet or outlet. Storage tanks should be sized based on projected use of chlorine. Sufficient capacity should exist to contain a four to six week supply (provided the tanks are opaque and are kept as cool as possible). Around the storage tank should be a containment basin of a volume at least 10% greater than that of the storage tank. This will prevent a spill to the environment should there be a problem with the integrity of the tank.

The tank outlet should be located off the bottom of the tank to allow some dead space for impurities to settle out. The tank outlet should be much larger than the pump suction and large enough to allow adequate draining during tank clean-out. All piping around the tank should be compatible with sodium hypochlorite. Most common is CPVC for cost-effectiveness, or plastic-lined steel pipe for long-term performance. Care should be taken to ensure that there are no threaded joints. Due to the low surface tension of sodium hypochlorite, any threaded joint will eventually leak. In place of threaded joints, the use of CPVC socket welded joints, O-ring sealed socket welded unions or flanged connections are preferred. Care should be taken to purchase certified pipe and fittings as well as the correct glue.

8.11.2 Chlorine Addition Equipment

Depending on the quantity required, sodium hypochlorite is added using either metering pumps or flow-controlled rotameter indicators. Because sodium hypochlorite solutions contain both chlorine and some excess caustic soda, the solutions are both corrosive and highly alkaline and require very carefully selected feeding equipment. The most successful metering pumps tend to be diaphragm style pumps, either motor driven for large volume systems or electric solenoid drives for small volumes. The metering pumps may have many additional features such as proportional dosing capability, low tank level alarm, or no flow alarm. The specifics you choose will depend on the system you are constructing. In all cases however, the pump should be equipped with teflon seats and seals or viton O-rings and it should be recommended for sodium hypochlorite service. *Avoid using stainless steel in sodium hypochlorite applications.*

The piping that takes sodium hypochlorite from the storage tank to the point of injection should be a fiberglass pipe with fiberglass joints as this material is most resistant to the hypochlorite solution but unfortunately this is rarely done. Alternatively, the use of well-supported CPVC pipe or plastic lined, flanged carbon steel pipe with plastic lined flanged fittings is recommended. The failure of many CPVC and PVC systems is due to improper pipe and joint preparation, incorrect glue, threaded connections and lack of proper pipe support, stressing pipe joints; causing nuisance leaks and a safety hazard.

Injection into the system takes place either into a pump well or another water body at atmospheric pressure or into a pressurized pipe at the pump discharge. The injection fitting, or diffuser, are an important feature of the system and should be designed by people with experience. Some facilities have reported "streaming" of sodium hypochlorite in common headers due to improper injection hardware. Sodium hypochlorite was not mixing evenly in the header and some systems that drew from points downstream were consistently not receiving as much chlorine as predicted.

It has been observed at many facilities that when the 12% sodium hypochlorite solution is injected into the raw water, precipitation of calcium carbonate can take place if the raw water contains high levels of calcium. This has become a problem at several facilities on the Great Lakes where the hypochlorite is mixed with the so-called carrier flow prior to the injection into the system. The piping where the initial mixing takes place tends to be of small diameter and calcium carbonate can quickly become a problem. It is also a problem for the associated valves and nozzles. If your facility is situated on a calcium-rich body of water, you need to take this problem into consideration during design.

Some hazards to be aware of: operators must wear rubber gloves when handling the chemical or maintaining the system; a splash shield should be worn and kept near the chemical area; an eyewash station is required in close proximity of the potential splash areas.

8.11.3 Chlorine Residual Analysis

In order to uphold the regulatory objectives for concentrations of residual chlorine in the discharge, one must be able to accurately analyze water samples for the presence of free and residual chlorine.

There are basically three methods for residual chlorine analysis: amperometric, colorimetric, and potentiometric. The most common method of analysis is the amperometric method. Amperometric titration, either on-line using an automatic chlorine analyzer or of discrete grab samples, is the de facto industry standard and it is the method used by most regulatory agencies to monitor chlorine residuals. It is beyond the scope of this book to discuss each method in detail. Excellent texts exist that can be consulted (Gordon et al., 1992). The following is just a very brief overview of the advantages and disadvantages as well as the detection limits of the individual methods.

8.11.3.1 Amperometric Titration

Electrochemical methods for the measurement of chlorine and chloramines in aqueous solutions were some of the earliest to be developed. Amperometry is an electrochemical technique in which, at a constant (external) potential, the current generated by the compound of interest is measured. The magnitude of the current is proportional to the concentration of that compound. It can be used for the measurement of both free and combined chlorine. The lower limit of detection appears to vary depending on the instrumentation used, operator using it, and the type of water sampled. Generally the detection limit lies in the 0.002–0.005 mg/L range (Gordon et al., 1987). A number of substances can interfere (give false readings) when samples are analyzed. Table 8.5 summarizes some of these.

On-line continuous analyzers were first introduced in the early 1950s and they continue to gain in popularity and in precision. The analyzers function on the basis of depolarization. When two electrodes of unlike material (e.g., silver and gold) are immersed in water, the ions in the water allow a DC current to be formed. In an on-line continuous analyzer, potassium iodide is used as a reagent. Any chlorine in the sample will liberate iodine and cause the current to increase. The analyzer is set to detect levels of free iodine. The end point is detected when further addition of potassium iodide does not cause any further change in the current reading. The analyzers do require periodic (typically once a season) maintenance. These devices are sensitive to sample flow and line pressure and require periodic (typically monthly or at least seasonal) calibration. It is advisable to verify the readings, from time to time, using an independent titration of the sample measured. Several automatic analyzers will output an analog signal for continuous monitoring, which can then be captured on a chart recorder or a datalogger.

Amperometry is the accepted method in the scientific and industrial community and it is the laboratory reference method used as the basis for comparisons of accuracy.

TABLE 8.5
Interferences in the Amperometric Titration of Free Chlorine

Interfering Substance	Apparent Chlorine mg/L (as Cl_2)
Iron, Fe^{3+}, 0.5 mg/L	Nil
Manganese, Mn(III), approx. 0.5 mg/L	0.02
Manganese, MnO^{4-}, 0.5 mg/L (as Cl_2)	0.35
Nitrite ion, 0.15 mg as N/L	Nil
Chlorite ion, 0.5 mg/L (as Cl_2)	0.03
Bromine, 0.5 mg/L (as Cl_2)	0.59
Iodine, 0.5 mg/L (as Cl_2)	0.65
Monochloramine, 0.5 mg/L (as Cl_2)	Nil
Dichloramine, 0.5 mg/L (as Cl_2)	Nil
Trichloramine, 0.5 mg/L (as Cl_2)	0.23

Source: After Gordon et al., 1992. Disinfectant residual measurement methods. Prepared for and published by the AWWA Research Foundation.

8.11.3.2 Colorimetric Methods

Over the years there have been numerous colorimetric methods developed for the determination of free and combined chlorine in aqueous solutions. Not many of these methods can be recommended. The most common and frequently preferred method is the *N,N*-diethyl-*p*-phenylenediamine (DPD) procedure. For detailed description of the chemical basis and analytical procedures, it is best to consult Standard Methods for Water Analysis. Colorimetric analyzers are similar to spectrophotometers. The DPD reagent is fed proportionally with the sample, into a measuring cell. A filtered light source shines a beam through the sample cell to a photodetector.

The colorimetric method has a number of weaknesses. Depending on the actual procedure used, the detection limit varies from 0.004 to 0.018 mg/L. The reagent DPD used in the colorimetric method is unstable in solution, which generally restricts its use to analysis of grab samples where the reagent is added as a tablet. The colorimetric method is not readily applied to on-line analysis. The required colorimetric calibration standards are difficult to prepare and maintain. By way of contrast, the amperometric technique is inherently an absolute method, so external standards may not be required; however, on-line analyzers in the field are usually calibrated with a standard solution concentration, which is determined in the laboratory by absolute amperometric titration.

8.11.3.3 Potentiometric Analysis

Oxidation–reduction potential analyzers utilize a platinum electrode in a sample paired with a reference cell to measure potential difference or voltage. When oxidizers are present, they exert an electromagnetic force on the platinum electrode, thus changing its electrical potential. This is then compared with the potential of the reference electrode. The potential difference is measured and displayed in millivolts. A calibration curve that compares potential difference (voltage) with known chlorine concentrations can be plotted. The calibration curve may vary from site to site and from one day to the next due to local water characteristics. Potentiometric TRC and FAC analyzers have high, on-going costs since regular replacement of the sensor is required, whereas some of the amperometric analyzers have built-in electrode cleaning systems. Also, the potentiometric methods can experience interference from chloramines, which can cause 1–50 times greater readings than the actual concentration when used for FAC analysis.

8.11.4 Dechlorination

The regulatory limits on the release of TRC to the environment tend to be very stringent. In Ontario, the objective is 0.002 mg/L TRC at the receiving waters. As this concentration is not detectable with the current equipment, the limit is set at 0.01 mg/L TRC in the discharge. Other regulators set different limits, but by and large, very little TRC or FAC is allowed in the discharge stream. Facilities that are not able to achieve the set discharge limits through dilution or sequential dosing of raw water systems have to neutralize the chlorine prior to discharge. Sodium sulfite, sulfur dioxide, or sodium bi or metabisulfite can be used. The most convenient is the use of a commercially available solution of sodium bisulfite "liquid sulfite."

The dosage requirement is 1.8–2.0 mg/L of sulfite per mg/L of residual chlorine. Reaction of sulfite with chlorine is almost instantaneous, although the reaction is slower in cold water. The sulfite can be fed into the discharge stream prior to the discharge reaching the natural waters.

8.12 SUMMARY

There are many advantages as well as disadvantages associated with the use of chlorine for macrofouling mussel control.

Advantages

1. Worldwide proven efficiency.
2. Works in most raw water systems.
3. Toxic at low concentrations with rapid loss of acute toxicity within the system treated.
4. No bioaccumulatory properties.
5. It is easy to measure TRO.
6. Costs are acceptable.
7. Chlorination systems are simple to construct and maintain.

Disadvantages

1. There are safety problems with transport and storage of liquefied chlorine.
2. Produces organo-halogen compounds in reaction with dissolved organic matter (trihalomethanes).
3. The cost of treatment varies with water quality rather than with the degree of fouling present.
4. Variations in chlorine demand could prevent the maintaining of constant TRO limits at discharge.
5. Insufficient analytical methodology for estimation of low concentrations (μg/L).

Overall, of all the chemical methods available, the use of chlorine is the best documented and most frequently used. For many facilities it is hard to justify switching from one chemical treatment to another when there are no clearly established benefits.

9 Mitigation of Aquatic Invasive Gastropods

9.1 INTRODUCTION

Chapter 7 describes control options for dreissenids, mytillids, and corbiculids. This chapter describes physical, chemical, and biological methods to mitigate impacts of aquatic invasive gastropods. By physical methods we mean both physical removal, for example by hand, and physical properties of water, for example temperature; in some cases, this section includes even mechanical removal methods. Chemical methods include use of dissolved oxygen, pH, salinity, oxidizing chemicals, nonoxidizing chemicals, and molluscicides to reduce or eradicate snail infestations. Biological control includes use of predators or parasites. Cowie (2001), after reviewing the literature, concluded that use of snails as biological controls has rarely had adequate prerelease testing of snails. The postrelease monitoring of nontarget impacts has always been incidental. Cowie (2001) stated that the use of nonnative snails for biocontrol purposes is poorly regulated; many introductions are unofficial and sometimes illegal. Cowie (2001) recommended that use of snails as biocontrol agents, if implemented, must be based on adequate prerelease testing, postrelease monitoring, and genuine concern for preservation of native biodiversity. Biological control using predators can usually only reduce pest numbers to acceptable levels, not eradicate the pest (Cowie, 2002). Howarth (1991), Civeyrel and Simberloff (1996), Simberloff and Stiling (1996a,b), and Williamson (1996), to mention a few, have made it increasingly clear that biological control can be as environmentally dangerous as the use of chemical pesticides. Also, it is probably not going to be possible to eradicate invasive gastropods from areas where they have become widely established.

Before describing various control options for the different species, it is important to echo the recommendations of Cowie (2002) that prevention of the spread of invasive species is the best way to avoid damage and the future costs of implementing management programs to control them. He recommends effective quarantine protocols at ports of entry and regulations restricting the raising, sale and purchase of snails, and movement of snails from infested to uninfested areas. "Implementing quarantine programs are highly cost-effective relative to the cost of damage once the snails have invaded" (Cowie, 2002). Public education and outreach programs are crucial to the success of attempts in limiting or even preventing the spread of invasive species.

9.2 GIANT RAMSHORN SNAIL, *MARISA CORNUARIETIS* (LINNAEUS, 1758)

9.2.1 Physical Control Options

Ferguson and Palmer (1958) state that the snail is not very successful during periods of drying and is not able to survive for long periods under wet, exposed vegetation, or in exposed mud. Therefore, control of populations of *Marisa* is possible by reducing water levels in systems where it exists. The species cannot tolerate cold water (<13°C) (Hunt, 1958) and could be eliminated under extended periods of cold temperatures.

9.2.2 Chemical Control Options

Marisa cannot tolerate salinities >9.5 ppt (Hunt, 1961). This species has been destroyed in ditches that are periodically treated with sodium pentachlorophenate (LC_{90} = 20–100 mg/L/h) or Bayluscide (LC_{90} = 3–8 mg/L/h), although the World Health Organization (1973) recommends the latter because it is less toxic to humans. Simmons (1998) states that Bayluscide meets all safety requirements by the American Environmental Protection Agency (EPA) and Environment Canada, including no long-term effects on the environment, human and animal life, no persistent residues, and does not react with other new chemicals to create hazardous substances. According to Perrett and Whitfield (1996), the main disadvantages of Bayluscide are its price and impacts on certain algae, aquatic plants, and fish.

9.2.3 Biological Control Options

There is no known biological method known for eradicating or even limiting population sizes of the giant ramshorn snail.

9.3 CHANNELED APPLESNAIL, *POMACEA CANALICULATA* (LAMARCK, 1822)

Of all the apple snail species, the channeled applesnail, also known as the golden apple snail (GAS), is most studied for management practices because of the damage it causes to rice and taro crops. However, due to taxonomic confusion, it is not entirely certain that all the studies are in fact on the GAS. Cowie (2002) states that pest management can only be successful against a background of clear understanding of the identity and biology of the pest species, which explains “why attempts to control apple snails has thus far proven extremely intractable.” He and Eversole (1992) and Halwart (1994a) recommend achieving a thorough understanding of the basic taxonomies, ecologies, and behaviors of all species before control measures can be more reliably developed and quarantine efforts can be more effective.

In general, the control options described below do not totally eradicate the snails, but most of the methods at least reduce their populations and the damage they cause. Cowie (2002) cites Olivares et al. (1992) and Litsinger and Estano (1993) who suggest that a combination of methods may eventually prove most effective, although no combination of treatments so far has proved entirely adequate. Furthermore, although many of the following measures appear to reduce snail numbers, at least to some extent, their impacts on yield losses are essentially unknown (Cowie, 2002). Rigorous quantitative assessments of this kind of integrated pest management approach are needed (Cowie, 2002).

Cowie (2002) cites van Dinther (1956), van Dinther and Stubbs (1963), Singh and Agarwal (1981), Madhu et al. (1982), Rao et al. (1983), Cheng (1989), Mochida (1991), Palis et al. (1996), and Schnorbach (1995) who have performed laboratory and/or field experiments to evaluate various pesticides against ampullariids. As a result of the ubiquitous use of pesticides in the environment and their potential harm to aquatic organisms, the literature on pesticides is enormous (Hill, 1985; Nimmo, 1985; Nimmo et al., 1987; Rand, 1995). However, Settle et al. (1996) and Cowie (2002) caution that adequate information on the environmental impacts of pesticides in rice paddy systems and their potential impacts on nontarget organisms are still largely unknown. Acosta and Pullin (1991) provide a bibliography specifically related to the control of invasive and nuisance ampullariid species. See also Cowie (2002), Levin and Kalo (2006), and Joshi and Sebastian (2006) for additional details on management options in different countries; many of the methods described below are from these references.

9.3.1 Physical/Mechanical Control Options

9.3.1.1 Handpicking

Physical removal by handpicking snails and eggs and then destroying them is an eradication option mainly for new infestations and it is the most effective nonchemical way to reduce snail numbers. Destruction of eggs can be facilitated by placing stakes in the paddy on which the snails oviposit; stakes with eggs are then readily removed (Cowie, 2002; Huynh, 2006). Eldredge (1994) and Cowie (1995) report successful eradication of *Pila conica* and some *Pomacea* sp. in Palau and Molokai (Hawaii), where all snails were manually collected from infested waters. Handpicking is a common initiative to control GASs in several countries, including the Philippines (Adalla and Magsino, 2006), China (Yin et al., 2006), Vietnam (Cuong, 2006), Japan (Wada, 2006), Indonesia (Hendarsih-Suharto, 2006), Suriname (Wiryareja and Tjoe-Awie, 2006), Taiwan (Yang et al., 2006), and Hawaii (Levin et al., 2006).

Although labor intensive, most large snails can be collected easily, and the juveniles that are missed do not grow and reproduce sufficiently quickly to become a serious problem until around harvest time (Cowie, 2002). Baskin (1996) and Naylor (1996) tout the economic feasibility of dealing with potential pest species in the early phase of invasion, prior to their becoming serious pests, rather than waiting until they are widely established and cause significant damage. The amount and frequency of effort required to achieve an acceptable level of control increases as grow rates and reproductive rates increase (Chunlin et al., 2006). Although not a delicacy for humans in Japan and Taiwan, handpicked GASs are eaten in mainland China (Yin et al., 2006) and Taiwan (Huynh, 2006). Joshi (2006) provides several recipes from the Philippines, Suriname, China, Cambodia, and Hawaii. Handpicked snails are also sold to restaurants (Yin et al., 2006). Dead snails are also processed as food for fish, ducks, pythons, crocodiles (Huynh, 2006), and shrimp (Bombeo-Tuburan et al., 1995), to mention a few.

9.3.1.2 Crop Rotation and Planting

Wada et al. (2004) and Wada (2006) state that crop rotation is mandatory in southern Japan and 30–40% of the rice fields must be used for planting soybean or plants other than rice each year. Although one year does not always result in eradication of the snails, densities are significantly reduced. Most of the new varieties of rice being developed for Asia are direct seeded rather than transplanted (Naylor, 1996; Wada, 1997).

Planting strategies and methods are considered especially important (Litsinger and Estano, 1993), and Cowie (2002) cites a couple of approaches: (i) Since susceptibility to damage declines with seedling age, planting out seedlings that are at least four to six weeks old has been recommended by Mochida (1991), Eversole (1992), Litsinger and Estano (1993), Halwart (1994a), Schnorbach (1995), Schnorbach et al. (2006), Naylor (1996), Wada et al. (2004), and Wada (2006); (ii) Increasing the seeding rate or the number of seedlings transplanted, or replacing damaged hills to offset yield loss, at least at low snail infestation levels (Halwart, 1994a).

9.3.1.3 Water Regulation

Lowering the water level or draining the paddy during the growing season will not kill the snails but it will drive the snails underground. Placing shallow trenches around the inside edges of a patch will help to concentrate the snails so they can be more easily removed (Levin et al., 2006). However, snail activity seems to be reduced if the water is shallower than their shell height (Wada, 1997). Also, periodic lowering of the water level may make the snails congregate in remaining areas of water, facilitating handpicking (above).

9.3.1.4 Raising or Lowering Water Temperature

Covering the paddy/patch with black plastic sheeting on sunny days has been suggested in Hawaii but is obviously impractical on a large scale. It can only be done during fallow periods, and would have to increase the water temperature probably to well over 45°C for an extended period in order to kill the snails (see temperature tolerance, above). The snails might be able to avoid the high temperatures by burying into the mud. Normal water temperatures in rice paddies probably rarely exceed 36°C (Halwart, 1994b). However, manipulating water temperature below lethal limits might impact snail activity, growth and reproduction, and deserves further research.

9.3.1.5 Grills and Metal Screens

Wire mesh grills are placed at the inlets and outlets to the rice paddies, taro patches, and so on (Adalla and Magsino, 2006; Cuong, 2006). The screens prevent at least the larger snails from moving between paddies, and snails that collect in the grills can be easily collected and destroyed. Snails are known to avoid strips of bare copper; placing copper barriers around taro patches has been suggested as a means of keeping snails out of uninfested small patches (Glover and Campbell, 1994; Levin and Kalo, 2004).

9.3.1.6 Baits

Cowie (2002) cites two studies: Glover and Campbell (1994) suggested the use of baits (sacks or nets filled with lettuce, cassava leaves, sweet potato leaves, taro leaves) to divert the snails from eating taro and facilitate handpicking of snails congregating at the baits. Eversole (1992) suggested the use of papaya leaves and promotion of the growth of Azolla. Levin and Kalo (2006) report the potential of using neem leaves with papaya extract to attract snails; once the plant is eaten, snails die within nine days. However, Cowie (2002) cautions that baits have to be significantly more attractive to the snails than the crop and it is possible that providing additional food as baits may enhance snail numbers. Indeed, neem leaves alone tend to increase the numbers of egg cases laid (Levin and Kalo, 2006).

9.3.1.7 Tilling Soil

In Japan, tillage or harrowing mechanically reduced the density of over-wintering snails, especially when tilled during winter because it crushed the snails and exposed them to the cold (Litsinger and Estano, 1993; Wada, 1997; Wada, 2006). Growing wheat as an off-season crop was also said to reduce snail numbers (Wada, 1997) or burning rice stalks after harvest kills snails near the surface of the mud, and the ash reportedly repels the snails (Eversole, 1992).

9.3.2 Chemical Control Options

Halwart (1994a) lists several synthetic chemicals that have been used against apple snails, including copper sulfate, calcium cyanamide, sodium pentachlorophenolate, niclosamide (Bayluscide), metaldehyde, various organo-tin compounds, endosulfan, cartap hydrochloride (Dimotin), and isazophos. Cowie (2002) states that Bayluscide, metaldehyde (Snailkil, Porsnail), Dimotin, and isazophos (Miral) are molluscicides that have become available in North America. In China, 6% (0.5–0.7 kg mixed with 5–10 kg fine sand scattered after rain fall) and 8% granules (22.5–30 kg/ha mixed with 9.5–110 kg fine soil and then directed at the base of plant) are used when GAS density reaches 2–3 m^{-2} (Chunlin et al., 2006); Wu and Xie (2006) achieved kill rates of 86% with 98% copper sulfate, 96% with 70% niclosamide, 98% with 98% cartap and 65% santobrite. Metaldehyde reduced snail damage significantly from 96% to 6% on Asian hills (Halwart, 1994a). Cheng and Kao (2006) state that metaldehyde is also used in Taiwan but it has slow reaction time and is highly temperature dependent and has poor kill rates in spring and winter months.

One of the problems with molluscicides is that more often than not, some snails can avoid exposure to the chemicals. Some individuals bury in the soil, others avoid exposure by crawling out of treated water on clay clumps or emergent vegetation (Halwart, 1994a; Van Dinther, 1973). These snails usually recover quickly from pesticide applications, partly because the half-life of the chemical is quite short. Also egg clutches are laid above the water and are sources of constant reinfestation (Halwart, 1994a). For this reason, a second application, perhaps a month after the first, is necessary to kill the newly hatched snails before they grow to reproductive maturity (Halwart, 1994a).

A more serious problem is the toxicity of some of the synthetic chemicals to nontarget organisms, including humans. Endosulfan is highly toxic to fish, shrimp, humans, birds, beetles, wasps, and so on and is now prohibited in Vietnam and Philippines (Huynh, 2006), and most rice-growing countries (Preap et al., 2006). Copper sulfate is highly toxic to most invertebrates and broad-scale application has potentially serious environmental consequences (Cowie, 1994). Copper sulfate is still used in Vietnam (Huynh, 2006) and was used in the 1950s in Surinam but has since been replaced by metaldehyde and fentin acetate, the only two pesticides currently legally registered there (Wiryareja and Tjoe-Aqie, 2006).

Among the least toxic are Bayluscide and metaldehyde because they provide effective control without negative impacts on either the environment or humankind and they degrade quickly (Schnorbach et al., 2006). Bayluscide is a lampricide and molluscicide, with an LC_{100} for mollusks, including snails that are intermediate hosts of trematode parasites of *Schistosoma* spp., at 1.0–5.0 mg/L (Schnorbach et al., 2006). However, the chemical is acutely toxic to fish at LC_{50} levels of 0.05 mg/L for rainbow trout and 0.23 mg/L for goldfish. The half-life of Bayluscide ranges from 0.8 to 2.5 days depending on the water system (Calumpang et al., 1995). Metaldehyde is highly selective for slugs and snails, has an LC_{100} of 2 mg/L, does not impact earthworms, beetles, or soil bacteria until 1000 mg/L and has a low toxicity to fish at LC_{50} values of 75 mg/L for rainbow trout and 100 mg/L for carp (Schnorbach et al., 2006). Metaldehyde completely degrades in water in 14 days and in soil within 22 days (Calumpang et al., 1995).

In laboratory bioassays, neem (cake, oil, and leaves) and papaya extracts containing the active pesticidal ingredient, azadirachtin, yielded 96% mortality within nine days; the snails stopped eating after one day of exposure (Levin and Kalo, 2006). Snails eating fresh neem leaves alone produced more egg cases than controls, suggesting neem leaves or cakes could be used to attract snails.

9.3.3 Biological Control Options

To date, there is no known specific parasite or predator of *Pomacea* snails that might be used for biological control in Asia. Exception to this is perhaps the predatory snail, *Marisa cornuarietis*, which feeds on *Pomacea* snails but which may also damage rice plants (Halwart, 1994a). It is unlikely that it would out-compete *Pomacea canaliculata*, or prey on it sufficiently to reduce its numbers (Cowie, 2002). In any case, it can be used only in countries where it already occurs. In this respect, *Melanoides tuberculata* has been used as biological control competitors in the Caribbean (Pointier and Guyard, 1992). However, *Melanoides tuberculata* already occurs in South-East Asia (and alongside ampullariids in Hawaii) and it is highly unlikely that it could displace the pest ampullariids (Cowie, 2002).

Debilitating diseases caused by viruses, bacteria, and fungi are rare in *Pomacea* species and are unlikely as biological control measures (Godan, 1983). Little is also known of microorganisms and of parasites that attack either the snails or their eggs. Most of the literature is on use of natural predators of GASs.

In Southeast Asia, various fish, birds (especially snail kites), rats, lizards, frogs, toads, beetles, and ants are known to feed on introduced apple snails or their eggs (Snyder and Snyder, 1970; Halwart, 1994a; Cazzaniga, 2006). Of these, only ducks and fish have attracted any serious consideration as potential control agents (Halwart, 1994a,b; Cuong, 2006; Huynh, 2006; Tamaru et al., 2006); whether any of these other predators could play a role in regulation of introduced apple snail populations is unknown. None of the predators of apple snails in their native ranges have been shown to play a significant role in snail population eradication. Some of these, especially rats, also cause serious damage to rice, and introduction or promotion of others as biocontrol agents may have unknown environmental consequences (Cowie, 2002).

Wong et al. (2008) proposed the use of common carp (*Cyprinus carpio*) as a biological control agent, based on an 8-week enclosure experiment; they quantified the impact of common carp on three species of aquatic macrophytes and nine species of snails, including the apple snail, in a shallow pond. The results showed that the apple snail or carp alone significantly reduced the plant biomass, although the apple snail had a stronger overall herbivorous effect than the carp. The carp completely removed juvenile apple snails, but had only a weak predatory effect on larger apple snails and no effect on the adults' oviposition frequency. Furthermore, the carp significantly reduced the populations of most species of other snails that occurred naturally in the pond. Wong et al. concluded that common carp can be an effective biological control agent against the invasive apple snail, but caution should be taken about its potential to reduce wetland floral and faunal diversity.

In Japan, Yusa (2006) examined the effectiveness and utilization of a wide variety of invertebrates (e.g., shrimp, crayfish, crabs, leeches, dragonfly larvae) and vertebrates (e.g., ducks, fish, frogs, turtles, rats) as biological control agents of both juvenile and adults GAS. He stated that the choice of predator candidates depends on three characteristics: effectiveness, cost, and negative impacts. Those eating juveniles may be able to regulate snail populations if used continuously; in this respect Yusa (2006) recommends carp, turtles, and ducks.

If using ducks, care must be taken that they do not damage the plants, especially new plantings (Kobayashi et al., 1993; Glover and Campbell, 1994). It is recommended that domestic ducks be released into the paddies prior to planting of seedlings, again once the seedlings have established (35–40 days after transplanting, otherwise the ducks may damage the seedlings), and finally, after harvest (Cowie, 2002). Significant reduction of both snail numbers and damaged rice hills can be achieved although the ducks prefer to eat juvenile snails (Halwart, 1994a; Huynh, 2006).

Carp (black and/or common) or tilapia are used as biological control agents in many countries including Indonesia (Hendarsih-Suharto et al., 2006), Taiwan (Mochida, 1988a,b, 1991; Yang et al., 2006), Vietnam (Cuong, 2006), China (Wu and Xie, 2006), Japan (Yusa, 2006), and Philippines (Halwart, 1994a,b; Halwart, 2006). Since the fish require sufficient water depth, they are generally used only at the start of each rice-growing season (Halwart, 1994b). Whether reducing snail numbers will translate into increased crop yield is unknown, but fishing enthusiasts use GAS as bait to catch black carp in Taiwan; this has created a new market for GAS and the fish can also be harvested before fields dry up and sold for additional profit (Yang et al., 2006).

9.4 TITAN APPLESNAIL, *POMACEA HAUSTRUM* (REEVE, 1856)

9.4.1 Physical Control Options

Very little is known about physical control options for limiting or eradicating populations of *Pomacea haustrum*. Handpicking probably has the most potential for reducing numbers of snails. One can also scrape off the egg masses from their attachment site and allow them to fall into the water since submersed eggs will not hatch.

9.4.2 Chemical Control Options

Elimination of titan apple snails by chemical means has not been reported in the literature, either because no effective chemical recommendation has been developed or the species' presence in North America is limited to Florida, and while the species feeds on all types of aquatic macrophytes, it has not required control. Rawlings et al. (2007) stated that *Pomacea haustrum* is currently of "relatively minor concern in the United States, given its failure to spread beyond Palm Beach County after 30 years or more in Florida." However, they cited Cox (2004) who cautioned that many species have maintained limited distributions, sometimes for decades, before becoming invasive.

9.4.3 Biological Control Options

The titan applesnail has been touted as a biological control agent of the planorbids, *Biomphalaria straminea*, *Biomphalaria tenagophila*, and *Biomphalaria glabrata* through the predation of their egg masses (Milward-de-Andrade, 1978; Guimares, 1983), but biological control of the applesnail is unknown.

9.5 ISLAND APPLESNAIL, *POMACEA INSULARUM* (D'ORBIGNY, 1839)

Rawlings et al. (2007) believe that some of the ecological and agricultural impacts in Asia associated with *Pomacea canaliculata* are almost certainly attributable to *Pomacea insularum*. They suggest that the species has not been explicitly acknowledged as a serious pest because of the confusion in identification of these two species, with most of the literature referring to *Pomacea canaliculata*. The species first appeared in the United States in Texas, probably in 1989 (Howells et al., 2006; Rawlings et al., 2007) and has since spread to Florida and Georgia. Given its rapid dispersal potential, the island applesnail may have impacts on aquatic ecosystems as significant as *Pomacea canaliculata*. Howells (2002) and Howells and Smith (2002) in particular expressed concern regarding the potential negative impacts that *Pomacea insularum* may have on natural aquatic habitats and rice agriculture in southeast Texas and other portions of the southern United States. In this scenario, many of the control options given for *Pomacea canaliculata* may likely apply to *Pomacea insularum*.

9.5.1 Physical Control Options

Handpicking adults and egg masses and then destroying them appears to be the most common method used for limiting population sizes of *Pomacea canaliculata* (see above), and probably applies here.

As part of a management strategy aimed at improving habitat conditions for fish populations, a drawdown of Lake Tohopekaliga, Florida, occurred between late 2003 and early 2004, and an intensive scraping treatment of littoral vegetation followed Kitchens et al. (2002). Prior to the drawdown and scraping treatment, native Florida applesnails dominated the littoral habitats on Lake Tohopekaliga, while the exotic Island applesnails were largely confined to the pelagic habitats on the interior side of the lake (Darby and Percival, 2000). The drawdown and scraping treatment, as well as the prolonged dry conditions that followed it, greatly suppressed the population of Florida applesnails in Lake Tohopekaliga (Darby et al., 2002, 2003). When the lake level rose, the island applesnail population spread out and invaded the littoral zone (Cattau, 2008). Had the island applesnail also been in the littoral zone, it may too have declined. As described above for channeled applesnails (see "Water regulation"), the lowered water levels at least concentrate the snails into small areas allowing for easier handpicking.

The chronic upper lethal temperature of *Pomacea insularum* is 37°C and the chronic lower lethal temperature is 15°C (Ramakrishnan, 2007). Karatayev et al. (2009) suggest that it is a higher tolerance to low winter temperatures that has allowed *Pomacea insularum* to colonize more water bodies than *Melanoides tuberculata*, *Tarebia granifera*, and *Marisa cornuarietis* in Texas. Levin and Kalo (2006) suggest using increased in-stream water flows, with subsequent drops in overall water temperature to help contain snail spread and to reduce the amount of available habitat for the snails, including lower stream reaches where water flows are severely diminished.

9.5.2 Chemical Control Options

Based on raw data from a salinity bioassay, Ramakrishnan (2007) found that *Pomacea insularum* could not tolerate salinities ≥13.6 ppt and adults did not attach to substrates at 10 ppt. Even if it was possible to manipulate salinity in rice fields to control *Pomacea insularum*, the germination, growth, and photosynthetic rate of rice are generally inhibited at ≥2–6‰ (Shannon et al., 1998). See *Pomacea canaliculata* for possible chemical options.

9.5.3 Biological Control Options

There are no published accounts of management using biological methods. The species has several predators, including snail kites (Cattau, 2008) but none appear to limit population sizes. Snail kites preferred the native *Pomacea paludosa* before the invasion by *Pomacea insularum*; in fact, Cattau (2008) found that adult kites were attracted to Lake Tohopekaliga, Florida, which has suffered a major invasion by *Pomacea insularum*, and that the relative contribution of the lake to the range-wide nesting effort increased from 6% to 33% after the invasion of the exotic snail. Cattau (2008) compared the energetics and behaviors of 15 snail kites when given the two snail species as food; he found that exotic snails required longer handling times, which led to increased drop rates, and resulted in depressed capture rates by the kites. However, he also found that exotic snails provided more energy than native snails. Consequently, the effects of the exotic snail on foraging behavior did not have negative energetic repercussions for the adult kites. Conversely, the effects of the exotic snail on juvenile foraging behavior can lead to insufficient daily energy balances and may suppress juvenile survival (Cattau, 2008), suggesting little value of kites as biological controls.

9.6 SPIKETOP APPLESNAIL, *POMACEA DIFFUSA* (BLUME, 1957)

Pomacea diffusa has generally been assumed to pose little threat in the United States and it is the only apple snail for which interstate transport is permitted but probably should not be permitted (Rawlings et al., 2007). Rawlings et al. (2007) state that the permit is not required because the species is perceived as innocuous (Gaston, 2006), presumably based on studies (e.g., Howells, 2002) and publications (e.g., Perera and Walls, 1996) that state that it feeds primarily on attached algae, not macrophytes. However, Rawlings et al. (2007) cite Aditya and Raut (2001) who state that, in addition to macrophytes, it feeds readily on animal carcasses, live worms, and the eggs of planorbid snails. Therefore, Rawlings et al. (2007) suggest that it may have direct effects on both aquatic vegetation and native snails and compete for food with native scavengers such as crayfish, shrimp, and fish.

9.6.1 Physical Control Options

Unknown or probably none required.

9.6.2 Chemical Control Options

Unknown or probably none required.

9.6.3 Biological Control Options

Unknown or probably none required.

9.7 FLORIDA APPLESNAIL, *POMACEA PALUDOSA* (SAY, 1829)

9.7.1 Physical Control Options

There is no need to control the species in Florida because it is native and is an important food item for endangered snail kites. In states where it is introduced (e.g., Hawaii) the most effective management methods are probably hand or mechanical removal of snails and egg masses, as is done for *Pomacea canaliculata*.

9.7.2 Chemical Control Options

Elimination of Florida applesnails by chemical means has not been reported in the literature, either because no effective chemical recommendation has been developed or the species has a very limited distribution in North America and appears to be of minor concern.

9.7.3 Biological Control Options

Conner et al. (2008) performed a laboratory experiment to determine the effect of densities of a new exotic snail, *Pomacea insularum* on growth and survival of native *Pomacea paludosa* in Florida. Growth of *Pomacea paludosa* decreased when native juvenile densities (without adults) increased from four to eight snails per 38 L aquarium; the presence of adults of either species decreased juvenile growth and also decreased juvenile survival of *Pomacea paludosa.* One *Pomacea insularum* adult had the equivalent effect of three to four native adults. Conner et al. (2008) concluded that based on the response of native juveniles to *Pomacea insularum* adults, populations of the native apple snail *Pomacea paludosa* could be negatively impacted (i.e., controlled) by expansion of the exotic snail *Pomacea insularum.*

9.8 QUILTED MELANIA, *T. GRANIFERA* (LAMARCK, 1822)

Pointier (2001) states that there appears to be no known control measures.

9.8.1 Physical Control Options

Temperature may be a control option. Laboratory thermal bioassays by Chaniotis et al. (1980b) resulted in a lower lethal temperature of 7°C, which may explain why this species is generally not found in water under 10°C (Abbot, 1952). Chaniotis et al. (1980c) reported total kills at water temperatures lower than 7°C and higher than 40°C.

9.8.2 Chemical Control Options

This species typically occurs in freshwaters and does not appear to tolerate salt water well. Chaniotis et al. (1980a) reported specimens burying themselves in response to rising salinity. It is able to survive pH between 9.1 and 9.5 (Chaniotis et al., 1980a) and salinity levels of at least 1 ppt (Oglesby, 1977). Some plants are toxic to snails, like American black nightshade, *Solanum nodiflorum*, the leaves of which are toxic to many pulmonate snails but not *T. granifera* (or *Melanoides tuberculata*) (Medina and Ritchie, 1980).

9.8.3 Biological Control Options

There is no known biological control.

9.9 RED-RIMMED MELANIA, *MELANOIDES TUBERCULATA* (MÜLLER, 1774)

Pointier (2001) states that there are no known control methods for the species. However, there appears to be few published attempts to control it, except with a molluscicide, as discussed below.

9.9.1 Physical Control Options

Low and high temperatures may be options for controlling infestations of *Melanoides tuberculata*; Duggan (2002) reports the minimum temperature for survival at 18°C and the maximum temperature for survival at 32.6°C. The species is typically found at temperatures of 18–25°C in the United States (Murray, 1971).

9.9.2 Chemical Control Options

Blay and Dongdem (1996) state that the species has little tolerance to low levels of oxygen saturation, suggesting low dissolved oxygen tensions may be an option for controlling infestations. Duggan (2002) found it is absent at saturation levels less than 47%. *Melanoides tuberculata* is not able to tolerate salinity levels grater than 3.3 ppt (Victor and Al-Farsi, 2001), suggesting that higher salinities could also be used to eliminate the species. Bayluscide (Bayer 73) has been used to control *Biomphalaria glabrata*, an intermediate host of trematodes causing Schistosomiasis and prey for *Melanoides tuberculata*. Francis-Lloyd et al. (1997) found that 24-h concentrations lethal to 50% of *Melanoides tuberculata* (LC_{50}) ranged from 0.062 to 0.085 mg/L, and 24-h concentrations lethal to 99% of snails (LC_{99}) ranged from 0.149 to 0.440 mg/L. Giovanelli et al. (2002) reported the LC_{50} and LC_{90} values for *Melanoides tuberculata* as 0.082 and 0.221 mg/L, respectively.

9.9.3 Biological Control Options

The species has a very hard shell and apparently has few predators that can break the shells. There appears to be no known methods of biological control of the species.

9.10 FAUCET SNAIL, *BITHYNIA TENTACULATA* (LINNAEUS, 1758)

9.10.1 Physical Control Options

Mitchell and Cole (2008) found that faucet snails exposed to heated water at 50°C for 1 min or longer eliminated all test snails. He considered the heated water treatment the most practical for use in fisheries applications.

9.10.2 Chemical Control Options

Mitchell and Cole (2008) found faucet snails to be highly resistant to ethanol, NaCl, formalin, Lysol, potassium permanganate, copper sulfate, Bacquacil, Virkon, household bleach, and to waters adjusted to pH values as low as 1 and as high as 13. He recommended two treatments, one with Hydrothol 191 at 20 mg/L or more for 24-h exposure, or exposure to 50°C for at least 1 min, as recommended above.

9.10.3 Biological Control Options

The faucet snail has numerous predators, including fish, waterfowl, crayfish, and leeches, but even all of them combined are unlikely to lower densities of the snail to the point of extirpation. Indeed, in the case of waterfowl, the faucet snail seems to be controlling duck populations. The faucet snail carries trematode parasites (e.g., *Leyogonimus polyoon*, *Cyathocotyle bushiensis*, and *Sphaeridiotrema globulus*) that when consumed is blamed for killing nearly 50,000 ducks in the last few years in the Upper Mississippi Wildlife Refuge, which borders Wisconsin, Minnesota, and Iowa; 22,000–26,000 birds, primarily American coots (*Fulica americana*) and lesser scaup (*Aythya affinis*) were killed in Navigation Pool 7 of the Upper Mississippi River near La Crosse, Wisconsin (Sauer et al., 2007).

9.11 NEW ZEALAND MUDSNAIL, *POTAMOPYRGUS ANTIPODARUM* (GRAY, 1843)

The New Zealand mudsnail (NZMS) is causing biofouling problems in utilities and industries (see Chapter 6).

The NZMS Management and Control Plan Working Group (2007) describe numerous pathway-specific management options that are summarized below.

1. *Close visual inspections of the following.*

   Fish hatcheries and aquaculture components: If NZMS is not yet present, inspect water systems, raceways, stocking equipment, as well as gut contents of imported fish. If NZMS is present use well water or other alternative uncontaminated sources; gear used in the field should not be used in the hatchery, and all gear should be stored in walk-in freezers, if available, treated with decontamination methods or thoroughly dried; release fish only at sites already contaminated by mudsnails. Preliminary investigations suggest that copper, carbon dioxide under pressure, and hydrocyclonic separators may prove useful in both decontaminating fish hatchery water supplies and preventing spread into uncontaminated areas of a hatchery.

   Recreational watercraft and trailers: Promote thorough drying of boats and trailers before launching at new locations.

   Transport by anchors, fishing gear and wet wells: Inspection and cleaning of anchors fouled with aquatic vegetation and mud; fishing gear (e.g., boots, waders—putting gear in a freezer for 6–8 h or in water at 50°C (120°F) for a few minutes) or dry gear at 29°C (84–86°F) for at least 24 h or at 40°C (104°F) for at least 2 h to kill all attached mudsnails (Medhurst and Herbst, 2003; Richards, 2004); and wet wells (use hot water values above, or see chemical control options).

Aquarium and aquatic plant trade and collections: Apply quarantine requirements for shipments of aquarium and water garden organisms, accompanied by effective requirements for disinfection or disposal of contaminated shipments.

Sand/gravel mining, extraction, and dredging: There do not appear to be regulatory requirements governing inspection or cleaning of associated equipment for these activities but thought should be given to address this.

2. *Provision of outreach and resources: Anglers, hunters, and natural resource management personnel:* If possible, provide resources and outreach programs to facilitate field decontamination of gear (e.g., wash stations at high use water access sites).
3. *Commercial shipping:* Most states and provinces with shipping ports should have programs in place at the state/provincial, federal, and/or international level to reduce the discharge of contaminated ballast water.
4. *Transport by fish, wildlife, and livestock:* Although impractical to limit in natural settings, netting and other predator avoidance methods typically used in fish hatcheries can reduce the ability for birds and other wildlife to spread NZMS from contaminated facilities. Livestock should be kept away from invaded streams and riparian areas.

9.11.1 Physical Control Options

NZMSs cannot withstand warm temperatures or low humidity situations (Dwyer et al., 2003; Richards et al., 2004). Richards et al. (2004) showed that *Potamopyrgus antipodarum* does not survive freezing or desiccation at high temperatures with low humidity. In bioassays, using seven length classes of snails, a variety of temperatures ranging from −3°C to 40°C, and two humidity levels (20–25% and 90–100%), they found that at all temperatures, larger *Potamopyrgus antipodarum* generally survived desiccation longer than smaller ones, and for all size classes mortality generally increased with increased exposure time. Richards et al. (2004) recommend thoroughly freezing or drying potentially contaminated equipment to limit the spread of *Potamopyrgus antipodarum* to uninfected aquatic ecosystems.

9.11.2 Chemical Control Options

The NZMS Management and Control Plan Working Group (2007) stated that Bayer 73, copper sulfate, and 4-nitro-3-trifluoromethylphenol sodium salt (TFM) are effective at killing NZMSs. They also stated that 100% mortality of the NZMS occurred after 48 exposure units of Bayluscide (1 ppm for 1 h) in a small spring creek along the lower Madison River.

Laboratory tests to solutions of benzethonium chloride, chlorine bleach, Commercial Solutions Formula 409 Cleaner Degreaser Disinfectant, Pine-Sol®, ammonia, and copper sulfate. With the exception of grapefruit seed extract, potassium permanganate and isopropyl alcohol, these materials all killed mudsnails within 5 min (Hosea and Finlayson, 2005). The most effective solutions for killing

NZMS, which can be used in the field, according to Hosea and Finlayson (2005) are copper sulfate, benzethonium chloride, and 50% Commercial Solutions Formula 409 Cleaner Degreaser Disinfectant. For copper sulfate, dissolve 3.785 g of copper sulfate pentahydrate crystals (99.1% purity) for each gallon of solution required; this will achieve a concentration of 252 mg/L of copper in the cleaning solution. For benzethonium chloride, dissolve 9.57 g of benzethonium chloride (97% purity) for each gallon of cleaning solution you want to make; this will achieve a concentration of 1.947 mg/L in the cleaning solution. For Commercial Solutions Formula 409 Cleaner Degreaser Disinfectant, dilute the commercially available solution 1:1 with clean water to achieve the needed concentration for the cleaning solution (i.e., one gallon of Formula 409 Disinfectant to one gallon of water) (NZMS Management and Control Plan Working Group, 2006).

9.11.3 Biological Control Options

Studies of the efficacy and specificity of a trematode parasite from the native range of NZMS in New Zealand as a biological control agent are promising (Dybdahl and Lively, 2005). Dybdahl and Lively (1998) and Lively and Dybdahl (2000) reported the parasite, *Microphallus* sp., to be highly specific in the native range, infecting the most common genotypes. Experimental infections have shown that populations of the parasite from the source range of western U.S. snail clones are very effective at infecting the genotype found in the western United States. Experimental infections and molecular genetic studies have also shown that these effective lineages of *Microphallus* sp. are highly specific (NZMS Management and Control Plan Working Group, 2006). However, the working group cautioned that if the biological control entails the introduction of another nonnative species, the costs of this have to be weighed against the costs of ecological damage caused by the NZMS.

10 Case Studies

10.1 ZEBRA AND QUAGGA MUSSELS IN THE INLAND WATERS OF EUROPEAN RUSSIA AND ADJACENT COUNTRIES

Marina I. Orlova

10.1.1 INTRODUCTION

Over the last two decades, the invasion and subsequent explosive expansion of freshwater invasive Ponto-Caspian dreissenids in Europe and North America have been focal areas for aquatic invasion research. Their vectors and mechanisms of dispersal, details of establishment and distribution, timing and pattern of paleo- and recent invasion, integration into living communities, predictive studies, as well as prevention and consequences of their biofouling have been examined. In the European portion of the former USSR, the invasions of Ponto-Caspian species have a long history. The causes and consequences of these invasions in a variety of ecosystems are well documented [e.g., waterways (since the last quarter of the nineteenth century), cascades of water reservoirs (since the second half of twentieth century), power generating installations (since the 1930s), and irrigation systems]. The last eight decades of this invasive history provide some lessons in coexistence and control of industrial fouling by dreissenids in a variety of facilities. At the "Periphyton and Fouling: Theory and Practice" conference held in St. Petersburg (October 22–25, 2008), it became clear that these mussels are currently a focal point for theoretical studies as well as the development of control strategies for beneficial and practical use (see www.zin.ru). At the same time, the south of Russia and adjacent countries are the native area for many invasive Ponto-Caspian species.

10.1.2 NONINVASIVE AND INVASIVE DREISSENIDS

In the recent past (prior to 1940–1970s), inland waters of the former USSR (in the Ponto-Caspian region and Aral and Balkhash Lakes' basins) possessed probably the highest number of fouling dreissenid* species and subspecies. Most of these have not been subject to invasive spread and have had their distribution limited to the giant brackish water basins of the Caspian and Aral Lakes. The following is a list of native brackish water Caspian dreissenids: *Dreissena elata* Andrusov and *Dreissena caspia caspia* Eichwald (which inhabited the coastal waters of the Caspian Lake

* All modern dreissenids are fouling organisms, while not all fossil dreissenids had attached mode of life, [see in Andrusov (1897), Baback (1983), Nuttall (1990), Starobogatov (1994), Geary et al. (2000) about paleontology of the family in ancient basins of Pratethys].

before the establishment of invasive Mediterranean mytilid, *Mytilaster lineatus* Gmelin*), deep-water dwelling *Dreissena rostriformis* Deshayes† (which is still present and inhabits the relatively deep waters of the Caspian Sea, which is free of *Mytilaster*) and *Dreissena polymorpha andrusovi* (Andrusov) that occurs in the eastern part of the Caspian Lake. In the Aral Lake basin, the existing dreissenids are freshwater oligohaline *Dreissena polymorpha aralensis* (Andrusov) (now inhabiting lakes and rivers of the northern part of the Aral Sea basin) and two brackish water taxa *Dreissena caspia pallasi* (Andrusov) and *Dreissena polymorpha obtusecarinata* (Andrusov) [once common in the Aral Lake proper but which is now extinct following salinity changes in the lake (Starobogatov and Andreeva, 1994)].

Of those mentioned above, at least one species *Dreissena rostriformis* [suggested to be the most probable ancestor of well-known invasive quagga mussel (Starobogatov, 1994; Orlova et al., 2005)] has a significant history of invasion in the past ("palaeoinvasions") in southern European ancient lakes and seas (Paratethys basin) as has been determined from paleontological records (Andrusov, 1897; Baback, 1983).

The invasive dreissenids zebra [*Dreissena polymorpha polymorpha* (Pallas)] and quagga mussels [*Dreissena rostriformis bugensis* (Andrusov)]‡ and the dark false mussel [*Mytilopsis leucophaeata* (Conrad)] are all characteristically found in the inland waters [lakes, rivers (especially large and those transformed into cascades of reservoirs) canals, estuaries] of European Russia, Ukraine, Moldova, Byelorussia, and Finland.

10.1.3 Invasive Dreissenids: Specific Features of Invasions and Modern Dispersal

Dreissena polymorpha has significantly expanded its range to the north of Europe from the Ponto-Caspian region (Figure 10.1) during the last two centuries by natural and men-mediated vectors (timber rafting, interbasin connections, and shipping) (Andrusov, 1897; Nowak, 1971; Karataev et al., 1998; Olenin et al., 1999; Bij de Vaate et al., 2002; Pollux et al., 2003). Now it occupies all of European Russia (up to 62°N), including the basins of the White (the Dvina River) and Baltic Seas (Figure 10.1). *Dreissena polymorpha* also has a paleolithic record of invasive spread: geological evidence confirms its occurrence during the last or even previous interglacials in the north-west portion of Russia (Starobogatov and Andreeva, 1994) and Western Europe

* *Mytilaster lineatus* is a component of industrial fouling in the south of the Caspian Sea and it is thought that *Mytilaster* outcompeted brackish water Caspian dreissenids by 1940–1960s (Logvinenko, 1965; Starobogatov, 1970; Starobogatov and Andreeva, 1994)

† The number of extant subspecies is unclear, but before the *M. lineatus* invasion there were four subspecies: indentified: *D. r. distincta* (Andrusov), *D. r. grimmi* (Andrusov), *D. r. pontocaspica* (Andrusov), *D. r. compressa* (Andrusov) (Starobogatov, 1994).

‡ Molecular resolution provides evidence of very close relationship between living Caspian *D. rostriformis distincta* and quagga mussel (Orlova et al., 2004; Therriault et al., 2004) and so far taxonomical status of quagga mussel is restored according to Andrusov (1890), who discovered and described this dreissenid as *Dreissena rostriformis*.

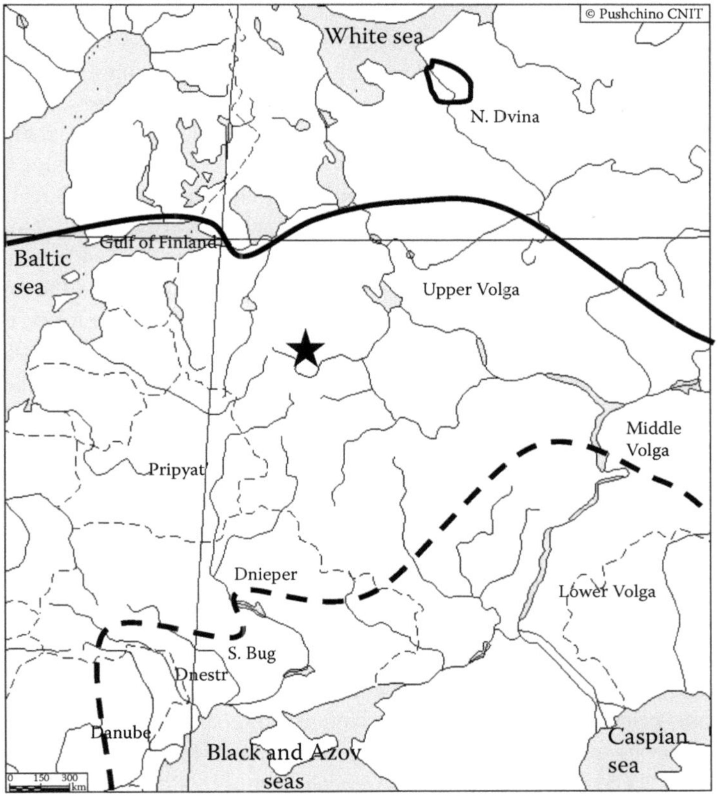

FIGURE 10.1 Early Holocene and recent range expansion of zebra mussel in European Russia and adjacent countries. Asterisk indicates lakes Udomlya and Pes'vo, where Kalininsk power plant is situated and monitoring of the zebra mussel is carried out.

(Kinzelbach, 1992). This means its recent spread to the north of Europe can be considered a recolonization of previously inhabited territory (Orlova et al., 2005).

Now, zebra mussels are present in fresh to oligohaline waters, in estuaries, pristine rivers (as a rule, in these rivers they form dense populations only in the lower reaches*), rivers transformed into cascades of reservoirs by damming (colonization occurs along all portions), and in lakes of different trophic status (mesotrophic lakes are most often colonized by zebra mussels) (Lyachnovich et al., 1994). It is important to note that unlike the other two invasive dreissenids, the zebra mussel also invades relatively small lakes [e.g., many lakes of Byelorussia and Valdai Hills (Russia)],

* Colonization of upper and middle reaches is often difficult for neo- and mesolimnic species that possess planktonic (drifting) larvae, as high velocity of downstream currents prevents settlement of juveniles near parental population and takes the larvae far downstream. Sometimes larvae settle hundreds of kilometers downstream of the place they were born. As for the parental population, situated far upstream, if there is no recruitment by multiple introductions, they gradually become extinct (Orlova et al., 2005).

including those serving as a source of water for the Kalininsk nuclear power plant (see Figure 10.1 and below). It has also been noticed that the species is extremely successful in cascading reservoirs and artificial systems: water supplies, service water systems and outer cooling circuits of power plants, irrigation systems south of Russia and Ukraine, and various canals (Zhadin and Gerd, 1961; Karatayev et al., 1994).

The quagga mussel (*Dreissena rostriformis bugensis*), previously maintained a very restricted distribution within its native region, the Dnieper-Bug Liman and lower Inguletz River between the 1890s [when it was first discovered and described by Andrusov (1890) and the 1940s (Zhuravel, 1951)]. From the 1940s to the present, *Dreissena bugensis* extended its range in the Ponto-Azov region: to the north of the Black Sea into the Dnieper River up to lower reaches of the Pripiat River, to the west and to the north-west into the Dniester (Kharchenko, 1995; Philippenko, 2008), Danube and then (via a channel) to the Main and Rhine Rivers (Van der Velde and Platvoet, 2007) and to the east to the Don and Manych Rivers (Zhulidov et al., 2004). Its invasion to the neighboring Volgo-Caspian basin happened only 40 years after the opening of the Volga–Don canal (opened in 1952), which linked the Ponto-Azov and Volgo-Caspian parts of the Ponto-Caspian region. *Dreissena bugensis* was first recorded in the Volga River in 1992 (Antonov, 1993), and in 1994 in the Caspian Sea delta and avan-delta of the Volga River (Orlova and Shcherbina, 2002). In the Volgo-Caspian area it has since expanded its range northward to the Sheksna reservoir (Figure 10.2) (Orlova and Shcherbina, 2002). Thus, the Volga River is now completely populated, wherever there is a cascade of hydroelectric power plants. No paleontological evidence exists, suggesting that *Dreissena bugensis* was ever present in areas other than its native range. The absence of paleoinvasion can be ascribed to the relatively young evolutionary age of *Dreissena bugensis* in relation to Quaternary time (Mordukhai-Boltovskoi, 1960; Orlova et al., 2005). Like *Dreissena polymorpha*, the recent distribution of the quagga mussel's invasionary region covers fresh to oligohaline waters, but unlike zebra mussels it is restricted by giant lakes, estuaries, large rivers, and some canals.* The quaaga mussel is especially abundant in large rivers transformed into cascades of reservoirs.

So far, the invasions of these freshwater dreissenids into Europe (zebra and quagga mussels) is much different in the timing and distributive patterns than in North America and have resulted in differences in the types of inland water bodies colonized.

Mytilopsis leucophaeata (Conrad) is a central American dreissenid that has been observed once in the south of Dniester Liman in 2004 (Therriault et al., 2004) but still there is no data of its further dispersal or other sightings from that area, while in other European regions it has greatly expanded its range in oligohaline and freshwaters (Marelli and Gray, 1985; Nuttall, 1990; Starobogatov and Andreeva, 1994; Darr and Zettler, 2000; Van der Velde et al., 2002; Therriault et al., 2004). Its northward location is in the vicinity (heated area) of the nuclear power plant in the Loviisa archipelago (Finland, Baltic Sea, Figure 10.3), where it exists as persistent population (Laine et al., 2006).

Despite extensive and regular monitoring of nonindigenous species in the eastern Gulf of Finland, carried out since 2004, there has been no record of dark false mus-

* Velocities higher than 2 m/s limit *D. r. bugensis* development in canals of the Ukraine, but *D. p. polymorpha* persists at these velocities (Kharchenko, 1995).

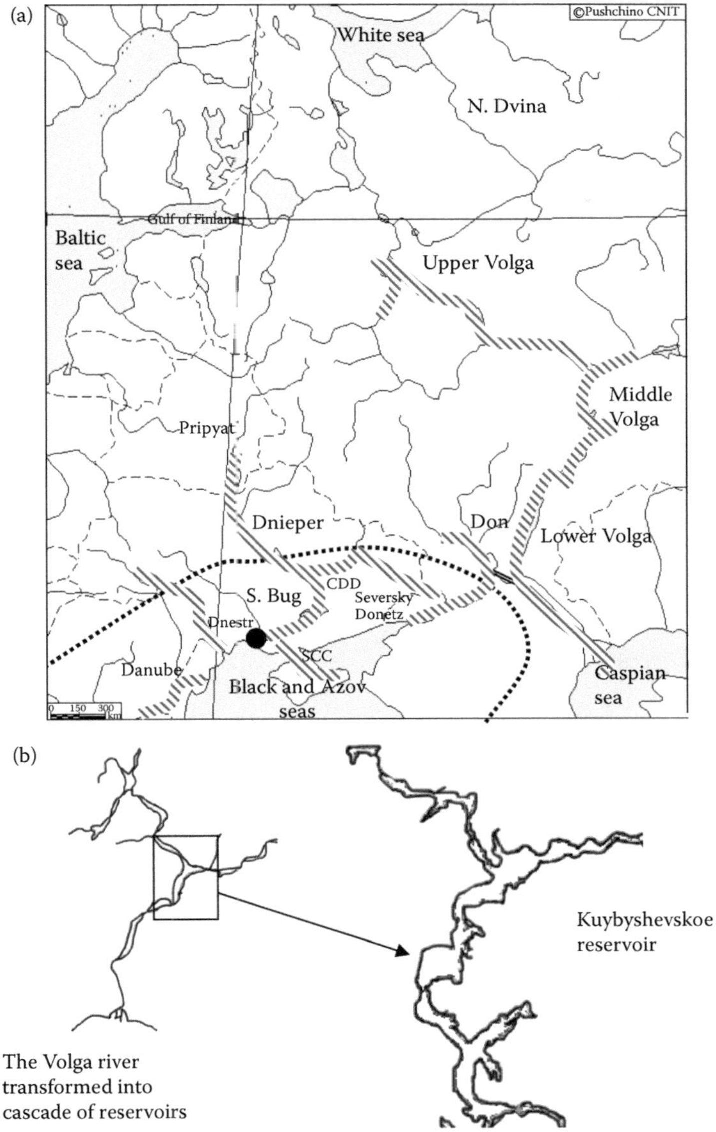

FIGURE 10.2 (a) Native (circle) and invaded area of quagga mussel in European Russia and adjacent countries; Ponto-Azov region is contoured with dotted line and Volga-Don Canal is indicated with double solid line. (b) An example of a river (The Volga River) transformed to a cascade of reservoir and the Volzhskaya hydroelectric power plant, mentioned in the text.

sels there and the quagga mussel has been found only once as have several other invasive species (Orlova et al., 2006; Alimov et al., 2008).

Thus, the three invasive dreissenid species inhabit a great variety of pristine and transformed ecosystems fitting their environmental requirements, mode of life, and

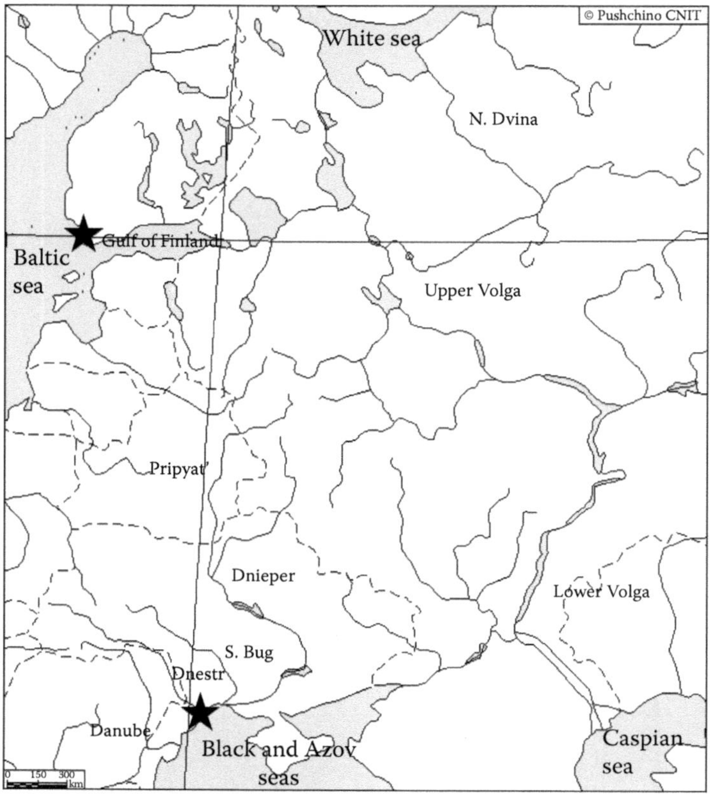

FIGURE 10.3 Records of dark false mussel in European Russia and adjacent countries.

requisite conditions for juvenile settlement.* Of these, the zebra mussel occurs in the most varying environmental conditions of inland waters, followed by the quagga mussel showing less variability and the dark false mussel showing the lowest level of adaptability.†

10.1.4 Man-Mediated Activity and Success of Zebra and Quagga Mussels Invasions and Recolonizations

Understanding what features of an ecosystem predispose it (this includes both assets that use an ecosystem as a source of water and those that are just situated there) to primary infestation and/or secondary recolonization by dreissenids is extremely important

* Colonization of upper and middle reaches is often difficult for neo- and mesolimnic species that possess planktonic (drifting) larvae, as high velocity of downstream currents prevents settlement of juveniles near parental population and takes the larvae far downstream. Sometimes larvae settle hundreds of kilometers downstream of the place they were born. As for the parental population, situated far upstream, if there is no recruitment by multiple introductions, they gradually become extinct (Orlova et al., 2005).

† When planning a strategy for fouling prevention or control, the diversity of fouling mussels, possessing various environmental requirements, and sensitivity to remedies should be taken into account.

in predicting future dispersal and planning a preventative strategy of control. Four or five factors appear to be most important for this prediction: (1) transformation of pristine ecosystems into cascades of reservoirs for hydroelectric facilities, shipping, and so on; (2) industry caused heating and change of water mass circulation in lakes used as a source of water or in parts of larger water bodies; (3) periodical disturbance with subsequent self-restoration of initial transparency conditions in the ecosystem; and probably to a lesser degree; (4) climate change; and (5) decrease of pollution following the industrial and agricultural recession at the beginning of the 1990s.

10.1.4.1 Transformation of Pristine Ecosystems into Cascades of Reservoirs

The Dnieper and Volga Rivers as was mentioned previously have significantly facilitated the dispersal of both invasive freshwater dreissenids that originate from estuarine portions of the Ponto-Caspian Rivers. *Dreissena bugensis* was not found in most colonized systems prior to the creation of the reservoir and although *Dreissena polymorpha* was common* there it formed abundant settlements only in the lower reaches of rivers and their estuaries as previously discussed. It is interesting to note here that in the Dnieper River system at least quagga mussel populations established themselves 5–10 years following the reservoir impoundment and several years after extensive colonization by the zebra mussel (Tzeyeb et al., 1966; Mills et al., 1996).

The most important prerequisite for facilitation of colonization is estuarization and limnification of the river system, which changes the conditions of a given river to those similar to the conditions in their native region.† Estuarization and limnification is a consequence of strong deceleration of river current velocities in comparison to the intact river [down by 70% (Butorin and Mordukhai-Boltovskoi, 1978)], combined with a change of riverine bed into reservoir floor (Figure 10.2b). This alteration prevents the downstream drift of dreissenid free-living larvae and facilitates persistent populations in all portions of the cascaded river. These changes also establish stratification (formation of a hypolimnion) of water masses, benefiting relatively cold-water invading species and provide a higher diversity of benthic habitats including the formation of habitats with artificial substrates and flooded forests. Stretches with relatively low velocities and zones of sedimentation are common in reservoirs and beneficial to fouling seston-feeding species, both native and nonindigenous. For example, by the mid-1960s it was noted that in the older, deeper reservoirs along the Dnieper River

* *D. polymorpha* was present in the Dnieper River before it was transformed into a cascade of reservoirs (Zhuravel, 1934) and the same is true for the Volga River (Zhadin and Gerd, 1961). However after the creation of cascade on a previously pristine river, the mussels increased in abundance more than some other fouling organisms. For example, the formation of the Volga River cascade of reservoirs was started in the 1940s and completed in 1981. Zebra mussels rapidly gained dominance within benthic communities shortly following creation of the first reservoir, Uglich Reservoir, in 1940 and were well established following impoundment of Rybinsk Reservoir in 1947 (Zhadin and Gerd, 1961).

† Analysis of intentional introductions' results of "liman opportunists" (*D. rostriformis bugensis* belongs to this faunistic group) to target ecosystems (Karpevich, 1975) have confirmed that ecosystems are only vulnerable to invasion by species whose tolerances match the environment to which they are being introduced in sufficient number. For example, establishment of *D. r. bugensis* in the Dnieper and Volga River basins only occurred following changes to the flow regime from a riverine to a more lacustrine/estuarine environment (these features are given in Table 3 in Orlova et al., 2005).

[e.g., Dnieprovskoe Reservoir (1934–1947) and Kakhovskoe Reservoir (1955)], *Dreissena bugensis* largely replaced *Dreissena polymorpha* and became the dominant dreissenid species (Tzeyeb et al., 1966). In contrast, the younger, shallower, and warmer reservoirs in this system contained extensive *Dreissena polymorpha* populations and limited *Dreissena rostriformis bugensis* populations (Tzeyeb et al., 1966). In both Dnieprovskoe and Kakhovskoe Reservoirs on the Dnieper river as well as in deep-water stretches of the Kuibyshevskoe reservoir on the Volga River, *Dreissena bugensis* occupies deeper habitats than *Dreissena polymorpha*, reaching depths of up to 28 m where temperatures are about 10°C lower than the epilimnion through the summer (Mills et al., 1996; Orlova et al., 2005). At most locations along the Volga River, *Dreissena bugensis* has become the more abundant species of *Dreissena* in the 2000s (Orlova et al., 2004). This confirms the hypothesis that given sufficient time, *Dreissena bugensis* will displace *Dreissena polymorpha*, in deeper, more oligotrophic habitats.* How does this impact cooling water systems? As larvae of both species are planktonic and *Dreissena bugensis* larvae generally prevail, recently many hydroelectric power plants have observed the quagga mussels becoming the dominant fouling species. This was found to be the case at the Volzhskaya hydroelectric power plant (Kablov et al., 2008; B. Bogush, presentation at conference,† N. Sokolova, personal communications).

10.1.4.2 Heating and Change of Water Circulation in Smaller Lakes

Based on Lakes Udomlya and Pes'vo, which serve as cooling ponds and water supply for the Kalininsk nuclear power plant, thermal recirculation can significantly accelerate zebra mussel population growth,‡ extend the breeding season and therefore the season for preventative treatment (Figure 10.4). For example, the breeding season in these lakes starts in March (2005–2007).§ This is due to the increase in circulation of surface waters and connection between the lakes (Figure 10.4). Veligers are recorded simultaneously in both a heated (Pes'vo) and cool (Udomlya, water source) sub-basin of the two lakes' system. In water bodies with natural temperatures at the upper temperate zone, the zebra mussel breeding season starts in the second half of June (Lvova and Makarova, 1994; Orlova et al., 2006). Despite the fact that the area of the lakes is only 18 km^2, one-third is covered by zebra mussel settlement with an estimated total biomass.

There is no doubt that establishment of the dark false mussel in cool coastal waters of Finland was only possible due to the hot waters discharge from Loviisa nuclear power plant (Laine et al., 2006).

* It seems intrinsic to carefully suggest that combating eutrophication, which along with biological invasions is considered as one of the most pronounced problems in inland waters (Millennium ecosystem Assessment report, 2005), could result in extension of quagga mussel range as a "by-product" of "oligotrophy."

† International scientific and practical conference "Periphyton and fouling: theory and practice," held on 22–25 October in St. Petersburg, Russia (see B. Bogush presentation, slide 21 at www.zin.ru).

‡ Species was unintentionally introduced with carps from Ivankov reservoir (Volga River) in the early 1980s.

§ The water temperature in March for the period of observation and times of sampling varied from 7°C to 14°C at stations, situated at different temperature zones, with the least in the second half of January (3–12°C) and highest in July (26–30°C).

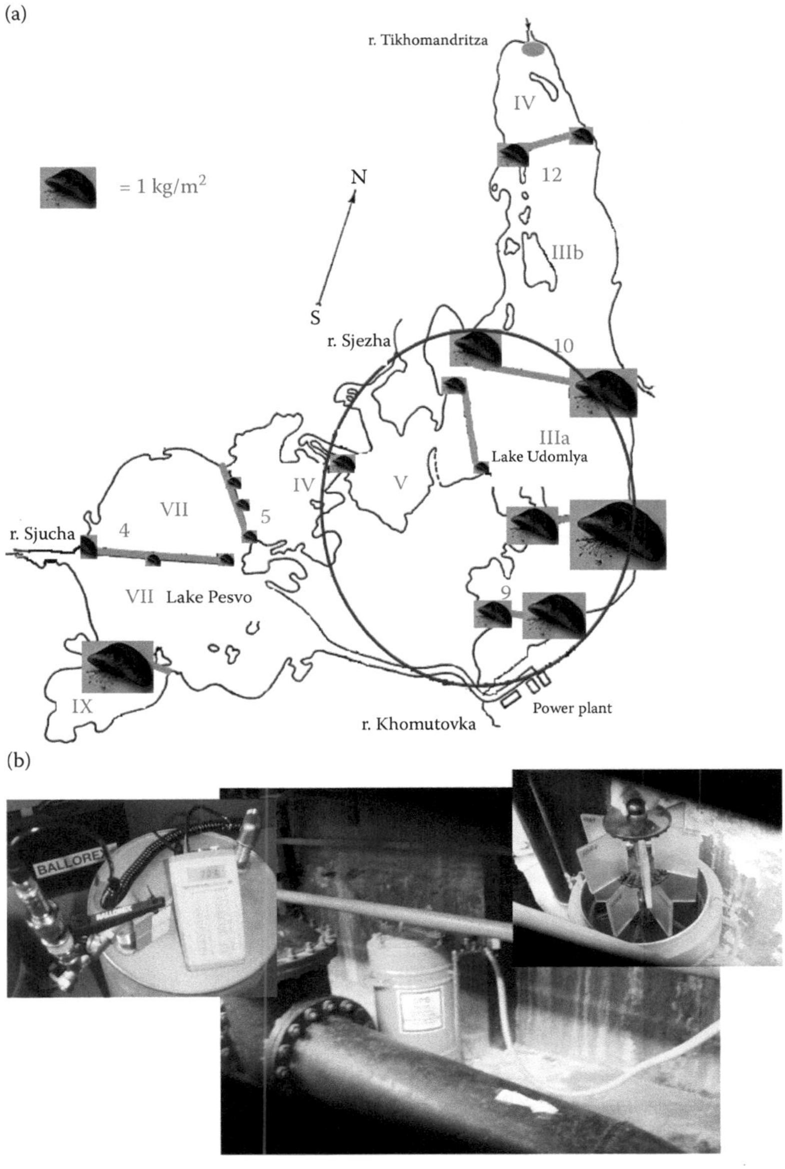

FIGURE 10.4 **(See color insert following page 286.)** Zebra mussel distribution in lakes and side-stream monitoring at Kalininsk power plants. (a) Zebra mussel biomass distribution in lakes Udomlya and Pes'vo (gray area—the zone of water circulation, Arabic numbers—numbers of transects for diving survey). (b) Example of a monitor installed at a pump house of the fire protection system at the Kalininsk power plant (monitors were constructed for both monitoring and experimentation; they are hermetic with regulated flow velocity and exchangeable interior equipment).

10.1.4.3 Disturbance and Subsequent Restoration of Normal Transparency Conditions in the Ecosystem

Disturbances, caused by dredging and in-wash activity, were observed in the eastern Gulf of Finland (Figure 10.5) in 2004–2008. Along with changes in transparency and other conditions, we observed changes in the quantitative abundance of zebra mussels and thereafter produced the following hypothesis: After a period of high turbidity [total suspended solids (TSS) levels of 6.2–25 mg/L, transparency (Secchi depth) of 0.3–0.7 m, contribution of inorganic TSS of 76–81%] in the area from 2006 to fall of 2007 (satellite image Figure 10.5a), a gradual but dramatic decrease of zebra mussel population at the most highly impacted portion of the Gulf of Finland, should be followed by a compensatory surge in the population (Figure 10.5b) just after the restoration of normal conditions (TSS of 4.3–11.6 mg/L, transparency of 0.9–2.1 m, percentage of inorganic fraction in TSS of 15–21%). Evidently, such rapid fluctuation and rejuvenation of populations (Figure 10.5c) should be thought of when choosing a control strategy, with current and seasonal variability of environmental conditions taken into consideration.

10.1.4.4 Climate Change

From our monitoring in the eastern Gulf of Finland (61°N), climate change is being expressed through more mild winters and an absence or a shorter duration of permanent ice cover period, which has evidently affected the timing of zebra mussel breeding and larvae appearance. For example, in 2008 after an iceless winter we first observed larvae in May, while they usually do not appear until the second half of June (2004–2007). These larvae were rather abundant throughout the whole warm season and successfully settled by October 2008 (Figure 10.5b, Report 2008*). These changes may also need to be taken into account in terms of prediction research and planning strategies.

10.1.4.5 Industrial and Agricultural Recession

There is no direct evidence that the industrial and agricultural recession that took place in the 1990s in the former USSR has led to a decrease in pollution by hazardous substances or a decrease in eutrophication in many regions, which in turn may have facilitated dispersal of the zebra and quagga mussels and other invasive species. However, this factor cannot be neglected. There has been a rapid increase or new findings of invasive species at least in the Gulf of Finland as of the late 1990s and early 2000s (Orlova et al., 2006). On the other hand, this increase in observation recording may be attributed to increased financial support of invasive species monitoring programs worldwide and in countries of the former USSR in particular.

10.1.5 Industrial Fouling

In the former USSR, zebra mussel fouling was first identified as a bio-obstacle in 1938 when hydroelectric plants were built on the Dnieper River. Along with the

* It should be noted that starting in the fall of 2007 the transparency has improved (see previous subsection) and probably we observed the combined effect of climate change and self-restoration of normal conditions after disturbance, rather then climate effect alone.

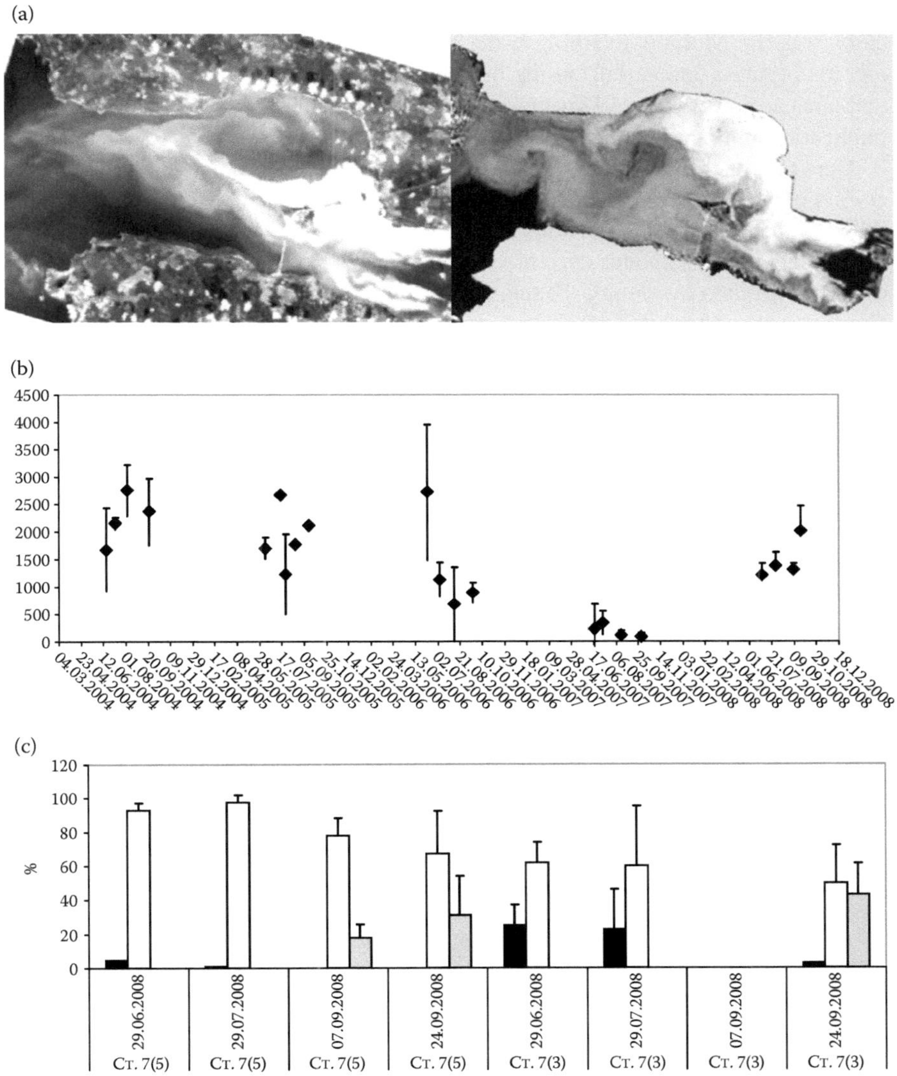

FIGURE 10.5 Results from invasive species monitoring of nonindigenous species in the eastern Gulf of Finland in 2004–2008. (a) Flumes of clay particles (whitish) induced by dredging and inwash in the Neva Bay in August and October 2006 (provided by Dr. L. Sukhacheva). (b) Dynamics of zebra mussel biomass (g/m^2). (c) Percentage of age cohorts in the population in 2008 [black areas—older than two years (no settlement observed in 2006), empty areas—one year old (settled in the fall of 2007 along after massive death of older cohorts and stop dredging and inwash), and gray areas—juveniles of 2008].

hydropower development industry, the concerns about fouling species were also identified as a problem for municipal drinking water, industrial cooling water, and irrigation systems in the basin of the Volga, Don, and Dniester Rivers. Those situated in the southern portion of the country were the most infested, while those situated in the north (in Leningrad region and in Karelia) were less so. One of the best known cases was the Moscow drinking water system (Sokolova, 1959).* Since the 1960s, records of heavy industrial fouling by zebra mussels were reported from fossil fuel and nuclear power plants. This period was also characterized by a rapid increase in publications on *Dreissena* (Limanova, 1964, 1978).

General conclusions from the published cases are that (1) enterprises using reservoirs created on large rivers as a water source are the most heavily infested, while those situated at pristine portions of rivers were less so†; (2) the degree of infestation is dependent on water current velocity, type of flow, and type of equipment in contact with the raw water‡; (3) the most vulnerable parts of industrial systems are similar to those reported by Claudi and Mackie (1993) and in the USACE Zebra Mussel Information System; (4) there are three sources of bio-obstacles produced by zebra mussels§; (5) fouling organisms are utilizing the resources and spaces within a facility and intake system according to their environmental requirements¶; (6) there is a need for the development of control methods and (7) utilization of removed shell material and deceased mussels; and (8) as mentioned there are geographical differences in infestation (between south and north).

10.1.6 Control: Current State and Future Perspectives

Modern programs give preference to strategies aimed at prevention of fouling development (*proactive*) rather than a *reactive* approach, as well as solving associated problems such as corrosion of materials. Control strategies should be planned

* Zebra mussels were found at an abundance of 19,000 ind/m^2 and biomass 15 kg/m^2.

† Volzhskaya hydroelectric plant has first reported zebra mussel infestation in 1958 (two years after the creation of Kuybyshevskoye reservoir), and then in 1961 it has reported that all underwater structures and equipment in contact with raw water were covered by fouling, which reached 8–10 cm thickness. During the same time period, water supply systems of industrial enterprises situated on the Volga River upstream of the reservoir had little or no infestation (Kirpichenko, 1963).

‡ On horizontally oriented surfaces and in water velocity of less than 1 m/s, the surface is densely settled by zebra mussels; increases of velocity greater than 1 m/s significantly limits settlement and settled juveniles occupy only burrs, cavities and depressions, and frontal surfaces of protrusions; at a velocity of 2 m/s the zebra mussel settles only in depressions, while at a velocity of 2–5 m/s they are absent even from horizontal surfaces. On vertical and sloping surfaces the infestation was the same as on horizontal surfaces only at velocity less than 0.6 m/s; in areas of turbulence zebra mussel settlement is heavier than in laminar flow (Karatayev et al., 1994).

§ In addition to (1) settling juveniles and (2) translocators, the (3) dead shells and adults drawn from water intake canal are important source of zebra mussel fouling at the facility. It was found that by the third to fourth year of power plant operation, the bottom of intake canals can be covered with a layer of dead shells with biomass of alive mussels as high as 16–19 kg/m^2; the collective biomass is critical as the material begins to be drawn into the system; at an initial velocity of 0.16–0.29 m/s small amount is drawn in, massive amounts are brought in (up to 56–71 kg of zebra mussel per second) when the velocity is 0.22–0.39 m/s (Afanasiev et al., 1990).

¶ In the forebay of Krivorozhkaya hydroelectric power plant (on the Dnieper river cascade) the zebra mussels and *Cordylophora caspia* (colonial Hydrozoa) form separate belts; within the facility a bryozoan *Plumatella emarginata* and zebra mussel demonstrate differential growth on equipment, situated in different temperature zones: at ambient temperatures and up to 30°C zebra mussel occurs with subsequent decrease of abundance, conversly *P. emartinata* has the highest biomass (18 kg/m^2) in the most heated zone, where temperature is 30–35°C (Protasov, 1994).

FIGURE 10.6 "Patio"-type artificial substrates (left photographs, general view) exposed in two zones of Udomlya Lake: in the zone of sedimentation (upper central photograph, no macrofouling organisms) and in the zone of circulation with zebra mussel (lower central photograph), *Plumatella emarginata* (Bryozoa) (upper right photograph), and *Epidaphya fluviatilis*, a colony and propagules under the colony (Porifera).

individually based on the specifications of a given facility, water source, and dynamics of fouling formation. Preference should be given to the most effective, ecologically sound methods with a reasonable cost. The application should be straightforward and easy to monitor. The selection is based and then verified by the results of thorough monitoring of macrofouling and biofilm target species and testing of the efficiency of the applied treatments (Orlova, 2008, see also her presentation at www.zin.ru). This monitoring is an important and initial constituent of the program. Such a program was developed for Kalininsk nuclear power plant (Orlova et al., 2007). It was implemented with the establishment of side-stream monitoring both within the facility (installation of monitors at observation and treatment points) and outside*; collection of weekly and biweekly planktonic samples; as well as periodic observation on fouling diversity and development on installed artificial substrates (Figure 10.6).

Generally, a reactive approach that addressed existing fouling was in wide use in the past, and is still used today. The most commonly applied remedies are mechanical cleaning, physical and chemical strategies (and their combination). The introduction of molluscivorous fish to the lakes is also still in use.† Of the chemical methods

* Monitoring in the lakes was aimed at both collection of basic information (biological variables, such as seasonal dynamics and structure of larval populations of target species; species composition, distribution and structure of benthic populations; dynamics of fouling formation on artificial substrates along with collection of standard material for experiments and tests) and the so-called shared information (environmental variables, that help to interpret results of biological sampling).

† For example, release of fishes into Udomlya Lake is carried out every year with no evidence of their effect on zebra mussel populations or even on population structure of introduced fishes. In summer 2007 we did a qualitative ichtiological survey and found no introduced fishes in catches. Among dissected native species, only roaches (*Rutilus rutilus*) of body length greater than 17 cm had zebra mussels as the main food item of diet.

available, chlorination by sodium hypochlorite and chlorine gas (Lubianov et al., 1965, 1977; Shevtzova and Kharchenko, 1986)* is the most commonly utilized strategies for cooling systems in power generation facilities, especially in cases of heavy infestation. In the Rosenergoatom company, periodic chlorination by sodium hypochlorite is obligatory at power plants infested by zebra mussels and it is considered as both a reactive and preventive remedy. The treatment regime needs to be optimized based on monitoring data as well as on the operability and safety of systems designed in the 1980–1990s (personal communications with staff of Kalininsk power plant).

Analyzing the available experience, technical reports, and literature collected in the 1990s, Karatayev et al. (1994) concluded that selection of treatment remedy (besides chlorination) is dependent on the type of industrial enterprise: at hydro facilities, there is no one specific remedy and a variety of available methods and approaches is often used, with preference given to regulation of flow velocity and retrofits of systems; fossil fuel and nuclear power plants often use thermal treatment; in irrigation systems, some mineral nutrients that act both as a biocide and a fertilizer are in use. Some industries use their own chemical production and wastage if they are permitted for use as biocides.

Due to the possible side effects of chemical treatments, extensive research on electric fields and acoustic technologies has been carried out in the former USSR. Many of the tested strategies were not found to be sufficiently efficient. Many were too costly, resource consuming, and difficult to operate but some were considered to be reasonably applicable: sedimentation with electric fields combined with the deceleration of flow velocity (Mikheev et al., 1969); use of ultrasound and cavitation effects (Elpiner and Feigina, 1957; Lubianov et al., 1965) and cathodic protection (Karatayev et al., 1994). However, recommended application of ultrasound, which was effective in laboratory tests, did not give the expected results when used on an industrial scale (a test case at the Volzhskaya hydroelectric power plant), at least in the design that has been in use there since the 1980s (Kablov et al., 2008; Dorefeev, personal communications).

For nonchemical preventive strategy, filtration and self-cleaning automated filters are recommended and are currently in use in many applications (Mikheev et al., 1969; Kablov et al., 2008). One of the most effective and simple but space-consuming technologies is sand–gravel filtration (Mikheev et al., 1969). Also, an increase in flow velocity by impellers was found to be effective in some portions of various systems (Tsimlyanskaya hydroelectric facility); substitution of fouling-sensitive materials with fouling-resistant plastics works well in some applications and antifouling coatings and cements with added biocides have also been found to be effective in certain situations (Karatayev et al., 1994; Protasov, 1994; Kablov

* The next regimes and concentrations are considered as effective (from Karatayev et al., 1994): in drinking water plants—2 times a year during 7–10 days keeping TRC as high as 3–5 mg/L; at hydroelectric facilities—pulse chlorination (50 min three times a day) with TRC concentration 0.5–0.6 mg/L measured at the output of system (case of Pridneprovskaya hydroelectric power plant); TRC of 1 mg/L is considered as sufficient for prevention of juvenile settling; 50 mg/L (no description of regime) as sufficient for killing adults.

et al., 2008). Some antifouling coatings were used successfully for up to four years (case of Volzhskaya hydroelectric power plant) (Karatayev et al., 1994). The use of a protective coating containing short-lived or species-specific biocides of biotic origin is also one of main novel directions of fouling protection research in marine systems today (Railkin, 2004).

Antifouling coatings utilizing fluorocarbon polymers were discussed at the previously mentioned conference (www.zin.ru) as being quite promising. These coatings, examined in laboratory and test facility situations, have demonstrated satisfactory resistance to abrasion, they are relatively cheap and effective in preventing settlement of zebra mussel juveniles by decreasing their adhesion to substrate (Kablov et al., 2008, tests were carried out at Volzhskaya hydroelectric power plant). Application of automatically regulated magnetic fields has also been considered as a possible method for fouling prevention. Results from laboratory tests and field observations on impulsive magnetic fields suggest that they have an impact on the mobility of aquatic organisms (Alexandrov et al., 2008) and that low frequency electromagnetic fields impact many groups of aquatic organisms, including nuisance species (Alexandrov, 2008).

10.1.7 Potential for Practical Use of *Dreissenids*

The freshwater dreissenids often inhabit relatively shallow areas and their dense and productive populations are often considered to be an easily available source of food for cultivated fish and agriculture (Lubianov, 1972). This is especially true for reservoirs, where mussels are easy to collect, with potential harvests of up to 60 tons of zebra mussels per hectare (in total wet body mass with shells) (Karatayev et al., 1994). According to Miroshnichenko (1990), in the relatively warm-water Tsymlianskoye reservoir, *Dreissena* has a fast growth rate and its annual production exceeds 1,000,000 tons with a biomass of 242,000 tons. The zebra mussel also has satisfactory nutrition values for many feedstock applications: 5000 tons of zebra mussel flesh contain about 500 tons of wet protein (Karatayev et al., 1994). Collection of *Dreissena* from storm casts is also considered to be useful. In the Curonial lagoon (Lithuania and Kaliningrad region of the Russian Federation), such mussels are used for forage of domestic ducks and pigs (Gasiunas, 1965). There are also known cases of using mussels collected directly from the mechanical cleaning of power plants for use in private ornamental gardens as mulch and fertilizer (personal communications).

Cultivation has been considered as an alternative to collection. A harvest of dreissenids cultivated on artificial substrates can be used as food for cultivated fishes, in agriculture (Kornobis, 1977), and for gardening. In cultivating blue mussels along the Baltic Sea coast, the additional and unexpected benefit of de-eutrophication occurred (Lindahl, 2008, see also his presentation at www.zin.ru). It is possible that cultivated zebra and quagga mussels could be used this way at the freshet areas of the Baltic Sea and in many reservoirs as well. Mobile modules for cultivation are considered to be preferable, as they allow movement along a water body, selecting the most propitious conditions for growth and harvest of mussels (Karatayev et al., 1994). Let us note that the same modules can be used as a barrier technology, preventing intake of mussel larvae into the facility. Recently, an experimental assessment of an installation for the cultivation of zebra mussels aimed at removing suspended matter from the effluent of a wastewater treatment plant (Kusserow et al., 2008) and de-eutrophication of highly eutrophic lagoons (Fenske, 2008) has been carried out.

Sturgeonids, salmonids, cyprinids, and other cultivated fish were successfully fed with live and cooked* *Dreissena*. Feeding peeled (free of shell), mashed (into spread), and cooked *Dreissena* is preferred by noncyprinid fishes, which do not possess pharyngeal teeth.

However, there are *two precautions* for use of dreissenids as a fish monodiet: (1) use of *Dreissena* with shell decreases its nutritional value, as contribution of shell and mantle liquid into total body mass is high in this mussel; (2) the presence of a high concentration of thiaminase in meat, which can induce thiamine deficit in fed fishes. To avoid this, it is recommended to alternate between feeding with raw mussels and with mussels where thiaminase is deactivated (by boiling or steaming) (Karatayev et al., 1994).

Being long-lived and sessile, zebra and quagga mussels are good biomonitors of hazardous substances, including radionuclides (Lubianov, 1972). Radioecological studies have shown that zebra and quagga mussels along with common native species of freshwater mussels (Unionidae) and snails (Viviparidae and Lymnaridae) can concentrate Sr-90 and other radionuclides (Lubianov, 1972; Orlova and Zuykov, 2008).

10.2 ZEBRA MUSSELS: REVIEW OF ECOLOGY AND IMPACTS SINCE INVASION IN IRELAND

Frances E. Lucy

In Ireland, the initial introduction of the zebra mussel is believed to have taken place in 1994, in the lower Shannon system (McCarthy et al., 1998; Minchin and Moriarty, 1998). This is the largest Irish navigable river system, flowing through 215 km of the Irish midlands (Figure 10.1). Imported second-hand leisure crafts were the most likely vector for their introduction. Live zebra mussels of English origin were found on the infested hulls of imported leisure craft on arrival in Ireland, over the period 1997–2001 and have also been genetically linked to English populations (Pollux et al., 2003; Astanei et al., 2005). Once established, significant zebra mussel settlement took place on native leisure craft and these mussels were carried to the upstream Shannon navigation via locks and swing bridges (Minchin et al., 2002). Since then the zebra mussel has spread within many other lakes in different parts of Ireland, with angling boats identified as the main vector of spread.

The most recent distribution map (Figure 10.7) includes many of these lakes but at the time of writing, more lakes have also been confirmed as having established zebra mussel populations. Most lakes utilized by anglers are in limestone waters, within the calcium and pH range suited to zebra mussels. While zebra mussels occur in some rivers

* One composition (percentage of each item) of fodder that facilitated fast growth of fishes is provided by Karatayev et al. (1994). It contains spread of boiled or raw Dreissena—15–20; minced boiled or raw fish—50–55; mixed fodder for livestock or bran—25; fodder yeasts—3–4; and spread from plants or vitamin premix—1–2.

FIGURE 10.7 Map of known Irish distribution of zebra mussels, 2008. (After Millane, 2009).

that are connected to lakes, river populations are inconsistent (Lucy et al., 2008). In fact, the combination of suitable environmental conditions, available access to angling waters, and lack of adequate legislation have led to rapid spread of large populations of the highly invasive zebra mussel within Irish lake waters in the last decade.

High densities (up to 120,000 m^2) and biomass (≤ 5 kg m^2) of zebra mussels have resulted in direct invasive impacts in infested lakes, both on the freshwater biota and on the human. In terms of lake biology, the most notable impact involves the extirpation of the native unionid clam, *Anodonta*, typically within four years of lake colonization (Lucy, 2005). This mirrors the experience in North America (Schloesser et al., 1993). The wide-scale removal of phytoplankton in lake waters has resulted in increased water clarity and hence textbook improvement of their OECD water quality status due to reduction in chlorophyll levels (OECD, 1982; Lucy et al., 2005). Phosphorus levels in several studies have not however shown a parallel reduction in terms of nutrient removal (Lucy, 2005; Millane, 2009) and so this improvement is scientifically dubious. Increased water clarity in the presence of nutrient availability has led to wide-scale increase in biomass of aquatic plants and benthic algae. In many cases this impedes navigation of leisure craft and angling activities. This has resulted in increased costs of waterway maintenance due to the need for mechanical weed-cutting. Lough Corrib, a world renowned trout lake, has recently been infested both with zebra mussel and curly African weed (*Lagarosiphon major*); there is no doubt that the increased transparency caused by the former invader will improve conditions for this highly invasive plant species to spread within the lake.

FIGURE 10.8 Fouled drinking water intake in Ireland.

The major impact on humans is the biofouling of water screens and pipework used for water abstraction (Figure 10.8). Many Irish towns and rural areas use zebra mussel colonized lakes for drinking water abstraction. These abstraction systems have not been designed to deal with fouling problems as they either predate the zebra mussel invasion or have not taken into account the necessary engineering required. For example, there may be no inspection hatches associated with the intakes and no parallel duplicate intake system. Additionally the nuts and bolts holding the screens in place quickly become corroded in the presence of zebra mussels, making them very difficult to remove.

In Ireland, removal of zebra mussels is carried out by divers, using physical cleaning operations on screens and pipework. In some cases, following diving operations, large stacks of decomposing zebra mussels have been left on lake shores. Zebra mussels have also biofouled water abstraction for fish hatcheries resulting in at least one fish kill and expensive recurrent damage to hatchery filtration systems. There is also a nuisance problem in one hydroelectric station, particularly in association with the fish pass system.

Despite the problems caused by this invasive species, zebra mussels can be advantageous to human, as they can be utilized as public health sentinels of human pathogens in watersheds used for drinking water abstraction and recreational waters. Multiple studies have shown that these filter feeders can harbor and concentrate environmentally derived protozoan parasites (Lucy et al., 2008 and references therein). When present in polluted waters, *Cryptosporidium*, *Giardia*, and human infectious microsporidia are retained by zebra mussels in their tissues. Zebra mussels are extremely useful biomonitors because of their widespread availability and abundance. This makes them easy to collect at different sites, from waters that need to be assessed for contamination.

Outbreaks of cryptosporidiosis occur regularly in many countries, due to consumption of contaminated drinking water, including the Milwaukee epidemic when over 400,000 people suffered from severe gastroenteritis due to *Cryptosporidium*, because the drinking water was contaminated with raw sewage.

In the laboratory, the presence, abundance, and viability of these pathogens in zebra mussels can be effectively assessed using fluorescence *in situ* hybridization (FISH) and immunofluorescent antibody (IFA) biomolecular techniques (Graczyk et al., 2004). By combining these lab results with environmental data, it is possible to source-track the origins of water contamination. The main environmental sources of these pathogens are human sewage and farm animal feces that find their way into local waters. Zebra mussels can therefore be used as an effective tool in the public health management of waters used for human resources, that is, drinking water abstraction and recreational usage.

There is little doubt that zebra mussels will continue to spread to all Irish lakes, with suitable water parameters, which are used for angling purposes. Unless lakes are effectively managed in the near future, via adequate scientific monitoring, new legislation, policy, and effective enforcement, there is no evidence that the rapid spread of this species in Irish waters will slow down until it has reached the biological limits of further expansion.

10.3 SUMMARY OF THE DISTRIBUTION AND IMPACT OF THE GOLDEN MUSSEL IN ARGENTINA AND NEIGHBORING COUNTRIES

Gustavo Darrigran

The first record of the golden mussel (*Limnoperna fortunei*) in South America was in September 1991 at Bagliardi Beach in the Rio de la Plata estuary in Argentina (Darrigran and Pastorino, 1995). Since the initial introduction, *Limnoperna fortunei* (Figure 10.9a) has spread throughout the Plata Basin at an average rate of 240 km/year (Darrigran, 2002), reaching the Paraguay river sometime between 1995 and 1996. In 2001, it was found in Uruguay, Paraguay, and Brazil. This means that it is now also found in Uruguay, Paraguay, and Brazil. Although not yet documented, it is probably present in Bolivia, as the golden mussel has been found in the Pantanal region, on the Bolivia–Brazil border.

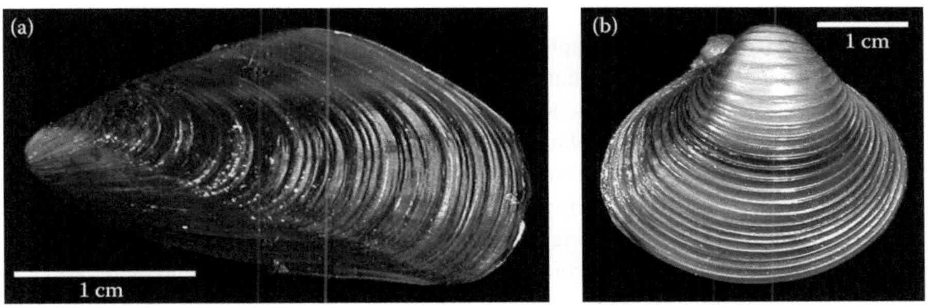

FIGURE 10.9 (a) *Limnoperna fortunei* (golden mussel) and (b) *Corbicula fluminea* (Asian clam).

In the initial years of the invasion, the original population of golden mussels increased in density and, in the following years, dispersed and extended its geographical distribution. This pattern continues to the present. The invasion of the golden mussel heads to the north of the continent, from a mild climate to a subtropical climate but not to the south, where the climate is cold.

At the beginning of 1994, *Limnoperna* caused the first known case of freshwater industrial macrofouling in South America (Darrigran, 1995). This event occurred at the intake for the drinking water plant of the city of La Plata. The infestation was controlled with the help of GIMIP (Grupo de Investigación sobre Moluscos Invasores/Plaga—Research Group on Invading Molluscs, UNLP, Argentina). Subsequent cases of macrofouling have occurred since 1994 as the mussel increased its range. Problems were encountered at the potable water intakes in Bernal and in the city of Buenos Aires (Darrigran and Ezcurra de Drago, 2000). These two drinking water plants supply water to more than half of the Argentine population.

In 1997, *Limnoperna* was found in the port of Asunción (Paraguay) on the Paraguay River. It was first recorded by the staff of the Paraguay Naval Prefecture when they found *Limnoperna* in the cooling systems of their ships. In 1998, *Limnoperna* was found in the Yacyretá Hydroelectric Power Plant (on the border of Argentina–Paraguay) in the upper Paraná and at the end of 1998, in the port of Posadas (Misiones) (Darrigran and Ezcurra de Drago, 2000).

Through a tributary of the Paraná River (the Carcarañá River, see Figure 10.10), the golden mussel invaded the center of Argentina. It was found at the Río Tercero Dam in 2002. It is from this water source that the Nuclear Plant Embalse draws its cooling water. *Limnoperna* reached Carcarañá River, Río Tercero River in Villa María, Río Tercero River in Río Tercero, Piedras Moras Dam, Segundo River in Despeñaderos, Los Molinos Dam, and San Roque Dam in Carlos Paz Village between 2002 and 2006 (Darrigran and Mansur, 2006; Darrigran et al., in press). As natural dispersion into some of these bodies of water is not possible due to a lack of interconnections, the presence of *Limnoperna* illustrates the influence that human activity has, acting as a vector for further dispersal of *Limnoperna fortunei* in South America.

10.3.1 Macrofouling Problems of Water Intakes

The impact of *Limnoperna* in South America, both on the native ecosystem and on the man-made infrastructure, is very similar to the impacts caused by the dreissenid mussels in the Northern Hemisphere.

Unlike freshwater bivalves native to South America, *Limnoperna fortunei* has an epifaunal mode of life and attaches to a wide variety of hard substrates, both natural (from tree trunks and aquatic plants to compact silt-sand) and artificial (docks, pipes, walls, etc). Freshwater macrofouling is a new economic and environmental problem for South America. The impacts of *Limnoperna fortunei* on numerous South American rivers include, among others, alteration of native benthic communities. For example, *Limnoperna fortunei* has modified the presence and abundance of several native macroinvertebrate species (Darrigran et al., 1998a,b; Silvester et al., 2007) and has negatively impacted several native species of molluscs (Darrigran and Ezcurra de Drago, 2000). *Limnoperna* settles on the shells of the native bivalve

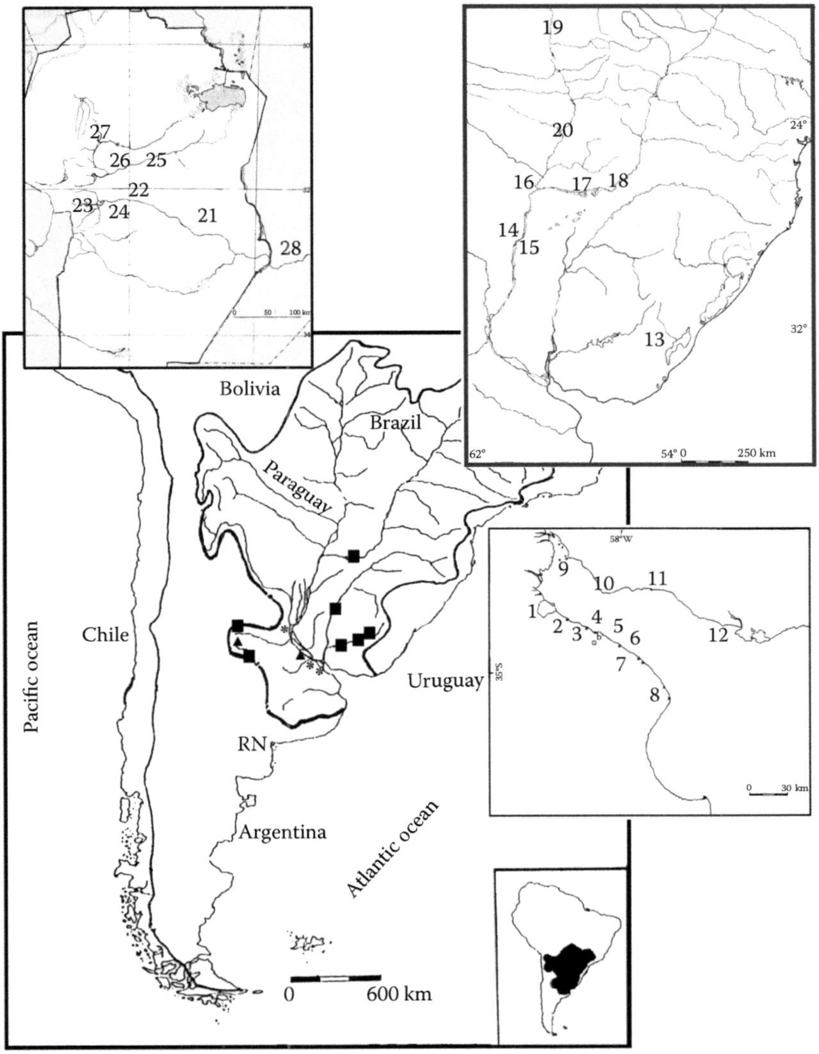

FIGURE 10.10 First localities where *Limnoperna fortunei* was present between 1991 and 2002 in the south of the Plata Basin. 1: Drinking water intake of Buenos Aires city; 2: drinking water intake of Bernal city; 3: Berazategui city; 4: Punta Lara beach; 5: drinking water intake of La Plata city; 6: Bagliardi beach; 7: Atalaya beach; 8: Punta; 9: Martín García Island; 10: Barracas de San Pedro; 11: Artilleros beach; 12: Pascual beach; 13: Laguna Merin lake, Tacuarí River; 14: Goya city; 15: Correntoso, San Javier, and Salado del Norte rivers; 16: Isla del Cerrito; 17: Yacyretá dam; 18: Posadas harbor; 19: Tamango canal; 20: Asunción harbor; 21: Río Tercero river in Villa María; 22: Río Tercero river in Río Tercero; 23: Río Tercero dam, Nuclear Power Station; 24: Piedras Moras dam; 25: Rio Segundo at Despeñaderos; 26: Los Molinos dam; 27: San Roque dam; and 28: Carcarañá river. RN: Rio Negro at Patagonia, Argentina (southeast distribution border in América of *Corbicula fluminea*); ■: Hydroelectric Power Plants with macrofouling; ▲: Nuclear Power Station with macrofouling; and *: Coal-fired Power Station with macrofouling.

Anodontites trapesialis (Lamarck) (Mycetopodidae) (Figure 10.11), the crab *Aegla platensis* Schmitt (Anomura, Aeglidae) and the shells of the introduced Asian clam *Corbicula fluminea* (Darrigran, 2002). The clams suffocate and crabs cannot move normally. Presence of *Limnoperna* has also caused an alteration of the native fish diet (Garcia and Protogino, 2005) and it now represents the main food source for *Leporinus obtusidens* Valenciennes (Anostomidea) in the Río de la Plata (Penchaszadeh et al., 2000).

The Argentine industrial facilities that draw water from the Paraná River, Uruguay River and the Rio de la Plata estuary (Figure 10.12a and b) have suffered macrofouling-related problems. The following list includes the most significant problems caused by *Limnoperna fortunei*:

1. Reduction in the diameter of pipes carrying raw water
2. Pipe blockages
3. Decreased water velocity and volume (due to the fact that the system is designed for laminar flow but the attachment of mussels to the walls causes turbulent flow)
4. Occlusion of orifices by accumulation of empty shells from upstream locations
5. Water contamination in the interior of the system by flesh of dead mussels
6. Filter and strainer blockages

Limnoperna fortunei also impacts many other human activities that take place in freshwater bodies. The impact on the fisheries industry, for example, could be seen in the sturgeon culture in República Oriental del Uruguay, where the mesh of fish cages is occluded (Figure 10.12c and d).

FIGURE 10.11 (See color insert following page 286.) This specimen of *Anodontites trapesialis* is used as a settlement substrate by *Limnoperna*, preventing the specimen from opening or closing its valves, causing eventual mortality.

(a) (c)

(b) (d)

FIGURE 10.12 **(See color insert following page 286.)** (a) Heat interchanger in a thermal power plant; (b) fish exclusion screens in a hydroelectric power plant; (c) cage for *Acipenser baeri* culture in Uruguay; and (d) detail of an aquaculture cage.

In central Argentina, the golden mussel has created significant problems for Cordoba City, one of the most important cities in the country. This city has a population of 1,400,000 inhabitants who have one of the highest standards of living in the country. The city also enjoys one of the highest economic growth levels per capita. On February 3, 2008, shells of the golden mussel accumulated in the cooling system of the San Roque power plant, causing the plant to be taken off-line. This loss of power in turn shut down the water distribution and processing facilities for the entire city, leaving Cordoba without water supply for 24 h. On the following morning, after mechanical cleaning of the system was completed, dense shell banks of *Limnoperna fortunei* were observed in the shallow water of the San Roque Dam (Figure 10.13a and b).

In mid-2008, the GIMIP was consulted on the first macrofouling case for an irrigation system in the region. This took place in the República Oriental del Uruguay on the Tacuarí River, a tributary of the Laguna Merin. The invasion of the golden mussel has now moved more than 50 km inland through this rice cultivation irrigation system (Figure 10.13c).

10.3.2 Prevention and Control in Argentina

In Argentina, the common practice is to control freshwater macrofouling problems from the intake inward. On this basis the GIMIP presents two criteria for the treatment based on Jenner et al. (1998):

1. Determine the biology of the target species, particularly the start and duration of the settling period.
2. Establish the growth rate of the target species in the particular environment of the facility in question. This information is crucial when designing a facility-specific control strategy.

Finally, as mentioned in Darrigran and Damborenea (2006), there is no single treatment option available for all facilities to mitigate *Limnoperna fortunei* fouling. What is required is a facility-specific combination of treatments suited to the environmental, practical, and economic situation at that location. This treatment is not only tailored spatially but also temporally in such a way that

- It causes the least possible impact to the environment.
- It does not alter the normal functioning of the plant.
- It is economical.

To design such a treatment, the following three steps should be taken:

- Diagnosis
- Generation of scientific action guidelines
- Monitoring

FIGURE 10.13 (See color insert following page 286.) (a) Shell bank of the golden mussel in the littoral of the San Roque Dam, Cordoba, Argentina; (b) width of the shell bank; and (c) intake on the Tacuarí River of the rice watering system (Laguna Merin Basin).

In parallel, public awareness must be raised regarding the matter of invasive aquatic species. Increasing public education and awareness on the topic of invasive species will increase pressure on governments to develop mechanisms to prevent further introductions and spread of these harmful organisms (Darrigran and Damborenea, 2006; Darrigran et al., 2008).

In almost all cases of industrial macrofouling control in Argentina, which GIMIP was involved in, the implementation of the Outreach and Education Program led to a change of public sentiment toward the problem of freshwater macrofouling. This was evident in Argentina, when in the process of designing two Thermal Energy Power Plants on the Paraná River (Figure 10.10) during 2006–2007, GIMIP was asked for advice on the plant's building plans and materials of construction in order to proactively prevent future fouling of the cooling systems. This is the first known instance of proactive preventative measures being taken to control for freshwater macrofouling in new facilities in South America.

Since the arrival of the golden mussel in 1991 to the present, much knowledge has accumulated on the biology and dispersion of the golden mussel in South America. It is essential to understand the dynamics of distribution of this invading species, explain its invasion process, and to have access to basic information which will help predict future distribution/dispersion patterns. To accomplish this, our laboratory is sharing *Limnoperna*-related information with a number of other individuals and institutions in South America.

10.4 RESPONSE OF A MAJOR BRAZILIAN UTILITY TO THE GOLDEN MUSSEL INVASION

Maria Edith Rolla and Hélen Regina Mota

10.4.1 Introduction

Most of the electrical energy in Brazil is produced by hydroelectric power plants. Cia. Energética de Minas Gerais (CEMIG) is one of the main producers of hydropower in Brazil. CEMIG provides 7% of Brazil's total generation capacity. It has 52 generating plants in three states: Minas Gerais, Santa Catarina, and Espírito Santo. The plants are located in 10 separate watersheds.

The arrival of the golden mussel has created new challenges for the CEMIG power-generating facilities in southern Brazil. Two years after finding the first golden mussel at the Itaipu Binacional Dam in the Paraná River and Lake Guaíba, the mussels have clogged intakes for all water treatment plants and industrial raw water users on Lake Guaíba, as well as the cooling systems of hydroelectrical plants north of the Paraná River. Studies have now verified that the growth and reproduction rates of the golden mussel in South America exceed the values found in their native region. While prior research literature was able to suggest numerous ways of controlling the mussels (with chlorination being considered to be the most efficient), practical considerations and site-specific concerns in many cases would not allow these control strategies. Alternative options had to be examined. The challenge lay in educating biologists about the needs of industrial facilities and educating engineers about the biology of golden mussels so that together they could develop the most efficient, site-specific control strategy with the least amount of impact on the utility's operations and the environment.

The Brazilian government initiated a National Task Force in 2003, which determined that education and outreach programs would be an effective method of slowing the spread of the mussels to other regions. These programs were aimed specifically at increasing awareness among the public to help slow the infestation. This was followed by the meetings of the Brazilian Generation Association and the Task Force of Minas Gerais State in 2005, which discussed methods to control the expansion of the golden mussel in local watersheds. The utility companies, the federal government, the research institutes, and universities cooperated in compiling information and drafting legislation to control the spread of the bivalve.

10.4.2 Response by CEMIG to *Limnoperna* Invasion

After its arrival in Argentina in 1991, the golden mussel quickly spread across the continent, following traditional boating routes. In 1998 it was found in the Yacyreta Hydro Electrical Plant, located on the Paraná River, where it invaded the cooling systems and other raw water circuits. The Itaipu Binacional Dam identified the presence of the mussel in their reservoir and cooling systems in April of 2001 (Zanella and Marenda, 2002) and the other dams of the Paraná River (Porto Primavera, Jupiá, and Ilha Solteira) detected the mussels in 2003. Since then, population densities at Itaipu have reached up to 180,000 individuals per square meter. *Limnoperna fortunei* is now causing problems in the water distribution systems at the Parana water treatment facility in Itaipu Lake.

Golden mussels were found by CEMIG in the Paranaíba River at the bridge of Porto Alencastro, situated on the state border between Minas Gerais and Mato Grosso do Sul. Larvae were also identified in plankton samples at the Paranaíba confluence with the Aporé and Claro Rivers, between Goiás and Minas Gerais in 2004. They did not reach the Sao Simao or any other hydroelectric dam maintained by CEMIG as of the end of 2008. It is thought that the slow rate of invasion is largely a result of a boat decontamination outreach program using chlorine at transitory harbors and the previously mentioned environmental education program initiated in 2001. Inspection of boats traveling on the Paranaíba River found no mussels on any boat hulls.

During the early stages of the *Limnoperna* invasion, CEMIG recruited the help of the Fundação Centro Tecnologico de Minas Gerais (CETEC) and the Federal University of Minas Gerais (UFMG) to provide assistance in tackling the problems associated with golden mussels. The agencies participated in two major projects. The first project was the "Research for the development of eco-control for *Limnoperna fortunei*" (2002–2004), with a budget of R$200,000. This project started with a bibliographic survey of previously conducted research on both invasive bivalves and *Limnoperna* specifically.

The second project was aimed at the "Development of methodologies and research on the ecosystem of plants and hydroelectric dams to control the Golden Mussel May/2005" (2008) and had a budget of R$1,200,000.

Physical characteristics of the environment often determine the ultimate success of a new invasive species. When environmental conditions are similar to those in its native habitat, the invasive species is able to quickly flourish in the absence of naturally occurring predators and parasites. Environmental variables such as temperature, pH, dissolved calcium concentration, and the presence of natural or artificial substrates to adhere to will all influence the environmental suitability for *Limnoperna*. In 2004, CEMIG organized the monitoring of six reservoirs in three different watersheds: Paranaíba, Grande, and Doce, which were considered "at risk" of infestation by the golden mussel. The Paranaíba river basin is one of the most important Brazilian basins. It is the second largest in total area and in hydroelectric potential. It also occupies a prime location (South-central Brazil), has large production of hydroelectricity, and is significant in terms of economic value (Coelho, 1996).

The first task was to determine the environmental suitability of these reservoirs using data available at Sisagua (the Information System for Monitoring the Quality of Water for Consumption Human) and the CEMIG database (which contains a record of physical water chemistry for the period December 1990 to June 2001). The parameters considered are temperature, dissolved oxygen, pH, conductivity, turbidity, ammonia-nitrogen, and total phosphate. Based on this information, all reservoirs were judged to have environmental suitability for *Limnoperna* invasion.

The reservoir monitoring program was set up and implemented between 2002 and 2004. The study included monitoring the reservoirs for the golden mussel as well as physical and chemical parameters. Water temperature, pH, electrical conductivity, dissolved oxygen, total phosphate, orthophosphate, ammonia-nitrogen, total nitrogen, and chlorophyll were recorded and collections of plankton were made. The data were collected every three months, more frequently if the area was considered to be a high risk.

An environmental suitability study of the Sao Simao reservoir (Table 10.1) suggested that habitat conditions at this site will allow *Limnoperna* to flourish. Evaluation of other reservoirs is still ongoing.

Along with the environmental suitability research and monitoring in the reservoirs, CEMIG initiated monitoring for the mussels in the hydroelectric plants. Five plants were chosen initially as they were considered most at risk: Emborcação and Sãn Simão on the Paranaíba River; Volta Grande and Igarapava on the Grande River; and Dona Rita on the Doce River. The first four plants are located in the Paraná Basin, which is considered "at risk." The last facility, Dona Rita, has a fish species present which came from the Pantanal region. Pantanal has been infested with the golden mussel since 2001 and there was concern that golden mussels may have been transferred at the same time as the fish. Each facility collected monthly plankton samples using a plankton net and sent the samples for analysis to detect the presence of larvae. Each facility also installed a biobox to visually check if any mussel settlement was occurring in the system.

To help slow the spread of *Limnoperna*, a public education program was initiated. A poster and booklet was prepared: "The golden mussel: A threat to Brazilian water and hydroelectric power" (Rolla et al., 2004) and distributed with the help of CEMIG to water users in the region. There were also three public education events

TABLE 10.1
Average Values of Physical and Chemical Parameters in São Simão Reservoir, Collected between 2006 and 2008

	Water Temperature (°C)	Dissolved Oxygen (mg/L)	Electrical Conductivity (µs/cm)	pH	Calcium (mg/L)	Alkalinity (mg/L)	Hardness (mg/L)	Total Phosphorus (mg/L)
Minimum	17.9	4.5	20	6.0	2.4	12.4	13.9	0.03
Maximum	32.1	9.6	159	10.5	16.1	68.4	74.3	1.07
Average	25.5	7.4	43	7.5	4.1	17.9	20.5	0.23

in the main regions of Minas Gerais, Goiás, and São Paulo, an area that contains the major reservoirs of CEMIG. The company continues to participate nationally and internationally in meetings regarding the golden mussel.

10.4.3 Research Results

10.4.3.1 *Limnoperna fortunei* and the Environment

A study by Fundação Centro Tecnologico de Minas Gerais (CETEC) in 2004 determined *in vitro* the tolerance limits for acidity and alkalinity of golden mussel adults taken from Itaipu Binacional. This study determined that in a normal temperature range, mussels were able to survive a pH ranging from 2.7 to 9. This study also investigated the effects of a range of temperatures on mortality (40°C, 37°C, 35°C, 32°C, 15°C, and 10°C). The bivalves were maintained in an incubation chamber at constant temperature and observations, including pH, temperature, attachment status, and mortality, were collected every 24 h. It was found that a 100% mortality rate was achieved in 24 h in the 40°C and 37°C conditions, 72 h at 35°C, and 96 h at 32°C. In the lower temperature conditions the mortality rate did not exceed 10% after the conclusion of the experiment. The ideal survival and reproduction temperature for the mussel was determined to be 19.5°C. The mortality rate was not observed to be affected by weight or length of the individuals and there was also no observable relation between filtration rate and temperature. Byssal attachment rates decreased in all temperature treatments over 15°C. Water temperature was also tested as a limiting factor for *Limnoperna* survival (Silva and Campos, 2004).

The golden mussels (*Limnoperna fortunei*) attach to substrate with a byssus-plaque system. It appears that adhesion to any material (organic, inorganic, biological, or synthetic) is a temperature-dependent property. We utilized the differential scanning calorimetric (DSC) technique in order to identify the phase transition temperature values for the byssus-plaque production in the golden mussel. We identified an endothermic peak temperature centered at 52.1°C with an enthalpy value of 577 J/g. This result and other recent experiments show that thermal treatments produce detachment in *Limnoperna fortunei* from substrates due to a permanent molecular alteration of the byssus subcomponents (bundles of collagens fibers). This alteration results in the loss of mechanical adhesiveness. Physiochemical experiments under way in our laboratory (not presented here) have shown that by combining thermal treatment with physicochemical treatment may contribute to the development of new antifouling strategies (CETEC, 2008).

In vitro experiments showed that *Limnoperna fortunei* is able to feed on the potentially toxic *Cyanobacteria* species *Microcystis viridis* and *Pseudoanabaena*. *Limnoperna* was able to filter an enormous amount of *Microcystis* in a short period of time. The observed filtration rates were as high as 17.20 mL/s or 24.52 mg/h. The filtration rates were similar to values found for the green algae *Selenastrum capricornutum*, which were observed to be as high as 11.91 mL/s. The ability of *Limnoperna* to feed on *Cyanobacteria* may lead to the bioaccumulation of cyanotoxins in this species. As the population densities of *Limnoperna* increase and the mussels are utilized as food by fish species normally used for human consumption, transfer of the toxin to higher trophic levels and ultimately to humans may become a problem (von Rückert et al., 2004).

Analysis of the diet of the native fish species *Leporinus friderici* (BLOCH, 1794) was done using stomach content to determine the extent to which golden mussels are fed on by this fish and by extension, if the mussels have entered the local food chain. Samples were collected at two locations: near the reservoir of São Simão and further downstream in the Paranaíba River. The first sample was taken in May (the dry season), and the second in October (the wet season). Stomach contents of the fish were analyzed for frequency of food occurrence and volumetric percentage. The stomach contents were categorized by frequency of occurrence and percentage of total volume (Hyslop, 1980; Table 10.2).

The golden mussel was found in all the sampled individuals from the Paranaíba River. This is consistent with other studies that found *Limnoperna fortunei* to be a significant component of the diet in many species of estuarine and freshwater fish. It appears likely that the population dynamics of some native species could be affected by the presence of these mussels in the ecosystem (Cataldo et al., 2002). No mussels were found in the stomachs of fish collected upstream of São Simão Reservoir (Mata and Silva, 2007).

Analysis of natural settlement substrates was carried out to determine if *Limnoperna* has a substrate preference in the Paranaíba River. Five substrates were identified at two sampling depths: rock, quartz, sandstone, silt, and conglomerate. The area of the sampled substrate was measured to verify the mussel settlement density on each substrate. It was determined that quartz completely deterred *Limnoperna* settlement. Rock was the most preferred substrate, followed by sandstone, conglomerate, and silt. The study was also able to conclude that mussel density on substrates at the depth of 1 m versus those from the surface showed no difference in mussel density (Souza and Mota, 2008).

TABLE 10.2
Percentage Composition of the Diet of *Leporinus friderici* in Paranaíba River

Food Items	Frequency of Occurrence (%)	Percentage Index Food (Hyslop, 1980) (%)
Plant matter	38.23	0.493
Seeds	29.41	0.314
Stems	14.70	0.003
Insects	8.82	0.013
Inorganic debris	20.58	0.070
Mussel remains	17.64	0.0025
Filamentous algae	8.82	0.005
Golden mussels	23.52	0.104
Chlorophytes	35.29	—
Cianofíceas	26.47	—
Diatoms	23.53	—

Studies conducted on the maximum levels of dissolved calcium tolerance of *Limnoperna fortunei* determined that any concentration higher than 100 mg/L produced 100% mortality (Mata and Campos, 2005).

Laboratory experiments confirmed the field observations of mussel settlement preference for wood, iron, aluminum, and fiber. Test plates made of copper or coated with antifouling paints were found to have much lower settlement of the golden mussel. Iron appeared as a suitable colonization medium in the laboratory experiments but when oxidation levels rose in the field, mussel mortality was observed (Fulgêncio and Mota, 2007).

10.4.3.2 Control Strategies

The golden mussel is of great concern to industrial water users in large part because of its ability as larva to penetrate all but the finest of filters, the ability of adults to form thick layers, and the capacity to produce prodigious amounts of tiny larvae. This is in addition to the tremendous adaptability of the mussels to various conditions. Most industries need to control *Limnoperna* populations in their facilities. They need tools to accomplish this control in an environmentally friendly manner and at a minimum economic cost.

Cataldo and Boltovskoy (1998) tested three commercial molluscicides alongside a chemical oxidant for *Limnoperna* control. They found that generally the required concentrations necessary to obtain mortality were much higher than has been previously found for other mussel species. The tests were performed using Clamtrol CT2, H130M, Bayluscide WP70, and chlorine as the oxidizing chemical. The maximum concentration of Clamtrol utilized to successfully control the zebra mussel is approximately 1 ppm. In *Limnoperna*, Clamtrol was found to have no significant effect at a concentration of 10 ppm for 24 h, had minimal effects at 50 ppm for 48 h, and required a time of 96 h to obtain a 100% mortality rate at 50 ppm. One of the main benefits claimed in using Clamtrol and H130M is that mussels are unable to detect it as a noxious substance and therefore do not close their valves, decreasing the time necessary to achieve mortality. Observation during the experimental treatment saw some valve closure taking place, suggesting that *Limnoperna* can sense these chemicals. Bayluscide was found to be generally as effective at the manufacturer's suggested treatment concentration and was determined to be successful when utilizing a treatment of 3–4 ppm for 48 h. After 24 h, a 90% mortality rate was found among the veligers but the adults required the full 2-day treatment for adequate control. The chlorine oxidant treatment confirmed that as with other mussel species, even high treatment concentrations (90 ppm) must be given sufficient time to produce adequate control results. As the mussels were found to close their valves on detecting chlorine and only attempted filtration when forced to, a lower dose concentration (1 ppm) was found to be just as successful as a high concentration after 14 days. The larval phases of *Limnoperna fortunei* were found to be more sensitive to chemical treatment than the adults. Lower concentrations of chemicals were required to achieve 100% mortality of the veligers.

An additional molluscicide, MXD100, was compared with sodium hypochlorite in laboratory tests for control of adult *Limnoperna* (Cataldo and Boltovskoy, 1998). The results showed that the MXD100 achieved mortality in adult mussels at lower concentrations than hypochlorite (Table 10.3) while having less of an impact on non-

Table 10.3
Values of CL50 Obtained with MXD100 and with Sodium Hypochlorite in *Daphnia similis* and *Limnoperna fortunei*

	MXD100 (ppm)	Sodium Hypochlorite (ppm)
CL50 in *Limnoperna fortunei*	0.05	5.5
CL50 in *Daphnia similis*	0.12	0.01

target species such as *Daphnia similis* than hypochlorite. They report that this may be due to the fact that the mussels do not sense MXD100.

Further studies using MXD100 are recommended, especially to verify performance under field conditions and long-term impact on the receiving environment.

Antifouling paint is one area of study that needs further exploration. Research on the mechanisms by which byssal attachment occurs and how it is affected by substrate material could aid in determining which substances will provide successful mussel fouling deterrent. Faria (2005) found no difference in the strength of mussel attachment between ceramic, tin, and plastic. Additionally, Faria confirmed that attachment to glass, iron, and aluminum is fairly strong whereas on copper, brass, and zinc the settlement level was zero, as these materials are generally toxic to most mussels.

The only nonchemical strategy investigated for *Limnoperna* control at this time was sand filtration or infiltration gallery. Other nonchemical methods that have been successful in controlling other forms of invasive bivalves would logically also be successful in controlling *Limnoperna*. However, until tested this is only an assumption.

A study performed in 2007 (Resende) found that *Limnoperna* infestation of various pipe diameters generally resulted in a loss of pump station energy efficiency between 50% and 300%. Considering that flow restrictions resulting from infestation by *Limnoperna fortunei* with the density of 0.5 cm^{-2} increase the energy use, there would be potential for much greater energy use in some of the areas where infestation densities of 10 individuals/cm^2 have already been found.

The results of the Resende study indicated that the loss of effective diameter due to infestation of piping is considerable. A pipe with a diameter of 4″, when infested by *Limnoperna fortunei* at a density of 0.5 cm^{-2}, will be equivalent to a clean pipe with a nominal diameter of 3″. Similarly, evaluating a pipe with a nominal diameter of 2″, it was observed that the infestation resulted in a reduction of its useful diameter to a diameter equal to 1.5″. In the case of industrial pipes infested by *Limnoperna fortunei*, it is necessary to make adjustments and interpolations for the estimated loss of flow.

The invasive nature of the golden mussel requires further research so that we can limit their spread to unaffected watersheds. We also need to prepare mitigation strategies for hydroelectric facilities which have not yet been affected, should the spread of *Limnoperna* continue. The research needed is an in-depth study of the organism's biology and ecology so that we can develop site-specific, efficient, environmentally sustainable, and low-cost mitigation strategies.

Glossary

Acute: Forming an angle of less than 90°.

Adult: Any individual that is more than half grown; it may or may not be gravid.

Anal siphon: The dorsal exhalant tube or opening derived from mantle-margin epithelia and located near the anus at the posterior end of the animal. Water currents flow from inside the mantle cavity through the anal siphon to the outside.

Angulate: Forming an angle; not rounded.

Annulate: Ringed; surrounded by a ring of a different color; formed in ring-like segments.

Annulation: Formation of ring-like parts or subdivisions of a segment.

Annulus: A ring or division.

Anterior: In the direction of the head; the opposite of posterior. In bivalve shells, the end with the foot, usually the narrower end.

Anterior end: The shorter end of the shell (from the beaks) in *Corbicula*, *Eupera*, *Musculium*, and *Sphaerium*; the longer end of the shell in *Pisidium*. The foot of the animal is directed toward this end.

Aperture: Opening, usually referring to the opening of a snail shell.

Apex: The tip or top of a snail shell.

Apical: Toward the tip of a structure or appendage.

Appendicula: A wrinkle on the beaks of some species of *Pisidium* forming a ridge or raised part (same as ridge).

Arch: A curved structure.

Asymmetrical: Lacking in symmetry; not capable of being divided into equal or corresponding parts.

Basal: At, or pertaining to, the base or point of attachment; nearest to the main body.

Base: The lowest part of a body or structure; the portion upon which a structure rests; the broadest portion of a cone-shaped structure.

Beak: The raised part on the dorsal margin of a shell. It is the oldest (embryonic) part of the shell around which the shell later develops.

Benthic: Of, or pertaining to, the bottom of a body of water or the organisms living at the bottom.

Bifid: Cleft or divided into two parts; two-pronged.

Bifurcate: Divided partly or forked into two.

Bilamellate: Divided into two lamellae or plates.

Bilobed: Having two lobes.

Bivalved: Having a shell consisting of two valves or valve-like parts.

Bivoltine: Producing two litters or broods each year.

Blood colorless: In most freshwater mollusks, the blood is a clear fluid containing hemocyanin.

Blood red: In some pulmonates, the blood is red in color due to the presence of hemoglobin.

Body whorl: The last, and largest, whorl of a snail shell.

Branchial siphon: The ventral inhalant tube or opening derived from mantle-margin epithelia. It is located at the posterior end of the animal and through it water carrying dissolved oxygen, food organisms, and male gametes flows outside.

Brood size: The number of embryos or larvae within a brood sac.

Byssus: A thread produced in the posterior portion of the foot for anchoring either the animal to the substratum (as in *Corbicula* and *Eupera)* or the shelled larvae to the parent's gill (as in *Musculium*).

Calcareous: Of the nature of, consisting of, or containing calcium carbonate.

Calyculate: Cup-like, as the beaks when separated from the rest of the shell by being distinctly raised above a sulcus, as in *Musculium.*

Canine-like: Refers to the shape of the cardinal teeth, in this case like the canine tooth of man (i.e., curved and somewhat tapered).

Cap: The raised portion of the beak above the sulcus in *Musculium.* Animals with caps are "calyculate."

Cardinal teeth: Lamellae, on the center of the hinge in the right and left valves, which function to stabilize the two valves against shearing forces. The opposing cardinal teeth of the two valves fit together in a complementary fashion. There are usually two cardinal teeth (C2 and C4) in the left valve and one (C3) in the right. Sometimes hinge anomalies and/or reversals (see this glossary) occur.

Carina: An elevated ridge or keel.

Chevron-shaped: Shaped like a wide-angled V, normally positioned or inverted.

Cilia: Fringes; series of moderate or thin hairs.

Cleft: Split, partly divided longitudinally.

Compressed: Having a low width/height ratio; flattened in such a way that the body is deep and narrow; the opposite of depressed.

Concave: Having a curved, depressed surface.

Concentric: A series of circles or ellipses having a common center.

Conical: Cone-shaped.

Contiguous: Adjacent to; in contact with; adjoining.

Convex: Having a curved rounded exterior, similar to that of a sphere.

Corneous: Hornlike; hardened proteinaceous layer.

Crenation: One of the rounded projections on a scalloped edge.

Crenulate: With small scallops, evenly rounded, and rather deeply curved.

Crest: A prominent ridge.

Crystalline style: A rod containing digestive enzymes in the stomach of bivalves.

Cusps: The highest elevations on the lateral teeth.

Deflected: Bent down or turned to one side.

Denticle: A small tooth.

Denticulate: Set with little teeth or notches.

Depressed: Flattened in such a way that the body is low and wide; having a low height/width ratio; the opposite of compressed.

Detritus: Loose material: rock fragments and organic material; can be suspended in a water column or part of benthos.

Dextral: To the right of the median line, or, for snails, coiled to the right.

Differentiated: Modified and specialized for the performance of specific functions.

Diffuse: Spread out; not localized.

Dioecious: Separate sexes, with some individuals being male and others female.

Discoidal: Disk-shaped; shaped like a round plate.

Dissections: Deeply cut divisions or lobes.

Distal: Near or toward the free end of any appendage; the part that is farthest from the body. Farthest from the beaks of the shell in an anterior–posterior axis.

Divergent: Spreading out from a common base.

Dorsal: Toward the upper surface when the body is in normal walking position; the opposite of ventral.

Dorsal margin: The upper edge of the shell, including the hinge.

Dyphyoidy: Nontaxonomic conchological feature of sphaeriids in which the shell shows a more or less developed transverse furrow, groove, or scar in the mesial part of each valve.

Elliptical: Shaped like an ellipse, with an oval body rounded at both ends.

Elongate: Lengthened; slender; longer than wide.

Emarginate: Having the margin notched or indented.

Embryo: The stages of marsupial development up to and including gastrulation.

Expanded: To open outwards or spread.

Exposed: Open to view; not shielded or protected.

Filament: In bivalves, a tube-like structure on the gill. Several filaments make up a gill lamella.

Fingernail clam: Common or vernacular name for members of the genera *Sphaerium* and *Musculium*.

Foot: Single in mollusks. It has creeping, crawling, or burrowing function.

Frontal: Of, or pertaining to, the anterior portion of the body or a part.

Fusiform: Spindle-shaped; tapered at each end.

Gill: Structure especially adapted for the exchange of dissolved gases between a mollusk and a surrounding liquid. In mollusks it is also called a demibranch.

Globose: Approaching the shape of a sphere.

Globule: A small spherical body.

Glochidium: A larval unionacean. Most are parasitic on fish.

Gravid: Carrying embryos or larvae.

Grooved: Having a narrow channel or depression.

Growth lines: In gastropods, raised lines running across the whorls; not to be confused with spiral striae.

Hatchet-shaped: Having an ax shape. Bivalve foot is hatchet-shaped.

Height: The greatest distance from the ventral edge of the shell to the top of the beak.

Height/Length (or H/L) ratio: The number or quotient obtained by dividing the greatest height of a clam shell by its greatest length.

Hermaphrodite: Has both male and female sexual organs in the same individual. Same as monoecious.

Hinge: The stabilizing lamellae (cardinal and lateral teeth) in the dorsal part of each valve. The opposing single lamella in the corbiculacean shell articulates with a pair of complementary lamellae in the opposing valve.

Hinge-plate: The part of the dorsal margin of the shell that is between the anterior and posterior lateral teeth of each valve; it does not include the lateral teeth.

Hinge teeth: Any of various ridges or thickenings on the dorsal side of the interior of a bivalve shell that aid in holding the two halves of the shell together.

Incised: Notched or deeply cut into.

Incision: The impressed line marking the juncture of two segments.

Incisor-like: Refers to the shape of the cardinal teeth, in this case like the incisor tooth of man (i.e., rather tall and chisel-like).

Inclined: To have a bend or slant.

Inflated: In a snail shell, refers to swollen, as if the whorls have been pumped full of air. In a bivalve, refers to laterally expanded, as opposed to compressed.

Irregular: Not uniform; not conforming to the usual pattern.

Iteroparous: Producing more than one litter or brood in the life span of an individual or the average life span of a cohort.

Juveniles: The born young who are less than half grown.

Keel: An elevated ridge; a carina.

Lamella: A sheet or thin plate. In bivalves, each gill is composed of an inner and outer lamella. Plural—lamellae.

Larvae: The stages of marsupial development where the shell is present.

Lateral: Toward the side of the body; the opposite of mesal.

Lateral teeth: The elongated lamellae at each end of the hinge-plate. They are designates AI (inside) and AIII for the anterior pair; PI (inside) and PIII for the posterior pair in the right valve; and AII for the anterior tooth and PII for the posterior tooth in the left valve.

Left valve: The shell half on the left side when the shell is placed with the hinge up and the anterior end forward.

Length: The greatest distance from the anterior end of the shell to the posterior end.

Lentic: Of, or pertaining to, standing water.

Ligament: The elastic material on the hinge by means of which the two valves articulate.

Limnetic: Pertaining to organisms inhabiting open waters of ponds, lakes, and inland seas.

Litter size: The number of embryos or larvae in a brood sac (same as brood size).

Lobate: Composed of, or possessing, lobes or divisions resembling lobes.

Lobe: A rounded division or projection of an organ.

Longitudinal: Lengthwise; extending along the long axis.

Lotic: Of, or pertaining to, flowing water.

Lucifugous: Avoiding light.

Macro: Being large, thick, or exceptionally prominent; of, involving, or intended for use with relatively large quantities or on a large scale.

Malleated: Having small flattened areas, as if having been hit with a hammer.

Mantle: In mollusks, a thin layer of tissue that secretes the shell and lines the shell's interior. The thin epithelial layer that hangs down like a skirt around all the soft parts of the animal.

Mantle edge: The growing edge of the mantle that usually bears three folds; the inner fold is large and muscular, the outer fold is elongate and secretes the periostracum and part of the shell, and the outer fold is elongate and sensory and helps to mold or form the growing edge of the shell.

Marsupium: The pouch used to contain the young. In Sphaeriidae, brood sacs produced from gill filaments of the inner demibranchs perform this service; in Unionidae and Corbiculidae, internal spaces within the inner gills serve this function.

Medial: Lying in the midline of the body.

Mesial: Pertaining to the middle, toward the middle; the opposite of lateral.

Monoecious: Has both male and female sexual organs in the same individual. Same as hermaphrodite.

Multivoltine: Producing more than two litters or broods each year.

Multispiral: Having numerous spirals, as on the operculum of Valvatidae.

Nacre: The pearly white or colored material covering the inside of a unionid shell.

Naiad: Common name for unionid shell.

Newborn: Both a plural and a singular noun that refers to those individuals that are newly born and have not added shell material since birth.

Nuclear whorl: The first whorl on the apex of a snail shell.

Obliquely: At, or from, an angle.

Obtuse: Forming an angle exceeding 90°.

Operculate: Having an operculum.

Operculum: A lid or cover. Present on the foot of all prosobranch snails.

Organic: Pertaining to, consisting of, or related to organs; pertaining to, originating in, or derived from, living organisms; pertaining to compounds that contain carbon.

Ovate: Somewhat oval in shape.

Ovoid: See ovate.

Parent: Any gravid individual or individual that was gravid.

Patelliform: Limpet-shaped.

Paucispiral: Marked by a spiral with a few turns, as on an operculum of hydrobiid snails.

Pea clam: Common or vernacular name for a member of the genus *Pisidium* (same as pill clam).

Periostracum: The colored, proteinaceous layer that covers the outside of mollusk shells.

Peripheral: Of, towards, or constituting an external surface or boundary; away from the center; toward the surface.

Pill clam: Common or vernacular name for a member of the genus *Pisidium* (same as pea clam).

Planktonic: Of, or pertaining to, aquatic organisms of fresh, brackish, or seawater that float passively or exhibit limited locomotor activity.

Posterior: In the direction away from the head; the opposite of anterior.

Posterior end: The longer end of the shell (from the beaks) in *Corbicula*, *Eupera*, *Sphaerium*, and *Musclium*; the shorter end of the shell in *Pisidium*. This is the end from which the siphons extend.

Process: An elongation of the surface, a margin, or an appendage; any prominent part of the body not otherwise definable.

Prosobranch: A group of snails that has an operculum, respires with gills, and are generally dioecious.

Proximal: Away from the tip of a structure or an appendage; the opposite of distal. Nearest to the beaks of the shell in an anterior–posterior axis.

Pulmonate: Having lungs, as in Physidae, Lymnaeidae, Planorbidae, and Ancylidae.

Pustule: A large projection lump.

Quadrate: Square, or nearly so.

Recess: A depression, fossa, cleft, or cavity.

Recurved: Bent backward.

Reduced: Diminished in size, amount, or number.

Respiratory: Of, or pertaining to, the interchange of gases between an organism and its environment.

Rest mark: A darker or thicker concentric part of the shell that is formed during a rest period in growth.

Reticulate: Covered with a network of lines.

Retractile: Capable of being drawn in or back.

Ridge: A crest or an elevation.

Ridge: A wrinkle or raised part on the beaks of some species of *Pisidium*.

Right valve: The shell half on the right side when the shell is placed with the hinge up and the anterior end forward.

Robust: Strongly formed or constructed; sturdy.

Rudimentary: Of, or pertaining to, an organ or a structure that is just beginning to develop or one in which development has been arrested; describing a vestige or the remains of a structure which was functional in an earlier stage of evolutionary development.

Scutellum: The projecting or pinched parts of the shell in front of the beaks.

Scutum: The projecting or pinched parts of the shell behind the beaks.

Secrete: To produce a substance discharged through a duct from a gland.

Sedentary: Permanently attached to the substratum; not free-moving; remaining in the same place; not migratory.

Segmentation: The state of being divided into segments or parts, which are marked off or separated from adjacent parts.

Semelparous: Producing only one litter or brood in the life span of an individual or in the average life span of a cohort.

Seminal receptacle: A sperm storage structure; spermatheca.

Serrate: Sawlike; with notched edges.

Serrated: Notched on the edge, like a saw blade.

Serrulate: Finely serrate; bearing minute teeth.

Sessile: Attached by the base; without stem or stalk.

Seta(e): Slender, hairlike appendage; hair.

Setiferous: Bearing setae.

Setose: Furnished or covered with setae or stiff hairs.

Shallow: Lacking in depth.

Sigmoid: S-shaped, curved in two directions.

Simple: Unmodified, not branched.

Sinistral: To the left of the median line, or for snails, coiled to the left; opposite of dextral.

Sinuous: Curving in and out.

Sinus: A cavity, hollow, recess, channel, or space.

Siphons: Fleshy tubes or openings at the posterior end through which inhalant or exhalant water passes into and out of the mantle cavity.

Slender: Small or narrow in circumference.

Solitary: Existing singly or alone.

Sparse: Not thickly grown; scattered.

Spatulate: Broad and rounded at the tip, more slender at the base; spoon-shaped.

Spherical: Having the form of a sphere or globe.

Spinous: With many spines.

Spinule: A very small spine.

Spire: The entire snail shell, excepting the body whorl.

Stria: A fine longitudinally impressed line. Plural—striae.

Striae: Concentric raised striations or lines on the exterior surface of the shell. They may vary from being very fine to very coarse, the latter often called as ribs.

Sub-: Prefix; more-or-less.

Subcentral: Not quite central in an anterior–posterior axis; off-centre.

Submersed: Covered with water; growing or adapted to grow underwater.

Succeeding: Coming next to one another in position; following in order.

Sulcus: A longitudinal furrow, one usually at each end of the hinge-plate of the right valve, which serves as a socket for the teeth of the opposing (left) valve, or any groove, furrow, or channel, as at the base of an umbonal cap.

Suture: A seam or an impressed line indicating the division of the distinct parts of the body wall, on a snail shell, separating the whorls.

Tapered: Becoming progressively smaller at one end.

Teeth: The opposing lamellae on the hinge that serve to stabilize the two valves against shearing forces. The central lamellae are called cardinal teeth and the elongate lamellae at each end of the hinge-plate are called lateral teeth.

Terminal: Pertaining to, placed at, or forming an end or extremity; being the last of a series; located at the extreme end of a body or structure.

Terminate: To come to, or form, a terminus or end.

Tooth: One of a pair of structures on the hinge of a lamellibranch shell which functions in locking the valves; a tooth-like or pointed process.

Transverse: At right angles to the long axis of the body.

Truncate: Having the end cut off more or less squarely.

Tubercle: A nodule or small eminence occurring on the surface of the shell or on a part of the shell (e.g., on the walls of the sulci of the lateral teeth).

Tubercular: Shape of a raised eminence or elevation that has a rounded top and vertical sides.

Tuberculate: Covered with tubercles or rounded knobs.

Tubular: Having the shape of a hollow cylindrical structure; composed of, or possessing, tubes.

Tubule: A small, elongate tube-like structure.

Tuft: A small cluster of flexible structures, such as hairs, feathers, or blades of grass, closely associated at their bases but with free ends spread apart.

Umbilicus: The small opening present in some snail shells behind the base of the aperture.

Umbo: The hump on the dorsal margin of a bivalve shell; also called the beak or umbone.

Umbone: The raised part on the dorsal margin of the shell (same as beak). Also called umbo.

Unequal: Of different sizes or proportions.

Uniform: Presenting an unvaried appearance of a surface, pattern, or color.

Univoltine: Producing only one litter or brood each year.

Valve: One of the opposing halves of the bivalve shell. The two halves articulate dorsally by an elastic ligament.

Veliger: A larval stage of dreissenid and corbiculid bivalves.

Ventral: Toward the lower surface when the body is in normal walking position; the opposite of dorsal.

Vertical: Of, or pertaining to, the highest point, namely summit or apex; upright; perpendicular to the horizontal plane; lengthwise or in the direction of the long axis.

Vestigial: Small or degenerate.

Weakly: Not strongly, not overly pronounced.

Whorl: One turn of the spiral of a snail shell.

Width: The greatest distance from one side of the shell to the other.

Width/Length (or W/L) ratio: The number or quotient obtained by dividing the greatest width of a clam shell by its greatest length.

References

References and further reading may be available for this article. To view references and further reading you must purchase this article.

Abbott, R.T., 1952. A study of an intermediate snail host (*Thiara granifera*) of the oriental lung fluke (*Paragonimus*). *Proceedings of the United States National Museum* 102: 71–116.

Abbott, R.T., 1958. *American Seashells*. Princeton: D. Van Nostrand Company.

Abbott, R.T., 1973. Spread of *Melanoides tuberculata*. *The Nautilus* 87: 29.

Abbott, R.T., 1974. *American Seashells. The Marine Mollusca of the Atlantic and Pacific Coasts of North America,* 2nd ed. New York: Van Nostrand Reinhold Co.

Abbott, R.T., 1975. Beware the Asiatic freshwater clam. *Tropical Fish Hobbyist* 23: 15.

Ackerman, J.D. and R. Claudi, 1991. The early life history of the zebra mussel: Overwintering juveniles and post-settlement movements. *Journal of Shellfish Research* 11: 217.

Ackerman, J.D., B. Sim, S.J. Nichols, and R. Claudi, 1994. A review of the early life history of zebra mussels (*Dreisssena polymorpha*): Comparisons with marine bivalves. *Canadian Journal of Zoology* 72: 1169–1179.

Ackerman, J.D., M.R. Loewen, and P.F. Hamblin, 2001. Benthic-pelagic coupling over a zebra mussel reef in western Lake Erie. *Limnology and Oceanography* 46: 892–904.

Acosta, B.O. and R.S.V. Pullin, 1991. Environmental impact of the golden snail (*Pomacea* sp.) on rice farming systems in the Philippines. Freshwater Aquaculture Center, Central Luzon State University, Munoz, Nueva Ecija: ICLARM.

Acres International, 1992. The study of centrifugal separators for zebra mussel control. Report P9895.29.01.

Adalia, C.B. and E.A. Magsino, 2006. Understanding the golden apple snail (*Pomacea canaliculata*): Biology and early initiatives to control the pest in the Philippines. In: *Global Advances in Ecology and Management of Golden Apple Snails*, R.C. Joshi and L.S. Sebastian, eds, pp. 199–214. Nueva Ecija: Philippine Rice Research Institute.

Adams, B.A., 1927. The lethal effects of various chemicals on *cyclops* and *daphnia*. *Water and Waste Engineering* 29: 361–364.

Aditya, G. and S.K. Raut, 2001. Food of the snail *Pomacea bridgesi*, introduced in India. *Current Science* 80: 919–921.

Afanasiev, S.A., A.V. Sidersky, and A.V. Shatokhina, 1990. Fouling of water supply canal of nuclear power plant as factor for bioobstacles. *Hydrobiological Journal* 26: 25–29 (in Russian).

Afanas'yev, S.A. and A.A. Protasov, 1987. Characteristics of a *Dreissena* population in the periphyton of a nuclear power plant cooling pond. *Hydrobiological Journal* 23: 42–49.

Ahlstedt, S., 1979. Recent mollusk transplants into the North Fork Holston River in southwestern Virginia. *The Bulletin of the North American Malacological Union Inc.* 1979: 21–23.

Akerlund G., 1974. Oxygen consumption in relation to environmental oxygen concentrations in the ampullariid snail *Marisa Cornuarietis* (L). *Comparative Biochemistry and Physiology* 47A: 1065–1075.

Albrecht, E.A., N.B. Carreño, and A. Castro-Vasquez, 1996. A quantitative study of copulations and spawning in South American apple-snail *Pomacea canaliculata* (Prosobranchia: Ampullariidae). *The Veliger* 39: 142–147.

Aldridge, A.D.C., P. Elliot, and G.D. Moggridge, 2006. Microencapsulated 'BioBullets' for the control of biofouling zebra mussels. *Environmental Science and Technology* 40: 975–979.

Aldridge, D.W. and R.F. McMahon, 1978. Growth, fecundity and bioenergetics in a natural population of the Asiatic clam, *corbicula manilensis* Philippi, from north central Texas. *Journal of Molluscan Studies* 44: 49–70.

Aldridge, D.W., B.S. Payne, and A.C. Miller, 1987. The effects of intermittent exposure to suspended solids and turbulence on three species of freshwater mussels. *Environmental Pollution* 45: 17–28.

Alexander, Jr. J.E, J.H. Thorp, and R.D. Fell, 1994. Turbidity and temperature effects on oxygen consumption in the zebra mussel (*Dreissena polymorpha*). *Canadian Journal of Fisheries and Aquatic Sciences* 51: 179–184.

Alexander, Y., A.Y. Karatayev, L.E. Burlakova, V.A. Karatayev, and D.K. Padilla, 2009. Introduction, distribution, spread, and impacts of exotic freshwater gastropods in Texas. *Hydrobiologia* 619: 181–194.

Alexandrov, B., A. Boltachev, T. Kharchenko, A. Lyashenko, M. Son, P. Tsarenko, and V. Zhukinsky, 2007. Trends of aquatic alien species invasions in Ukraine. *Aquatic Invasions* 2: 215–242.

Alexandrov, V.V. and B.V. Alexandrov, 2004. Earth electromagnetic fields and aquatic ecosystems. *Periphyton and Fouling: Theory and Practice*. Book of Abstracts, St. Petersburg: Russian Academy of Sciences, (in Russian with English summary).

Alexandrov, V.V., E.S. Gorshkov, V.V. Ivanov, and L.B. Lider, 2008. Transitory influence by impulsive magnetic fields on physiological reactions of biological objects *in vivo*. *Periphyton and Fouling: Theory and Practice*. Book of Abstracts, St. Petersburg: Russian Academy of Sciences, pp. 152–154 (in Russian with English summary).

Alimov, A.F., 1975. The rate of metabolism in freshwater bivalve molluscs. *Soviet Journal of Ecology* 5: 6–13.

Alimov, A.F., M.I. Orlova, T.M. Florinskaya, and N.D. Sorokin, 2008. Development of system of monitoring of non-indigenous species. In: *Ecosystem of the Neva River Estuary: Biological Diversity and Ecological Problems*, A.F. Alimov and S.M. Golubkov, eds. Moskow: KMK Press.

Allhands, M., 2003. State fish hatcheries and invasive species don't mix. In: *Proceedings of the 12th International Conference on Aquatic Invasive Species*, Windsor, Canada, June 9–12.

Alonso, A. and P. Castro-Díez, 2008. What explains the invading success of the aquatic mud snail *Potamopyrgus antipodarum* (Hydrobiidae, Mollusca)? *Hydrobiologia* 614: 107–116.

Anderson, B., 1993. The Philippine snail disaster. *The Ecologist* 23: 70–72.

Anderson, R., 2005. An annotated list of the non-marine mollusca of Britain and Ireland. *Journal of Conchology (London)* 38: 607–638.

Andrews, E.B., 1964. The functional anatomy of the mantle cavity, kidney and blood system of some pilid gastropods. *Journal of Zoology* 146: 70–94.

Andrews, E.B., 1965. The functional anatomy and histology of the reproductive system of some pilid gastropod molluscs. *Proceedings of the Malacological Society of London* 36: 121–140.

Andrews, J., 1981. *Texas Shells: A Field Guide*. Austin: University of Texas Press. Revised 1992 as *A Field Guide to Shells of the Texas Coast*. Houston: Gulf Publishing Company.

Andrusov, N.I., 1890. *Dreissena rostriformis* Desh. in the Bug River. *Vestnik Estestvoznaniya* 6: 261–262.

Andrusov, N.I., 1897. Fossil and recent Dreissenidae of Eurasia. Trudy Sankt-Peterburgskago Obschestva Estestvoispitatelei. *Department of Geology and Mineralogy* 25: 1–683 (in Russian with German summary).

Angarano, M.B., R.F. McMahon, D.L. Hawkins, and J.A. Schetz, 2007. Exploration of structure antifouling relationships of capsaicin-like compounds that inhibit zebra mussel (*Dreissena polymorpha*) macrofouling. *Biofouling* 23: 295–305.

Anonymous, 1993. *Zebra Mussel Research Program.* U.S. Army Engineer Waterways Experiment Station, Technical Notes, Section 4, Miscellaneous. Technical Note ZMR-4-03.

Anonymous, 1994. Heat trials to control ballast water introductions. *Marine Pollution Bulletin* 28: 647.

Antonov, P.I., 1993. About invasion of bivalve *Dreissena bugensis* (Andr.) into the Volga River Reservoirs. Ecological problems of large rivers basins. *Abstract from the International Conference, September 6–10, Togliatti, Russia (in Russian).*

Appleton, R.D. and Palmer, A.R., 1988. Water-borne stimuli released by predatory crabs and damaged prey induce more predator-resistant shells in a marine gastropod. *Proceedings of the National Academy of Sciences* 85: 4387–4391.

Araujo, R., D. Moreno, and M.A. Ramos, 1993. The Asiatic clam *Corbicula fluminea* (Müller, 1774) in Europe. *American Malacological Bulletin* 10: 39–49.

Archambault-Guezou, J., 1982. Comparaison microstructurale des tests de diverses especes actuelles des genres *Dreissena* et *Con-geria* (Dreissenidae, Mollusca Bivalvia). *Malacologia* 22: 325–332.

Armengol, J., L. Caputo, M. Comerma, C. Feijoó, J.C. Garcíal, R. Marcé, E. Navarro, and J. Ordoñez, 2003. Sau reservoir's light climate: Relationships between Secchi depth and light extinction coefficient. *Limnetica* 22: 195–210.

Armistead, D.C., 1995. *Tolerances* of Zebra Mussels to various temperatures in the Mississippi and Ohio Rivers, 1988–1992. Zebra Mussel Research Technical Note ZMR-1–32, U.S. Army Corps of Engineers, Waterways Experiment Station, Vicksburg.

Armitage, K.B., 1958. Lagos volcanicos de El Salvador. *Community Institute of Tropical Investigations* 7: 39–48.

Armonies, W. and D. Hartke, 1995. Floating of mud snails *Hydrobia ulvae* in tidal waters of the Wadden Sea, and its implications in distribution patterns. *Helgoland Marine Research* 49: 529–538.

Arnott, D.L. and M.J. Vanni, 1996. Nitrogen and phosphorus recycling by the zebra mussel (*Dreissena polymorpha*) in the western basin of Lake Erie. *Canadian Journal of Fisheries and Aquatic Sciences* 53: 646–659.

Arthur, J.W. and J.W. Eaton, 1971. Chloramine toxicity to the amphipod, *Gammarus pseudolimnaeus* and the fathead minnow (*Pimephales promelas*). *Journal of the Fisheries Research Board of Canada* 28: 1841–1845.

Ashby, S.L., W.A. Boyd, and R.H. Kennedy, 1998. Assessing the potential for zebra mussel habitat at U.S. Army Corps of Engineers Water Resources Projects using GIS techniques and water quality data. In: *Proceedings of the Eighth International Zebra Mussel and Aquatic Nuisance Species Conference*, Sacramento, USA, March 16–19.

ASI Group, 2005. *Proof of Principle Evaluation of an Innovative Ballast Water Treatment Methodology: Phase I Final Report.* ASI Group Project M20115. Ottawa: ASI Group.

Astanei, I., E. Gosling, J. Wilson, and E. Powell, 2005. Genetic variability and phylogeography of the invasive zebra mussel, *Dreissena polymorpha* (Pallas). *Molecular Ecology* 14: 1655–1666.

Aufderheide, J., R. Warbritton, N. Pounds, S. File-Emperador, C. Staples, N. Caspers, and V. Forbes, 2006. Effects of husbandry parameters on the life-history traits of the apple snail, *Marisa cornuarietis:* Effects of temperature, photoperiod, and population density. *Invertebrate Biology* 125: 9–20.

AWWA (American Water Works Association), 2006. *Water Chlorination/Chloramination Practices and Principles*, 2nd ed. American Water Works Association, Denver, Colorado.

Baback, E.V., 1983. The Pliocene and Quaternary Dreissenidae of the Euxinian Basin. In: *Proceedings of the Palaeontological Institute Academy of Science of USSR* (in Russian), Moscow: Russian Academy of Sciences.

Babinec, J., 2003. Copper ion treatment for zebra mussel mitigation in house service water systems. *Power Plant Chemistry* 5: 539–547.

Bailey, R.C., E.H. Anthony, and G.L. Mackie, 1983. Environmental and taxonomic variation in fingernail clam (Bivalvia: Pisidiidae) shell morphology. *Canadian Journal of Zoology* 61: 2781–2788.

Baker, F.C., 1902. The Mollusca of the Chicago area, pt. II. *Chicago Academy of Science, Natural History Survey Bulletin III*, pp. 131–418.

Baker, F.C., 1928. The American *Bithynia* not wholly an introduced species. *Transactions of the Illinois State Academy of Science* 20: 56–62.

Baker, P. and S. Baker, 1993. Criteria for estimating zebra mussel risk for non-invaded regions. *Dreissena polymorpha Information Review* (Zebra Mussel Information Clearinghouse, New York Sea Grant) 4: 4–8.

Baker, P., S. Baker, and R. Mann, 1993a. Criteria for predicting Zebra Mussel invasions in the Mid-Atlantic Region. School of Marine Science, College of William and Mary, Gloucester Point, VA (not seen, cited by Cohen, 2001).

Baker, P., S. Baker, and R. Mann, 1993b. Potential range of the zebra mussel, *Dreissena polymorpha*, in and near Virginia. School of Marine Science, College of William and Mary, Gloucester Point, VA.

Baker, S.M. and D.J. Hornbach, 1997. Acute physiological effects of zebra mussel (*Dreissena polymorpha*) infestation on two unionid mussels, *Actinonaias ligamentina* and *Amblema plicata*. *Canadian Journal of Fisheries and Aquatic Sciences* 54: 512–519.

Baker, S.M., J.S. Levinton, J.P. Kurdziel, and S.E. Shumway, 1998. Selective feeding and biodeposition by zebra mussels and the relation to changes in phytoplankton composition and seston load. *Journal of Shellfish Research* 17: 1207–1213.

Balcom, N.C., 1994. *Aquatic Immigrants of the Northeast, No. 4: Asian Clam, Corbicula fluminea*, Connecticut Sea Grant College Program. Available at: http://www.sgnis.org/publicat/nespp_4.htm. Accessed March 2, 2009.

Baldwin, B.S., M. Black, O. Sanjur, R. Gustafson, R.A. Lutz, and R.C. Vrijenhoek, 1996. A diagnostic molecular marker for zebra mussels *(Dreissena polymorpha)* and potentially co-occurring bivalves: Mitochondrial *COI. Molecular Marine Biology and Biotechnology* 5: 9–14.

Baldwin, B.S., P. Filippetti, and S. Sanderson, 1997. A test of the potential spread of zebra and quagga mussels from the St. Lawrence River to inland waters of northern New York (abstract). Presented at the *2nd Northeast Conference on Nonindigenous Aquatic Nuisance Species*, Burlington, USA.

Baldwin, B.S., M.S. Mayer, J. Dayton, N. Pau, J. Mendilla, M. Sullivan, A. Moore, A. Ma, and E.L. Mills, 2002. Comparative growth and feeding in zebra and quagga mussels (*Dreissena polymorpha* and *Dreissena bugensis*): Implications for North American lakes. *Canadian Journal of Fisheries and Aquatic Sciences* 59: 680–694.

Bamber, R.N. and J.D. Taylor, 2002. The brackish water mussel *Mytilopsis leucophaeata* (Conrad, 1831) (Bivalvia: Dreissenidae) in the river Thames. *Journal of Conchology* 37: 403–404.

Barber, B.J., 1992. Preliminary investigation of the salinity tolerance of zebra mussels, *Dreissena polymorpha* implications for Chesapeake Bay. (Abstract) *Journal of Shellfish Research* 11: 218.

Barbiero, R.P. and M.L. Tuchman, 2004. Long-term dreissenid impacts on water clarity in Lake Erie. *Journal of Great Lakes Research* 30: 557–565.

Barbiero, R.P., D.C. Rockwell, G.J. Warren, and M.L. Tuchman, 2006. Changes in spring phytoplankton communities and nutrient dynamics in the eastern basin of Lake Erie since the invasion of *Dreissena* spp. *Canadian Journal of Fisheries and Aquatic Sciences* 63: 1549–1563.

Barbiero, R.P., M.L. Tuchman, and E.S. Millard, 2006. Post-dreissenid increases in transparency during summer stratification in the offshore waters of Lake Ontario: Is a reduction in whiting events the cause? *Journal of Great Lakes Research* 32: 131–141.

Bargar, T.A., 1991. *The effects of the molluscicide H-130 (didecyl-dimethyl ammonium chloride) and bentonite clay upon Hexagenia limbata, Pimephelas promelas and Ceriodaphnia dubia.* MSc thesis, Ohio State University, Columbus, OH.

Barnes, D.K., M.S. Alix, and J.W. Camp, 1998. The effects of magnetic water treatment on zebra mussels in laboratory and industrial settings. Paper presented at the *8th International Zebra Mussel and Other Nuisance Species Conference*, March 16–19, Sacramento, USA.

Barnes, M.A., R.K. Fordham, and R.L. Burks, 2008. Fecundity of the exotic applesnail, *Pomacea insularum. Journal of the North American Benthological Society* 27: 38–745.

Bartell, S.M., Y. Wu, S.K. Nair, J. Orr, and J. Ragland, 2007. Risk assessment and decision analysis support for invasive mussel management for the St. Croix Basin and adjacent Upper Mississippi River. Final Report for U.S. Army Corps of Engineers, St. Paul District, Contract No. W912ES-05-D-0002, Task Order 07.

Barton, L.K., 1993. Control program for zebra mussels (*Dreissena polymorpha*) at the Perry nuclear power plant, Lake Erie. In: *Zebra Mussels Biology, Impacts, and Control*, T.F. Nalepa and D.W. Schloesser, eds, pp. 555–562. Boca Raton: Lewis Publishers.

Baskin, Y., 1996. Curbing undesirable aliens. *BioScience* 46: 732–736.

Bass, M.L. and A.G. Heath, 1975. Physiological effects of intermittent chlorination on fish. *American Zoologist* 15: 818.

Bass, M.L. and A.G. Heath, 1977. Cardiovascular and respiratory changes in rainbow trout, *Salmo Gairdneri* exposed intermittently to chlorine. *Water Resources* 11: 497.

Bastviken, D.T.E, N.F. Caraco, and J.J. Cole, 1998. Experimental measurements of zebra mussel (*Dreissena polymorpha*) impacts on phytoplankton community composition. *Freshwater Biology* 39: 375–386.

BAT-cooling: European IPPC bureau Sevilla., Document on the application of best available techniques to industrial cooling systems, November 2000 (http://eippcb.jrc.es) BREF (11.00) Cooling systems.

Baxter, R.M., 1977. Environmental effects of dams and impoundments. *Annual Review* of *Ecology and Systematics* 8: 255–283.

Beasley, C.R., C.H. Tagliaro, and W.B. Figueiredo, 2003. The occurrence of the Asian clam *Corbicula fluminea* in the Lower Amazon Basin. *Acta Amazonica* 33: 317–324.

Beauchamp, W.M. 1880. *Bithynia tentaculata* Linn. *American Naturalist* 14: 523.

Beaver, R.J., T.L. Crisman, and R.J. Brock, 1991. Grazing effects of an exotic bivalve (*Corbicula fluminea*) on hypereutrophic lake water. *Lake and Reservoir Management* 7(1): 45–51.

Bedë, L.C., 1992. Dinâmica populacional de *Melanoides tuberculata* (Prosobranchia: Thiaridae) no reservatório da Pampulha, Belo Horizonte, M.G., Brasil. Dissertação de Mestrado. Universidade Federal de Minas Gerais, Belo Horizonte (not seen, cited by Paz et al., 1995).

Behrendt, A., 2007. Photos. Available at: http://www.allesumdieschneck.de/html/marisa_cornuarietis_english.html. Accessed December 6, 2008.

Beissinger, S.R., T.J. Donnay, and R. Walton, 1994. Experimental analysis of diet specialization in the snail kite: The role of behavioural conservatism. *Oecologia* 100: 54–65.

Belanger, S.E., 1991. The effect of dissolved oxygen, sediment, and sewage treatment plant discharges upon growth, survival and density of Asiatic clams. *Hydrobiologia* 218: 113–126.

Belanger, S.E., D.S. Cherry, J.L. Farris, K.G. Sappington, and J. Cairns, Jr., 1991. Sensitivity of the Asiatic clam to various biocidal control agents. *Journal of the American Water Works Association* 83: 79–87.

Belluati M., L. Bartole, and G. Bressan, 1997. In chlorine dioxide and disinfection. In: *Proceedings of the First European Symposium on Chlorine Dioxide and Disinfection,* Rome, Italy, November 7–8.

Bennike, O.J., B. Jensen, W. Lemke, A. Kuijpers, and S. Lomholt, 2008. Late- and postglacial history of the Great Belt, Denmark. *Boreas* 33: 18–33.

Benson, A.J., 2008a. Zebra mussel sightings distribution. Available at: http://nas.er.usgs.gov/taxgroup/mollusks/zebramussel/zebramusseldistribution.asp. Accessed January 20, 2009.

Benson, A.J., 2008b. Quagga mussel sightings distribution. Available at: http://nas.er.usgs.gov/taxgroup/mollusks/zebramussel/quaggamusseldistribution.assp. Accessed January 20, 2009.

Benson, A.J., 2008c. *Pomacea insularum.* USGS Nonindigenous Aquatic Species Database, Gainesville, FL. Available at: http://nas.er.usgs.gov/queries/FactSheet.asp?speciesID=2599 Revision Date: 8/14/2007. Accessed October 30, 2008.

Benson, A.J., P.L. Fuller, and C.C. Jacono, 2001. Summary Report of Nonindigenous Aquatic Species in U.S. Fish and Wildlife Service Region 4. Report prepared for U.S. Fish and Wildlife Service, Arlington, Virginia.

Beran, L. and M. Horsák, 2007. Distribution of the alien freshwater snail *Ferrissia fragilis* (Tryon, 1863) (Gastropoda: Planorbidae) in the Czech Republic. *Aquatic Invasions* 2: 45–54.

Berezina, N.A., 2001. Influence of ambient pH on freshwater invertebrates under experimental conditions. *Russian Journal of Ecology* 32: 343–351.

Berezina, N.A., 2007. Expansion of the North American amphipod *Gammarus tigrinus* Sexton, 1939 to the Neva Estuary (easternmost Baltic Sea). *Oceanologia* 49: 129–135.

Bergstrom, P., 2004. An introduction to dark false mussels. Available at: http://www.chesapeakebay.net/pubs/calendar/LRSC_09-30-04_Presentation_1_5352.pdf. Accessed February 28, 2009.

Berkman, P.A., M.A. Haltuch, E. Tichich, D.W. Garton, G.W. Kennedy, J.E. Gannon, S.D. Mackey, J.A. Fuller, and D.L. Liebenthal, 1998. Zebra mussels invade Lake Erie muds. *Nature* 393: 27–28.

Berry, A.J. and A.B. Haji Kadri, 1974. Reproduction in the Malayan freshwater cerithiacean gastropod *Melanoides tuberculata. Journal Zoology London* 172: 369–381.

Berry, E.G. 1943. *The Amnicolidae of Michigan: Distribution, ecology and taxonomy.* Museum Zoological Publication No. 57, University of Michigan.

Bertram, P. 1993. Total phosphorus and dissolved oxygen trends in the central basin of Lake Erie. *Journal of Great Lakes Research* 19: 224–236.

Beukema, J.J. and J. de Vlas, 1989. Tidal-current transport of thread-drifting postlarval juveniles of the bivalve *Macoma balthica* from the Wadden Sea to the North Sea. *Marine Ecology Progress Series* 52: 193–200.

Bierman, Jr. V.J., D.W. Dilks, J. Kaur, J.V. DePinto, and T.J. Feist, 2005. Modeling the role of zebra mussels in the proliferation of blue-green algae in Saginaw Bay, Lake Huron. *Journal of Great Lakes Research* 31: 32–55.

Bij de Vaate, A., 1989. Occurrence and population dynamics of the zebra mussel, *Dreissena polymorpha* (Pallas, 1771), in the Lake IJsselmeer area (the Netherlands). In: Paper presented at *New York Sea Grant Conference on Zebra Mussels in the Great Lakes,* November 22–23, New York, USA.

Bij de Vaate, A., 1991. Colonization of the River Rhine by the Asiatic clam, *Corbicula fluminea* Müller, 1774 (Pelecypoda, Corbiculidae). *Bulletin Zoologisch Museum Universiteit van Amsterdam* 13: 13–16.

Bij de Vaate, A., K. Jazdzewski, H.A.M. Ketelaars, S. Gollasch, and G. van der Velde, 2002. Geographical patterns in range extension of Ponto-Caspian macroinvertebrate species in Europe. *Canadian Journal of Fisheries and Aquatic Sciences* 59: 1159–1174.

Bilos, C., J.C. Colombo, and M.J.R. Presa, 1998. Trace metals in suspended particles, sediments and Asiatic clams (*Corbicula fluminea*) of the Rio de la Plata Estuary, Argentina. *Environmental Pollution* 99: 1–11.

Bilton, D.T., J.R. Freeland, and B. Okamura, 2001. Dispersal in freshwater invertebrates. *Annual Review of Ecological Systems* 32: 159–181.

Binnie, N.E., 2009. Overloaded? Mussels, biofouling and material condition observations for the *Hamilton* and *Scourge* Shipwreck Site. In: *ACUA Underwater Archaeology Proceedings 2009,* E. Laanela and J. Moore, eds. Advisory Council on Underwater Archaeology, manuscript in preparation.

Binnie, N.E., P. Engelbert, L.D. Murdocka, and J. Moore, 2000. "Shipwrecks, archaeology and zebra mussels: Is mussel attachment a threat to our submerged cultural resources?" In: *Proceedings of the 10th International Aquatic Nuisance Species and Zebra Mussel Conference*, Toronto, Ontario, pp. 121–131, February 2000.

Birger, T.I., A.Y. Malarevskaja, O.M. Arsan, V.D. Solomatina, and Y.M. Gupalo, 1978. Physiological aspects of adaptations of mollusks to abiotic and biotic factors due to blue-green algae. *Malacological Review* 11: 100–102.

Blackburn, R.D., T.M. Taylor, and D.L. Sutton, 1971. Temperature tolerance and necessary stocking rates of *Marisa cornuarietis* L. for aquatic weed control. In: *Proceedings of the European Weed Resource Council, Third International Symposium on Aquatic Weeds,* Oxford: European Weed Resource Council.

Blanck, C.A., D.F. Mead, and D.J. Adams, 1996. Effective control of Zebra Mussels using a high molecular weight polymer. Presented at the *Sixth International Zebra Mussel and Other Aquatic Nuisance Species Conference,* March, Dearborn, USA.

Blay, Jr. J. and F. Dongdem, 1996. Preliminary observations on the benthic macrofauna of a polluted coastal lagoon in Ghana (West Africa). *Tropical Ecology* 37: 127–133.

Block, R.M., 1977. Physiological responses of estuarine organisms to chlorine. *Chesapeake Science* 18: 158.

Block, R.M., J.C. Rhoderick, and S.R. Gullans, 1977. Physiological response of white perch (*Morone americana*) to chlorine. *Association of the Southeastern Biologists Bulletin* 24: 37.

Blok, J.W. and M. Tan-Maas, 1977. Function of byssus threads in young postlarval mytilus. *Nature* 267: 558.

Blossey, B., D. Schroeder, S.D. Hight, and R.A. Malecki, 1994. Host specificity and environmental impact of two leaf beetles (*Galerucella calmariensis* and *G. pusilla*) for biological control of purple loosestrife (*Lythrum salicaria*). *Weed Science* 42: 134–140.

Blume, W.J., P.C. Fraleigh, and W.R. van Cott, 1994. Evaluation of copper ions and aluminum floc for preventing settlement of zebra mussels. In: *Proceedings of the Fourth International Zebra Mussel Conference*, Madison, USA, March.

Bobat, A., M.O. Hengürmen, and W. Zapletal, 2004. Zebra mussel and fouling problems in the Euphrates Basin. *Turkish Journal of Zoology* 28: 161–177.

Bobeldyk, A.M., J.M. Bossenbroek, M.A. Evan-White, D.M. Lodge, and G.A. Lamberti, 2005. Secondary spread of zebra mussels (*Dreissena polymorpha*) in lake-stream systems. *Ecoscience* 12: 414–421.

Boegman, L., M.R. Loewen, P.F. Hamblin, and D.A. Culver, 2008. Vertical mixing and weak stratification over zebra mussel colonies in western Lake Erie. *Limnology and Oceanology* 53: 1093–1110.

Boelman, S.F., F.M. Neilson, E.A. Dardeau, and T. Cross, 1997. Zebra mussel (*Dreissena polymorpha*) control handbook for facility operators. Zebra mussel Research Program, Miscellaneous paper EL-97-1, U.S. Army Corps. of Engineers, Washington. Available at: http://el.erdc.usace.army.mil/zebra/zmis/zmishelp3/zebra_mussel_dreissena_polymorpha_control_handbook_for_facility_operators_first_edition.htm. Accessed February 27, 2009.

Boettger, C.R., 1933. Über die Ausbreitung der Muschel *Congeria cochleata* Nyst in europäischen Gewässern und ihr Auftreten im Nordostseekanal. *Zoologischer Anzeiger* 101: 43–48.

Bøggild, O.B., 1930. The shell structure of the mollusks. K. Dan. *Vidensk. Selsk. Skr. Naturvidensk. Math. Afd* 2: 232–235.

Boltovskoy, D. and D. Cataldo, 1999. Population dynamics of *Limnoperna fortunei* an invasive fouling mollusk in the lower Paraná river (Argentina). *Biofouling* 14: 255–263.

Boltovskoy, D., N. Correa, D. Cataldo, and F. Sylvester, 2006. Dispersion and ecological impact of the invasive freshwater bivalve *Limnoperna fortunei* in the Rio de la Plata watershed and beyond. *Biological Invasions* 8: 947–963.

Bombeo-Tuburan, I., S. Fukumoto, and E.M. Rodriguez, 1995. Use of the golden apple snail, cassava, and maize as feeds for the tiger shrimp, *Penaeus monodon,* in ponds. *Aquaculture* 131: 91–100.

Bondesen, P. and E.W. Kaiser, 1949. *Hydrobia (Potamopyrgus) jenkinsi* (Smith) in Denmark illustrated by its ecology. *Oikos* 1: 252–281.

Borcherding, J., 1991. The annual reproductive cycle of the freshwater mussel *Dreissena polymorpha* (Pallas) in lakes. *Oecologia* 87: 208–218.

Bossenbroek, J.M., C.E. Kraft, and J.C. Nekola, 2001. Prediction of long-distance dispersal using gravity models: Zebra mussel invasion of inland lakes. *Ecological Applications* 11: 1778–1788.

Bossenbroek, J.M., L.E. Johnson, B. Peters, and D.M. Lodge, 2007. Forecasting the expansion of zebra mussels in the United States. *Conservation Biology* 21: 800–810.

Botts, P.S., B.A. Patterson, and D.W. Schloesser, 1996. Zebra mussel effects on benthic invertebrates: Physical or biotic? *Journal of the North American Benthological Society* 15: 179–184.

Bourne, G.R. and J.A. Berlin, 1982. Predicting *Pomacea dolioides* (Reeve) (Prosobranchia: Ampulariidae) weights from measurements of their shells. *The Veliger* 24: 367–370.

Bowler, P.A., 1991. The rapid spread of the freshwater hydrobiidae snail *Potamopyrgus antipodarum* (Gray) in the Middle Snake River, southern Idaho. *Proceedings of the Desert Fishes Council* 21: 173–182.

Bowler, P.A. and T.J. Frest, 1992. The non-native snail fauna of the Middle Snake River, Southern Idaho. *Proceedings of the Desert Fishes Council* 23: 28–44.

Bowman, M.F. and R.C. Bailey, 1998. Upper pH tolerance limit of the zebra mussel (*Dreissena polymorpha*). *Canadian Journal of Zoology* 76: 2119–2123.

Brady, Jr. R.F., 2001. A fracture mechanical analysis of fouling release from nontoxic antifouling coatings. *Progress in Organic Coatings* 43: 188–192.

Brady, T., 2007. The use of ozone to control zebra mussels, Ontario power generation, Lennox generating station. In: *Proceedings of the 15th International Conference on Aquatic Invasive Species*, Nijmegen, the Netherlands, September 23–27.

Brady, T.J., J.E. Van Benschoten, and J.N. Jensen, 1996. Chlorination effectiveness for zebra and quagga mussels. *Journal of the American Water Works Association* 88: 107–110.

Brand, C.J., S.M. Schmitt, R.M. Duncan, and T.M. Cooley, 1988. An outbreak of Type E botulism among common cloons (*Gavia immer*) in Michigan's upper peninsula. *Journal of Wildlife Diseases* 24: 471–476.

Breitig, G., 1965. *The use of ultrasound in the eradication of larvae*. PhD dissertation, Mathematisch-Naturwissenschaftliche Fakultät, University Greifswald.

Brendelberger, H., 1997. Contrasting feeding strategies of two freshwater gastropods, *Radix peregra* (Lymnaeidae) and *Bithynia tentaculata* (Bithynidae). *Archiv fur Hydrobiologie* 140: 1–21.

Brendelberger, H. and S. Jürgens, 1993. Suspension feeding in *Bithynia tentaculata* (Prosobranchia, Bithyniidae), as affected by body size, food and temperature. *Oecologia* 94: 36–42.

Briers, R.A., 2003. Range size and environmental calcium requirements of British freshwater gastropods. *Global Ecology and Biogeography* 12: 47–51.

Britton, J.C. and B. Morton, 1979. *Corbicula* in North America: The evidence reviewed and evaluated. In: *Proceedings of the First International Corbicula Symposium*, J.C. Britton, ed., pp. 249–287. Fort Worth, TX: Texas Christian University Research Foundation.

Britton, J.C. and B. Morton, 1982. A dissection guide, field and laboratory manual for the introduced bivalve *Corbicula fluminea. Malacological Review Supplem*ent 3: 1–82.

Britton, J.C. and B. Morton, 1986. Polymorphism in *Corbicula fluminea* (Bivalvia: Corbiculoidea) from North America. *Malacological Review* 19: 1–43.

Britton, J.C. and B. Morton, 1989. *Shore ecology of the Gulf of Mexico*. Austin, TX: University of Texas Press.

Britton, J.C., D.R. Coldiron, L.P. Evans, Jr., C. Golightly, K.D. O'kane, and J.R. Teneyck, 1979. Reevaluation of the growth pattern in *Corbicula fluminea* (Muller). In: *Proceedings of the First International* Corbicula *Symposium*, J.C. Britton, ed., pp. 177–192. Fort Worth, TX: Texas Christian University.

Britton, J.C., ed, 1979. *Proceedings of the First International Corbicula Symposium*. Fort Worth, TX: Texas Christian University Research Foundation.

Brohmer, P., 1962. Die Weichtiere Mitteleuropas. In: *Die Tierwelt Mitteleuropas*, Band 2, 1. Ergänzung. Leipzig: Quelle & Meyer.

Brook, A.J. and A.L. Baker, 1972. Chlorination at power plants: Impact on phytoplankton productivity. *Science* 176: 1414–1415.

Brooks, G. and G. Matisoff, 1993. Chlorine dioxide control of adult zebra mussels. In: *Third International Zebra Mussel Conference, Agenda and Abstracts,* Toronto, Canada, February.

Brown, C.D., M.V. Hoyer, R.W. Bachmann, and D.E. Canfield, Jr., 2000. Nutrient–chlorophyll relationships: An evaluation of empirical nutrient–chlorophyll models using Florida and northern temperate lake data. *Canadian Journal of Fisheries and Aquatic Sciences* 57: 1574–1583.

Brown, D.S., 1994. *Freshwater Snails of Africa and their Medical Importance*, 2nd ed. Bristol: Taylor & Francis.

Brown, K.M., B. Lang, and K.E. Perez, 2008. The conservation ecology of North American pleurocerid and hydrobiid gastropods. *Journal of the North American Benthological Society* 27: 484–495.

Brugnoli, E., J. Clemente, L. Boccardi, A. Borthagaray, and F. Scarabino, 2005. Golden mussel *Limnoperna fortunei* (Bivalvia: Mytilidae) distribution in the main hydrological basins of Uruguay: Update and predictions. *Anais da Academia Brasileira de Ciências* 77: 235–244.

Brungs, W.A., 1973. Effects of residual chlorine on aquatic life. *Journal of the Water Pollution Control Federation* 45: 2180–2193.

Brungs, W.A., 1977. General considerations concerning the toxicity to aquatic life of chlorinated condenser effluent. In: *Biofouling Control Procedures, Technology and Ecological Effects,* L.D. Jensen, ed., pp. 109–113. New York: Marcel Dekker.

Bryan, D.C., 1990. Apple snail densities at Alexander Springs, Lake County, and observations on snail ecology. *Florida Scientist* 53: 13.

Buchan, L.A.J. and D.K. Padilla, 1999. Estimating the probability of long-distance overland dispersal of invading aquatic species. *Ecological Applications* 9: 254–265.

Buchanan, A.C., 1987. Die-off impacts on the mussel fauna of selected reaches of the Bourbeuse and Meramec rivers, Missouri. In: *Proceedings of the Workshop on Die-offs of Freshwater Mussels in the United States*, R.J. Neves, ed., pp. 44–54. U.S. Fish and Wildlife Servive, Upper Mississippi River Conservation Committee, and Virginia Polytechnic Institute and State University (Blacksburg).

Buckley, J.A., 1977. Heinz body hemolytic anemia in coho salmon (*Oncorhynchus kisutch*) exposed to chlorinated wastewater. *Journal of the Fisheries Resource Board of Canada* 34: 215.

Bunt, C.M., H.J. MacIsaac, and W.G. Sprules, 1993. Pumping rates and projected filtering impacts of juvenile zebra mussels (*Dreissena polymorpha*) in western Lake Erie. *Canadian Journal of Fisheries and Aquatic Sciences* 50: 1017–1022.

Burch, J.B., 1982. North American freshwater snails. *Walkeriana* 1: 217–365.

Burch, J.B., G-M. Park, and E-Y. Chung, 1998. Michigan's polyploidy clams. *Michigan Academician* 30: 351–352.

Burch, J.Q., 1944. Checklist of West North American marine Mollusca from San Diego, California to Alaska. *Minutes of the Conchological Club of Southern California* 38: 11–18.

Burky, A.J., 1983. Physiological ecology of freshwater bivalves. In: *The Mollusca,* Vol. 6, *Ecology,* W.D. Russell-Hunter, ed., pp. 281–327. Orlando: Academic Press, Inc.

Burky, A.J. and P.A. White, 1989. Seasonal low and high temperature tolerance of *Corbicula fluminea* from a northern habitat. *Bulletin of the North American Benthological Society* 6: 59.

Burlakova, L.E., A.Y. Karatayev, and D.K. Padilla, 2003. Functional changes in benthic freshwater communities after *Dreissena polymorpha* (Pallas) invasion and consequences for filtration. In: *Proceedings of the NATO Advanced Research Workshop on the Comparative Roles of Suspension-Feeders in Ecosystems Nida,* R.F. Dame and S. Olenin, eds, pp. 263–276. Lithuania, October 4–9, *Earth and Environmental Science* 47: 1–359.

Burlakova, L.E., A.Y. Karatayev, and D.K. Padilla, 2006. Changes in the distribution and abundance of *Dreissena polymorpha* within lakes through time. *Hydrobiologia* 571: 133–146.

Burlakova, L.E., A.Y. Karatayev, D.K. Padilla, L.D. Cartwright, and D.N. Hollas, 2008. Wetland restoration and invasive species: Apple snail (*Pomacea insularum*) feeding on native and invasive aquatic plants. *Restoration Ecology* 16: 1–8.

Burns, M.D.M., A.M. Garcia, M.A. Bemvenuti, J.P. Vieira, D.M.L. Motta Marques, A. Moresco, M. Vinicius, and L. Condini, *Notes on Geographic Distribution Check List* 2006: 2.

Butorin, N.V. and F.D. Mordukhai-Boltovskoi, eds, 1978. *The Volga and its life.* Leningrad: Nauka Press (in Russian).

Cada, C., 2003. Effects of *Potamopyrgus antipodarum* on trout and fish diets and growth. Presented paper at *New Zealand Mudsnail in the Western USA Conference,* August.

Cada, C., 2004. *Interactions between the invasive New Zealand mud snail, Potamopyrgus antipodarum, baetid mayflies, and fish predators.* MSc thesis, Montana State University, Bozeman.

Calheiros, D.F., 2003. Influencia do pulso de inundação na composição isotópica ($\delta^{13}C$ and $\delta^{15}N$) das fontes primárias de energia na planície de inundação do rio Paraguai (Pantanal-MS). Ph.D. dissertation, Centro de Energia Nuclear na Agricultura, Univ. São Paulo.

Calheiros, D.F. and S.K. Hamilton, 1998. Limnological conditions associated with natural fish kills in the Pantanal Wetland of Brazil. *Verh Internat Verein Limnol* 26: 2189–2193.

Calumpang, S.M.F., M.B.J. Median, A.W. Tejada, and J.R. Medina, 1995. Environmental impact of two molluscicides: Niclosamide and mealdehyde in a rice paddy ecosystem. *Bulletin of Environmental Contamination and Toxicology* 55: 494–501.

Canfield, Jr. D.E. and L.M. Hodgson, 1983. Prediction of Secchi disc depths in Florida lakes: Impact of algal biomass and organic color. *Hydrobiologia* 99: 51–60.

Capítoli, R.R. and C.E. Bemvenuti, 2004. Distribuição do mexilhão dourado *Limnoperna fortunei* (Dunker, 1857) na área estuarina da Lagoa dos Patos e Canal São Gonçalo. *Anais do VI Simpósio de Ecossistemas Brasileiros* 1–8.

Caraco, N.F., J.J. Cole, P.A. Raymond, D.L. Strayer, M.L. Pace, S.E.G. Findlay, and D.T. Fischer, 1997. Zebra mussel invasion in a large, turbid river: Phytoplankton response to increased grazing. *Ecology* 78: 588–602.

Caraco, N.F., J.J. Cole, and D.L. Strayer, 2006. Top-down control from the bottom: Regulation of eutrophication in a large river by benthic grazing. *Limnology and Oceanography* 51: 664–670.

Carlson, R., 1980. More complications in the chlorophyll-Secchi disk relationship. *Limnology and Oceanography* 25: 379–382.

Carlson, R.E., 1977. A trophic state index for lakes. *Limnology and Oceanography* 22: 361–369.

Carlsson, N.O.L., 2006. Invasive golden apple snails are threatening natural ecosystems in southeast Asia. In: *Global Advances in Ecology and Management of Golden Apple Snails*, R.C. Joshi, and L.S. Sebastian, eds, pp. 61–72. Nueva Ecija: Philippine Rice Research Institute.

Carlsson, R., 2000. The distribution of the gastropods *Theodoxus fluviatilis* (L.) and *Potamopyrgus antipodarum* (Gray) in lakes on the Åland Islands, southwestern Finland. *Boreal Environment Research* 5: 187–195.

Carlsson, R., 2006. Freshwater snail assemblages of semi-isolated brackish water bays on the Åland Islands, SW Finland. *Boreal Environmental Research* 11: 371–382.

Carlton, J.T., 1985. Transoceanic and interoceanic dispersal of coastal marine organisms: The biology of ballast water. *Oceanography and Marine Biology Annual Review* 23: 313–371.

Carlton, J.T., 1992. Introduced marine and estuarine mollusks of North America: An end-of-The-20th century perspective. *Journal of Shellfish Research* 11: 489–505.

Carlton, J.T., 1993. Dispersal mechanisms of the zebra mussel *Dreissena polymorpha*. In: *Zebra Mussels: Biology, Impact, and Control*, T.E. Nalepa and D.W. Schloesser, eds, pp. 677–697. Ann Arbor: Lewis (CRC Press).

Carpenter, S.R., J.F. Kitchell, and J.R. Hodgson, 1985. Cascading trophic interactions and lake productivity. *BioScience* 35: 634–639.

Carr, J.F. and J.K. Hiltunen, 1965. Changes in the bottom fauna of western Lake Erie from 1930 to 1961. *Limnology and Oceanography* 10: 551–569.

Carter, G.S., T.F. Nalepa, and R.R. Rediske, 2006. Status and trends of benthic populations in a coastal drowned river mouth lake of Lake Michigan. *Journal of Great Lakes Research* 32: 578–595.

Cary, D.M., 1985. *Climatological and environmental factors effecting the foraging behavior and ecology of Everglade Kites*. Masters thesis, University of Miami, Coral Gables.

Caspers, N., C. Staples, U. Friederich, E. Mihaich, D. Thomas, and S. Hentges, 2008. Effects of 'bisphenol A' on reproduction and development in the gastropod, *Marisa cornuarietis* and use of the results in environmental risk assessment. SETAC Europe Annual Meeting 2008.

Castagna, M. and P. Chanley, 1973. Salinity tolerance of some marine bivalves from inshore and estuarine environments in Virginia waters on the western mid-Atlantic coast. *Malacologia* 12: 47–96.

Castillo, L.V. and C.M.V. Casal, 2006. Golden apple snail utilization in small-scale aquaculture in the Philippines. In: *Global Advances in Ecology and Management of Golden Apple Snails*, R.C. Joshi and L.S. Sebastian, eds, pp. 475–482. Nueva Ecija: Philippine Rice Research Institute.

Cataldo, D. and D. Boltovskoy, 1998. Simposio internacional sobre gestao de recursos hidricos, instituto de pesquisas hidraulicas, universidade federal de rio grande do sul, porto alegre/gramado, Brasil, Octubre 1998. Trabajo presentado: *Evaluación de la efectividad de moluscidas seleccionados para el control de limnoperna fortunei (bivalvia)*.

Cataldo, D. and D. Boltovskoy, 1999. Population dynamics of *Corbicula fluminea* (Bivalvia) in the Paraná River Delta (Argentina). *Hydrobiologia* 380: 153–163.

Cataldo D., D. Boltovskoy, V. Marini, and N. Correa, 2002. Limitantes de *limnoperna fortunei* en lacuenca del plata: predación por peces. *Terceira Jornada Sobre Conservação da Fauna Ictica do Rio Uruguai, caru, paysandú.*

Cataldo, D., D. Boltovskoy, J.L. Hermosa, and C. Canzi, 2005. Temperature-dependent rates of larval development *Limnoperna fortunei* (bivalvia: mytilidae). *Journal of Molluscan Studies* 71: 41–46.

Cattau, C., 2008. *Effects of the invasive exotic apple snail (Pomacea insularum) on the snail kite (Rostrhamus sociabilis plumbeus) in Florida*. MSc thesis, University of Florida.

Cawein, J.L., B.G. Lewis, and J.E. Marsden, 1993. Zebra mussel aggregation: Physical stimulus or chemical cue? In: *Third International Zebra Mussel Conference, Agenda and Abstracts,* Toronto, Canada, February.

Cazzaniga, N.J., 2002. Old species and new concepts in the taxonomy of *Pomacea* (Gastropoda: Ampullariidae). *Biocell*, 26: 71–81.

Cazzaniga, N.J., 2006. *Pomacea canaliculata*: Harmless and useless in its natural realm (Argentina). In: *Global Advances in Ecology and Management of Golden Apple Snails*, R.C. Joshi and L.S. Sebastian, eds, pp. 37–60. Nueva Ecija: Philippine Rice Research Institute.

Cañete, R., M. Yong, J. Sánchez, L. Wong, and A. Gutiérrez, 2004. Population dynamics of intermediate snail hosts of *Fasciola hepatica* and some environmental factors in San Juan y Martinez Municipality, Cuba. *Memoires Institute de Oswaldo Cruz* 99: 257–262.

Cecala, R.K., C.M. Mayer, K.L. Schulz, and E.L. Mills, 2008. Increased benthic algal primary production in response to the invasive zebra mussel (*Dreissena polymorpha*) in a productive ecosystem, Oneida Lake, New York. *Journal of Integrative Plant Biology* 50: 1452–1466.

Cedeño-León, A. and J.D. Thomas, 1983. The predatory behaviour of *Marisa cornuarietis* on eggs and neonates of *Biomphalaria glabrata*, the snail host of *Schistosoma mansoni*. *Malacologia* 24: 289–297.

Cejka, T., 1994. First record of the New Zealand mollusc *Potamopyrgus antipodarum* from the Slovak section of the Dunaj River. *Biologia, Bratislava* 49: 657–658.

Cerutti, R., 1998. An infestation of *Pomacea canaliculata* (Lamarck, 1804) in Lake Miramar, San Diego, California. *Festivus* 30: 25–27.

CETEC., 2003. Pesquisa para o desenvolvimento de ecotecnologias de prevenção e controle de *limnoperna fortunei*; estudo de caso:reservatório de volta grande. *Relatório parcial*, M.C.S. Campos, ed. Belo horizonte: CETEC.

CETEC., 2008. In: *Desenvolvimento de metodologias e pesquisas no ecossistema e em plantas de usinas hidrelétricas para controle de mexilhão dourado*, A.C. Silva, ed. Relatório parcial. Belo horizonte: fundação centro tecnológico de minas gerais.

Chalker-Scott, L., R. Carnevale, K. Smith, and J. Scott, 1993. Comparison of acute and chronic mid-range ultraviolet radiation (280–320 nm) effects on adult zebra mussels. In: *Third International Zebra Mussel Conference, Agenda and Abstracts*, Toronto, Canada, February.

Chaniotis, B.N.C., J.M. Butler, Jr., F. Ferguson, and W.R. Jobin, 1980a. Bionomics of *Tarebia granifera* (Gastropoda: Thiaridae) in Puerto Rico, an Asiatic vector of *Paragonimiasis westermani*. *Caribbean Journal of Science* 16: 81–90.

Chaniotis, B.N.C., J.M. Butler, Jr., F. Ferguson, and W.R. Jobin, 1980b. Presence of males in Puerto Rican *Thiara (Tarebia) granifera* (Gastropoda: Thiaridae), a snail thought to be parthogenic. *Caribbean Journal of Science* 16: 95–97.

Chaniotis, B.N.C., J.M. Butler, Jr., F. Ferguson, and W.R. Jobin, 1980c. Thermal limits, desiccation tolerance, and humidity reactions of *Thiara (Tarebia) granifera mauiensis* (Gastropoda: Thiaridae) host of the Asiatic lung fluke disease. *Caribbean Journal of Science* 16: 91–93.

Charlton, M.N., 1980. Oxygen depletion in Lake Erie: Has there been any change? *Canadian Journal of Fisheries and Aquatic Sciences* 37: 72–81.

Chaudhurti, J.C., N. Bhargava, and A.D. Purohit, 1965. Conductivity and its relation to total dissolved solids in ground water of Rajasthan Desert. Def. *Science Journal* 15: 125–130.

Chavaud, L., J.K. Thompson, J.E. Cloern, and G. Thouzeau, 2003. Clams as CO_2 generators: The *Potamocorbula amurensis* example in San Francisco Bay. *Limnology and Oceanography* 48: 2086–2092.

Chen, K., 2004. *A preliminary study of the reproductive ecology of the snail Tarebia granifera (Lamarck, 1822) Prosobranchia: Thiaridae in Jinlun River, Southeastern Taiwan.* MSc thesis, Nation Sun Yat-Sen University, Taiwan.

Cheng, E.Y., 1989. Control strategy for the introduced snail *Pomacea lineata*, in rice paddy. In: *Slugs and Snails in World Agriculture*, I.F. Henderson, ed., pp. 69–75. Farnham, UK: British Crop Protection Council Monograph 41.

Cheng, E.Y. and C.-H. Kao, 2006. Control of golden apple snail, *Pomacea canaliculata* (lamarck), in Taiwan. In: *Global Advances in Ecology and management of Golden Apple Snails*, R.C. Joshi and L.S. Sebastian, eds, pp.155–168. Nueva Ecija: Philippine Rice Research Institute.

Chernin, E., E.H. Michelson, and D.L. Augustine, 1956. Studies on the biological control of Schistosome-bearing snails. *American Journal of Tropical Medicine and Hygiene* 5: 297–307.

Cherry, D.S., J. Rodgers, R. Graney, and J. Cairns, 1980. Dynamics and control of the Asiatic clam in the New River, Virginia. *Virginia Water Resources Research Center Bulletin* No. 123.

Cherry, D.S., S.R. Larrick, K.L. Dickson, R.C. Hoehn, and J. Cairns, 1977. Significance of hypochlorous acid in free residual chlorine to the avoidance response of spotted bass (*Micropterus punctulatus*) and rosyface shiner (*Notropis rubellus*). *Journal of the Fisheries Research Board of Canada* 34: 1365–1372.

Cherry, D.S, J.L. Scheller, N.L. Cooper, and J.R. Bidwell, 2005. Potential effects of Asian clam (*Corbicula fluminea*) die-offs on native freshwater mussels (Unionidae) I: Water-column ammonia levels and ammonia toxicity. *Journal of the North American Benthological Society* 24: 369–380.

Chick, J.H., C.F. Jordan, J.P. Smith, and C.C. McIvor, 1992. A comparison of four enclosure traps and methods used to sample fishes in aquatic macrophytes. *Journal of Freshwater Ecology* 7: 353–361.

Choi, S.S. and J.S. Kim, 1985. Studies on the metamorphosis and the growth of larva in *Limnoperna fortunei. Korean Journal of Malacology* 1: 13–18.

Christensen, V.G., A.C. Ziegler, and J. Xiaodong, 2001. Continuous turbidity monitoring and regression analysis to estimate total suspended solids and fecal coliform bacteria loads in real time. USGS Government Report. Available at: http://ks.water.usgs.gov/pubs/reports/vgc.fisc.pdf. Accessed February 12, 2009.

Chunlin, W., W. Lifeng, and C. Jun, 2006. Golden apple snail prevention and management in China. In: *Global Advances in Ecology and Management of Golden Apple Snails*, R.C. Joshi and L.S. Sebastian, eds, pp. 281–298. Nueva Ecija: Philippine Rice Research Institute.

Ciereszko, A., K. Dabrowskil, B. Piros, M. Kwasnik, and J. Glogowski, 2001. Characterization of zebra mussel (*Dreissena polymorpha*) sperm motility: Duration of movement, effects of cations, pH and gossypol. *Hydrobiologia* 452: 225–232.

Civeyrel, L. and D. Simberloff, 1996. A tale of two snails: Is the cure worse than the disease? *Biodiversity and Conservation* 5: 1231–1252.

Clarke, A.H., 1981. *The Freshwater Molluscs of Canada.* National Museum of Natural Sciences, National Museums of Canada, Ottawa.

Clarke, A.H., 1984. Freshwater molluscan survey of the Roanoke, Tar and Neuse river systems, N.C. *American Malacological Bulletin* 3: 104–105.

Clarke, Jr. A.H., 1973. The freshwater mollusca of the Canadian interior basin. *Malacologia* 13: 1–509.

Clarke, Jr. A.H., 1981. *Corbicula fluminea*, in Lake Erie. *The Nautilus* 95: 83–84.

Clarke, K.B., 1952. The infestation of waterworks by *Dreissena polymorpha*, a freshwater mussel. *Journal of the Institution of Water Engineers* 6: 370–378.

Clarke, M. and R.F. McMahon, 1993. Tissue freezing and time to complete mortality on exposure to freezing air temperatures in the zebra mussel (*Dreissena polymorpha*) with special reference to dewatering during freezing conditions as a mitigation strategy. In: *Third International Zebra Mussel Conference, Agenda and Abstracts*, Toronto, Canada, February.

Clarke, M. and R.F. McMahon, 1996a. Effects of temperature on byssal thread production by the freshwater mussel, *Dreissena polymorpha* (Pallas). *American Malacological Bulletin* 13: 105–110.

Clarke, M. and R.F. McMahon, 1996b. Effects of current velocity on byssal-thread production in the zebra mussel (*Dreissena polymorpha*). *Canadian Journal of Zoology* 74: 63–69.

Clarke, M. and R.F. McMahon, 1996c. Effects of hypoxia and low-frequency agitation on byssogenesis in the freshwater mussel, *Dreissena polymorpha* (Pallas). *Biological Bulletin* 191: 413–420.

Claudi, R., 1992. Zebra mussels and their impact. Paper Presented at the *North American Lake Management Society Meeting,* Cincinnati, USA, November.

Claudi, R. and D. Lowther, 1990. Effect of sodium hypochlorite on adult zebra mussels at an electrical generating facility—case study. Paper and Abstract Presented at the *Annual North American Benthological Society Meetings,* Blacksburg, USA, May.

Claudi, R. and D.W. Evans, 1993. Chemical addition strategies for zebra mussel (*Dreissena polymorpha*) control in once-through service water systems. In: *Zebra Mussel Biology, Impacts and Control*, T.F. Nalepa and D.W. Schloesser, eds, pp. 563–573. Boca Raton: Lewis Publishers.

Claudi, R. and F. Spencer, 1993. Summary of the 1992 zebra mussel research at Ontario Hydro. Ontario Hydro Report N-IR-07015-061-0001.

Claudi, R. and G. Mackie, 1994. *Zebra Mussel Monitoring and Control.* Boca Raton: Lewis Publishers, Inc.

Claudi, R., T.H. Prescott, and A.C. Taraborelli, 2009. Assessment of the 'ballast safe' filter performance in removing quagga mussel veligers from the raw water of lake Havasu. Prepared for Sigma Design Company LLC, Springfield, NJ.

Claudi, R. and A.J. Van Oostrom, 2009. Mitigation of biofouling in once-through cooling systems: An overview and case study on treatment optimization. In: *The Zebra Mussel in Europe*, G. Van der Velde, S. Rajagopal, and A. Bij de Vaate, eds. pp. 429–441, Weikersheim: Backhuys Publishers, Leiden/Margraf Publishers.

Claxton, W.T., A. Martel, R.M. Dermott, and E.G. Boulding, 1997. Discrimination of field-collected juveniles of two introduced dreissenids (*Dreissena polymorpha* and *Dreissena bugensis*) using mitochondrial DNA and shell morphology. *Canadian Journal of Fisheries and Aquatic Sciences* 54: 1280–1288.

Claxton, W.T. and E.G. Boulding, 1998. A new molecular technique for identifying field collections of zebra mussel (*Dreissena polymorpha*) and quagga mussel (*Dreissena bugensis*) veliger larvae applied to eastern Lake Erie, Lake Ontario,and Lake Simcoe. *Canadian Journal of Zoology* 76: 194–198.

Claxton, W.T. and G.L. Mackie, 1998. Seasonal and depth variations in gametogenesis and spawning of *Dreissena polymorpha* and *Dreissena bugensis* in eastern Lake Erie. *Canadian Journal of Zoology* 76: 2010–2019.

Claxton, W.T., G.L. Mackie, and E.G. Boulding, 1998. A genetic and morphological comparison of deep and shallow water populations of the introduced *Dreissena polymorpha* and *Dreissena bugensis*. *Canadian Journal of Zoology* 76: 1269–1276.

Clench, W.J., 1966. *Pomacea bridgesi* (Reeve) in Florida. *The Nautilus* 79: 105.

Clench, W.J., 1969. *Melanoides tuberculata* (Muller) in Florida. *The Nautilus* 83: 72.

Coelho, M.A., 1996. *Geografia do Brasil*, 4th ed. São Paulo: Companhia Editora Nacional.

Cohen, A.N., 2001. Zebra mussel's calcium threshold and implications for its potential distribution in North America, San Francisco Estuary Institute Richmond, CA.

Cohen, A.N., 2005. A review of zebra mussels' environmental requirements. A report for the California Department of Water Resources, Sacramento CA. San Francisco Estuary Institute, Oakland, CA.

Cohen, A.N., 2007. Potential distribution of zebra mussels (*Dreissena polymorpha*) and quagga mussels (*Dreissena bugensis*) in California. Phase 1 Report for the California Department of Fish and Game. San Francisco Estuary Institute, Oakland, CA.

Cohen, A.N. and J.T. Carlton, 1998. Accelerating invasion rate in a highly invaded estuary. *Science* 279: 555–558.

Cohen, A.N. and A. Weinstein, 1998. *The Potential Distribution and Abundance of Zebra Mussels in California*. San Francisco Estuary Institute, Richmond, CA.13pp. Available at: http://www.sfei.org/bioinvasions/Reports/1998-ZebraMusInCalif225.pdf. Accessed February 4, 2009.

Cohen, A.N. and A. Weinstein, 2001. Zebra mussel's calcium threshold and implications for its potential distribution in North America. A report for the California Sea Grant College Program, La Jolla, CA, and the Department of Energy, National Energy Technology Center, Morgantown WV. San Francisco Estuary Institute, Richmond, CA.

Cohen, R.H., P.V. Dresler, E.J.P. Phlips, and R.L. Cory, 1984. The effect of the Asiatic clam, *Corbicula fluminea,* on phytoplankton of the Potomac River, Maryland. *Limnology and Oceanography* 29: 170–180.

Cohn, A., 1996. Zebra mussels and their impact on historic shipwrecks, Technical Report No. 15, Lake Champlain Maritime Museum, January 1996.

Cole, G.A., 1994. *Textbook of Limnology,* 4th ed. Prospect Heights: Waveland Press, Inc.

Cole, R.A. and M. Friend, 1999. Miscellaneous Parasitic Diseases. In: *Field Manual of Wildlife Diseases: General Field Procedures and Diseases of Birds*, Chapter 35, M. Friend, J.C. Franson, and E.A. Ciganovich, eds. Biological Resources Division. Information and Technology Report 1999–001, USGS, Madison.

Collier, K.J., R.J. Wilcock, and A.S. Meredith, 1998. Influence of substrate type and physico-chemical conditions on macroinvertebrate faunas and biotic indices in some lowland Waikato, New Zealand, streams. *New Zealand Journal of Marine and Freshwater Research* 32: 1–19.

Conn, D.B., G.C. LeTendre, S. Cluett, and S.R. LaPan, 1991. Colonization of navigational buoys by zebra mussels (*Dreissena polymorpha*) in the St. Lawrence River. *Dreissena polymorpha Information Review* (New York Sea Grant Extension) 2: 5.

Conn, D.B., R.A. Lutz, Y.P. Hu, and V.S. Kennedy, 1993. Guide to identification of larval and postlarval stage of zebra mussels *Dreissena polymorpha* and the dark false mussel, *Mytilopsis leucophaeata*. New York Sea Grant Institute, Stonybrook.

Conn, D.B., S.R. LaPan, and G.C. LeTendre, 1992. Range expansion of zebra mussels (*Dreissena polymorpha*) and quagga mussels (*Dreissena* sp.) in the St. Lawrence River and the eastern basin of Lake Ontario. *Dreissena polymorpha Information Review* (New York Sea Grant Extension) 3: 1.

Conn, D.B. and D.A. Conn, 2004. Review and prospective of a long-term monitoring program for assessing the invasion, establishment, and population trends of dreissenid mussels in the Upper St. Lawrence River and Eastern Lake Ontario, 1990–2003. *Aquatic Invaders* 15: 1–7.

Conner, S.L., C.M. Pomory, and P.C. Darby, 2008. Density effects of native and exotic snails on growth in juvenile apple snails Pomacea paludosa (Gastropoda: Ampullariidae): A laboratory experiment. *Journal of Molluscan Studies* 74: 355–362.

Conrad, T.A., 1831. Description of fifteen new species of recent, and three of fossil shells, chiefly from the coast of the United States. *Journal of the Academy of Natural Sciences of Philadelphia* 6: 256–268.

Conroy, J.D. and D.A. Culver, 2005. Do dreissenid mussels affect Lake Erie ecosystem stability processes? *The American Midland Naturalist* 153: 20–32.

Conroy, J.D., W.J. Edwards, R.A. Pontius, D.D. Kane, H.Y. Zhang, J.F. Shea, J.N. Richey, and D.A. Culver, 2005. Soluble nitrogen and phosphorus excretion of exotic freshwater mussels (*Dreissena* spp.): Potential impacts for nutrient remineralisation in western Lake Erie. *Freshwater Biology* 50: 1146–1162.

Cooksey, C., J. Hyland, D. Aurin, D. Bearden, and L. Balthis, 2001. MERHAB Florida Monitoring Program. FY01 Annual Report for the CCMA—Charleston Team's Component of the MERHAB Florida Monitoring Program: Benthic Monitoring and System-Wide Data Management. *NOAA National Ocean Service, National Centers for Coastal Ocean Science*, Charleston.

Coon, G.T., H.W. Belcher, D.J. Hartson, and Y. Zai-Chun, 1993. Potential infestation of subirrigation systems in saginaw bay by zebra mussels and evaluation of control options. In: *Third International Zebra Mussel Conference, Agenda and Abstracts*, Toronto, Canada, February.

Cooper, N.L., J.R. Bidwell, and D.S. Cherry, 2005. Potential effects of Asian clam (*Corbicula fluminea*) die-offs on native freshwater mussels (Unionidae) II: Pore-water ammonia. *Journal of the North American Benthological Society* 24: 381–394.

Cordeiro, J.R. and S. MacWilliams, 1999. Occurrence of the Asian Clam *Corbicula fluminea* (Müller, 1774) (Bivalvia: Sphaeriacea: Corbiculidae) in Colorado. *The Veliger* 42: 278–280.

Cordeiro, J.R., A.P. Olivero, and J. Sovell, 2007. Corbicula fluminea (Bivalvia: Sphaeriacea: Corbiculidae) in Colorado. *The Southwestern Naturalist* 52: 424–445.

Corrao, N.M., P.C. Darby, and C.M. Pomory, 2006. Nitrate impacts on the Florida apple snail, *Pomacea paludosa. Hydrobiologia* 568: 135–143.

Cortes Cabanero M.A., J.A. Martinez, R.G. Miguelez, R.C. Iparraguirre, J.A.F. Garcia, R.H. Martinez, R. Claudi, and T. Prescott, 2007. Proactive approach to zebra mussel treatment at a Spanish nuclear power plant: Case study. Presentation at the *15th International Conference on Aquatic Invasive Species*, Nijmegen, Netherlands, September 23–27.

COSEWIC., 2003a. COSEWIC Assessment and Status Report on the Kidneyshell, *Ptychobranchus fasciolaris*, in Canada. Committee on the Status of Endangered Wildlife in Canada, Ottawa.

COSEWIC., 2003b. COSEWIC Assessment and Status Report on the Round Hickorynut, *Obovaria subrotunda*, in Canada. Committee on the Status of Endangered Wildlife in Canada.

COSEWIC., 2004. COSEWIC assessment and status report on the Round Pigtoe *Pleurobema sintoxia* in Canada. Committee on the Status of Endangered Wildlife in Canada, Ottawa.

COSEWIC., 2006a. COSEWIC Assessment and Status Report on the Rainbow mussel, *Villosa iris*, in Canada. Committee on the Status of Endangered Wildlife in Canada, Ottawa.

COSEWIC., 2006b. COSEWIC Assessment and Status Report on the Mapleleaf, *Quadrula quadrula*, in Canada. Committee on the Status of Endangered Wildlife in Canada, Ottawa.

Costa, R., P. Elliott, P.M. Saraiva, D. Aldridge and G.D. Moggridge, 2008. Development of sustainable solutions for zebra mussel control through chemical product engineering. *Chinese Journal of Chemical Engineering* 16: 435–440.

Costil, K., G.B.J. Dussart, and J. Daquzan, 2001. Biodiversity of aquatic gastropods in the Mont St. Michel basin (France) in relation to salinity and drying of habitats. *Biodiversity and Conservation* 10: 1–18.

Counts, C.L., 1986. The zoogeography and history of the invasion of the United States by *Corbicula fluminea* (Bivalvia: Corbiculidae). In: *Proceedings of the Second International Corbicula Symposium*, J.C. Britton, ed., *American Malacological Bulletin Special Edition,* No. 2.

Courchene, J.E. and J.D. Chapman, 1975. Algae control in northwest reservoirs. *Journal of the American Water Works Association* 67: 127–130.

Cowie, R.H., 1994. Freshwater snail surveys in relation to eradication of apple snails with copper sulphate on Kauai. *Bishop Museum Technical Report* 5. 26 pp.

Cowie, R.H., 1995. Identity, distribution and impacts of introduced Ampullariidae and Viviparidae in the Hawaiian Islands. *Journal of Medical and Applied Malacology* 59: 61–67.

Cowie, R.H., 1998. Patterns of introduction of non-indigenous non-marine snails and slugs in the Hawaiian Islands. *Biodiversity and Conservation* 7: 349–368.

Cowie, R.H., 2001. Can snails ever be effective and safe biocontrol agents? *International Journal of Pest Management* 47: 23–40.

Cowie, R.H., 2002. Apple snails as agricultural pests: Their biology, impacts, and management. In: *Mollusks as Crop Pests*, B.M. Barker, ed., pp.145–192. Wallingford: CAB International.

Cowie, R.H. and S.C. Thiengo, 2003. The apple snails of the Americas (Mollusca: Gastropoda: Ampullariidae: *Asolene, Felipponea, Marisa, Pomacea, Pomella*): A nomenclatural and type catalog. *Malacologia* 45: 41–100.

Cowie, R.H., K.A. Hayes, and S.C. Thiengo, 2006. What are apple snails? Confused taxonomy and some preliminary resolution. In: *Global Advances in ECOLOGY and Management of Golden Apple Snails*, R.C. Joshi and L.S. Sebastian, eds, pp. 3–23. Nueva Ecija: Philippine Rice Research Institute.

Cox, G.W., 2004. *Alien Species and Evolution.* Washington: Island Press.

Cox, T.J. and J.C. Rutherford, 2000. Thermal tolerances of two stream invertebrates exposed to diurnally varying temperature. *New Zealand Journal of Marine and Freshwater Research* 34: 203–208.

Coyne, K.J., X-Xia Qin, and J.H. Waite, 1997. Extensible collagen in mussel byssus: A natural block copolymer. *Science Reports* 277: 1830–1832.

Crisp, D.J., 1974. Factors influencing the settlement of marine inveretbrtae larvae. In: *Chemoreception in Marine Organisms*, P.T. Grant and A.M. Mackie, eds, pp. 177–253. New York: Academic press.

Croop, J.D., 1996. A quantitative comparison of sampling techniques and habitat use for the apple snail (*Pomacea paludosa*) in the Florida Everglades. MSc thesis. University of Florida, Gainesville.

Cuker, B.E., 1983. Competition and coexistence among the grazing snail *Lymnaea*, Chironomidae, and microcrustacea in an arctic epilithic lacustrine community. *Ecology* 64: 10–15.

Cuong, D.N., 2006. The golden apple snail in Vietnam. In: *Global Advances in ecology and Management of Golden Apple Snails*, R.C. Joshi and L.S. Sebastian, eds, pp. 243–254. Nueva Ecija: Philippine Rice Research Institute.

Custer, C.M. and T.W. Custer, 1997. Occurrence of zebra mussels in near-shore areas of western Lake Erie. *Journal of Great Lakes Research* 23(1): 108–115.

Čejka T., L. Dvoák, and V. Košel, 2008: Present distribution of *Potamopyrgus antipodarum* (Gray, 1843) (Mollusca: Gastropoda) in the Slovak Republic. *Malacologica Bohemoslovaca* 7: 21–25.

Dame, R.F. and S. Olenin, 2003. The comparative roles of suspension-feeders in ecosystems. In: *Proceedings of the NATO Advanced Research Workshop on The Comparative Roles of Suspension-Feeders in Ecosystems Nida, Lithuania*, October 4–9. *Earth and Environmental Science* 47: 1–359.

Darby, P.C., P.L. Valentine-Darby, R.E. Bennetts, J.D. Croop, H.F. Percival, and W.M. Kitchens, 1997. Ecological studies of apple snails (*Pomacea paludosa,* Say). *South Florida Water Management District Special and St. Johns River Water Management District* Publication SJ98-SP6.

Darby, P.C. and H.F. Pervical, 1999. Dry down tolerance of the Florida apple snail (*Pomacea paludosa*, Say). *Journal of Molluscan Studies* 65: 195–208.

Darby, P.C., R.E. Bennetts, J.D. Croop, P.L. Valentine-Darby, and W.M. Kitchens, 2000. A comparison of sampling techniques for quantifying abundance of the Florida apple snail (*Pomacea paludosa,* Say). *Journal of Molluscan Studies* 65:195–208.

Darby, P.C., P.L. Valentine-Darby, H.F. Percival, and W.M. Kitchens, 2001. Collecting Florida applesnails (*Pomacea paludosa*) from wetland habitats using funnel traps. *Wetlands* 21: 308–311.

Darby, P.C., R.E. Bennetts, S.J. Miller, and H.F. Percival, 2002. Movements of Florida apple snails in relation to water levels and drying events. *Wetlands* 22: 489–498.

Darby, P.C., P.L. Valentine-Darby, and H.F. Percival, 2003. Dry season survival in a Florida apple snail (*Pomacea paludosa,* Say) population. *Malacologia* 45: 179–184.

Darby, P.C., P.L. Valentine-Darby, P.H. Franklin, and W.M. Kitchens, 2004. Florida apple snail (*Pomacea paludosa,* Say) responses to lake habitat restoration activity. *Archiv für Hydrobiologie* 161: 561–575.

Darling, J.A. and M.J. Blum, 2007. DNA-based methods for monitoring invasive species: A review and prospectus. *Biological Invasions* 9: 751–765.

Darr, A. and M.L. Zettler, 2000. Erstnachweis von *Congeria leucophaeata* (Conrad, 1831) in Mecklenburg-Vorpommern (Bivalvia: Dreissenidae). *Malakologische Abhandlungen* 20: 197–200.

Darra, B. and K.M. Mehta, 1963. *Journal of Indian Society of Soil Science* 8: 287 (not seen, cited by Chaudhurti et al., 1965).

Darrigran, G., 1995. *Limnoperna fortunei*: ¿un problema para los sistemas naturales de agua dulce del mercosur? *Revista del museo* 5: 85–87.

Darrigran, G., 1997. Invading bivalves in South America. *Aliens* 5: 3–4.

Darrigran, G. and G. Pastorino, 1993. Bivalvos invasores en el Río de la Plata, Argentina. *Com. Soc. Malac. Urug* 7: 309–313.

Darrigran, G., M.C. Damborenea, and P. Penchaszadeh, 1998a. A case of hermaphroditism in the freshwater invading bivalve *Limnoperna fortunei* (Dunker, 1857) (Mytilidae) from its first American habitat (Rio de la Plata, Argentina). *Iberus* 16: 99–104.

Darrigran, G., S.M. Martin, B. Gullo, and L. Armendariz, 1998b. Macroinvertebrates associated with *Limnoperna fortunei* (Dunker, 1857) (Bivalvia, Mytilidae) in Rio de la Plata, Argentina. *Hydrobiologia* 367: 223–230.

Darrigran, G., P. Penchaszadeh, and C. Damborenea, 1999. The reproductive cycle of *Limnoperna fortunei* (Dunker, 1857) (Mytilidae) from a Neotropical temperate locality. *Journal of Shellfish Research* 18: 361–365.

Darrigan, G. and I. Ezcurra de Drago, 2000. Invasion of *Limnoperna fortunei* (Dunker, 1857) (Bivalvia: Mytilidae) in America. *Nautilus* 2: 69–74.

Darrigran, G., M.C. Damborenea, P. Penchaszadeh, and C. Taraborelli, 2003. Reproductive stabilization of *Limnoperna fortunei* (Bivalvia Mytilidae) after ten years of invasion in the Americas. *Journal of Shellfish Research* 22: 141–146.

Darrigran, G. and C. Damborenea, 2005. El mejillon dorado *Limnoperna fortunei* (Dunker, 1857) en la Cuenca del Plata. In *Invasores. Invertebrados bentonicos introducidos en el Rio de la Plata y region costera marina aledana*, P.E. Penchaszadeh, ed., pp. 39–102. Buenos Aires: Eudeba UDEBA.

Darrigran, G. and M.C. Mansur, 2006. Distribución, abundancia y dispersión. En: *Bio-invasión del mejillón dorado en el continente Americano*, G. Darrigran and M.C. Damborenea, eds. Edulp, la plata: Argentina.

Darrigran, G., C. Damborenea, and N. Greco, 2007. An evaluation pattern for antimacrofouling procedures: *Limnoperna fortunei* larvae study in a hydroelectric power plant in South America. *Ambio* 36: 575–579.

Darrigran, G., A. Vilches, and T. Legarralde, 2008. Desinterés del pasado, decisiones del futuro: Educación para prevenir las invasiones biológicas. *Revista Educación en Biología* 11: 39–52.

Darrigran, G., W. Boeger, C. Damborenea, and M. Maroñas, In press. Evaluation of sampling and analysis techniques for early detection of *Limnoperna fortunei* (mytilidae) in limit areas of its distribution. *Brazilian Journal of Biology* 69.

Darrigran, G.A., 2000. Invasive freshwater bivalves of the neotropical region. *Dreissena* 11(2): 7–13.

Darrigran, G.A., 2002. Potential impact of filter-feeding invaders on temperate inland freshwater environments. *Biological Invasion* 4: 145–156.

Darrigran, G.A. and G. Pastorino, 1995. The recent introduction of a freshwater Asiatic bivalve, *Limnoperna fortunei* (Mytilidae) into South America. *The Veliger* 38: 171–175.

Davidson, T.M., V.E.F. Brenneis, C. de Rivera, R. Draheim, and G.E. Gillespie, 2008. Northern range expansion and coastal occurrences of the New Zealand mud snail *Potamopyrgus antipodarum* (Gray, 1843) in the northeast Pacific. *Aquatic Invasions* 3: 349–353.

De Lafontaine, Y. and B. Cusson, 1997. Veligers of zebra mussels in the Richelieu River: An intrusion from Lake Champlain? In: *Proceedings of the Second Northeast Conference on Nonindigenous Aquatic Nuisance Species*, N.C. Balcom, ed., pp. 30–40. Connecticut Sea Grant College Program, April 18–19.

De Oliveira, M.D., L.F. de Barros, and L.C.R. da Silva, 2004. Density of *Limnoperna fortunei* in Paraguay River, Brazil. Abstract from the *13th International Conference on Aquatic Invasive Species.* Ennis, County Clare, Ireland, September 20–24, 2004.

De Stasio, Jr. B.T. and S. Richman, 1998. Phytoplankton spatial and temporal distributions in Green Bay, Lake Michigan, prior to colonization by the zebra mussel (*Dreissena polymorpha*). *Journal of Great Lakes Research* 24: 620–628.

De Stasio, B.T., M.B. Schrimpf, A.E. Beranek, and W.C. Daniels, 2008. Increased chlorophyll *a*, phytoplankton abundance, and cyanobacteria occurrence following invasion of Green Bay, Lake Michigan by dreissenid mussels. *Aquatic Invasions* 3: 21–27.

DePinto J.V., V.J. Bierman, T.J. Feist, and J. Kaur, 1999. Conceptualization of an aquatic ecosystem model for integrated management of Lake Erie. In: Paper Submitted for Presentation *at IJC-Council of Great Lakes Research Managers' Great Lakes Modeling Summit IAGLR '99 Annual Conference,* Cleveland, USA, May 25–28.

Death, R., 1991. *Environmental stability: Its effects on stream benthic communities.* PhD dissertation, University of Canterbury (not seen, cited in NZMS Management and Control Plan Working Group, 2007).

Death, R.G., B. Baillie, and P. Fransen, 2003. Effect of *Pinus radiata* logging on stream invertebrate communities in Hawke's Bay, New Zealand. *New Zealand Journal of Marine and Freshwater Research* 37: 507–520.

Deaton, L.E., J.G.S. Derby, N. Subhedar, and M.J. Greenberg, 1989. Osmorregulation and salinity tolerance in two species of bivalve mollusk: *Limnoperna fortunei* and *Mytilopsis leucophaeta. Journal of Experimental Marine Biology and Ecology* 133: 67–79.

Demian, E.S. and A.M. Ibrahim, 1971. The egg mass, egg laying and mating behaviour of the snail *Marisa cornuarietis* (L.). *Bulletin of the Zoological Society of Egypt* 23: 1–12.

Demian, E.S. and A.M. Ibrahim, 1972. Tolerance of the snail *Marisa cornuarietis* (L.) to extreme temperatures under laboratory conditions. *Bulletin of the Zoological Society of Egypt* 24: 20–30.

Demian, E.S. and F. Yousif, 1973. Embryonic development and organogenesis in the snail *Marisa cornuarietis* (Mesogastropoda: Ampullariidae) I. General outlines of development. *Malacologia* 12: 123–150.

Denson, D.R. and S.Y. Wang, 1994. Morphological differences between zebra and quagga mussel spermatozoa. *American Malacological Bulletin* 11: 79–81.

Depew, D.C., S.J. Guildford, and R.E.H. Smith, 2006. Nearshore-offshore comparison of chlorophyll a and phytoplankton production in the dreissenid-colonized eastern basin of Lake Erie. *Canadian Journal of Fisheries and Aquatic Sciences* 63: 1115–1129.

Dermott, R., 2001. Sudden disappearance of the amphipod *Diporeia* from eastern Lake Ontario, 1993–1995. *Journal of Great Lakes Research* 27: 423–433.

Dermott, R. and M. Munawar, 1993. Invasion of Lake Erie offshore sediments by *Dreissena*, and its ecological implications. *Canadian Journal of Fisheries and Aquatic Sciences* 50: 2298–2304.

Dermott, R., J. Mitchell, I. Murray, and E. Fear, 1993. Biomass and production of zebra mussels (*Dreissena polymorpha*) in shallow waters of northeastern Lake Erie. In: *Zebra Mussels: Biology, Impacts, and Control*, T.F. Nalepa and D.W. Schloesser, eds, pp. 399–413. Boca Raton: CRC Press.

Dermott, R. and Kerec, D., 1997. Changes in the deepwater ben-thos of eastern Lake Erie since the invasion of *Dreissena*: 1979–1993. *Canadian Journal of Fisheries and Aquatic Sciences* 54: 922–930.

Desy, J.C., J.-F. Archambault, B. Pinel-Alloul, J. Hubert, and P.G.C. Campbell, 2000. Relationships between total mercury in sediments and methyl mercury in the freshwater gastropod prosobranch *Bithynia tentaculata* in the St. Lawrence River, Quebec. *Canadian Journal of Fisheries and Aquatic Sciences* 57: 164–173.

Devercelli, M. and E. Peruchet, 2008. Trends in chlorophyll-*a* concentration in urban water bodies within different man-used basins. *Annales de Limnologie—International Journal of Limnology* 44: 75–84.

Dias, P.J., L. Sollelis, E.J. Cook, S.B. Piertney, I.M. Davies, and M. Snow, 2008. Development of a real-time PCR assay for detection of *Mytilus* species specific alleles: Application to a sampling survey in Scotland. *Journal of Experimental Marine Biology and Ecology* 15: 253–258.

Dickson, K.L., A.C. Hendricks, J.S. Crossman, and J. Cairns, 1974. Effects of intermittently chlorinated cooling tower blowdown on fish and invertebrates. *Environmental Science & Technology* 8: 845–849.

Diers, J.A., J.J. Bowling, S.O. Duke, S. Wahyuono, M. Kelly, and M.T. Hamann, 2006. Zebra mussel antifouling activity of the marine natural product aaptamine and analogs. *Marine Biotechnology* 8: 366–372.

Dillon, P.J. and F.H. Rigler, 1974. The phosphorous-chlorophyll relationship in lakes. *Limnology and Oceanography* 19: 767–763.

Dillon, P.J. and F.H. Rigler, 1975. A simple method for predicting the capacity of a lake for development based on trophic status. *Journal Fisheries Research Board of Canada* 32: 1519–1531.

Dillon, R.T., 2000. *Ecology of Freshwater Mollusks*. Cambridge: Cambridge University Press.

Dillon, R.T., A.R. Wethington, J.M. Rhett, and T.P. Smith, 2002. Populations of the European freshwater pulmonate *Physa acuta* are not reproductively isolated from American *Physa heterostropha* or *Physa integra*. *Invertebrate Biology* 12: 226–234.

Dmitry P., D.P. Filippenko, and M.O. Son, 2008. The New Zealand mud snail *Potamopyrgus antipodarum* (Gray, 1843) is colonizing the artificial lakes of Kaliningrad City, Russia (Baltic Sea Coast). *Aquatic Invasions* 3: 345–347.

Dobson, E. and G.L. Mackie, 1998. Biodeposition and accumulation of polychlorinated biphenyls and Cd by zebra mussels in Lake Erie. *Canadian Journal of Fisheries and Aquatic Sciences* 55: 1131–1139.

Doll, B., 1997. Zebra Mussel Colonization: North Carolina's Risks. Sea Grant North Carolina, University of North Carolina, Raleigh, NC (UNC SG-97-01).

Domm, S., R.W. McCauley, E. Kott, and J.D. Ackerman, 1993. Physiological and taxonomic separation of two dreissenid mussels in the Laurentian Great Lakes. *Canadian Journal of Fisheries and Aquatic Sciences* 50: 2294–2297.

Donnay, T.J. and S.R. Beissinger, 1993, Apple snail (*Pomacea dolioides*) and freshwater crab (*Dilocarcinus dentatus*) population fluctuations in the llanos of Venezuela. *Biotropica* 25: 206–214.

Donskoy, D.M. and M.L. Ludyanskiy, 1995. Low frequency sound as a control measure for zebra mussel fouling. Presented at the *5th International Zebra Mussel and Other Aquatic Nuisance Organisms Conference*, Toronto, Canada, February 14–21.

Dorgelo, J., 1987. Density fluctuations in populations (1982–1986) and biological observations of *Potamopyrgus jenkinsi* in two trophically differing lakes. *Hydrobiological Bulletin* 21: 95–110.

Dorgelo, J., 1991. Growth, food and respiration in the prosobranch snail *Potamopyrgus jenkinsi* (E.A. Smith) Hydrobiidae, Mollusca. *Verh Internat Verein Limnol* 24: 2947–2953.

Dorgelo, J. and M. Gorter, 1984. Preliminary data on size composition and settlement of *Dreissena polymorpha* (Pallas) (Molluska: Bivalvia) in lakes differing in trophic state. *Hydrobiology Bulletin* 18: 159–163.

Dorlego, J., 1993. Growth and population structure of the zebra mussel (*Dreissena polymorpha*) in Dutch lakes differing in trophic state. In: *Zebra Mussels Biology, Impacts, and Control*, T.F. Nalepa and D.W. Schloesser, eds, pp. 79–94. Boca Raton: Lewis Publishers.

Dorgelo, J., H. Meester, and C. van Velzen, 1995. Effects of diet and heavy metals on growth rate fertility in the deposit-feeding snail *Potamopyrgus jenkinsi* (Smith) (Gastropoda: Hydrobiidae). *Hydrobiologia* 316: 199–210.

Dos Santos, C.P., N.L. Würdig and M.C.D. Mansur, 2005. Fases larvais do mexilhão dourado *Limnoperna fortunei* (Dunker) Mollusca, Bivalvia, Mytilidae na do Guaiba, Rio Grande do Sul, Brasil. *Revista Brasileira de Zoologia* 22(3): 702–708.

Dos Santos, S.B., I.C. Miyahira, and L.E. Macedo de Lacerda, 2007. First record of *Melanoides tuberculatus* (Müller, 1774) and *Biomphalaria tenagophila* (d´Orbigny, 1835) on Ilha Grande, Rio de Janeiro, Brazil. *Biota Neotropica* 7: 362–364.

Drake, J.M. and J.M. Bossenbrooek, 2004. The potential distribution of zebra mussels in the United States. *BioScience* 54: 931–941.

Dreier, H. and J.A. Tranquilli, 1981. Reproduction, growth, distribution, and abundance of *Corbicula* in an Illinois cooling lake. *Illinois Natural listing Sunday Bulletin* 32: 378–393.

Dresler, P.V. and R.L. Cory, 1980. The Asiatic clam, *Corbicula fluminea* (Muller), in the tidal Potomac River, Maryland. *Estuaries* 3: 150–151.

Duggan, I.C., 2002. First record of a wild population of the tropical snail *Melanoides tuberculata* in New Zealand natural waters. *New Zealand Journal of Marine and Freshwater Research* 36: 825–829.

Duggan, I.C., K.J. Collier, P.D. Champion, G.F. Croker, R.J. Davies-Colley, P.W. Lambert, J.W. Nagels, and R.J. Wilcock, 2002. Ecoregional differences in macrophyte and macroinvertebrate communities between Westland and Waikato: Are all New Zealand lowland streams the same? *New Zealand Journal of Marine and Freshwater Research* 36: 831–845.

Duggan, I.C., S.A. Bailey, R.I. Colautti, D.K. Gray, J.C. Makarewicz, and H.J. MacIsaac, 2003. Biological invasions in Lake Ontario: Past, present and future. In: *State of Lake Ontario (SOLO)—Past, Present and Future*, M. Munawar, ed., pp. 1–13. Ecovision World Monograph Series, Aquatic Ecosystem Health and Management Society.

Duggan, I.C., I.K.G. Boothroyd, and D.A. Speirs, 2007. Factors affecting the distribution of stream macroinvertebrates in geothermal areas: Taupo Volcanic Zone, New Zealand. *Hydrobiologia* 592: 235–247.

Duncan, A. and R.Z. Klekowski, 1967. The influence of salinity on the survival, respiratory rate and heart beat of young *Potamopyrgus jenkinsi* (Smith) Prosobranchiata. *Comparative Biochemistry and Physiology* 22: 495–505.

Dundee, D.S., 1974. Catalog of introduced molluscs of eastern North America (north of Mexico). *Sterkiana* 55: 1–37.

Dundee, D.S. and A. Paine, 1977. Ecology of the snail *Melanoides tuberculata* (Müller), intermediate host of the human liver fluke (*Opisthorchis sinensis*) in New Orleans, Louisiana. *The Nautilus* 91: 17–20.

Durfor, C.N. and E. Becker, 1964. Public water supplies of the 100 largest cities in the United States, 1962: U.S. Geological Survey Water-Supply Paper 1812.

Dwyer, W., 1999. The New Zealand mud snail (*Potamopyrgus antipodarium*), its introduction, biology and vectors. In: *Proceedings of the Ninth International Conference on Zebra Mussels and Aquatic Nuisance Species*, Duluth, USA, April 26.

Dwyer, W.P., B.L. Kerans, and M.M. Gangloff, 2003. Effect of acute exposure to chlorine, copper sulfate, and heat on the survival of New Zealand mud snails. *Intermountain Journal of Sciences* 9: 53–58.

Dybdahl, M.F. and C.M. Lively, 1995a. Diverse, endemic and polyphyletic clones in mixed populations of a freshwater snail (*NZ mudsnail*). *Journal of Evolutionary Biology* 8: 385–398.

Dybdahl, M.F., and C.M. Lively, 1995b. Host-parasite interactions: Infection of common clones in natural-populations of a freshwater snail (*Potamopyrgus antipodarum*). *Proceedings of the Royal Society of London Series B* 260: 99–103.

Dybdahl, M.F. and C.M. Lively, 1996. The geography of coevolution: Comparative population structures for a snail and its trematode parasite. *Evolution* 50: 2264–2275.

Dybdahl, M.F. and C.M. Lively, 1998. Host-parasite coevolution: Evidence for rare advantage and time-lagged selection in a natural population. *Evolution* 52: 1057–1066.

Dybdahl, M.F. and A.C. Krist, 2004. Genotypic versus condition effects on parasite-driven rare advantage. *Journal of Evolutionary Biology* 17: 967–973.

Eaton, J.W., C.F. Kolpin, H.S. Swofford, C.M. Kjellestrand, and H.S. Jacob, 1973. Chlorinated urban water: A cause of dialysis-induced hemolytic anemia. *Science* 181: 463.

Eckblad, J.W., 1975. The Asian clam *Corbicula* in the Upper Mississippi River. *The Nautilus* 89: 4.

Edmondson, W.T., 1980. Secchi disk and chlorophyll. *Limnology and Oceanography* 25: 378–379.

Edwards, W.J., C.R. Rehmann, E. Mcdonald, and D.A. Culver, 2005. The impact of a benthic filter feeder: Limitations imposed by physical transport of algae to the benthos. *Canadian Journal of Fisheries and Aquatic Sciences* 62: 205–214.

Effler, S.W., S.R. Boone, C.A. Siegfried, L. Walrath, and S.L. Ashby, 1997. Mobilization of ammonia and phosphorus by zebra mussels (*Dreissena polymorpha*) in the Seneca River, New York. *Zebra mussels and aquatic nuisance species,* F. Diltri, ed., pp. 187–207. Ann Arbor: Ann Arbor Press.

Effler, S.W. and C. Siegfried, 1998. Tributary water quality feedback from the spread of zebra mussels: Oswego River, New York. *Journal of Great Lakes Research* 24: 453–463.

Ekholm, M., 2000. US Filter/Johnson Screens. Paper presented at the *10th International Aquatic Nuisance Species and Zebra Mussel Conference,* Toronto, Canada, February 13–17.

Ekroat, L.E., E.C. Mastellar, J.C. Schaffer, and L.M. Steele, 1993. The byssus of the zebra mussel (*Dreissena polymorpha*): Morphology, byssal thread formation, and detachment. In: *Zebra Mussels Biology, Impacts, and Control*, T.F. Nalepa and D.W. Schloesser, eds, pp. 239–263. Boca Raton: Lewis Publishers.

Elderkin, C.L. and P.L. Klerks, 2005. Variation in thermal tolerance among three Mississippi River populations of the zebra mussel *Dreissena polymorpha. Journal of Shellfish Research* 24: 221–226.

Eldredge, L.G., 1994. Perspectives in aquatic exotic species management in the Pacific islands. Vol. I. *Introductions of Commercially Significant Aquatic Organisms to the Pacific Islands.* South Pacific Commission, Noumea, New Caledonia, 127 pp.

Electric Power Research Institute (EPRI), 1980. Review of open literature on effects of chlorine on aquatic organisms. Report to the Electric Power Research Institute, Palo Alto, California, by Oak Ridge National Laboratory, Oak Ridge.

Elexova, E., 1998. Interaction of the Danube River and its left side tributaries in Slovak stretch from benthic fauna point of view. *Biologia, Brataslavia* 53: 621–632.

Eliopoulos, C. and P. Stangel, 1997. Lake Champlain 1996 zebra mussel monitoring program, Final Report. Vermont Department of Environmental Conservation, Waterbury, VT.

Eliopoulos, C. and P. Stangel, 1998. Lake Champlain 1997 zebra mussel monitoring program, Final Report. Vermont Department of Environmental Conservation, Waterbury, VT.

Eliopoulos, C. and P. Stangel, 1999. Lake Champlain 1998 zebra mussel monitoring program, Final Report. Vermont Department of Environmental Conservation, Waterbury, VT.

Eliopoulos, C. and P. Stangel, 2000. Lake Champlain 1999 zebra mussel monitoring program, Final Report. Vermont Department of Environmental Conservation, Waterbury, VT.

Eliopoulos, C. and P. Stangel, 2001. Lake Champlain 2000 zebra mussel monitoring program, Final Report. Vermont Department of Environmental Conservation, Waterbury, VT.

Eliopoulos, C. and P. Stangel, 2003. Lake Champlain 2001–2002 zebra mussel monitoring program, Final Report. Vermont Department of Environmental Conservation, Waterbury, VT.

Elliott, P. and P.S.E. zu Ermgassen, 2008. The Asian clam (*Corbicula fluminea*) in the River Thames, London, England. *Aquatic Invasions* 3(1): 54–60.

Ellis, A.E., 1978. *British Freshwater Bivalve Mollusca. Keys and Notes for the Identification of the Species*. Synopses of the British fauna, No. 11. London: Academic Press.

Elpiner, I.E. and Z.S. Feigina, 1957. Ultrasound vawes in control of hydrobionts. *Water Supply and Sanitary Engineering* 8: 14–16 (in Russian).

Elzinga, W.J. and T.S. Butzlaf, 1994. Carbon dioxide as a narcotizing pre-treatment for chemical control of *Dreissena polymorpha*—Environmental Science and Engineering, Inc. In: *Proceedings of the Fourth International Zebra Mussel Conference,* Madison, USA, March.

Elzinga, W.J. and T.S. Butzlaf, 1998. Carbon dioxide as a narcotizing pre-treatment for chemical control of *Dreissena polymorpha.*Environmental Science and Engineering, Inc. In: *Proceedings of the Fourth International Zebra Mussel Conference,* Madison, USA, March.

Emerson, W.K. and M.K. Jacobson, 1976. *The American Museum of Natural History Guide to Shells, Land, Freshwater and Marine, from Nova Scotia to Florida*, 1st ed. New York: Alfred A Knopf Inc.

Empire State Electric Energy Research Corporation (ESEERCO), 1992. Evaluation of ultrasound for zebra mussel mitigation. Research Report EP 91-17.

Eng, L.L., 1979. Population dynamics of the Asiatic clam, *Corbicula fluminea* (Muller), in the concrete lined Delta-Mendota Canal of central California. In: *Proceedings of the First International Corbicula Symposium*, J.C. Britton, ed., pp. 39–68. Fort Worth, TX: Texas Christian University.

Epa, U.P.K., 2006. Alien invasive snail, *Pomacea sp.,* in Sri Lanka. In: *Global Advances in Ecology and Management of Golden Apple Snails*, R.C. Joshi and L.S. Sebastian, eds, pp. 319–324. Nueva Ecija: Philippine Rice Research Institute.

Escot, C., A. Basanta, F. Cobo, and M.A. Gonzalez, 2003. Sobre la presencia de *Mytilopsis leucophaeata* (Conrad, 1831) (Bi-valvia, Dreissenacea, Dreissenidae) en el rio Guadalquivir (sur de la Peninsula Iberica). *Graelisia* 59: 91–94.

Estebenet, A.L., 1995. Food and feeding in *Pomacea canaliculata* (Gastropoda: Ampullariidae). *Veliger* 38: 277–283.

Estebenet, A.L. and N. Cazzaniga, 1992. Growth and demography of *Pomacea canaliculata* under laboratory conditions. *Malacalogical Review* 25: 1–12.

Estebenet, A.L., P.R. Martín, and S. Burela, 2006. Conchological variation in *Pomacea canaliculata* and other South American Ampullariidae (Caenogastropoda, Architaenioglossa). *Biocell* 30: 329–335.

Estoy, Jr. G.F., Y. Yusa, T. Wada, H. Sakurai, and K. Tsuchida, 2002. Size and age at first copulation and spawning of the apple snail, *Pomacea canaliculata* (Gastropoda: Ampullariidae). *Applied Entomology and Zoology* 37: 199–205.

Evans, D.W., 1991. Analysis of nanticoke zebra mussels for toxic substances. Ontario Hydro Report C91-04-K.

Evans, D.W. and B.F. Sim, 1993. Optimizing the use of hypochlorite in zebra mussel control. In: *Third International Zebra Mussel Conference, Agenda and Abstracts,* Toronto, Canada, February.

Evans D.W., D. Lewis, E. Whitby, and G. McKenna, 1995. UV light for zebra mussel control. *US Army Corps of Engineers Zebra Mussel Workshop,* New Orleans, USA, November 25–30.

Eversole, A.G., 1992. Golden snail problem on rice: Report for short-and long-term steps for management. Unpublished Report. IRRI, Los Baños (not seen, cited by Cowie, 2002).

Ezhova, E., L. Żmudziński, and K. Maciejewska, 2005. Longterm trends in the macrozoobenthos of the Vistula Lagoon, southeastern Baltic Sea. Species composition and biomass distribution. *Bulletin of the Sea Fisheries Institute in Gdynia* 164: 55–73.

Fahnenstiel, G., G.A. Lang, T.F. Nalepa, and T.H. Johengen, 1995. Effects of zebra mussel (*Dreissena polymorpha*) colonization on water quality parameters in Saginaw Bay, Lake Huron. *Journal of Great Lakes Research* 21: 435–448.

Faria, E.A., 2005. Caracterização de superfícies antiincrustantes para o *limnoperna fortunei*—mexilhão dourado, dissertação (pós-graduação em engenharia de materiais). *Rede Temática Em Engenharia De Materiais,* UFOP, CETEC, UEMG.

Faria, E.A., J.R.T. Branco, M.C.S. Campos, M.D. Oliveira, and M.E. Rolla, 2006. Estudo das características antiincrustantes de materiais. *Revista da Escola de Minas* 59(2): 233–238.

Farrapeira, C.M.R., A.V. De Oliveira, M. De Melo, D.F. Barbosa, and K.M. Euzebio Da Silva, 2007. Ship hull fouling in the port of Recife, Pernambuco. *Brazilian Journal of Oceanography* 55(3): 207–221.

Fast, A.W., 1971. The invasion and distribution of the Asiatic clam (*Corbicula fluminea*) in a southern California reservoir. *Bulletin of Southern California Academy of Science* 70: 91–98.

Fava, J.A. and C.F. Tsai, 1975. The avoidance reaction of blacknose dace, *Rhinichtys atratulus* (Hermann), to domestic sewage effluents and their toxic constituents. *Transactions of the American Fisheries Society* 105: 430–441.

Fears, C. and G.L. Mackie, 1994. Use of low level electric current (AC) for controlling quagga mussels. In: *Proceedings of the Fourth International Zebra Mussel Conference,* Madison, USA, March 7–10.

Fears, C.D. and G.L. Mackie, 1995. Efficacy of low level electric current (AC) for controlling quagga mussels in the Welland Canal. Presentation at the Fifth International *Zebra Mussel and Other Aquatic Nuisance Organisms Conference,* Toronto, Canada, February 14–21.

Fears, C. and G.L. Mackie, 1996a. Use of low level electric current (AC) to prevent settlement of zebra/quagga mussels on concrete and steel panels at Nanticoke T.G.S. In: *Zebra Mussels and Aquatic Nuisance Species,* F.M. D'Itri, ed., Chapter 26, pp. 407–417. Ann Arbor: Ann Arbor Press.

Fears, C. and G.L. Mackie, 1996b. Efficacy of low level electric current (AC) for controlling quagga mussels on concrete and steel panels at Nanticoke T.G.S. In: *Proceedings of the Sixth International Zebra Mussel and Other Nuisance Organisms Conference,* Dearborn, USA, March 14–21.

Fears, C.D. and G.L. Mackie, 1997. Effectiveness of low level electric current (AC) for preventing settlement of zebra/quagga mussels on concrete and steel panels after one year exposure in Nanticoke T.G.S. Forebay. Presentation at the *Seventh International Zebra Mussel and Aquatic Nuisance Species Conference,* New Orleans, USA, January 28–31.

Feminella, J.W. and C.P. Hawkins, 1995. Interactions between stream herbivores and periphyton: A quantitative analysis of past experiments. *Journal of the North American Benthological Society* 14: 465–509.

Fenske, C., 2008. Water quality improvement in coastal waters—can fouling organisms help? In: *Periphyton and Fouling: Theory and Practice*. Book of Abstracts, St. Petersburg, pp. 184–185.

Ferguson, F.F. and J.R. Palmer, 1958. Biological notes on *Marisa cornuaurietis*, a predator of *Australorbis glabratus*, the snail intermediate host of schistosomiasis in Puerto Rico. *American Journal of Tropical Medicine and Hygiene* 7: 640–642.

Filippenko, D.P. and M.O. Son, 2008. The New Zealand mud snail *Potamopyrgus antipodarum* (Gray, 1843) is colonising the artificial lakes of Kaliningrad City, Russia (Baltic Sea Coast). *Aquatic Invasions* 3: 345–347.

Fisher, S.A., S.W. Fisher, and K.R. Polizotto, 1993. Field tests of the molluscistatic activity of potassium chloride on zebra mussel veligers. In: *Third International Zebra Mussel Conference, Agenda and Abstracts,* Toronto, Canada, February.

Fisher, S.W., 1991. The toxicity of potassium chloride to zebra mussel veligers and select nontarget organisms. Paper and Abstract Presented at the *Second Annual International Zebra Mussel Conference,* Rochester, USA.

Flessas, C., Y. Couillard, B. Pinel-Alloul, L. St-Cyr, and P.G.C. Campbell, 2000. Metal concentrations in two freshwater gastropods (Mollusca) in the St. Lawrence River and relationships with environmental contamination. *Canadian Journal of Fisheries and Aquatic Sciences* 57: 126–137.

Foe, C. and A. Knight, 1985. The effect of phytoplankton and suspended sediment on the growth of *Corbicula fluminea* (Bivalvia). *Hydrobiologia* 127: 105–115.

Forbes, V.E., R. Warbritton, J. Aufderheide, N. Van Der Hoeven, and N. Caspers, 2008. Effects of 'bisphenol A' on fecundity, egg hatchability, and juvenile growth of *Marisa cornuarietis*. *Environmental Toxicology and Chemistry* 27: 2332–2340.

Forester, J.S., R.H. Kay, and R.L. Walter, 1993. Zebra mussel monitoring in the lower Mississippi River, 1991–1992. *Dreissena polymorpha Information Review, Zebra Mussel Information Clearinghouse* 4: 3–5.

Foster, M., P. Fuller, A. Benson, S. Constant, and D. Raikow, 2008. *Corbicula fluminea*. USGS nonindigenous aquatic species database, Gainesville, FL. Available at: http://nas.er.usgs.gov/queries/FactSheet.asp?speciesID=92. Accessed March 2, 2009.

Fox, R.O., 1969. The *Corbicula* story: A progress report. Second Annual Meeting of the Western Society of Malacologists held at Stanford University, Stanford, CA, June 19, 1969. 11 p. California Academy of Sciences, Department of Invertebrate Zoology, San Francisco, CA.

Fox, R.O., 1970a. The *Corbicula* story: Chapter two. Third Annual Meeting of the Western Society of Malacologists held at Stanford University, Stanford, CA, June 24, 1970. 10 pp. California Academy of Sciences, Department of Invertebrate Zoology, San Francisco, CA.

Fox, R.O., 1970b. *Corbicula* in Baja California. *The Nautilus* 83: 145.

Fox, R.O., 1971. The *Corbicula* Story: Chapter 3. Fourth Annual Meeting, Western Society of Malacologists.

Fox, R.O., 1972. The *Corbicula* story: Chapter four. Fifth Annual Meeting of the Western Society of Malacologists held at the University of Redlands, June 19, 1972. 5 pp. California Academy of Sciences, Department of Invertebrate Zoology, San Francisco, CA.

Fraleigh, P.C., W.R. Van Cott, M.E. Wenning, and J.A. DeKam, 1993. Effects of hypochlorite, permanganate, chlorine dioxide and chloramine on zebra mussel settling. In: *The Third International Zebra Mussel Conference, Agenda and Abstracts,* Toronto, Canada, February.

Francis-Floyd, R.J., J. Gildea, P. Reed, and R. Klinger, 1997. Use of Bayluscide (Bayer 73) for snail control in fish ponds. *Journal of Aquatic Animal Health* 9: 41–48.

Franzen, A., 1983. Ultrastructural studies of the spermatozoa in 3 bivalve species with notes on evolution of elongated sperm nucleus in primitive spermatozoa. *Gamete Research* 7: 199–214.

Freitas, J.R., L.C. Bede, P. de Marco, L.A. Rocha, and M.B.L. Santos, 1987. Population dynamics of aquatic snails in Pampulha reservoir. *Memorias do Instituto Oswaldo Cruz, Rio de Janeiro* 82: 299–305.

French, J.R.P. and D.W. Schloesser, 1991. Growth and overwinter survival of the Asiatic clam, *Corbicula fluminea*, in the St. Clair River, Michigan. *Hydrobiologia* 219: 165–170.

French, J.R.P. and D.W. Schloesser, 1991. Growth and overwinter survival of the Asiatic clam, *Corbicula fluminea*. In: *The St. Clair River, Michigan, and Environmental Assessment and Habitat Evaluation of the Upper Great Lakes Connecting Channels. Developments in Hydrobiology*, 65, M. Munawar and T. Edsall, eds, pp. 165–170. Dordrecht: Kluwer Publishers.

French, J.R.P. and D.W. Schloesser, 1996. Distribution and winter survival health of Asian clams, *Corbicula fluminea*, in the St. Clair River, Michigan. *Journal of Freshwater Ecology* 11: 183–192.

Frest, T. and E. Johannes, 2000. An annotated checklist of Idaho land and freshwater mollusks. *Journal of Idaho Academy of Science* 36(2): 1–51.

Frieburg, M.W. and D.H. Hazelwood, 1977. Oxygen consumption of two amphibious snails: *Pomacea paludosa* and *Marisa cornuarietis* (Prosobranchia: Ampullariidae). *Malacologia* 16: 541–548.

Frischer, M.E., A.S. Hansen, J.A. Wyllie, J. Wimbush, J. Murray, and S.A. Nierzwicki-Bauer, 2002. Specific amplification of the 18S rRNA gene as a method to detect zebra mussel (*Dreissena polymorpha*) larvae in plankton samples. *Hydrobiologia* 487: 33–44.

Fujio, Y., E. Von Brand, and M. Kobayashi, 1991. Apparent differential hatchabilities associated with degrees of heterozygosity at leucine aminopeptidase isozyme loci in the applesnail *Pomacea canaliculata. Nippon Suisan Gakkaishi* 57: 459–461 (not seen, cited by http://nis.gsmfc.org/nis_factsheet.php?toc_id=154. Accessed October 31, 2008.

Fujio, Y., H. Kurihara, and E. von Brand, 1991a. Differences metric traits among three strains of apple snail, *Pomacea canalicualta. Tohuku Journal of Agricultural Research* 41: 61–68.

Fulgencio, M.Q. and H.R. Mota, 2007. Preferência de *limnoperna fortunei* (Dunker, 1857) a diferentes tipos de materiais. *Relatório Final Bolsa Iniciação Científica* CNPQ/CETEC.

Gale, W.F., 1969. *Bottom fauna of pool 19, Mississippi River, with emphasis on the life history of the fingernail clam,* Musculium transversum. PhD dissertation, Iowa State University, Ames. University Microfilm No. 69-20642, 12p., Ann Arbor, MI.

Gandhi, B., 2001. Efficacy assessment of a medium pressure UV iradiation system for zebra mussel control. OPG Report N-08131.02T07, 11p.

Gandhi, B., 2001. High flow, small pore filter for zebra mussel control at Nanticoke GS. OPG Report N-08131.02T07.

Garcia, A.M., M.A. Bemvenuti, J.P. Vieira, D.M.L. Motta Marques, M.D.M. Burns, A. Moresco, and M.V. Condini, 2006. Checklist comparison and dominance patterns of the fish fauna at Taim Wetland, South Brazil. *Neotropical Ichthyology* 4: 261–268.

Garcia, M.L. and L.C. Protogino, 2005. Invasive freshwater molluscs are consumed by native fishes in South America. *Journal of Applied Ichthyology* 21: 34–38.

Garland, E.D. and C.A. Zimmer, 2002. Techniques for the identification of bivalve larvae. *Marine Ecology Program Series* 225: 299–310.

Garton, D.W. and W.R. Haag, 1991. Heterozygosity, shell length and metabolism in the zebra mussel, *Dreissena polymorpha,* from a recently established population in Lake Erie. *Comparative Biochemistry and Physiology* 99A: 45–48.

Garton, D.W. and W.R. Haag, 1993. Seasonal reproductive cycles and settlement patterns of *Dreissena polymorpha* in western Lake Erie. In: *Zebra Mussels: Biology, Impacts, and Control,* T.F. Nalepa and D.W. Schloesser, eds, pp. 111–128. Boca Raton: CRC Press.

Gasiunas, I.I., 1965. Distribution and practical meaning of mollusk *Dreissena polymoprha* (Pallas) in water-bodoies of Lithuania. In: *Molluscs: Questions of Theretical and Applied Malacology, II Conference on Mollusk Research.* Book of abstracts. Leningrad (in Russian).

Gaston, E.E., 2006. Aquatic snails; permit requirements for importation and interstate movement. *Federal Register* 71: 16973–16975.

Gauthier, C. and G. Nicolaides, 1993. Remedial cleaning methods to remove zebra mussels. In: *The Third International Zebra Mussel Conference, Agenda and Abstracts,* Toronto, Canada, February.

Gaygusuz. Ö., Ç.G. Gaygusuz, A.S. Tarkan, H. Acipinar, and Z. Türer, 2007. Preference of zebra mussel, *Dreissena polymorpha* in the diet and effect on growth of gobiids: A comparative study between two different ecosystems. *Ekologi* 17: 1–6.

Geary, D.H., I. Magyar, and P. MuËller, 2000. Ancient Lake Pannon and its endemic molluscan fauna (Central Europe; Mio-Pliocene). *Advances in Ecological Research* 31: 463–482.

Gerard, C., A. Blanc, and K. Costil, 2003. *Potamopyrgus antipodarum (*Mollusca: Hydrobiidae) in continental aquatic gastropod communities: Impact of salinity and trematode parasitism. *Hydrobiologia* 493: 167–172.

Gerking, S.D., 1957. A method of sampling the littoral macrofauna and its application. *Ecology* 38: 219–266.

Giamberini, L., N. Czembor, and J.C. Phian, 1998. Effects of Mexel 432 on the settling, detachment and mortality of adult zebra mussels. In: *Proceedings of the Fourth International Zebra Mussel Conference, Madison, USA.*

Gillespie, D.M. and C.D.J. Brown, 1966. A quantitative sampler for macroinvertebrates associated with aquatic macrophytes. *Limnology and Oceanography* 11: 404–406.

Gillis, P., 1989. Gametogenesis in *Dreissena polymorpha.* An unpublished report. Department of Zoology, University of Guelph.

Gillis, P. and G.L. Mackie, 1992b. The effect of the epizoic *Dreissena polymorpha* on the filtration activities of the host unionid, *Proptera alata.* In: *Biology and Impact of Zebra Mussels in Lake St.Clair and Lake Erie*, G.L. Mackie, D. Pathy, P. Gillis, and E. Dobson, eds., pp. 59–93. Final Report to the Ministry of the Environment, Project 443G.

Gillis, P.L. and G.L. Mackie, 1994. The impact of *Dreissena polymorpha* on populations of Unionidae in Lake St. Clair. *Canadian Journal of Zoology* 72: 1260–1271.

Giovanelli, A., C.L.P.A.C. da Silva, L. Medeiros, and M.C. de Vasconcellos, 2002. The molluscicidal activity of niclosamide (Bayluscide WP70®) on *Melanoides tuberculata* (Thiaridae), a snail associated with habitats of Biomphalaria glabrata (Planorbidae). *Memoirs des Instituto Oswaldo Cruz, Rio de Janeiro* 97: 743–745.

Gittenberger, E. and A.W. Janssen, eds, 1998. De Nederlandse zoetwatermollusken. Recente en fossiele weekdieren uit zoet en brak water. Naturalis, Leiden (not seen, cited by Wolff, 2005).

Gittenberger, E., A.W. Jansen, W.J. Kuiper, J.G.J. Kuiper, T. Meijer, G. Van der Velde, J.N. de Vries, and G.A. Peeters, 1998. De Nederlandse zoetwatermollusken; recente en fossiele weekdieren uit zoet en brak water. Nederlandse fauna 2. *Nationaal Natuurhistorisch Museum, KNNV uitgeverij and European Invertebrate Survey*, Leiden (not seen, cited by Verween et al. 2006a).

Glaubrecht, M., T. v. Rintelen, and A.V. Korniushin, 2003. Toward a systematic revision of brooding freshwater Corbiculidae in southeast Asia (Bivalvia, Veneroida): Oon shell morphology, anatomy and molecular phylogenetics of endemic taxa from islands in Indonesia. *Malacologia* 45: 1–40.

Glover, N. and C. Campbell, 1994. Apple snails in wetland taro. Pacific Islands Farm Manual, *Taro Pest & Disease Leaflet* 5: 1–4.

Glöer. P.C.A., C. Albrecht, and T. Wilke, 2007. Enigmatic distribution patterns of the Bithyniidae in the Balkan Region (Gastropoda: Rissooidea). *Mollusca* 25: 13–22.

Godan, D., 1983. *Pest Slugs and Snails. Biology and Control.* New York: Springer.

Gordon, G., W.J. Cooper, R.G. Rice, and G.E. Pacey, 1987. Disinfectant residual measurement methods. Prepared for and published by the AWWA Research Foundation.

Gordon, G., W.J. Cooper, R.G. Rice, and G.E. Pacey, 1992. *Disinfectant Residual Measurement Methods.* Second Edition. AWWARF and AWWA, Denver, CO.

Graczyk, T.K., D.B. Conn, D.J. Marcogliese, H. Graczyk, and Y. de Lafontaine, 2003. Accumulation of human waterborne parasites by zebra mussels (*Dreissena polymorpha*) and Asian freshwater clams (*Corbicula fluminea*). *Parasitology Research* 89: 107–112.

Graczyk, T.K., D.B. Conn, F. Lucy, D. Minchin, L. Tamang, L.N.S. Moura, and A.J. DaSilva, 2004. Human waterborne parasites in zebra mussels (*Dreissena polymorpha*) from the Shannon River drainage area, Ireland. *Parasitology Research* 93: 385–391.

Graney, R.L., D.S. Cherry, J.H. Rodgers, Jr., and J. Cairns, Jr, 1980. The influence of thermal discharge and substrate composition on the population structure and distribution of the Asiatic clam, *Corbicula fluminea*, in the New River, Virginia. *The Nautilus* 94: 130–135.

Green, D.E. and P.K. Stumpf, 1946. The mode of action of chlorine. *Journal of the American Water Works Association* 38: 1301–1305.

Green, R., K. Murray, T. Salvango, M. Pencak, F. Bartolone, and J. Lacy, 1993. The establishment of a successful non-oxidizing biocide zebra mussel control strategy at nine mile point unit 2. In: *Third International Zebra Mussel Conference, Agenda and Abstracts,* Toronto, Canada, February.

Greenshields, F. and J.F. Ridley, 1957. Some researches on the control of mussels in water pipes. *Journal of the Institute of Water Engineers* 11: 300–306.

Gregg, B.C., 1974. *The effects of chlorine and heat on selected stream invertebrates.* PhD dissertation, Virginia Polytechnic Institute and State University.

Griffiths, J.S., 1992. Control of zebra mussels by thermal shock. Ontario Hydro Report 92-65-K.

Griffiths, R.W., 1987. Environmental quality assessment of Lake St. Clair in 1983 as reflected by the distribution of benthic macroinvertebrate communities. Report prepared for the Ontario Ministry of the Environment by Aquatic Ecosystems Limited.

Griffiths, R.W., 1993. Effects of zebra mussels (*Dreissena polymorpha*) on the benthic fauna of Lake St.Clair. In: *Zebra Mussels: Biology, Impacts, and Control*, T.F. Nalepa and D.W. Schloesser, eds, pp. 415–437. Ann Arbor: Lewis Publishers.

Griffths R.W., D.W. Schloesser, J.H. Leach, and W.P. Kovalak, 1991. Distribution and dispersal of the zebra mussel *Dreissena polymorpha* in the Great Lakes region. *Canadian Journal of Fisheries and Aquatic Sciences* 48: 1381–1388.

Grigorovich, I.A., A.V. Korniushin, D.K. Gray, I.C. Duggan, R.I. Colautti, and H.J. MacIsaac, 2003. Lake Superior: An invasion coldspot? *Hydrobiologia* 499: 191–210.

Grigorovich, I.A., J.R. Kelly, J.A. Darling, and C.W. West, 2005. The quagga mussel invades the Lake Superior Basin. *Journal of Great Lakes Research* 34: 342–350.

Gruszka, P., 1999. The River Odra estuary as a gateway for alien species immigration to the Baltic Sea Basin. *Acta Hydrochimica et Hydrobiologica* 27: 374–382.

GSMFC (Gulf States Marine Fisheries Commission), 2009. Available at: http://nis.gsmfc.org/nis_records.php?gss_id=377. Accessed January 24, 2009.

Guimaraes, C.T., 1981. Algumas observaoes de laboratoria sobre biologia e ecologia de *Pomacea haustrum* (Reeve, 1856). *Memorias do Instituto Oswaldo Cruz, Rio de Janeiro* 76: 33–46.

Guimares, C.T., 1983. Biological control: *Pomacea haustrum* Reeve, 1856 (Mollusca Pilidae) over planorbids under laboratory conditions. *Revista Sa de Publica* 17: 138–147.

Gunther, A.J., J.A. Davis, D.D. Hardin, J. Gold, D. Bell, J.R. Crick, G.M. Scelfo, J. Sericano, and M. Stephenson, 1999. Long-term bioaccumulation monitoring with transplanted bivalves in the San Francisco estuary. *Marine Pollution Bulletin* 38: 170–181.

Gutierrez, A., G. Perera, M. Yong, and J.A. Fernandez, 1997. Relationships of the prosobranch snails *Pomacea paludosa, Tarebia granifera and Melanoides tuberculata* with the abiotic environment and freshwater snail diversity in the central region of Cuba. *Malacological Review* 30: 39–44.

Haag, W.R., D.J. Berg, D.W. Garton, and J.L. Farris, 1993. Reduced survival and fitness in native bivalves in response to fouling by the introduced zebra mussel (*Dreissena polymorpha*) in western Lake Erie. *Canadian Journal of Fisheries and Aquatic Sciences* 50: 13–19.

Haas, G., M. Brunke, and B. Streit, 2002. Fast turnover in dominance of exotic species in the Rhine River determines biodiversity and ecosystem function: An affair between amphipods and mussels. In: *Invasive Aquatic Species of Europe: Distribution, Impacts and Management*, E. Leppäkoski, S. Gollasch, and S. Olenin, eds, pp. 426–432. Dordrecht: Kluwer Academic Publishers.

Hakenkamp, C.C. and M.A. Palmer, 1999. Introduced bivalves in freshwater ecosystems: The impact of *Corbicula* on organic matter dynamics in a sandy stream. *Oecologia* 119: 445–451.

Hakenkamp, C.C., S.G. Ribblett, M.A. Palmer, C.M. Swan, J.W. Reid, and M.R. Goodison, 2001. The impact of an introduced bivalve (*Corbicula fluminea*) on the benthos of a sandy stream. *Freshwater Biology* 46: 491–501.

Hall, L.W., D.T. Burton, and L.H. Liden, 1981. An interpretive literature analysis evaluating the effects of power plant chlorination on freshwater organisms. *CRC Critical Reviews in Toxicology* 9: 1–20.

Hall, R.O., M.C. Vanderloop, and M.F. Dybdahl, 2002. Production of New Zealand mudsnails and native invertebrates in Firehole River, Gibbon River, and Polecat Creek. Final Report: *The Invasiveness of an Exotic Snail in the Greater Yellowstone Ecosystem and Energy Flow Under Ambient Conditions,* B.L. Kerans, M. Dybdahl, and R. Hall, eds (not seen, cited in NZMS Management and Control Plan Working Group, 2007).

Hall, R.O., J.L. Tank and M.F. Dybdahl, 2003. Exotic snails dominate carbon and nitrogen cycling in a highly productive stream. *Frontiers in Ecology and the Environment* 1: 408–411.

Hall, R.O. Jr., M.F. Dybdahl, and M.C. Vanderloop, 2006. Extremely high secondary production of introduced snails in rivers. *Ecological Applications* 16(3): 1121–1131.

Hall, S.R., C.P. Schneider, S.J. Lary, F. Arrhenius, N.K. Pauliukonis, E.L. Mills, and L.G. Rudstam, 2003. A comparison of total phosphorus, chlorophyll *a*, and zooplankton in embayment, nearshore, and offshore habitats of Lake Ontario. *Journal of Great Lakes Research* 29: 54–69.

Hallac, D.E. and J.E. Marsden, 2001. Comparison of conservation strategies for unionids threatened by zebra mussels (*Dreissena polymorpha*): Periodic cleaning vs. quarantine and translocation. *Journal of the North American Benthological Society* 20: 200–210.

Halwart, M., 1994a. The golden apple snail *Pomacea canaliculata* in Asian rice farming systems: Present impact and future threat. *International Journal of Pest Management* 40: 199–206.

Halwart, M., 1994b. The ecology of fish in rice fields. *Annals of Tropical Research* 10: 1–15.

Halwart, M., 2006. Fish as biological control agents of golden apple snails in Philippine rice fields. In: *Global Advances in Ecology and Management of Golden Apple Snails*, R.C. Joshi and L.S. Sebastian, eds, pp. 363–374. Nueva Ecija: Philippine Rice Research Institute.

Hamilton, S.K., S.J. Sippel, D.F. Calheiros, and J.M. Melack, 1997. An anoxic event and other biogeochemical effects of the Pantanal wetland on the Paraguay River. *Limnological Oceanography* 42: 257–272.

Hammarlund, D., S. Björck, B. Buchardt, C. Israelson, and C.T. Thomsen, 2003. Rapid hydrological changes during the Holocene revealed by stable isotope records of lacustrine carbonates from Lake Igelsjön, southern Sweden. *Quaternary Science Reviews* 22: 353–370.

Hann, B.J., 1999. A prairie coastal wetland (Lake Manitoba's delta marsh): Organization of the invertebrate community. In: *Invertebrates in Freshwater Wetlands of North America: Ecology and Management,* D.P. Batzer, R.B. Rader, and S.A. Wissinger, eds, pp. 1013–1039. New York: Wiley.

Hanna, G.D., 1966. Introduced mollusks of western North America. *Occasional Papers, California Academy of Sciences* 48: 1–108.

Hanning, G.W., 1979. *Aspects of reproduction in Pomacea paludosa* (Mesogastropoda: Pilidae). MS thesis, Florida State University, Tallahassee.

Hanning, G.W. and W.S. Leedom, 1978. Schistosome dermatitis from *Pomacea paludosa* (Say) (Prosobranchia: Pilidae). *The Nautilus* 92(3): 105–106.

Hanovia Ltd., http://www.hanovia.com

Hansen, A.S., S.A. Nierzwicki-Bauer, and M.E. Frisher, 1998. Presence, growth and survival of zebra mussels in Lake George, New York. Presentation at the *Eighth International Zebra Mussel and Aquatic Nuisance Species Conference,* Sacremento, USA, March 16–19.

Harada, E. and M. Nishino, 1995. Differences in inhalant siphonal papillae among the Japanese species of *Corbicula* (Mollusca: Bivalvia). *Publications of Seto Marine Biology Laboratory* 36: 389–408.

Harman, W., 1968. Replacement of pleurocerids by *Bithynia* in polluted waters of central New York. *The Nautilus* 81: 77–83.

Harman, W.N., 1974a. Clams and mussels (Mollusca: Bivalvia). In: *Pollution Ecology of Freshwater Invertebrates,* C. Hart and S. Fuller, eds, pp. 215–273. New York: Academic Press.

Harman, W.N., 1974b. Snails (Mollusca: Gastropoda). In: *Pollution Ecology of Freshwater Invertebrates*, C. Hart and S. Fuller, eds, pp. 275– 312. New York: Academic Press.

Harman, W.N., 2000. Diminishing species richness of mollusks in Oneida Lake, New York State, USA. *The Nautilus* 114: 120–126.

Harper, F., 1936. The distribution of the limpkin and its staple food *Pomacea. The Nautilus* 51: 37–40.

Harrington, D.K., J. Van Benschoten, J.N. Jensen, T. Brady, D. Lewis, J. Sferrazza, and E. Neuhauser, 1993. Combined use of heat and oxidants for controlling zebra mussels. In: *Third International Zebra Mussel Conference, Agenda and Abstracts,* Toronto, Canada, February.

Harry, H.W. and D.V. Aldrich, 1958. The ecology of *Australorbis glabratus* in Puerto Rico. *Bulletin of The World Health Organization* 18: 819–832.

Hasloop, U., 1992. Establishment of the Asiatic Clam *Corbicula C. fluminalis* in the tidal Weser River (N. Germany). *Archiv für Hydrobiologie* 126: 175–180.

Havel, J.E., C.E. Lee, and M.J. Vander Zanden, 2005. Do reservoirs facilitate invasions into landscapes? *BioScience* 55: 518–525.

Hawkins, C.P. and J.K. Furnish, 1987. Are snails important competitors in stream ecosystems? *Oikos* 49: 209–220.

Hayes, K.A., R.C. Joshi, S.C. Thiengo, and R.H. Cowie, 2008. Out of South America: Multiple origins of non-native apple snails in Asia. *Diversity and Distributions* 14: 701–712.

Haynes, A.B., J.R. Taylor, and M.E. Varley, 1985. The influence of the mobility of *Potamopyrgus jenkinsi* (Smith, E.A.) (Prosobranchia: Hydrobiidae) on its spread. *Archiv für Hydrobiologie* 103: 497–508.

Haynes, J.M., N.A. Trisch, C.M. Mayer, and R.S. Rhyne, 2005. Benthic macroinvertebrate communities in southwestern Lake Ontario following invasion of *Dreissena* and *Echinogammarus*: 1983–2000. *Journal of the North American Benthological Society* 24: 148–167.

Hayward, D. and E. Estevez, 1997. Suitability of Florida Waters to Invasion by the Zebra Mussel, *Dreissena polymorpha.* Florida Sea Grant College Program, Mote Marine Laboratory, Sarasota. Technical Report No. 495.

Heath, A.G., 1979. Toxicity of intermittent chlorination to rainbow trout, coho salmon and carp. *Association Southeastern Biological Bulletin* 23: 65.

Heath, R.T., G.L. Fahnenstiel, W.S. Gardner, J.F. Cavaletto, and S.-J. Hwang, 1995. Ecosystem-level effects of zebra mussels (*Dreissena polymorpha*): An enclosure experiment in Saginaw Bay, Lake Huron. *Journal of Great Lakes Research* 21: 501–516.

Hebert, P.D.N., B.W. Muncaster, and G.L. Mackie, 1989. Ecological, and genetic studies on *Dreissena polymorpha* (Pallas): A new mollusc in the Great Lakes. *Canadian Journal of Fisheries and Aquatic Sciences* 46: 1587–1591.

Hebert. P.D.N., C.C. Wilson, M.H. Murdoch, and R. Lazar, 1991. Demography and ecological impacts of the invading mollusc *Dreissena polymorpha*. *Canadian Journal of Zoology* 59: 405–409.

Heiler, K.C.M., P.V. von Oheimb, K. Ekschmitt, and C. Albrechct, 2008. Studies on the temperature dependence of activity and on the diurnal activity rhythm of the invasive *Pomacea canaliculata* (Gastropoda: Ampullariidae). *Mollusca* 26: 73–81.

Heinson, G.E., 1958. *Life history and ecology of the freshwater clam, Corbicula fluminea*. MSc thesis, University of California, Berkeley (not seen, cited by Britton and Morton, 1979).

Hendarsih-Suharto, R., M. Marwoto, M. Heryanto, and S.S. Siwi, 2006. The golden apple snail, *Pomacea* spp., in Indonesia. In: *Global Advances in Ecology and Management of Golden Apple Snails*, R.C. Joshi and L.S. Sebastian, eds, pp. 231–242. Nueva Ecija: Philippine Rice Research Institute.

Herbst, D.B., M.T. Bogan, and R.A. Lusardi, 2008. Low specific conductivity limits growth and survival of the New Zealand mud snail from the Upper Owens River, California. *Western North American Naturalist* 68: 324–333.

Hibberd, J. and C. Sferrazza, 1993. The use of chlorine residual analyzers and controllers on effluent streams: Present problems and future needs. In: *Third International Zebra Mussel Conference, Agenda and Abstracts,* Toronto, Canada, February.

Higgins, T.M., J.M. Grennan, and K.T. McCarth, 2008. Effects of recent zebra mussel invasion on water chemistry and phytoplankton production in a small Irish lake. *Aquatic Invasions* 3: 14–20.

Hill, I.R., 1985. Effects on non-target organisms in terrestrial and aquatic environments. In: *The Pyrethroid Insecticides*, J.P. Leahey, ed., pp. 151–262. Philadelphia: Taylor & Francis.

Hillis, D.M. and J.C. Patton, 1982. Morphological and electrophoretic evidence for two species of *Corbicula* (Bivalvia: Corbiculidae) in North American. *American Midlands Naturalist* 108: 74–80.

Hillis, D.M. and R.L. Mayden, 1985. Spread of the Asiatic clam *Corbicula* (Bivalvia: Corbiculacea) into the new world tropics. *Southwest Naturalist* 30: 454–456.

Hincks, S.S., 1994. *Environmental factors limiting the success of zebra mussels in Ontario.* MSc thesis, University of Guelph, Guelph.

Hincks, S.S. and G.L. Mackie, 1993. The effects of calcium and alkalinity on the growth and reproductive success of *Dreissena polymorpha*. In: *Third International Zebra Mussel Conference, Agenda and Abstracts,* Toronto, Canada, February 23–26.

Hincks, S.S. and G.L. Mackie, 1997. Effects of pH, calcium, alkalinity, hardness, and chlorophyll on the survival, growth and reproductive success of zebra mussels (*Driessena polymorpha*) in Ontario lakes. *Canadian Journal of Fisheries and Aquatic Sciences* 54: 2049–2057.

Ho, K.T. and S.P. Gloss, 1987. Distribution and persistence of the lampricide Bayer 73 following a control application. *Canadian Journal of Fisheries and Aquatic Sciences* 44: 112–119.

Holland, R.E., 1993. Changes in planktonic diatoms and water transparency in Hatchery Bay, Bass Island area, western Lake Erie since the establishment of the zebra mussel. *Journal of Great Lakes Research* 19: 617–624.

Holliday, C.P., T.C. Rasmussen, and W.P. Miller, 2003. Establishing the relationship between turbidity and total suspended sediment concentration. In: *Proceedings of the 2003 Georgia Water Resources Conference, University of Georgia*, K.J. Hatcher, ed. Institute of Ecology, Athens, Georgia, April 23–24. Available at: http://cms.ce.gatech.edu/gwri/uploads/proceedings/2003/Holliday,%20Rasmussen%20and%20Miller.PDF. Accessed February 12, 2009.

Hollingsworth, R.G. and R.H. Cowie, 2006. Apple snails as disease vectors. In: *Global Advances in Ecology and Management of Golden Apple Snails*, R.C. Joshi and L.S. Sebastian, eds, pp. 37–60. Nueva Ecija: Philippine Rice Research Institute.

Holmes, N., 1970. Marine fouling power stations. *Marine Pollution Bulletin* 1: 105–106.

Holomuzki, J.R. and B.J.F. Biggs, 1999. Distributional responses to flow disturbance by a stream–dwelling snail. *Oikos* 87: 36–47.

Holomuzki, J.R. and B.J.F. Biggs, 2000. Taxon–specific responses to high–flow disturbances in streams: Implications for population persistence. *Journal of the North American Benthological Society* 19: 670–679.

Holt, D.A. and E. Ryan, 1997. Chlorine dioxide—adult zebra mussel eradication at Dofasco steel mill. In: *Proceedings of the Seventh International Zebra Mussel and Aquatic Nuiscance Species Conference*, New Orleans, USA, January 28.

Hoorfar, J., B. Malorny, A. Abdul-Mawjood, N. Cook, M. Wagner, and P. Fach, 2004. Practical considerations in design of internal amplification controls for diagnostic PCR assays. *Journal of Clinical Microbiology* 42: 1863–1868.

Hopkins, G.J., 1990. The zebra mussel, *Dreissena polymorpha*: A photographic guide to the identification of microscopic veligers. Available from Queen's Printer for Ontario.

Horgan, M.J. and E.L. Mills, 1997. Clearance rates and filtering activity of zebra mussel (*Dreissena polymorpha*): Implications for freshwater lakes. *Canadian Journal of Fisheries and Aquatic Sciences* 54: 249–255.

Hornbach, D.J., 1985. A review of metabolism in the Pisidiidae with new data on its relation to life history traits in *Pisidium casertanum. American Malacological Bulletin* 3: 187–200.

Horne, A.J. and C.R. Goldman, 1994. *Limnology*, 2nd ed. New York: McGraw-Hill.

Horne, F.R., 1979. Comparative aspects of estivating metabolism in the gastropods, *Marisa cornuarietis. Comparative Biochemistry and Physiology* 64A: 309–311.

Horne, F.R. and McIntosh, S., 1979. Factors influencing distribution of mussels in the Blanco River of central Texas. *The Nautilus* 94: 119–133.

Horning, W.B. and L. Keup, 1964. Decline of the Asiatic clam in the Ohio River. *The Nautilus* 78: 29–30.

Hosea, R.C. and B. Finlayson, 2005. Controlling the spread of New Zealand mudsnails on wading gear. California Department of Fish and Game, Rancho Cordova.

Howarth, F.G., 1991. Environmental impacts of classical biological control. *Annual Review of Entomology* 36: 485–509.

Howe, P.H., P. Redmon, R. Atteberry, and E. Masters, 1993. A summary of federal regulations related to use of FIFRA registered biocides and region V U.S. EPA concerns related to use of these biocides for zebra mussel control. In: *Third International Zebra Mussel Conference, Agenda and Abstracts*, February, Toronto, Canada.

Howell, E.T., C.H. Marvin, R.W. Bilyea, P.B. Kauss, and K. Somers, 1996. Changes in environmental conditions during *Dreissena* colonization of a monitoring station in eastern Lake Erie. *Journal of Great Lakes Research* 22: 744–756.

Howells, R.G., 1992. Annotated list of introduced non-native fishes, mollusks, crustaceans and aquatic plants in Texas waters. Texas Parks and Wildlife Department, Management Data Series 78, Austin.

Howells, R.G., 2001. The channeled applesnail (*Pomacea canaliculata*) invasion: A threat to aquatic ecosystems and the price of rice crispies. *American Conchologist* 29: 8–10.

Howells, R.G., 2002. Comparative feeding of two species of apple snails (*Pomacea*). *Ellipsaria* 4: 14–16.

Howells, R.G., R.W. Neck, and H.D. Murray, 1996. *Freshwater Mussels of Texas*. Austin: Texas Parks and Wildlife Press.

Howells, R.G. and J.W. Smith, 2002. Status of the applesnail *Pomacea canaliculata* in the United States. In: *The Seventh International Congress on Medical and Applied Malacology*, T. Wada, Y. Yusa, R.C. Joshi, and A.I. S. Ghesquiere, eds, pp. 86–96. SEAMEO Regional Centre for Graduate Study and Research in Agriculture, Los Baños, Laguna, Philippines.

Howells, R.G., L.F. Burlakova, A.Y. Karatayev, P.K. Marfurt, and R.L. Burks, 2006. Native and introduced Ampullariidae in North America: History, status, and ecology. In: *Global Advances in the Ecology and Management of Golden Apple Snails*, R.C. Joshi, and L.S. Sebastian, eds, pp. 73–112. Philippine Rice Research Institute.

Hoyer, M.V. and J.R. Jones, 1983. Factors affecting the relation between phosphorus and chlorophyll *a* in midwestern reservoirs. *Canadian Journal of Fisheries and Aquatic Sciences* 40: 192–199.

Hoyer, M.V., K.F. Thomas, S.K. Notestein, and D.E. Canfield, Jr., 2003. Nutrient, chlorophyll, and water clarity relationships in Florida's nearshore coastal waters with comparisons to freshwater lakes. *Canadian Journal of Fisheries and Aquatic Sciences* 59: 1024–1031.

Hoyle, J.A., T. Shaner, J.M. Casselman, and R. Dermott, R., 1999. Changes in the lake whitefish (*Coregonus clupeaformis*) stocks in eastern Lake Ontario following *Dreissena* mussel invasion. *Great Lakes Research Reviews* 4: 5–10.

Hromada M., V. DudiÁák, and R. Yosef, 2000. An inside-out perspective of the true shrikes, a review of the helminthofauna. *Ring* 22: 185–204.

Hunt, B.P., 1958. Introduction of *Marisa* into Florida. *The Nautilus* 72: 53–55.

Hunt, B.P., 1961. Tolerance of a freshwater snail *Marisa cornuarietis* (L) to sea water. *Quarterly Journal of the Florida Academy of Science* 23: 278–284.

Hutchinson, G.E., 1957. *A Treatise on Limnology*. New York: Wiley.

Huynh, N.K., 2006. Golden apple snails in Vietnam. In *Global Advances in Ecology and Management of Golden Apple Snails*, R.C. Joshi and L.S. Sebastian, eds, pp. 255–266. Nueva Ecija: Philippine Rice Research Institute.

Hylleberg, J. and H.R. Siegismund, 1987. Niche overlap in mud snails (Hydrobiidae): Freezing tolerance. *Marine Biology* 94: 403–407.

Hyslop, E.J., 1980. Stomach contents analysis—a review of methods and their application. *Journal of Fish Biology* 17: 411–29.

Idrisi, N., E.L. Mills, L.G. Rudstam, and D.J. Stewart, 2001. Impact of zebra mussels (*Dreissena polymorpha*) on the pelagic lower trophic levels of Oneida Lake, New York. *Canadian Journal of Fisheries and Aquatic Sciences* 58: 1–12

Imlay, M.J., 1973. Effects of K^+ on survival and distribution of fresh water mussels. *Malacologia* 12: 97–113.

Ingols, R.S., H.A. Wyckoff, T.W. Kethley, H.W. Hodgden, E.L. Fincher, J.C. Hildebrand, and J.E. Mandel, 1953. Bactericidal studies of chlorine. *Industrial Engineering Chemistry* 45: 996–1000.

Isom, B.G., 1968. The naiad fauna of Indian Creek, Madison County, Alabama. *American Midland Naturalist* 79: 514–516.

Isom, B.G., 1986. Historical review of Asiatic clam (*Corbicula*) invasion and biofouling of waters and industries in the Americas. *American Malacological Bulletin, Special Edition No. 2* 1986: 1–5.

Ito, K., 2002. Environmental factors influencing overwintering success of the golden apple snail, *Pomacea canaliculata* (Gastropoda: Ampullariidae), in the northernmost population of Japan. *Applied Entomology and Zoology* 37: 655–661.

Ito, K., 2003. Expansion of the golden apple snail, *Pomacea canaliculata*, and features of its habitat. *Food and Fertilizer Technology Center, Extension Bulletin* 2003–11–01.

Ituarte, C.F. 1981. Primera noticia acerca de la introduccion de pelecipods Asiaticos en al area Rioplatense. *Neotropica* 27: 79–82.

Ituarte, C.F., 1984. Aspectos biologicos de las poblaciones de *Corbicula largillierti* (Philippi, 1844) (Mollusca Pelecypoda) en el Rio de La Plata. *Revista del Museo de La Plata (nueva serie)* 13: 231–247.

Ituarte, C.F., 1994. *Corbicula* and *Neocorbicula* (Bivalvia: Corbiculidae) in the Paraná, Uruguay, and Rio de La Plata basins. *The Nautilus* 107: 129–135.

Iwanyzki, S. and R.W. McCauley, 1993. Upper lethal temperatures of adult zebra mussels (*Dreissena polymorpha*). In: *Zebra Mussels: Biology, Impacts, and Control*, Nalepa, T.F. and D.W. Schloesser, eds, pp. 667–673. Boca Raton: Lewis Publishers.

Iwasaki, K., 1997. Climbing behaviour and tolerance to aerial exposure of freshwater mussel, *Limnoperna fortunei. Venus* 56(1): 15–25.

Iwasaki, K. and Y. Uriu, 1998. Life cycle of a freshwater mytilid mussel, *Limnoperna fortunei*, in Uji River, Kyoto. *Venus* 57: 105–113.

Jacobsen, R. and V.E. Forbes, 1997. Clonal variation in life–history traits and feeding rates in the gastropod, *Potamopyrgus antipodarum*: Performance across a salinity gradient. *Functional Ecology* 11: 260–267.

Jacobson, M.K., 1953. *Congeria leucophaeta* (Conrad) in the Hudson River. *The Nautilus* 66: 125–127.

Jacobson, M.K., 1975. The freshwater prosobranch, *Tarebia granifera*, in Oriente, Cuba. *The Nautilus* 89: 106.

Jacobson, M.K., 1978. *Tarebia* (Prosobranchia: Thiaridae) in Cuba. *The Nautilus* 92: 54–55.

James, M., 1985. Changes in the faunal composition of two thermal streams near Taupo, New Zealand. *New Zealand Journal of Marine and Freshwater Research* 19: 439–443.

James, W.F., J.W. Barko, and H.L. Eakin, 1997. Nutrient regeneration by the zebra mussel (*Dreissena polymorpha*). *Journal of Freshwater Ecology* 12: 209–216.

James, W.F., J.W. Barko, and H.L. Eakin, 2001a. Phosphorus recycling by zebra mussels in relation to density and food resource availability. *Hydrobiologia* 455: 55–60.

James, W.F., J.W. Barko, and H.L. Eakin, 2001b. Enhanced phosphorus recycling by zebra mussels at high density levels in relation to food supply. Report to US Army Corps of Engineers, ERDC WQTN-PD-09.

James, W.G., 1967. Mussel fouling and use of exomotive chlorination. *Chemical Industry* 1: 994–996.

Janech, M.G. and R.D. Hunter, 1995. *Corbicula fluminea* in a Michigan River: Implications for low temperature tolerance. *Malacological Review* 28: 119–124.

Janik, J., 1997. Zebra mussels and potential impacts in California. Paper presented at *American Fisheries Society, 127th Annual Meeting*, Monterey, California, August 24–28 (not seen, cited by Cohen, 2001).

Jantz, B. and D. Neumann, 1992. Shell growth and population dynamics of Dreissena polymorpha in the River Rhine. In: *The Zebra Mussel Dreissena polymorpha: Ecology, Biological Monitoring and First Applications in the Water Quality Management*, D. Neumann and H.A. Jenner, eds, pp. 49–66. Stuttgart: Gustav Fisher.

Jarvis, P., J. Dow, R. Dermott, and R. Bonnell, 2000. Zebra (*Dreissena polymorpha*) and quagga mussel (*Dreissena bugensis*) distribution and density in Lake Erie, 1992–1998. Canadian Technical Report of Fisheries and Aquatic Sciences No. 2304.

Jenner, H. and H.J.G. Polman, 1998. Pulse-Chlorination: A new way of reducing chlorine for mitigation of mussel fouling. KEMA Power Generation & Environmental Technology. In: *Proceedings of the Eighth International Zebra Mussel and Other Nuisance Species Conference,* Sacramento, USA, March 16–19.

Jenner, H. and H.J.G. Polman, 2003. New antifouling technique by combined carbon dioxide and sodium hypochlorite dosing. In: *Proceedings of the 13th International Conference on Aquatic Invasive Species*, Ennis, Ireland, September 20–24.

Jenner, H.A., 1978. Chlorine minimization in macrofouling control in the Netherlands. In: *Water Chlorination: Environmental Impact and Health Effects, Proceedings of the Third Conference on Water Chlorination*, Colorado Springs, USA, October 28 to November 2.

Jenner, H.A. and J.P.M. Janssen-Mommem, 1993. Monitoring and control of *Dreissena polymorpha* and other macrofouling bivalves in the Netherlands. In: *Zebra Mussels: Biology, Impacts, and Control*, T.F. Nalepa, T.F., and D.W. Schloesser, eds, pp. 537–553. Boca Raton: Lewis Publishers.

Jenner, H.A., J.W. Whitehouse, C.J.L. Taylor, and M. Khalanski, 1998. Cooling water management in European power stations. Biology and control of fouling. *Hydroécologie appliquée* 10: 1–225.

Jenner, H.A., H.J.G. Polman, and R. van Wijck, 2003. Four years experience with a new chlorine dosing regime against macrofouling. VGB Chemie im Kraftwerk, Essen, October 22–23, 2003.

Jobin W.R., 1970. Population dynamics of aquatic snails in three farm ponds of Puerto Rico. *American Journal of Tropical Medicine and Hygiene* 19: 1038–1048.

Johnson, L.E., 1995. Enhanced early detection and enumeration of zebra mussel (*Dreissena* spp.) veligers using cross-polarized light. *Canadian Journal of Fisheries and Aquatic Science* 312: 139–146.

Johnson, L.E. and D.K. Padilla, 1996. Geographic spread of exotic species: Ecological lessons and opportunities from the invasion of the zebra mussel *Dreissena polymorpha. Biological Conservation* 78: 23–33.

Johnson, L.E. and J.T. Carlton, 1992. Counter-productive public information: The "Noah fallacy" and mussel myths. *Dreissena polymorpha Information Review* 3: 2–6.

Johnson, L.E. and J.T. Carlton, 1996. Post-establishment spread in large-scale invasions: Dispersal mechanisms of the zebra mussel, *Dreissena polymorpha. Ecology* 77: 1686–1690.

Johnson, P.D. and R.F. McMahon, 1998. Effects of temperature and chronic hypoxia on survivorship of the zebra mussel (*Dreissena polymorpha*) and Asian clam (*Corbicula fluminea*). *Canadian Journal of Fisheries and Aquatic Sciences* 55: 1564–1572.

Jokinen, E., 1992. *The Freshwater Snails (Mollusca: Gastropoda) of New York State*. The University of the State of New York, the State Education Department, New York State Museum, Albany.

Jones, A.T., 2008. Can we reposition the preferred geological conditions necessary for an infiltration gallery? The development of a synthetic infiltration gallery. *Desalination* 221: 598–601.

Jones, L.A. and A. Ricciardi, 2005. Influence of physicochemical factors on the distribution and biomass of invasive mussels (*Dreissena polymorpha* and *Dreissena bugensis*) in the St. Lawrence River. *Canadian Journal of Fisheries and Aquatic Sciences* 62: 1953–1962.

Jordan, F., H.L. Jelks, and W.M. Kitchens, 1994. Habitat use by the fishing spider *Dolmoides triton* in a northern Everglades wetland. *Wetlands* 14: 239–242.

Jordan, F., K.J. Babbitt, C.C. McIvor, and S.J. Miller, 1996. Spatial ecology of the crayfish *Procambarus alleni* in a Florida wetland mosaic. *Wetlands* 16: 134–142.

Jordan, P.J. and L.E. Deaton, 1999. Osmotic regulation and salinity tolerance in the freshwater snail *Pomacea bridgesi* and the freshwater clam *Lampsilis teres. Comparative Biochemistry and Physiology Part A* 122: 199–205.

Jordan, R.A. and C.E. Sutton, 1984. Oligohaline benthic invertebrate communities at two Chesapeake Bay power plants. *Estuaries* 7: 192–212.

Joshi, R.C. 2006. Golden apple snail recipes. In: *Global Advances in Ecology and Management of Golden Apple Snails*, R.C. Joshi, and L.S. Sebastian, eds, pp. 537–560. Nueva Ecija: Philippine Rice Research Institute.

Joshi, R.C. and L. Sebastian, eds, 2006. *Global Advances in Ecology and Management of Golden Apple Snails*. Nueva Ecija: Philippine Rice Research Institute.

Joshi, R.C., E.C. Martin, T. Wada, and L.S. Sebastian, 2006. Role of golden apple snails in organic rice cultivation and weed management. In: *Global Advances in Ecology and Management of Golden Apple Snails*, R.C. Joshi and L.S. Sebastian, eds, pp. 483–487. Nueva Ecija: Philippine Rice Research Institute.

Joy, J.E. and L.E. McCoy, 1975. Comparisons of shell dimensions and viscera mass weights in *Corbicula manilensis* (Philippi, 1844). *The Nautilus* 89: 51–54.

Kahabka, J. and A. Talgo, 1993. Filtration as a control method for zebra mussels: A small hydroelectric plant application. In: *Third International Zebra Mussel Conference, Agenda and Abstracts,* Toronto, Canada, February.

Karatayev, A., 1995. Factors determining the distribution and abundance of *Dreissena polymorpha* in lakes, dam reservoirs and channels. In: *Proceedings of the Fifth International Zebra Mussel and Other Aquatic Nuisance Organisms Conference*, Toronto, Canada, February.

Karatayev, A.Y., V.P. Mikheev, S.A. Afanasyev, M.Y. Kirpichenko, F.F. Protasov, L.V. Shevtsova, and T.G. Kharchenko, 1994. Practical meaning, use and methods of control of fouling of hydrotechnical constructions. In: *Dreissena polymorpha* (Pall.) (Bivalvia, Dreissenidae) *Taxonomy, Ecology and Practical Meaning*, Y.I. Starobogatov, ed., pp. 206–221. Moskow: Nauka Press (in Russian).

Karatayev, A.Y. and L.E. Burlakova, 1995. The role of *Dreissena* in lake ecosystems. *Russian Journal of Ecology* 26: 207–211.

Karatayev, A.Y., L.E. Burlakova, and D.K. Padilla, 1997. The effects of *Dreissena polymorpha* (Pallas) invasion on aquatic communities in Eastern Europe. *Journal of Shellfish Research* 16: 187–203.

Karateyev, A., L. Burlakova, and D. Padilla, 1998. Physical factors that limit the distribution and abundance of *Dreissena polymorpha* (Pall.). *Journal of Shellfish Research* 17: 1219–1235.

Karatayev, A.Y., L.E. Burlakova, and D.K. Padilla, 2002. Impacts of zebra mussels on aquatic communities and their role as ecosystem engineers. In: *Invasive Aquatic Species of Europe: Distribution, Impacts, and Management*, E. Leppäkoski, S. Gollasch, and S. Olenin, 433–446. Dordrecht: Kluwer Academic Publishers.

Karatayev, A.Y., L.E. Burlakova, and D.K. Padilla, 2003a. Contrasting distribution and impacts of two exotic suspension feeders, *Dreissena polymorpha* and *Corbicula fluminea*. In: *Proceedings of the NATO Advanced Research Workshop on the Comparative Roles of Suspension-Feeders in Ecosystems*, R.F. Dame and S. Olenin, eds, pp. 239–262. Nida, Lithuania October 4–9. *Earth and Environmental Science* 47: 1–359.

Karatayev, A.Y., L.E. Burlakova, D.K. Padilla, and L.E. Johnson, 2003b. Patterns of spread of the zebra mussel *Dreissena polymorpha* (Pallas): The continuing invasion of Belarussian lakes. *Biological Invasions* 5: 213–221.

Karatayev, A.Y., L.E. Burlakova, and D.A. Padilla, 2006. Growth rate and longevity of *Dreissena polymorpha* (Pallas): A review and recommendations for future study. *Journal of Shellfish Research* 25: 23–32.

Karatayev, A.Y., D. Boltovskoy, D.K. Padilla, and L.E. Burlakova, 2007a. The invasive bivalves *Dreissena polymorpha* and *Limnoperna fortunei*: Parallels, contrasts, potential spread, and invasion impacts. *Journal of Shellfish Research* 26: 205–213.

Karatayev, A.Y., D.K. Padilla, D. Minchin, D. Boltovskoy, and L.E. Burlakova, 2007b. Changes in global economies and trade: The potential spread of exotic freshwater bivalves. *Biological Invasions* 9: 161–180.

Karatayev, A.Y., L.E. Burlakova, V.A. Karatayev, and D.K. Padilla, 2009. Introduction, distribution, spread, and impacts of exotic freshwater gastropods in Texas. *Hydrobiologia* 619: 181–194.

Karpevich, A.F., 1947. The adaptability of metabolism in North Caspian mussels (genus *Dreissena*) to variations in the salinity regime. *Zoologicheskij Zhurnal* 26: 331–338 (not seen, cited in Strayer and Smith, 1993).

Karpevich, A.F., 1955. Some data on formation in the bivalved mollusks. *Zoologicheskij Zhurnal* 34: 6–67.

Karpevich, A.F., 1975. *Theory and Practice of Acclimatization of Aquatic Organisms*. Moscow: 'Pishchevaya promyshlennost' publishers (in Russian).

Karski, W., 2001. Efficacy assessment of a MABOS intermittent ozone injection system at Bruce A Nuclear Power Station OPG Report N-08131.02 T07.

Katz, B.M. and U.M. Harder, 1976. The attraction of fish. *Journal of Water Pollution Control Federation* 48: 592.

Kauffman, E.G., 1969. Form, function and evolution. In: *Treatise on Invertebrate Palaeontology. Part N. Mollusca 6*. R.C. Moore, ed., New York: Geological Society of America, pp. 129–205.

Kavanaugh, K.B., 2007. Hagg Lake Project monitoring report, 1999–2005. Prepared by K.B.K. of Flow Science Incorporated for the Tualatin River Flow Management Technical Committee.

Keer, J.T. and L. Birch, 2003. Molecular methods for the assessment of bacterial viability. *Journal of Microbiological Methods* 53: 175–183.

Keevin, T.M., R.E. Yarbrough, and A.C. Miller, 1992. Long-distance dispersal of zebra mussels (*Dreissena polymorpha*) attached to hulls of commercial vessels. *Journal of Freshwater Ecology* 7: 437.

Keillor, P., 1991. Sand filter intakes could safeguard vital water-supply systems from zebra mussels. Wisconsin Sea Grant Institute WIS-SG-91-428-13.

Keller, R.P. 2006. Ecological and bioeconomic risk assessment for invasive species. Ph.D. dissertation, University of Notre Dame, Indiana. 158 pp.

Keller, R.P., J.M. Drake, and D.M. Lodge, 2007. Fecundity as a basis for risk assessment of nonindigenous freshwater molluscs. *Conservation Biology* 21(1): 191–200.

Kennedy, A.J., R.N. Millward, J.W. Lynn, K.D. Perry, and J.A. Steevens, 2006. Generation of life stage sensitivity data to optimize chemical control strategies. Presentation at *the 14th International Conference Aquatic Invasive Species*, Key Biscayne, USA, May 14–19.

Kennedy, V.S. and L. van Huekelem, 1985. Gametogenesis and larval production in a population of the introduced Asiatic clam, *Corbicula* sp. (Bivalvia: Corbiculidae), in Maryland. *Biological Bulletin* 168: 50–60.

Kennedy, V.S., S.C. Fuller, and R.A. Lutz, 1991. Shell and hinge development of young *Corbicula fluminea* (Muller) (Bivalvia: Corbiculoidea). *American Malacological Bulletin* 8: 107–111.

Kennedy, W.J., J.D. Taylor, and A. Hall, 1969. Environmental and biological controls on bivalve shell mineralogy. *Biological Reviews of the Cambridge Philosophical Society* 44: 499–530.

Kerans, B.L., 2003. Models of interactions of two species: Lessons for invasion of *Potamopyrgus antipodarum*. Paper presented at the *Third Annual New Zealand Mud Snail in the Western USA Conference* (not seen, cited in NZMS Management and Control Plan Working Group, 2007).

Kerans, B.L. and C. Cada, 2002. Foraging behavior and interference competition. Final Report: *The invasiveness of an exotic snail in the Greater Yellowstone Ecosystem and Energy Flow under Ambient Conditions*, B.L. Kerans, M. Dybdahl, and R. Hall, eds (not seen, cited in NZMS Management and Control Plan Working Group, 2007).

Kerans, B.L., M.F. Dybdahl, M.M. Gangloff, and J.E. Jannot, 2005. *Potamoprygus antipodarum*: Distribution, abundance, and effects on native macroinvertebrates in the Greater Yellowstone Ecosystem. *Journal of the North American Benthological Society* 24: 123–138.

Kerney, M, 1999. *Atlas of the Land and Freshwater Molluscs of Britain and Ireland*. Colchester: Harley Books.

Kew, K.W., 1893. *The Dispersal of Shells.* Kegan Paul French, Trubner and Co., Ltd., London. 291 p.

Keyes, D., and J. Moore, 2009. "The *Hamilton & Scourge* Shipwreck Site Condition Survey 2008: Rationale, organization and objectives." In: *ACUA Underwater Archaeology Proceedings 2009,* Laanela, E. and J. Moore, eds. Advisory Council on Underwater Archaeology, manuscript in preparation.

Khalanski, M., 1993. Testing of five methods for the control of zebra mussels in cooling circuits of power plants located on the Mosselle River. In: *Proceedings of the Third International Zebra Mussel Conference,* Toronto, Canada.

Khalanski, M. and F. Bordet, 1979. Effects of chlorination on marine mussels. In: *Water Chlorination: Environmental Impact and Health Effects. Proceedings of the Third Conference on Water Chlorination,* Colorado Springs, USA, October 28–November 2.

Kharchenko, T.A., 1995. *Dreissena*: Range, ecology, fouling impacts. *Hydrobiological Journal* 31: 3–20 (in Russian).

Khin, O., O.O. Thein, M. Thwin, and N.N. Yin, 2006. Efficacy of selected insecticides against the golden apple snail, *Pomacea canaliculata* (Lamarck), in rice in Myanmar. In: *Global Advances in Ecology and Management of Golden Apple Snails*, R.C. Joshi, and L.S. Sebastian, eds, pp. 315–318. Nueva Ecija: Philippine Rice Research Institute.

Kijviriya, V., E.S. Upatham, V. Viyanant, and D.S. Woodruff, 1991. Genetic studies of Asiatic clams, *Corbicula*, in Thailand: Allozymes of 21 nominal species are identical. *American Malacological Bulletin* 8: 97–106.

Kilgour, B.W. and G.L. Mackie, 1993. Colonization of different construction materials by the zebra mussel, *Dreissena polymorpha* (Bivalvia: Dreissenidae). In: *Zebra Mussels: Biology, Impact and Control*, T. Nalepa and D.W. Schloesser, eds, pp. 167–174. Boca Raton: CRC Publications.

Kilgour, B.W. and R. Kepple, 1993. Effects of salinity on the survival of zebra mussel veliger larvae. In: *Third International Zebra Mussel Conference, Agenda and Abstracts,* Toronto, Canada, February.

Kilgour, B.W., M. Baker, and G.L. Mackie, 1993. Effects of salinity on growth and condition of adult zebra mussels (*Dreissena polymorpha*). *Estuaries* 17: 385–393.

Kim, I., 1996. Occurrence of the planktonic crustaceans in relation to pH in the rice fields of eastern Korea. *Korean Journal of Limnology* 29: 233–240.

Kimura, T. and M. Tabe, 1997. Large genetic differentiation of the mussels *Limnoperna fortunei fortunei* (Dunker) and *Limnoperna fortunei kikuchii* Habe (Bivalvia: Mytilidae). *Venus* (Japanese Journal of Malacology) 56(1): 27–34.

King, P.B. and H.M. Beikman, 1974. *Geologic Map of the United States* (*Exclusive of Alaska and Hawaii*). Reston, VA: U.S. Geological Survey, three sheets, scale 1:2,500,000.

Kinzelbach, R., 1992. The main features of the phylogeny and dispersal of the zebra mussel *Dreissena polymopha. Limno-lology Aktuell* 4: 5–17.

Kipp, R.M. and A. Benson, 2008. *Bithynia tentaculata.* USGS Nonindigenous Aquatic Species Database, Gainesville, FL. Available at: http://nas.er.usgs.gov/queries/FactSheet.asp?speciesID=987. Revision Date: 2/28/2007. Accessed December 12, 2008.

Kirpichenko, M.Y., 1963. Peculiarities of dispersal of *Dreissena* in conditions of regulation of flow of the Volga River. *Proceedings of IBIW AS USSR* 10: 153–158 (in Russian).

Kitchens, W.M., R.E. Bennetts, and D.L. DeAngelis, 2002. Linkages between the snail kite population and the wetland dynamics in a highly fragmented South Florida Hydroscape. In: *The Everglades, Florida Bay, and Coral Reefs of the Florida Keys*, K.G. Porter, ed., pp. 183–203. Boca Raton: CRC Press.

Kiyota, H. and K. Sogawa, 1996. Ecology and management of the apple snail in Kyushu, Japan. In: *Proceedings of the International Workshop on the Pest Management Strategies in Asian Monsoon Agroecosystems* (*Kumamoto 1995*), pp. 187–195.

Kjeldsen, K., 1996. Regulation of algal biomass in a small lowland stream: Field experiments on the role of invertebrate grazing, phosphorus and irradiance. *Freshwater Biology* 36: 535–546.

Klerks, P.L., P.C. Fraleigh, and R.C. Stevenson, 1993. Controlling zebra mussel (*Dreissena polymorpha*) veligers with three oxidizing chemicals: Chlorine, permanganate, and peroxide + iron. In: *Zebra Mussels Biology, Impacts, and Control*, T.F. Nalepa, and D.W. Schloesser, eds, pp. 621–642. Boca Raton: Lewis Publishers.

Klerks, P.L., P.C. Fraleigh, and J.E. Lawniczak, 1996. Effects of zebra mussel (*Dreissena polymorpha*) on seston levels and sediment deposition in western Lake Erie. *Canadian Technical Report of Fisheries and Aquatic Sciences* 53: 2284–2291.

Kluza, D.A. and M. McNyset, 2005. Ecological niche modeling of aquatic invasive species. *Aquatic Invaders* 16: 1–7.

Knoll, L.B., O. Sarnelle, S.K. Hamilton, C.E.H. Kissman, A.E. Wilson, J.B. Rose, and M.R. Morgan, 2008. Invasive zebra mussels (*Dreissena polymorpha*) increase cyanobacterial toxin concentrations in low-nutrient lakes. *Canadian Journal of Fisheries and Aquatic Sciences* 65: 448–455.

Knox, W.E., P.K. Stumpf, D.E. Green, and V.H. Auerbach, 1948. The inhibition of sulfhydryl enzymes as the basis of the bactericidal action of chlorine. *Journal of Bacteriology* 55: 541–548.

Kobayashi, A., Y. Tanaka, and M. Shimada, 2003. Genetic variation of sex allocation in the parasitoid wasp *Heterospilus prosopidis*. *Evolution* 57: 2659–2664.

Kobayashi, W., R.H. Cowie, and N. Glover, 1993. Escargot has got to go. *The Taro Tattler* 5: 2.

Koch, L.M., 1989. *Mytilopsis leucophaeta* (Conrad, 1831) from the upper Mississippi River (Bivalvia: Dreissenidae). *Malacology Data Net* 2: 153–154.

Kok, L.T., T.J. McAvoy, R.A. Malecki, S.D. Hight, J.J Drea, and J.R. Coulson, 1992. Host specificity tests of *Galerucella calmariensis* (L.) and *G. pusilla* (Duft.) (Coleoptera: Chrysomelidae), potential biological control agents of purple loosestrife, *Lythrum salicaria* L. (Lythraceae). *Biological Control* 2: 282–290.

Komaru, A. and K. Konishi, 1996. Ultrastructure of biflagellate spermatozoa in the freshwater clam, *Corbicula leana* (Prime). *Invertebrate Reproduction and Development* 29: 193–197.

Komaru, A. and K. Konishi, 1999. Non-reductional spermatozoa in three shell color types of the freshwater clam *Corbicula fluminea* in Taiwan. *Zoological Science* 16: 105–108.

Komaru, A., K. Konishi, I. Nakayama, T. Kobayashi, H. Sakai, and K. Kawamura, 1997. Hermaphroditic freshwater clams in the genus *Corbicula* produce non-reductional spermatozoa with somatic DNA content. *Biological Bulletin* 193: 320–323.

Komaru, A., K. Konishi, K. Kawamura, and H. Sakai, 1998. Morphological remarks on a *Corbicula* species collected in Saga Prefecture, Japan. *Bulletin of the National Research Institute of Aquaculture* 27: 37–41.

Komaru, A., K. Ookubo, and M. Kiyomoto, 2000. All meiotic chromosomes and both centrosomes at spindle pole in the zygotes discarded as two polar bodies in clam *Corbicula leana*: Unusual polar body formation observed by antitubulin immunofluorescence. *Development Genes Evolution* 210: 263–269.

Kononova, M.M., 1966. *Soil Organic Matter*. Oxford: Pergamon Press Ltd.

Koopmans, R. and R.W. Hughes, 1993. Mechanical filtration as a control measure for zebra mussels, *Third International Zebra Mussel Conference, Agenda and Abstracts*, Toronto, Canada, February.

Kornobis, S., 1977. Ecology of *Dreissena polymoprha* (Pall.) (Dreissenidae, Bivalvia) in lakes received heated water discharges. *Polish Archives of Hydrobiology* 24: 531–545.

Kornuishin A.V. and M. Glaubrecht, 2003. Novel reproductive modes in freshwater clams: Brooding and larval morphology in Southeast Asian taxa of *Corbicula* (Mollusca, Bivalvia, Corbiculidae). *Acta Zoologica* 84: 293–315.

Kornuishin, A.V., 2006. Non-unionid freshwater bivalves. In: *The Mollusks. A Guide to Their Study, Collection and Preservation*, C.F. Sturm, T.A. Pearce, and A. Valdes, eds, pp. 327–338. American Malacological Society, Sheridan Press, Hanover, PA.

Korotneva, N.V. and I.N. Dregol'skaya, 1992. Effect of the elevated temperature in the habitat of fresh water mollusk *Bithynia tentaculata* L. on its oogenesis. *Tsitologiya* 34(2): 30–36.

Kotrla, M.B., 1975. New geographical location for *Philophthalmus* sp. In: *Thiarid Snails and Waterfowl in Texas. Bulletin of the American Malacological Union, Inc.*, pp. 41–42.

Koutnik, M. and D.K. Padilla, 1994. Predicting the spatial distribution of *Dreissena polymorpha* (zebra mussel) among inland lakes of Wisconsin: Modeling with a GIS. *Canadian Journal of Fisheries and Aquatic Sciences* 51: 1189–1196.

Kovalak, W., 1992. Control of zebra mussel fouling by coatings. *Journal of Shellfish Research* 11: 237.

Kovalak, W.P., G.D. Longton, and R.D. Smithee, 1993. Infestation of power plant systems by the zebra mussel (*Dreissena polymorpha*). In: *Zebra Mussels Biology, Impacts, and Control*, T.F. Nalepa and D.W. Schloesser, eds, pp. 359–380. Boca Raton: Lewis Publishers.

Kowalewski, J. and P. Patrick, 1993. Acoustics as a possible mitigation strategy against zebra mussel settling. In: *Third International Zebra Mussel Conference, Agenda and Abstracts,* Toronto, Canada, February.

Kozlowski, S., C. Page, and J. Whetstone, 2002. *Zebra Mussels in South Carolina: The Potential Risk of Infestation*. South Carolina Department of Natural Resources, South Carolina Sea Grant Consortium and Clemson University.

Kraemer, L.R., 1978. *Corbicula fluminea* (Bivalvia: Sphaeriacea): The functional morphology of its hermaphroditism. *Bulletin American Malacological Union* 1978: 40–49.

Kraemer, L.R., 1979. Juvenile *Corbicula*: Their distribution in Arkansas River benthos. In: *Proceedings of the First International Corbicula Symposium*, J.C. Britton, ed., pp. 89–97. Fort Worth, TX: Texas Christian University Research Foundation.

Kraemer, L.R. and D.S. Lott, 1977. Microscopic anatomy of the visceral mass of *Corbicula* (Bivalvia: Sphaeriacea). *Bulletin American Malacological Union* 1977: 48–55.

Kraemer, L.R. and M.L. Galloway, 1986. Larval development of *Corbicula fluminea* (Muller) (bivalvia: corbiculacea): An appraisal of its heterochrony. *American Malacololgical Bulletin* 4: 61–79.

Kraft, C., 1994. *Zebra Mussel Update #21*. University of Wisconsin, Wisconsin Sea Grant Institute, Madison.

Kraft, C.E. and L.E. Johnson, 2000. Regional differences in rates and patterns of North American inland lake invasions by zebra mussels (*Dreissena polymorpha*). *Canadian Journal of Fisheries and Aquatic Sciences* 57: 993–1001.

Kraft, C.E., P.J. Sullivan, Y.A. Karatayev, L.E. Burlakova, J.C. Nekola, L.E. Johnson, and D.K. Padilla, 2002. Landscape patterns of an aquatic invader: Assessing dispersal extent from spatial distributions. *Ecological Applications*, 12: 749–759.

Kreis, Jr. R.G., M.D. Mullin, R. Rossmann, and L.L. Wallace, 1991. organic contaminants and heavy metal concentration in zebra mussel tissue from western Lake Erie. Abstract and Paper Presented at *the Second International Conference,* Rochester, USA.

Kreiser, B.R. and J.B. Mitton, 1995. The evolution of cold tolerance in *Corbicula fluminea* (Bivalvia: Corbiculidae). *The Nautilus* 109: 111–112.

Krieger, K.A., 1985. Snail distribution in Lake Erie, USA, Canada; the influence of anoxia in the southern central basin nearshore zone. *Ohio Journal of Science* 85: 230–244.

Kuijper, W., 2000. De weekdieren van de Nederlandse brakwatergebieden (Mollusca). *Ned. Faun. Meded* 12: 41–120 (not seen, cited by Wolff, 2005).

Kushlan, J.A., 1975. Population changes of the applesnail, *Pomacea paludosa*, in the Southern Everglades. *The Nautilus* 89: 21–23.

Kushlan, J.A., 1978. Predation on applesnail eggs (*Pomacea*). *The Nautilus* 92: 57–58.

Kushlan, J.A. and M.S. Kushlan, 1975. Food of the white ibis in Southern Florida. *Florida Field Naturalist* 3: 31–38.

Kusserow, R., U. Schröter-Bobsin, and J. Mählmann, 2008. Zebra-Mussel-Biofilters in waste water treatment plants—how efficient is their filtration under difficult conditions. *Periphyton and Fouling: Theory and Practice*. Book of Abstracts, St. Petersburg.

Lachner, E.A., C.R. Robins, and W.R. Courtenay, Jr., 1970. Exotic fishes and other aquatic organisms introduced into North America. *Smithsonian Contributions to Zoology* 59: 1–29.

Laine, A.O., J. Mattila, and A. Lehikoinen, 2006. First record of the brackish water dreissenid bivalve *Mytilopsis leucophaeata* in the northern Baltic Sea. *Aquatic Invasions* 1: 38–41.

Lane, D.J.W., J.A. Nott, and D.J. Crisp, 1982. Enlarged stem glands in the foot of the post-larval mussel, *Mylilus edulis*: Adaptation for bysso-pelagic migration. *Journal of the Marine Biology Association* (U.K.) 62: 809–818.

Lane, D.J.W., A.R. Beaumont, and J.R. Hunter, 1985. Byssus drifting and the drifting threads of young post-larval mussel *mylilus edulis*. *Marine Biology* 84: 301–308.

Lange, C.L., G.A. Smythe, J.F. Doyle, and P.M. Sawyko, 1993. Application of low voltage electric fields to deter attachment of zebra mussels to structures. Phase II. In: *Third International Zebra Mussel Conference, Agenda and Abstracts*, Toronto, Canada, February.

Larsen, J.B., M.E. Frischer, L.J. Rasmussen, and B.W. Hansen, 2005. Single step nested multiplex PCR to differentiate between different bivalve larvae. *Marine Biology* 146: 1119–1129.

Larson, G.L., F.E. Hutchins, and L.P. Lamperti, 1977. Laboratory determinations of acute and sublethal toxicities of inorganic chloramines to early life stages of coho salmon. *Transactions of the American Fisheries Association* 106: 268–277.

Lassen, H.H., 1978. The migration potential of freshwater snails exemplified by the dispersal of *Potamopyrgus jenkinsi*. *Natura Jutlandica* 20: 237–241.

Lauritsen, D.D., 1986. Filter-feeding in *Corbicula fluminea* and its effects on seston removal. *Journal of the North American Benthological Society* 5: 165–172.

Lavrentyev, P.J., W.S. Gardner, J.F. Cavaletto, and J.R. Beaver, 1995. Effects of the zebra mussel (*Dreissena polymorpha* Pallas) on protozoa and phytoplankton from Saginaw Bay, Lake Huron. *Journal of Great Lakes Research* 21: 545–557.

LePage, W.L., 1993. The impact of *Dreissena polymorpha* on waterworks operations at monroe, michigan: A case history. In: *Zebra Mussels Biology, Impacts, and Control*, T.F. Nalepa, and D.W. Schloesser, eds, pp. 333–354. Boca Raton: Lewis Publishers.

Leach, J.H. 1993. Impacts of the zebra mussel (*Dreissena polymorpha*) on water quality and fish spawning reefs in western Lake Erie. In: *Zebra Mussels: Biology, Impacts, and Control*, T.F. Nalepa and D.W. Schloesser, eds, pp. 381–397. Boca Raton: Lewis Publishers.

Lee, C.E. and M.A. Bell, 1999. Causes and consequences of recent freshwater invasions by saltwater animals. *Tree* 14: 284–288.

Lee, T., S. Siripattrawan, C.F. Ituarte, and D. O'Foighil, 2005. Invasion of the clonal clams: Corbicula lineages in the new world. *American Malacological Bulletin* 20: 113–122.

Lee, V., 1973. Some common snails of vegetable fields in Hong Kong. *Agriculture Hong Kong* 1: 123–129.

Leff, L.G., J.L. Burch, and J.V. McArthur, 1990. Spatial distribution, seston removal, and potential competitive interactions of the bivalves *Corbicula fluminea* and *Elliptio complanata*, in a coastal plain stream. *Freshwater Biology* 24: 409–416.

Leitch, E.G., 1993. Evaluation of coatings for zebra mussels. In: *Third International Zebra Mussel Conference, Agenda and Abstracts,* Toronto, Canada, February.

Leppäkoski, E. and S. Olenin, 2000. Non-native species and rates of spread: Lessons from the brackish Baltic Sea. *Biological Invasions* 2: 151–163.

Leppäkoski, E., S. Gollasch, and S. Olenin, 2002. *Invasive Aquatic Species of Europe: Distribution, Impacts, and Management.* Springer.

Leung, B., J.M. Drake, and D.M. Lodge, 2004. Predicting invasions: Propagule pressure and the gravity of Allee effects. *Ecology* 85: 1651–1660.

Leung, B., J.M. Bossenbroek, and D.M. Lodge, 2006. Boats, pathways, and aquatic biological invasions: Estimating dispersal potential with gravity models. *Biological Invasions* 8: 241–254.

Levin, P., and O.N.H. Kalo, 2006. Statewide strategic control plan for apple snail (*Pomacea canaliculata*) in Hawaii. Final Report submitted by The Hawaii Land Restoration Institute, Wailaku.

Levri, E.P., A.A. Kelly, and E. Love, 2007. The invasive New Zealand mud snail (*Potamopyrgus antipodarum*) in Lake Erie. *Journal of Great Lakes Research* 33: 1–6.

Lewandowski. K., 1982. The role of early developmental stages in the dynamics of *Dreissena polymorpha* (Pall.) (Bivalvia) populations in lakes; II. Settling of larvae and the dynamics of number of settled individuals *Polish Journal of Ecology* 30: 223–286.

Lewandowski, K. and J. Ejsmont-Karabin, 1983. Ecology of planktonic larvae of *Dreissena polymorpha* (pall.) in lakes of different degrees of heating. *Polish Archives of Hydrobiology* 30: 89–102.

Lewis, D. and G.E. Whitby, 1993. Use of ultraviolet radiation for zebra mussel control. In: *Third International Zebra Mussel Conference, Agenda and Abstracts,* Toronto, Canada, February.

Lewis, D. and R.L. Pawson, 1993. Use of cathodic protection as a means of control and removal of zebra mussels. In: *Third International Zebra Mussel Conference, Agenda and Abstracts,* Toronto, Canada, February.

Lewis, D., J.E. Vanbenschoten, and J.N. Jensen, 1993. A study to determine effective ozone dose at various temperatures for inactivation of adult zebra mussels. Unpublished Report for Ontario Hydro.

Lewis, D.P., G.E. Whitby, and G. McKenna, 1996. Evaluation of low and medium UV for zebra mussel control. Presentation at the *Sixth International Zebra Mussel and Other Aquatic Nuisance Species Conference,* Dearborn, USA, March.

Lewis, D.P., J. Piontkowski, E. Neuhauser, and J. Knowlton, 1996. Chronic exposure of adult and larval zebra mussels at low level potassium concentrations; laboratory studies. In: *Proceedings of the Sixth International Conference on Zebra Mussels and Other Aquatic Nuisance Species,* Dearborn, USA, March 5–7.

Lewis, D.P. and G.E. Whitby, 1996. Control of zebra mussels with medium and low pressure ultraviolet radiation: Optimization study. *Aquatic Sciences,* Project M9295.

Lewis, D.P. and W. Cairns, 1998. Ultraviolet control for zebra mussels—the evolution of a non-chemical alternative. Presentation at the *8th International Zebra Mussel and Aquatic Nuisance Species Conference,* Sacramento, USA, March 16–19.

Lewis, E.L., 1980. The Practical Salinity Scale 1978 and its antecedents. *IEEE Journal of Oceanic Engineering* OE-5: 3–8.

Limanova, N.A., 1964. Dreissena: Bibliography. *Biology of Dreissena and its control. Proceedings of IBIW of AS USSR* 7: 83–136 (in Russian).

Limanova, N.A., 1978. Dreissena: *Bibliographica Guide.* Moskow (in Russian).

Lindahl, O., 2008. Mussel farming for coastal water quality remediation. *Periphyton and Fouling: Theory and Practice.* Book of Abstracts, St. Petersburg.

Litsinger, J.A. and D.B. Estano, 1993. Management of the golden apple snail *Pomacea canaliculata* (Lamarck) in rice. *Crop Protection* 12: 363–370.

Lively, C.M. and M.F. Dybdahl, 2000. Parasite adaptation to locally common host genotypes. *Nature* 405: 679–681.

Lively, C.M. and J. Jokela, 2002. Temporal and spatial distribution of parasites and sex in a freshwater snail. *Evolutionary Ecology Research* 4: 219–226.

Livi, S., C. Cordisco, C. Damiani, and M. Romanelli, 2006. Identification of bivalve species at an early developmental stage through PCR-SSCP and sequence analysis of partial 18S rDNA. *Mar. Biol.* 149: 1149–1161.

Livshits, G. and L. Fishelson, 1983. Biology and reproduction of the freshwater snail *Melanoides tuberculata* (Gastropoda: Prosobranchia) in Israel. *Israel Journal of Zoology* 32: 21–35.
Livshits, G., L. Fishelson, and G.S. Wise, 1984. Genetic similarity and diversity of parthenogenic and bisexual populations of the freshwater snail *Melanoides tuberculata* (Gastropoda: Prosobranchia). *Biological Journal of the Linnean Society* 23:41–54.
Logvinenko, B.M., 1965. About changes in fauna of Caspian mollusks of genus *Dreissena* abter invasion of *Mytilaster lineatus* (Gmel.). *Biological Sciences* 4: 1–19 (in Russian).
Logvinenko, B.M. and Y.A. Starobogatov, 1968. Phyllum Mollusca. In: *Atlas of the Invertebrates of the Caspian Sea.* Moscow, 308–413 (in Russian).
Lohner, R.N., V. Sigler, C.M. Mayer, and C. Balogh, 2007. A comparison of the benthic bacterial communities within and surrounding *Dreissena* Clusters in Lakes. *Microbial Ecology* 54: 469–477.
Lorenzen, M.W., 1980. Use of chlorophyll–Secchi disk relationships. *Limnology and Oceanography* 25: 371–372.
Lovell, S.J., F. Stone, and L. Fernandez, 2005. The economic impacts of aquatic invasive species: A review of the literature. U.S. EPA, National Center for Environmental Economics, Working Paper No. 05-02.
Lowe, R.L. and R.W. Pillsbury, 1995. Shifts in benthic algal community structure and function following the appearance of zebra mussels (*Dreissena polymorpha*) in Saginaw Bay, Lake Huron. *Journal of Great Lakes Research* 21: 558–566.
Lozano, S.J., J.V. Scharold, and T.F. Nalepa, 2001. Recent declines in benthic macroinvertebrate densities in Lake Ontario. *Canadian Journal of Fisheries and Aquatic Sciences* 58: 518–529.
Lubianov, I.P., 1972. *Biological background for protection of hydrotechnical constructions from faunal of fouling biocoenosis in Dneprodserzhinskoe and Zaporozhskoe reservoirs.* Abstract of PhD dissertation, Dnepropetrovsk (in Russian).
Lubianov, I.P., Y.M. Norokha, M.M. Bogolubova, and A.K. Dyga, 1965. Questions of technical hydrobiology and approaches to their solution in relation to defence of water supply systems of power plants and industrial enterprises from biological fouling. *Ist Congress of Hydrobiological Society of USSR.* Book of Abstracts, p. 257 (in Russian).
Lubianov, I.P., Y.M. Norokha, A.K. Dyga, and M.M. Bogolubova, 1977. Biological fouling in circulation pipelines of a fossil fuel plant and remedies of control. Control of pollution of turbine condenser and other tracts of technical service water systems of fossil fueled power plants. Moskow, pp. 211–223 (in Russian).
Lucy, F., 2005. *The dynamics of zebra mussel (Dreissena polymorpha) populations in Lough Key, Co. Roscommon, 1998–2003.* PhD dissertation, Institute of Technology, Sligo.
Lucy, F., 2006. Early life stages of *Dreissena polymorpha* (zebra mussel): The importance of long-term datasets in invasion ecology. *Aquatic Invasions* 1: 171–182.
Lucy, F. and P. Timpson, 1998. Zebra mussel (*Dreissena polymorpha*) monitoring programme in Lough Allen. A report for Enterprise Ireland.
Lucy, F., M. Sullivan, and D. Minchin, 2005. Nutrient levels and the zebra mussel population in Lough Key. ERTDI Report Series No. 34. EPA, Wexford. Available at: www.epa.ie/EnvironmentResearch/ReportsOutput/FileUpload,8806,en.pdf. Accessed April 14, 2009.
Lucy, F.E., D. Minchin, and R. Boelens, 2008. From lakes to rivers: Downstream larval distribution of *Dreissena polymorpha* in Irish river basins. *Aquatic Invasions* 3: 297–304.
Lucy, F.E., T.K. Graczyk, D. Minchin, L. Tamang, and A. Miraflor, 2008. Biomonitoring of surface and coastal water for *Cryptosporidium*, *Giardia* and human virulent microsporidia using molluscan shellfish. *Parasitology Research* 103: 1369–1375.
Luschützky, E., 2005. Bioaccumulation of heavy metals in the zebra mussel (*Dreissena polymorpha*) in the rivers Danube and Drau and its role as a bioindicator organism. *Umweltwissenschaften und Schadstoff-Forschung* 17: 68–76.

Lvova, A.A. and G.E. Makarova, 1994. Gametogenesis and reproductive cycle. In: Dreissena *polymorpha (Pall.) (Bivalvia, Dreissenidae): Taxonomy, Ecology and Practical Meaning*, Y.I. Starobogatov, ed., pp. 138–149. Moscow: Nauka Press (in Russian).
Lyachnovich, V.Y., A.Y. Karatayev, S.M. Lyakhov, N.I. Andreev, S.A. Afanasiev, A.K. Dyga, V.P. Zakusky, V.I. Zolotareva, A.A. Lvova, M.Y. Nekrasova, V.F. Osadchikh, Y.V. Pligin, A.A. Protasov, and G.M. Tishchikov, 1994. Conditions in habitats. In: *Dreissena polymorpha (Pall.) (Bivalvia, Dreissenidae): Taxonomy, Ecology and Practical Meaning*, Y.I. Starobogatov, ed., pp. 109–119. Moscow: Nauka Press (in Russian).
Lyashenko, A.V. and T.A. Karchenko, 1989. Annual dynamics of energy metabolism in *Dreissena. Hydrobiological Journal* 25: 34–41.
MacIsaac, H.J., 1994a. Size-selective predation on zebra mussels (*Dreissena polymorpha*) by crayfish (*Orconectes propinquus*). *Journal of the North American Benthological Society* 13: 206–216.
MacIsaac, H.J., 1994b. Comparative growth and survival of *Dreissena polymorpha* and *Dreissena bugensis*, exotic molluscs introduced to the Great Lakes. *Journal of Great Lakes Research* 20: 783–790.
MacIsaac, H.J., W.G. Sprules, and J.H. Leach, 1991. Ingestion of small-bodied zooplankton by zebra mussels (*Dreissena polymorpha*): Can cannibalism on larvae influence population dynamics? *Canadian Journal of Fisheries and Aquatic Sciences* 48: 2051–2060.
MacIsaac, H.J., W.G. Sprules, O.E. Johannsson, and J.H. Leach, 1992. Filtering impacts of larval and sessile zebra mussels (*Dreissena polymorpha*) in western Lake Erie. *Oecologia* 92: 30–39.
MacIsaac, H.J., J.V.M. Borbely, J.R. Muirhead, and P.A. Graniero, 2004. Backcasting and forecasting biological invasions of inland lakes. *Ecological Applications* 14: 773–783.
Mackie, G.L., 1978a. Are sphaeriid clams ovoviviparous or viviparous? *The Nautilus* 92: 135–146.
Mackie, G.L., 1978b. Shell structure in freshwater Sphaeriacea (Bivalvia: Heterodonta). *Canadian Journal of Zoology* 56: 1–6.
Mackie, G.L., 1979. Dispersal mechanisms in Sphaeriidae (Mollusca: Bivalvia). *Bulletin of the American Malacological Union* 1979: 17–21.
Mackie, G.L., 1984. Bivalves. In: *The Mollusca*, K.M. Wilbur, ed., Chapter 5, pp. 351–418. *Reproduction*, A.S. Tompa, N.H. Verdonk, and J.A.M. van de Biggelaar. New York: Academic Press.
Mackie, G.L., 1984. Reproduction in Bivalvia. In: *The Mollusca*. Vol. VII, *Reproduction*, A.S. Tompa, N.H. Verdonk, and J.A.M. van de Biggelaar, Chapter 5, pp. 351–418. Academic Press, New York.
Mackie, G.L., 1990. Protocol for monitoring and control of zebra mussels. Unpublished manual. Available from Mackie & Associates Water Systems Analysts Inc., 381 Elmira Rd., Guelph, Ontario, Canada.
Mackie, G.L., 1990. Zebra mussels: The Great Lakes under siege, again. *Rotunda* 23: 19–26.
Mackie, G.L., 1991. Monitoring, sampling, and identification techniques for adult and veliger zebra mussels. In: *Resources, Engineering and Operations for the New Decade. Annual Conference Proceedings*, American Water Works Association, Philadelphia, USA.
Mackie, G.L., 1991a. Biology of the exotic zebra mussel, *Dreissena polymorpha*, relative to native bivalves and its potential impact in Lake St. Clair. In: *Environmental Assessment and Habitat Evaluation of the Connecting Channels*, M. Munawar and T. Edsall, eds. *Hydrobiologia* 219: 251–268.
Mackie, G.L., 1991b. Monitoring, sampling and identification techniques for adult and veliger zebra mussels. In *Annual Conference Proceedings, American Water Works Association, Resources, Engineering and Operations for the New Decade*, Philadelphia, USA, June 23–27.

Mackie, G.L., 1993. Early biological and life history attributes of the zebra mussel, *Dreissena polymorpha* (Bivalvia: Dreissenidae) and impacts on native bivalves in Lake St. Clair. In: *Zebra Mussels Biology, Impacts, and Control*, T.F. Nalepa and D.W. Schloesser, eds, pp. 153–165. Boca Raton: Lewis Publishers.
Mackie, G.L., 1995. Adaptations of exotic Mollusca for life in regulated rivers and their potential impacts. Keynote Speaker. In: *Proceedings of the Fifth Annual Symposium on Natural History of Lower Tennessee and Cumberland River Valleys,* Murray State University, March 3–4.
Mackie, G.L., 2000a. Chapter 9: Mollusc introductions through aquarium trade. In: *Non-indigenous Freshwater Organisms: Vectors, Biology and Impacts*, R. Claudi and J. Leach, eds, pp. 135–150. Boca Raton: Lewis Publishers.
Mackie, G.L., 2000b. Chapter 15: Ballast water introductions of Mollusca. In: *Non-indigenous Freshwater Organisms: Vectors, Biology and Impacts*, R. Claudi and J. Leach, eds, pp. 255–272. Boca Raton: Lewis Publishers.
Mackie, G.L., 2000c. Introduction of molluscs through the import for live food. In: *Non-indigenous Freshwater Organisms: Vectors, Biology and impacts,* R. Claudi, and J. Leach, eds, Chapter 21, pp. 305–314. Boca Raton: Lewis Publishers.
Mackie, G.L., 2001. *Flora and Fauna of the Speed River Watershed, Including Exotic and Endangered Species*. Water Systems Analysts, Guelph, Ontario.
Mackie, G.L., 2002. Traits of endangered and invading freshwater molluscs in North America. In: *Alien Invaders in Canada's waters, Wetlands, and Forests*, R. Claudi, P. Nantel, and E. Muckle-Jeffs, eds, pp. 187–198. Ottawa: Publications of Canadian Forest Service, Natural Resources Canada.
Mackie, G.L., 2004. *Applied Aquatic Ecosystem Concepts, 2nd ed.* Dubuque: Kendall/Hunt Publishing Company.
Mackie, G.L., 2007. Biology of freshwater corbiculid and sphaeriid clams of North America. *Ohio Biological Survey, Bulletin New Series* 15.
Mackie, G.L. and S.U. Qadri, 1971. A quantitative sampler for phytomacrofauna. *Journal of the Fisheries Research Board of Canada* 28: 1322–1324.
Mackie, G.L., D.S. White, and T.W. Zdeba, 1980. A guide to freshwater mollusks of the Laurentian Great Lakes with special emphasis on the genus Pisidium. Environmental Research Laboratory, Office of Research and Development, U.S. Environmental Protection Agency, Duluth.
Mackie, G.L. and L.A. Flippance, 1983a. Calcium sources for growth of *Musculium securis* (Bivalvia: Pisidiidae). *Canadian Journal of Zoology* 61: 874–878.
Mackie, G.L. and L.A. Flippance, 1983b. Intra- and interspecific variations in calcium content of freshwater Mollusca in relation to calcium content of the water. *Journal of Molluscan Studies* 49: 204–211.
Mackie, G.L. and L.A. Flippance, 1983c. Relationships between buffering capacity of water and the size and calcium content of freshwater mollusks. *Freshwater Invertebrate Biology* 2: 48–55.
Mackie, G.L. and J.M. Topping, 1988. Historical changes in the unionid fauna of the sydenham river watershed and downstream changes in shell morphometrics of three common species. *Canadian Field-Naturalist* 102: 617–626.
Mackie, G.L., W.N. Gibbons, B.W. Muncaster and I.M. Gray, 1989. The zebra mussel, *Dreissena polymorpha*: A synthesis of european experiences and a preview for North America. Report prepared for Water Resources Branch, Great Lakes Sect. Available from Queen's Printer for Ontario, ISBN 0-7729-5647-2.
Mackie, G.L. and B.W. Kilgour 1992a. Effects of salinity on growth and survival of zebra mussels (*Dreissena polymorpha*). A report prepared for ESSERCO.
Mackie, G.L. and B.W. Kilgour, 1992b. Toxicity of sodium metabisulfite and formation of sulfur compounds in soft tissues of adult zebra mussels. Final Report to Tennessee Valley Authority, manuscript in review by *Water Research.*

Mackie G.L. and B.W. Kilgour, 1993. Efficacy and role of alum in removal of zebra mussel veliger larvae from raw water supplies. In: *Third International Zebra Mussel Conference, Agenda and Abstracts,* Toronto, Canada, February.

Mackie, G.L. and C.A. Wright, 1994. Ability of the zebra mussel, *Dreissena polymorpha*, to biodeposit and remove phosphorous and B.O.D. from diluted activated sewage sludge. *Water Research* 28: 1123–1130.

Mackie, G.L. and B.W. Kilgour, 1995. Efficacy and role of alum in removal of zebra mussel veliger larvae from raw water supplies. *Water Research* 29: 731–744.

Mackie, G.L. and D.W. Schloesser, 1996. Comparative biology of zebra mussels in Europe and North America: An overview. *American Zoologist* 36: 244–258.

Mackie, G.L., P. Lower, and C. Cooper, 2000. Plasma pulse technology to control zebra mussel biofouling, *Zebra Mussel Technical Notes Collection* (ERDC TN-ZMR-2-22), U.S. Army Engineer Research and Development Center, Vicksburg. Available at: www.wes.army.mil/el/elpubs/zebtnote.html. Accessed April 8, 2009.

Mackie, G.L. and L.E. Taylor, 2007. *The Cottage Bible*. Erin: The Boston Mills Press.

MacNeill, D.B., 1991a. Identification of juvenile *Dreissena* and *Mytilopsis leucophaeata*. In: *Dreissena polymorpha Information Review*, S.G. Moore, ed., pp. 1–2. New York Zebra Mussel Information Clearinghouse, New York Sea Grant Extension.

MacNeill, D.B., 1991b. Potential overlap of *Dreissena* and *Mytilopsis* in the Hudson River. In: *Dreissena polymorpha Information Review*, S.G. Moore, ed., pp. 2–2. New York Zebra Mussel Information Clearinghouse, New York Sea Grant Extension.

MacNeill, D.B., 1991c. Ecological perspectives on functional morphologies of *Dreissena polymorpha* and *Mytilopsis leucophaeata*. In: *Dreissena polymorpha Information Review,* S.G. Moore, ed., pp. 2–5. New York Zebra Mussel Information Clearinghouse, New York Sea Grant Extension.

Madenjian, C.P., 1995. Removal of algae by the zebra mussel (*Dreissena polymorpha*) population in western Lake Erie: A bioenergetics approach. *Canadian Journal of Fisheries and Aquatic Sciences* 52: 381–390.

Madhu, C., K.J. Rao, K.S. Rao., and K.V.R. Rao, 1982. Comparative evaluation of the toxicity of lindane and sumithion on freshwater snail, *Pila globosa* (Swainson). *Geobios* 9: 181–183.

Madon, S.P., D.W. Schneider, J.A. Stoekel, and R.E. Sparks, 1998. Effects of inorganic sediment and food concentrations on energetic processes of the zebra mussel, *Dreissena polymorpha*: Implications for growth in turbid rivers. *Canadian Journal of Fisheries and Aquatic Sciences* 55: 401–413.

Magara, Y., Y. Matsui, Y. Gotto, and A. Yuasa, 2001. Invasion of the non-indigenous nuisance mussel, *Limnoperna fortunei*, into water supply facilities in Japan. *Journal of Water Supply: Research and Technology—AQUA* 50: 113–124.

Malecki, R.A., B. Blossey, S.D. Hight, D. Schroeder, L.T. Kok, and J.R. Coulson, 1993. Biological control of purple loosestrife: A case for using insects as control agents, after rigorous screening, and for integrating release strategies with research. *Bioscience* 43: 680–686.

Malek, E.A. and T.C. Cheng, 1974. *Medical and Economic Malacology*. London: Academic Press.

Mangin, S., 1998. The 100th meridian initiative: A strategic approach to prevent the westward spread of zebra mussels and other aquatic nuisance species. U.S. Fish and Wildlife.

Manguin, S., R. White, B. Blossey, and S.D. Hight, 1993. Genetics, taxonomy, and ecology of certain species of *Galerucella* (Coleoptera: Chrysomelidae). *Annals of the Entomological Society of America* 86: 397–410.

Mansur, M.C.D., L.M.Z. Richinitti, and C.P. Dos Santos, 1999. *Limnoperna fortunei* (Dunker, 1857) molusco bivalve invasor na bacia do Guaíba, Rio Grande do Sul, Brasil. *Biociências* 7: 147–149.

Mansur, M.C.D., C.P. Dos Santos, G. Darrigran, I. Heydrich, C.T. Callil, and F.R. Cardoso, 2003. Primeiros dados quali-quantitativos do mexilhão-dourado, *Limnoperna fortunei* (Dunker), no Delta do Jacuí, no Lago Guaíba e na Laguna dos Patos, Rio Grande do Sul, Brasil e alguns aspectos de sua invasão no novo ambiente. *Revista Brasileira de Zoologia* 20: 75–84.

Mansur, M.C.D., C.T. Callil, F.R. Cardoso, and J.A.A. Ibarra, 2004. Uma retrospectiva e mapeamento da invasão de espécies de *Corbicula* (Molusca, Bivalvia, Veneroida, Corbiculidae) oriundas do Sudeste Asiático, na América do Sul. In: *Água de lastro e bioinvasão*, J.S.V. Silva and R.C.C.L. Souza, eds. Rio de Janeiro: Interciência.

Marelli, D.C., and S. Gray, 1983. Conchological redescriptions of *Mytilopsis sallei* and *Mytilopsis leucophaeata* of the brackish Western Atlantic (Bivalvia: Dreissenidae). *Veliger* 25: 185–193.

Marelli, D.C. and S. Gray, 1985. Comments on the status of recent members of the genus *Mytilopsis* (Bivalvia: Dreissenidae). *Malacological Reviews* 18: 117–122.

Marfurt, R.K. and R.L. Burks, 2005. Invaders from the south: Applesnail (*Pomacea canaliculata*) ecology and life history. Presentation at the *Freshwater Mollusk Conservation Society, Biennial Symposium,* St. Paul, Minnesota, May 15–18 (not seen, cited by Howells et al., 2006).

Marsden, E., 1991. Status of zebra mussels in Lake Michigan—temporal and spatial distribution of veligers and adults. In: Paper Presented at the *Second International Zebra Mussel Conference,* Rochester, USA, November.

Marsden, J.E., A.P. Spidle, and B. May, 1996. Review of genetic studies of *Dreissena* spp. *American Zoologist* 36: 259–270.

Marsh, C.P., et al., 1999. *Robotic Underwater Corrosion Inspection/Assessment of Sheet Pile Along Two Rivers at Cleveland, Ohio.* Construction Engineering Research Lab (Army) Champain, IL, Report ADA364913, May.

Marshall, W.B., 1933. *Bulimus tentaculata* (L.) (*Bithynia tentaculata*) living in the Potomac River. *The Nautilus* 47: 141–142.

Martel, A. Recruitment rates of drifting post-metamorphic zebra mussels on off-bottom collectors in Lake Erie: A correlation with wind speed and direction. In: *Second International Zebra Mussel Conference, Agenda and Abstracts,* Toronto, Canada, February.

Martel, A., 1995. Demography and growth of the exotic zebra mussel (*Dreissena polymorpha*) in the Rideau River (Ontario). *Canadian Journal of Zoology* 73: 2244–2450.

Martin, I.D., G.L. Mackie, and M. Baker, 1992a. Control of the biofouling bivalve, *Dreissena polymorpha*, with sodium hypochlorite and with two polyquaternary ammonia and benzothiazole compounds. *Archives of Environmental Contamination and Toxicology* 24: 381–388.

Martin, I.D., G.L. Mackie, and M. Baker, 1992b. Acute toxicity and pulse-dosed delayed mortality at 12 and 22°C with sodium hypochlorite, bulab 6002, and hydrogen peroxide on the biofouling mollusk, *Dreissena polymorpha. Archives of Environmental Contamination and Toxicology* 24: 389–398.

Martín, P.R., A.L. Estebenet, and N.J. Cazzaniga, 2001. Factors affecting the distribution of *Pomacea canaliculata* (Gastropoda: Ampullariidae) along its southernmost natural limit. *Malacologia* 43: 13–23.

Martín, P.R. and A.L. Estebenet, 2002. Interpopulation variation in life-history traits of *Pomacea canaliculata* (Gastropoda: Ampullariidae) in southwestern Buenos Aires Province, Argentina. *Malacologia* 44: 153–163.

Mata, A.R. and A.C. da Silva, 2007. Mudanças sazonais na dieta de *leporinus friderici* (Bloch, 1794), na área de influência do reservatório de são simão, bacia do rio paranaíba, brasil. *Relatório Final Bolsa Iniciação Científica* FAPEMIG/CETEC.

Mata, F.R and M.S. Campos, 2004. Preferência e tolerância de *limnoperna fortunei* (dunker, 1857) a variáveis abióticas. *Relatório Final Bolsa Iniciação Científica* CNPQ/CETEC.

Mata, F.R. and M.C.S. Campos, 2005. Limites de tolerância de adultos da espécie *limnoperna fortunei* (Dunker, 1857) a variáveis físico-químicas da. *Relatório Final Bolsa Iniciação Científica* CNPQ/CETEC.

Matthews, R.F., 1998. Extremely low frequency electromagnetism: An effective nonchemical method for control of zebra mussel infestation. *Proceedings of the Eighth International Zebra Mussel and Other Nuisance Species Conference*, Sacramento, USA, March 16–19.

Matthews, M.A. and R.F. McMahon, 1994. The survival of zebra mussels (*Dreissena polymorpha*) and Asian clams (*Corbicula fluminea*) under extreme hypoxia. In: *Proceedings of the Fourth International Zebra Mussel Conference,* Madison, USA, March.

Matthews, M.A. and R.F. McMahon, 1995. Survival of zebra mussels (*Dreissena polymorpha*) and Asian clams (*Corbicula fluminea*) under extreme hypoxia. Technical Report EL-95-3 to US Army Corps of Engineers.

Mattice, J.S., 1970. *Trophic biology of a natural population of Bithynia tentaculata in terms of ecological energetics*. PhD dissertation, Syracuse University, Syracuse.

Mattice, J.S. and L.L. Dye, 1976. Thermal tolerance of Asiatic clam. In: *Thermal Ecology*, G.W. Esch and R.W. McFarlane, eds, pp. 130–135. CONF-750452, U.S. Energy Research and Development, National Technological Information Service, Department of Commerce, Springfield.

Mattison, R.G., T.S. Dunn, R.E.B. Hanna, W.A. Nizami, and Q.M. Ali, 1995. Population dynamics of freshwater gastropods and epidemiology of their helminth infections with emphasis on larval parmphistomes in northern India. *Journal of Helminthology* 69: 125–138.

May, B. and J.E. Marsden, 1992. Genetic identification and implications of another invasive species of Dreissenid mussel in the Great Lakes. *Canadian Journal of Fisheries and Aquatic Sciences* 49: 1501–1506.

Mayer, C.M., A. VanDeValk, J.L. Forney, L.G. Rudstam, and E.L. Mills, 2000. Response of yellow perch (*Perca flavescens*) in Oneida Lake, New York, to the establishment of zebra mussels (*Dreissena polymorpha*). *Canadian Journal of Fisheries and Aquatic Sciences* 57: 742–754.

Mayer, C.M., R.A. Keats, L.G. Rudstam, and E.L. Mills, 2002. Scale-dependent effects of zebra mussels on benthic invertebrates in a large eutrophic lake. *Canadian Journal of Fisheries and Aquatic Sciences* 21: 616–633.

McCarter, N.H., 1986. Food and energy in the diet of brown and rainbow trout from Lake Benmore, New Zealand. *New Zealand Journal of Marine and Freshwater Research* 20: 551–559.

McCarthy, T.K., J. Fitzgerald, and W. O'Connor, 1998. The occurrence of the zebra mussel *Dreissena polymorpha* (Pallas, 1771), an introduced biofouling freshwater bivalve in Ireland. *Irish Naturalists' Journal* 25: 413–415.

McCauley, R.W. and E. Kott, 1993. Lethal effects of hydrogen ion on adult zebra mussels, *Dreissena polymorpha*, in relation to calcium concentration of the surrounding water (abstract). In: *Third International Zebra Mussel Conference, Agenda and Abstracts*, Toronto, Canada, February 23–26.

McClary, A., 1964. Surface inspiration and ciliary feeding in *Pomacea palludosa. Malacologia* 2: 87–104.

McDermott, R.M., 1993. Distribution and ecological impact of quagga mussels in the lower Great Lakes. In: *Proceedings of the Third International Zebra Mussel Conference,* J.L. Tsou and Y.G. Mussalli, eds, pp. 2-1–2-21. EPRI TR-102077, Electric Power Research Institute, Palo Alto, California.

McGoldrick, D.J., J.L. Metcalfe-Smith, M.T. Arts, D.W. Schloesser, T.J. Newton, G.L. Mackie, E.M. Monroe, J. Biberhofer, and K. Johnson, 2009. Characteristics of a refuge for native freshwater mussels (Bivalvia: Unionidae) in Lake St. Clair. *Journal of Great Lakes Research* 35: 137–146.

McKay, S., 1991. Control of zebra mussels by electroshocking. Ontario Hydro Report 91-171-K.

McLeod, M.J., 1986. Electrophoretic variation in North American *Corbicula*. *American Malacological Bulletin Special Edition* 2: 125–132.

McMahon, R.E., T.A. Ussery, and M. Clarke, 1993. Use of emersion as a zebra mussel control method. Technical Report No. EL-93-1, U.S. Army Corps of Engineers, Waterways Experiment Station, Vicksburg.

McMahon, R.F. and A.E. Bogan, 2001. Mollusca: Bivalvia. In: *Ecology and Classification of North American Freshwater Invertebrates*, J.H. Thorp and A.P. Covich, eds, 2nd ed, Chapter 10, pp. 297–330. San Diego: Academic Press.

McMahon, R.F. and C.J. Williams, 1986a. Upper thermal limit and downstream colonization rates in a natural population of *Corbicula fluminea* (Müller) receiving thermal effluents. *American Malacological Bulletin* 2: 231–239.

McMahon, R.F. and C.J. Williams, 1986b. A reassessment of growth rate, life span, life cycles and population dynamics in a natural population and field cages individuals of *Corbicula fluminea* (Müller) (Bivalvia: Corbiculacea). *American Malacological Bulletin* 2: 151–166.

McMahon, R.F. and C.J. Williams, 1994. A unique respiratory adaptation to emersion in the introduced freshwater clam, *Corbicula fluminea* (Müllller) (Lamellibranchia: Corbiculacea). *Physiological Zoology* 57: 274–279.

McMahon, R.F. and J.L. Tsou, 1990. Impact of European zebra mussel infestation to the electric power industry. *Proceedings of the American Power Conference* 52: 988–997.

McMahon, R.F., 1977. Shell size-frequency distributions of *Corbicula manilensis* Philippi from a clam-fouled steam condenser. *The Nautilus* 91: 54–59.

McMahon, R.F., 1979. Response to temperature and hypoxia in the oxygen consumption of the introduced Asiatic freshwater clam, *Corbicula fluminea* (Miiuller). *Comparitive Biochemical and Physiology* 63A: 383–388.

McMahon, R.F., 1982. The occurrence and spread of the introduced Asiatic freshwater clam, *Corbicula fluminea* (Muller), in North America: 1924–1982. *The Nautilus* 96: 134–141.

McMahon, R.F., 1983. Ecology of an invasive pest bivalve: *Corbicula fluminea*. In: *The Molluscs, Vol. 6, Ecology*, W.D. Russell-Hunter, ed., pp. 505–561. Orlando: Academic Press.

McMahon, R.F., 1991. Mollusca: Bivalvia. In: *Ecology and Classification of North American Freshwater Invertebrates*, J.H. Thorp and A.P. Covich, eds, pp. 315–399. New York: Academic Press.

McMahon, R.F., 1992. The zebra mussel—the biological basis for its macrofouling and potential for distribution in North America. In: *Corrosion '92: Proceedings of the 47th NACE Annual Conference*, Reprint No. 342. National Association of Corrosion Engineers, Houston, Texas.

McMahon, R.F., 1996. The physiological ecology of the zebra mussel, *Dreissena polymorpha*, in North America and Europe. *American Zoologist* 36: 339–363.

McMahon, R.F., 1999. Invasive characteristics of the freshwater bivalve, *Corbicula fluminea*. In: *Nonindigenous Freshwater Organisms: Vectors, Biology, and Impacts*, R. Claudi and J.H. Leach, eds, pp. 315–343. Boca Raton: Lewis Publishers.

McMahon, R.F., 2000. Invasive characteristics of the freshwater bivalve *Corbicula fluminea*. In: *Non-indigenous Freshwater Organisms: Vectors, Biology and Impacts*, R. Claudi and J. Leach, eds, pp. 315–343. Boca Raton: Lewis Publishers.

McMahon, R.F., 2002. Evolutionary and physiological adaptations of aquatic invasive animals: Selection versus resistance. *Canadian Journal of Fisheries and Aquatic Sciences* 59: 1235–1244.

McMahon, R.F., 2007. Analysis of the shell length distributions of quagga mussel (*Dreissena rostriformis bugensis*) taken from lakes Mead and Mohave on the Colorado River Drainage during January–March 2007. Report published by Department of Biology, The University of Texas at Arlington, Arlington.

McMahon, R.F., M.A. Matthews, L.R. Shaffer, and P.D. Johnson, 1995. Effects of elevated carbon dioxide concentrations on survivorship in zebra mussels (*Dreissena polymorpha*) and Asian clams (*Corbicula fluminea*). Presentation at the *Fifth International Zebra Mussel and Other Aquatic Nuisance Organisms Conference,* Toronto, Canada, February.

McMahon, R.F., M.A. Matthews, T.H. Ussery, R. Chase, and M. Clarke, 1994. Further studies of heat tolerance in zebra mussels: Effects of temperature acclimation and chronic exposure to lethal temperatures. In: *Proceedings of the Fourth International Zebra Mussel Conference*, Madison, USA.

McMahon, R.F., T.A. Ussery, A.C. Miller, and B.S. Payne, 1993. Thermal tolerance in zebra mussels (*Dreissena polymorpha*) relative to rate of temperature increase and acclimation temperature. In: *Proceedings of the Third International Zebra Mussel Conference*, Toronto, Canada, March.

McMahon, R.F., T.A. Ussery, and A.C. Miller, 1992. Temperature and relative humidity effects on emersion tolerance in zebra mussels *Dreissena polymorpha*: Is dewatering a mitigation option? In: *2nd International Zebra Mussel Conference, Agenda and Abstracts, Toronto, Canada, February.*

McPherson, M.J. and S.G. Moller, 2006. *PCR: The Basics*. Taylor and Francis: New York.

McTighe, T.C., F.J. Ritz, and J.R. Amend, 1993. Responding to the zebra mussel threat a case history. In: *Third International Zebra Mussel Conference, Agenda and Abstracts,* Toronto, Canada, February.

Medhurst, R.B. and D.B. Herbst, 2003. An alternative method for decontamination: Bleach toxicity in New Zealand mudsnails from upper Owens River. In: *NZ Mudsnail in the Western USA: Conference 2003. Minutes of the Third Annual Conference*. August 26 and 27, 2003. Chavez, ed., Montana State University, Bozeman, Montana.

Medina, F.R. and L.S. Rirchie, 1980. Moluscicidal activity of the Puerto Rican weed, *Solanum nodiflorum*, against snail hosts of *Fasciola hepatica*. *Economic Botany* 34: 368–375.

Meisenheimer, J., 1901. Entwicklungsgeschichte von *Dreissena polymorpha* Pall. *Zeitschr. Wissensch. Zool. Bd* 69: 1–137.

Mellina, E. and J.B. Rasmussen, 1994. Patterns in the distribution and abundance of zebra mussel (*Dreissena polymorpha*) in rivers and lakes in relation to substrate and other physiochemical factors. *Canadian Journal of Fisheries and Aquatic Sciences* 51: 1024–1036.

Mellina, E., J.B. Rasmussen, and E.L. Mills, 1995. Impact of zebra mussel (*Dreissena polymorpha*) on phosphorus cycling and chlorophyll in lakes. *Canadian Journal of Fisheries and Aquatic Sciences* 52: 2553–2573.

Menzel, D.B. and M.O. Amdur, 1986. Toxic responses of the respiratory system. In: *Cassaret and Doull's Toxicology: The Basic Science of Poisons*, C.D. Klassen, M.O. Amdu, and J. Doull, eds, pp. 330–358. Toronto: Collier MacMillan.

Metcalf, A.L., 1966. *Corbicula manilensis* in the Mesilla Valley of Texas and New Mexico. *The Nautilus* 80: 16–20.

Metcalf, A.L. and R. Smartt, 1972. Records of introduced mollusks: New Mexico and western Texas. *The Nautilus* 85: 144–145.

Metcalfe-Smith, J.L., S.K. Staton, G.L. Mackie, and E.L. West, 1998. Assessment of the current status of rare species of freshwater mussels in southern Ontario. Environment Canada, National Water Research Institute, Burlington, ON, NWRI Contribution No. 98-019.

Metcalfe-Smith, J.L., S.K. Staton, G.L. Mackie, and I.M. Scott, 1999. Range, population stability and environmental requirements of rare species of freshwater mussels in southern Ontario. Environment Canada, National Water Research Institute, Burlington, ON, NWRI Contribution No. 99-058.

Metcalfe-Smith, J.L., G.L. Mackie, J. Di Maio, and S. Staton, 2000. Changes over time in the diversity and distribution of freshwater mussels (Unionidae) in the Grand River, southwestern Ontario. *Journal of Great Lakes Research* 26: 445–459.

Metcalfe-Smith, J.L., J. Di Maio, S.K. Staton, and S.R. De Solla, 2003. Status of the freshwater mussel communities of the Sydenham River, Ontario, Canada. *American Midland Naturalist* 150: 37–50.

Metcalfe-Smith, J.L., D.J. McGoldrick, M. Williams, D.W. Schloesser, J. Biberhofer, G.L. Mackie, M.T. Arts, D.T. Zanatta, K. Johnson, P. Marangelo, and T.D. Spencer, 2004. Status of a refuge for native freshwater mussels (Unionidae) from the impacts of the exotic zebra mussel (*Dreissena polymorpha*) in the delta area of Lake St. Clair. Environment Canada. National Water Research Institute, Burlington, Ontario. Technical Note No. AEI-TN-04-001.

Michelson, E.H., 1961. On the generic limits in the family Pilidae (Prosobranchia: Mollusca). *Breviora* 133: 1–10.

MIE Consulting Engineers Limited, 1994. Pitting corrosion of steel beneath zebra mussels, underwater investigations. A report prepared for Public Works and Government Services Canada.

Mihuc, T.B., J.M. Battle, J.R. Mihuc, and C.F. Bryan, 1999. Zebra mussel (*Dreissena polymorpha*) seasonal colonization patterns in a sub-tropical floodplain river. *Hydrobiologia* 392: 121–128.

Mikheev, V.P., 1964. Mortality rate of *Dreissena* in anaerobic conditions. In: *Biology and Control of Dreissena,* B.K. Shtegman, ed., pp. 65–68. Moskow, Russia: Academy of Sciences of the U.S.S.R., Institute of Biology of Inland Waters.

Mikheev, V.P., V.F. Dudnikov, and E.P. Shtern, 1969. *Defense of Hydrotechnical Installations from Mussel Fouling*. Moskow: "Energy" publishers (in Russian).

Milke, L.M. and V.S. Kennedy, 2001. *Mud crabs (Xanthidae) in Chesapeake Bay: Claw characteristics and predation on epifaunal bivalves. Invertebrate Biology* 120: 67–77.

Millane, M., 2009. *Impact of the zebra mussel Dreissena polymorpha (Pallas, 1771) on the ecological integrity of Lough Sheelin*, Ireland. PhD dissertation, University College, Dublin.

Miller, A.C., B.S. Payne, and R.F. McMahon, 1992. *The Zebra Mussel: Biology, Ecology, and Recommended Control Strategies*. Zebra Mussel Research Technical Note ZMR-1-01, U.S. Army Corps of Engineers, Waterways Experiment Station, Vicksburg.

Miller, A.C., F. Neilson, and K. Gunnison, 1993. *Susceptibility of Navigation Locks to Zebra Mussel Infestations*, Zebra Mussel Research Technical Note ZMR-4-03, U.S. Army Corps of Engineers, Waterways Experiment Station, Vicksburg.

Miller, A.C., J. Lei, and J. Tom, 1994. Shell strength of the nonindigenous zebra mussel *Dreissena polymorpha* (Pallas) in comparison to two other freshwater bivalve species. *Veliger* 37: 319–321.

Miller, A.C. and P. Lowry, 2000. Controlling zebra mussels, quagga mussels, and biofilm growth with the plasma sparker. *Zebra Mussel Technical Notes Collection* (ERDC TN-ZMR-2-23), U.S. Army Engineer Research and Development Center, Vicksburg, MS. Available at: www.wes.army.mil/el/elpubs/zebtnote.html. Accessed March 16, 2009.

Miller, G.H. and F.A. McClure, 1931. The fresheater clam industry of the Peral River. *Lingnan Science Journal* 10: 307–322.

Mills, E.L., J.H. Leach, J.T. Carlton, and C.L. Secor, 1993a. Exotic species in the Great Lakes: A history of biotic crises and anthropogenic introductions. *Journal of Great Lakes Research* 19: 1–54.

Mills, E.L., R.M. Dermott, E.F. Roseman, D. Dustin, E. Mellina. D.B. Conn, and A.P. Spidle, 1993b. Colonization, ecology, and population structure of the "Quagga" mussel (Bivalvia: Dreissenidae) in the lower Great Lakes. *Canadian Journal of Fisheries and Aquatic Sciences* 50: 2305–2314.

Mills, E.L., G. Rosenberg, A.P. Spidle, M. Ludyanskiy, Y. Pligin, and B. May, 1996. A review of the biology and ecology of the quagga mussel (*Dreissena bugensis*), a second species of freshwater Dreissenid introduced to North America. *American Zoologist* 36: 271–286.

Mills, E.L., J.R. Chrisman, B. Baldwin, R.W. Owens, R.O. Gorman, T. Howell, E.F. Rosemand, and M.K. Raths, 1999. Changes in the Dreissenid community in the lower Great Lakes with emphasis on southern Lake Ontario. *Journal of Great Lakes Research* 25: 187–197.

Milward-de-Andrade, R., O.S. Carvalho, and C.T. Guimarães, 1978. Some bioecological data regarding *Pomacea haustrum* (Reeve, 1856) predator–competitor of intermediate hosts of *Schistosoma mansoni* Sambon, 1907. *Revista Sa de Publica* 12: 78–89.

Minchin, D. and C. Moriarty, 1998. Zebra mussels in Ireland. *Fisheries Leaflet* No. 177, Marine Institute, Dublin.

Minchin, D., F. Lucy, and M. Sullivan, 2002. Zebra mussel: Impacts and spread. In: *Invasive Aquatic Species of Europe: Distribution, Impacts, and Management*, E. Leppäkoski, S. Gollasch, and S. Olenin, eds, pp. 267–275. Dordrecht: Kluwer Academic Publishers.

Miroshnichenko, M.P., 1990. Growth, production and role of *Dreissena polymoprha* (Pall.) in forage reserve of Tsymlianskoe water reservoir. In: *Species in its Range: Biology, Ecology and Productivity of Aquatic Invertebrates,* pp. 141–146. Minsk: "Nauka i technika" publishers (in Russian).

Mitchell, A.J. and T.M. Brandt, 2003. Temperature tolerance of red-rimmed melania, *Melanoides tuberculata* (Gastropoda: Prosobranchia: Thiaridae). In: *Abstracts, World Aquaculture Society Meeting,* February 19.

Mitchell, A.J. and T. Brandt, 2005. Temperature tolerance of the red-rim melania *Melanoides tuberculatus*, an exotic aquatic snail established in the United States. *American Fisheries Society Transaction* 134: 126–131.

Mitchell, A.J. and R. Cole, 2008. Survival of the faucet snail Bithynia tentaculata after chemical disinfection, pH extremes, and heated water bath treatments. *North American Journal of Fisheries Management* 28: 1597–1600.

Mitchell, R. and J. Maki, 1993. Use of microorganisms for unconventional control treatments of zebra mussels. In: *Third International Zebra Mussel Conference, Agenda and Abstracts,* Toronto, Canada, February.

Miyahara, Y., Y. Hirai, and S. Oya, 1986. Oviposition and hatching rate of *Ampullarius insularus* (D'Orbigny) in Kyushu. In: *Proceedings of the Association for Plant Protection of Kyushu* 32: 96–100.

Mochida, O., 1988a. Nonseedborne rice pests of quarantine importance. In: *Rice Seed Health.* IRRI, Los Banos, pp. 117–129.

Mochida, O., 1988b. The rice water weevil (*Lissorhoptrus pryzophilus*) and the freshwater sanil (*Pomacea canaliculata* (Lamarck)) as important pests of crops for plant quarantine in Asia. *Asean Planti.* 1988: 71–80.

Mochida, O., 1991. Spread of freshwater *Pomacea* snails (Pilidae, Mollusca) from Argentina to Asia. *Micronesica, Supplement* 3: 51–62.

Molina, F.R. and S. José de Paggi, 2008. Zooplankton in the Paraná River floodplain (South America) before and after the invasion of *Limnoperna fortunei* (Bivalvia). *Wetlands* 28: 695–702.

Mondadori, A., ed., 1980. *The Macdonald Encyclopedia of Shells*. London: Macdonald & Co.

Moore, J. and D. Keyes, 2009. Initial visual results from the *Hamilton and Scourge Shipwreck* Site Condition Survey 2008. In: *ACUA Underwater Archaeology Proceedings 2009,* E. Laanela and J. Moore, eds. Advisory Council on Underwater Archaeology, manuscript in preparation.

Mordukhai-Boltovskoi, F.D., 1960. *Caspian Fauna in Azov and Black Sea Basin*. Moskow: Publishers of the Academy of Sciences (in Russian).

Morgan, D.E., M. Kesser, J.F. Foertch, and J.M. Vozarik, 1992. The Asiatic clam, *Corbicula fluminea* extends its North American distribution to the Connecticut River (abstract only), Page 523. In: *Aquaculture and Environmental Stewardship: Milford Shellfish Biology Seminar*. Blogoslowski, V.J. ed., *Environmental Management* 16: 521–529.

Morgan, D.E., M. Keser, J.T. Swenarton, and J.F. Foertch, 2003. Population dynamics of the Asiatic calm, *Corbicula fluminea* (Müller) in the lower Conneticut River: Establishing a foothold in New England. *Journal of Shellfish Research* 22: 193–203.

Morley, N.J., M.E. Adam, and J.W. Lewis, 2004. The role of *Bithynia tentaculata* in the transmission of larval digeneans from a gravel pit in the Lower Thames Valley. *Journal of Helminthology* 78: 129–135.

Morris, C., 2008. Offutt base lake zebra mussel eradication memorandum. interim report, Nebraska, USA. Available at: http://www.aquaticnuisance.org/wordpress/wp-content/uploads/2009/01/Interim%20Report.pdf. Accessed April 13, 2009.

Morris, P.A., 1975. *A Field Guide to Shells of the Atlantic and Gulf Coasts and the West Indies*, 3rd ed. Boston: Houghton Mifflin Professional Publishers.

Morrison, J.P.E., 1954. The relationships of old and New World melanians. In: *Proceedings of the United States National Museum* 103: 357–393.

Morton, B., 1969a. Studies on the biology of *Dreissena polymorpha* Pall.: I. General anatomy and morphology. *Proceedings of the Malacological Society of London* 38: 301–321.

Morton, B., 1969b. Studies on the biology of *Dreissena polymorpha* Pall.: III. Population dynamics. *Proceedings of the Malacological Society of London* 38: 471–482.

Morton, B., 1970. The evolution of the heteromyarian condition in the Dreissenacea (Bivalvia). *Palaeontology London* 13: 563–572.

Morton, B., 1971. Studies on the biology of *Dreissena polymorpha* Pall. Some aspects of filter-feeding and the effect of micro-organisms upon the rate of filtration. In: *Proceedings of the Malacological Society of London* 39: 289–301.

Morton, B., 1973. Some aspects of the biology and functional morphology of the organs of feeding and digestion of *Limnoperna fortunei* (Dunker) (Bivalvia: Mytilacea). *Malacologia* 12: 265–281.

Morton, B., 1975. The colonization of Hong Kong's water supply system by *Limnoperna fortunei* (Dunker, 1857) (Bivalvia: Mytilacea) from China. *Malacological Review* 8: 91–105.

Morton, B., 1977. The population dynamics of *Limnoperna fortunei* (Dunker, 1857) (Bivalvia: Mytilicea) in Plover Cove Reservoir, Hong Kong. *Malacologia* 16: 165–182.

Morton, B., 1980. *Mytilopsis sallei* (Recluz) (Bivalvia: Dreissenacea) recorded from Hong Kong: An introduction by Vietnamese refugees? *Malacological Review* 13: 90–92.

Morton, B., 1982. The reproductive cycle in *Limnoperna fortunei* (Dunker, 1857) (Bivalvia: Mytilidae) fouling Hong Kong's raw water supply system. *Chinese Journal of Oceanology and Limnology* 13: 312–325.

Morton, B., 1983. The sexuality of *Corbicula fluminea* in lentic and lotic waters in Hong Kong. *Journal of Molluscan Studies* 49: 81–83.

Morton, B., 1986. *Corbicula* in Asia—an updated synthesis. *American Malacological Bulletin* 2: 113–124.

Morton, B., 1989. Life-history characteristics and sexual strategy of *Mytilopsis sallei* (Bivalvia: Dreissenacea), introduced into Hong Kong. *Zoology London* 219: 469–485.

Morton, B., C.S. Au, and W.W. Lam, 1976. Control of *Limnoperna fortune. Journal of the Institution of Water Engineers and Scientists* 30: 147–156.

Morton, B. and K.Y. Tong, 1985. The salinity tolerance of *Corbicula fluminea* (Bivalvia: Corbiculoidea) from Hong Kong. *Malacological Review* 18: 91–95.

Mota, H.R., 2006. Análise da influência da geoquímica do ambiente e das características do substrato na estruturação da população de *corbicula fluminea*, Muller, 1794 (mollusca, bivalvia) no reservatório da usina elétrica de volta grande—mg/sp, dissertação (mestrado)—universidade federal de ouro preto. Escola de Minas. Departamento de Geologia. *Área De Concentração: Geologia Ambiental E Conservação de Recursos Naturais.*

Motta Marques, D.M.L., C. Tucci, C. Calazans, V.L.M. Callegaro, and A. Villanueva, 2002. O Sistema Hidrológico do Taim—site 7. In: *Os Sites e o programa Brasileiro de pesquisas ecológicas de longa duração, Belo Horizonte*, U. Seeliger, C.V. Cordazzo, and F. Barbosa, eds, pp. 125–144. MCT-CNPq.

Mouthon, J., 1981. Sur la presence en France et en Portugal de *Corbicula* (Bivalvia: Corbiculidae) originaire d'Asie. *Basteria* 45: 109–116.

Mouthon, J., 2000. Repartition du genre *Corbicula* Megerle von Mühlfeld (Bivalvia: Corbiculidae) en France a l'aube du 21e siecle [Distribution of *Corbicula* Megerle von Mühlfeld (Bivalvia; Corbiculidae) in France at the beginning of the 21st century]. *Hydroecologie Appliquee* 12: 135–146 [French with English summary].

Mouthon, J., 2001. Life cycle and population dynamics of the Asian clam *Corbicula fluminea* (Bivalvia: Corbiculidae) in the Saone and Rhone rivers (France). *Annales De Limnologie-International Journal of Limnology* 39: 15–25.

Muchmore, D. and D. Epel, 1973. The effects of chlorination of wastewater on fertilization in some marine invertebrates. *Marine Biology* 19: 93–95.

Murray, H.D., 1964. *Tarebia granifera* and *Melanoides tuberculatus* in Texas. *Bulletin of the American Malacological Union Inc.* 1964: 15–16.

Murray, H.D., 1971. The introduction and spread of thiarids in the United States. *The Biologist* 53: 133–135.

Murray, H.D., 1975. *Melanoides tuberculata* (Müller), Las Morras Creek, Bracketville, Texas. *Bulletin of the American Malacological Union, Inc.* 1975: 43.

Murray, H.D. and A.J. Stewart, 1968. Establishment of a trematode cycle in *Tarebia granifera* (Lamarck) in Texas. *The American Malacological Union, Inc.* 1968: 17–18.

Murray, H.D. and D. Haines, 1969. *Philophthalmus* sp. (Trematoda) in *Tarebia granifera* and *Melanoides tuberculatus* in South Texas. Annual Reports of The American Malacological Union, pp. 44–45.

Murray, T.E., P.H. Rich, and E.H. Jokinen, 1993. Invasion potential of the zebra mussel *Dreissena polymorpha* (Pallas) in Connecticut: Predictions from water quality data. Connecticut Institute of Water Resources, University of Connecticut, Storrs, CT. Special Report No. 36.

Naddafi, R., K. Pettersson, and P. Eklöv, 2007. The effect of seasonal variation in selective feeding by zebra mussels (*Dreissena polymorpha*) on phytoplankton community composition. *Freshwater Biology* 52: 823–842.

Naegel, L.C.A. and J.I.M. Alvarez, 2005. Biological and chemical properties of the secretion from the hypobranchial gland of the purple snail *Plicopurpura pansa*. *Journal of Shellfish Research* 24: 421–428.

Nagaya, K., Y. Matsui, H. Ohira, A. Yuasa, H. Yamamoto, K. Ohkawa, and Y. Magara, 2001. Attachment strength of an adhesive nuisance mussel, *Limnoperna fortunei*, against water flow. *Biofouling* 17: 263–274.

Nalepa, T.F., 1994. Decline of native bivalves (Unionidae: Bivalvia) in Lake St. Clair after infestation by the zebra mussels *Dreissena polymorpha*. *Canadian Journal of Fisheries and aquatic Sciences* 51: 2227–2233.

Nalepa, T.F. and J.M. Gauvin, 1988. Distribution, abundance, and biomass of freshwater mussels (Bivalvia: Unionidae) in Lake St. Clair. *Journal of Great Lakes Research* 14: 411–419.

Nalepa, T.F., B.A. Manny, J.C. Roth, S.C. Mozley, and D.W. Schloesser, 1991. Long term decline in freshwater mussels (Bivalvia: Unionidae) of the western basin of Lake Erie. *Journal of Great lakes Research* 17: 214–219.

Nalepa, T.F. and D.W. Schloesser, eds, 1993. *Zebra Mussels: Zebra Mussels Biology, Impact and Control*. Boca Raton: Lewis Publishers.

Nalepa, T.F., D.J. Hartson, G.W. Gostenik, D.L. Fanslow, and G.A. Lang, 1996. Changes in the freshwater mussel community of Lake St. Clair: from Unionidae to *Dreissena polymorpha* in eight years. *Journal of Great Lakes Research* 22: 354–369.

Nalepa, T.F., D.J. Hartson, D.L. Fanslow, G.A. Lang, and S.J. Lozano, 1998. Declines in benthic macroinvertebrate populations in southern Lake Michigan, 1980–1993. *Canadian Journal of Fisheries and Aquatic Sciences* 11: 2402–2413.

Nalepa, T.F., D.L. Fanslow, M.B. Lansing, and G.A. Lang, 2002. Trends in the benthic macroinvertebrate community of Saginaw Bay, Lake Huron, 1987 to 1996: Responses to phosphorus abatement and the zebra mussel, *Dreissena polymorpha*. *Journal of Great Lakes Research* 29(1): 14–33.
National Native Mussel Conservation Committee, 1998. National strategy for the conservation of native freshwater mussels. *Journal of Shellfish Research* 17: 1419–1428.
National Research Council Canada (NRCC), 1985. TFM and Bayer 73: Lamppricides in the aquatic environment. NRCC Associate Committee on Scientific Criteria for Environmental Quality. No. 22488. Ottawa, Canada.
Naylor, R., 1996. Invasions in agriculture: Assessing the cost of the golden apple snail in Asia. *Ambio* 25: 443–448.
Neary, B.P., P.J. Dillon, J.R. Munro, and B.J. Clark, 1990. The acidification of Ontario lakes: An assessment of their sensitivity and current status with respect to biological damage. Queen's Printer for Ontario, Toronto, Ontario. ISBN 0-7729-6550-1.
Neary, B.P. and J.H. Leach, 1992. Mapping the potential spread of the zebra mussel (*Dreissena polymorpha*) in Ontario. *Canadian Journal of Fisheries and Aquatic Sciences* 49: 406–415.
Neck, R.W., 1984. Occurrence of the striped ram's horn snail, *Marisa cornuarietis*, in central Texas (Ampullariidae). *The Nautilus* 98: 119–120.
Neck, R.W., 1985. *Melanoides tuberculata* in extreme Southern Texas. *Texas Conchologist* 21: 150–152.
Neck, R.W. and J.G. Schultz, 1992. First record of a living channeled applesnail, *Pomacea canaliculata* (Pilidae), from Texas. *Texas Journal of Science* 44: 115–116.
Nedeau, E.J., M.A. McCollough, and B.I. Swartz, 2000. The freshwater mussels of maine. *Maine Department of Inland Fisheries and Wildlife*. Augusta, ME. pp. 118.
Negovetic, S. and J. Jokela, 2000. Food choice behaviour may promote habitat specificity in mixed populations of clonal and sexual *Potamopyrgus antipodarum*. *Experimental Ecology* 60: 435–441.
Neitzel, D.A., K.I. Johnson, T.L. Page, J.S. Young, and P.M. Dalling, 1984. Correlation of bivalve biological characteristics and service water design. Bivalve fouling of nuclear power plant service water systems. Vol 1. NUREG/CR-4070, PNL-5300, Vol 1. US Regulatory Commission, Washington, DC.
Neuhauser, E.F., M.A. Rhode, J.J. Knowlton, R.J. Wahanik, M. Borden, D. P. Lewis, and G.L. Mackie, 1993. Thermal treatment to control zebra mussels at the Dunkirk Steam Station. In: *Proceedings of the Third Internaternational Zebra Mussel Conference,* Toronto, Canada, February 23–26.
Neumann, D. and H.A. Jenner, eds, 1992. *The Zebra Mussel Dreissena polymorpha.* New York: Gustav Fischer.
Neumann, D., J. Borcherding, and B. Jantz, 1993. Growth and seasonal reproduction of *Dreissena polymorpha* in the Rhine River and adjacent waters. In: *Zebra Mussels: Biology, Impacts, and Control,* T.F. Nalepa and D.W. Schloesser, eds, pp. 95–109. Boca Raton: CRC Press.
Nichols, S.J. and M.G. Black, 1994. Identification of larvae: The zebra mussel (*Dreissena polymorpha*), quagga mussel (*Dreissena rosteriformis bugensis*), and Asian clam (*Corbicula fluminea*). *Canadian Journal of Zoology* 72: 406–417.
Nichols, S.J., 1993. Maintenance of the zebra mussel (*Dreissena polymorpha*) under laboratory conditions. In: *Zebra mussels: Biology, Impacts, and Control*, T.F. Nalepa and D.W. Schloesser, eds, pp. 733–747. Boca Raton: Lewis Publishers.
Nichols, S.J., 1996. Variations in the reproductive cycle of *Dreissena polymorpha* in Europe, Russia, and North America. *American Zoologist* 36: 311–325.
Nicholls, K.H. and G.J. Hopkins, 1993. Recent changes in Lake Erie (north shore) phytoplankton: Cumulative impacts of phosphorus loading reductions and the zebra mussel introduction. *Journal of Great Lakes Research* 19: 637–647.

Nierzwicki-Bauer, S.A., B.R. McGrath, A.S. Hansen, and J. Wimbush, 2000. Bioassay for predicting survival of zebra mussel veligers. Presentation at the *10th International Aquatic Nuiscance Species and Zebra Mussel Conference,* Toronto, Canada, February 13–17.

Nimmo, D.R., 1985. Pestcides. In: *Fundamentals of Aquatic Toxicology*, G.M. Rand and S.R. Petrocelli, eds, pp. 335–373. New York: Hemisphere.

Nimmo, D.R., D.L. Coppage, Q.H. Pickering, and D.J. Hansen, 1987. Assessing the toxicity of pesticides to aquatic organisms. In: *Silent Spring Revisited*, G.J. Marco, R.M. Hollingsworth, and W. Durham, eds, pp. 49–70. American Chemical Society, Washington, DC.

Nocker, A., C. Cheung, and A.K. Camper, 2006. Comparison of propidium monoazide with ethidium monoazide for differentiation of live vs. dead bacteria by selective removal of DNA from dead cells. *Journal of Microbiological Methods* 67: 310–320.

Noonburg, E.G., B.J. Shuter, and P.A. Abrams, 2003. Indirect effects of zebra mussels (*Dreissena polymorpha*) on the planktonic food web. *Canadian Journal of Fisheries and Aquatic Sciences* 60: 1353–1368.

Nowak, E., 1971. The range expansion of animals and its cause. *Zeszyty Naukowe* 3: 1–255. Translation by the Smithsonian Institution and National Science Foundation, Washington, DC. Foreign Science Publication Department of the National Center for Science, Technology, and Economic Information, Warsaw, Poland, 1975.

Nuttall, C.P., 1990. Review of the Caenozoic heterodont bivalve superfamily Dreissenacea. *Palaeontology London* 33: 707–737.

NZMS Management and Control Plan Working Group, 2007. National management and control plan for the New Zealand Mudsnail (*Potamopyrgus antipodarum*). Prepared for the Aquatic Nuisance Species Task Force by the New Zealand Mudsnail Management and Control Plan Working Group, May 2007. 100 p. Available online at: http://www.anstaskforce.gov/Documents/NZMS_MgmtControl_Final.pdf. Accessed August 2009.

OECD, 1982. *Eutrophication of Waters Monitoring Assessment and Control.* Paris: Organisation for Economic Co-operation and Development.

Odum, E.P., 1969. The strategy of ecosystem development. *Science* 164: 262–270.

Ogawa, T. and T.H. Prescott, 2002. Use of intermittent ozone for zebra mussel control in industrial settings. In: *Proceedings of the 11th International Conference on Aquatic Invasive Species,* Alexandria, USA, February 25.

Oglesby, L.C., 1977. A newly introduced, brackish-water snail in the Salton Sea Basin, California. *California Fish and Game* 63: 180–182.

Okamoto, A. and B. Arimoto, 1986. Chromosomes of *Corbicula japonica*, *C. sandai* and *C.* (*Corbiculina*) *leana* (Bivalvia: Corbiculidae). *Venus* 45: 194–202.

Okland, J., 1990. *Lakes and Snails. Environment and Gastropoda in 1500 Morwegian Lakes.* The Netherlands: Universal Book Service.

Okland, J. and A. Okland, 1992. Lakes and ponds in Norway—site evaluation for the protection of invertebrates, with special reference to species of Mollusca. *Fauna* 45: 124–149.

Okuma, M., K. Tanaka, and S. Sudo, 1994a. Feeding habit of apple snail (*Pomacea canaliculata*) to paddy weeds and damage avoidance to rice seedlings. *Weed Research* (*Japan*) 39: 109–113.

Okuma, M., K. Tanaka, and S. Sudo, 1994b. Weed control method using apple snail (*Pomacea canaliculata*) in paddy fields. *Weed Research* (*Japan*) 39: 114–119.

Olenin, S., M. Orlova, and D. Minchin, 1999. *Dreissena polymorpha* (Pallas, 1771). In: *Case Histories on Introduced Species: Their General Biology, Distribution, Range Expansion and Impact*, S. Gollasch, D. Minchin, H. Rosenthal, and M. Voigh, eds, pp. 38–42. Kiel: University of Kiel.

Olivares, Jr. F.M., J.R. Medina, Jr., and S.R. Obien, 1992. Integrated pest management activities on the golden snail in the Philippines: A progress report. Unpublished report presented to the Crop and Resource management Networks Joint Meeting of INSURF, ARFSN, and IPMN, Vietnam, October 12–17, 1993 (not seen, cited by Cowie, 2002).

Oliveira, M., A. Takeda, L. Barros, B. Domingos, and E. Resende, 2006. Invasion by *Limnoperna fortunei* (Dunker, 1857) (Bivalvia, Mytilidae) of the Pantanal Wetland, Brazil. *Biological Invasions* 8: 97–104.

Oliveira, M.D., 2003. Ocorrência e impactos do Mexilhão dourado *Limnoperna fortunei* (Dunker, 1857) no Pantanal Mato-Grossense. *EMBRAPA, Corumbá, MS: Circular Técnica* 38.

Oliver, P.G., A.M. Holmes, and C. Mettam, 1998. *Mytilopsis leucophaeata* (Conrad, 1831) (Bivalvia: Dreissenoidea). A species new to the British fauna. *Journal of Conchology* 36: 13–18.

Oliver-Gonzales, J., P.M. Bauman, and A.S. Benenson, 1956. Effect of the snail, *Marisa comuarietis* on *Australorbis glabratus* in natural bodies of water in Puerto Rico. *Journal of Agriculture* (University of Puerto Rico) 46: 241–242.

O'Neill, C., 1990. Misidentification of mussels in the Hudson River prompts development of identification guide. In: *Dreissena Polymorpha Information Review*, S.G. Moore, ed., Vol. 1, No. 2, pp. 9–10. Brockport: New York Zebra Mussel Information Clearinghouse, New York Sea Grant Extension.

O'Neill, Jr. C.R., 1997. Economic impact of zebra mussels—results of the 1995 National zebra mussel Information Clearinghouse Study. *Great Lakes Research Review* 3: 35–44.

Opresko, D.M. and E.H. Hannon, 1979. Chemical effects of power plant cooling waters: An annotated bibliography. EPRI EA-1072. Electric Power Research Institute, Palo Alto.

Orensanz, J.M., E. Schwindt, G. Pastorino, A. Bortolus, G. Casas, G. Darrigran, R. Elias, J.J. Lopez Gappa, S. Obenat, M. Pascual, P. Penchaszadeh, M. Luz Piriz, F. Scarabino, E.D. Spivak, and E.A. Vallarino, 2002. No longer the pristine confines of the world ocean: A survey of exotic marine species in the southwestern Atlantic. *Biological Invasions* 4: 115–143.

Orlova, M.I., 2008. Background of a program for prevention of fouling at a power plant. *Periphyton and Fouling: Theory and Practice*. Book of Abstracts, St. Petersburg, pp. 164–168.

Orlova, M.I. and G.H. Shcherbina, 2002. About range expansion of *Dreissena bugensis* Andr. (Dreissenidae, Bivalvia) in Upper Volga Reservoirs. *Zoological Journal* 81: 515–520 (in Russian).

Orlova, M.I., J.R. Muirhead, P.I. Antonov, G.K.H. Shcherbina, Y.I. Starobogatov, G.I. Biochino, T.W. Therriault, and H.J. MacIsaac, 2004. Range expansion of quagga mussels *Dreissena rostriformis bugensis* in the Volga River and Caspian Sea basin. *Aquatic Ecology* 38: 561–573.

Orlova, M.I., T.W. Therriault, P.I. Antonov, and G. Kh. Shcherbina, 2005. Invasion ecology of quagga mussels (*Dreissena rostriformis bugensis*): A review of evolutionary and phylogenetic impacts. *Aquatic Ecology* 39: 401–418.

Orlova, M.I., I.V. Telesh, N.A. Berezina, A.E. Antsulevich, A.A. Maximov, and L.F. Litvinchuk, 2006. Effects of nonindigenous species on diversity and community functioning in the eastern Gulf of Finland (Baltic Sea). *Helgoland Marine Research* 60: 98–105.

Orlova, M.I. and M.A. Zuykov, 2008. Accumulation of ^{241}Am by fouling mussel *Dreissena polymoprha* Pallas (Dreissenidae, bivalvia) and perspective of practical use. *Periphyton and Fouling: Theory and Practice*. Book of Abstracts, St. Petersburg, pp. 192–197.

Owens, R.W. and D.E. Dittman, 2003. Shifts in the diets ofslimy sculpin (*Cottus cognatus*) and lake whitefish (*Coregonus clupeaformis*) in Lake Ontario following the collapse of the burrowing amphipod (*Diporeia*). *Aquatic Ecosystem Health Management* 6: 311–323.

Oya, S., Y. Hirai, and Y. Miyahara, 1987. Overwintering of the apple snail, *Pomacea canaliculata* Lamarck, in North Kyushu. *Japanese Journal of Applied Entomology and Zoology* 31: 206–212.

Paavola, M., A.O. Laine, M. Helavuori, and P. Kraufvelin, 2007. Profiling four brackish-water harbours: Zoobenthic composition and invasion status. *Boreal Environmental Research* 12: 159–175.

Packman, J.J., K.J. Comings, and D.B. Booth, 1999. Using turbidity to determine total suspended solids in urbanizing streams in the Puget lowlands. Research Report, University of Washington, Seattle. Available at: http://water.washington.edu/Research/Reports/tssturb.pdf. Accessed February 12, 2009.

Padilla, D.K., M.A. Chotkowski, and L.A.J. Buchan, 1996. Predicting the spread of zebra mussels (*Dreissena polymorpha*) to inland waters using boater movement patterns. *Global Ecology and Biogeography Letters* 5: 353–359.

Page, T.L., A. Neitzel, M.A. Simmons, and P.F. Hayes, 1986. Biofouling of power plant service systems by *Corbicula*. In: *Proceedings of second International Corbicula Symposium, Special Edition No. 2.*, pp. 41–45.

Pain, T., 1950. *Pomacea* (Ampullariidae) of British Guiana. *Proceedings of the Malacological Society of London* 28: 63–74.

Pain, T., 1960. *Pomacea* (Ampullariidae) of the Amazon River system. *Journal of Conchology* 24: 421–432.

Palis, F.V., R.F. Macatula, and L. Browning, 1996. Niclosamide, an effective molluscicide for the golden apple snail (*Pomacea canaliculata* Lamarck) control in Philippine rice production systems. In: *British Crop Protection Council Symposium Proceedings* 66: 213–230 (not seen, cited by Cowie, 2002).

Pareschi, D.C., T. Matsumura-Tundisi, G.R. Medeiros, A.P. Luzia, and J.G. Tundisi, 2008. First occurrence of *Limnoperna fortunei* (Dunker, 1857) in the Rio Tietê watershed (São Paulo State, Brazil). *Brazilian Journal of Biology* 68: 1107–1114.

Parker, S.P., ed, 1994. *McGraw-Hill Dictionary of Scientific and Technical Terms*, 5th ed. New York: McGraw-Hill.

Parsons M.G. and R.W. Harkins, 2002. Full-scale particle removal performance of three types of mechanical separation devices for the primary treatment of ballast water. *Marine Technology* 39: 211–222.

Pastorino, G., G. Darrigran, S. Martín, and L. Lunaschi, 1993. *Limnoperna fortunei* (Dunker, 1857) (Mytilidae), nuevo bivalvo invasor en aguas del río de la Plata. *Neotrópica* 34(39): 101–102.

Pathy, D.A. and G.L. Mackie, 1993. Shell features of *Dreissena polymorpha*, the quagga mussels, and *Mytilopsis leucophaeata* (Bivalvia: Dreissenidae) in North America. *Canadian Journal of Zoology* 71: 1012–1023.

Patil, J.G., R.M. Gunasekera, B.E. Deagle, and N.J. Bax, 2005. Specific detection of Pacific oyster (*Crassostrea gigas*) larvae in plankton samples using nested polymerase chain reaction. *Marine Biotechnology* 7: 11–20.

Patterson, C.G., 1983. Effect of aggregation on the respiration rate of the freshwater unionid bivalve, *Elliptio complanata* (Solander). *Freshwater Invertebrate Biology* 2: 139–146.

Pavlichenko, V.I., 1977. The role of *Hydropsyche angustipennis* (Trichoptera: Hydropsychidae) larvae in destroying black flies in flowing reservoirs of the Zaporozyhe oblast, USSR. *Ekologiya (Moscow)* 1: 104–105.

Payne, B.S. and D. Lowther, 1992. Larval monitoring in a chlorine treatment program to prevent zebra mussel settlement in hydropower facilities. Zebra Mussel Research Program, U.S. Army Engineer Waterways Experiment Station, Technical Note ZMR-1-02.

Payne, B.S., A. Miller, G. Adams, R.F. McMahon, and M.A. Matthews., 1998. Army Engineer Waterways Experiment Station Vicksburg, Report No. A060063.

Paz, R.J., T. Watanabe, M. Priscila, M. Dijck, and F.J.P. Abílio, 1995. First record of melanoides tuberculata (müller, 1774) (gastropoda: prosobranchia: thiaridae) in the state of paraíba (Brazil) and its possible ecological implications. *Revista nordestina de biologia* 10: 79–84.

Peebles, C.R., J. Oliver-González, and F.F. Ferguson, 1972. Apparent adverse effect of *Marisa cornuarietis* upon *Lymnaea columella* and *Biomphalaria glabrata* in an ornamental pond in Puerto Rico. In: *Proceedings of the Hawaiian Entomological Society* 21: 247–256.

Penalver. L.M., 1950. Infeccion experimental de moluscos fluviales Cubanos con *Schistosoma mansoni. Arch. Venezol. Patol. Trop. Y Parasitol. Med.* 2: 297–308.

Penchaszadeh, P., G.A. Darrigran, C. Angulo, A. Averbuj, N. Brignoccoli, M. Brögger, A. Dogliotti, and N. Pírez, 2000. Predation on the invasive freshwater mussel *Limnoperna fortunei* (Dunker, 1857) (mytilidae) by the fish *Leporinus obtusidens* valenciennes, 1846 (anostomidae) in the río de la plata, Argentina. *Journal of Shellfish Research* 19: 229–231.

Penny-Levin, P., R.H. Cowie, J.M. Taylor, K.A. Hayes, K.M. Burnett, and C.A. Ferguson, 2006. Apple snail invasions and the slow road to control: Ecological, economic, agricultural, and cultural perspectives in Hawaii. In: *Global Advances in Ecology and Management of Golden Apple Snails*, R.C. Joshi and L.S. Sebastian, eds, pp. 325–336. Nueva Ecija: Philippine Rice Research Institute.

Perera, G. and J. Walls, 1996. Apple Snails in the Aquarium. 121 pages, T.F.H. Publishing Ltd. England: Waterlooville.

Pérez-Bote, J.L. and J. Fernández, 2008. First record of the Asian clam *Corbicula fluminea* (Müller, 1774) in the Guadiana River Basin (southwestern Iberian Peninsula). *Aquatic Invasions* 3: 87–90.

Perrett, S. and P.J. Whitfield, 1996. Currently available molluscicides. *Parasitology Today* 12: 156–159.

Peterson, R.T., 1980. *A Field Guide to the Birds*, 4th ed. Boston: Houghton Mifflin.

Petrie, S.A. and R.W. Knapton, 1999. Rapid increase and subsequent decline of zebra and quagga mussels in Long Point Bay, Lake Erie: Possible influence of waterfowl predation. *Journal of Great Lakes Research* 25: 772–782.

Petrille, J.C. and S.D. Miller, 2000. Efficacy of hydrogen peroxide for control of adult zebra mussels, *Dreissena polymorpha* and Asiatic clams, *corbicula fluminea.* Presentation at *10th International Aquatic Nuiscance Species and Zebra Mussel Conference,* Toronto, Canada, February 13–17.

Pfenninger, M., F. Reinhardt, and B. Streit, 2002. Evidence for cryptic hybridization between different evolutionary lineages of the invasive clam genus *Corbicula* (Veneroida, Bivalvia). *Journal of Evolutionary Biology* 15: 818–829.

Phelps, H.L., 1994. The Asiatic clam (*Corbicula fluminea*) invasion and system-level ecological change in the Potomac River Estuary near Washington, D.C. *Estuaries* 17: 614–621.

Philippenko, S.I., 2008. Dynamic processes in the development of population on the zebra mussel (*Dreissena polymoprha*) in the Kuchurgan water storage basin-cooler of the Moldavian state district power station. *Periphyton and Fouling: Theory and Practice.* Book of Abstracts, St. Petersburg, pp. 106–108 (in Russian with English summary).

Pickles, S.B., 2000. Use of ultraviolet radiation for zebra mussel control at Ontario power generation. Presentation at the *10th International Aquatic Nuiscance Species and Zebra Mussel Conference,* Toronto, Canada, February 13.

Piesik, Z., 1983. Biology of *Dreissena polymorpha* settling of stylon nets and the role of this mollusk in eliminating the seston and the nutrients from the water course. *Polish Archives of Hydrobiology* 30: 353–362.

Pilsbry, H.A., 1932. *Bithynia tentaculata* in the Potomac. *The Nautilus* 48: 69.

Pimentel, D., 2002. *Biological Invasions: Economic and Environmental Costs of Alien Plant, Animal, and Microbe Species.* Boca Raton: CRC Press.

Pimentel, D., R. Zuniga, and D. Morrison, 2005. Update on the environmental and economic costs associated with alien-invasive species in the United States. *Ecological Economics* 52: 273–288.

Pinel-Alloul, B. and E. Magnin, 1971. Cycle vital et croissance de *Bithynia tentaculata* L. (Mollusca: Gastropoda: Prosobranchia) du Lac St. Lois pre de Montreal. *Canadian Journal of Zoology* 49: 759–766.

Pointier, J.-P., 1993. The introduction of *Melanoides tuberculata* (Mollusca: Thiaridae) to the Island of Saint Lucia (West Indies) and its role in the decline of *Biomphalaria glabrata*, the snail intermediate host of *Schistosoma mansoni*. *Acta Tropica* 54:13–18.

Pointier, J.-P., 2001. Invading freshwater snails and biological control in Martinique Island, French West Indies. *Rio de Janeiro* 96: 67–74.

Pointier, J.-P., A. Théron, and D. Imbert-Establet, 1988. Decline of a sylvatic focus of *Schistosoma mansoni* in Guadeloupe (French West Indies) following competitive displacement of the snail host *Biomphalaria glabrata* by *Ampullaria glauca*. *Oecologia* 75: 38–43.

Pointier, J.-P., C. Balzan, P. Chrosciechowski, and R.N. Incani, 1991a. Limiting factors in biological control of the snail intermediate hosts of *Schistosoma mansoni* in Venezuela. *Journal of Medical and Applied Malacology* 3: 53–67.

Pointier, J.-P., A. Théron, D. Imbert-Establet, and G. Borel, 1991b. Eradication of a sylvatic focus of *Schistosoma mansoni* using biological control by competitor snails. *Biological Control* 1: 244–247.

Pointier, J.-P., B. Delay, J.L. Toffart, M. Lefèvre, and R. Romero-Alvarez, 1992. Life history traits of three morphs of *Melanoides tuberculata* (Gastropoda: Thiaridae), an invading snail in the French West Indies. *Journal of Molluscan Studies* 58: 415–423.

Pointier, J.-P. and A. Guyard, 1992. Biological control of the snail intermediate hosts of *Schistosoma mansoni* in Martinique, French West Indies. *Tropical Medical Parasitology* 43: 98–101.

Pointier, J.-P., A. Théron, and G. Borel, 1993a. Ecology of the introduced snail *Melanoides tuberculata* (Gastropoda: Thiaridae) in relation to *Biomphalaria glabrata* in the Marshy Forest Zone of Guadeloupe, French West Indies. *Journal of Molluscan Studies* 59: 421–428.

Pointier, J.-P., L. Thaler, A.F. Pernot, and B. Delay, 1993b. Invasion of the Martinique Island by the parthenogenetic snail *Melanoides tuberculata* and the succession of morphs. *Acta Oecologica* 14: 33–42.

Pointier, J.-P., R.N. Incani, P. Chrosciechowski, and S. Prypchan, 1994. Invasion of the rivers of the Littoral Central Region of Venezuela by *Thiara granifera* and *Melanoides tuberculata* (Mollusca: Prosobranchia: Thiaridae) and the absence of *Biomphalaria glabrata*, snail host of *Schistosoma mansoni*. *The Nautilus* 107:124–128.

Polewski, B. and C. Gauthier, 1990. Impact of zebra mussels on the fire protection system at nanticoke TGS. Abstract and Paper Presented at the *EPRI International Macrofouling Sypmosium,* Orlando, USA, December.

Pollux, B., D. Minchin, G. Van der Velde, T. Van Alen, S.Y. Moon-Van Der Staay, and J. Hackstein, 2003. Zebra mussels (*Dreissena polymorpha*) in Ireland, AFLP-fingerprinting and boat traffic both indicate an origin from Britain. *Freshwater Biology* 48: 1127–1139.

Polman, H.J.G. and H.A. Jenner, 2002. Pulse-Chlorination® the best available technique in macrofouling mitigation using chlorine. *Power Plant Chemistry* 4: 93–97.

Ponder, W.F., 1988. *NZ mudsnail*, a Molluscan colonizer of Europe and Australia. *Journal of Molluscan Studies* 54: 271–286.

Pontius, R.A., 2000. *The impact of zebra mussels (Dreissena polymorpha) on pelagic food webs*. PhD dissertation, Ohio State University, Columbus.

Pontius, R.A. and D.A. Culver, 2001. Estimating zebra mussel impact on pelagic food webs: The role of size-specific grazing rates. *Verhandlungen Internationale Vereinigung fur Theoretische und Angewandte Limnologie* 27: 3025–3028.

Popa, O.P. and L.O. Popa, 2006. The most westward European occurrence point for *Dreissena bugensis* (Andrusov, 1897). *Malacologica Bohemoslovaca* 5: 3–5.

Porte, D., R. Verzola, and N. Uphoff, 2006. Using golden apple snails for weed control. In: *Global Advances in Ecology and Management of Golden Apple Snails*, R.C. Joshi, and L.S. Sebastian, eds, pp. 457–513. Nueva Ecija: Philippine Rice Research Institute.

Porter, A.E. and J.E. Marsden, 2008. Adult zebra mussels (*Dreissena polymorpha*) avoid attachment to mesh materials. *Northeastern Naturalist* 15: 589–594.

Pothoven, S.A., T.F. Nalepa, P.J. Schneeberger, and S.B. Brandt, 2001. Changes in diet and body condition of lake whitefish in southern Lake Michigan associated with changes in benthos. *North American Journal of Fisheries Management* 21: 876–883.

Preap, V., P. Chanty, K. Bunnarith, and K. Hin, 2006. The golden apple snail situation in Cambodia. In: *Global Advances in Ecology and Management of Golden Apple Snails*, R.C. Joshi, and L.S. Sebastian, eds, pp. 267–274. Nueva Ecija: Philippine Rice Research Institute.

Prentice, M.A., 1983. Displacement of *Biomphalaria glabrata* by the snail *Thiara granifera* in field habitats in Santa Lucia, West Indies. *Annals of Tropical Medicine and Parasitology* 77: 51–59.

Prescott, D.R.C. and M.M. Curteanu, 2004. Survey of aquatic gastropods in the central Parkland subregion of Alberta. *Alberta Species at Risk Report* 92: 1–50.

Prezant, R.S. and K. Chalermwat, 1984. Flotation of the bivalve *Corbicula fluminea* as a means of dispersal. *Science* 225: 1491–1493.

Prokopovich, N.P. and D.J. Hebert, 1965. Sedimentation in the Delta-Mendota Canal. *Journal of American Water Works Association* 57: 375–382.

Prosser, C.L. and F.A. Brown, 1962. *Comparative Animal Physiology*. Philadelphia: W.B. Saunders.

Protasov, A.A., 1994. *Freshwater peiphyton.* Kiev: Naukova dumka publishers (in Russian with Ukrainian summary).

Purchon, R.D., 1977. *The Biology of the Mollusca.* New York: Pergamon Press.

Qiu A., A. Shi, and A. Komaru, 2001. Yellow and brown shell color morphs of *Corbicula fluminea* (Bivalvia: Corbiculidae) from Sechuan Province, China, are triploids and tetraploids. *Journal of Shellfish Research* 20: 323–328.

Quigley, M.A., W.S. Gardner, and W.M. Gordon, 1993. Metabolism of the zebra mussel (*Dreissena polymorpha*) in Lake St. Clair of the Great Lakes, Chap, 18. In: *Zebra Mussels: Biology, Impacts, and Control*, T.F. Nalepa and D.W. Schloesser, eds, pp. 295–306. Boca Raton: Lewis Publishers, Inc.

Quinn, J.M., G.L. Steele, C.W. Hickey, and M.L. Vickers, 1994. Upper thermal tolerances of twelve New Zealand stream invertebrate species. *New Zealand Journal of Marine and Freshwater Research* 28: 391–397.

Radea, C., I. Louvrou, and A. Economou-Amilli, 2008. First record of the New Zealand mud snail *Potamopyrgus antipodarum* J.E. Gray, 1843 (Mollusca: Hydrobiidae) in Greece—notes on its population structure and associated microalgae. *Aquatic Invasions* 3: 341–344.

Radke, H.G., L.E. Ritchie, and F.F. Ferguson, 1961. Demonstrated control of *Australorbis glabratus* by *Marisa cornuarietis* under field conditions in Puerto Rico. *American Journal of Tropical Medicine and Hygiene* 10: 370–373.

Raikow, D.F., O. Sarnelle, A.E. Wilson, and S.K. Hamilton, 2004. Dominance of the noxious cyanobacterium *Microcystis aeruginosa* in low-nutrient lakes is associated with exotic zebra mussels. *Limnology and Oceanography* 49: 482–487.

Railkin, A.I., 2004. *Marine Biofouling Colonization Processes and Defenses*. Boca Raton: CRC Press.

Railkin, A.I., M.I. Orlova, and S.Z. Chikadze, 2008. Modern problems of protection from marine and aquatic fouling. *Periphyton and Fouling: Theory and Practice*. Book of Abstracts, St. Petersburg, pp. 170–173.

Rajagopal, S., J. Azariah, J., and K.V.K. Nair, 1994. Heat treatment as a fouling control method for Indian coastal power plants. In: *Recent Developments in Biofouling Control*, M.F. Thompson, R. Nagabhushanam, R. Sarojini, and M. Fingerman, eds, pp. 391–396. New Delhi: Oxford and IBH.

Rajagopal, S., G. Van der Velde, and H.A. Jenner, 1995. Aangroeibestrijding van de brakwatermossel *Mytilopsis leucophaeata* bij de kerncentrales Hemweg en Velsen. Deel 1: Biologie en klepbewegingsgedrag met behulp van de MosselMonitor. KEMA-rapport (not seen, cited by Verween et al. 2006).

Rajagopal, S., G. Van der Velde, H.A. Jenner, M. Van Der Gaag, and A.J. Kempers, 1996. Effects of temperature, salinity and agitation on byssus thread formation of zebra mussel, *Dreissena polymorpha. Aquatic Ecology* 30: 187–195.

Rajagopal, S., G. Van der Velde, and H.A. Jenner, 1997. Response of zebra mussel, *Dreissena polymorpha,* to elevated temperatures in the Netherlands. In: *Zebra Mussels and Aquatic Nuisance Species,* F.M. D'Itri, ed., pp. 257–273. Chelsea: Ann Arbor Press.

Rajagopal, S., G. Van der Velde, and A. Bij de Vaate, 2000. Reproductive biology of the Asiatic clams *Corbicula fluminalis* and *Corbicula fluminea* in the river Rhine. *Archiv für Hydrobiologie* 149: 403–420.

Rajagopal S, M. Van der Gaag, G. Van der Velde, and H.A. Jenner, 2002a. Control of brackish water fouling mussel, *Mytilopsis leucophaeata* (Conrad) with sodium hypochlorite. *Archives of Environmental Contamination and Toxicology* 43: 296–300.

Rajagopal, S., G. van der Velde, M. van der Gaag, and H.A. Jenner, 2002b. Sublethal responses of zebra mussel, *Dreissena polymorpha* to low-level chlorination: An experimental study. *Biofouling* 18: 95–108.

Rajagopal, S., G. Van der Velde, M. Van der Gaag, and H.A. Jenner, 2005a. Factors influencing the upper temperature tolerances of three mussel species in a brackish water canal: Size, season and laboratory protocols. *Biofouling* 21: 87–97.

Rajagopal, S., M. Van der Gaag, G. Van der Velde, and H.A. Jenner, 2005b. Upper temperature tolerances of exotic brackish-water mussel, *Mytilopsis leucophaeata* (Conrad): An experimental study. *Marine Environmental Research* 60: 512–530.

Ram, J. and S.M. Palazzolo, 2008. Globalization of an aquatic pest: Economic costs, ecological outcomes, and positive applications of zebra mussel invasions and expansions. *Geography Compass* 2: 1755–1776.

Ramakrishnan, V., 2007. *Salinity, pH, temperature, desiccation and hypoxia tolerance in the invasive freshwater apple snail Pomacea insularum.* PhD dissertation, University of Texas at Arlington.

Ramcharan, C.W., D.K. Padilla, and S.I. Dodson, 1992. Models to predict potential occurrence and density of the zebra mussel, *Dreissena polymorpha. Canadian Journal of Fisheries and Aquatic Sciences* 49: 2611–2620.

Ranamukhaarachchi, S.L. and S. Wickramasinghe, 2006. Golden apple snails in the world: Introduction, impact, and control measures. In: *Global Advances in Ecology and Management of Golden Apple Snails,* R.C. Joshi and L.S. Sebastian, eds, pp. 133–151. Nueva Ecija: Philippine Rice Research Institute.

Rand, G.M., 1995. *Fundamentals of Aquatic Toxicology. Effects, Environmental Fate and Risk Assessment,* 2nd ed. Washington: Taylor & Francis.

Rao, M.R., K.S.P. Rao, M.S. Reddy, K.R.S.S. Rao, and N.M. Reddy, 1983. Toxicity of phenthoate and changes in the organic constituents of the snail, *Pila globosa* (Swainson) under sublethal and lethal impacts. *Geobios* 10: 250–253.

Rawlings, T.A., K.A. Hayes, R.H. Cowie, and T.M. Collins, 2007. The identity, distribution, and impacts of non-native apple snails in the continental United States. *BMC Evolutionary Biology* 7: 97.

Reed, T., A.L. Fettes, S.J. Wielgus, A.K. Barnes, and J.J. Schiefelbein, 2004. Refugia and local controls: Benthic invertebrate dynamics in lower Green Bay, Lake Michigan following zebra mussel invasion. *Journal of Great Lakes Research* 30: 390–396.

Reed-Andersen, T., E.M. Bennett, B.S. Jorgensen, G. Lauster, D.B. Lewis, D. Nowacek, J.L. Riera, B.L. Sanderson, and R. Stedman, 2000. Distribution of recreational boating across lakes: Do landscape variables affect recreational use? *Freshwater Biology* 43: 439–448.

Reed-Andersen, T., S.R. Carpenter, D.K. Padill, and R.C. Lathrop, 2000. Predicted impact of zebra mussel (*Dreissena polymorpha*) invasion on water clarity in Lake Mendota. *Canadian Journal of Fisheries and Aquatic Sciences* 57: 1617–1626.

Reeders, H.H., A. Bij de Vaate, and F.J. Slim, 1989. The filtration rate of *Dreissena polymorpha* (Bivalvia) in three Dutch lakes with reference to biological water quality management. *Freshwater Biology* 22: 133–141.

Reeders, H.H. and A. Bij de Vaate, 1990. Zebra mussels *(Dreissena polymorpha):* A new perspective for water quality management. *Hydrobiologia* 200/201: 437–450.

Reeders, H.H., A.B. bij De Vaate, and R. Noordhuis, 1993. Potential of the zebra mussel Dreissena polymorpha for water quality management. In: *Zebra Mussels: Biology, Impacts and Control.* Nalepa, T.F. and D.W. Schloesser (eds). pp. 439–452, Lewis Publishers, Boca Raton, FL.

Rehder, H.A., 1937. Congeria leucophaeta (Con.) in the Hudson River. *The Nautilus* 50: 143.

Reise, K., S. Gollasch, and W.J. Wolff, 1999. Introduced marine species of the North Sea coasts. *Helgolaender Meeresuntersuchungen* 52: 219–234.

Remane, A. and C. Schlieper, 1971. *Biology of Brackish Water.* New York: Wiley.

Renner, M., 2005. *State of the World 2005: Redefining Global Security.* World Watch Institute. W.W. Norton and Company.

Resende, M.F., 2007. Variação das características das características hidraúlicas em condutos forçados devido à infestação pelo *limnoperna fortunei.* Belo Horizonte, Escola de Engenharia da UFMG.

Ricciardi, A., 1998. Global range expansion of the Asian mussel *Limnoperna fortunei* (Mytilidae): Another fouling threat to freshwater systems. *Biofouling* 13: 97–106.

Ricciardi, A., 2001. Facilitative interactions among aquatic invaders: Is an "invasional meltdown" occurring in the Great Lakes? *Canadian Journal of Fisheries and Aquatic Sciences* 58: 2513–2525.

Ricciardi, A., 2003. Predicting the impacts of an introduced species from its invasion history: An empirical approach applied to zebra mussel invasions. *Freshwater Biology* 48: 972–981.

Ricciardi, A., R. Serrouya, and F.G. Whoriskey, 1995. Aerial exposure tolerance of zebra and quagga mussels (Bivalvia: Dreissenidae): Implications for overland dispersal. *Canadian Journal of Fisheries and Aquatic Sciences* 52: 1470–1477.

Ricciardi, A., F.G. Whoriskey, and J.B. Rasmussen, 1997. The role of the zebra mussel (*Dreissena polymorpha*) in structuring macroinvertebrate communities on hard substrata. *Canadian Journal of Fisheries and Aquatic Sciences* 54: 2596–2608.

Ricciardi, A., R. Neves, and J.B. Rasmussen, 1998. Impending extinctions of North American freshwater mussels (Unionoida) following the zebra mussel (*Dreissena polymorpha*) invasion. *Journal of Animal Ecology* 67: 613–619.

Ricciardi, A. and F.G. Whoriskey, 2004. Exotic species replacement: Shifting dominance of dreissenid mussels in the Soulanges Canal, upper St. Lawrence River, Canada. *Journal of the North American Benthological Society* 23: 507–514.

Richards, D.C., 2002a. The New Zealand mudsnail in the western USA. Available at: http://www2.montana.edu/NZ mudsnail/. Department of Ecology, Montana State University, Bozeman.

Richards, D.C., 2002b. Greeting and overview presentation. In: *New Zealand Mudsnail in the Western USA: Conference*, August, 2002 (not seen, cited in NZMS Management and Control Plan Working Group, 2007).

Richards, D.C., 2002c. The New Zealand mudsnail invades. *Aquatic Nuisance Species Digest* 4: 42–44.

Richards, D.C., 2003. Competition between *NZ mudsnail* and threatened Bliss Rapids snail. Paper Presented at the *New Zealand Mudsnail in the Western USA: Conference,* August (not seen, cited in NZMS Management and Control Plan Working Group, 2007).

Richards, D.C., 2004. *Competition between the threatened Bliss Rapids Snail, Taylorconcha serpenticola (Hershler et al.) and the invasive, aquatic snail, NZ mudsnail* (Gray). PhD dissertation, Montana State University, Bozeman.

Richards, D.C., L.D. Cazier, and G.T. Lester, 2001. Spatial distribution of three snail species, including the invader *Potamopyrgus antipodarum,* in a freshwater spring. *Western North American Naturalist* 61: 375–380.

Richards, D.C., P. O'Connell, and D.C. Shinn, 2004. Simple control method to limit the spread of New Zealand mudsnail *Potamopyrgus antipodarum. North American Journal of Fisheries Management* 24: 114–117.

Richards, D.C., C.M. Falter, G.T. Lester, and R. Myers, 2005. Listed Mollusks. Responses to FERC additional information request AR-2. Hells Canyon Project. FERC No. P-1971-079 (not seen, cited in NZMS Management and Control Plan Working Group, 2007).

Richter, B.D., D.P. Braun, M.A. Mendelson, and L.L. Master, 1997. Threats to imperiled freshwater fauna. *Conservation Biology* 11: 1081–1093.

Rigden, B., 1993. Use of small pore filters in an Ontario hydro facility. In: *Third International Zebra Mussel Conference, Agenda and Abstracts,* Toronto, Canada, February.

Riley, L.A., 2003. *Exotic species impact: Exploitative competition between stream snails*? MS thesis, Washington State University, Pullman, Washington.

Riley, L.A., M.F. Dybdah, and R.O. Hall, 2002. Invasive species impact: Resource competition between stream snails within the Greater Yellowstone Ecosystem. In: *The Invasiveness of an Exotic Snail in the Greater Yellowstone Ecosystem and Energy Flow under Ambient Conditions*, B.L. Kerans, M. Dybdahl, and R. Hall, eds, Final Report (not seen, cited in NZMS Management and Control Plan Working Group, 2007).

Rinne, J.N., 1974. The introduced Asiatic clam, *Corbicula* in central Arizona reservoirs. *The Nautilus* 88: 56–61.

Robertson, I.C.S. and C.L. Blakselee, 1948. The Mollusca of the Niagara frontier region. *Bulletin of the Buffalo Society of Natural Sciences* 19(3): 1–191.

Robins C.H., 1971. Ecology of the introduced snail, *Marisa cornuarietis* (Ampullariidae) in Dade County, Florida. *The Biologist* 53: 136–152.

Rodgers, J.H., Jr., D.S. Cherry, K.L. Dickson, and J. Cairns, Jr., 1979. Invasion, population dynamics and elemental accumulation of Corbicula fluminea in the New River at Glen Lyn, Virginia. In: *Proceedings, First International Corbicula Symposium,* Texas Christian University Research Foundation, Fort Worth, Texas. Britton, J.C (ed). pp. 99–110.

Roditi, H.A., N.F. Caraco, J.J. Cole, and D.L. Strayer, 1996. Filtration of the Hudson River water by the zebra mussel (*Dreissena polymorpha*). *Estuaries*, 19: 824–832.

Roe, S.L. and H.J. MacIsaac, 1997. Deepwater population structure and reproductive state of quagga mussels (*Dreissena bugensis*) in Lake Erie. *Canadian Journal of Fisheries and Aquatic Science* 54(10): 2428–2433.

Roessler, M.A., G.L. Beardsley, and D.C. Tabb, 1977. New records of the introduced snail, Melanoides tuberculata (Mollusca: Thiaridae) in south Florida. *Florida Scientist* 40: 87–94.

Rolla, M.E., N.D.C. Barbosa, M.C.S. Campos, M.V. Junqueira, G.H.O. von Ruckert, 2004. O mexilhão dourado: Uma ameaça às águas e hidrelétricas brasileiras. *Belo Horizonte*: CEMIG.

Romano, M.A., A.D. Byers, R.V. Anderson, and N.C. Krumanocker, 1997. Characterization of dreissenid mussels from the Mississippi River using specific DNA markers from mitochondrial cytochrome C oxidase. Triannual Unionid Report No. 13, p. 2.

Roper, J.M., D.S. Cherry, J.W. Simmers, and H.E. Tatem, 1996. Bioaccumulation of toxicants in the zebra mussel, *Dreissena polymorpha*, at the Times Beach confined disposal facility, Buffalo, New York. *Environmental Pollution* 94(2): 117–119.

Roper, J.M., D.S. Cherry, J.W. Simmers, and H.E. Tatem, 1997. Bioaccumulation of PAHs in the Zebra Mussel at Times Beach, Buffalo, New York. *Environmental Monitoring and Assessment* 46: 267–277.

Ross, L.T., A. Olsen, and L. Groenvelde, 1982. The relationship between Jackson Turbidity Units and Nephelometric Turbidity Units, as determined from various Florida sediment samples. A report to Florida Department of Environmental Regulation. Available at: ftp://ftp.dep.state.fl.us/pub/labs/biology/miscpubs/ross82.pdf. Accessed February 12, 2009.

Rothe, M., 2007. Control of golden mussel by ozonation of cooling water in power plants. In: *Proceedings of the 15th International Conference on Aquatic Invasive Species*, Nijmegen, Netherlands, September 23–27.

Ruiz, G.M., P.W. Fofonoff, J.T. Carlton, M.J. Wonham, and A.H. Hines, 2000. Invasion of coastal marine communities in North America: Apparent patterns, processes, and biases. *Annual Review of Ecological Systems* 31: 481–531.

Russo, T.N., 1974. Discovery of the gastropod snail *Melanoides (Thiara) tuberculata* (Miiller) in Florida. *Florida Scientist* 36: 212–213.

Rusznak, L., N. Smolik, L. Hale, and S. Freymark, 1995. Examination of the potential of chlorine dioxide for use in zebra mussel veliger control. In: *Proceedings of the Fifth International Zebra Mussel and Other Aquatic Nuisance Organisms Conference*, Toronto, Canada, February 14–21.

Ryan, M.F., 1998. Extremely low frequency electromagnetism: An effective nonchemical method for control of zebra mussel infestation. Presentation at the *8th International Zebra Mussel and Other Nuisance Species Conference*, Sacramento, USA, March 16–19.

Ryan, P., 1982. Energy contents of some New Zealand freshwater animals. *New Zealand Journal of Marine and Freshwater Research* 16: 283–287.

Rzepecki, L.M. and J.H. Waite, 1993. Zebra mussel byssal glue: An unwelcome bond between man and mollusk? In: *Third International Zebra Mussel Conference, Agenda and Abstracts,* Toronto, Canada, February.

Sakamoto, M., 1966. Primary production by phytoplankton community in some Japanese lakes and its dependence on lake depth. *Archives fur Hydrobiologia* 62: 1–28.

Sambrook, J. and D. Russell, 2001. *Molecular Cloning: A Laboratory Manual.* CSHL Press, Cold Spring Harbor.

Santos, C.A.Z., C.H.S. Penteado, and E.G. Mendes, 1987. The respiratory responses of an amphibious snail *Pomacea lineata* (Spix, 1827), to temperature and oxygen tension variations. *Comparative Biochemistry and Physiology* 86A: 409–415.

Sarina E.S.E. Loo, S.E., R. Mac Nally, and P.S. Lake, 2007. Forecasting New Zealand mudsnail invasion range: Model comparisons using native and invaded ranges. *Ecological Applications* 17: 181–189.

Sarnelle, O., D.F. Raikow, A.E. Wilson, S.K. Hamilton, and L.B. Knoll, 2005. Complex interactions between the zebra mussel, *Dreissena polymorpha*, and the harmful phytoplankter, *Microcystis aeruginosa. Limnology and Oceanography* 50: 896–904.

Satake, K., 1980. Limnological studies on inorganic acid lakes in Japan. *Japanese Journal of Limnology* 41: 41–50.

Satuito, C.G., I. Katuyama, and S. Den, 2000. Effect of electric fields on the planktonic-staged larvae of fouling organisms. Paper presented at the *10th International Aquatic Nuisance Species and Zebra Mussel Conference*, Toronto, Canada, February 13–17.

Sauer, J.S., R.A. Cole, and J.M. Nissen, 2007. Finding the exotic faucet snail (*Bithynia tentaculata*): Investigation of waterbird die-offs on the Upper Mississippi River National Wildlife and Fish Refuge. U.S. Geological Survey Open-File Report 2007-1065.

Sawada, H., H. Saito, M. Hosoi, and H. Toyohara, 2008. Evaluation of PCR methods for fixed bivalve larvae. *Journal of Marine Biological Association, U.K.* 88: 1441–1449.

Scavia, D. and S.C. Chapra, 1977. Comparison of an ecological model of Lake Ontario and phosphorous loading models. *Journal Fisheries Research Board of Canada* 34: 286–290.

Schaefer, R. and R. Claudi, 2004. Development of an efficient low-cost sparker technology for controlling zebra mussels. Presentation at the *13th International Conference on Aquatic Invasive Species,* Ennis, Ireland, September 20–24.

Schaner, T., 1991. Detection of zebra mussel veligers in plankton samples using sugar solution. Lake Ontario Fisheries Unit Annual Report, LOA 91.1.

Scheller, J.L., 1997. *The effect of die-offs of Asian clams (Corbicula fluminea) on native freshwater mussels (Unionidae).* MSc thesis, Virginia Polytechnic Institute and State University, Blacksburg.

Schlesch, H., 1937. Bemerkungen über die Verbreitung der Süsswasser- und Meeresmollusken im östlichen Ostseege-biete. *Tartu Loodusuurijate Seltsi Aruanded* 43: 37–64.

Schloesser, D.W. and W. Kovalak, 1991. Infestation of unionids by *Dreissena polymorpha* in a power plant canal in Lake Erie. *Journal of Shellfish Research* 10: 355–359.

Schloesser, D.W. and T.F. Nalepa, 1994. Dramatic decline of unionid bivalves in offshore waters of western Lake Erie after infestation by the zebra mussel, *Dreissena polymorpha. Canadian Journal of Fisheries and Aquatic Sciences* 51: 2234–2242.

Schloesser, D.W., T.F. Nalepa, and G.L. Mackie, 1996. Review of the impacts of the zebra mussel on Unionidae in North America. *American Zoologist* 36: 300–310.

Schloesser, D.W., J.L. Metcalfe-Smith, W.P. Kovalak, G.D. Longton, and R.D. Smithee, 2006. Extirpation of freshwater mussels (Bivalvia: Unionidae) following the invasion of dreissenid mussels in an interconnecting river of the Laurentian Great Lakes. *American Midland Naturalist* 155: 307–320.

Schmeck, E.H., 1942. *Viviparus malleatus* in Niagara River. *The Nautilus* 55: 102–103.

Schneider, R., 1997. Generation of chlorine dioxide. In: Chlorine dioxide and disinfection. *Proceedings of the First European Symposium on Chlorine Dioxide and Disinfection*, Rome, Italy, November 7–8.

Schneider, D.W., C.D. Ellis, and K.S. Cummings, 1998a. A transportation model assessment of the risk to native mussel communities from zebra mussel spread. *Conservation Biology* 12: 788–800.

Schneider, D.W., S.P. Madon, J.A. Stoeckel, and R.E. Sparks, 1998b. Seston quality controls zebra mussel (*Dreissena polymorpha*) in turbid rivers. *Oecologia* 117: 331–341.

Schnorbach, H.-J., 1995. The golden apple snail (*Pomacea canaliculata* Lamarck), an increasingly important pest in rice, and methods of control with Bayluscide. *Pflanzenschutz-Nachrichten Bayer* 48: 313–346.

Schnorbach, H.-J., H-W. Rauen, and M. Bieri, 2006. Chemical control of the golden apple snail, *Pomacea caniculata.* In: *Global Advances in Ecology and Management of Golden Apple Snails*, R.C. Joshi and L.S. Sebastian, eds, pp. 419–438. Nueva Ecija: Philippine Rice Research Institute.

Schoenbach, K.H., A. Abou-Ghazala, R.W. Alden, R. Turner, and T.J. Fox, 1997. Biofouling prevention with pulsed electric fields. Paper presented at the *Seventh International Zebra Mussel and Aquatic Nuisance Species Conference*, New Orleans, USA, January 28–31.

Schreiber, E.S.G., A. Glaister, G.P. Quinn, and P.S. Lake, 1998. Live history and population dynamics of the exotic snail *NZ mudsnail* (Prosobranchia: Hydrobiidae) in Lake Purrumbete, Victoria, Australia. *Australian Journal of Marine and Freshwater Research* 49: 73–78.

Schreiber, E.S.G., P.S. Lake, and G.P. Quinn, 2002. Facilitation of native stream fauna by an invading species: Experimental investigations of the interaction of the snail, *NZ mudsnail* (Hydrobiidae) with native benthic fauna. *Biological Invasions* 4: 317–325.

Schreiber, E.S.G., G.P. Quinn, and P.S. Lake, 2003. Distribution of an alien aquatic snail in relation to flow variability, human activities and water quality. *Freshwater Biology* 48: 951–961.

Schruben, P.G., R.E. Arndt, and W.J. Bawiek, Compilers, display software by A. Russell and R.A. Ambroziak, 1998. Geology of the conterminous United States at 1:2,500,000 scale—a digital representation of the 1974 P.B. King, and H.M. Beikman Map: U.S. Geological Survey Digital Data Series DDS-11, Release 2 (CD-ROM).

Scott-Wasilk, J., G.G. Downing, and J.S. Lietzow, 1983. Occurrence of the Asiatic clam *Corbicula fluminea* in the Maumee River and western Lake Erie. *Journal of Great Lakes Research* 9: 9–13.

Seaman, D.E. and W.A. Porterfield, 1964. Control of aquatic weeds by the snail *Marisa cornuarietis. Weeds* 12: 87–92.

Seegert, G.L., A.S. Brooks, and D.L. Latimer, 1977. The effects of a 30-minute exposure of selected lake michagan fishes and invertebrates to residual chlorine. In: *Biofouling Control Procedures, Technology and Ecological Effects*, L.D. Jensen, ed., pp. 91–99. New York: Marcel Dekker.

Selck, H., J. Aufderheide, N. Pounds, C. Staples, N. Caspers, and V. Forbes, 2006. Effects of food type, feeding frequency, and temperature on juvenile survival and growth of *Marisa cornuarietis* (Mollusca: Gastropoda). *Invertebrate Biology* 125: 106–116.

Selvamani, M.J.P., S.M. Degnan, and B.M. Degnan, 2001. Microsatellite genotyping of individual larvae: Parentage assignment in aquaculture. *Marine Biotechnology* 3: 478–485.

Settle, W.H., H. Ariawan, E.T. Astuti, W. Cahyana, A.L. Hakim, D. Hindayana, A.S. Lestari, and P. Sartanto, 1996. Managing tropical rice pests through conservation of generalist natural enemies and alternative prey. *Ecology* 77: 1975–1988.

Shankland, E.L. and R.L. Lowe, 2005. Effects of a recent zebra mussel invasion on the phytoplankton community in Lake Leelanau in Northern Michigan. Research Articles, University of Michigan Research Forum, 2005, Issue 2: 33–36. Available at: http://www.umich.edu/~umforum/Issue2/Issue2-p32.pdf. Accessed February 17, 2009.

Shannon, M.C., J.D. Rhoades, J.H. Draper, S.C. Seardaci, and M.D. Spyres, 1998. Assessment of salt tolerance in rice cultivars in response to salinity problems in California. *Crop Science* 38: 394–398.

Shapiro, J., 1957. Chemical and biological studies on the yellow organic acids of lake water. *Limnology and Oceanography* 2: 161–179.

Sharfstein, B. and A.D. Steinman, 2001. Growth and survival of the Florida apple snail (*Pomacea paludosa*) fed 3 naturally occurring macrophyte assemblages. *Journal of the North American Benthological Society* 20: 84–95.

Shevtsova, L.V., 1971. Determination of the weight of *Dreissena polymorpha* Pall. and *D. bugensis* Andr. from their size. *Hydrobiological Journal* 7: 103–105 (English translation of Gidrobiologicheskii Zhurnal).

Shevtsova, L.V. and T.A. Kharchenko, 1986. Technology of removing of zebra mussel fouling from pipelines of irrigation systems. Kiev: Naukova dumka publishers (in Russian).

Shkorbatov, G.L., 1986. Interspecific differentiation and integrity of species as a system. In: *System of Integration of Species*. Volskis, R.S. ed., pp. 118–137. Vilnius: Akademiia Nauk SSSR.

Shkorbatov, G.L., A.F. Karpevich, and P.I. Antonov, 1994. Ecological physiology. *Dreissena polymorpha* (Pall.) (Bivalvia: Dreissenidae). *Systematics, Ecology and Practical Significance.* Moscow: Nauka (not seen, cited by Karatayev, 1995).

Sickel, J.B., 1976. *An ecological study of the Asiatic clam, Corbicula manilensis (Philippi, 1841), in the Altamaha River, Georgia, with emphasis on population dynamics, productivity and control methods* PhD dissertation, Emory University, Atlanta.

Sickel, J.B., 1979. Population dynamics of *Corbicula* in the Altamaha River, Georgia. In: *Proceedings, First International Corbicula Symposium,* J.C. Britton, ed., pp. 69–80.

Sickel, J.B. and M.W. Heyn, 1980. Decline of the Asiatic Clam, *Corbicula fluminea*, in the Lower Tennessee and Cumberland Rivers. *Bulletin of the American Malacological Union* 1980: 24–26.

Siddall, S.E., 1980. Early development of *Mytilopsis leucophaeata* (Bivalvia, Dreissenacea). *Veliger* 22: 378–379.

Sigurdson, J.B., C.W. Titman, and P.A. Davies, 1976. The dispersal of young post-larval bivalve molluscs by byssus threads. *Nature* 262: 386–387.

Silva, F.A. and M.S. Campos, 2004. Temperatura como fator limitante ao desenvolvimento de *limnoperna fortunei* (Dunker, 1857). *Relatório Final Bolsa Iniciação Científica* CNPQ/CETEC.

Simberloff, D. and P. Stiling, 1996a. Risks of species introduced for biological control. *Biological Conservation* 78: 185–192.

Simberloff, D. and P. Stiling, 1996b. How risky is biological control? *Ecology* 77: 1965–1974.

Simmons, J., 1998. Algal control and destratification at Hanningfield Reservoir. *Water Science and Technology* 17: 309–316.

Simpson, H.J., S.N. Chillrud, R.F. Bopp, E. Shuster, and D.A. Chaky, 2006. Major ion geochemistry and drinking water supply issues in the Hudson River basin. In: *The Hudson River Estuary,* J.S. Levinton and J.R. Waldman, eds, pp. 79–98. Cambridge: Cambridge University Press.

Singh, O. and R.A. Agarwal, 1981. Toxicity of certain pesticides to two economic species of snails in northern India. *Journal of Economic Entomology* 74: 568–571.

Siripattrawan, S., J.-K. Park, and D. O'Foighil, 2000. Two lineages of the introduced Asian freshwater clam *Corbicula* occur in North America. *Journal of Molluscan Studies* 66: 423–429.

Siver, P., 1993. Inferring the specific conductivity of lake water with scaled chrysophytes. *Limnology and Oceanography* 38: 1480–1492.

Skubinna, J.P., T.G. Coon, and T.R. Batterson, 1995. Increased abundance and depth of submersed macrophytes in response to decreased turbidity in Saginaw Bay, Lake Huron. *Journal of Great Lakes Research* 21: 476–488.

Smirnova, N.F., 1973. Reaction of *Dreissena* to salinity and petrol. *Biologiya Vnutrennikh Vod* 18: 37–39.

Smirnova, N.F. and G.A. Vinogradov, 1990. Biology and ecology of *Dreissena polymorpha* from the European USSR. Presentation at the U.S. Environmental Protection Agency Workshop, *Zebra Mussels and Other Introduced Aquatic Nuisance Species*, Saginaw, USA, September 26–28.

Smirnova, N.F., G.I. Biochino, and G.A. Vinogradov, 1993. Some aspects of the zebra mussel (*Dreissena polymorpha*) in the former European USSR with morphological comparisons to Lake Erie. In: *Zebra Mussels: Biology, Impacts, and Control*, T.F. Nalepa and D.W. Schloesser, eds, pp. 217–226. Boca Raton: Lewis Publishers.

Smit, H., A. Bij de Vaate, H.H. Reeders, E.H. van Ness, and R. Noordhuis, 1993. Colonization, ecology and positive aspects of zebra mussels (*Dreissena polymorpha*) in the Netherlands. In: *Zebra Mussels Biology, Impacts, and Control*, T.F. Nalepa, and D.W. Schloesser, eds, pp. 55–78. Boca Raton: Lewis Publishers.

Smith, D.G. and K.J. Boss, 1996. The occurrence of *Mytilopsis leucophaeta* (Conrad, 1831) (Veneroida: Dreissenidae) in southern New England. *Veliger* 39: 359–360.

Smith, V.H., 1982. The nitrogen and phosphorus dependence of algal biomass in lakes: An empirical and theoretical analysis. *Limnology and Oceanography* 27: 1101–1112.

Smithson, J.A., 1986. Development of a *Corbicula* control treatment at the Baldwin power station. *American Malacological Bulletin, Special Edition No. 2* 1986: 63–67.

Smythe, A.G., C.L. Lange, D.C. Steuernagel, and R.L. Tuttle, 1993. Efficiency evaluations for in-line high speed agitators in controlling zebra mussel veligers. In: *Third International Zebra Mussel Conference, Agenda and Abstracts,* Toronto, Canada, February.

Smythe, A.G. and C.L. Lange, 1999. Application of pulse-power electric fields as a control for settlement and planktonic-stage zebra mussels: Phase II. Presentation at the *Ninth International Zebra Mussel and Aquatic Nuisance Species Conference,* Duluth, USA, April 26–30.

Smythe, A.G. and A.C. Miller, 2003. Pulse-power: A possible alternative to chemicals for zebra mussel control; Summary of 2000 field studies. *ANSRP Technical Notes Collection* (ERDC TN-ANSRP-03–2), U.S. Army Engineer Research and Development Center, Vicksburg, MS. Available at: www.wes.army.mil/el/ansrp. Accessed February 22, 2009.

Smythe, G., C.L. Lange, and T.M. Short, 1993. Efficiency evaluations for in-line automatic backwash filters in controlling zebra mussel veligers. In: *Third International Zebra Mussel Conference, Agenda and Abstracts,* Toronto, Canada, February.

Smythe, G., C.L. Lange, and P.M. Sawyko, 1997. Evaluation of an in-line magnetic device as a control method for adult and settlement-stage zebra mussels. Paper presented at the *Seventh International Zebra Mussel and Aquatic Nuisance Species Conference,* New Orleans, USA, January 28–31.

Smythe, G. and E.A. Dardeau, Jr, 1999. Overview of electrical technologies for controlling dreissenids, with emphasis on pulse-power systems, *Zebra Mussel Technical Notes collection* (ZMR-3–22), U.S. Army Engineer Research and Development Center, Vicksburg, USA. Available at: *www.wes.army.mil/el/elpubs/zebtnote.html.* Accessed April 6, 2009.

Smythe, G.A., C.L. Lange, and G.H. Keipper, 1998. A study to assess the efficacy of pH to reactively control adult zebra and quagga mussels in an industrial setting. Acres International Corporation. In: *Proceedings of the Eighth International Zebra Mussel and Other Nuisance Species Conference,* Sacramento, USA, March 16–19.

Snyder, N.F.R. and H.A. Snyder, 1969. A comparative study of mollusc predation by limpkins, Everglade kites, and boat-tailed grackles. *Living Bird* 8: 177–223.

Snyder, N.F.R. and H.A. Snyder, 1971. Defenses of the Florida apple snail *Pomacea paludosa. Behaviour* 40: 175–215.

Sokolova, N.Y., 1959. To the fauna of fouling in water reservoirs and water bodies. In: *Proceedings of the Conference on Problems of Inland Waters Wiology*. AS USSR publishers (in Russian).

Son, M.O., 2007. Native range of the zebra mussel and quagga mussel and new data on their invasions within the Ponto-Caspian region. *Aquatic Invasions* 2: 169–179.

Son, M.O., 2008. Rapid expansion of the New Zealand mud snail *Potamopyrgus* antipodarum (Gray, 1843) in the Azov-Black Sea Region. *Aquatic Invasions* 3: 335–340.

Son, M.O., M.V. Nabozhenko, and I.V. Shokhin, 2008. The Don River Basin is a new stage of expansion of *Potamopyrgus jenkinsi* (Smith, 1889) (Gastropoda, Hydrobioidea) in Europe. *Doklady Biological Sciences* 419: 129–130.

Sorba, E.A. and D.A. Williamson, 1997. Zebra mussel colonization potential in Manitoba, Canada. Water Quality Management Section, *Manitoba Environment*, Report No. 97-07.

Sousa, R., S. Dias, and C. Antunes, 2006a. Spatial subtidal macrobenthic distribution in relation to abiotic conditions in the Lima estuary, NW of Portugal. *Hydrobiologia* 559: 135–148.

Sousa, R., C. Antunes, and L. Guilhermino, 2006b. Factors influencing the occurrence and distribution of *Corbicula fluminea* (Müller, 1774) in the River Lima estuary. *International Journal of Limnology* 42: 165–171.

Sousa, R., M. Rufino, M. Gaspar, C. Antunes, and L.C. Guilhermino, 2008. Abiotic impacts on spatial and temporal distribution of Corbicula fluminea (Müller, 1774) in the River Minho Estuary, Portugal. *Aquatic Conservation: Marine and Freshwater Ecosystems* 18: 98–110.

Souza, B.B. and H.R. Mota, 2008. Preferência de *limnoperna fortunei* (Dunker, 1857) a diferentes substratos naturais da bacia do rio paranaíba. *Relatório Final Bolsa Iniciação Científica* FAPEMIG/CETEC.

Souza, J.R.B., C.M.C. Rocha, and M.P.R. Lima, 2005. Ocorrência do bivalve exótico *Mytilopsis leucophaeta* (Conrad) (Mollusca, Bivalvia), no Brasil. *Revista Brasileira de Zoologia* 22: 1204–1206.

Spidle, A., 1994. *A comparison of exotic bivalves, the quagga mussel (Dreissena Bugensis Andrusov) and the zebra mussel (D. Polymorpha* Pallas), *using genetic variation and tolerance to temperature and salinity*. MS thesis, Cornell University, Ithaca.

Spidle, A.P., E.P. Marsden, and B. May, 1994. Identification of the Great Lakes quagga mussel as *Dreissena bugensis* from the Dnieper River, Ukraine, on the basis of allozyme variation. *Canadian Journal of Fisheries and Aquatic Science* 51: 1485–1489.

Spidle, A.P., E.L. Mills, and B. May, 1995. Limits to tolerance of temperature and salinity in the quagga mussel *(Dreissena bugensis)* and the zebra mussel (*Dreissena polymorpha*). *Canadian Journal of Fisheries and Aquatic Science* 52: 2108–2119.

Sprague, J.B. and D.E. Drury, 1969. Avoidance reactions of salmonid fish to representative pollutants. *Water Resources* 2: 23–24.

Sprung, M., 1987. Ecological requirements of developing *Dreissena polymorpha* eggs. *Archives of Hydrobiology Supplement* 79: 69–86.

Sprung, M., 1989. Field and laboratory observations of *Dreissena polymorph*a larvae: Abundance, growth, mortality and food demands. *Archives of Hydrobiology Supplement* 115: 537–561.

Sprung, M., 1993c. The other life: An account of present knowledge of the larval phase of *Dreissena polymorpha.* In: *Zebra Mussels: Biology, Impacts, and Control*, T.F. Nalepa, and D.W. Schloesser, eds, pp. 39–53. Boca Raton: Lewis Publishers.

Sprung, M., 1995a. Physiological energetics of the zebra mussel *Dreissena polymorpha* in lakes. I. Growth and reproductive effort. *Hydrobiologia* 304: 117–132.

Sprung, M., 1995b. Physiological energetics of the zebra mussel *Dreissena polymorpha* in lakes. II. Food uptake and gross growth efficiency. *Hydrobiologia* 304: 133–146.

Sprung, M. and U. Rose, 1988. Influence of food size and food quantity on the feeding of the mussel *Dreissena polymorpha. Oecologia* 77: 526–532.

St. Petersburg Government, 2008. Monitoring of non-indigenous species in 2008 year in the Gulg of Finland. Technical Report, St. Petersburg (in Russian).

Stanczykowska, A., 1964. On the relationship between abundance, aggregations and "condition" of *Dreissena polymorpha* Pall. in 36 Masurian lakes. *Polish Journal of Ecology* 12A: 653–690.

Stanczykowska, A., 1975. Ecosystem of the Mikolajskie Lake, Poland regularities of the *Dreissena polymorpha* bivalvia occurrence and its function in the lake. *Polish Archives of Hydrobiology* 22: 73–78.

Stanczykowska, A., 1977. Ecology of *Dreissena polymorpha* (Pall.) (Bivalvia) in lakes. *Polish Archives of Hydrobiology* 24: 461–530.

Stanczykowska, A., 1984. The effect of various phosphorus loadings on the occurrence of *Dreissena polymorpha* (Pall.). *Limnologica* (*Berlin*) 15: 535–539.

Stanczykowska, A., W. Lawacz, J. Mattice, and K. Lewandowski, 1976. Bivalves as a factor affecting the circulation of matter in Lake Mikolajskie (Poland). *Limnologica* 10: 347–352.

Stanczykowska, A. and W. Lewandowski, 1993. Thirty years of studies of *Dreissena polymorpha* ecology in Mazurian lakes of northeastern Poland. In: *Zebra Mussels Biology, Impacts, and Control*, T.F. Nalepa and D.W. Schloesser, eds, pp. 3–38. Boca Raton: Lewis Publishers.

Stangel, P., 2004. Lake Champlain 2003 Zebra Mussel Monitoring Program, Final Report. Vermont Department of Environmental Conservation, Waterbury, VT.

Stangel, P. and A. Shambaugh, 2005. Lake Champlain 2004 zebra mussel monitoring program, Final Report. Vermont Department of Environmental Conservation, Waterbury, VT.

Starobogatov, Y.I., 1970. *Molluscan Fauna and Geographic Zoning of Inland Waters of the Globe*. Leningrad: Nauka Press (in Russian).

Starobogatov, Y.I., 1994. Taxonomy and palaeontology. In: *Dreissena polymorpha (Pall.) (Bivalvia, Dreissenidae). Taxonomy, Ecology and Practical Meaning*, Y.I. Starobogatov, ed., pp. 18–46. Moskow: Nauka Press (in Russian).

Starobogatov, Y.I. and S.I. Andreeva, 1994. Range. In: *Dreissena polymorpha (Pall.) (Bivalvia, Dreissenidae). Taxonomy, Ecology and Practical Meaning*, Y.I. Starobogatov, ed. pp. 47–53. Moskow: Nauka Press (in Russian).

Staton, S.K., J.L. Metcalfe-Smith, and E.L. West, 2000a. Status of the Northern riffleshell, *Epioblasma torulosa rangiana* (Bivalvia: Unionidae), in Ontario and Canada. *The Canadian Field-Naturalist* 114: 224–235.

Stearns, S.C., 1976. Life-history tactics: A review of the ideas. *Quarterly Review of Biology* 51: 3–47.

Stepien, C.A., C.D. Taylor, and K.A. Dabrowska, 2002. Genetic variability and phylogeographical patterns of a nonindigenous species invasion: A comparison of exotic vs. native zebra and quagga mussel populations. *Journal of Evolutionary Biology* 15: 314–328.

Stepien, C.A., C.D. Taylor, I.A. Grigorovich, S.V. Shirman, R. Wei, A.V. Korniushin, and K.A. Dabrowska, 2003. DNA and systematic analysis of invasive and native dreissenid mussels: Is *Dreissena bugensis* really *D. rostriformis*? *Aquatic Invaders* 14(2): 1–10.

Sterki, V., 1911. Civilization and snails. *The Nautilus* 24(9): 98–101.

Stevens, A.J., Z.C. Welch, P.C. Darby, and H.F. Percival, 2002. Temperature effects on Florida Applesnail activity: Implications for Snail Kite foraging success and distribution. *Wildlife Society Bulletin* 30: 75–81.

Steward, K.K. and W.H. Ornes, 1975. The autoecology of sawgrass in the Florida Everglades. *Ecology* 56: 162–171.

Stewart, T.W. and J.M. Haynes, 1994. Benthic macroinvertebrate communities of southwestern Lake Ontario following invasion of *Dreissena*. *Journal of Great Lakes Research* 20: 479–493.

Stoeckmann, A., 2003. Physiological energetics of Lake Erie *Dreissenid mussels:* A basis for the displacement of *Dreissena polymorpha* by *Dreissena bugensis*. *Canadian Journal of Fisheries and Aquatic Sciences* 60: 126–34.

Strayer, D.L., 1991. Projected distribution of the zebramussel, *Dreissena polymorpha*, in North America. *Canadian Journal of Fisheries and Aquatic Sciences* 48: 1389–1395.

Strayer, D.L., 1999. Effects of alien species on freshwater molluscs in North America. *Journal of the North American Benthological Society* 18: 74–98.

Strayer, D.L. and L.C. Smith, 1996. Relationships between zebra mussels (*Dreissena polymorpha*) and unionid clams during the early stages of the zebra mussel invasion of the Hudson River. *Freshwater Biology* 36: 771–779.

Strayer, D.L., L.C. Smith, and D.C. Hunter, 1998. Effects of the zebra mussel (*Dreissena polymorpha*) invasion on the macrobenthos of the freshwater tidal Hudson River. *Canadian Journal of Zoology* 76: 419–425.

Strayer, D.L., N.F. Caraco, J.J. Cole, S. Findlay, and M.L. Pace, 1999. Transformation of ecosystems by bivalves. *Bioscience* 49: 19–27.

Strayer, D.L. and H.M. Malcom, 2006. Long-term demography of a zebra mussel *(Dreissena polymorpha)* population. *Freshwater Biology* 51: 117–130.

Strayer, D.L. and H.M. Malcom, 2007. Shell decay rates of native and alien freshwater bivalves and implications for habitat engineering. *Freshwater Biology* 52: 1611–1617.

Strzelec, M., 1993. Snails (Gastropoda) of anthropogenic water environments in Silesian Upland. *Prace Naukowe Uniwersytetu S´la¸skiego* no 1358: 1–104 (in Polish with English summary).

Strzelec M., 1999. Effect of artificially elevated water temperature on growth and fecundity of *Potamopyrgus antipodarum* (Gray) in anthropogenic water bodies in Southern Poland (Gastropoda: Prosobranchia: Hydrobiidae). *Malakologische Abhandlungen* 19: 266–272.

Strzelec, M. and M. Krodkiewska, 1994. The rapid expansion of *Potamopyrgus jenkinsi* (E.A. Smith, 1889) in Upper Silesia (Southern Poland) (Gastropoda: Prosobranchia: Hydrobiidae). *Malakologische Abhandlungen Staatliches Museum fur Tierkunde Dresden* 17: 83–86.

Stuckey, R.L. and D.L. Moore, 1995. Return and increase in abundance of aquatic flowering plants in Put-in-Bay harbor, Lake Erie, Ohio. *Ohio Journal of Science* 95: 261–266.

Subda Rao, N.V. and S.C. Mitra, 1982. Bioecology of two melaniid snails (Mollusca: Gastropoda) in ponds near Calcutta. *Journal of the Zoological Society of India* 34: 21–32.

Sugiura, N. and T. Wada, 1999. Cold tolerance of the apple snail, *Pomacea canaliculata* (Lamarck), collected from various locations. *Proceedings of the Association for Plant Protection of Kyushu* 45: 59–62 (in Japanese) (not seen, cited by Ito, 2002).

Suren, A.M., 2005. Effects of deposited sediment on patch selection by two grazing stream invertebrates. *Hydrobiologia* 549: 205–218.

Sverdrup, H.U., M.W. Johnson, and R.H. Fleming, 1942. *The Oceans: Their Physics, Chemistry and General Biology*. Englewood Cliffs, NJ: Prentice-Hall.

Swinnen, F., M. Leynen, R. Sablon, L. Duvivier, and R. Vanmaele, 1998. The Asiatic clam *Corbicula* (Bivalvia: Corbiculidae) in Belgium. Bulletin del l' Institut Royal des Sciences Naturelles de Belgique. *Biologie* 68: 47–53.

Sylvester, F., J. Dorado, D. Boltovskoy, A. Juárez, and D.l. Cataldo, 2004. Filtration rates of the invasive pest bivalve *Limnoperna fortunei* as a function of size and temperature. Presentation at the *13th International Conference on Aquatic Invasive Species,* Ennis, Ireland, September 20–24.

Sylvester, F., J. Dorado, D. Bolstovskoy, A. Juárez, and D. Cataldo, 2005. Filtration rate of the invasive pest bivalve *Limnoperna fortunei* as a function of size and temperature. *Hydrobiologia* 534: 71–80.

Sylvester, F., D. Boltovskoy, and D.H. Cataldo, 2007. The invasive bivalve *Limnoperna fortunei* promotes growth of benthic invertebrates in South American floodplain rivers. *Hydrobiologia* 589: 15–27.

Syobu, S., H. Mikuriya, J. Yamaguchi, and M. Matsuzaki, 2001. Estimating the overwinter mortality of the apple snail in paddy field of southern Japan using temperature data. *Japanese Journal of Applied Entomology and Zoology* 45: 203–207.

Takashi Wada, T., K. Ichinose, Y. Yusa, and N. Sugiura, 2004. Decrease in density of the apple snail *Pomacea canaliculata* (Lamarck) (Gastropoda: Ampullariidae) in paddy fields after crop rotation with soybean, and its population growth during the crop season. *Applied Entomology and Zoology* 39: 367–372.

Tamaru, C.S., 1996. Control of the apple snail (Pomacea canaliculata), planning project. *Six month report, University of Hawaii Sea Grant College Program,* Honolulu, 59 p.

Tamaru, C.S. and F. Hun, 1996. Survey of apple snails for Hui Na'auao. Unpublished report, University of Hawaii Sea Grant Program, Honolulu, 13 p.

Tamaru, C.S., A.H. Ako, and C.-T.C. Tamaru, 2006. Control of the Apple Snail, *Pomacea canaliculata,* in Hawai'i: Challenge or opportunity? In: *Global Advances in Ecology and Management of Golden Apple Snails*, R.C. Joshi and L.S. Sebastian, eds, pp. 459–474. Nueva Ecija: Philippine Rice Research Institute.

Tammi, K.A., L.T. Green, E. Herron, M. Kerr, and M. Rice, 1995a. Likelihood of invasion of the aquatic exotic zebra mussel, *Dreissena polymorpha* (Pallas), into Rhode Island freshwater systems. *Proceedings of the Northeastern Conference on Nonindigenous Aquatic Nuisance Species*, Connecticut Sea Grant College Program, Groton (not seen, cited by Cohen, 2001).

Tammi, K.A., L.T. Green, M. Kerr, and M. Rice, 1995b. Likelihood of invasion of the aquatic exotic zebra mussel, *Dreissena polymorpha* (Pallas), into Rhode Island freshwater systems: 1995 assessment of boating activity at potential zebra mussel habitats for monitored areas in Rhode Island. Rhode Island Cooperative Extension Publication No. 3076, Kingston: University of Rhode Island (not seen, cited by Cohen, 2001).

Tanaka, K., T. Watanabe, H. Higuchi, K. Miyamoto, Y. Yusa, T. Kiyonaga, H. Kiyota, Y. Suzuki, and T. Wada, 1999. Density-dependent growth and reproduction of the apple snail, *Pomacea canaliculata*: A density manipulation experiment in a paddy field. *Researches on Population Ecology* 41: 253–262.

Tashiro, S.J. and S.D. Colman, 1982. Filter-feeding in the freshwater prosobranch snail *Bithynia tentaculata*: Bioenergetic partitioning of ingested carbon and nitrogen. *American Midland Naturalist* 107: 114–132.

Taxson, A. and A. Taxson, 1980. At Sky Lake, North Miami. New York Shell Club Notes No. 259: 6 (not seen, cited by Darby et al. 1997).

Therriault, T.W., M.F. Docker, M.I. Orlova, D.D. Heath, and H.J. MacIsaac, 2004. Molecular resolution of the family Dreissenidae (Mollusca: Bivalvia) with emphasis on Ponto-Caspian species, including first report of Mytilopsis leucophaeata in the Black Sea basin. *Molecular Phylogenetics and Evolution* 30: 479–489.

Thiengo, S.C., 1987. Observations on the morphology of *Pomacea lineata* (Spix, 1827) (Mollusca, Ampullariidae). *Memorias de Instituto Oswaldo Cruz* 82: 563–570.

Thomas, K.J., 1975. Biological control of *Slavinia* by the snail *Pila globosa* Swainson. *Biological Journal of the Linnean Society* 7: 243–247.

Thomas, N.A. and K.M. Mackenthum, 1964. Asiatic clam infestation at Charleston, West Virginia. *The Nautilus* 78: 28–29.

Thompson, F.G., 1984. *The Freshwater Snails of Florida: A Manual for Identification.* Gainesville: University of Florida Press.

Thorp, J.H., J.E. Alexander, Jr., B.L. Bukaveckas, G.A. Cobbs, and K.L. Bresko, 1998. Responses of Ohio River and Lake Erie dreissenid molluscs to changes in temperature and turbidity. *Canadian Journal of Fisheries and Aquatic Sciences* 55: 220–229.

Todd, M.E., 1964. Osmotic balance in *Hydrobia ulvae* and *Potamopyrgus jenkinsi* (Gastropoda: Hydrobiidae). *Journal of Experimental Biology* 41: 665–677.

Toetz, D., L.P. Varga, and M. Pierce, 1977. Effects of chlorine and chloramine on uptake of inorganic nitrogen by phytoplankton. *Water Resources* 11: 253–258.

Toledo, R., C. Munoz-Antoli, M. Perez, and J.G. Esteban, 1998. Larval trematode infections in freshwater gastropods from the Albufera Natural Park in Spain. *Journal of Helminthology* 72: 79–82.

Torres-Orozco, R.B., and E. Revueltas-Valle, 1996. New southernmost record of the Asiatic clam *Corbicula fluminea* (Bivalvia: Corbiculidae), in Mexico. *Southwestern Naturalist* 41: 60–61.

Truchan, J.G., 1977. Toxicity of residual chlorine to freshwater fish: Michigan's experience. In: *Biofouling Control Procedures, Technology and Ecological Effects*, L.D. Jensen, ed., pp. 79–89. New York: Marcel Dekker.

Tsai, C.F., 1968. Effects of chlorinated sewage effluents of fishes in upper Patuxent River, Maryland. *Chesapeake Science* 9: 83–91.

Tsai, C.F., 1970. Changes in fish populations and migration in relation to increased sewage pollution in Little Patuxent River, Maryland. *Chesapeake Science* 11: 34–41.

Tsai, H.-C., S.S.-J. Lee, C.-K. Huang, C.-M. Yen, E.-R. Chen, and Y.-C. Liu, 2004. Outbreak of eosinophilic meningitis associated with drinking raw vegetable juice in southern Taiwan. *American Journal of Tropical Medicine and Hygiene* 71: 222–226.

Tsai, Y.-L. and P.A. Rochelle, 2001. Extraction of nucleic acids from environmental samples. In: *Environmental Molecular Microbiology: Protocols and Applications*, P.A. Rochelle, ed., pp. 15–30, Wymondham: Horizon Scientific Press.

Tucker, J.K., 1994. Colonization of unionid bivalves by the zebra mussel, *Dreissena polymorpha*, in pool 26 of the Mississippi River. *Journal of Freshwater Ecology* 9: 129–134.

Turner, R.L., 1996. Use of stems of emergent vegetation for oviposirion by the Florida apple snail (*Pomacea paludosa*), and implications for marsh management. *Florida Scientist* 59: 34–49.

Tzeyeb, Y.Y., A.M. Almazov, and V.I. Vladimirov, 1966. Regularities in the changes of the hydrological, hydrochemical and hydrobiological regimes of the Dniepr River on flow-off regulation and their effect on fish biology and sanitary state of the reservoirs. *Hydrobiological Journal* 2: 3–18 (in Russian).

UNESCO, 1981a. The Practical Salinity Scale 1978 and the International Equation of State of Seawater 1980. *Technical Papers on Marine Science* 36: 25pp.

UNESCO, 1981b. Background papers and supporting data on the Practical Salinity Scale 1978. *Technical Papers on Marine Science* 37: 144pp.
U.S. Army Corps of Engineers, Technical Note ZMR-4-02 A method for calculating the total mass of zebra mussels, Zebra Mussel Research Program, U.S. Army.
USACE (U.S. Army Corps of Engineers Waterways) Experiment Station, 1993. Zebra Mussel Research Technical Notes. Section 4, Miscellaneous, Technical Note ZMR-4-03 (Prepared and published by the Zebra Mussel Research Program, U.S. Army Engineer Waterways Experiment Station, Vicksburg, MS 39180-6199).
U.S. Bureau of Reclamation, 2009a. Preliminary evaluation of quagga mussel fouling potential for Cu–Ni (90-10) wedge wire screen. *Zebra and Quagga Mussel Research Program Research Note* ZQMRP-2009-RN-01. Available at: http://www.usbr.gov/mussels/research/docs/researchnotes/ZQMRP-2009-RN-01-CU-NI.pdf. Accessed April 12, 2009.
U.S. Bureau of Reclamation, 2009b. Quagga mussel control for pipes using water jetting. Zebra & quagga mussel research program research note ZQMRP-2009-RN-03. Available at: http://www.usbr.gov/mussels/research/docs/researchnotes/ZQMRP-2009-RN-03-Water-Jetting.pdf. Accessed April 12, 2009.
USGS (United States Geological Survey), 2008. *Pomacea bridgesii*. USGS Nonindigenous Aquatic Species Database, Gainesville, FL. Available at: http://nas.er.usgs.gov/queries/FactSheet.asp?speciesID=982 Revision Date: 11/23/2004. Accessed October 30, 2008.
USGS (United States Geological Survey), 2009a. (*Marisa cornuarietis*) Avaialble at: http://nas.er.usgs.gov/queries/collectioninfo.asp?NoCache=1%2F23%2F2009+1%3A06%3A39+PM&SpeciesID=981&State=&County=&HUCNumber= Accessed January 2009.
USGS (United States Geological Survey), 2009b. (*Pomacea canaliculata*) Avaialble at: http://nas.er.usgs.gov/queries/collectioninfo.asp?NoCache=1%2F23%2F2009+12%3A32%3A56+PM&SpeciesID=980&State=&County=&HUCNumber= Accessed January 2009.
USGS (United States Geological Survey), 2009c. (*Pomacea haustrum*) Available at: http://nas.er.usgs.gov/queries/collectioninfo.asp?SpeciesID=2600 Accessed January 2009.
USGS (United States Geological Survey), 2009d. (*Pomacea insularum*) Available at: http://nas.er.usgs.gov/queries/collectioninfo.asp?SpeciesID=2599. Accessed January 23, 2009.
USGS (United States Geological Survey), 2009e. (*Pomacea diffusa*) Available at: http://nas.er.usgs.gov/queries/collectioninfo.asp?NoCache=1%2F23%2F2009+12%3A32%3A56+PM&SpeciesID=2662&State=&County=&HUCNumber= Accessed January 2009.
USGS (United States Geological Survey), 2009f. (*Pomacea paludosa*) Avaialble at: http://nas.er.usgs.gov/queries/collectioninfo.asp?NoCache=1%2F23%2F2009+12%3A32%3A56+PM&SpeciesID=985&State=&County=&HUCNumber= Accessed January 2009.
USGS (United States Geological Survey), 2009g. (*Tarebia granifera*) Available at: http://nas.er.usgs.gov/queries/collectioninfo.asp?NoCache=1%2F23%2F2009+12%3A32%3A56+PM&SpeciesID=1039&State=&County=&HUCNumber= Accessed January 2009.
USGS (United States Geological Survey), 2009h. (*Melanoides tuberculata*) Available at: http://nas.er.usgs.gov/queries/collectioninfo.asp?NoCache=1%2F23%2F2009+1%3A06%3A39+PM&SpeciesID=1037&State=&County=&HUCNumber= Accessed January 2009.
USGS (United States Geological Survey), 2009i. (*Bithynia tentaculata*) Avaialble at: http://nas.er.usgs.gov/queries/collectioninfo.asp?NoCache=1%2F23%2F2009+1%3A12%3A23+PM&SpeciesID=987&State=&County=&HUCNumber= Accessed January 2009.
USGS (United States Geological Survey), 2009j. (*Potamopyrgus antipodarum*) Available at: http://nas.er.usgs.gov/queries/collectioninfo.asp?NoCache=1%2F23%2F2009+1%3A15%3A27+PM&SpeciesID=1008&State=&County=&HUCNumber=Accessed January 2009.
Vaillancourt, G. and E. Lafarriere, 1983. Relationship between the quality of the environment and the benthic groupings in the littoral zone of the St. Lawrence River, Canada. *Naturaliste Canadien (Quebec)* 110: 385–396.

Valentine-Darby, P.I., P.C. Darby, R.E. Bennetts, W.M. Kitchens, and H.F. Percival, 2008. The use of mark-recapture to estimate Florida apple snail (*Pomacea paludosa* say) density in wetland habitats. *Florida Scientist* 71(2): 115–127.

van Benschoten, J.E., J.N. Jensen, D. Lewis, and T.J. Brady, 1993. Chemical oxidants for controlling zebra mussels (*Dreissena polymorpha*): A synthesis of recent laboratory and field studies. In: *Zebra Mussels Biology, Impacts, and Control*, T.F. Nalepa and D.W. Schloesser, eds, pp. 599–620. Boca Raton: Lewis Publishers.

van Benthem Jutting, T., 1936. Brakwatermollusken. In: *Flora en fauna der Zuiderzee. Monografie van een brakwatergebied*, H.C. Redeke, ed., Suppl., pp. 137–144. De Boer, Den Helder (not seen, cited by Wolff, 2005).

van Benthem Jutting, T., 1943. Mollusca. C. Lamellibranchia—*Fauna van Nederland* 12: 1–477 (not seen, cited by Wolff, 2005).

Van Cott, W.R., P.C. Fraleigh, M.E. Wenning, and J.A. DeKam, 1993. Zebra mussel settling and infestation patterns. In: *Third International Zebra Mussel Conference, Agenda and Abstracts,* Toronto, Canada, February.

van der Velde, G., R. Van der Gaag, S. Rajagopal, and H.A. Jenner, 1998. Where exotic mussels *Dreissena polymorpha* and *Mytilopsis leucophaeata* meet in the brackish Noordzeekanaal, the Netherlands. Abstracts from the Eighth International Zebra Mussel and Other Nuisance Species Conference, Sacramento California, March 16–19, 1998.

van der Velde, G. and D. Platvoet, 2007. Quagga mussel *Dreissena rostriformis bugensis* (Andrusov, 1897) in the Main River (Germany). *Aquatic Invasions* 2: 261–264.

van der Velde, G., I. Nagelkerken, S. Rajagopal, and A. bij de Vaate, 2002. Invasions by alien species in inland freshwater bodies in Western Europe: The Rhine Delta. In: *Invasive Aquatic Species of Europe: Distribution, Impacts and Management*, Leppakoski E., S. Gollash, and S. Olenin, eds, pp. 501–506. Dordrecht: Kluwer Academic Publishers.

van Dinther, J.B.M., 1956. Control of Pomacea (Ampullaria) snails in rice fields. Agricultural Experiment Station, Paramaribo, *Surinam Bulletin* 68: 1–20 (not seen, cited by Cowie, 2002).

van Dinther, J.B.M. and R.W. Stubbs, 1963. Summary of research on the control of rice snails in Surinam. Agricultural Experiment Station, Paramaribo, *Surinam Bulletin* 82: 415–420 (not seen, cited by Cowie, 2002).

Van Dinther, J., 1973. Mollusks and their control. *World Crops* 25(6): 282–286.

Vanderploeg, H.A., T.H. Johengen, G.L. Fahnenstiel, T.F. Nalepa, J.R. Liebig, W.W. Carmichael, and M.A. Agy, 2001. Zebra mussel (*Dreissena polymorpha*) selective filtration promoted toxic *Microcystis* blooms in Saginaw Bay (Lake Huron) and Lake Erie. *Canadian Journal of Fisheries and Aquatic Sciences* 58: 1208–1221.

Vanderploeg, H.A., T.F. Nalepa, D.J. Jude, E.L. Mills, K.T. Holeck, J.R. Liebig, I.A. Grigorovich, and H. Ojaveer, 2002. Dispersal and emerging ecological impacts of Ponto-Caspian species in the Laurentian Great Lakes. *Canadian Journal of Fisheries and Aquatic Sciences* 59: 1209–1228.

Vaughn, C.C. and C.C. Hakenkamp, 2001. The functional role of burrowing bivalves in freshwater ecosystems. *Freshwater Biology* 46: 1431–1446.

Verelst, L., F. Ollevier, R. Vanmaele, L. Duvivier, P. Girboux, and J.N. Delanay, 1999. The use of ozone for biofouling control in auxiliary circuit of cooling water system for a power plant. In: *Proceedings of the Ninth International Conference on Zebra Mussels and Aquatic Nuisance Species,* Duluth, USA, April 26.

Verween, A., S. Degraer, and M. Vincx, 2002. Reducing the economic impact of an invasive bivalve, Mytilopsis leucophaeata, in the harbour of Antwerp VLIZ Young Scientists' Day, Brugge, Belgium 13 March 2002: Book of abstracts. VLIZ Special Publication, 7: pp. 53.

Verween, A., M. Vincx, J. Mees, and S. Degraer, 2005. Seasonal variability of *Mytilopsis leucophaeata* larvae in the harbour of Antwerp: Implications for ecologically sound and economically sound biofouling control. *Belgian Journal of Zoology* 135: 91–93.

Verween, A., M. Vincx, and S. Degraer, 2006. Growth patterns of *Mytilopsis leucophaeata*, an invasive biofouling bivalve in Europe. *Biofouling* 22: 221–231.

Verween, A., M. Vincx, and S. Degraer, 2007. The effect of temperature and salinity on the survival of *Mytilopsis leucophaeata* larvae (Mollusca, Bivalvia): The search for environmental limits. *Journal of Experimental Marine Biology and Ecology* 348: 111–120.

Victor, R. and A.A.I. Al-Farsi, 2001. Water quality and invertebrate fauna of farm wells in an area affected by salinization in Oman. *Journal of Arid Environments* 48: 419–428.

Vincent, B., G. Vaillaincourt, and M. Harvey, 1981. Cycle de development. Croissance effectifs, biomasse et production de *Bithynia tentaculata* L. (Gastropoda: Prosobranchia) dans le Saint-Laurent (Quebec). *Canadian Journal of Zoology* 59: 1237–1250.

Vincent, B., H. Rioux, and M. Harvey, 1981. Factors affecting the structure of epiphytic gastropod communities in the St. Lawrence River (Quebec, Canada). *Hydrobiologia* 220: 57–71.

Vinson, M.R. and M.A. Baker, 2008. Poor growth of Rainbow Trout fed New Zealand mud snails *Potamopyrgus antipodarum. North American Journal of Fisheries Management* 28: 701–709.

Vitousek, P.M., C.M. D'Antonio, L.L. Loope, M. Rejmánek, and R. Westbrooks, 1996. Introduced species: A significant component of human-caused global change. *New Zealand Journal of Ecology* 21: 1–16.

Vitousek, P.M., C.M. D'Antonio, L.L. Loope, and R. Westbrooks, 1996. Biological invasions as global environmental change. *American Scientist* 84: 468–478.

Voelz, N.J., J.V. McArthur, and R.B. Rader, 1998. Upstream mobility of the Asiatic clam *Corbicula fluminea*: Identifying potential dispersal agents. *Journal of Freshwater Ecology* 13: 39–45.

Von Rückert, G., M.C.S. Campos, and M.E. Rolla, 2004. Alimentação de *limnoperna fortunei* (Dunker, 1857): Taxas de filtração com ênfase ao uso de cyanobacteria. *Acta Scientiarum Maringá* 26: 421–429.

Vreenegoor, S.M., R.M. Block, J.C. Rhoderick, and S.R. Gullans, 1977. The effects of chlorination on the osmoregulatory ability of the blue crab, *Callinectes sapidus. Association of Southeastern Biologists Bulletin* 24: 93.

von Proschwitz, T., 1997. *Bithynia tentaculata* (L.) in Norway—a rare species on the edge of its western distribution, and some notes on the dispersal of freshwater snails. *Fauna (Oslo)* 50: 102–107.

Wada, T., 1997. Introduction of the apple snail *Pomacea canaliculata* and its impact on rice agriculture. In: *Proceedings on International Workshop on Biological Invasions of Ecosystems by Pests and Beneficial Organisms.* National Institute of Agro-Environmental Sciences, 170–180. Ministry of Agriculture, Forestry and Fisheries, Tsukuba.

Wada, T., 2006. Impact and control of introduced Apple Snail, *Pomacea canaliculata* (Lamarck), in Japan. In: *Global advances in Ecology and Management of Golden Apple Snails*, R.C. Joshi and L.S. Sebastian, eds, pp. 181–198. Nueva Ecija: Philippine Rice Research Institute.

Wada, T., Y. Ichinose, Y. Yusa, and N. Sigiura, 2004. Decrease in density of the apple snail *Pomacea canaliculata* (Lamarck) (Gastropoda: Ampulariidae) in paddy fields after crop rotation with soybean, and its population growth during the crop season. *Applied Entomology and Zoology* 39: 367–372.

Wainman, B., S. Hincks, N. Kaushik, and G.L. Mackie, 1996. Biofilm and substrate preference in dreissenid larvae in Lake Erie. *Canadian Journal of Fisheries and Aquatic Sciences* 53: 134–140.

Waller, D.L., J.J. Rach, W.G. Cope, L.L. Marking, S.W. Fisher, and H. Dabrowska, 1993. Toxicity of candidate molluscicides to zebra mussels (*Dreissena polymorpha*) and selected nontarget organisms. *Journal of Great Lakes Research* 19: 695–702.

Walton, W.C., 1993. The invasion of the Hudson River estuary by the zebra mussel *Dreissena polymorpha,* and its subsequent range overlap with the dark false mussel, *Mytilopsis leucophaeata*. Final Report, Polgar Fellowship Program.

Walton, W.C., 1996. Occurrence of zebra mussel (*Dreissena polymorpha*) in the oligohaline Hudson River, New York. *Estuaries and Coasts* 19: 612–618.

Walz, N., 1978. The energy balance of the freshwater mussel *Dreissena polymorpha* in the laboratory and in Lake Constance, Switzerland. III. Growth under standard conditions. *Archives of Hydrobiology Supplement* 55: 121–141.

Ward, R.W. and G.M. DeGraeve, 1978. Residual toxicity of several disinfectants in domestic wastewater. *Journal of the Water Pollution Control Federation* 50: 46–60.

Watanabe, I. and W. Ventura, 1990. Management practices to control golden apple snail (*Pomacea* sp.) in lowland rice. *International Rice Research Newsletter* 15: 33.

Watson, E.T., J.L. Metcalfe-Smith, and J. Di Maio, 2000a. Status of the Snuffbox*, Epioblasma triquetra*, in Canada. COSEWIC Status Report, Ottawa, Ontario.

Watson, E.T., J.L. Metcalfe-Smith, and J. Di Maio, 2000b. Status of the Mudpuppy Mussel, *Simpsonaias ambigua*, in Canada. COSEWIC Status Report, Ottawa, Ontario.

Watzin, M.C., A.B. Cohn, M.M. Lescaze, and M. Miranda, 2000. Zebra mussels, shipwrecks and the environment, Lake Champlain Maritime Museum and University of Vermont School of Natural Resources Rubenstein Ecosystem Science Laboratory, Final Report, January 31, 2000.

Watzin, M.C., A.B. Cohn, and B.P. Emerson, 2001. "Zebra mussels, shipwrecks and the environment, Final Report," Lake Champlain Maritime Museum and University of Vermont School of Natural Resources Rubenstein Ecosystem Science Laboratory, May 2001.

Way, C.M., D.J. Hornbach, C.A. Miller-Way, B.S. Payne, and A.C. Miller, 1990. Dynamics of filter-feeding in *Corbicula fluminea* (Bivalvia: Corbiculidae). *Canadian Journal of Zoology* 68: 115–120.

Weatherhead, M.A. and M.R. James, 2001. Distribution of macroinvertebrates in relation to physical and biological variables in the littoral zone of nine New Zealand lakes. *Hydrobiologia* 462: 115–129.

Webb, S.C., A. Fidler, and T. Renault, 2007. Primers for PCR-based detection of ostreid herpes virus-1 (OsHV-1): Application in a survey of New Zealand molluscs. *Aquaculture* 272: 126–139.

Welch, C.R., J.W. Johnson, H. Blake, A. Miller, T. Dardeau, and C. Lunderman, 2000. Water shock pressure fields generated by the sparktec plasma sparker. Paper presented at the *10th International Aquatic Nuisance Species and Zebra Mussel Conference*, Toronto, Canada, February 13–17.

Werner, S. and K-O. Rothhaupt, 2008. Mass mortality of the invasive bivalve *Corbicula fluminea* induced by a severe low-water event and associated low water temperatures. *Hydrobiologia* 613: 143–150.

West, E.L., J.L. Metcalfe-Smith, and S.K. Staton, 2000. Status of the Rayed Bean, *Villosa fabalis* (Bivalvia: Unionidae), in Ontario and Canada. *The Canadian Field-Naturalist* 114: 248–258.

Wetzel, R.G., 1975. *Limnology*. Philadelphia: W.B. Saunders Co.

White, D.S., M.H. Winnell, and D.D. Jude, 1984. Discovery of the Asiatic clam, *Corbicula fluminea*, in Lake Michigan. *Journal of Great Lakes Research* 10: 329–331.

White, G.C., 1986. *The Handbook of Chlorination.* New York: Van Nostrand Reinhold Company.

Whittier T.R., A.T. Herlihy, and S.M. Pierson, 1995. Regional susceptibility of northeast lakes to zebra mussel invasion. *Fisheries* 20: 20–27.

Whittier, T.R., P.L. Ringold, A.T. Herlihy, and S.M. Pierson, 2008. A calcium-based invasion risk assessment for zebra and quagga mussels (*Dreissena* spp.). *Frontiers in Ecology and the Environment* 6: 180–184.

Wiancko, P. and D.R. Lowther, 1990. Options for underwater mechanical cleaning of zebra mussels. In: *Paper and Abstract Presented at the EPRI International Macrofouling Symposium, December, Orlando, USA.*

Widdows, J., P. Fieth, and C.M. Worrall, 1979. Relationships between seston, available food and feeding activity in the common mussel *Mytilus edulis. Marine Biology* 50: 195–207.

Wiktor, L., 1969. The biology of *Dreissena polymorpha* (Pall.) and its ecological importance in the firth of szczecin. *Stud. Mater. Morsk. Inst. Ryb. Gdynia,* Series 5A: 1–88 (cited in Stanczykowska, 1977).

Wilber, C.G., 1983. *Turbidity in the Aquatic Environment: An Environmental Factor in Fresh and Oceanic Waters.* Springfield: Charles C. Thomas Publisher.

Wilbur, K.M. and C.M. Yonge (Chief eds), 1983a. *The Mollusca, Vol. 1.* New York: Academic Press.

Wilbur, K.M. and C.M. Yonge (Chief eds), 1983b. *The Mollusca, Vol. 2. Environmental Biochemistry and Physiology*, P.W. Hochachka, ed. New York: Academic Press.

Wilbur, K.M. and C.M. Yonge (Chief eds) 1983c. *The Mollusca, Vol. 3. Development*, A.S. Verdonk, J.A.M. Van den Biggelaar, and A.S. Tompa, eds. New York: Academic Press.

Wilbur, K.M. and C.M. Yonge (Chief eds), 1983d. *The Mollusca, Vol. 4. Physiology*, Part I, A.S.M. Saleuddin and K.M. Wilbur, eds. New York: Academic Press.

Wilbur, K.M. and C.M. Yonge (Chief eds), 1983e. *The Mollusca, Vol. 5. Physiology*, Part 2, A.S.M. Saleuddin, and K.M. Wilbur, eds. New York: Academic Press.

Wilbur, K.M. and C.M. Yonge (Chief eds), 1983f. *The Mollusca, Vol. 6. Ecology*, W.D. Russel-Hunter, ed. New York: Academic Press.

Wilbur, K.M. and C.M. Yonge (Chief eds), 1983g. *The Mollusca, Vol. 11, Form and Function*, E.R. Trueman and M.R. Clarke, eds. New York: Academic Press.

Wilbur, K.M., 1964. Shell formation and regeneration. In: *Physiology of Mollusca. Vol. 1*, K.M. Wilbur and C.M. Yonge, eds, pp. 243–282. New York: Academic Press.

Wildridge, P.J., R.G. Werner, F.G. Doherty, and E.F. Neuhauser, 1998. Acute effects of potassium on filtering rates adult zebra mussels *Dreissena polymorpha. Journal of Great Lakes Research* 21: 629–636.

Wilhelm, S.W., G.S. Bullerjahn, M.L. Eldridge, J.M. Rinta-Kanto, L. Poorvin and R.A. Bourbonniere, 2006. Seasonal hypoxia and the genetic diversity of prokaryote populations in the central basin hypolimnion of Lake Erie: Evidence for abundant cyanobacteria and photosynthesis. *Journal of Great Lakes Research* 32: 657–671.

Williams, C.J. and R.F. McMahon, 1985. Seasonal variation in oxygen consumption rates, nitrogen excretion rates and tissue organic carbon:nitrogen ratios in the introduced freshwater bivalve, *Corbicula fluminea* (Müller) (Lamellibranchia: Corbiculidae). *American Malacological Bulletin* 3: 267–268.

Williams, C.J. and R.F. McMahon, 1986. Power station entrainment of *Corbicula fluminea* (Müller) in relation to population dynamics, reproductive cycle and biotic and abiotic variables. In: *Proceedings of the Second International Corbicula Symposium*, J.C. Britton, ed., pp. 99–111. *American Malacological Bulletin Special Edition,* No. 2.

Williams, C.J. and R.F. McMahon, 1989. Annual variation of tissue biomass and carbon and nitrogen-content in the fresh-water bivalve *Corbicula fluminea* relative to downstream dispersal. *Canadian Journal of Zoology* 67: 82–90.

Williams, J.D., A.E. Bogan, and J.T. Garner, 2008. *Freshwater Mussels of Alabama and the Mobile Basin in Georgia, Mississippi and Tennessee.* Tuscaloosa: The University of Alabama Press.

Williamson, M.H., 1996. *Biological Invasions.* London: Chapman & Hall.

Winner, B.E., 1989. A comparison of three species of *Pomacea* and their spawn. *American Conchologist* 17: 15–16.

Winner, B.E., 1991. *A Field Guide to Molluscan Spawn. II.* North Palm Beach: Privately published.

Winner, B.E., 1996. A comparison of four species of *Pomacea* and their spawn. *Hawaiian Shell News* 44: 5–7.

Winter, J.E., 1978. A review of the knowledge of suspension-feeding in lamellibranchiate bivalves, with special reference to artificial aquaculture systems. *Aquaculture* 13: 1–33.

Winterbourn, M.J., 1968. The faunas of thermal waters in New Zealand. *Tuatara 16:* 111–122.

Winterbourn, M.J., 1969. Water temperature as a factor limiting the distribution of *Potamopyrgus antipodarum* (Gastropoda-Prosobranchia) in the New Zealand thermal region. *New Zealand Journal of Marine and Freshwater Research* 3*:* 453–458.

Winterbourn, M.J., 1970. The New Zealand species of *Potamopyrgus* (Gastropoda: Hydrobiidae). *Malacologia* 10: 283–321.

Winterbourn, M.J., 1974. Larval Trematoda parasitising the New Zealand species of *Potamopyrgus* (Gastropoda: Hydrobiidae). *Mauri Ora* 2: 17–30.

Winterbourn, M.J. and A. Fegley, 1989. Effects of nutrient enrichment and grazing on periphyton assemblages in some spring-fed, South Island streams. *New Zealand Natural Sciences* 16: 57–65.

Wiryareja, S. and J.R. Tjoe-Awie, 2006. Golden Apple Snail: Its Occurrence and Importance in Suriname's Rice Ecosystem. In *Global Advances in Ecology and Management of Golden Apple Snails*, R.C. Joshi, and L.S. Sebastian, eds, pp. 337–342. Nueva Ecija: Philippine Rice Research Institute.

Wisniewski, R., 1990. Shoals of *Dreissena polymorpha* as bioprocessor of seston. *Hydrobiologia* 200/201: 451–458.

Wolff, W.J., 1969. The mollusca of the estuarine region of the Rivers Rhine, Meuse and Scheldt in relation to the hydrography of the area. II. The Dreissenidae. *Basteria* 33: 93–103.

Wolff, W.J., 1999. Exotic invaders of the meso-oligohaline zone of estuaries in The Netherlands: Why are there so many? *Helgolaender Meeresuntersuchungen*, 52.

Wolff, W.J., 2005. Non-indigenous marine and estuarine species in the Netherlands. *Zoologische Mededelingen* 79: 1–116.

Wong, P.K., K.L. Kwong, and J-W. Qiu, 2008. Complex interactions among fish, snails and macrophytes: implications for biological control of an invasive snail. *Biological Invasions* 10(8): 230–240.

World Health Organization, 1973. Schistosomiasis control. *Technical Report Series WHO* 515: 47.

Wright, D.A., E.M. Setzler-Hamilton, J.A. Magee, V.S. Kennedy, and S.P. McIninch, 1996. Effect of salinity and temperature on survival and development of young zebra (*Dreissena polymorpha*) and quagga (*Dreissena bugensis*) mussels. *Estuaries and Coasts* 19: 619–628.

Wright, D.A., J.A. Magee, E.M. Setzler-Hamilton, L. Chalker-Scott, and G.L. Morgan, 1998. Use of high energy monochromatic UV light to kill dreissenid larvae. In: *Proceedings of the Sixth International Zebra Mussel and Other Aquatic Nuisance Species Conference,* Dearborn, USA, March.

Wright, D.A., R. Dawson, and M. Wildgust, 2000. Excimer UV light as a potential treatment for ballast water. Presented at the *10th Annual Aquatic Nuisance Species and Zebra Mussel Conference,* Toronto, Canada, February 13–17.

Wright D.A. and R. Dawson, 2002. *Shipboard Trial of Primary and Secondary Ballast Water Treatment.* Systems Ballast Water Treatment R&D Directory. London: Program Coordination Unit, Global Ballast Water Management Program, International Maritime Organization.

Wu, M. and V. Xie, 2006. The Golden Apple Snail *(Pomacea canaliculata)* in China. In: *Global Advances in Ecology and Management of Golden Apple Snails*, R.C. Joshi and L.S. Sebastian, eds, pp. 285–298. Nueva Ecija: Philippine Rice Research Institute.

Yacoubi, B., A. Zekhnini, D. Rondelaud, P. Vignoles, G. Dreyfuss, J. Cabaret, and A. Moukrim, 2007. Habitats of *Bulinus truncatus* and *Planorbarius metidjensis*, the intermediate hosts of urinary schistosomosis, under a semiarid or an arid climate. *Parasitology Research* 101: 311–316.

Yahaya, H., M. Nordin, M.N.M. Hisham, and A. Sivapragasam, 2006. Golden apple snails in Malaysia. In: *Global Advances in Ecology and Management of Golden Apple Snails*, R.C. Joshi and L.S. Sebastian, eds, pp. 215–230. Nueva Ecija: Philippine Rice Research Institute.

Yang, P.-S., E.Y.Y.-H. Chen, W.-C. Lee, and Y.-H. Chen, 2006. Golden apple snail management and prevention in Taiwan. In: *Global Advances in Ecology and Management of Golden Apple Snails*, R.C. Joshi, and L.S. Sebastian, eds, pp. 169–180. Nueva Ecija: Philippine Rice Research Institute.

Yin, Y., H. Jiang, D. Guo, and Z. Wang, 2006. Golden apple snails in China. In: *Global Advances in Ecology and Management of Golden Apple Snails*, R.C. Joshi and L.S. Sebastian, eds, pp. 299–314. Nueva Ecija: Philippine Rice Research Institute.

Yong, M. 1990. Seasonal studies on *Pomacea paludosa* in Cuba. *Florida Scientist* 53: 15.

Yonge, C.M. and J.I. Campbell, 1968. On the heteromyarian condition in the Bivalvia with special reference to zebra mussels and certain Mytilacea. *Transactions of the Royal Society of Edinburgh* 68: 21–43.

Yusa, Y., 2004. Brood sex ratio in the apple snail *Pomacea canaliculata* is determined genetically and not by environmental factors. *Journal of Molluscan Studies* 70: 269–275.

Yusa, Y., 2006. Genetics of sex-ratio variation inferred from parent-offspring regressions and sib correlations in the apple snail *Pomacea canaliculata*. *Heredity* 96: 100–105.

Zanatta, D.T., G.L. Mackie, J.L. Smith, and D.A. Woolnough, 2002. A refuge for native freshwater mussels (Bivalvia: Unionidae) from impacts of the exotic zebra mussel (*Dreissena polymorpha*) in Lake St. Clair. *Journal of Great Lakes Research* 28: 479–489.

Zanella, O. and L.D. Marendad, 2002. Ocorrência de *limnoperna fortunei* na central hidrelétrica de itaipu. In: *V Congresso Latinoamericano De Malacologia*. Instituto Butantan/ Instituto De Biociências—Usp, São Paulo.

Zaranko, D.T., D.G. Farara, and F.G. Thompson, 1997. Another exotic mollusk in the Laurentian Great Lakes: The New Zealand native *Potamopyrgus antipodarum* (Gray, 1843) (Gastropoda, Hydrobiidae). *Canadian Journal of Fisheries and Aquatic Sciences* 54: 809–814.

Zeitoun, I.H., 1979. The effect of chlorine toxicity on certain blood parameters of adult rainbow trout (*Salmo gairdneri*). *Environmental Biology of Fishes* 1: 189.

Zeitoun, I.H., L.D. Hughes, and D.E. Ullrey, 1977. Effect of shock exposures of chlorine on the plasma electrolyte concentrations of adult rainbow trout (*Salmo gairdneri*). *Journal of the Fisheries Resource Board of Canada* 34: 1034.

Zettler, M.L. and D.D. Daunys, 2007. Long-term macrozoobenthos changes in a shallow boreal lagoon: Comparison of a recent biodiversity inventory with historical data. *Limnologica* 37: 170–185.

Zhadin, V.I., 1965. Mollusks of fresh and brackish waters of the USSR. *Keys to the Fauna of the USSR*, No. 46, Acad. Nauk USSR, Moscow.

Zhadin, V.I. and S.V. Gerd, 1961. *Rivers, Lakes and Reservoirs of the USSR: Their Fauna and Flora*. Moskow: Uchpedgis Publishers (in Russian).

Zhang, H., 2006. *Ecological modeling of the lower trophic levels of Lake Erie*. PhD dissertation, Ohio State University, Columbus.

Zhang, H., D.A. Culver, and L. Boegman, 2008. A two-dimensional ecological model of Lake Erie: Application to estimate dreissenid impacts on large lake plankton populations. *Ecological Modelling* 214: 219–241.

Zhou, W.C., Y.F. Wu, and J.Q. Yang, 2006. Viability of the ampullaria snail in China. *Fujian Journal of Agricultural Sciences* 18: 25–28.

Zhu, B., D.G. Fitzgerald, C.M. Mayer, L.G. Rudstam, and E.L. Mills, 2006. Alteration of ecosystem function by zebra mussels in Oneida Lake: Impacts on submerged macrophytes. *Ecosystems* 9: 1017–1028.

Zhulidov, A.V., D.F. Pavlov, T.F. Nalepa, G.H. Scherbina, D.A. Zhulidov, and T.Y. Gurtovaya, 2004. Relative Distributions of *Dreissena bugensis* and *Dreissena polymorpha* in the Lower Don River System, Russia. *International Review of Hydrobiology* 89: 326–333.

Zhulidov, A.V., D.A. Zhulidov, D.F. Pavlov, T.F. Nalepa, and T.Y. Gurtovaya, 2005. Expansion of the invasive bivalve mollusk *Dreissena bugensis* (quagga mussel) in the Don and Volga River Basins: Revisions based on archived specimens. *Ecohydrobiology* 5: 127–133.

Zhulidov, A.V., T.F. Nalepa, A.V. Kozhara, D.A. Zhulidov, and T.Y. Gurtovaya, 2006. Recent trends in relative abundance of two dreissenid species, *Dreissena polymorpha* and *Dreissena bugensis* in the Lower Don River system, Russia. *Arhiv fur Hydrobiologia* 165: 209–220.

Zhuravel, P.A., 1934. Some notes on changes in fauna of rapid region of the Dniepr River in connection with Dnieproges. *Priroda* 8: 50–56 (in Russian).

Zhuravel, P.A., 1951. About *Dreissena bugensis* (Mollusca) from the system of the Dnieper River and about it's recent appearance in Dneprovskoye Reservoir. *Zoological Journal* 30: 186–188 (in Russian).

Zhuravel, P.A., 1967. About range expansion of *Dreissena bugensis* in artificial water bodies. *Hydrobiological Journal* 3: 87–90 (in Russian).

Zolotareva, V.I., 1976. Length-weight characteristics of *Dreissena bugensis* from the Zaporozh'ye Reservoir. *Hydrobiological Journal* 12: 94–95.

Zolotareva, V.I., A.V. Makhonina, and A.K. Dyga, 1978. The filtration ability of *Dreissena, bugensis* Andrusov. *Malacological Review* 11: 96–97.

Index

B

C

D

E

F

K

L

M

Q